W9-CRY-886

Applications of Infrared, Raman, and Resonance Raman Spectroscopy in Biochemistry

Frank S. Parker

Department of Biochemistry
New York Medical College
Valhalla, New York

PLENUM PRESS • NEW YORK AND LONDON

1983

Library of Congress Cataloging in Publication Data

Parker, Frank S., 1921–
Applications of infrared, raman, and resonance raman spectroscopy in biochemistry.

Bibliography: p.
Includes index.
1. Spectrum analysis. 2. Biological chemistry—Technique. I. Title. [DNLM: 1. Spectrophotometry, Infrared. 2. Spectrum analysis, Raman. 3. Biochemistry. QD 272.S57 P239a]
QP519.9.S6P37 1983 574.19′285 83-11012
ISBN 0-306-41206-3

A Division of Plenum Publishing Corporation
233 Spring Street, New York, N.Y. 10013

Printed in the United States of America

To Gladys and Judith
in memory of our George
December 7, 1951–May 13, 1981

PREFACE

This book is the outgrowth of a review I wrote several years ago for *Applied Spectroscopy* at the invitation of Dr. John R. Ferraro, who was then editor of the journal and a scientist at the Argonne National Laboratories. In that review I discussed biochemical applications of infrared and Raman spectroscopy, including resonance Raman applications. In this book I have greatly expanded the scope of that original paper [*Applied Spectroscopy* **29,** 129–147 (1975)].

This book is not intended to be a basic text in the several types of spectroscopy. Rather, I have sought to describe applications that would be interesting to a variety of people: advanced undergraduate chemistry and biology students, graduate students and research workers in several disciplines, and spectroscopists. With this aim in mind there was no intent to have exhaustive coverage of the literature.

This book may also be considered an extension and expansion of my earlier one, *Applications of Infrared Spectroscopy in Biochemistry, Biology, and Medicine* (Plenum Press, New York, 1971). The responses from various scientists to the publication of that work were partly responsible for my setting out to do this one. Also, encouragement by the late Dr. Ward Pigman to "do the next one" was also instrumental in my attempting it.

It is a pleasure to thank all those authors who kindly sent me reprints of their papers.

F.S.P.

ACKNOWLEDGMENTS

I thank Ruth Prunty, Juanita O'Brien, and Ann Fusek for typing the manuscript and for being helpful at all times. It is appropriate here to remember the kind cooperation and help given me by the late Alice Cross. I also thank Anne Meagher for her help in obtaining permissions.

It is a pleasure to acknowledge the help of my friends on the staff of Westchester Medical Center Library at New York Medical College: Luiza Balthazar, Maureen Czujak, Danine L'Hommedieu, Arlene Miller, Judith Myers, and Martha Taylor. They have given their time and assistance willingly and cheerfully. May libraries everywhere be staffed with people like them! I thank my personal librarian, Gladys Parker, for her assistance with the index and other parts of this book, and Evelyn Roberts for her continual interest and friendship.

I thank Steven Zatrak for his excellent drawings and his constant cooperation.

With pleasure I thank Professor Michael S. Feld and Professor Ramachandra Dasari for the opportunity to work at the Regional Laser Center in the Spectroscopy Laboratory of the Massachusetts Institute of Technology, and for their warm hospitality during my stay there.

I want to remember here Professor Harold D. Appleton, who encouraged me in the writing of this book. He was a fine friend and colleague.

I thank Blossom and Paul Licht, Belle and Sid Romash, and Joyce and Herb Williams for their beautiful, dependable friendships.

To my brother-in-law, Herbert Baker, and my sister-in-law, Rebecca, I give my heartfelt thanks for their constant support and love, especially during our most difficult days.

To my daughter Judith I give my love and deepest gratitude for being who she is and what she is. I thank Rick Silberman, who has also given us much support and love.

To two other persons whose friendship over the years I have valued highly, I give my sincere thanks: Dr. J. Logan Irvin and Dr. Murray Lieberman. Each in his own way has helped to make this book possible.

I also wish to acknowledge help from my late parents, Jennie and Louis, and my late sister, Evelyn, who gave me their love, and also made my early college education possible.

I remember with deep affection Rose and Bernard Baker—warm, generous people, great respecters of education and the world of books. They were the parents of my very special wife.

To my wife, Gladys, always devoted and loving, always tolerant of my work habits, who has patiently and steadfastly shared with me the vicissitudes of life, I owe the deepest gratitude. I can never thank her enough.

F.S.P.

CONTENTS

Chapter 1

INTRODUCTION AND BRIEF THEORY

There are many texts available that treat the theory of molecular vibrations and infrared and Raman spectra. Detailed treatments are given in the books by Herzberg (1945), Wilson *et al*. (1955), Szymanski (1967), Colthup *et al*. (1975), Long (1977), Tobin (1971), Pinchas and Laulicht (1971), Nakanishi and Solomon (1977), and Nakamoto (1978). Warshel (1977) and Van Wart and Scheraga (1978) discuss the theory of resonance Raman spectroscopy. Other useful texts and review articles are referred to in various sections of this book.

Pinchas and Laulicht (1971) have discussed the isotope effect on frequencies, deuterated and tritiated compounds, labeled saturated hydrocarbons, deuterated unsaturated hydrocarbons, oxygenated organic compounds, deuterated nitrogen and other deuterated organic compounds, carbon-, nitrogen-, and oxygen-18-labeled compounds; and analytical applications.

It is not necessary to discuss the various aspects of IR spectroscopy as it has been performed for many years with dispersion instruments. Many biochemical examples occur in the literature (Parker, 1971). Recent advances in which IR spectroscopy has given us new biochemical pieces of information in a variety of fields have come from the use of Fourier transform IR (FTIR) instruments. Thus, it is important that discussion of some theory and instrumentation of FTIR spectroscopy be given. Biochemical applications of FTIR are given in this chapter and elsewhere in this book.

The main disadvantage of prism or grating IR spectrometers is the monochromator. These instruments have narrow slits at the entrance and the exit of the monochromator, which limit the frequency interval of the radiation arriving at the detector to one frequency of radiation at any given measurement instant.

In order to collimate the IR radiation beam, one uses a very limited area of the radiation emanating from the source. Much of the original radiation falls on the jaws of the exit slit and is lost, i.e., the detector never “sees” it. As one narrows the entrance and exit slits, the resolution of the instrument becomes greater; but this process results in a greater loss of the total radiation from the source, and consequently the detector sees less energy. For a grating spectrometer having a resolution of 1 cm^{-1}, the proportion of radiation reaching the detector is only 0.03%.

Thus, by using a monochromator while trying to obtain good resolution over a broad range of frequencies, we are not recording the spectra in an efficient way.

Fourier transform IR (FTIR) spectroscopy uses an interferometer, derived from an optical instrument invented by Albert Abraham Michelson (1852–1931), the first American to receive the Nobel Prize in physics (1907). This instrument was the basis of the famous Michelson–Morley experiment, by which the constancy of the speed of light, irrespective of the motion of a light source or of the observer, was established. Mated to the interferometer (for FTIR work) are the mathematical principle of the French mathematician, Baron Jean Baptiste Fourier (1768–1830), and a dedicated digital computer. Bates (1976) has reviewed the basic principles and given some current applications of FTIR spectroscopy.

Fellgett (1949) seems to have made the first modern application in Fourier transform spectroscopy by calculating a spectrum from an interferogram based on astronomical measurements. (An interferogram is a complex summation of cosine waves.) The increased speed of measurement by this method, compared to that when a monochromator is used, is called *Fellgett's advantage* (or the multiplex advantage).

The term *throughput* is defined as the product of the area and solid angle of the beam from the source accepted by the instrument. *Jacquinot's advantage* (1954) is the increased spectral signal-to-noise ratio that results from the increased signal at the detector, the consequence of a greater throughput of radiation than is obtained when a monochromator measures the same spectrum; but this advantage is not as important as that of Fellgett, since the throughput of most FTIR spectrometers being used is not very different from that of grating instruments made to perform the same measurements (Griffiths *et al.*, 1977). We shall give a brief outline of the theory.

Figure 1.1 shows a schematic diagram of the operation of a Michelson interferometer. The apparatus has two mirrors, M_1 and M_2, and a beam splitter. The beam splitter transmits half of all incident radiation from the source to M_1, a moving mirror, and reflects half to the fixed mirror M_2. The components reflected by M_1 and M_2 return to the beam splitter,

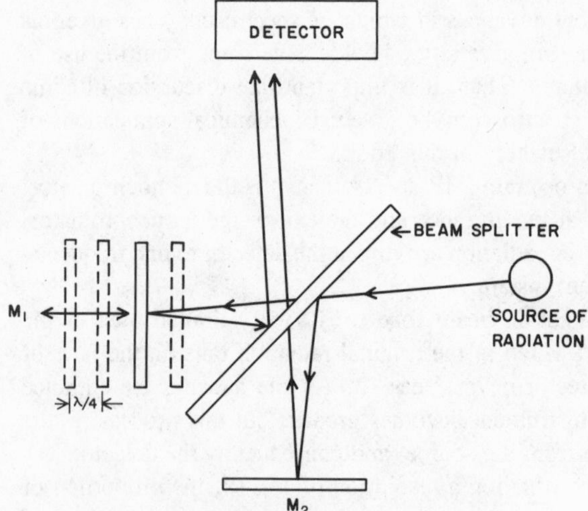

Fig. 1.1. Optical diagram of operation of a Michelson interferometer. (Kagel and King, 1973.) Beam angles at M_1 and M_2 are exaggerated for purposes of explanation. See text.

where they recombine. (The angles of the reflected beams are drawn in exaggerated fashion in Fig. 1.1.) If M_1 and M_2 are equidistant from the beam splitter, the amplitudes of the waves of light combine constructively. If the source gives monochromatic radiation of wavelength λ, and mirror M_1 is moved a distance of $\lambda/4$, the exiting beam is the net result of two beams 180° out of phase recombining destructively. The detector sees a cosine signal, the amplitude of which depends on the position of the mirror.

If the interferometer is illuminated with a polychromatic source of light, the output is more complicated. It is measured as a signal intensity, I, as a function of the optical path difference, δ, of the two paths of light waves (Bates, 1976). The detector receives a signal (the interferogram) which is a summation of all the interferences resulting from constructive or destructive interaction between each wavelength component and all the others. A cosine Fourier transform [Eq. (1.1)] relates the intensity of the interferogram as a function of mirror travel, $I(x)$, and the intensity of the source as a function of light frequency, $I(\nu)$:

$$I(x) = \int_{-\infty}^{\infty} I(\nu) \cos(2\pi x\nu)\, d\nu \tag{1.1}$$

The inverse transform is Eq. (1.2):

$$I(\nu) = \int_{-\infty}^{\infty} I(x) \cos(2\pi \nu x)\, dx \tag{1.2}$$

It relates the interferogram to the optical spectrum. Once the interferogram is digitized and stored by use of a computing system, the digitized equivalent of a complex Fourier transform can be performed on the interferogram to decode the multiplexed information to obtain a complex spectrum.

Some review articles (Griffiths, 1974; Griffiths *et al.*, 1972; Becker and Farrar, 1972; Bates, 1976) and books (Bell, 1972; Griffiths, 1975) describe the principles of Fourier transform spectroscopy and give examples of applications. Mme. Janine Connes (1961) has reviewed in detail the theory of Fourier transforms.

The Digilab Model 296 Interferometer uses the following beam splitters, depending on the frequency range to be observed: For the range 10,000–3,300 cm^{-1}, a coating of Fe_2O_3 on a quartz substrate; 6,000–1,700 cm^{-1}, Fe_2O_3 on CaF_2; 3,800–400 cm^{-1}, Ge on KBr; 400–10 cm^{-1}, mylar substrate.

The Nicolet Instrument Corporation puts out a very useful brochure that gives a technical description and specifications of their 7000 Series FTIR spectrometer.

The Transept Interferometer—An Innovation

The Transept optical system used by Analect Instruments, a division of Laser Precision Corporation (Irvine, California 92714), is the first fundamental change in FTIR interferometers. We have seen above the system that had been used in commercial FTIR

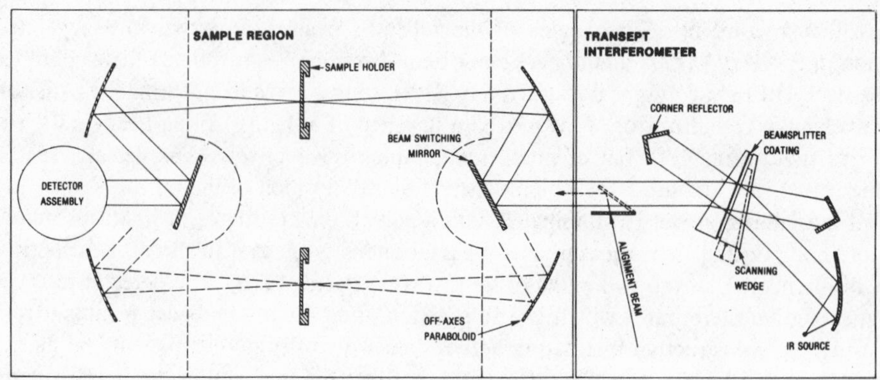

Fig. 1.2. fx-6200 Optical configuration. (Courtesy of Analect Instruments.)

instruments until 1979, the Michelson interferometer. The Analect Instruments interferometer shown schematically in Fig. 1.2 differs from the Michelson version in three significant ways: (1) path length scanning is accomplished by a moving wedge of transparent (refractive) material instead of a moving mirror; (2) the interferometer reflectors are cube corners instead of plane mirrors; (3) the interferometer arms are not mutually perpendicular. Features (1) and (2) enhance the ruggedness and reliability of the instrument by significantly reducing the criticalness of optical alignment and motion control. Feature (3) eliminates displacement of the optical beam during scanning.

In the Analect interferometer, path length scanning is done by moving one of the two matched wedges so as to change effectively the thickness of a uniform layer of transparent material in one arm. The result is a change in optical path length expressed as

$$\Delta p = \Delta y(n - 1) \sin \alpha$$

where Δy is an increment of wedge translation, n is the refractive index of the wedge, and α is its apex angle.

Refractive scanning amplifies the change of optical path length by an order of magnitude, thus substantially reducing the required position repeatability. Also, the effect of any inadvertent tilt of the wedge will be just a lateral displacement of the beam rather than a change in angular alignment. Analect reports that this distinction, compared to the moving mirror design, leads to a factor of 300 reduction in the required angular stability during scanning.

The beam splitter substrate in the Transept system is identical to the scanning wedge and is symmetrically positioned in the second arm of the interferometer. The resulting mutual compensation, along with proper orientation of the two arms, gets rid of any degradation due to dispersion, astigmatism, or spherical aberration. Corner reflectors return light rays on a path parallel to their direction of incidence, thus compensating for any dispersive effects that might come as a result of incomplete matching of the fixed and moving wedges.

Because the instrument is rugged, it can usually be moved without need of realignment of the optical system. Realignment, if necessary, can be done by untrained workers.

The Analect fx-6000 series of FTIR spectrometers reverse the processing order used in the conventional approach to FTIR data processing. A fast Fourier transform (FFT) is performed on every scan by using a high-speed (0.2-sec transform) dedicated processor. The output of the FFT processor is then coadded to obtain the desired signal-to-noise ratio by means of a second processor. A third processor controls the standard readout functions. Oscilloscope display of a complete spectrum is said to occur within 12 sec. The use of such a system provides the following: periodic updating of the display during coadding, access to interactive display functions during data collect, and greater flexibility for the investigator.

Comparison of Fourier Transform and Grating IR Spectrophotometers

Griffiths *et al.* (1977), in an excellent discussion, have compared FT and grating IR spectrophotometers from a theoretical viewpoint in terms of the multiplex advantage of FT spectrometers, the relative optical throughputs, the comparative performance of the detectors used with each type of system, and other more minor operating conditions, e.g., for difference spectroscopy. These authors have compared the signal-to-noise ratios of spectra measured on commercial FT (Digilab FTS-14) and grating instruments (Beckman 4240 and Perkin–Elmer 180). They discussed optimum applications for each type of instrument. The use of an FTIR spectrometer is strongly favored by Griffiths *et al.* (1977) for most measurements in which the time to acquire data is limited by the type of experiment [e.g., on-line gas chromatography–infrared (GC–IR) or absorption spectrometry of transient species, or in situations where the time to acquire data on a grating spectrometer is very long, as with absorption spectrometry of samples with very high absorbance]. Yet, there is still much application to be made without an FT instrument, e.g., the use of a grating spectrometer with a minicomputer-based data system. The choice of the type of instrument is best decided by the users to satisfy their particular needs.

Compensation of Solvent

Griffiths (1975) and Koenig (1975) have discussed how to compensate for solvent (water) by the digital subtraction of solvent spectra from the spectra of the biological solute in FTIR spectroscopy. The strong water band at 1640 cm^{-1} which is usually found in ordinary IR aqueous spectra is readily subtracted and its interference removed.

FTIR spectra of such substances as ribonuclease, α-casein, bovine serum albumin, hemoglobin, and β-lactoglobulin have been recorded by Tabb (1974), who found that the amide I band has practically the same frequency in water and in the solid state for these proteins. Amide I of α-casein (much random coil) is at 1655 cm^{-1}; ribonuclease displays two amide I bands: in the solid state, 1653 and 1647 cm^{-1}; in solution, 1656 and 1646 cm^{-1}. The latter band was caused by irregularly formed β-sheet and random structure. Hemoglobin and serum albumin, having much helical content, show amide I bands at 1656 cm^{-1}. β-Lactoglobulin displays its amide I at 1632 cm^{-1}, which is evidence of antiparallel β-sheet structure (Krimm, 1962; Susi, 1969).

Thus, by means of FTIR spectroscopy one can easily subtract the spectrum of water from that of biological samples. This practice permits the study of those vibrational features that are obscured by the water absorption bands. Nevertheless, one should observe two main precautions when performing such operations. The water spectrum that is subtracted must be recorded at the same temperature as that of the biological sample, since the spectrum of water is temperature dependent. (See Fig. 5 in Cameron *et al.*, 1979, where it is shown that the effect of temperature is especially important in the study of weak C—^{2}H stretching modes between 2100 and 2200 cm^{-1} or bands near 1640 cm^{-1} where the water fundamental is frequently as wide as that of other bands being examined.) The second caveat is that the windows of the cells employed must be of the same substance, and their path lengths must also be the same. Otherwise, one may be recording artifacts due to anomalous dispersion (which varies with refractive index) and changes in absorption due to path length changes (Young and Jones, 1971).

Time-Lapse FTIR

A recent industrial application of FTIR spectroscopy might conceivably be used to study biochemical films, e.g., protein. Time-lapse IR spectroscopy has been used by Hartshorn (1979) to study the progress of film formation in a manner similar to time-lapse photography. This was an industrial application of the method to study a film sample of 60% soya oil pentaerythritol alkyd, to which was added 0.05% cobalt drier. The film was cast on a KBr plate. The FTIR method involves sequential subtraction of the initial spectrum from each later spectrum, or for example, subtraction of a 2-h spectrum from a 4-h spectrum. A word of caution is necessary, however. Hartshorn shows that when small differences between two very large numbers are to be measured, any small statistical errors in the photometric accuracy can and do yield relatively large, erratic errors. Care must be taken not to misinterpret false peaks produced by the subtraction.

FTIR Use with Purple Membrane and Photoreceptor Membrane

Recent applications of FTIR spectroscopy to purple membrane in films and solution (Rothschild and Clark, 1979a,b) and to bovine photoreceptor membrane (Rothschild *et al.*, 1980) have demonstrated that this technique is a sensitive tool for investigating membrane structure. FTIR studies yield information on the three-dimensional structure of membrane lipids and proteins in oriented membranes. For example, the average spatial orientation of α-helices of bacteriorhodopsin has been obtained by Rothschild and Clark (1979a) from measurement of the IR dichroism of oriented purple membranes (see p. 741).

Rothschild *et al.* (1982) have used FTIR difference spectroscopy to examine bacteriorhodopsin in both dark and illuminated purple membrane films. On the basis of IR evidence and isotopic labeling they concluded that the Schiff base is protonated in bR570 (see Chapter 4) in agreement with resonance Raman results.

FTIR Advantage for Kinetic Studies

Until recently, if the concentration of any species was to be observed by the use of IR absorption bands, the spectrometer was not scanned, but the intensity of one band was monitored continuously. Since a measurement time for each interferogram of less than one second is achievable, a whole spectrum can be stored in the computer memory, and one can measure several species simultaneously instead of measuring only one band of one component (Bell, 1972).

Gas Chromatography–FTIR

Griffiths (1978) has given detailed information about the use of FTIR spectroscopy for the on-line measurement of the IR spectra of compounds separated by gas chromatography, a process commonly called GC–IR. He has presented material on sampling, design of light pipe, single-beam and dual-beam GC–IR, software for GC–IR, and applications. Among the applications he has cited are the analyses of cigarette smoke and the trace organic compounds in water.

Krishnan *et al.* (1981) have discussed recent developments in GC–FTIR spectroscopy: software and hardware and examples to illustrate these developments. Krishnan and Ferraro (1981) have recently reviewed the GC–IR technique.

Peaks in capillary-column gas chromatograms have been identified at the nanogram level by dual-beam FTIR spectrometry (Kuehl *et al.*, 1980). These preliminary results indicated that on-line GC–FTIR measurements of peaks eluted from support-coated open tubular columns appeared to be quite feasible. "On-the-fly" detection limits of 1 ng of injected sample were anticipated.

Rossiter (1982) has described recent developments in FTIR–GC and GC–IR. This is an interesting short paper that discusses several facets of these techniques. Rossiter pointed out that the availability of modestly priced ratio-recording IR spectrophotometers and IR data stations has significantly changed the capabilities of dispersive IR spectroscopy. The new low-cost fast FTIR spectrophotometers make on-line real-time FTIR–GC more widely accessible. Techniques that he described allow the slower scanning dispersive instruments to perform effective GC–IR. He also mentioned a cell that is not designed to work in the trapping mode, but is optimized for on-line work with high-quality dispersive IR spectrophotometers. One complete GC–IR system that he discussed uses a Perkin–Elmer 681 ratio-recording IR instrument and a data station with an Accuspec model 2 GC–IR interface and a bypass valve connected to a compact GC.

Vibrational Circular Dichroic Spectra

An FTIR spectrometer was recently used to observe vibrational circular dichroic spectra of (+)-camphor and (−)-camphor (Nafie *et al.*, 1979). The method employed in this work was the first demonstration of high-frequency modulated, differential FTIR spectroscopy. The FTIR spectrometer system was said to have fundamental advantages over a dispersive grating instrument both in signal quality, owing to increased throughput—Jacquinot's advantage (Griffiths, 1975)—and in shorter measurement time, owing to spectral multiplexing—Fellgett's advantage (Griffiths, 1975). However, in this work the advantage of signal quality had not yet been achieved. The recording of vibrational circular dichroism (VCD) by FTIR methods offers a new horizon in VCD spectroscopy, which will undoubtedly be broadened as further developments are made.

Minicomputer Use

Foskett (1977) has discussed the rationale behind the choice of a minicomputer. He compared the needs of a dispersive IR spectrophotometer with an FTIR interferometer. His discussion complements that by Mattson and Smith (1977), where the same family of minicomputers is interfaced to a dispersive IR spectrophotometer. The latter authors have discussed the application of internal reflection spectroscopy with a germanium prism to study the amide I band of a protein (fibrinogen) in aqueous solution. Ninety-five percent of the band intensity comes from the 775-nm thin region at the germanium–solution interface. By using such a thin sampling region in combination with signal-to-noise enhancing techniques (spectrum averaging and mathematical smoothing) and by subtracting the solvent spectrum (phosphate buffer) from the spectrum of the protein solution, Mattson and Smith were able to separate the obscured amide I band from the overlying water band at 1639 cm^{-1}.

A microcomputer-controlled IR analyzer for multicomponent analysis has been described by Telfair *et al.* (1976). The applications given to demonstrate the instrument's capability were the analysis of a five-component liquid solvent mixture and the measurement of eight contaminants in aviator's breathing oxygen.

The entire computer system is on a single printed circuit board. The 8080 A Intel microprocessor is used for the computer central processor unit. Eight thousand eight-bit bytes of programmable read-only memory and 1024 bytes of random access memory comprise the memory of the computer.

Linear matrix algebra is used to handle multicomponent analyses. An inversion function is required in the analysis of the five-component mixture, and this step is done by use of an off-line macrocomputer.

Of greater interest to this author is the measurement of trace impurities in a pilot's oxygen supply. The contaminants examined in a preliminary study (Telfair *et al.*, 1976) were CH_4, C_2H_2, C_2H_4, C_2H_6, *n*-butane, CO_2, N_2O, and CO. Table 1.1 shows the analysis of synthetic gas mixtures in oxygen at 9 atm pressure and a path length of 20.25 m.

Computer Programs

Jones (1977) has discussed modular computer programs for infrared absorption spectrophotometry. All fifty programs were written in FORTRAN IV. They appear in Bulletin Nos. 11–17 of the National Research Council of Canada.

Table 1.1 Analysis of Synthetic Gas Mixtures in Oxygen at 9 atm and 20.25-m Path[a]

Printout	20 ppm CH_4	8 ppm C_2H_6	5 ppm C_2H_4	10 ppm C_4H_{10}	2 ppm C_2H_2	5 ppm CO_2	3 ppm N_2O	5 ppm CO	5 ppm C_2H_6 5 ppm C_4H_{10}
CH_4	19.7	0.1	0.3	0.8	0.7	0.6	0.4	1.3	0.1
C_2H_6	0.2	8.2	0.1	−0.4	0	−0.1	0	−0.1	5.3
C_2H_4	0.1	0.2	4.7	0.3	0.2	0.1	0	0.1	0
C_4H_{10}	0.3	−0.1	0.2	10.3	0	0.1	0	0	4.3
C_2H_2	0	0	0	0	1.6	0.1	0	0	0
CO_2	0.1	0.3	0	2.1	0.1	4.9	0.1	0.1	1.1
N_2O	0	0	0	−0.1	0	−0.1	3.2	−0.2	0
CO	0.1	0.1	0.2	0.4	0.2	0.2	−0.1	5.3	0

[a] Telfair *et al.* (1976).

Raman Spectroscopy

For short, concise reviews of Raman spectroscopy, see Schrader (1974), O'Bremski (1971), Tobin (1972), Delhaye and Merlin (1975), Thomas (1971), and Spiro (1978).

When an atom with a spherical, symmetric electron cloud is placed between the plates of a charged condenser, the electrons are pulled toward the positive plate and the protons toward the negative plate. The atom is said to be polarized now, and it has an induced dipole moment. Representing the vector of the electric force of the external field as **E,** and the induced dipole moment oriented parallel to the direction of **E** as μ, we may say, $\mu = \alpha\mathbf{E}$, where α is the polarizability of the atom. Using Cartesian coordinates to resolve the electric field, we may rewrite this equation as

$$\mu_x = \alpha\mathbf{E}_x$$

$$\mu_y = \alpha\mathbf{E}_y$$

$$\mu_z = \alpha\mathbf{E}_z$$

Since we rarely deal with a symmetric molecule, and generally are presented with non-symmetric ones, α may be different for the x, y, and z directions, i.e., the molecule is anisotropic. For the general situation we have the following equations:

$$\mu_x = \alpha_{xx}\mathbf{E}_x + \alpha_{xy}\mathbf{E}_y + \alpha_{xz}\mathbf{E}_z$$

$$\mu_y = \alpha_{yx}\mathbf{E}_x + \alpha_{yy}\mathbf{E}_y + \alpha_{yz}\mathbf{E}_z$$

$$\mu_z = \alpha_{zx}\mathbf{E}_x + \alpha_{zy}\mathbf{E}_y + \alpha_{zz}\mathbf{E}_z$$

where $\alpha_{xx},\alpha_{xy},\alpha_{xz}, \ldots ,\alpha_{zz}$ are proportionality constants between μ_x and $\mathbf{E}_x$,μ_x and $\mathbf{E}_y$, μ_x and $\mathbf{E}_z$ . . . μ_z and $\mathbf{E}_z$, etc. A system of such constants characterizes the total polarizability, and it describes a linear relationship between vectors. Such a system is called a tensor (Margenau and Murphy, 1943).

For a symmetric tensor (the polarizability tensor), this last set of equations becomes

$$\mu_{x'} = \alpha_{x'x'}\mathbf{E}_{x'}$$

$$\mu_{y'} = \alpha_{y'y'}\mathbf{E}_{y'}$$

$$\mu_{z'} = \alpha_{z'z'}\mathbf{E}_{z'}$$

where the x',y', and z' coordinates are assumed such that only $\alpha_{x'x'}$,$\alpha_{y'y'}$, and $\alpha_{z'z'}$ are different from zero [all terms containing $\alpha_{x'y'}$, $\alpha_{x',z',}$, and $\alpha_{y',z'}$ are equal to zero (Brand and Speakman, 1960)]. The axes $\mu_{x'}$, $\mu_{y'}$, $\mu_{z'}$ are three mutually perpendicular directions in the molecule, whose induced moments are parallel to the electric field.

The polarizability ellipsoid is the surface generated by a plot of $1/\sqrt{\alpha}$ in any direction from the origin. Its axes are x',y', and z'. For a totally anisotropic molecule, $\alpha_{x'x'}$,$\alpha_{y'y'}$, and $\alpha_{z'z'}$ are not equal to each other, so there are three unequally long axes for the ellipsoid. When two of the three axes are equal, the ellipsoid is a rotational ellipsoid. Its polarizability is identical for the x' and y' directions. An isotropic molecule has three equal axes, i.e., equal polarizability in all three directions.

When light of frequency ν_0 impinges on a molecule, the electronic system of the molecule develops an induced frequency. An induced dipole moment vibrates at frequency ν_0, and its amplitude is proportional to the polarizability of the molecule. As a result, the molecule emits Rayleigh radiation, the frequency of which is ν_0.

The polarizability of the molecule (see Colthup *et al.*, 1964; Szymanski, 1967) depends on its size, shape, and orientation; and it can be viewed as a polarizability ellipsoid. Figure 1.3 shows vibrational and polarizability ellipsoids of SO_2, a simple molecule.

The polarizability ellipsoid may be modified as a result of change in the shape of the molecule due to atomic nuclear vibrations. Consequently, modulation of the radiation emitted by the molecule will occur. As a result of these processes, radiation from the molecule contains not only ν_0, the exciting frequency, but also the sum and difference of the exciting and the vibrational frequency ν_s, i.e., $\nu_R^+ = \nu_0 + \nu_s$ and $\nu_R^- = \nu_0 - \nu_s$. Figure 1.4 shows graphically the exciting radiation ν_0, the molecule with vibration frequency ν_s, and the effect on ν_0 by ν_s. The frequency sum, $\nu_0 + \nu_s$, and the frequency difference, $\nu_0 - \nu_s$, are also shown. The frequencies (really frequency "shifts") ν_R^+ and ν_R^- are called anti-Stokes and Stokes Raman lines (or bands), respectively. The Stokes Raman radiation is of higher frequency than the anti-Stokes. In the usual trace of a Raman spectrum, only the more intense Stokes spectrum is recorded.

Polarization measurements by means of Raman spectroscopy are quite useful for structure determination, as in infrared work. In the case of liquids, gases, and single crystals, two spectra are recorded in order to obtain the depolarization ratio, ρ : one with the polarizer oriented parallel to the incident beam and the other oriented perpendicular to the incident beam (Fig. 1.5). ρ is expressed mathematically by

$$\rho = I_{\perp}/I_{\parallel}$$

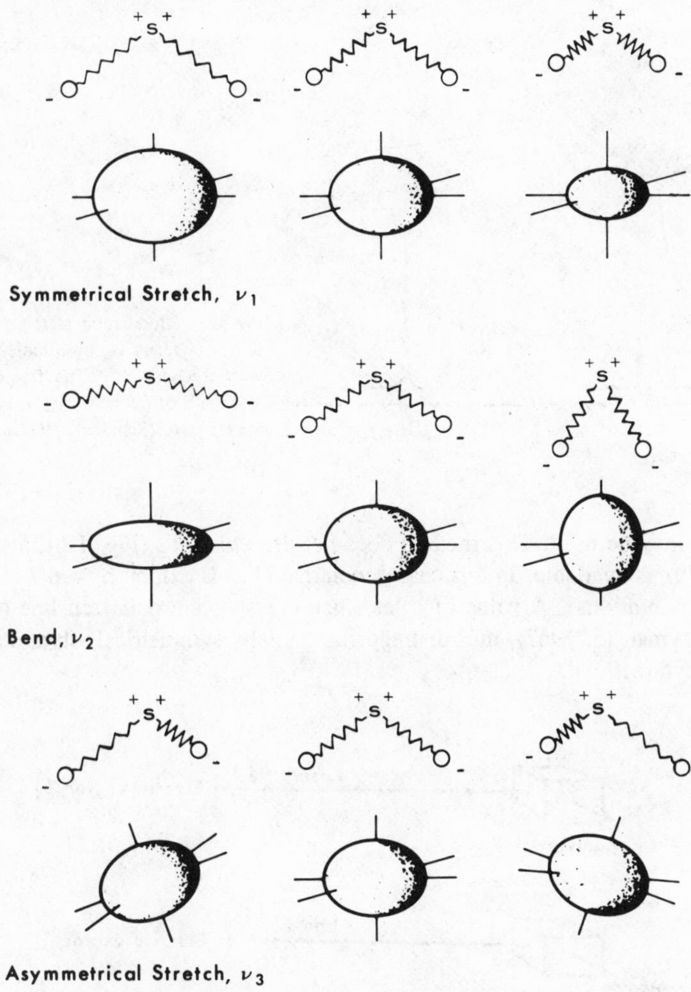

Fig. 1.3. Vibrational modes and polarizability ellipsoids of SO_2. (O'Bremski, 1971.)

where $I_\perp$ and $I_\parallel$ are the scattered light intensities observed perpendicular and parallel, respectively, to the incident beam. For totally symmetric vibration modes $\rho = 3/4$, depolarized (abbreviated as *dp*) in nonresonance spectra. In the case where vibration modes are not totally symmetric, $\rho < 3/4$, polarized (abbreviated as *p*). Three more types of Raman line (or band) can appear in resonance Raman spectra: (1) for vibration modes that are not totally symmetric, $\rho \geqq 3/4$; (2) ρ is greater than 3/4 and less than infinity—denoted as anomalous polarization (abbreviated as *ap*); (3) ρ approaches infinity, inversely polarized (abbreviated as *ip*) for antisymmetric modes, such as that found for heme proteins by Spiro and Strekas (1972).

If the recorded Raman spectra are weak, another method of measuring the depolar-

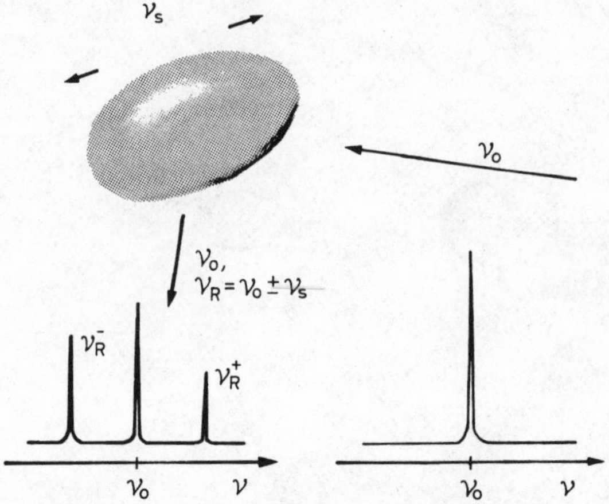

Fig. 1.4. Modulation of the exciting radiation, frequency ν_0, by a molecule vibrating with frequency ν_s. The frequency ν_0 and sidebands of frequency $\nu_0 + \nu_s$ and $\nu_0 - \nu_s$ are emitted. (Schrader, 1974.)

ization ratio may be used. This method does not use Polaroids (Fig. 1.6), and therefore more light flux is available. In this case depolarized bands exhibit $\rho = 6/7$ (non-totally-symmetrical vibrations). A value of ρ less than 6/7 shows a polarized line (Colthup *et al.*, 1964; Szymanski, 1967), thus distinguishing totally symmetrical vibrations from the non-totally symmetrical.

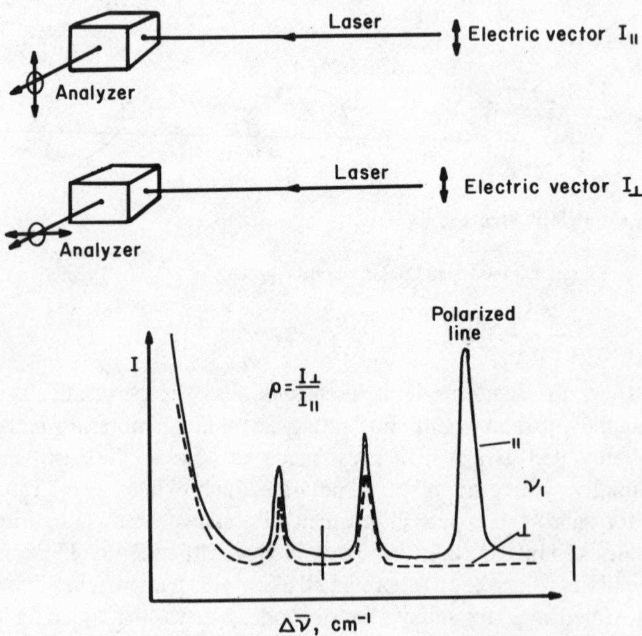

Fig. 1.5. Measurement of depolarization ratios. (Hendra, 1974.)

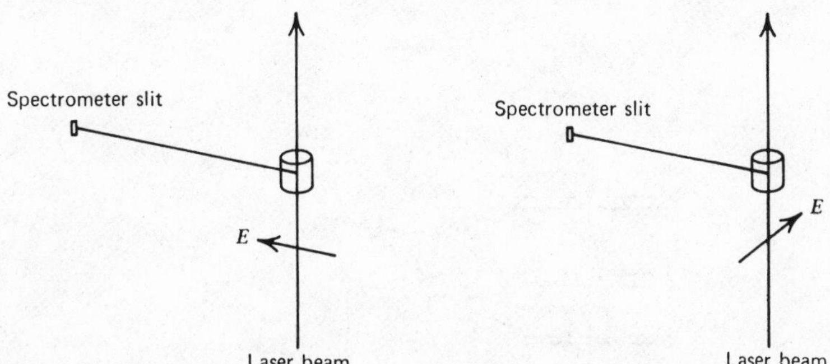

Fig. 1.6. One way of measuring depolarization ratios of Raman bands. (Tobin, 1971.)

Raman Instrumentation

A good example of an illumination system devised to optimize the intensity of the Raman spectrum while eliminating spurious light is given here. The basic principles involved have been discussed (*SPEX Speaker,* 1966 and 1967).

The first component (Fig. 1.7) is the dielectric mirror assembly with which any laser beam of visible light can be deflected through 90°. The next component is the removable polarization rotator, an 8-mm-diameter polished crystal quartz disk cut along its *c* axis to the length needed to rotate the beam polarization exactly 90°; it is obligatory for the study of oriented single crystals. The beam expander, sustaining collimation, increases the beam area 16-fold, fills the microscope objective lens, optimizing its ability to focus. It includes a rotatable polarizer to attenuate the beam again without ruining the attenuating element.

The beam expander also illuminates the Claassen filter, a high-dispersion prism monochromator whose function is to separate out a given laser frequency to produce "pure" excitation not contaminated by plasma and adjacent lasing frequencies, e.g., the weak frequency at 488.9 nm, which is very near to the much more intense, much used 488.0-nm frequency of an Ar laser. The Claassen filter is tunable and has a very narrow bandpass compared to interference filters; thus, it replaces them.

The other components in the sequence of the optical system have the following functions: (a) guide the light upward to the sample and (b) focus the beam on the sample (microscope objective).

A silvered scatter collection mirror behind the sample and another silvered multipass excitation mirror above the sample cause at least a twofold increase in the Raman scatter. Next, the Raman beam proceeds to the *f*/1 collecting optics and then to the polarizing analyzer, which can rotate through 90°, and swing out of the way when not required. On the same mount is a crystal quartz compensated scrambler (which is essential for accurate measurements of depolarization ratios), and provision is made for an optional long-pass filter (different for each excitation frequency) which is used for highly scattering substances.

The RAMALOG 4 (Spex Industries, Metuchen, New Jersey) features a double monochromator. A third monochromator is available, which is a grating–mirror interchange

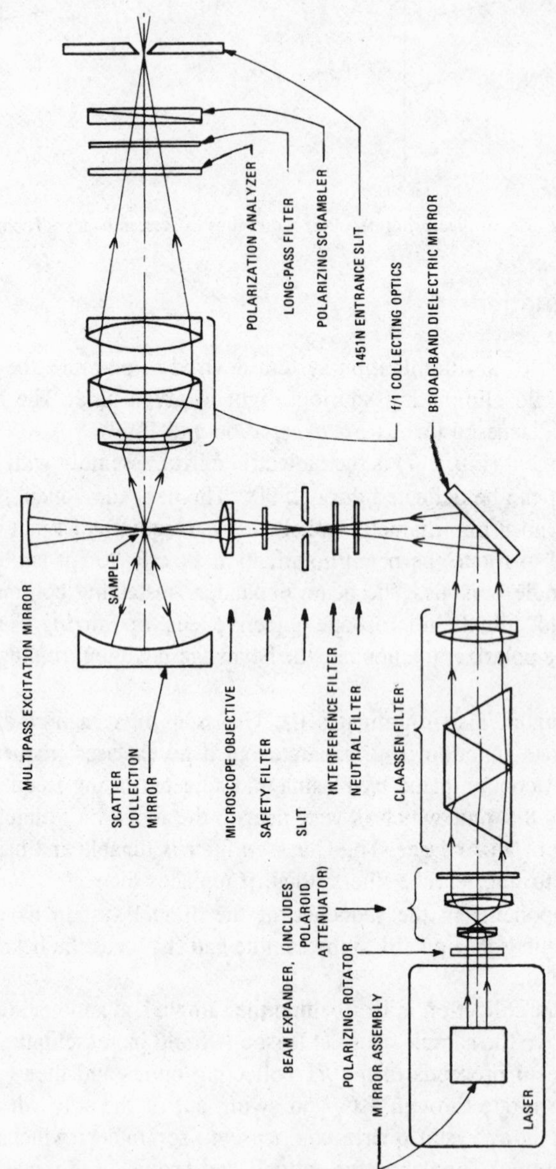

Fig. 1.7. Spex Raman illuminating system. (Courtesy of Spex Industries.)

device that affords triple monochromator performance in obtaining otherwise unachievable spectral frequencies that are very close to the excitation frequency. The third monochromator also decreases scattered light very efficiently.

The detection system uses a GaAs photomultiplier (to maximize signal-to-noise ratio throughout the visible region), thermoelectrically cooled to −30°C. The recorder has 12 chart drive speeds. The RAMACOMP is a computerized Raman system consisting basically of the RAMALOG 4, a dedicated computer, interactive teletype, and pertinent software. The digitized signal from the photon counter is processed by a scaling register.

Biomembranes are complex and have, therefore, been studied by a recording technique that involves computer assistance. For example, in one laboratory (Wallach *et al.*, 1979), the following system and conditions have been used with a Spex Industries' RAMALOG 4 Raman spectrometer. The spectrometer is interfaced with a computer. The excitation source is an argon ion laser, usually tuned at 488 nm. For concentrated liposomes the slit setting is 50 μm and the excitation has 800 mW of power; for membranes the slit setting is 200 μm, with 300 mW of power; but for resonance Raman work the power is 50 mW.

A thermoelectrically cooled photomultiplier detects the Raman scattering perpendicular to the laser beam. The scatter is measured as photons/sec and has a magnitude of 10^3 to 10^5 per second. Scanning is performed through the computer (Interdata Model 70). Wallach *et al.* (1979) give the following typical specifications for scanning: (a) maximum time and minimum time for each recorded point, 1.5 and 0.5 sec, respectively (longer with smaller slits), except for resonance Raman work where 0.5 and 0.1 sec, respectively, are used; (b) photon counts are 10^4–10^5 maximum and 100 minimum. No counts are recorded between data points. During scanning, the computer memory stores photon counts (1–4 scans; about 300). The stored spectra are averaged and smoothed by a computer-programed least-squares procedure, and are finally plotted on the RAMALOG recorder with proper suppressions of background and use of appropriate scale expansions. To measure accurately the locations of some bands, first- and second-derivative spectra are computed and recorded. When necessary, integrated intensities can be computed.

Luminescence of some biomembranes, which is a problem for various reasons, may be eliminated by using an argon-ion laser to pump a tunable dye laser and using the latter as the exciting source (Wallach *et al.*, 1979). This approach avoids the commonly used "burning off" procedure to get rid of luminescent properties, a process which is not allowable for undenatured biological specimens.

Laser-Raman Microprobes

Abraham and Etz (1979) have used a laser-Raman microprobe to identify microscopic inclusions of a silicone polymer in standard paraffin sections of lymph node. They believe that their microprobe system (Fig. 1.8) has much potential for many applications in toxicology, pathology, forensics, and environmental studies. Figure 1.9 shows Raman spectra of the cytoplasm of a giant cell of lymph node, a foreign body located within such a cell, and a small particle of silicone elastomer from a joint prosthesis. The authors noted that micro-Raman spectra sometimes yield more detailed information than the

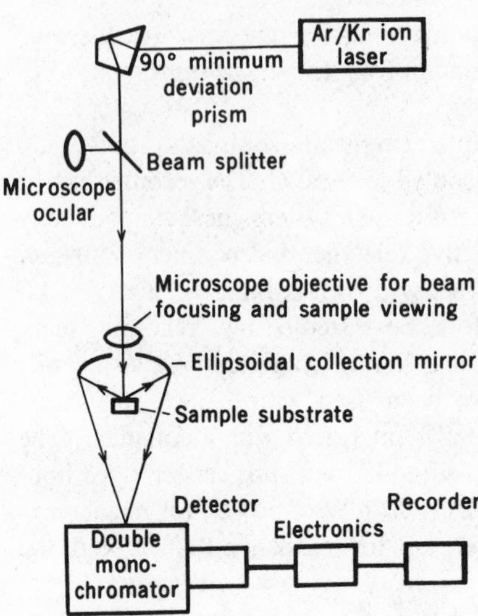

Fig. 1.8. Schematic diagram of the laser-Raman microprobe developed at the National Bureau of Standards. Any one of several laser wavelengths in the visible region of the spectrum is used to excite the micro-Raman spectrum. Nonlasing plasma lines are removed by use of a predispersing prism. The radiation scattered by the sample is collected over a large solid angle in 180° backscattering geometry. Lateral spatial resolution of the probe measurement is determined by the spot size of the laser on the sample and a spatial filter (exit pinhole, not shown) placed in the path of the collected scattered light. Depth resolution is several micrometers (but less than ~12 μm), depending on the optical transparency and surface topography of the sample. Typical measurement parameters employed in the microanalysis of thin sections of biological soft tissue are: laser wavelength, 514.5 nm (green) and 647.1 nm (red); laser power, 5 to 60 mW (at sample); laser spot diameter, 6 to 20 μm; time constant, 1 to 5 sec; scan rate, 50 to 10 cm^{-1} per minute; and spectral slit width, 3 cm^{-1}. (Abraham and Etz, 1979.)

corresponding bulk Raman spectra recorded from crystals or powders of the same materials.

Another technique has been developed that combines a microscope and a Raman spectrometer. The instrument used is called a laser-Raman molecular microprobe. Some of its applications have been described by Dhamelincourt *et al.* (1979). In the microprobe, photons from a laser excite the sample and cause the production of Raman bands of different components. The Raman bands are used to detect, identify, and then locate each component by making a micrographic image that gives the "map" of its distribution in the sample. The microprobe gives an image under many conditions such as in the presence of water, air, and at high temperature.

Figure 1.10 shows spherocrystals (1–12 μm) (Ballan-DuFrancais, 1975) of fat body of *Blatella germanica L. (Insecta Dictyoptera)*. Purine catabolism is important in such an organism. Figure 1.11 shows the spectra obtained from reference samples of potassium urate and uric acid, and from fat body spherocrystals. The comparison shows that fat body is composed mainly of uric acid and a small amount of urate.

Resonance Raman Spectroscopy

Spiro (1974) has written an introductory paper on resonance Raman spctroscopy, which includes several applications in biochemistry. He and Gaber (1977) have presented an excellent review of applications of resonance Raman and Raman spectroscopy to the study of proteins. Gaber (1977) has discussed biological applications of Raman spectroscopy, including resonance Raman. Lewis and Spoonhower (1974) have described

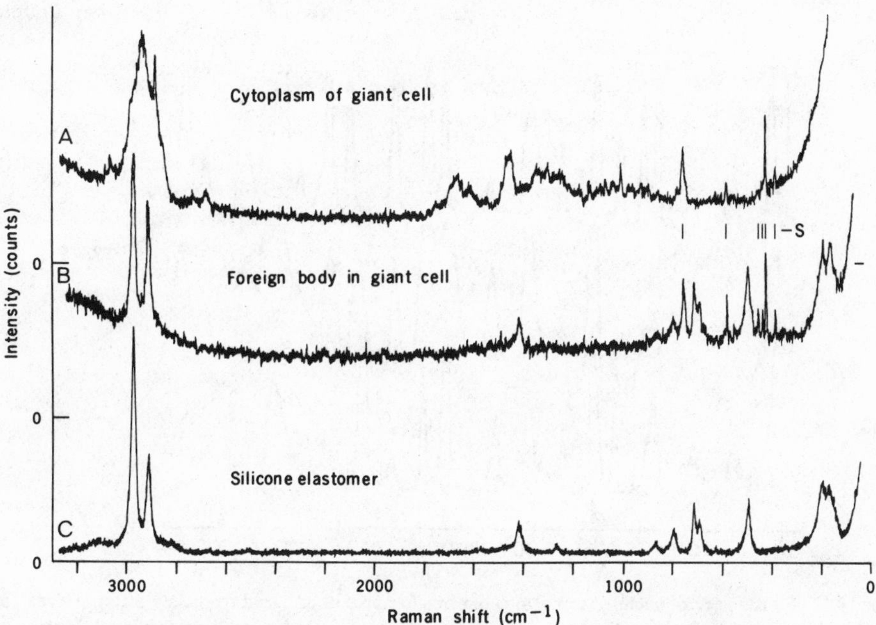

Fig. 1.9. Spectra recorded in the Raman probe microanalysis of a deparaffinized standard 5-μm section of lymph node, mounted on a sapphire (α-Al_2O_3) substrate. Measurement parameters common to each spectrum are: excitation, 514.5 nm; laser spot diameter, 16 μm; exit pinhole spatial filter, 140 μm in diameter; and spectral slit width, 3 cm^{-1}. (A) Spectrum of the cytoplasm away from foreign bodies. Laser power was 40 mW (at sample); time constant, 5.0 sec; and scan rate, 20 cm^{-1} per minute. (B) Spectrum of a foreign body (size ~24 μm) located within a giant cell. Measurement conditions were the same as for (A). (C) Spectrum of a small (~60 μm) particle of silicone elastomer from a joint prosthesis. Laser power was 60 mW (at sample); time constant, 0.5 sec; and scan rate, 100 cm^{-1} per minute. The bands marked *S* on spectra (A) and (B) are contributed by the sapphire substrate. The vertical scale (scattered light intensity) of each spectrum extends from 0 to 1000 counts, with the strongest peaks at ~3000 cm^{-1} having nearly full-scale intensity. Zero intensity is indicated for each spectrum, allowing differences in background signal levels to be noted. (Abraham and Etz, 1979.)

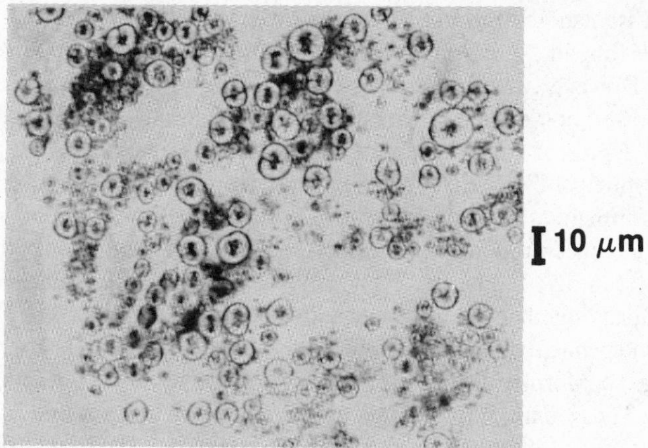

Fig. 1.10. Micrographic image (× 300) of part of a histological section of *Blatella germanica L.* (Dhamelincourt *et al.*, 1979.) Reprinted with the permission of the American Chemical Society. Copyright 1979.

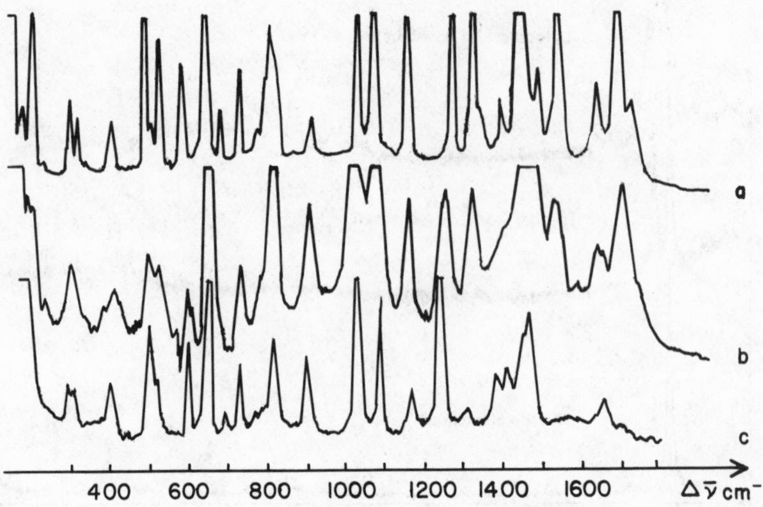

Fig. 1.11. Raman spectra obtained from (a) commercial sample of uric acid (particle ~5 μm in size), (b) fat body spherocrystal (~5 μm in diameter), (c) commercial sample of potassium urate (particle ~5 μm in size). (Dhamelincourt *et al.*, 1979.) Reprinted with the permission of the American Chemical Society. Copyright 1979.

biological applications of tunable laser resonance Raman spectroscopy, particularly with respect to heme proteins, nonheme materials (e.g., vitamin B_{12}, antibody active sites, binding sites of proteins), and the visual process. Lewis (1976a,b) has given interesting discussions of resonance Raman spectroscopy and its application to studies of rhodopsin, the principal light absorber in the eye. Warshel (1977) has presented resonance Raman theory and discussions of visual pigments, metalloporphyrins, and enzymic reactions. Parker (1975) has presented a comprehensive review of biochemical applications of Raman, resonance Raman, and infrared spectroscopy. Van Wart and Scheraga (1978) have discussed Raman and resonance Raman spectroscopy.

Resonance Raman spectroscopy and normal Raman spectroscopy use the same instrumentation. However, the former uses certain precautions due to accompanying side effects of the absorption of light. The characteristic property of resonance Raman spectroscopy is the degree of enhancement of chromophoric vibrations to the point that they are the main features of the spectrum, and parts of the molecule that are not chromophoric do not give information. With the resonance Raman effect one can obtain spectra from chromophores at concentrations of 10^{-4} *M* or less. The chromophore itself may be a site of biochemical activity; therefore, knowledge of its resonance Raman spectrum may indicate accompanying changes in structure.

The region spanned is approximately from 200 to 2000 cm^{-1}, and in this range one can follow, e.g., conformational changes, charge effects, bond distortions, and chemical rearrangement. The resonance Raman labeling technique (to be discussed later) and normal Raman methods give essentially complementary information. The former gives data about localized sites, while normal Raman scattering can indicate facts about conformation, e.g., backbone structure in nucleic acids, or secondary structure of proteins, such as α-helix or random coil. Resonance Raman studies require solutions of much lower concentrations of solute than normal Raman work. Solutions for the latter technique must be one or two orders of magnitude more concentrated, on a molar basis.

Qualitatively, the origin of the resonance Raman phenomenon has been described (Behringer, 1974; Tang and Albrecht, 1970; Carey and Schneider, 1978; Van Wart and Scheraga, 1978).

The normal Raman effect is Raman scattering induced by excitation far removed from any electronic transitions. If, however, the laser frequency ν_{laser} is allowed to be near or to coincide with an electronic absorption ν_{abs}, quite a marked enhancement of intensity may result such that

$$\text{intensity} \propto (\nu_{\text{laser}} - \nu_{\text{vib}})^4 \frac{(\nu_{\text{abs}}^2 + \nu_{\text{laser}}^2)^2}{(\nu_{\text{abs}}^2 - \nu_{\text{laser}}^2)^2}$$

This is the resonance Raman effect (Gaber, 1977), and the advantage for applications to biological systems is in its great sensitivity and selectivity as a tool for investigating chromophore structure, because only vibrational modes directly associated with the chromophore have their intensities enhanced (Hirakawa and Tsuboi. 1975). Normal modes with a large shift in equilibrium geometry upon electronic excitation produce intense resonance Raman features. Intensity enhancement can originate from two mechanisms: one, dependent on coupling to a single excited electronic state; and the other involving more than one excited state.

Kiefer (1974) has summarized techniques used for resonance Raman spectroscopy.

Tunable Laser Raman Spectroscopy

Tunable laser resonance Raman spectroscopy employs a high-powered laser that is tunable in intervals of less than 1/100 Å (depending on conditions) from about 280 nm to about 1,500 nm with the proper choices of dyes, temperature-tuned frequency doublers, and the application of a dye laser to pump a dye laser. For a resonance Raman experiment, one tunes the laser into the electronic transition of some molecular entity (e.g., the porphyrin ring vibrations of heme proteins), and one observes in the scattered radiation collected off the sample the enhancement of the sensitive vibrational spectrum above the background of vibrations in the nearby environment that are produced by the protein.

CARS (Coherent Anti-Stokes Raman Scattering) Spectra

For a discussion of the CARS technique and applications to biophysical studies, see B.S. Hudson [*Ann. Rev. Biophys. Bioeng.* **6,** 135 (1977)]. Another excellent source of information is A.B. Harvey, “Coherent Anti-Stokes Raman Spectroscopy: CARS” [*Analyt. Chem.* **50**(9), 905A (1978)].

Coherent anti-Stokes emission is the result of the interaction, through the third-order molecular polarizability, of two laser photons of frequency ω_1 and one laser photon of frequency ω_2, yielding a fourth photon, ω_3, of frequency $2\omega_1 - \omega_2$. There is a Raman emission whenever

$$\omega_1 - \omega_2 = \Delta\omega$$

where $\Delta\omega$ is a Raman active vibrational frequency. Begley *et al.* (1974) have given the scheme of the energy level describing this process. Also, conservation of momentum requires the wave vector equality

$$2k_1 - k_2 = k_3$$

where $k_i = \omega_i n/c$, n is the index of refraction of the interaction medium, and c is the velocity of light. This requirement is attained if the two incident beams intersect at a small angle in the solution. The anti-Stokes beam that is produced, ω_3, is located differently in space and can readily be separated from the other two. Enhancement of the CARS conversion efficiency in dilute absorbing solutions is thought to occur when the ω_1 and/or ω_3 frequencies are in resonance with electronic transitions of the molecule.

Nestor *et al.* (1976) have described experimental conditions for obtaining good spectra on aqueous solutions. They found, for example, that for cytochrome *c* in a cell with a path length of 1 mm the optimum concentration is about 1 mM. Their equipment has been described previously (Chabay *et al.*, 1976). An advantage of the CARS spectrum is that fluorescence is eliminated.

Among the lasers presently obtainable, the pulsed nitrogen-tunable dye system has the advantage of the broadest tuning range, which permits the systematic investigation of resonance effects.

Infrared and Raman Group Frequencies

Tables 1.2 and 1.3 present lists of infrared and Raman group frequencies, respectively, that investigators of organic compounds are likely to encounter. It should be noted that although frequencies usually are found within the limits indicated, they may, in special cases, lie outside these ranges. Therefore, when the correlations are used, all other available evidence should also be considered.

The Complementary Use of IR and Raman Spectroscopy

Some examples are given below (Washburn, 1978) of how infrared and Raman spectroscopy complement each other in the elucidation of structures. In general (of course, exceptions exist), the more unsymmetrically substituted a particular bond (or the greater the change of dipole moment for a given vibration), the stronger the IR band and the

Table 1.2 Characteristic Infrared Bands of Various Groups[a]

Range (cm^{-1})	Intensity[b]	Group	Type of vibration[c]	Comments
4505–4200	w	C—H	str	Aliphatic (combination)
4255–4000	w	C—H	str	Aromatic (combination)
3650–3500	var	O—H	str	Free OH, oxime
3640–3623	m (sh)	O—H	str	Free OH, alcohols
3600–3100	m	O—H	str	Water of crystallization
3590–3425	var (sh)	O—H	str	Intramolec. bonded OH
3550–3500	m	O—H	str	Free OH, carboxylic acid (very dilute solution)
3550–3450	var (sh)	O—H	str	Intermolec. bonded OH (dimeric)
3550–3195	w	C═O	str	Carbonyl (first overtone)
~3520	s	N—H	str	Primary amide (free)
~3500	m	N—H	str (asym)	Primary amine, free NH (dilute solution)
3500–3300	m	N—H	str	Secondary amine,, free NH
3500–3060	m	N—H	str	Associated NH, amine or amide
~3400	s	N—H	str	Primary amide (free)
~3400	m	N—H	str (sym)	Primary amine, free NH (dilute solution)
3400–3225	s (br)	O—H	str	Intermolec. bonded OH (polymeric)
~3380	m	NH_3^+	str	Amine salt (soln.)
3355–3145	m	NH_3^+	str	Amine salt (solid); several bands
~3350	m	N—H	str	Primary amide (bonded)
~3300	s	C—H	str	≡C—H, acetylenes
3300–2500	w (vbr)	O—H	str	H-bonded carboxylic acid dimers
~3280	m	NH_3^+	str	Amine salt (soln.)
~3175	m	N—H	str	Primary amide (bonded)
3155–3050	w	C—H	str	—CH═C—O— and —C═CH—O—
3095–3075	m	C—H	str	RCH═CH_2, olefin
3075–3030	w–m	C—H	str	C—H of aromatic ring
3050–2995	w	C—H	str	Of epoxide (shifts to 3040–3030 if ring strain increases)
3040–3010	s, m	C—H	str	>C—H; RCH═CH_2, RCR═CHR′ (*cis* or *trans,*) RCR′═CHR″, olefin
~2960	s	C—H	str (asym)	*C*-Methyl
~2925	s	C—H	str (asym)	>CH_2, methylene, Ar—CH_3
2900–2880	w	C—H	str	C—H, methine
2900–2705 (two)	w	C—H	str	—C(═O)H, aldehyde
2900–2300 (several)	w	N—H	str	Quaternary amine salt, bonded
~2875	s	C—H	str (sym)	*C*-Methyl
~2850	s	C—H	str (sym)	>CH_2, methylene

Continued

Table 1.2 *(Continued)*

Range (cm^{-1})	Intensity[b]	Group	Type of vibration[c]	Comments
2835–2815		C—H	str	*O*-methyl
~2825	m	C—H	str	$-CH(OCH_2-)_2$, alkyl acetal
~2780		C—H	str	$-O-CH_2-O-$
2705–2560	w (br)	P—OH	str	Phosphoric ester, H-bonded
2705–2300	s	NH_2^+, NH^+	str	(May be several bands)
~2580	w	S—H	str	Thiol, free
~2400	w	S—H	str	Thiol, H-bonded
~2270	vs	N=C=O	asym str	Isocyanate
2260–2240	w	—C≡N	str	Saturated nitrile
2260–2190	var	C≡C		RC≡CR′; acetylenes
2230–2215	s	—C≡N	str	Unsaturated conjugated nitrile
2200–2050	vs	C=S	asym str	—N—C=S, isothiocyanate (2 or more bands)
2200–2000	s			Cyanide, thiocyanate, cyanate
2180–2120		C≡N	str	$R-N^+\equiv C^-$
2160–2120	s	N≡N	str	Azide
2140–2100	w	C≡C		RC≡C—H; acetylenes
~1810	s	C=O	str	—COCl, aliphatic acid chloride
1780–1740	s	C=O	str	—O—(C=O)—O—, carbonate
~1770	s	C=O	str	γ-Lactone
1745–1735	s	C=O	str	Saturated esters
~1740	s	C=O	str	δ-Lactone
1740–1720	s	C=O	str	—C(=O)H, aldehyde
~1725	s	C=O	str	Formic ester
1725–1705	s	C=O	str	Ketone
~1720	s	C=O	str	Benzoic ester
1720–1700	s	C=O	str	—COOH; aliphatic carboxylic acid (dimer)
1700–1670	s	C=O	str	—CONHR, secondary amide, free (dil. soln.): Amide I
1690–1670	s	C=O	str	$-CONH_2$, primary amide, free (dil. soln.): Amide I
1680–1630	s	C=O	str	Secondary amide (solid)
1680–1620	var	C=C	str	Nonconjugated C=C
1678–1668		C=C		*trans* Olefin; RHC=CHR′
~1675	s	C=S	str	Thioester
~1675	s	C=O	str	Thioester
~1670	w	C=N	str	Aliphatic oxime
1670–1620	s	C=O	str	Primary amide (solid), H-bonded, 2 bands: Amide I
1662–1652		C=C		*cis* Olefin; RHC=CHR′
1658–1648	m	C=C		Terminal olefin; $RR'C=CH_2$
1650–1620	s	N—H	def	Primary amide (solid): Amide II band
1650–1600	s	NO_2	asym str	$-O-NO_2$, nitrate

Continued

Table 1.2 *(Continued)*

Range (cm^{-1})	Intensity[b]	Group	Type of vibration	Comments
1650–1580	m–s	N—H	def	NH_2; primary amine
1650–1550	w	N—H	def	NHR; secondary amine
1648–1638	var	C═C	str	Terminal olefin; $RHC{=}CH_2$
~1625	s	C═C	str	Ph-Conjugated C═C
1625–1585	m	C═C	sk, i-p	Aromatic C═C
1620–1590	s	N—H	def	Primary amide (dil. soln.)
1620–1560	m–s	NH_2^+	def	
1610–1540	vs	C—O	asym str	$—COO^-$, carboxylate
~1600	s	C═C	str	CO or C═C conjugated with C═C
~1585	m	NH_3^+	asym def	Amine salt
1580–1520	m	C═N (plus C═C)	int eff	Pyrimidines
1570–1515	s	N—H	def	Secondary amide (solid): Amide II band
1550–1510	s	N—H	def	Secondary amide (dilute solution)
~1500	var	C═C	sk, i-p	Aromatic C═C
1500–1470	s	C═S	str	—N—C═S
1500–1300	m	NH_3^+	sym def	Amine salt
~1468	s	C—H	sc	Alkane $—CH_2—$
~1460	m	C—H	bend (asym)	$—CH_3$
1460–1400	s	C—O	sym str	$—COO^-$, carboxylate
~1455	s	C—H	sc	Alicyclic $—CH_2—$
1450–1400	w	—N═N—	str	azo
1440–1395	w	C—O	str (plus OH def)	Carboxylic acid
1440–1350	s	S═O	str	$(RO)_2SO_2$, sulfuric ester
1440–1325	m	C—C		Aliphatic aldehyde
1420–1406	w	C—H	i-p bend	$C{=}CH_2$
1420–1330	s	S═O	str	$ROSO_2R'$, sulfonic ester
1418–1400	m	C—N	str	Primary amide
~1410	w	C—N	str	Aliphatic amine
1390–1360 doublet	m	C—H	bend (sym)	*gem*-Dimethyl
1385–1375	m	C—H	bend (sym)	$—CH_3$
1370–1250		C—O	str	Lactone
~1340	w	C—H	bend	Alkane C—H
1340–1280	s	S═O	sym str	R_2SO_2, sulfone
1340–1180	w	N≡N	str	Azide
1320–1210	s	C—O	str	Carboxylic acid
1310–1250	s	C—O	str	Benzoic ester, phthalic ester
1305–1200	m	N—H	def	Secondary amide, Amide III band
1300–1250	s	NO_2	sym str	$—O—NO_2$, nitrate
1300–1200	s	P═O	str	Phosphoric ester, free P═O
1270–1150	s	C—O	str	—(O═)C—O—R in esters
1256–1232	s	C—O	str	CH_3COOR, acetic ester
~1250		C—O	str	Methylene acetal
~1250		C—O	str	Epoxide
~1250	vs	$Si—CH_3$	sym CH_3 def	$Si(CH_3)_3$, trimethylsilyl

Continued

Table 1.2 *(Continued)*

Range (cm^{-1})	Intensity[b]	Group	Type of vibration	Comments
1250–1150	vs	P═O	str	Phosphoric ester, H-bonded P═O
1235–1212	s	C═S	str	$(RO)_2C$═S, thioketone
1230–1150	s	S═O	str	$(RO)_2SO_2$, sulfuric ester
1225–1175 1125–1090 1070–1000 (two)	w	C—H	i-p bend	*p*-Substituted phenyl
1220–1020	m	C—N	str	Aliphatic amine
1200–1145	s	S═O	str	$ROSO_2R'$, sulfonic ester
1200–1040		C—O	str	C—O—C—O—C, cyclic acetal (4–5 bands)
1200–1170	s	C—O	str	Propionic and higher esters
1200–1000	s	C—OH	str	Alcohols
1185–1175	s	C—O	str	Formic ester
1175–1165 1170–1140	s	C—H	sk	$(CH_3)_2C<$, isopropyl
1175–1125 1110–1070 1070–1000	w	C—H	i-p bend	Unsubstituted phenyl
1150–1100	s	S═O	asym str	R_2SO_2, sulfone
1150–1100	s	C—O	str	Benzoic ester, phthalic ester
1150–1070	s	C—O—C	asym str	Aliphatic ether
~1120	s	C═S	str	—NH—(C═S)—, thioamide
1110–1000	s	C—F	str	Monofluoro derivatives
1090–1030	vs	P—O—C		Phosphoric ester
1090–1020	vs	Si—O	str	Si—O—C, trimethylsilyl
1058–1053	s	C═S	str	$(RS)_2C$═S, trithiocarbonate
1050–1020	s	S═O	str	>S═O, sulfoxide
~1040		C—O	str	Methylene acetal
1005–990 915–910	vs	C—H	bend	C═C—H, vinyl
995–985 910–905		C—H	o-o-p bend	RCH═CH_2
980–965		C—H	o-o-p bend	*trans* RHC═CHR′
970–940	br	P—O—P		Pyrophosphate
965–960 945–940	s	C—H	bend	Vinyl ether
960–930		N—O	str	Oxime
950–810		C—O	str	Epoxide
~948	s	C—H	bend	Vinyl ester
~925		C—O	str	Methylene acetal
895–885		C—H	o-o-p bend	RR′C═CH_2
~840	vs	Si—CH_3	str	$Si(CH_3)_3$, trimethylsilyl
840–790		C—H	o-o-p bend	RR′C═CHR″
840–790	m	C—H	sk	$(CH_3)_2C<$, isopropyl
840–750		C—O	str	Epoxide
833–810	vs	C—H	o-o-p bend	*p*-substitued phenyl
~800	w	NH_3^+	rock	Amine salt

Continued

Table 1.2 *(Continued)*

Range (cm^{-1})	Intensity[b]	Group	Type of vibration	Comments
~800	w	NH_2^+	rock	
770–730 710–690	s	C—H	o-o-p bend	Unsubstituted phenyl
~755	vs	Si—CH_3	str	$Si(CH_3)_3$, trimethylsilyl
750–700	s	C—Cl	str	Monochloro derivatives
~720	m(br)	N—H	def	Secondary amide, bonded: Amide V band
705–570	w	C—S	str	Thiol, sulfide
~690		C—H	o-o-p bend	*cis* RHC═CHR′
~650	s	C—Br	str	Bromo derivatives
600–480	s	C—I	str	Iodo derivatives
550–450	vw	S—S	str	Disulfide

[a] Parker (1971).
[b] br, broad; m, medium; s, strong; sh, sharp, v, very; var, variable; w, weak.
[c] asym, asymmetrical; def, deformation; i-p, in-plane; int eff, interaction effects; o-o-p, out-of-plane; sc, scissoring; sk, skeletal; str, stretching; sym, symmetrical.

Table 1.3. Characteristic Raman Frequencies[a]

Frequency (cm^{-1})	Vibration	Compound
3400–3330	Bonded antisymmetric NH_2 stretch	Primary amines
3380–3340	Bonded OH stretch	Aliphatic alcohols
3374	CH stretch	Acetylene (gas)
3355–3325	Bonded antisymmetric NH_2 stretch	Primary amides
3350–3300	Bonded NH stretch	Secondary amines
3335–3300	≡CH stretch	Alkyl acetylenes
3300–3250	Bonded symmetric NH_2 stretch	Primary amines
3310–3290	Bonded NH stretch	Secondary amides
3190–3145	Bonded symmetric NH_2 stretch	Primary amides
3175–3154	Bonded NH stretch	Pyrazoles
3103	Antisymmetric ═CH_2 stretch	Ethylene (gas)
3100–3020	CH_2 stretches	Cyclopropane
3100–3000	Aromatic CH stretch	Benzene derivatives
3095–3070	Antisymmetric ═CH_2 stretch	C═CH_2 derivatives
3062	CH stretch	Benzene
3057	Aromatic CH stretch	Alkyl benzenes
3040–3000	CH stretch	C═CHR derivatives
3026	Symmetric ═CH_2 stretch	Ethylene (gas)
2990–2980	Symmetric ═CH_2 stretch	C═CH_2 derivatives
2986–2974	Symmetric NH_3^+ stretch	Alkyl ammonium chlorides (aqueous solution)
2969–2965	Antisymmetric CH_3 stretch	*n*-Alkanes
2929–2912	Antisymmetric CH_2 stretch	*n*-Alkanes
2884–2883	Symmetric CH_3 stretch	*n*-Alkanes

Continued

Table 1.3. *(Continued)*

Frequency (cm^{-1})	Vibration	Compound
2861–2849	Symmetric CH_2 stretch	*n*-Alkanes
2850–2700	CHO group (2 bands)	Aliphatic aldehydes
2590–2560	SH stretch	Thiols
2316–2233	C≡C stretch (2 bands)	$R—C\equiv C—CH_3$
2301–2231	C≡C stretch (2 bands)	$R—C\equiv C—R'$
2300–2250	Pseudoantisymmetric N═C═O stretch	Isocyanates
2264–2251	Symmetric C≡C—C≡C stretch	Alkyl diacetylenes
2259	C≡N stretch	Cyanamide
2251–2232	C≡N stretch	Aliphatic nitriles
2220–2100	Pseudoantisymmetric N═C═S stretch (2 bands)	Alkyl isothiocyanates
2220–2000	C≡N stretch	Dialkyl cyanamides
2172	Symmetric C≡C—C≡C stretch	Diacetylene
2161–2134	$\overset{+}{N}\equiv\overset{-}{C}$ stretch	Aliphatic isonitriles
2160–2100	C≡C stretch	Alkyl acetylenes
2156–2140	C≡N stretch	Alkyl thiocyanates
2104	Antisymmetric N═N═N stretch	CH_3N_3
2094	C≡N stretch	HCN
2049	Pseudoantisymmetric C═C═O stretch	Ketene
1974	C≡C stretch	Acetylene (gas)
1964–1958	Antisymmetric C═C═C stretch	Allenes
1870–1840	Symmetric C═O stretch	Saturated 5-membered ring cyclic anhydrides
1820	Symmetric C═O stretch	Acetic anhydride
1810–1788	C═O stretch	Acid halides
1807	C═O stretch	Phosgene
1805–1799	Symmetric C═O stretch	Noncyclic anhydrides
1800	C═C stretch	$F_2C═CF_2$ (gas)
1795	C═O stretch	Ethylene carbonate
1792	C═C stretch	$F_2C═CFCH_3$
1782	C═O stretch	Cyclobutanone
1770–1730	C═O stretch	Halogenated aldehydes
1744	C═O stretch	Cyclopentanone
1743–1729	C═O stretch	Cationic α-amino acids (aqueous solution)
1741–1734	C═O stretch	*O*-Alkyl acetates
1740–1720	C═O stretch	Aliphatic aldehydes
1739–1714	C═C stretch	$C═CF_2$ derivatives
1736	C═C stretch	Methylene cyclopropane
1734–1727	C═O stretch	*O*-Alkyl propionates
1725–1700	C═O stretch	Aliphatic ketones
1720–1715	C═O stretch	*O*-Alkyl formates
1712–1694	C═C stretch	RCF═CFR
1695	Nonconjugated C═O stretch	Uracil derivatives (aqueous solution)
1689–1644	C═C stretch	Monofluoroalkenes
1687–1651	C═C stretch	Alkylidene cyclopentanes
1686–1636	Amide I band	Primary amides (solids)
1680–1665	C═C stretch	Tetraalkyl ethylenes
1679	C═C stretch	Methylene cyclobutane
1678–1664	C═C stretch	Trialkyl ethylenes

Continued

Table 1.3. *(Continued)*

Frequency (cm^{-1})	Vibration	Compound
1676–1665	C═C stretch	*trans*-Dialkyl ethylenes
1675	Symmetric C═O stretch (cyclic dimer)	Acetic acid
1673–1666	C═N stretch	Aldimines
1672	Symmetric C═O stretch (cyclic dimer)	Formic acid (aqueous solution)
1670–1655	Conjugated C═O stretch	Uracil, cytosine, and guanine derivatives (aqueous solution)
1670–1630	Amide I band	Tertiary amides
1666–1652	C═N stretch	Ketoximes
1665–1650	C═N stretch	Semicarbazones (solid)
1663–1636	Symmetric C═N stretch	Aldazines, ketazines
1660–1654	C═C stretch	*cis*-Dialkyl ethylenes
1660–1650	Amide I band	Secondary amides
1660–1649	C═N stretch	Aldoximes
1660–1610	C═N stretch	Hydrazones (solid)
1658–1644	C═C stretch	$R_2C{=}CH_2$
1656	C═C stretch	Cyclohexene, cycloheptene
1654–1649	Symmetric C═O stretch (cyclic dimer)	Carboxylic acids
1652–1642	C═N stretch	Thiosemicarbazones (solid)
1650–1590	NH_2 scissors	Primary amines
1649–1625	C═C stretch	Allyl derivatives
1648–1640	N═O stretch	Alkyl nitrites
1648–1638	C═C stretch	$H_2C{=}CHR$
1647	C═C stretch	Cyclopropene
1638	C═O stretch	Ethylene diothiocarbonate
1637	Symmetric C═C stretch	Isoprene
1634–1622	Antisymmetric NO_2 stretch	Alkyl nitrates
1630–1550	Ring stretches (doublet)	Benzene derivatives
1623	C═C stretch	Ethylene (gas)
1620–1540	Three or more coupled C═C stretches	Polyenes
1616–1571	C═C stretch	Chloroalkenes
1614	C═C stretch	Cyclopentene
1596–1547	C═C stretch	Bromoalkenes
1581–1465	C═C stretch	Iodoalkenes
1575	Symmetric C═C stretch	1,3-Cyclohexadiene
1573	N═N stretch	Azomethane (in solution)
1566	C═C stretch	Cyclobutene
1560–1550	Antisymmetric NO_2 stretch	Primary nitroalkanes
1555–1550	Antisymmetric NO_2 stretch	Secondary nitroalkanes
1548	N═N stretch	1-Pyrazoline
1545–1535	Antisymmetric NO_2 stretch	Tertiary nitroalkanes
1515–1490	Ring stretch	2-Furfuryl group
1500	Symmetric C═C stretch	Cyclopentadiene
1480–1470	OCH_3, OCH_2 deformations	Aliphatic ethers
1480–1460	Ring stretch	2 Furfurylidene or 2-furoyl group
1473–1446	CH_3, CH_2 deformations	*n*-Alkanes
1466–1465	CH_3 deformation	*n*-Alkanes
1450–1400	Pseudoantisymmetric N═C═O stretch	Isocyanates
1443–1398	Ring stretch	2-Substituted thiophenes
1442	N═N stretch	Azobenzene
1440–1340	Symmetriic CO_2^- stretch	Carboxylate ions (aqueous solution)

Continued

Table 1.3. *(Continued)*

Frequency (cm^{-1})	Vibration	Compound
1415–1400	Symmetric CO_2^- stretch	Dipolar and anionic α-amino acids (aqueous solution)
1415–1385	Ring stretch	Anthracenes
1395–1380	Symmetric NO_2 stretch	Primary nitroalkanes
1390–1370	Ring stretch	Naphthalenes
1385–1368	CH_3 symmetric deformation	*n*-Alkanes
1375–1360	Symmetric NO_2 stretch	Secondary nitroalkanes
1355–1345	Symmetric NO_2 stretch	Tertiary nitroalkanes
1350–1330	CH deformation	Isopropyl group
1320	Ring vibration	1,1-Dialkyl cyclopropanes
1314–1290	In-plane CH deformation	*trans*-Dialkyl ethylenes
1310–1250	Amide III band	Secondary amides
1310–1175	CH_2 twist and rock	*n*-Alkanes
1305–1295	CH_2 in-phase twist	*n*-Alkanes
1300–1280	CC bridge bond stretch	Biphenyls
1282–1275	Symmetric NO_2 stretch	Alkyl nitrates
1280–1240	Ring stretch	Epoxy derivatives
1276	Symmetric N═N═N stretch	CH_3N_3
1270–1251	In-plane CH deformation	*cis*-Dialkyl ethylenes
1266	Ring “breathing”	Ethylene oxide (oxirane)
1230–1200	Ring vibration	Para-disubstituted benzenes
1220–1200	Ring vibration	Mono- and 1,2-dialkyl cyclopropanes
1212	Ring “breathing”	Ethylene imine (aziridine)
1205	C_6H_5—C vibration	Alkyl benzenes
1196–1188	Symmmetric SO_2 stretch	Alkyl sulfates
1188	Ring “breathing”	Cyclopropane
1172–1165	Symmetric SO_2 stretch	Alkyl sulfonates
1150–950	CC stretches	*n*-Alkanes
1145–1125	Symmetric SO_2 stretch	Dialkyl sulfones
1144	Ring “breathing”	Pyrrole
1140	Ring “breathing”	Furan
1130–1100	Symmetric C═C═C stretch (2 bands)	Allenes
1130	Pseudosymmetric C═C═O stretch	Ketene
1112	Ring “breathing”	Ethylene sulfide
1111	NN stretch	Hydrazine
1070–1040	S═O stretch (1 or 2 bands)	Aliphatic sulfoxides
1065	C═S stretch	Ethylene trithiocarbonate
1060–1020	Ring vibration	Ortho-disubstituted benzenes
1040–990	Ring vibration	Pyrazoles
1030–1015	In-plane CH deformation	Monosubstituted benzenes
1030–1010	Trigonal ring “breathing”	3-Substituted pyridines
1030	Trigonal ring “breathing”	Pyridine
1029	Ring “breathing”	Trimethylene oxide (oxetane)
1026	Ring “breathing”	Trimethylene imine (azetidine)
1010–990	Trigonal ring “breathing”	Mono-, meta-, and 1,3,5 substituted benzenes
1001	Ring “breathing”	Cyclobutane
1000–985	Trigonal ring “breathing”	2- and 4-Substituted pyridines
992	Ring “breathing”	Benzene

Continued

Table 1.3. *(Continued)*

Frequency (cm^{-1})	Vibration	Compound
992	Ring "breathing"	Pyridine
939	Ring "breathing"	1,3-Dioxolane
933	Ring vibration	Alkyl cyclobutanes
930–830	Symmetric COC stretch	Aliphatic ethers
914	Ring "breathing"	Tetrahydrofuran
906	ON stretch	Hydroxylamine
905–837	CC skeletal stretch	*n*-Alkanes
900–890	Ring vibration	Alkyl cyclopentanes
900–850	Symmetric CNC stretch	Secondary amines
899	Ring "breathing"	Pyrrolidine
866	Ring "breathing"	Cyclopentane
877	OO stretch	Hydrogen peroxide
851–840	Pseudosymmetric CON stretch	*O*-Alkyl hydroxylamines
836	Ring "breathing"	Piperazine
835–749	Skeletal stretch	Isopropyl group
834	Ring "breathing"	1,4-Dioxane
832	Ring "breathing"	Thiophene
832	Ring "breathing"	Morpholine
830–720	Ring vibration	Para-disubstituted benzenes
825–820	C_3O skeletal stretch	Secondary alcohols
818	Ring "breathing"	Tetrahydropyran
815	Ring "breathing"	Piperidine
802	Ring "breathing"	Cyclohexane (chair form)
785–700	Ring vibration	Alkyl cyclohexanes
760–730	C_4O skeletal stretch	Tertiary alcohols
760–650	Symmetric skeletal stretch	*tert*-Butyl group
740–585	CS stretch (1 or more bands)	Alkyl sulfides
735–690	"C=S stretch"	Thioamides, thioureas (solid)
733	Ring "breathing"	Cycloheptane
730–720	CCl stretch, P_C conformation	Primary chloroalkanes
715–620	CS stretch (1 or more bands)	Dialkyl disulfides
709	CCl stretch	CH_3Cl
703	Ring "breathing"	Cyclooctane
703	Symmetric CCl_2 stretch	CH_2Cl_2
690–650	Pseudosymmetric N=C=S stretch	Alkyl isothiocyanates
688	Ring "breathing"	Tetrahydrothiophene
668	Symmetric CCl_3 stretch	$CHCl_3$
660–650	CCl stretch, P_H conformation	Primary chloroalkanes
659	Symmmetric CSC stretch	Pentamethylene sulfide
655–640	CBr stretch, P_C conformation	Primary bromoalkanes
630–615	Ring deformation	Monosubstituted benzenes
615–605	CCl stretch, S_{HH} conformation	Secondary chloroalkanes
610–590	CI stretch, P_C conformation	Primary iodoalkanes
609	CBr stretch	CH_3Br
577	Symmetric CBr_2 stretch	CH_2Br_2
570–560	CCl stretch, T_{HHH} conformation	Tertiary chloroalkanes
565–560	CBr stretch, P_H conformation	Primary bromoalkanes
540–535	CBr stretch, S_{HH} conformation	Secondary bromoalkanes
539	Symmetric CBr_3 stretch	$CHBr_3$

Continued

Table 1.3. ***(Continued)***

Frequency (cm^{-1})	Vibration	Compound
525–510	SS stretch	Dialkyl disulfides
523	CI stretch	CH_3I
520–510	CBr stretch, T_{HHH} conformation	Tertiary bromoalkanes
510–500	CI stretch, P_H conformation	Primary iodoalkanes
510–480	SS stretch	Dialkyl trisulfides
495–485	CI stretch, S_{HH} conformation	Secondary iodoalkanes
495–485	CI stretch, T_{HHH} conformation	Tertiary iodoalkanes
484–475	Skeletal deformation	Dialkyl diacetylenes
483	Symmetric CI_2 stretch	CH_2I_2
459	Symmetric CCl_4 stretch	CCl_4
437	Symmetric CI_3 stretch	CHI_3 (in solution)
425–150	"Chain expansion"	*n*-Alkanes
355–335	Skeletal deformation	Monoalkyl acetylenes
267	Symmetric CBr_4 stretch	CBr_4 (in solution)
200–160	Skeletal deformation	Aliphatic nitriles
178	Symmetric CI_4 stretch	CI_4 (solid)

[a] Dollish *et al.* (1974).

weaker the Raman. The more symmetrically substituted a particular bond (or the greater its polarizability), the stronger the Raman intensity and the weaker the IR. Such different characteristics produce strong Raman scattering for bonds like S—S, N═N, C—C, and C≡C and usually weak IR absorptions, while bonds like C═O, C—O, O—H, and N—H produce strong IR bands and generally weak Raman scattering.

The spectra of dibromocyanomethyl methyl sulfone (Fig. 1.12) show the differences between the appearance of the C≡N stretching mode in the IR and Raman presentations. In the Raman, the C≡N presents a prominent line at about 2230 cm^{-1}; in the IR, the corresponding band is almost invisible. The SO_2 stretching vibrations (1348 cm^{-1}, *antisym,* and 1160 cm^{-1}, *sym*) are strong in the IR; in the Raman, the 1350 cm^{-1} stretch is very weak and the 1160 cm^{-1} very strong.

Figure 1.13 shows the spectra of 3a,7a-dichloro-3a,4,7,7a-tetrahydro-*N*-phenyl-1H-isoindole-1,3(2H)-dione. The ring C═C stretching mode, although completely obscured in the IR by the nearby strong carbonyl, is clearly present at 1640 cm^{-1} in the Raman. The monosubstituted benzene is much more prominent in the Raman than in the IR. The imide C═O grouping displays strong IR absorption and very weak Raman intensity at 1735 cm^{-1}. This grouping also shows a weak band in the IR, but a relatively stronger one in the Raman at 1800 cm^{-1}. These are, respectively, antisymmetric and symmetric C═O stretching modes. This reversal of band intensity strengths between Raman and IR spectra offers a fine way to confirm anhydride, imide, or other groupings in which carbonyl coupling occurs.

Figure 1.14 presents spectra of pentobarbital, a derivative of barbituric acid. The IR recording suggests a barbiturate by its complex carbonyl spectrum. The Raman spectrum supports this finding by displaying an extremely intense band at ca. 630 cm^{-1}, owing

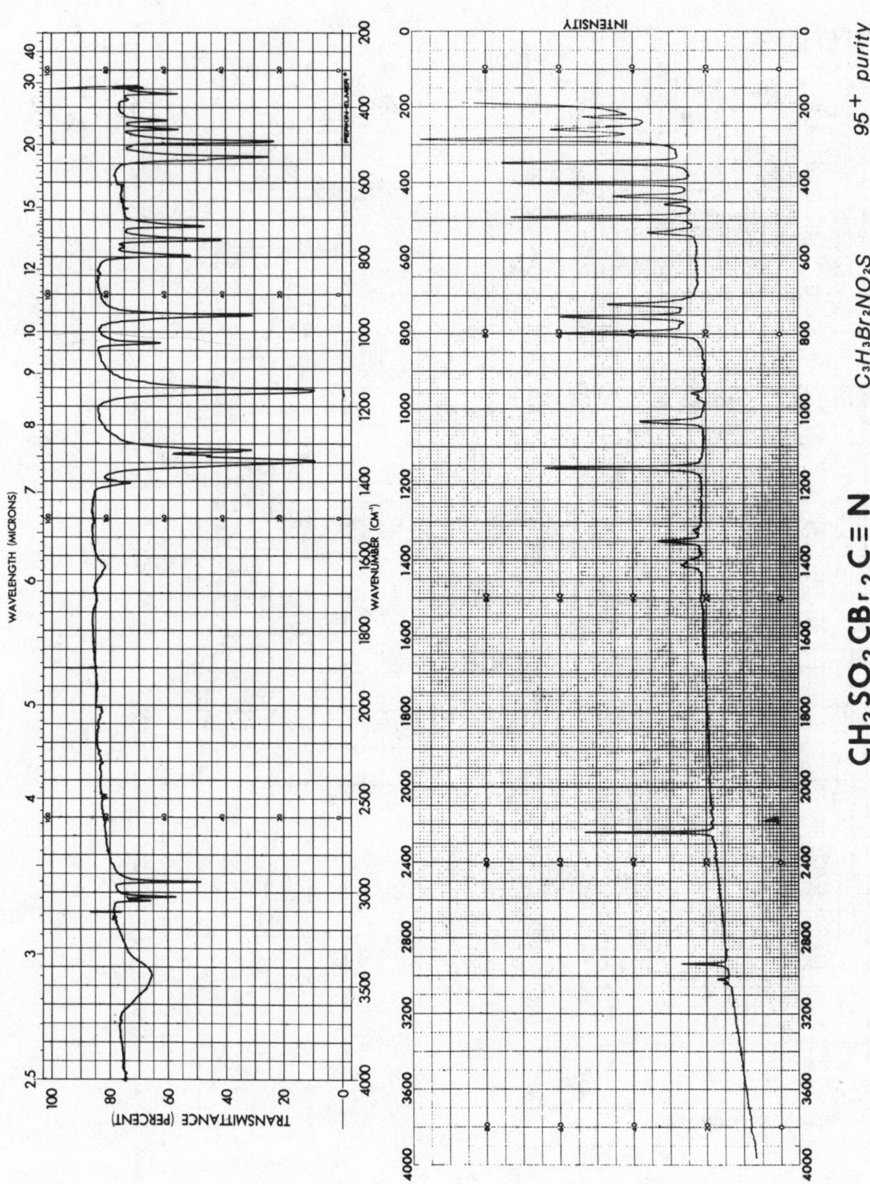

Fig. 1.12. Dibromocyanomethyl methyl sulfone. (Washburn, 1978.)

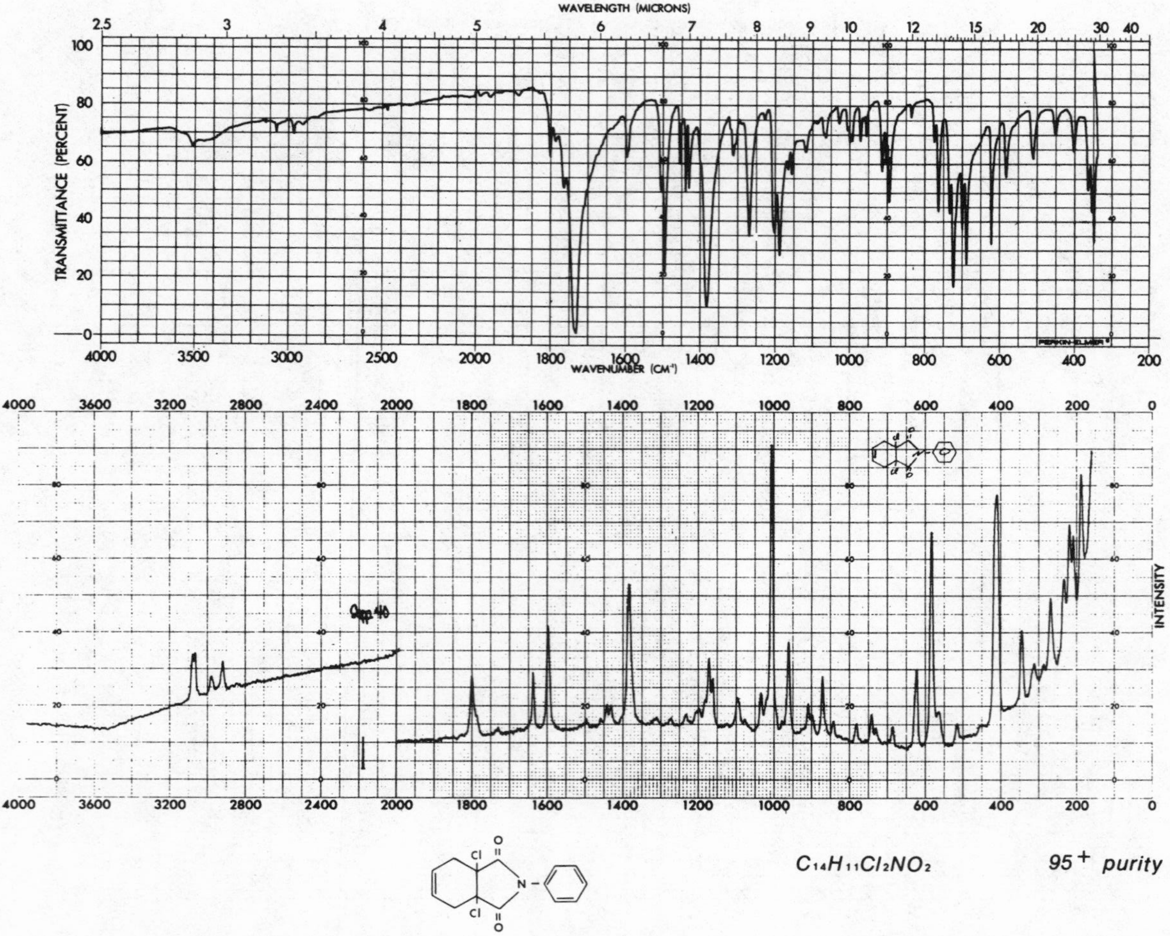

Fig. 1.13. 3*a*,7*a*-Dichloro-3*a*,4,7,7*a*-tetrahydro-*N*-phenyl-1*H*-isoindole-1,3(2*H*)-dione. (Washburn, 1978.)

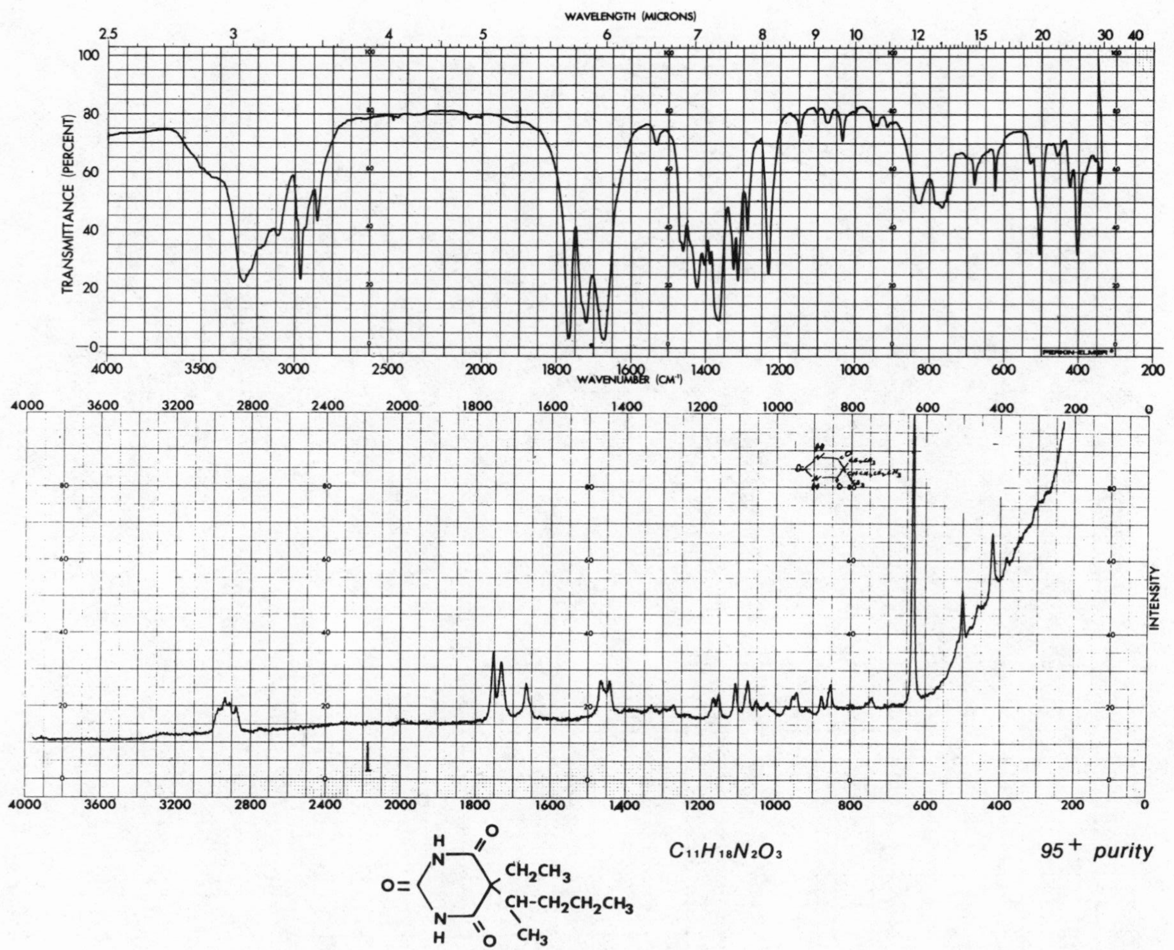

Fig. 1.14. Pentobarbital. (Washburn, 1978.)

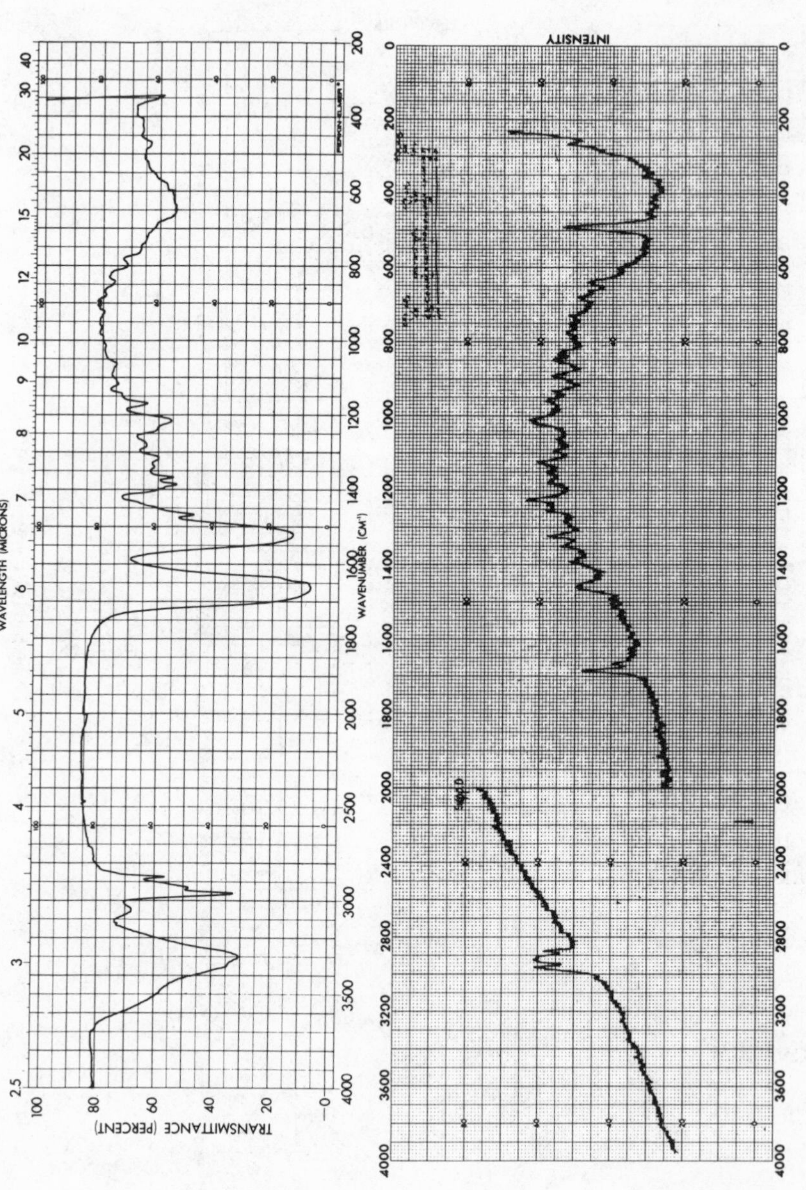

Fig. 1.15. Cyclo Val-D-Cys-Cys-Val-D-Leu. (Washburn, 1978.)

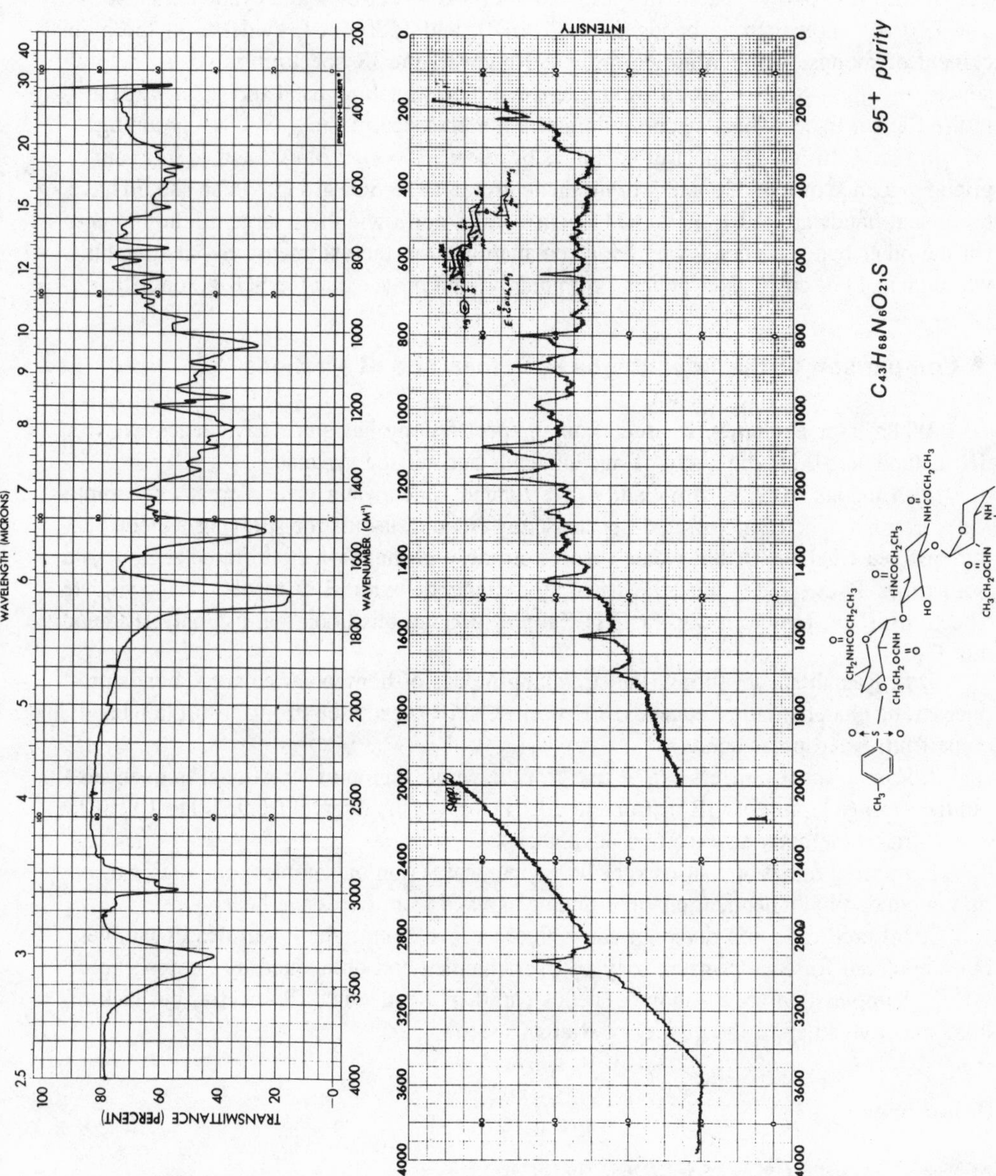

Fig. 1.16. 3′Tosylpercarboethoxyseldomycin 5. (Washburn, 1978.)

to a ring-breathing vibration that characterizes barbiturate compounds. The corresponding band, since it originates from a highly symmetric vibrational frequency, is very weak in the IR spectrum.

Figure 1.15 shows spectra of cyclo Val-D-Cys-Cys-Val-D-Leu, a cyclic tetrapeptide. The IR trace shows strong bands at 3300, 1670, and 1530 cm^{-1}, evidence of the high content of monosubstituted amide groups. However, the IR spectrum is not so helpful concerning the —S—S— group. An intense disulfide stretching band appears at 490 cm^{-1} in the Raman trace, which strongly indicates the presence of the —S—S— grouping.

In Fig. 1.16, the IR spectrum displays bands at 3350(s), 1700(s), and 1530(s) cm^{-1}, giving evidence of a high content of urethane groups. The SO_3 group, which should have prominent bands around 1350 and 1150 cm^{-1}, does not, owing to a large dilution effect. On the other hand, a very strong band, emanating from the symmetric S—O stretching vibration at 1170 cm^{-1}, does appear prominently, and proves that the SO_3 group is there.

A Comparison of the Advantages of Raman and IR Methods

When laser excitation is used, Raman spectroscopy has some advantages over the IR method, e.g., for the study of nucleic acids and related systems:

1. Aqueous solutions are readily investigated, since water gives a weak, uncomplicated Raman spectrum. Infrared spectroscopy is quite useful for studying aqueous solutions (see Chapter 2), but water absorbs strongly in much of the infrared region, and sometimes D_2O (^2H_2O) must be used in conjunction with H_2O to obtain a complete spectrum. (See, however, various FTIR applications in this book that use only H_2O and not D_2O.)

2. In Raman spectroscopy, the stretching modes of homopolar covalent bonds produce strong characteristic frequencies. However, in IR spectra the strong absorption bands come from heteropolar bonds.

3. Raman spectrometers can be used for the whole region of vibrational frequencies, but the average laboratory IR instrument does not cover the interval from 200 to 10 cm^{-1}.

Infrared methods have certain advantages:

1. Infrared radiation causes very little chemical action on samples, but a laser beam may produce photochemical changes and/or damage from excessive heat.

2. Infrared radiation does not cause fluorescence from trace impurities in samples, but a potential Raman spectrum may be overwhelmed and obliterated by fluorescence.

3. Samples that are not homogeneous can display adequate IR spectra, but lack of homogeneity can spoil the quality of Raman spectra.

References

Abraham, J. L., and Etz, E. S., *Science* **206,** 716 (1979).

Ballan-DuFrancais, C., doctoral thesis, Paris, France, 1975; quoted in Dhamelincourt *et al.*, 1979.

Bates, J. B., *Science* **191,** 31 (1976). An excellent concise review for the beginner.

Becker, E. D., and Farrar, T. C., *Science* **178,** 361 (1972).

Begley, R. F., Harvey, A. B., and Byer, R. L., *Appl. Phys. Lett.* **25,** 387 (1974).

Behringer, J., *Mol. Spectrosc.* **2,** 100 (1974).

Bell, R. J., *Introductory Fourier Transform Spectroscopy,* Academic Press, New York, 1972.

Brand, J. C. D., and Speakman, J. C., *Molecular Structure: The Physical Approach,* Edward Arnold, London, p. 175, 1960.

Cameron, D. G., Casal, H. L., and Mantsch, H. H., *J. Biochem. Biophys. Methods* **1,** 21 (1979).

Carey, P. R., and Schneider, H., *Acc. Chem. Res.* **11,** 122 (1978).

Chabay, I., Klauminzer, G., and Hudson, B., *Appl. Phys. Lett.* **28,** 27 (1976).

Colthup, N. B., Daly, L. H., and Wiberley, S. E., *Introduction to Infrared and Raman Spectroscopy,* Academic Press, New York, p. 31, 1964.

Colthup, N. B., Daly, L. H., and Wiberley, S. E., *Introduction to Infrared and Raman Spectroscopy,* 2nd Ed., Academic Press, New York, 1975. Presents 624 IR spectra and 36 Raman spectra with interpretations of bands. Also, gives 24 IR unknowns for interpretation as an exercise in functional group analysis; answers are also given. Excellent discussions of infrared and Raman theory.

Connes, Mme. Janine, *Rev. Opt. Théor. Instrum.* **40,** 45, 116, 171, 233 (1961). The Clearinghouse for Federal Scientific and Technical Information, Cameron Station, Virginia, has an English translation in Document AD 409869.

Delhaye, M., and Merlin, J.-C., *Biochimie* **57,** 401–415, (1975) (in French).

D'Esposito, L., and Koenig, J. L., in *Fourier Transform Infrared Spectroscopy, Applications to Chemical Systems,* Vol I, (J. R. Ferraro and L. J. Basile, eds.) Academic Press, New York, 1978. These authors have discussed biological macromolecules.

Dhamelincourt, P., Wallart, F., Leclercq, M., N'Guyen, A. T., and Landon, D. O., *Anal. Chem.* **51,** 414A (1979).

Dollish, F. R., Fateley, W. G., and Bentley, F. F., *Characteristic Raman Frequencies of Organic Compounds,* John Wiley and Sons, New York, 1974.

Fellgett, P. D., (1949), Cited in *Aspen Int. Conf. Fourier Spectrosc.,* 1970 (G. A. Vanasse, A. T. Stair, and D. J. Baker, eds.), AFCRL-71-0019.

Foskett, C. T., "The Laboratory Computer as an Information Channel," in *Infrared, Correlation, and Fourier Transform Spectroscopy* (J. S. Mattson, H. B. Mark, Jr., and H. C. MacDonald, Jr., eds.), p. 191, Marcel Dekker, Inc., New York, 1977.

Gaber, B. P., *Am. Lab.,* **March,** 1977, p. 15.

Griffiths, P. R., *Anal. Chem.* **46,** 645A (1974).

Griffiths, P. R., *Chemical Infrared Fourier Transform Spectroscopy,* John Wiley and Sons, New York, 1975.

Griffiths, P. R., in *Fourier Transform Infrared Spectroscopy, Applications to Chemical Systems,* Vol. I, (J. R. Ferraro and L. J. Basile, eds.), Chapter 4, p. 143, Academic Press, New York, 1978.

Griffiths, P. R., Sloane, H. J., and Hannah, R. W., *Appl. Spectrosc.* **31,** 485 (1977).

Griffiths, P. R., Foskett, C. T., and Curbelo, R., *Appl. Spectrosc. Rev.* **6,** 31 (1972).

Hartshorn, J. H., *Appl. Spectrosc.* **33,** 111 (1979).

Hendra, P. J., in *Polymer Spectroscopy* (D. O. Hummel, ed.), Verlag Chemie, Weinheim/Bergstr. and Academic Press, New York, 1974.

Herzberg, G., *Molecular Spectra and Molecular Structure: Infrared and Raman Spectra of Polyatomic Molecules,* D. Van Nostrand, New York, 1945.

Hirakawa, A. Y., and Tsuboi, M., *Science* **188,** 359 (1975).

Jacquinot, P., *17e Congres du GAMS,* Paris (1954).

Jones, R. N., "Modular Computer Programs for Infrared Spectrophotometry," in *Infrared, Correlation, and Fourier Transform Spectroscopy* (J. S. Mattson, H. B. Mark, Jr., and H. C. MacDonald, Jr., eds.), Marcel Dekker, Inc., New York, 1977.

Kagel, R. O., and King, S. T., *Ind. Res.,* **October,** 1973.

Kiefer, W., *Appl. Spectrosc.* **28,** 115 (1974).

Koenig, J. L., *Appl. Spectrosc.* **29,** 293 (1975).

Krimm, S., *J. Mol. Biol.* **4,** 528 (1962).

Krishnan, K., Brown, R. H., Hill, S. L., Simonoff, S. C., Olson, M. L., and Kuehl, D., *Am. Lab.,* **March,** 1981, p. 122.

Krishnan, K., and Ferraro, J. R., in *FTIR Spectroscopy,* Vol. 3 (J. R. Ferraro and L. J. Basile, eds.), Academic Press, New York, 1981.

Kuehl, D., Kemeny, G. J., and Griffiths, P. R., *Appl. Spectrosc.* **34,** 222 (1980).

Lewis, A., *The Spex Speaker,* **21**(2), June, 1976, Spex Industries, Metuchen, New Jersey 08840 (1976a).

Lewis, A., *Fed. Proc.* **35,** 51 (1976b).

Lewis, A., and Spoonhower, J. in *Spectroscopy in Biology and Chemistry* (S.-H. Chen and S. Yip, eds.) p. 347, Academic Press, New York, 1974.

Long, D. A., *Raman Spectroscopy,* McGraw-Hill International, New York, 1977.

Margenau, H., and Murphy, G. M., *The Mathematics of Physics and Chemistry,* Van Nostrand, New York, 1943.

Mattson, J. S., and Smith, C. A., "An On-Line Minicomputer System for Infrared Spectrophotometry," in *Infrared, Correlation, and Fourier Transform Spectroscopy* (J. S. Mattson, H. B. Mark, Jr., and H. C. MacDonald, Jr., eds.), p. 71, Marcel Dekker, Inc., New York, 1977.

Nafie, L. A., Diem, M., and Vidrine, D. W., *J. Am. Chem. Soc.* **101,** 496 (1979).

Nakamoto, K., *Infrared and Raman Spectra of Inorganic and Coordination Compounds,* 3rd Ed., John Wiley and Sons, New York, 1978.

Nakanishi, K., and Solomon, P. H., *Infrared Absorption Spectroscopy,* 2nd Ed., Holden–Day, Inc., San Francisco, 1977. Contains a chapter on laser Raman spectroscopy with many examples and interpretations of spectra. Also contains 100 problems in IR spectroscopy with answers worked out.

Nestor, J., Spiro, T. G., and Klauminzer, G., *Proc. Natl. Acad. Sci., USA* **73,** 3329 (1976).

O'Bremski, R. J., *Introduction to Raman Spectroscopy,* Beckman Instruments, Inc., Fullerton, California, pp. 1–52, 1971.

Parker, F. S., *Appl. Spectrosc.* **29,** 129 (1975).

Pinchas, S., and Laulicht, I., *Infrared Spectra of Labelled Compounds,* Academic Press, New York, 1971.

Rossiter, V., *Am. Lab.*, p. 144, **February** 1982.

Rothschild, K. J., and Clark, N. A., *Biophys. J.* **25,** 473 (1979a).

Rothschild, K. J., and Clark, N. A., *Science* **204,** 311 (1979b).

Rothschild, K. J., DeGrip, W. J., and Sanches, R., *Biochim. Biophys. Acta* **596,** 338 (1980).

Rothschild, K. J., Zagaeski, M., and Cantore, W. A., Abstracts of the 26th Annual Meeting, *Biophys. J.* **37,** No. 2 Part 2, 229a, February 1982.

Schrader, B., "Raman Spectroscopy," in *Methodicum Chimicum,* Vol. 1, Part A (F. Korte, ed.), pp. 290–307, Academic Press, New York, 1974.

SPEX Speaker, Vol. XI, No. 4, December, 1966; Vol. XII, No. 2, June, 1967.

Spiro, T. G., *Acc. Chem. Res.* **7,** 339 (1974).

Spiro, T. G., "Resonance Raman Spectra of Hemoproteins," in *Methods in Enzymology,* Vol. LIV, *Biomembranes,* Part E, (S. Fleischer and L. Packer, eds.), pp. 233–249, Academic Press, New York, 1978.

Spiro, T. G., and Gaber, B. P., *Annu. Rev. Biochem.* **46,** 553 (1977).

Spiro, T. G., and Strekas, T. C., *Proc. Natl. Acad. Sci. U.S.A.* **69,** 2622 (1972).

Susi, H., in *Structure and Stability of Biological Macromolecules* (S. N. Timasheff and G. D. Fasman, eds.), Chap. 7, p. 575, Marcel Dekker, Inc., New York, 1969.

Szymanski, H. A., ed., *Raman Spectroscopy, Theory and Practice,* Plenum Press, New York, 1967.

Tabb, D. L., Ph.D. thesis, Case Western Reserve University, 1974; quoted in D'Esposito and Koenig (1978).

Tang, J., and Albrecht, A. C. in *Raman Spectroscopy, Theory and Practice,* Vol. 2 (H. A. Szymanski, ed.), Plenum Press, New York, 1970.

Telfair, W. B., Gilby, A. C., Syrjala, R. J., and Wilks, P. A., Jr., *Am. Lab.*, **November** 1976, p. 91.

Thomas, G. J., Jr., "Infrared and Raman Spectroscopy," in *Physical Techniques in Biochemical Research,* Vol. IA, 2nd Ed. (G. Oster, ed.), pp. 277–346, Academic Press, New York, 1971.

Tobin, M. C., *Laser Raman Spectroscopy,* Wiley-Interscience, New York, 1971.

Tobin, M. C., "Raman Spectroscopy," in *Methods in Enzymology,* Vol. 26, Part C (C. H. W. Hirs and S. N. Timasheff, eds.), pp. 473–497, 1972.

Van Wart, H. E., and Scheraga, H. A., in *Methods in Enzymology,* Vol. XLIX, Part G (C. H. W. Hirs and S. N. Timasheff, eds.), pp. 67–149, Academic Press, 1978.

Wallach, D. F. H., Verma, S. P., and Fookson, J., *Biochim. Biophys. Acta* **559,** 153 (1979).

Warshel, A., "Interpretation of Resonance Raman Spectra of Biological Molecules," in *Ann. Rev. Biophys. Bioeng.* **6,** 273–300 (1977).

Washburn, W. H., *Am. Lab.* **10,** 47 (November, 1978).

Wilson, E. B., Decius, J. C., and Cross, P. C., *Molecular Vibrations,* McGraw-Hill, New York, 1955.

Young, R. P., and Jones, R. N., *Chem. Rev.* **71,** 219 (1971).

Chapter 2

SAMPLING METHODS

Raman

The size of the sample is in the milligram—microgram region, but samples of 1 nl from gas chromatographic columns have been used. For crystals, Raman spectra have been observed on samples of less than 1 μg.

One of the most commonly mentioned advantages of Raman spectroscopy for biochemical applications is that water can readily be used as a solvent, since it shows only a very weak Raman spectrum. Infrared spectroscopy of aqueous solutions can be difficult in certain spectral regions of interest, but there are useful spectral "windows" in the IR region where water can be used; and there are cells available that do not dissolve in water, e.g., Irtran-2. (See discussion of aqueous samples in this chapter.) Raman spectroscopy uses glass cells for aqueous solutions, e.g., melting-point capillaries, and this factor is a decided advantage.

For Raman work with aqueous solutions 0.2 ml of, for example, 5% protein or nucleic acid solution is usually more than enough; a minimum of ~0.02 ml is required. However, resonance Raman experimentation requires much less concentrated solutions, on the order of those used for obtaining visible or UV absorption spectra.

Powders can be sampled by direct scattering off their surface, or they may be packed in melting-point capillaries and illuminated either horizontally or vertically. Small crystals can be contained in capillary tubes for sampling. If a crystal is large enough to hold, it can be held directly in the laser beam. For solid materials one does not have to grind the sample and consequently destroy its integrity. Sample thickness is not a problem, since Raman spectroscopy is essentially akin to an emission process rather than an absorption phenomenon.

A problem with sampling in Raman spectroscopy is the fluorescence one may produce when exposing a sample to laser excitation. For example, the disadvantage of the argon ion laser is in the greater probability of fluorescence and sample decomposition from shorter-wavelength excitation. However, using a He–Ne laser source (red excitation), Lord and Yu (1970) found no serious fluorescence problems in the examination of the enzyme lysozyme and its constituent amino acids.

Sometimes the fluorescence seen in Raman spectra of biochemical substances is not caused by the materials themselves but rather by organic impurities. These impurities are often "removed" by long exposure to the source. Commercial samples are sometimes rendered spectroscopically acceptable by treatment with activated charcoal.

Filtering solutions through fritted glass disks or centrifuging prior to filling sample cells minimizes Tyndall scattering, which is caused by dust, suspended material, colloids, bubbles, or some other undissolved material with particle size about the same as or greater than the excitation wavelength.

Absorption of the excitation wavelength by the sample and fluorescence caused by this wavelength are to be avoided, since these processes make it difficult to obtain adequate Raman spectra. Also, conditions should be optimized for obtaining as strong Raman scattering as possible, since the Raman effect is basically a weak phenomenon. Bailey *et al.* (1967) and Hawes *et al.* (1967) have described several spatial arrangements for excitation of relatively small amounts of sample, and these arrangements produce efficient scattering.

Figure 2.1 shows an axial excitation/transverse viewing cell (Bailey *et al.*, 1967), in which 0.04 μl is the smallest sample size. The focused laser beam is sent down the axis of a cylinder and the scattered radiation is viewed horizontally, perpendicular to the main axis of the cylinder. Figure 2.2 shows a transverse excitation/transverse viewing cell (Pez, 1968), in which 0.008 μl is the smallest sample. Freeman and Landon (1968) readily obtained excellent depolarization measurements on such quantities.

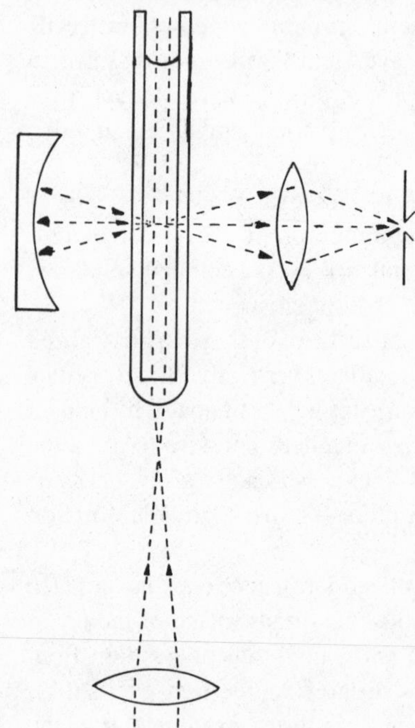

Fig. 2.1. Axial excitation/transverse viewing cell of Bailey *et al.* (1967) in which minimum sample size is 0.04 μl. Reprinted with the permission of the American Chemical Society. Copyright 1967.

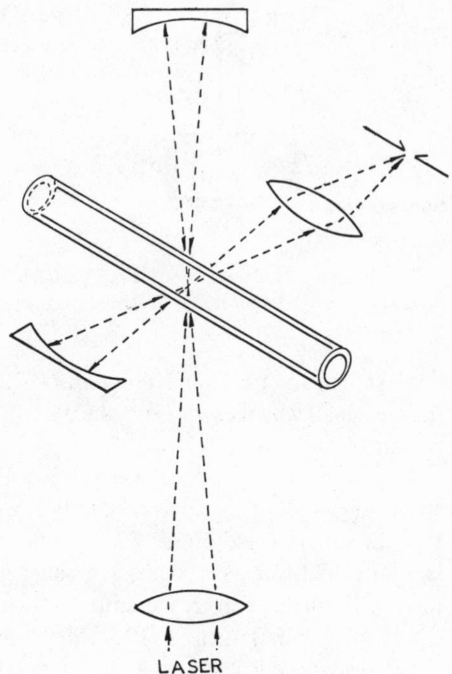

Fig. 2.2. Transverse excitation/transverse viewing cell of Pez in which minimum sample size is 0.008 μl. (Pez, 1968).

Raman spectra have been obtained on liquid samples as small as 8 nl, and 30-nl quantities have been examined routinely (Freeman and Landon, 1968). Solids can be run as conveniently as liquids with samples as small as 0.1 mg. These workers obtained signal intensities with the transverse/transverse capillary system at least twice as great as those obtained from the axial/transverse technique, and their laboratory uses this technique routinely. Fused silica capillaries were entirely free of objectionable fluorescence, but such tubes require very high sealing temperatures and cost much more than the standard Kimax 1-mm capillaries. The latter show less fluorescence than many silica glasses. The Toronto arc which was used as a light source years ago, before the advent of laser spectroscopy, was run under conditions that generated much heat. The short wavelength of excitation, 4358 Å, produced fluorescence from many samples, often obscuring the weaker Raman bands. The sample often decomposed. The presently used heatless 80-mW He–Ne laser emits a monochromatic wavelength so long that samples seldom fluoresce or decompose. The energy of the laser beam today is roughly 10,000 times greater than that of the strongest Toronto arc. Use of such focused high-density energy allows excitation of very small sample volumes with no loss of scattering energy. In addition, the Ar^+ and Kr^+ lasers give one a choice for selecting a particular excitation frequency, thus avoiding the chances of decomposition, absorption, and fluorescence.

The methods used to handle lyophilized powders and aqueous solutions in order to obtain their Raman spectra have been described by Yu *et al.* (1972) in an investigation of the conformation of insulin and proinsulin.

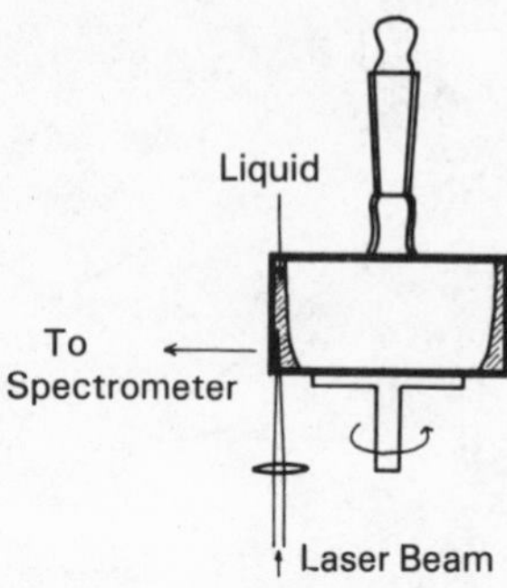

Fig. 2.3. Cell for resonance Raman excitation in liquids and its optical arrangement. (Kiefer and Bernstein, 1971.)

Yu and Jo (1973) and Yu *et al*. (1972a) have stated that freeze-drying a protein may induce small changes in both the backbone and side-chain conformations of globular proteins.

Fox and Tu (1979) have described a sample-heating apparatus for use with Raman spectroscopy. It is useful at relatively high temperatures (+130°C). These workers measure the amide I and amide III regions of 10% solutions of lysozyme at 24°, 50°, 55°, and 60°C and showed progressive changes from predominantly α-helical to random coil form with increase in temperature.

Kiefer and Bernstein (1971) have described a simple, convenient rotating cell for use in resonance Raman excitation with lasers in liquids. Use of the cell (Fig. 2.3) avoids thermal decomposition of the sample due to local heating; it also avoids having to compromise between absorption and emission of the scattered light. The cell is designed for a Raman spectrometer with 90° viewing. Improved spectra were found by these authors, including more overtones than previously found by another laboratory for I_2 in an organic solvent and an increase in band intensities by a factor of ~10. Aqueous solutions of the MnO_4^- ion also showed spectra that were improved over those in a static cell.

Photolysis, photoreduction, or thermal decomposition may be caused by laser irradiation. To overcome this problem, the sample has been circulated (Woodruff and Spiro, 1973; Ondrias *et al.*, 1980), spun (Kiefer and Bernstein, 1971), and scanned at its surface (Koningstein and Gächter, 1973). By use of a back-scattering arrangement (Vergoten *et al.*, 1976) the Raman spectra of samples ordinarily decomposed by laser radiation can be observed at high or low temperatures.

Anderson and Kincaid (1978) have developed a strictly anaerobic, self-contained, circulating flow cell of small volume (less than 1 ml), and have applied it to the resonance Raman investigation of a HbCO complex, and to cytochrome *c*. The cell was effective in reducing photodissociation of light-sensitive compounds.

Infrared

The methods for handling samples to record infrared spectra in a biochemical laboratory are essentially those used in the everyday testing by the organic chemist, with certain exceptions.

Solids

One can readily record the spectra of many solid compounds as mulls and as alkali halide pellets (or even without the use of a suspending medium, as in films, or by means of attenuated total reflection). The solid particles should be extremely fine (5μm or less) or excessive loss of energy due to scattering of light will result. One can minimize scattering losses by dispersing the solid in a medium having a similar refractive index, but scattering losses can be considerable even when the solid is dispersed in mineral oil (as a mull) or in alkali halide (as a disk).

Crystalline samples sometimes produce spectra with distorted band shapes, an effect known as the Christiansen effect [See Potts (1963)]. Also polymorphic forms of the same substance frequently show differences in infrared spectra. An example is *N*-benzoyl-2,3,4,6-tetra-*O*-benzoyl-β-D-glucosylamine, a compound that exists in a form with melting point 113–115°C which, when heated to 117–120°C and allowed to crystallize from the melt, gives a form with melting point 184°C having a somewhat different spectrum in Nujol (Tipson, 1968). Also, different crystal habits (same melting point) of a compound may display partially differing spectra, especially if examined as mulls, in which little pressure is applied. Shifts of up to 20 cm^{-1} for certain bands have been observed (Barker *et al.*, 1956) for crystalline and amorphous forms of some carbohydrates. In all such instances, however, spectra of samples of each of the forms, recorded after dissolution in the same solvent, or as a molten substance, are identical.

The Mull Technique. The proper preparation of a mull or paste is an excellent way to get a sample ready for qualitative infrared spectroscopy. One grinds vigorously about 3 to 10 mg of substance with a hard pestle in a hard and smooth mortar (e.g., agate) for 1 to 5 minutes until the powder is so fine that its caked surface takes on a "glossy" appearance. Then *a small drop* of mulling fluid is added, and the vigorous grinding continued until the slurry has the consistency of cold cream. Another small drop of mulling fluid may be added if the slurry seems too dry, and grinding is continued just beyond the point at which the last substance caked on the mortar becomes dispersed in the slurry. A rod with a rubber tip, known to the chemist as a "policeman," is then used to scrape the material and transfer it to a *flat* sodium chloride or potassium bromide plate. Another *flat* salt plate is placed on the slurry, and the slurry is squeezed to form a thin uniform film by a gentle rotary motion. The "sandwich" formed is placed in a holder, as shown in Fig. 2.4, for mounting in the spectrophotometer. After recording, plates are easily cleaned by rubbing on a flat polishing cloth wetted with acetone, and then dried by rubbing on a dry part of the cloth.

Mulling agents commonly used are Nujol (mineral oil), perfluorokerosene (Fluorolube), and hexachlorobutadiene. Figure 2.5 shows spectra of the first two substances. Fluorolube is a polymer of —(CF_2—CFCl)— units. One can obtain a spectrum essentially without interfering bands from mulling agents by recording the spectrum of a Fluorolube mull from 4000 to ~1330 cm^{-1} and then recording from 1330 to 400 cm^{-1} with a Nujol mull. Hexachlorobutadiene absorbs strongly in the regions 1640–1510 cm^{-1}, 1200–1140 cm^{-1}, and 1010–760 cm^{-1}. Nujol is not usable in the near-infrared region because of interference from C—H overtone and combination bands, and has no appreciable absorption bands in the far-infrared region.

Frequently, it is quite difficult to grind up plastic or rubbery substances, and such materials lend themselves more readily to casting as a film on a sodium chloride or

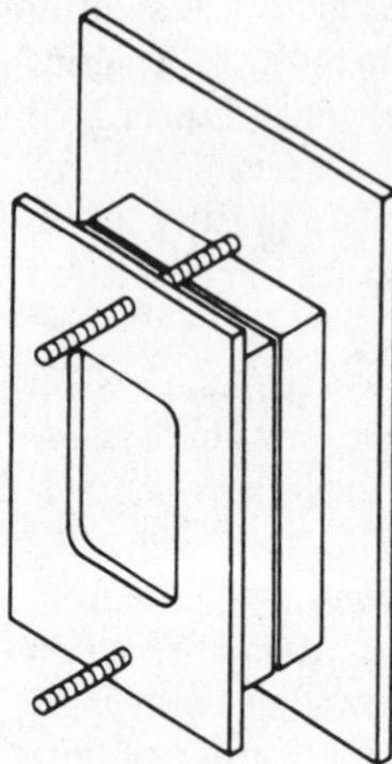

Fig. 2.4. A simple device for holding a capillary film between salt plates. (Potts, 1963.)

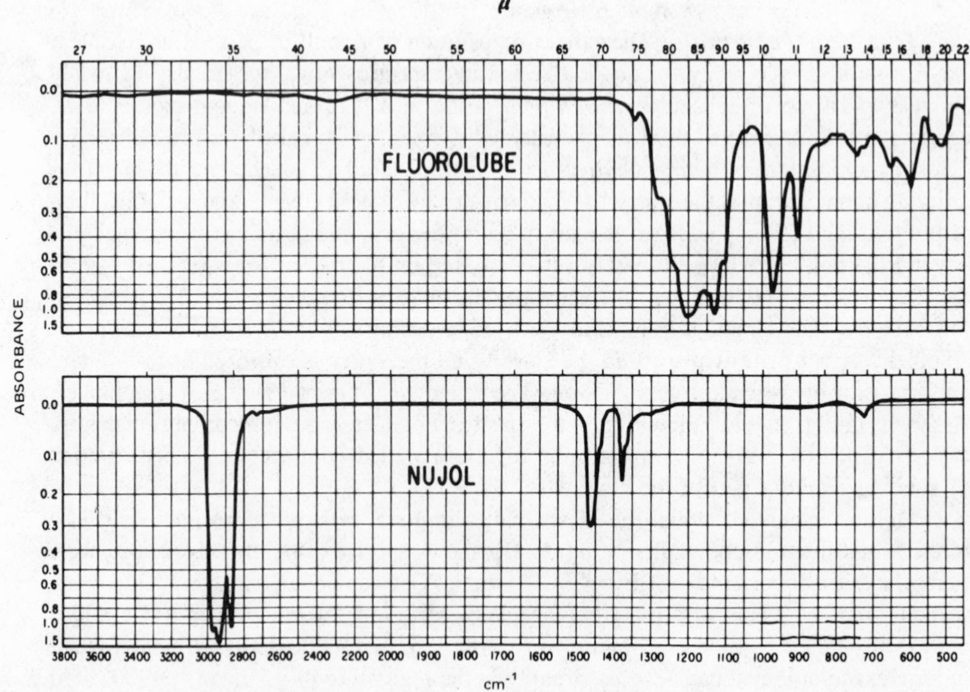

Fig. 2.5. Spectra of Fluorolube and Nujol (capillary films). (Potts, 1963.)

potassium bromide plate. This is done by dissolving the sample in a volatile solvent, pouring the solution onto a salt plate, evaporating off the solvent, and mounting the plate in a holder, as shown in Figure 2.4.

Biological polymers such as fibrous proteins, cellulose, resins, and other natural materials are not soluble in volatile solvents, but certain types of proteins and polysaccharides have frequently been cast as films from aqueous solutions onto silver chloride plates. To obtain mull spectra, however, one must use other methods. For example, a little finely powdered sodium chloride (particles $<2\mu m$ size) admixed with the sample in the mortar can act as an abrasive. If the material is quite hard, a fingernail file may be used to produce a powder which can be collected in the mortar. The powdered form often makes mulls readily. One should use a "dry box" or some other enclosed space purged of moisture for making mulls of hygroscopic materials, since they will otherwise produce poor spectra with prominent absorption by water.

Quantitative analysis by the mull technique is difficult, but internal standards can reduce such problems. Internal standards such as calcium carbonate and lead thiocyanate have been used in quantitative analysis with mulls (Barnes *et al.*, 1947; Bradley and Potts, 1958).

The Pellet Technique. The use of alkali halide pellets in which the organic sample is uniformly suspended has become very popular for recording spectra. The method of forming the pellets consists of grinding a few milligrams of the sample with about a gram of an alkali halide, placing the mixture in a die, evacuating air to remove moisture, and compressing the mixture to form a transparent disk about 1 mm thick under a pressure of about 80,000 lb/in.^{2}. (Special dies and presses can be obtained from manufacturers.) The disk is then removed from the die, placed in a holder of suitable dimensions, and scanned directly in the spectrophotometer. Micropellets ($\sim$5 mm^2) have also been prepared by various methods. A common way is to place a piece of blotting paper with a small central hole into the die, put the desired amount of mixture into the hole, and apply pressure. The microdisk can then be examined while supported by the paper in a holder. A beam condenser may be used for working with micropellets to ensure that adequate energy impinges on the sample. The very pure nonhygroscopic matrix material must be stable and have high transmittance in the spectral range examined. It must also have a low sintering pressure and the proper refractive index. The best matrix material down to 400 cm^{-1} is potassium bromide. Sodium chloride, potassium chloride, potassium iodide, and cesium bromide are also used, the last being suitable down to 290 cm^{-1}.

Owing to moisture in the atmosphere, pellet spectra often show weak hydroxyl bands when there is no hydroxyl group present in the sample. Care should be used when one interprets spectra of pellets in the O—H stretching region. A check on this problem is to record the spectrum of a disk of the alkali halide alone.

A very convenient and good method for grinding and mixing the sample with the matrix is to use a mechanical device, such as the "Wig-L-Bug" amalgamator, which is sold by the Crescent Dental Manufacturing Co., Chicago, Illinois. The powder and several steel balls are put into a stainless steel capsule and then placed in the machine, where they are shaken vigorously for a timed period (timing helps to achieve reproducibility in the spectra). This process yields satisfactory mixing and grinding, and the technique can be used for quantitative work.

There are other ways of mixing sample with matrix material. Usually they involve hand grinding with a mortar and pestle or a ball mill. The sample is sometimes ground separately in the presence of a highly volatile solvent to obtain improved distribution of particle size. A solution of the sample in a volatile solvent may be added to the matrix powder. During the grinding of the matrix powder, the solvent evaporates and leaves a fine dispersion of the sample on the matrix particles. A difficulty encountered with methods involving evaporation of solvent is that moisture is absorbed during the process.

A very useful technique for preparing a mixture of a biochemical substance and alkali halide is to freeze rapidly a solution of the mixture and remove the water under vacuum (lyophilization). Although this method is not used in many organic and spectroscopic laboratories because of the need for special apparatus and the length of time required for sample preparation, it is frequently used to advantage in biochemistry laboratories, most of which own lyophilization equipment.

When used properly, the mull technique and the pellet method both yield good results for solid-state spectra. The pellet technique does, however, have certain advantages: interfering bands are absent; concentration and homogeneity of sample are easier to control; small samples are easier to handle; and samples can be stored for future use. On the other hand, the pellet technique has the disadvantage of chemical interaction between certain samples and the alkali halide, thereby producing spectra not characteristic of the sample. Such anomalous spectra may come from chemical and physical alterations produced by grinding the sample and halide together. One often observes such changes for polar and ionic compounds. Frequently, the pellet spectrum of a compound is not identical with the mull spectrum, and these anomalies in pellet spectra are caused by exchange of ions, formation of solid solutions, orientation effects, relaxation of stress (effects of pressure), and crystallization (polymorphism).

When certain crystalline sugars are suspended in alkali halide pellets they are changed to amorphous forms (Farmer, 1957 and 1959). The progressive changes in the spectra of some sugars, for example α-D-glucose, prepared as a pellet, are due to complex formation with trace amounts of sodium bromide or sodium iodide present as contaminants of the corresponding potassium halide (Farmer, 1959). Moist alkali halide may allow the sugar to mutarotate or form a hydrate while the pellet is stored (Barker *et al.*, 1954).

Tipson and Isbell found that, even though specimens were examined immediately after preparation of pellets, 8 of 24 aldopyranosides (1960) had interacted with the halide, and 6 of 27 free sugars (1962) displayed spectra with different characteristics in potassium iodide and in Nujol.

Other effects which may cause differences between pellet and mull spectra are particle size effects, differences in the refractive index of the medium, lattice energies of the sample and the matrix material, adsorption of the sample on the matrix, and atmospheric interactions with the matrix (Rao, 1963). It should be kept in mind that the crystal lattice is distorted during grinding and pelleting. When pressure is removed the distorted structure undergoes degrees of reversion to the original structure with time, and as a result the pellet spectrum at various intervals of time after the removal of pressure shows gradual changes during the recrystallization of the sample from a disordered to an ordered state. Attention has been called to the effect of the matrix material (KBr, KI, CsBr, CsI) on band contours in the pellet technique (Schiele, 1966), and to the effect of this type of sample preparation on the distribution of rotational isomers (Park and Wyn-Jones, 1966).

Use of Solid Films. Solid films have been examined frequently for infrared analysis in biochemical work, for example, in structural studies of proteins, polypeptides, and polysaccharides. Such films have been of particular value for studying polarization spectra of macromolecules in intact films and in oriented ones (stretched, rolled, or stroked), thereby permitting knowledge to be gained concerning spatial arrangements within the molecule and conformational effects among molecules. A few workers have discussed the film technique (Lecomte, 1948; Randall *et al.*, 1949; Hacskaylo, 1954).

To prepare a film one allows a solution of the sample to evaporate slowly on the face of a suitable window material. The plate may be gradually heated if necessary. For many compounds this means dissolution in a volatile organic solvent, and an alkali halide plate is used. Films of such substances as proteins and mucopolysaccharides have frequently been cast from aqueous solution on a silver chloride plate. IRTRAN-2 (ZnS) and KRS-5 (thallium bromoiodide) plates can also be used.

To obtain uniform films the plate may be placed on a platform that can be leveled properly. Thickness is controlled by adjusting the concentration of the solution. Poor spectra are obtained from layers of grainy crystals. The orientation of molecules in some preferred alignment leads to difficulties in film spectra. Interference fringes found frequently in the spectra of films cause difficulties with analysis. The spectrum of a polystyrene film, for example, shows interference fringes in the 4000–2000 cm^{-1} region. These fringes are absent in the spectrum determined by the attenuated total reflection technique. Interference fringes can be prevented by placement of a film of Nujol between the sample film and the window material (Lutinski, 1958). Melts of compounds can be deposited on windows, but gradual warming of the windows is also necessary to prevent cracking. The examination of powders rubbed into polyethylene film has been described (Schwing and May, 1966).

Solution Techniques

The biochemist is quite familiar with ultraviolet and visible spectroscopy, in which a compound is frequently dissolved in an aqueous solution, a good spectrum is obtained, and quantitative analysis can be readily applied. In the case of infrared spectroscopy a common method of obtaining a spectrum is to dissolve the sample in an appropriate solvent, place the solution in a suitable cell, and record the spectrum. Certainly the solvent must have reasonable transparency to infrared radiation in the region to be used. This method is used widely in qualitative analysis, and is the most commonly used method in quantitative analysis.

If it is possible to use a solution for recording the spectrum of a substance, there are definite advantages to be had by doing so. It is easiest to interpret a spectrum when the molecule is in the simplest, least complicated, and most reproducible environment. With such conditions established, one can more facilely make useful correlations between vibrational frequencies and molecular structure. The best way to maintain materials in similar and simple surroundings is to dissolve them in dilute solutions in an inert nonpolar solvent.

Because water dissolves alkali halides and has intense interfering absorption bands, it has not been used much by spectroscopists for infrared work; but the easy availability of heavy water (D_2O) has made it possible to use both forms of water in conjunction to obtain adequate spectra for many biochemical substances in cells which are insoluble in water. A discussion of the use of water as a solvent is given later under *Aqueous Solutions*.

When studying spectra of solutions one should be aware that interactions between solute and solvent have effects on band position and intensity, but it seems preferable to put up with the relatively mild interactions between solute and nonpolar solvent than to try to identify frequencies from the much stronger interactions between adjacent polar molecules in a pure liquid or the even stronger interactions between juxtaposed molecules in a crystal lattice. Moreover, in a pure nonpolar liquid the interactions of adjacent molecules are just as important as interactions between these molecules and nonpolar molecules of solvent in a dilute solution.

Without dilution, many of the absorption bands of pure substances are so intense that spectra must be recorded from very thin layers to obtain bands of appropriate intensity. It is difficult to reproduce very thin optical-path lengths in cells for thin layers of sample.

Hydrogen bonding between compound and solvent may cause band shifts or alterations. In solution, molecules of a compound may form intermolecular hydrogen bonds. With the concentration less than 0.005 *M*, very little intermolecular hydrogen bonding occurs for solutions in such solvents as carbon tetrachloride (Kuhn, 1952, 1954).

Using standard solution techniques allows one to develop a sense of the relationship of band intensity to molecular structure and to apply this information in determining the structure. The problem that arises with the solution technique is that a nonpolar solvent does not exist which is transparent to infrared radiation throughout the ranges of interest and which dissolves every organic substance. However, several good solvents exist which have adequate transparency in useful ranges of the infrared region. A very satisfactory method commonly used for recording spectra of solutions is to use one solvent in certain regions and another solvent in other regions. The two solvents most often used because of their transparent properties and because many organic compounds dissolve in them are carbon tetrachloride and carbon disulfide (Fig 2.6). The useful range for carbon tetrachloride is 4000 to ~1330 cm^{-1}, except for a narrow range between 1475 and 1600 cm^{-1}, where tetrachloroethylene may be used. The useful range for carbon disulfide is ~1330 to 450 cm^{-1}. Various concentrations of solute may be used in a variety of cells with different thicknesses (optical paths), the best compromise for having adequate concentration of solute, satisfactory transparency of solvent, and ease in cleaning cells being the use of 10% solutions (weight per volume) in cells with 0.1 mm optical path.

Many compounds are not soluble enough in either of these two solvents, for example, organic and inorganic substances of ionic nature, highly polar substances, and most polymers. Other solvents can be used for dissolving these materials, but most of these solvents have strong infrared absorption or low transmission, and they are usually not satisfactory for use in the identification of compounds or for structural determinations. Therefore, spectra of polar substances for such uses are recorded by techniques mentioned earlier. Many solvents are available, however, which dissolve polar substances and do have adequate transmission in some regions, so that it is still possible to do quantitative analysis by the preferred solution methods.

A chart showing the useful regions of solvents in the 5000–625 cm^{-1} region is presented in Fig. 2.7. Figure 2.8 shows the useful regions of solvents from 667 to 286 cm^{-1} (Szymanski, 1964).

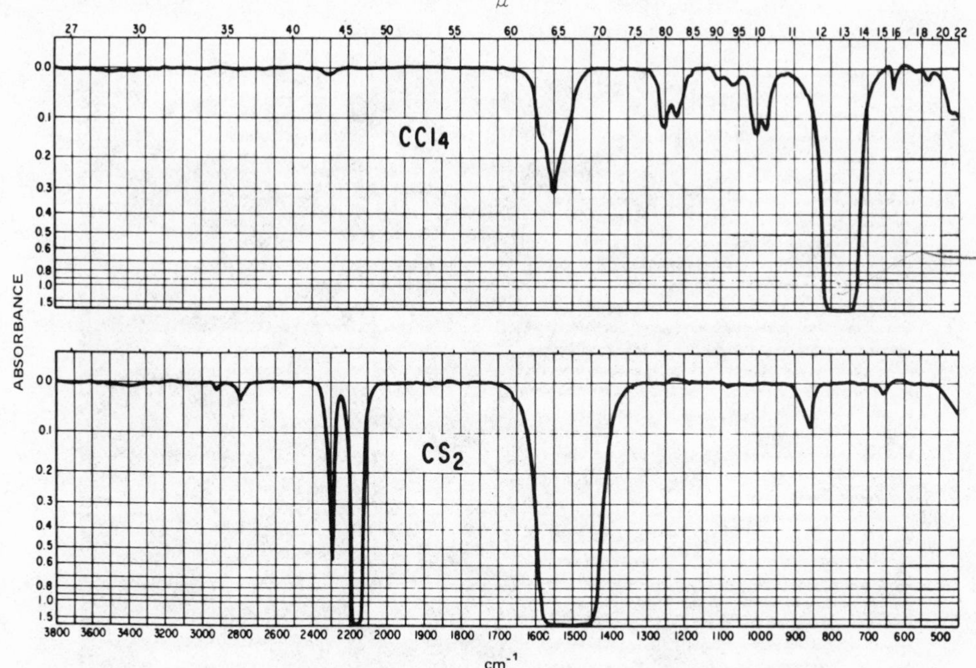

Fig. 2.6. The spectra of carbon tetrachloride and carbon disulfide liquids, cell length ~ 0.1 mm. (Potts, 1963.)

Chloroform has transmission properties such that it is not quite as useful as carbon tetrachloride or carbon disulfide, but it dissolves polar substances better. Methylene chloride is also a much better solvent for polar compounds and its transmission properties are almost as good as those of chloroform.

A wide range of polar materials is soluble in acetone, acetonitrile, dioxane, dimethyl formamide, and nitromethane, all of which have some spectral regions free of absorption where measurements are feasible.

McNiven and Court (1970) have published the spectra of twelve solvents in their deuterated and undeuterated forms, along with a chart showing regions of transmittance over 10% (Fig. 2.9) for all of them. These deuterated solvents will provide additional "windows" for studying the infrared-frequency shifts of various functional groups in solvents [see reviews by Rao (1963) and Hallam (1961)]. The transparency in certain regions will also allow the deuterated solvents to be used for quantitative analyses of dissolved solute mixtures where absorption bands are obscured by nondeuterated solvents. Data for several solvents in the near-infrared region are presented in Fig. 2.10 (Goddu and Delker, 1960).

Compensation of Solvent Absorption Bands. It is very advantageous, of course, to be able to "remove" solvent absorption bands in a spectrum. In double-beam infrared spectrophotometers, atmospheric absorption bands are readily "blanked out." These instruments can also be used to blank out solvent absorption bands, thus producing

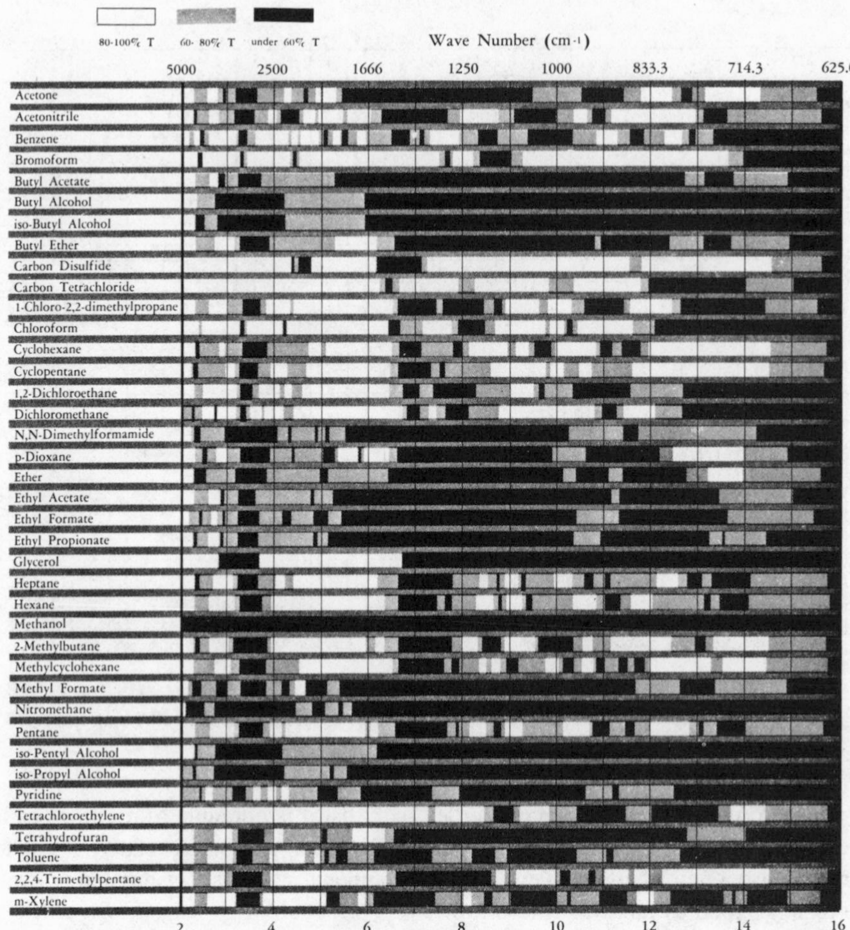

Fig. 2.7. Solvents for the 5000–625-cm^{-1} region. Maximum transmittance in the infrared is shown in three ranges, below 60% transmittance, 60%–80%, and 80%–100% transmittance, obtained with a 0.10-mm cell path. (Courtesy of Matheson Coleman and Bell Division of the Matheson Company, Inc.)

an unperturbed baseline. This is done by placing in the reference beam of the spectrophotometer another cell of matched optical path containing pure solvent. The low-intensity bands of carbon tetrachloride (Fig. 2.6) in a 0.1-mm cell are blanked out easily in this way. This is true for other solvents having some bands of low intensity (e.g., acetone at ~2600 and 2000 cm^{-1}, and dimethyl formamide at ~1950 cm^{-1}, in 0.1-mm cells).

One must avoid attempting to cancel out any solvent band having *total* absorption, for example, the broad band in the region 1600–1400 cm^{-1} in the carbon disulfide spectrum (Fig. 2.6). If one attempts to blank out this band and to obtain the spectrum of a solute that absorbs in this region, no band(s) would be recorded, since in this region there is

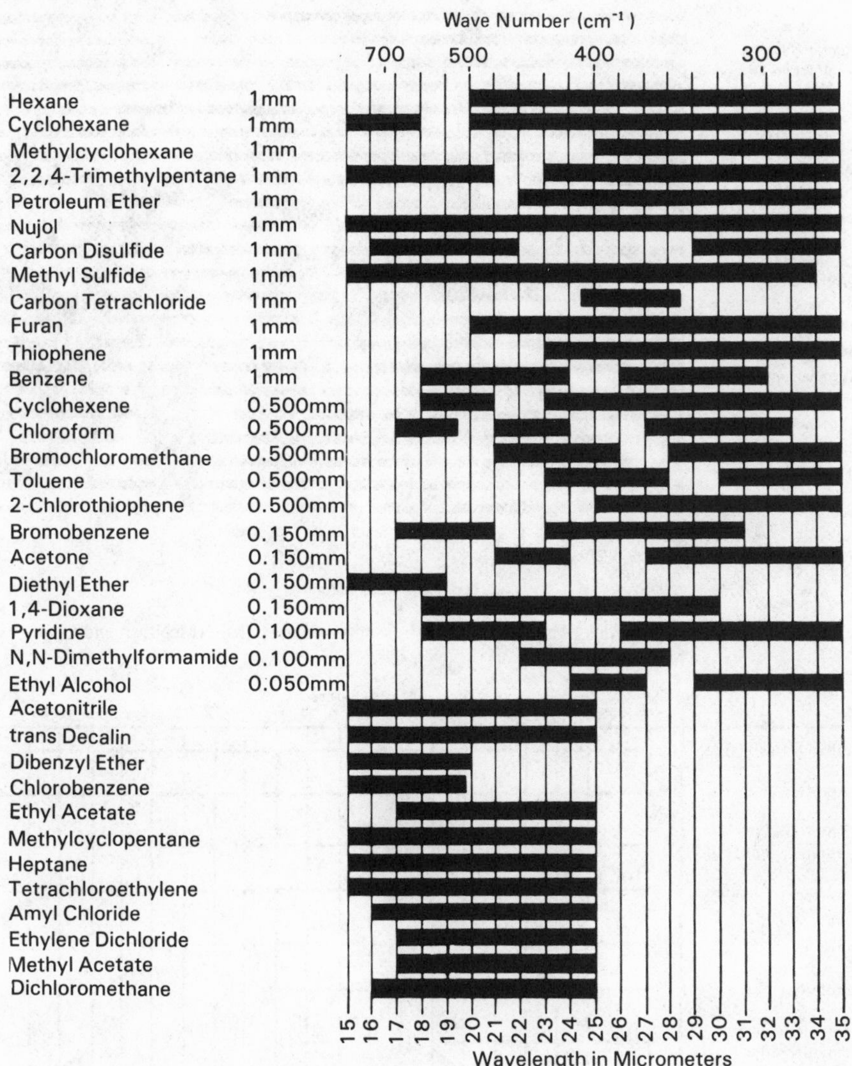

Fig. 2.8. Solvents for the 667–286 cm^{-1} region. The black lines represent useful regions. (Szymanski, 1964.)

no infrared radiation at all impinging on the detector (except perhaps scattered light), and the servo system is completely inactive. Thus, the region 1600–1400 cm^{-1} would display a meaningless straight line in the spectrum, with perhaps some upward or downward movement of the pen due to drift of the servo system. A difficulty of a similar nature will also be observed if compensation is attempted on a strong but not totally absorbing band. One will obtain misinformation or very inaccurate data if he tries to do

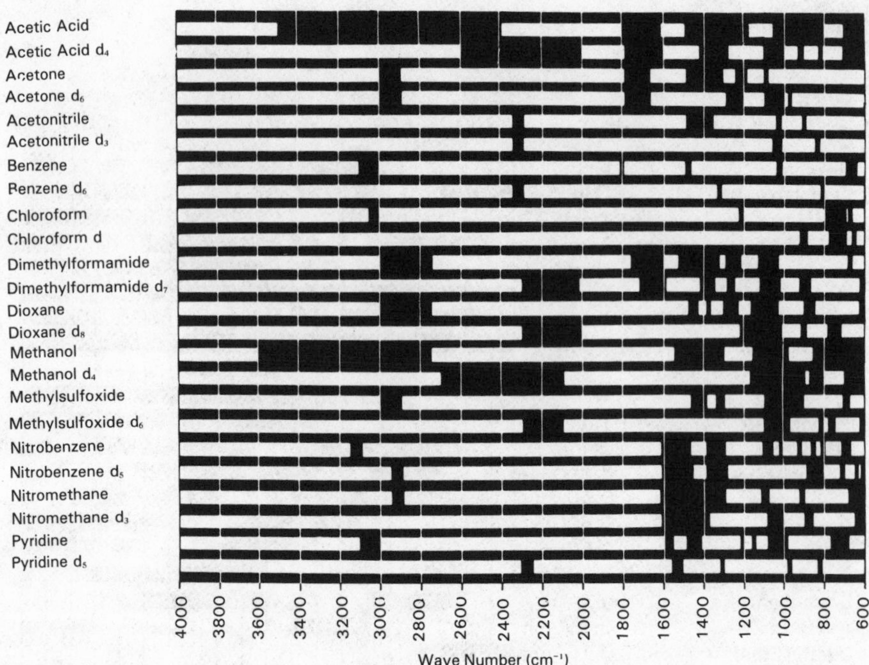

Fig. 2.9. Solvents showing regions (white) of infrared transmittance over 10%. (McNiven and Court, 1970.)

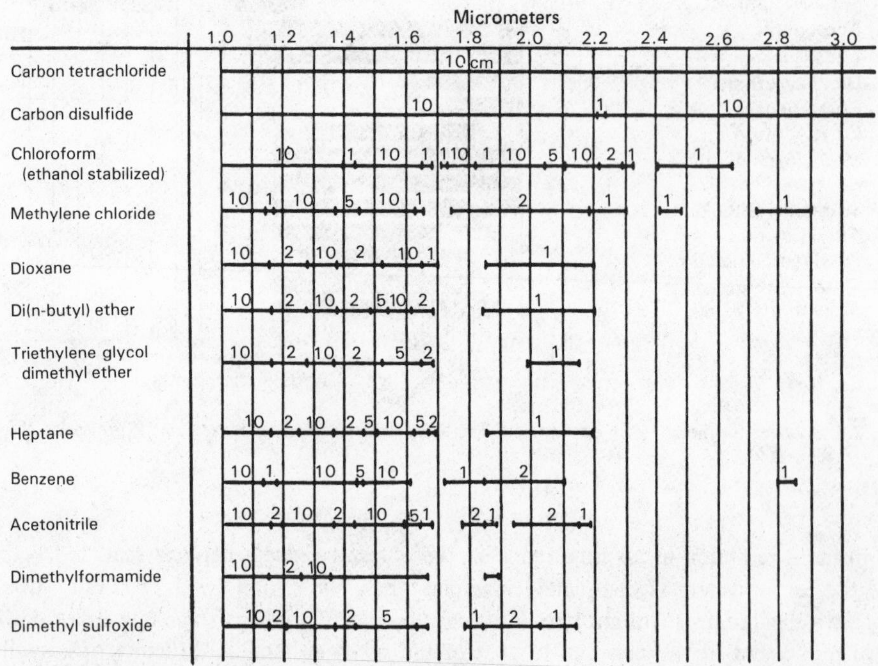

Fig. 2.10. Data on solvents in the near-infrared region. Maximum path length (cm) which can be tolerated in qualitative work in each case is also indicated. (Goddu and Delker, 1960.) Reprinted with the permission of the American Chemical Society. Copyright 1960.

quantitative measurements on a solute in the region of such a solvent absorption band, or even to interpret it correctly. The reader is referred to more detailed discussions of such problems in Potts (1963).

Many aliphatic and aromatic hydrocarbons are useful as solvents in the 700 to 290 cm^{-1} range of the far-infrared region. For example, there are 1,4-dioxane, furan, thiophene, benzene, hexane, cyclohexane, and 2,4,4-trimethylpentane. Benzene has adequate transparency (>40%) in the 555–285 cm^{-1} region and dioxane has greater than 60% transmittance from 555 to 322 cm^{-1}. Acetone transmits well in the regions from ~475 to 425 and ~375 to 300 cm^{-1}. Ether transmits well in the regions from ~700 to ~525, ~415 to ~390, and 350 to 325 cm^{-1}. By using a combination of acetone, ether, and dioxane one can examine the whole cesium bromide region (Bentley *et al.*, 1958). Cells of optical path as large as 0.2 mm are satisfactory for polar solvents such as acetone.

Cells for the Use of Liquids. Many types of cells are available from manufacturers for use with liquids. Various window materials are available (NaCl, KBr, CsBr, AgCl, CaF_2, BaF_2, IRTRAN-2, thallium bromoiodide, etc.) in both the demountable and the sealed cells. For these cells there are available spacers of various thicknesses of lead, Teflon, polyethylene, etc. Since the fixed-thickness sealed cells often contain lead spacers cemented to the windows by a mercury amalgam (and these metals are biological poisons) it is wise to consider whether the use of such cells is advisable in a particular instance (e.g., an aqueous enzymic system in one such cell). Cells are available for semimicro, micro, and ultramicro work. Variable path cells of the micrometer or wedge type are available, and these are particularly useful for solvent compensation.

Figure 2.11 shows one type of completely demountable cell for liquids. The parts are laid out in order of assembly as follows: cell frame, supporting gasket, window spacer(s), window with holes, gasket, metal plate with holes for syringes and Teflon stoppers, O rings, and the screw-on cover for tightening the cell.

Obviously, any cell material should be insoluble in the solvent(s) being used. Most nonpolar solvents like carbon disulfide, carbon tetrachloride, chloroform, tetrachloroethylene, benzene, etc., can be employed in cells of any material, while polar solvents like methyl alcohol cannot be used in NaCl cells.

Aqueous Solutions. Coblentz (1905) had used water as an infrared solvent as early as 1905, and Gore *et al.* (1949) had studied aqueous solutions of several amino

Fig. 2.11. Complete demountable cell in order of assembly. (Courtesy of Barnes Engineering Co.)

acids in 1949. Blout has published many infrared spectra of biochemical polymers in water and D_2O solution, examples of which can be found in Blout and Lenormant (1953) and Blout (1957). Figure 2.12 (Blout, 1957) shows absorption spectra of water and D_2O (with and without compensation) of 0.025 mm thickness in the region 4000–600 cm^{-1}. It can be seen that D_2O transmits where water absorbs and vice versa, thus making the combination of these solvents useful for examining aqueous solutions. The O—H deformation modes of water are present between 1700 and 1600 cm^{-1} and the O—D deformation of D_2O lies at ~1200 cm^{-1}. Except for these regions the spectra show better than 40% transmittance and satisfactory compensation is readily obtained in a double-beam spectrometer.

Maxwell and Caughey (1976) have studied hemoglobin solutions in D_2O 4.9 m*M* in heme in CaF_2 cells with a path length of 0.26 mm, and found it difficult to detect a weak band due to Hb-bound $^{15}N^{16}O$ at 1587 cm^{-1}. The difficulty here was that percent transmittance was recorded for hemoglobin solutions versus air. But when they used difference spectroscopy they were able to obtain useful spectra of hemoglobin solutions 2.7 m*M* in D_2O phosphate buffer (see below), and had no problem detecting the 1587 cm^{-1} band.

Reviews have appeared on the subject of aqueous solution infrared spectroscopy, each giving emphasis to different applications (Nachod and Martini, 1959; Goulden, 1959; Jencks, 1963; Parker, 1962, 1967). Table 2.1 shows some infrared frequency assignments in D_2O solution (see Jencks, 1963, for details).

FTIR experiments are discussed in other chapters for cases in which the spectrum

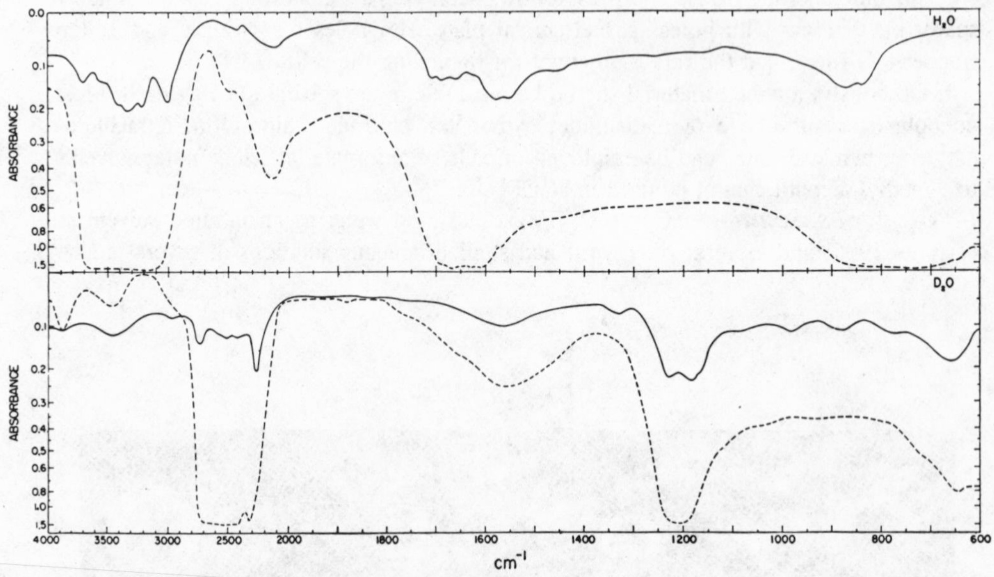

Fig. 2.12. Absorption curves of H_2O and D_2O in 0.025-mm cells: Broken curves, spectra with no compensation in the reference beam. Solid curves, spectra showing balance obtainable in a double-beam spectrometer. (Blout, 1957.)

Table 2.1. Some Infrared Assignments in Aqueous Solution[a]

Bond	Frequency[b] (cm^{-1})
	A. X—H
O—H of H_2O (solvent)	2800–3800, 2100, 1600–1800
O—H of HDO	3380, 1455(b)
O—D of D_2O (solvent)	2200–2850, 1550(w), 1150–1250
CH_3	2950–2980(m), 1420–1470(m), 1365–1390(m)
CH_2	2915–2950(m), 1430–1480
CH_2, α to C═O	2915–2950(m), 1405–1435
C—H, aromatic and olefinic	3010–3130(m)
	B. C═O (all strong or very strong)
Esters	
Normal	1710
Ethyl carbamate	1681
Acetylcholine chloride	1735
Alanine ethyl ester hydrochloride	1743
Ethyl trifluoroacetate	1786
Acids	
Normal	1710
α-Keto and amino acids	1720–1730
Acyl phosphates	
Acetyl phosphate dianion	1713
Acetyl phosphate monoanions	1735–1750
Carbamyl phosphate	1670
Thiol esters	1670–1680
Amides	
Normal	1625–1680
Urea	1604
Acetylimidazole	1740
Aldehydes and ketones	1665–1730
Carboxylate ions	
Normal	1560–1590, 1405–1420
α-Keto and amino acids	1610–1630, 1400–1410
Bicarbonate	1630, 1363
Carbamate	1540, 1441
Carbonate	1416
	C. C—O
Esters and acids	1100–1350(s)
Alcohols and ethers	1050–1200(s)
	D. C═C
General	1600–1680(w—m)
Aromatic	~1600, ~1500, ~1450 (variable intensity)

Continued

Table 2.1. *(Continued)*

Bond	Frequency[b] (cm^{-1})
	D. C═C *(continued)*
Nucleotides (principal ring absorptions at neutrality)	
AMP	1626(s)
UMP	1645–1658(s), 1678–1692(s)
IMP	1674(s)
CMP	1493–1505(s), 1610(s), 1649(s)
TMP	1620(s), 1650(s)
	E. Phosphates
Dianions	975–985(s), 1070–1090(s), 1105–1120 (variable)
Monoanions	1050–1090(s), 1205–1240(s), 915–940(?)

[a] Jencks (1963).
[b] b = broad, w = weak, m = medium, s = strong.

of water is readily subtracted out, thus making aqueous solution spectroscopy quite easy to perform.

The reader will find many references in this book to spectroscopy done on biochemical substances in aqueous solution. Examples of only a few of the types of compounds conveniently studied in water (H_2O) in the region between ~1550 and 950 cm^{-1} are: organic acids of the tricarboxylic acid cycle (Parker, 1958), amino acids (Parker and Kirschenbaum, 1950; Goulden, 1959), and carbohydrates (Goulden, 1959; Parker, 1960).

In a study of NO bonding to heme B and hemoglobin A Maxwell and Caughey (1976) used path lengths of approximately 0.025 mm for solutions of hemoglobin and 0.1 mm for solutions of hemes. For careful work it was necessary to use sample and reference cells of almost identical path lengths in order to cancel out protein absorption bands, thus producing a nearly flat baseline. They obtained such matching by overlapping the interference fringe patterns of the empty CaF_2 cells. Superimposability of the wave patterns for the two cells was attained by the use of spacers of nearly identical thickness and by adjusting carefully the pressures put on the spacers. This procedure of matching path lengths allowed them to obtain satisfactory difference spectra of protein solutions in the N—O stretching region, provided that the protein concentrations were also closely matched in the sample and reference cells.

It is worth noting that, although either H_2O or D_2O may be used as the solvent for the C—O region near 1950 cm^{-1} or the O—O region near 1100 cm^{-1}, only D_2O is satisfactory for the N—O region (between 1575 and 1700 cm^{-1}), since this solvent has a window (Fig. 2.12). The reader should consult Maxwell and Caughey (1976) for a discussion of the problem involved in the detection of a relatively weak band due to hemoglobin-bound $^{15}N^{16}O$ at 1587 cm^{-1} (in the presence of strong amide bands), and how they avoided this difficulty by matching Hb—$^{15}N^{16}O$ in the sample beam with Hb—CO in the reference beam. The difference spectrum readily showed the band due to $^{15}N^{16}O$.

Cells for Aqueous Solutions. Jencks (1963) has discussed the types of cell materials available for work with aqueous solutions in the mid-infrared region. Calcium fluoride and barium fluoride cells are frequently used. The former is transparent down to ~1111 cm^{-1} and the latter to ~769 cm^{-1}. These materials are unstable to acid and concentrated ammonium salts, and are quite expensive, with BaF_2 nearly twice the cost of CaF_2. Zinc sulfide (IRTRAN-2) has reasonably good transmission from 10,000 to ~769 cm^{-1}, is resistant to dilute acids, moderately concentrated bases, and organic solvents, and cells made of it cost less. An example has been given of the use of a zinc sulfide cell to study an enzymic reaction (Parker, 1964).

Very thin cells made of other water-insoluble windows such as zinc selenide and cadmium telluride have been used by Cameron *et al.* (1979) in hydrated systems, e.g., in the study of model and natural membranes.

Inexpensive silver chloride cells can be constructed (Nachod and Martini, 1959). Silver chloride is transparent down to 400 cm^{-1} and has been used in many studies, but it deteriorates with exposure to light, is attacked by amines, and is too pliable to give cells of reproducible path length. Cells made of polyethylene supported on sodium chloride plates have been described (Antikainen, 1958). Other cells made of polyethylene (which is both inert and cheap) have also been mentioned (Robinson, 1959; Gordon, 1963). Teflon-on-NaCl cells were used to study acidic and basic solutions (Fogelberg and Kaila, 1957), and adequate spectra were obtained except in the 1200-cm^{-1} region where there is strong absorption from the Teflon. Various other types of polyethylene and Teflon cells have been used, but their main difficulties are that they are flexible, it is hard to distribute fluid evenly, and it is difficult to obtain matched cells.

For cases in which it is not necessary to use a matching cell in the reference beam, an attenuator (supplied by manufacturers) may be used in the reference beam, or a screen may be employed, in order to obtain adequate spectra from the available energy.

Calibration of Infrared Absorption Cells

For quantitative work it is essential to obtain accurately the length (optical path) of each cell so that comparisons can be made between substances not examined in identical cells. The interference-fringe method is used for this purpose. It is based on the interference of light which is directly transmitted through a cell with light that has been twice reflected internally in the cell. The fringe method is carried out in the following manner:

1. Place the dry, empty cell in the spectrophotometer in the usual sample beam. Do not place a cell in the reference beam.
2. Operate the spectrophotometer as near as possible to the 100% transmittance line on the paper to produce fringes of the greatest amplitude, and obtain a spectrum containing from 20 to 50 fringes as in Fig. 2.13.
3. Calculate the cell thickness by the following equation:

$$b(\text{in millimeters}) = \frac{n\lambda_1\lambda_2}{(\lambda_2 - \lambda_1)(2000)}$$

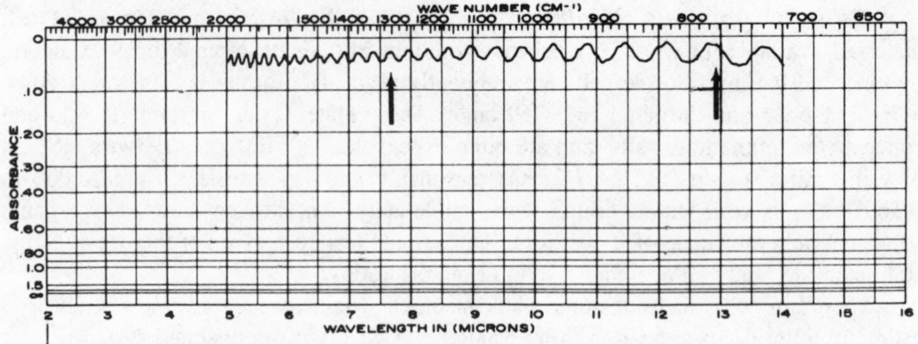

Fig. 2.13. Fringes obtained with a 0.1-mm sealed cell. (Courtesy of Barnes Engineering Co.)

where b is the cell thickness in millimeters, λ_1 is the starting wavelength and λ_2 the final wavelength (in micrometers), and n is the number of fringe maxima between λ_1 and λ_2. When measurements are recorded in wavenumbers, the following equation is used:

$$b(\text{in millimeters}) = \frac{5n}{f_1 - f_2}$$

where f_1 and f_2 are the starting and finishing wavenumber values. In Fig. 2.13 there are 11 maxima between 1305 and 775 cm^{-1}. A cell thickness of 0.104 mm is then calculated.

Cells for Gases

There are fewer molecules in a given volume of gas than in the same volume of liquid, and a greater sample thickness is needed. A commonly used cell has a 10-cm thickness, which yields a reasonable absorbance level for most gases and vapors at convenient values of partial pressure (for example, 5–40 mm Hg). Figure 2.14 shows a simple gas cell with a 5-cm path length. A side-arm is put on the cell so that a small, convenient amount of sample may be frozen out for later expansion into the cell chamber. Windows are made of NaCl or KBr.

Cells of different path lengths and windows can be purchased. Cells with multiple reflections have effectively long path lengths and these are used to permit the study of dilute gases and for measurement of both strong and weak absorption bands in the same sample. Several types of multiple-pass gas cells have been described (White and Alpert, 1958). A good discussion of the use of a 40-m gas cell for atmospheric studies has included lower detection limits for many gases (Hollingdale-Smith, 1966). The variable-path gas cell contained in the MIRAN gas analyzer (Fig. 2.15) has a readily adjustable path length controlled by a knob, so that one can produce any path length between 0.75 and 20.25 m in 1.5-m increments.

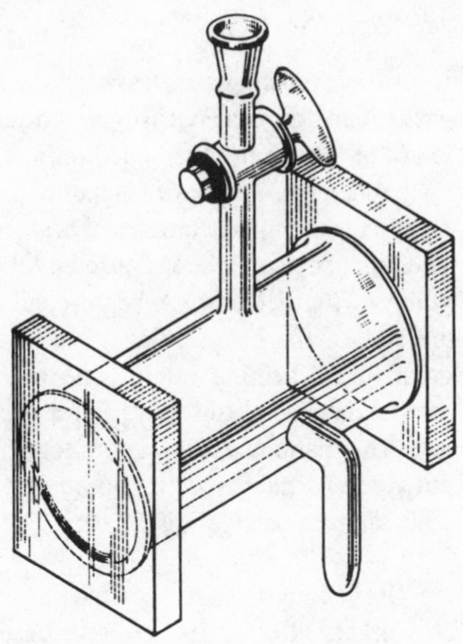

Fig. 2.14. A simple gas absorption cell. (Potts, 1963.)

Infrared spectroscopy can be used to perform qualitative analysis of gases. Pierson *et al* (1956) have given a chart to be used for this purpose. Also, infrared spectroscopy and gas chromatography are a useful combination for characterizing volatile substances. Sometimes fractions from a vapor fractometer are collected as vapors in a gas cell. White *et al*. (1959) have described a microgas cell of 1 m path for use with a beam condenser to observe the spectra of gas chromatographic fractions less than 0.05% of the total charge.

There are many devices used for trapping gaseous fractions from conventional gas chromatographs, to be transferred to the spectrometer (Anderson, 1959; Senn and Drushel, 1961; Flett and Hughes, 1963; Chang *et al.*, 1961). Many descriptions of liquid traps which can be used in conjunction with gas chromatographic instruments are available in the literature. Stewart (1958) has discussed the use of a low-temperature microcell for the study of condensed gases.

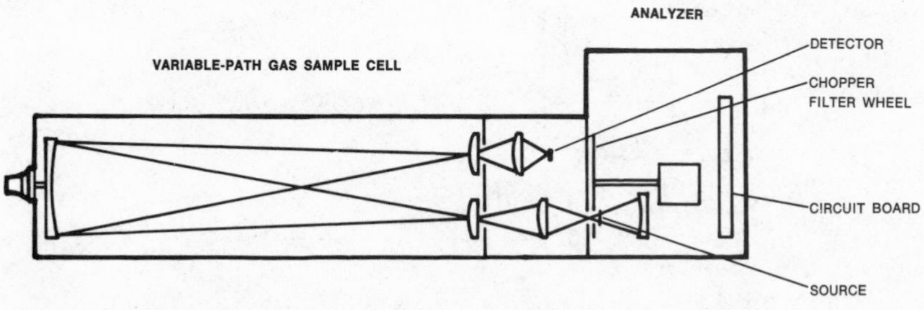

Fig. 2.15. Variable-path gas cell. (Courtesy of Foxboro Co.)

Micromethods

It is sometimes necessary to work with very small samples of solids, liquids, or gases, and for such occasions various types of cells and equipment have been devised. many of which are available commercially.

When the sample size is reduced, the area or thickness of the sample is also decreased. A thinner sample decreases the absorbance, and ordinate-scale expansion is required. The reduced sample area caused by the use of a small cell decreases the amount of energy going to the detector. A beam-condensing device overcomes this problem by reducing the size of the beam going through the sample. When the beam size must be reduced greatly, a reflecting microscope or microilluminator is used. Workers have described several devices of this kind; Barer *et al*. (1949) used one of the first of such attachments.

The Nanometrics Corporation (Sunnyvale, California 94086) has produced a computerized IR microspectrophotometer which gives spectra in the range 4000 to 690 cm^{-1} with samples as small as 20 μm. The optical diagram of this instrument is given in Fig. 2.16.

Reflecting microscopes have been used for samples of 0.1 μg or less. Anderson and co-workers (1953a; 1953b) have described another beam condenser with silver chloride

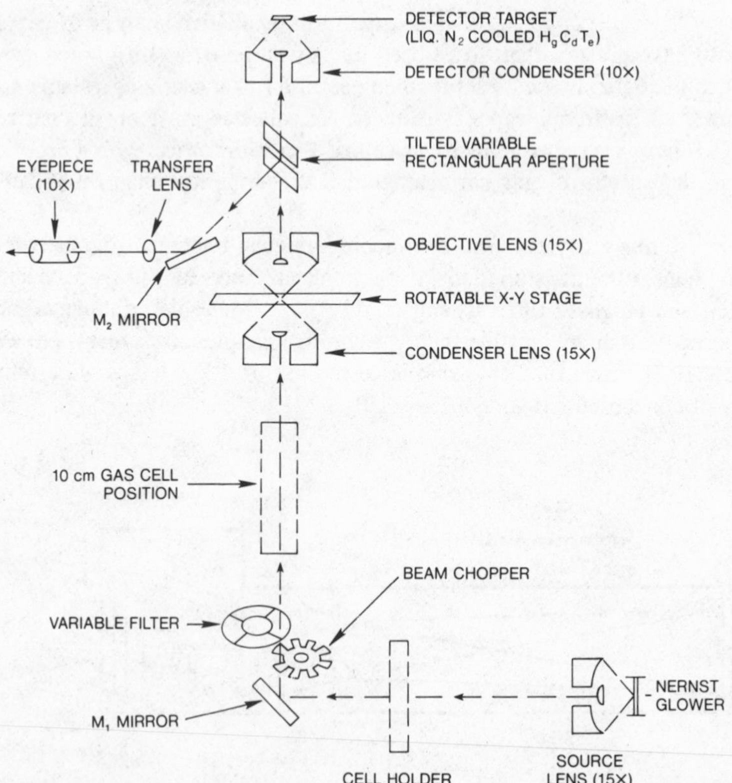

Fig. 2.16. Infrared microspectrophotometer. (Courtesy of Nanometrics Corp.)

lenses that can illuminate 10–100 μg, and White *et al.* (1958) have presented a better model with potassium bromide lenses. Infrared spectra of very small samples can be obtained with reflecting microscopes, but a larger amount of sample is needed since there are losses in preparation and handling. Special cells and methods have been used to lessen these difficulties. Micromethods have their greatest application with solid samples. The use of micromulls (Lohr and Kaier, 1960) and micropellets (Anderson and Miller, 1953a, White *et al.*, 1958; Clark and Boer, 1958; Bisset *et al.*, 1959; Dinsmore and Edmondson, 1959; Resnik *et al.*, 1957) has been discussed. One to 20 μg of sample can be used to make pellets of cross section 0.5–5 mm, weighing from a fraction of a milligram to a few milligrams of the pelleting material.

Many microsamples come from thin-layer chromatographic (TLC) separations. The sample is separated by chromatography, the adsorbent removed from the area of the chromatogram containing the separated material, the sample eluted with a suitable solvent, filtered to remove the suspended adsorbent, and finally mixed with KBr. Garner and Packer (1968) describe a time-saving technique. By using a porous triangle of pressed KBr ("Wick-Stick," Harshaw Chemical Co., Cleveland) in a small glass vial capped so that evaporation is restricted to the center of the vial, the filtration of adsorbent and deposition of the sample on KBr can be accomplished in a single step. The adsorbent containing the sample is scraped from the TLC plate and transferred to a glass vial containing a Wick-Stick by means of a thin-stemmed funnel to prevent the adsorbent from dusting on the top half of the Wick-Stick. A suitable eluting solvent is added and a vented cap is placed on the vial.

The pressed KBr triangle is 2.5 cm high, 0.8 cm wide at its base, and 0.2 cm thick. The vial is 3.5 cm high with a 1.0-cm inside diameter. The stainless-steel cap has a vent hole 0.3 cm in diameter. A stainless steel spring clip holds the KBr triangle upright and centered in the vial. The solvent climbs the KBr by capillary action and evaporation takes place preferentially at the apex of the triangle, depositing the sample (Fig 2.17).

Next, 1 to 2 mm of the tip is cut off with a sharp scalpel, mashed on a clean metal surface, and pressed into a transparent disk with a microdie. Satisfactory spectra for qualitative analysis are obtained from 10 to 50 μg of sample when a 1.5-mm-diameter micro-KBr-die and a beam-condensing unit are used.

Figure 2.18 shows a type of fixed-thickness microcell used for liquids. One can use as little as ~0.02 ml of liquid in cells of regular design. Cells holding ~2 μl and little dead space have been made with 0.1-mm spacers. Demountable cells holding 0.1–0.5 μl can be used with nonvolatile liquids. As little as 0.02 μl of material can be used for recording spectra if slits are masked and spacers reduced to 0.02 mm. The spectrum of the 727-cm^{-1} band of toluene has been recorded by means of 90–100% ordinate scale expansion (Stewart, no date) on only 86×10^{-12} liter in a cell of 0.02 mm thickness and 1 mm^2 area.

Figure 2.19 (Potts, 1963) shows a "cavity cell" which is convenient for microsamples, and although the machining process used to produce the cavity in a small solid block of sodium chloride does not yield a cell of precise optical path length, it does not matter for qualitative work with microsamples. A mounting device is used for placing the cell in a beam-condensing system.

A microcell has been made by drilling holes in a salt crystal (Price *et al.*, 1967).

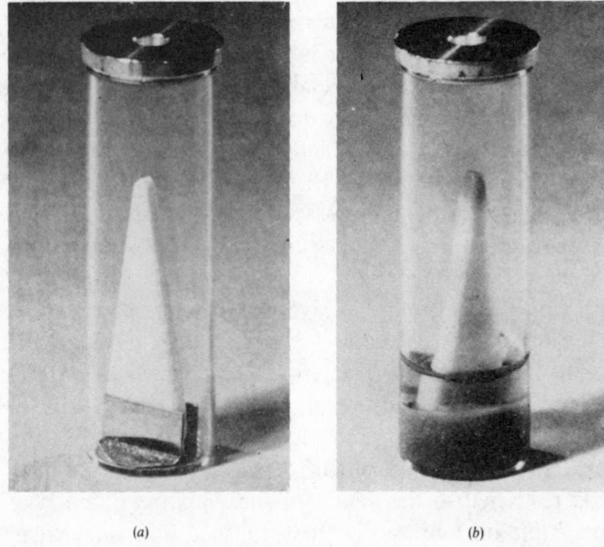

Fig. 2.17. (a) Wick-Stick is shown here ready for use. The triangular-shaped KBr Wick-Stick is placed in a holder clip and set inside the glass vial. Solvent and sample can now be carefully put into the vial and solvent allowed to evaporate from the tip of the Wick-Stick. (b) Sample has migrated to the tip of the Wick-Stick. (Note darkened tip area.) When the solvent has evaporated, the tip of the Wick-Stick containing the sample is sliced off, placed in a KBr die and pressed into a micropellet. (Courtesy of Harshaw Chemical Co.)

Fig. 2.18. Infrared microcell. The item at the left is an adapter to fit the instrument. Filling takes place through the needle by capillarity or suction. (Courtesy of Perkin–Elmer Corp.)

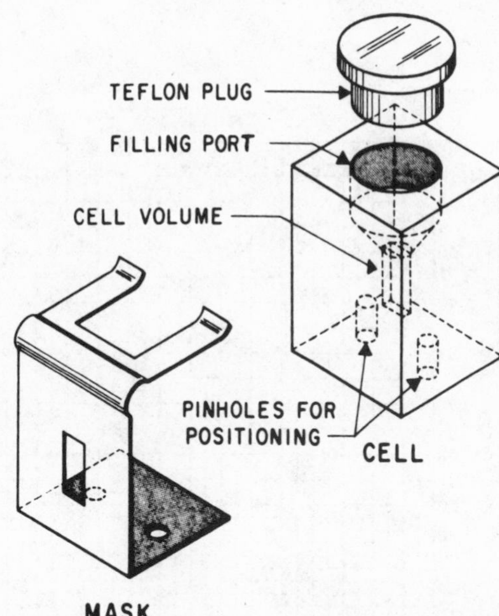

Fig. 2.19. A cavity cell for obtaining spectra of microsamples. (Potts, 1963.)

The spectrum of 5.04 μg of benzyl acetate run in the microcell compared favorably with the spectrum of 2283 μg run in a macrocell.

Low-Temperature Work

There have been a few thorough studies of the temperature dependence of the infrared spectra of simple compounds, but not many such studies with solids have been done. The effect of temperature on vibrational spectra has been discussed by several workers (Walsh and Willis, 1950; Richards and Thompson, 1945; Hainer and King, 1950). Generally, if no change in phase occurs, the effects of the lowering of temperature are: a slight narrowing of the absorption bands, a slight increase in the intensity of bands, and sometimes, some band splitting (Dows, 1963). Hainer and King (1950) have reported the effect of temperature on a portion of the spectrum of cholesterol. Zhbankov *et al.* (1966) studied the spectra of some sugars at liquid-nitrogen temperature, but did not study the whole spectral region. These two studies covered only the region 1500–1400 cm^{-1}, in which the effects of temperature are not so pronounced as in other regions. Katon *et al.* (1969) have discussed critically what they consider to be erroneous conclusions of Zhbankov *et al.* on low-temperature effects.

Caspary (1968a) indicated the usefulness of low temperature for the identification of materials having room-temperature spectra which are similar to each other [e.g., *n*-hexyl bromide and *n*-heptyl bromide; *n*-butylamine and *n*-hexylamine; and sebacic acid di-*n*-butyl ester and azelaic acid di-*n*-butyl ester (Fig. 2.20)]. He compared other com-

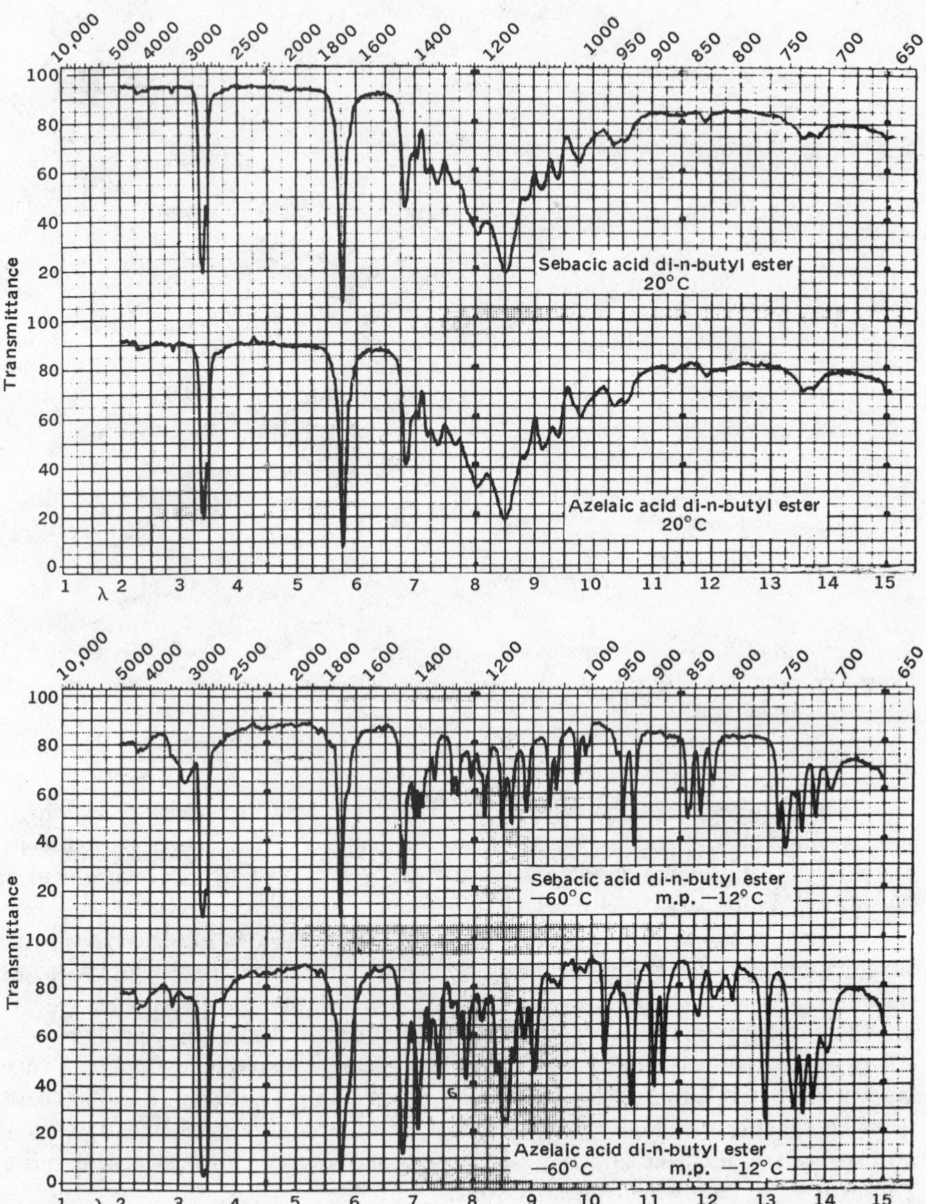

Fig. 2.20. Infrared spectra of sebacic acid di-*n*-butyl ester and azelaic acid di-*n*-butyl ester in the liquid state (top) and solid state, −60°C (bottom). (Caspary, 1968a.)

pounds at room temperature and low temperature, such as various ketones (1968b) and aldehydes (1965). These substances, all of which are liquids at room temperature, show many differences at the low temperature.

Omori and Kanda (1967) published similar spectra of several carbonyl compounds at about 100 K, e.g., aldehydes, acid halides, and ketones. They found that the carbonyl vibration decreased markedly in frequency—about 11 to 25 cm^{-1} in aldehydes and acid halides, and about 17 to 50 cm^{-1} in ketones—upon changing from the vapor to the solid state.

These effects and those reported by Caspary, which involve differences in the spectra due to phase change from liquid to solid (or gas to solid) are not the same as those referred to by Katon *et al.* (1967, 1969), in which all materials are solids at room temperature, and in which no phase change occurs as the temperature is lowered.

Katon *et al.* (1967; 1969) used liquid-nitrogen temperature to investigate the detailed structure of crystalline sugars and the usefulness of infrared spectroscopy in the differentiation of sugars. Some of their spectra cover the range from 4000 to 33 cm^{-1}. Figure 2.21 shows the infrared spectrum of α,α-trehalose dihydrate at both room temperature and 113 K recorded from a Nujol mull. (Similar results are obtained when the KBr-pellet technique is used, but this method can give unreproducible results.) There is marked improvement in band definition as well as the appearance of new bands. X-ray diffraction patterns of this compound at these two temperatures show that there is no gross phase change, but the low-temperature diffraction pattern shows a striking line intensification. The absence of a phase change is confirmed by the fact that the infrared spectrum changes continuously and smoothly as the temperature is lowered.

These facts can be explained by an order–disorder transition involving the hydrogen atoms bonded to the oxygen atoms. A similar situation occurs in ammonium chloride and is caused by the hindered rotation of the ammonium ion in the crystal lattice (Garland and Schumaker, 1967). In sugars the disorder can be due to internal rotation about the C—O bond, a motion which is hindered by a potential barrier of the order of magnitude of that in ammonium chloride (5–6 kcal/mole). The continuous nature of the temperature dependence is then due to the fact that the number of molecules possessing sufficient energy to surmount this barrier and lead to disorder follows the usual Boltzmann distribution and is a smooth function of temperature. It is therefore expected that further cooling of the sample will lead to further reduction of background of the spectrum due to further ordering of the hydrogen atoms.

Figure 2.22 shows the spectra of lactose monohydrate at three temperatures—298, 113, and ~20 K (temperature of boiling hydrogen). The spectrum at 20 K has the best resolution and the greatest band intensities. Figure 2.23 shows spectra for the same compound at higher frequencies, and it is seen that, on cooling, the bonded-OH band breaks up and shows much fine structure. The monomer-OH stretching frequency appears at ~3525 cm^{-1}. Apparently, one hydroxyl group of lactose is not hydrogen-bonded in the crystal, as is the case with sucrose, which has one hydroxyl group unbonded.

The effects seen above with sugars are to be expected with other complex molecules which possess internal motions that can be described by a potential function possessing more than one minimum and which are hindered by a relatively small potential barrier in the solid state.

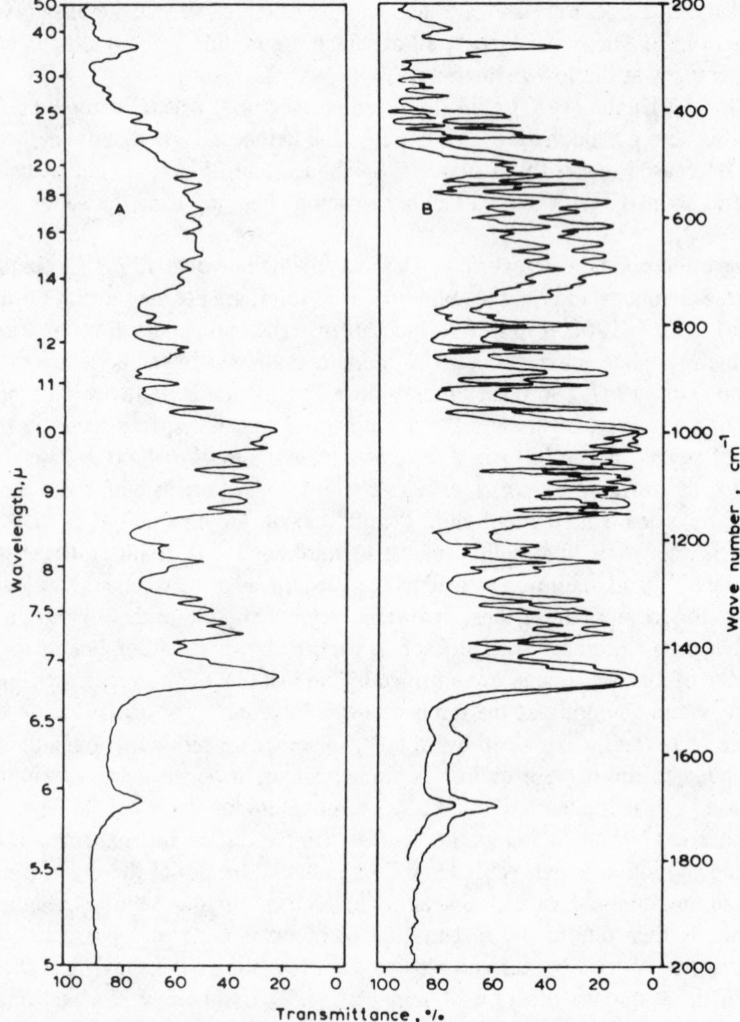

Fig. 2.21. Part of the infrared spectrum of a Nujol mull of α,α-trehalose dihydrate. (A), at room temperature, (B), at ~110 K. (Katon *et al.*, 1969.)

Katon *et al.* (1968) have extended their low-temperature work to other types of compounds and have given a description of the low-temperature cell they used. Other carbohydrates were examined (e.g., raffinose, sucrose, fructose, arabinose, xylose, lactose, mannose, maltose, galactose, rhamnose, cellobiose, and melibiose) as well as a noncrystalline trypsin (little or no change at the lower temperature), urea, diphenyl, stearoyl chloride, cholesterol (Fig. 2.24), cholesteryl acetate, serotonin creatinine sulfate, sodium creatinine phosphate hexahydrate, daunomycinone, carnosine, and others. Some

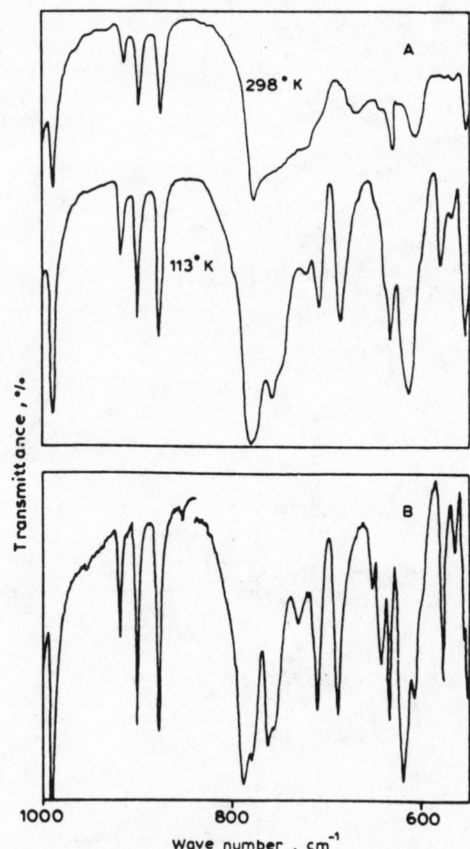

Fig. 2.22. The infrared spectrum of a Nujol mull of lactose monohydrate in the 1000–500-cm^{-1} region. (A), at 298 K and 113 K; (B), at ~20 K. (Katon *et al.*, 1969.)

of the spectra (lactose and sucrose) were run at liquid-*hydrogen* temperatures at 20 K by E.R. Lippincott. Most of the low-temperature spectra have very much improved resolution and often extra absorption bands. These striking results obtained on lowering the temperature have only been found with compounds that can exhibit hydrogen bonding.

Michell (1970) has reported the low-temperature spectra of some polysaccharides, including cellulose. McCall *et al.* (1971) have studied the low-temperature spectra of algal cellulose, hydrocellulose I (from cotton), and hydrocellulose II (from regenerated cellulose). The changes which were observed on cooling were associated with hydrogen bonding effects.

Bradbury and Elliott (1963), in an interesting conformational study of peptide group interactions, examined the infrared spectra of orthorhombic *N*-methylacetamide with the electric vector of the normally incident radiation along each of the three crystal axes in turn. Their measurements were made at room temperatures (when the crystal structure was not completely ordered) and at temperatures down to that of liquid nitrogen. The use of the low temperatures to study peptide group interactions in the unit cell allowed

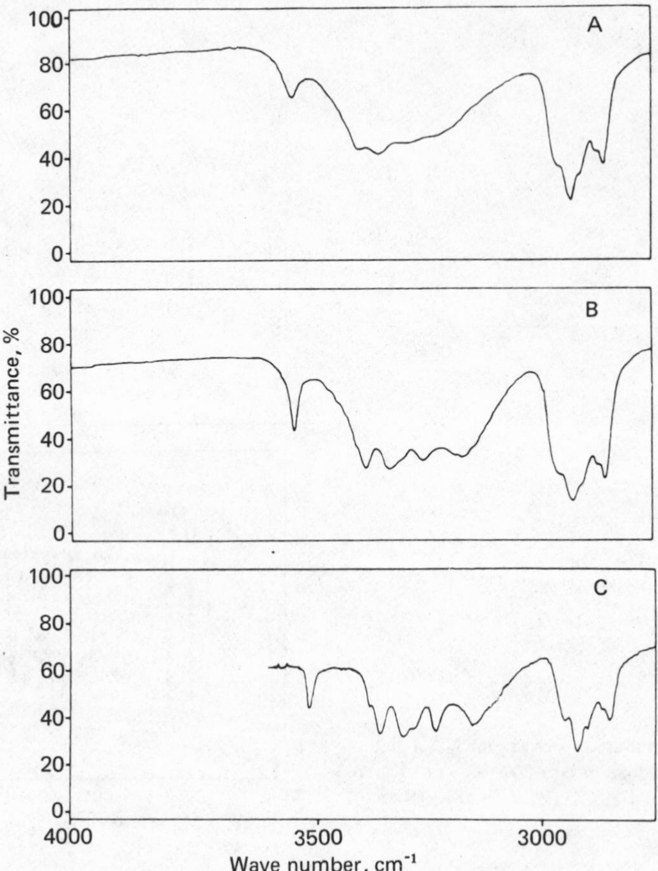

Fig. 2.23. The infrared spectrum of a Nujol mull of lactose monohydrate in the 3600–2800-cm^{-1} region. (A), at room temperature; (B), at ~110 K; (C), at ~20 K. (Katon *et al.*, 1969.)

them to reach different conclusions regarding some of the assignments of Miyazawa *et al.* (Miyazawa *et al.*, 1956, 1958; Miyazawa, 1960a, 1960b). Also, the liquid-nitrogen temperatures enhanced the fine structures of the NH and ND stretching bands.

Various types of low-temperature cells are available commercially. Rochkind (1968) has presented a low-temperature (20 K) technique, which provides a practical and sensitive method of infrared quantitative analysis of all infrared-absorbing gases and volatile liquids. The method, called pseudomatrix isolation spectroscopy (PMI), also provides a tool for the analysis of complex gas mixtures. The PMI method distinguishes between molecular isotopes; for example Fig. 2.25 shows a PMI spectrum of a mixture of isotopic d_2-ethylenes (condensed on a 20 K CsI window).

Hermann *et al.* (1969) have reviewed the subject of infrared spectroscopy at subambient temperatures, including: an extensive literature review with over 600 references;

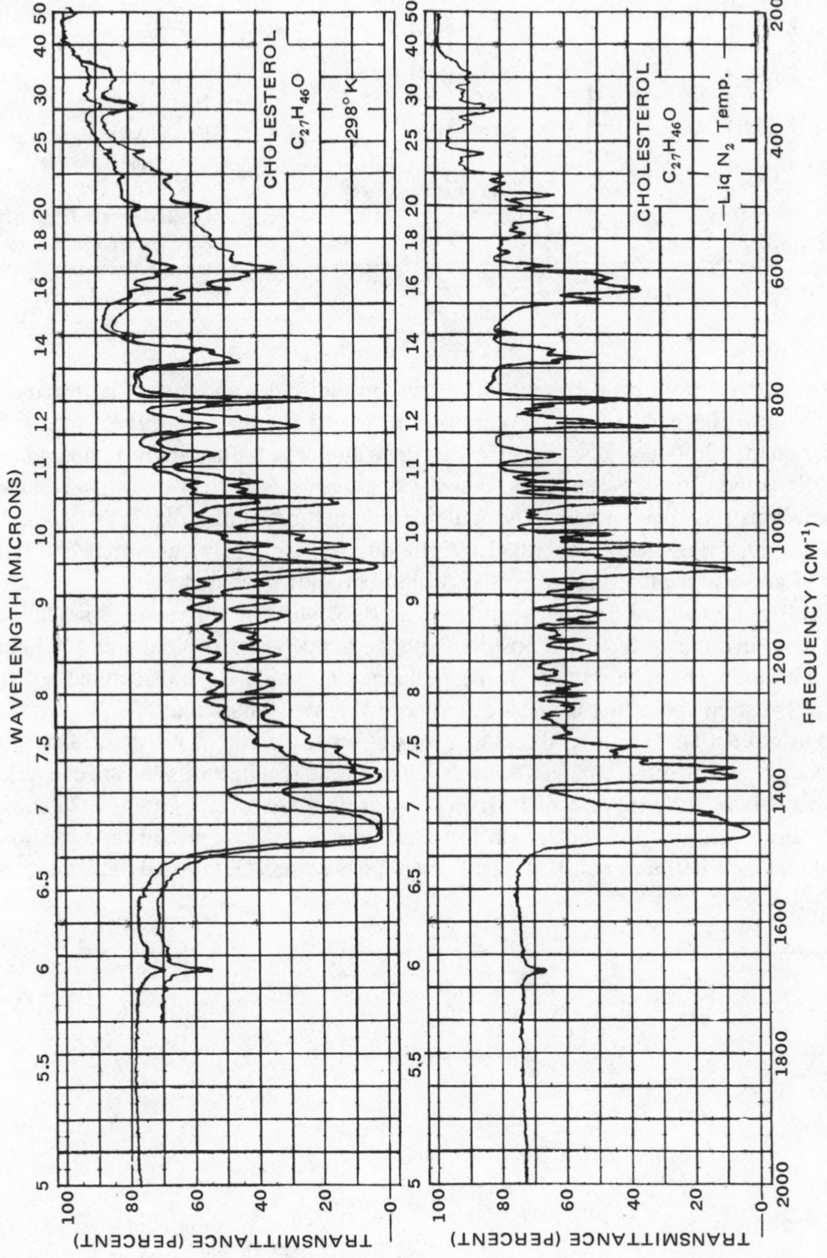

Fig. 2.24. The partial infrared spectrum of a Nujol mull of cholesterol at 298 K and at about liquid-nitrogen temperature. (Katon *et al.*, 1968.)

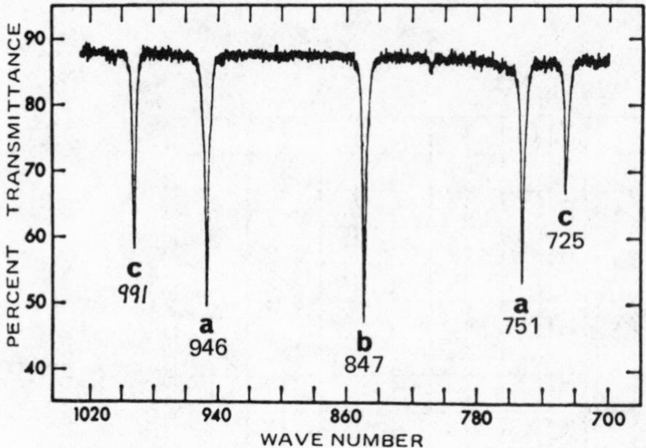

Fig. 2.25. Infrared PMI spectrum, 20 K, of a mixture of d_2-ethylenes at 1% in nitrogen (700–1025 cm^{-1}). (a) (1,1)ethylene-d_2; (b) *cis*-(1,2) ethylene-d_2; (c) *trans*-(1,2)ethylene-d_2. Twelve micromoles of ethylene mixture was deposited. Spectral slit, ~0.9 cm^{-1}. (Rochkind, 1968.)

and discussions of work on pure molecules, and on molecules and molecular fragments within matrices. Hermann and co-workers (Hermann and Harvey, 1969; Hermann *et al.*, 1969; Hermann, 1969) have also discussed the design of cells for use at low temperatures. These cells belong in four categories: conventional transmission cells; matrix isolation cells; pseudomatrix isolation cells; and multiple internal reflection cells (ATR).

Ford *et al.* (1969) have described a technique for accurately calibrating the temperature of low-temperature infrared cells. A liquid of known melting point is introduced into the cell as a capillary film between NaCl or AgCl windows. The cell is assembled with the thermocouple in the usual position, and the temperature lowered until the liquid freezes. The monochromator is set at the frequency at which the transmittance of the liquid and the solid shows the greatest difference. The transmittance at this frequency is recorded as a function of time as the sample warms up at the rate of 1.5 to 3.3 deg/min. Thermocouple emf readings are marked on the chart at frequent intervals and are converted later to temperatures by the use of thermocouple calibration tables. Figure 2.26 shows that the transmittance of the sample (in this case, ice) is nearly constant until the solid begins to melt and becomes constant again when it has completely melted. The midpoint

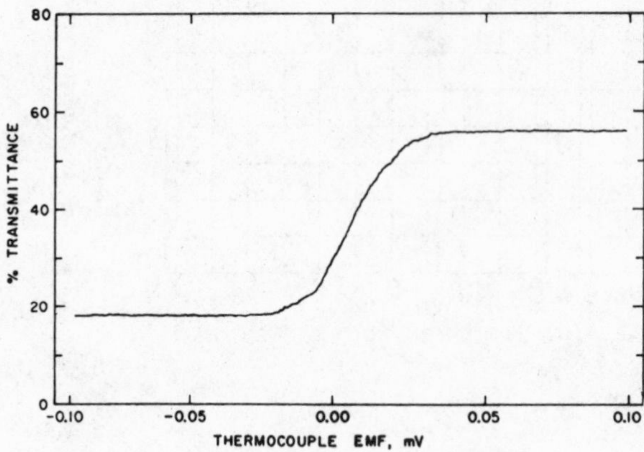

Fig. 2.26. Changes of transmittance at 2650 cm^{-1} during the melting of a thin film of ice. Recorded on a Perkin–Elmer model 521 spectrophotometer. (Ford *et al.*, 1969.)

of the melting curve (which usually extends over a range of 1° or less) is taken as the observed melting point. The melting points of the following sharply melting substances have been useful for calibration: *n*-butyl ethyl ether (−124°C); methanol (−97.8°C); ethyl acetate (−83.6°C); anisole (−37.3°C); H_2O (0.0°C); and D_2O (4.0°C).

Attenuated Total Reflection

Sometimes the sample to be studied transmits infrared radiation poorly, or the material is difficult to prepare by the usual methods, or one does not want to change the specimen in any way before studying it. The technique known as attenuated total reflection or ATR (also called multiple internal reflection, MIR, and FMIR, frustrated multiple internal reflection) can be applied in such cases. It was developed independently by Fahrenfort (1961) and Harrick (1960). To produce a useful ATR spectrum the basic requirement is that the infrared radiation enter a crystal of high refractive index, strike a sample of lower refractive index one or more times at an angle above the critical angle at the reflecting interface, and emerge through the crystal into the monochromator. The spectrum obtained is very much like that of a conventional spectrum produced by transmission techniques, and is entirely independent of the sample thickness. The depth of penetration of the radiation beam into the sample is 5 μm or less. Figure 2.27 is an optical diagram of a basic internal reflection accessory. Commonly used crystals are made of thallium-bromide–thallium-iodide (KRS-5), and a typical set of dimensions for a crystal is approximately 50 × 20 × 2 mm. One may use a micro MIR accessory and accomplish analyses less expensively. For example, in a micro unit the dimensions of the crystal may be 10 × 5 × 0.5 mm, and still 20 reflections occur within the crystal at a 45° angle of incidence. The smaller dimensions of the micro crystal require smaller sample area for a similar number of reflections, and lead to lower cost. The diagram shows a sample placed on both sides of the trapezoidal crystal.

Figure 2.28 presents the optical diagram of an MIR accessory that shows a variety of crystals with different incident angles in position. Each of these crystals (KRS-5, ZnSe, AgCl, etc.) is used with a sample holder that can be positioned by means of locating pins so that 30°, 45°, or 60° angles of incidence can be employed. This variability allows the user to vary penetration depth and the number of reflections. Liquid, powder, solid, and smear samples and biological tissues can be examined.

Some useful crystals for ATR work are thallium bromoiodide (KRS-5), silver chlo-

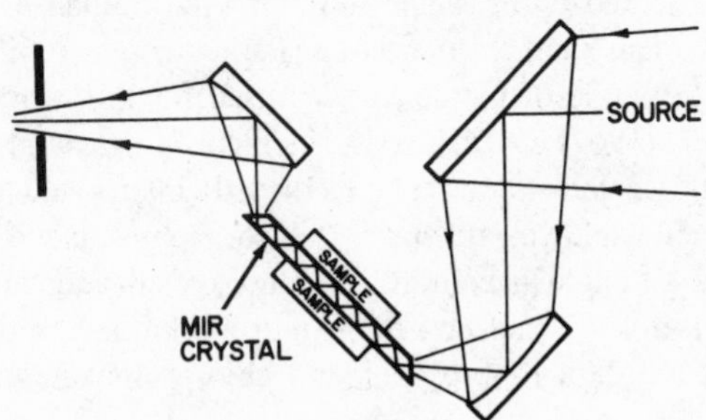

Fig. 2.27. Optical diagram of basic internal reflection accessory, top view. (Courtesy of Perkin–Elmer Corp.)

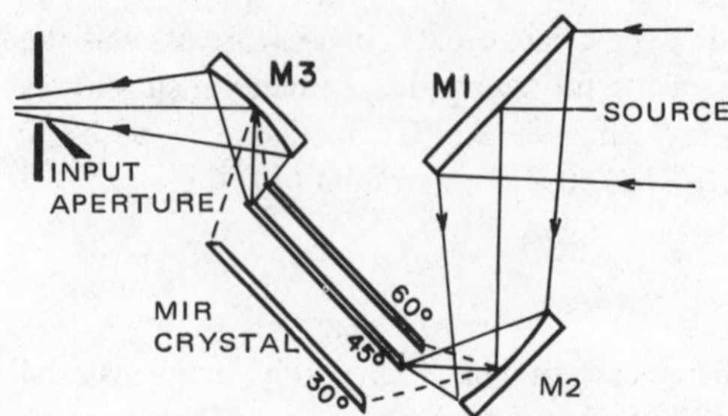

Fig. 2.28. Optical diagram, multiple internal reflection accessory. (Courtesy of Perkin–Elmer Corp.)

ride, zinc sulfide (IRTRAN-2), germanium, ZnSe, and AgBr. A variety of instrumentation and ATR accessories are available from Foxboro Company, Perkin–Elmer Corporation, Barnes Engineering Company, and Harrick Scientific Corporation.

Harrick's (1967) book is a comprehensive study of internal reflection spectroscopy. Polchlopek (1966) has discussed some theoretical aspects, instrumentation, sampling techniques, and applications of the method. Wilks and Hirschfeld (1967) have discussed advantages of the ATR method, materials for crystals, sampling techniques, quantitative analysis, microtechniques, and numerous applications. Pawlak *et al.* (1967) have also reviewed the subject of ATR.

Fujiyama *et al.* (1970), in a discussion of some systematic errors in infrared absorption spectrophotometry of liquid samples, have concluded that workers interested in serious study of integrated intensities or of band shapes would do well to avoid reflection–interference troubles by using the ATR technique, or to take careful and detailed account of these effects in any conventional absorption spectrophotometry.

A misleading statement that ATR of water solutions requires concentrations of 20% or higher was made in the review by Pawlak *et al.* (quoting a paper by Katlafsky and Keller, 1963). Parker (1963) has shown that suitable ATR spectra can be obtained on 5% solutions of glycyl-L-alanine, and evidence points to the detection of other substances at concentrations much less than 20% (Hermann, 1965b).

The author of this book has had considerable experience with the ATR technique and finds it particularly useful for work with biological specimens. The optical properties of tissues are such that they must be very thinly sectioned to obtain *transmission* spectra, whereas the ATR method does not depend on sample thickness. Useful spectra have been recorded from normal and diseased human arterial tissue (Fig. 2.29) and heart tissue (chicken, rat, calf) (Parker and Ans, 1967). Examinations of the biochemical lesions in diseased corneas and lenses of the human eye (Parker and D'Agostino, 1976) and in human xanthelasma tissue (Parker *et al.*, 1969) have been carried out by use of ATR. A 45° angle of incidence was used for this work. This fact is mentioned since Pawlak *et al.* (1967) say that a 45° incidence angle unit cannot be used with water solutions, and this statement might be incorrectly interpreted to mean that water-containing tissues would not display useful spectra. Also, Robinson and Vinogradov (1964) have successfully used a 45° angle in examining amino acids in aqueous solutions. Hermann (1965b) has recorded useful ATR spectra of rat muscle, kidney, and stomach tissues. In one case he claims to have shown the presence of chlorpromazine or a metabolite of it in tissue.

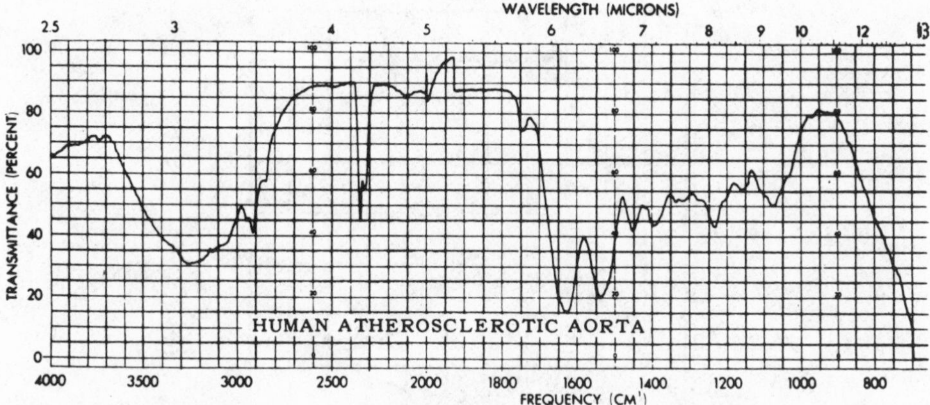

Fig. 2.29. Single reflection spectrum of adult human aorta containing atherosclerotic plaque. The bands near 2320 and 2350 cm^{-1} were caused by CO_2. (Parker and Ans, 1967.)

The fact that useful spectra can be obtained from polymers in various forms, from fibers which cannot be studied by transmission techniques, from other intractable materials, from aqueous solutions, etc., makes this technique useful in many disciplines. The ATR technique has been used for analysis of bacterial cultures (Johnson, 1966) and in forensic science (Denton, 1965). It has also been applied to a great variety of substances: molecular species present at electrode interfaces (Hansen *et al.*, 1966; Mark and Pons, 1966); carbohydrates (Parker and Ans, 1966); a single crystal of pentaerythritol (Tsuji *et al.*, 1970); cosmetics on the skin (Wilks Scientific Corp., 1966); pesticidal traces (Hermann, 1965a); water–alcohol mixtures (Malone and Flournoy, 1965); nitrate ion (Wilhite and Ellis, 1963); leather (Pettit and Carter, 1964); and blood spectra from within the human circulatory system (Kapany and Silbertrust, 1964). The last-mentioned application requires special equipment.

The migration (both directed and random) and adhesion of phagocytes of the peripheral circulation have been investigated frequently. This process is particularly important in order to initiate phagocytosis (Van Oss and Gillman, 1972), as it is required for the ingestion and final destruction of invading microorganisms. DeBari and Needle (1977) have used a ZnSe substrate for studying the adhesion of leukocytes to a surface, and examined the contact angles made between extracellular fluid and leukocytes, an important physical determination, since it is thought that for adhesion to occur the contact angle with the extracellular fluid of the surface under "attack" must be greater than that of the phagocytes, resulting in the squeezing out of water at the interface (Andrade *et al.*, 1973; Van Oss *et al.*, 1972).

The method employed to study leukocyte adhesion onto ZnSe was multiple internal reflection IR spectroscopy with a micro sampling attachment. These workers found a linear relationship between the number of phagocytes in the peripheral blood specimen and the relative change in transmittance (ΔT_{a-r}) between 1700 and 1620 cm^{-1}, where T_a and T_r (1700 cm^{-1}) are the transmittance at the analytical and reference wavelengths,

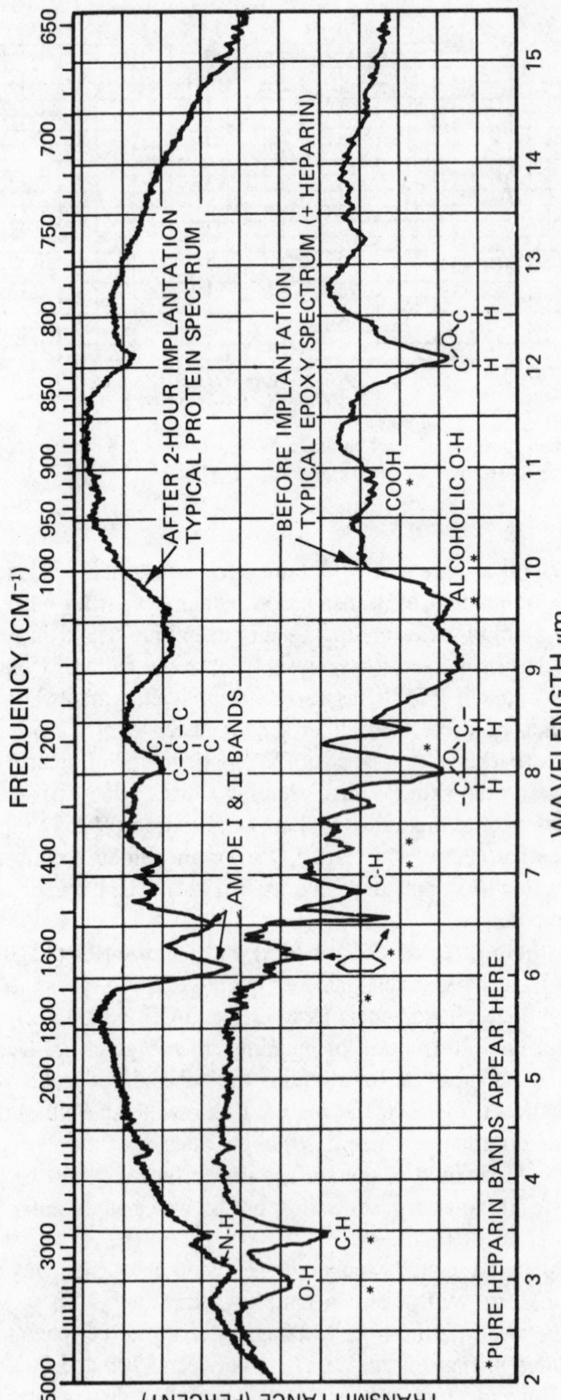

Fig. 2.30. MIR infrared spectra of an epoxy before and after implant in vein. (Baier, 1970.)

respectively. High concentrations of platelets and heparin appear not to affect this relationship.

Research on blood compatibility of materials has shown that interaction between a nonbiological surface and flowing blood is initiated by the spontaneous adsorption of a thin film of proteinaceous substance (Baier *et al.*, 1971; Lyman and Kim, 1971; Lyman *et al.*, 1974). The adsorption of fibrinogen, a plasma protein, to a polymer surface (a type of polyurethane) was studied by means of ATR infrared spectroscopy (Stupp *et al.*, 1977). Adsorption was much greater on polymer surfaces containing a higher relative concentration of polyether than aromatic segments.

Stellite artificial-heart valves have a tightly bound, hard-to-remove organic coating exposing only closely packed CH_3 groups from waxy constituents (stearic and palmitic acid chains) of the polishing agents used on Stellite (an alloy of cobalt, chromium, molybdenum, and tungsten). The presence of the waxy coating was confirmed (Baier, 1970) by surface chemical analyses and MIR spectroscopy. Another point worth mentioning is that the Stellite valves, when exposed to flowing blood, acquire a strongly adherent protein film within seconds of exposure. In addition, Fig. 2.30 shows the differences between an epoxy surface (plus heparin) before its implantation in a dog's jugular vein and the protein layer deposited on the epoxy surface after it has been in place in the vein for two hours.

Submicrogram quantities of organic phosphorus compounds in air have been monitored by ATR spectroscopy (Prager and LaRosa, 1968). Adsorptive platinum films (50–80 nm thick) on germanium crystals (57 × 14 × 2 mm) were used in order to increase the sensitivity of the method. Adsorption was reversible by air flowing through the germanium cell, thus permitting repeated use of the platinum-coated crystals. At the optimum thickness and temperature it was possible to detect 0.1 μg of dimethylmethylphosphonate per liter of air in 30 sec.

Parker (1969) has shown ATR spectra for erythritol (Fig. 2.31) and DL-phenylalanine (Fig. 2.32), at room temperature and at liquid-nitrogen temperature. The resolution increases in both cases at the lower temperature.

The reader will find discussions of additional work in which ATR was used in other sections of this book.

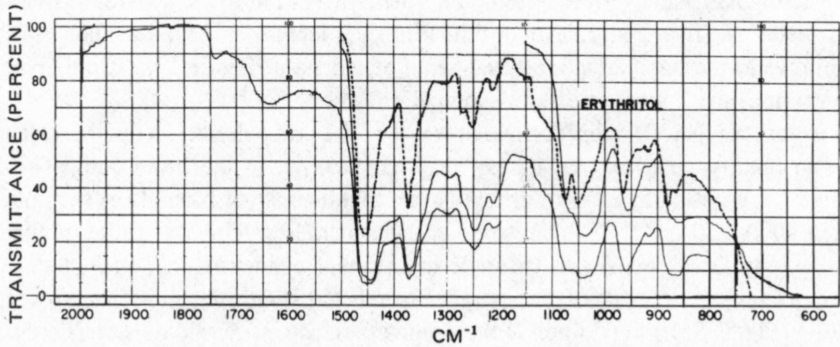

Fig. 2.31. ATR spectra of erythritol (Nujol mull). Solid line, room temperature; dotted line, liquid-nitrogen temperature, on 1-mm KRS-5 plate. 1375- and 1450-cm^{-1} bands are caused by Nujol. (Parker, 1969.)

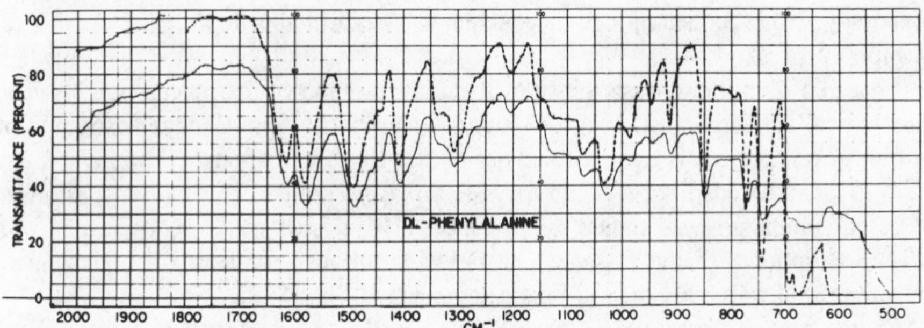

Fig. 2.32. ATR spectra of DL-phenylalanine (powder, run on 1-mm KRS-5 plate directly). Solid line, room temperature; dotted line, liquid-nitrogen temperature. (Parker, 1969.)

The Use of Polarized Infrared Radiation and the Measurement of Dichroism

Polarized infrared radiation is used to obtain information about the direction of transition moments of normal modes of vibration in solid oriented compounds. If one knows the molecular orientation in a solid, he can use polarization studies in making band assignments. The measured direction of the transition moment of the vibration producing a band must coincide with the direction deduced from the structure if the assignment is correct. On the other hand, knowing the band assignment but not the molecular orientation in the solid, one can deduce some knowledge of the molecular orientation.

In the crystalline state the normal modes of the unit cell, not the molecule, interact with radiation impinging on it. The normal modes of a unit cell are comprised of the normal modes of the individual molecules which exist in as many different phases as the number of molecules in a unit cell. For example, in a unit cell consisting of two molecules, each having a C═O group, the two C═O stretching modes are comprised of in-phase and out-of-phase stretching of the two C═O groups.

Polarizers. To obtain a beam of polarized infrared radiation one can use silver chloride (Newman and Halford, 1948) or selenium (Elliott *et al.*, 1948a) polarizers. Several plates of silver chloride or a thin film of selenium are tilted at the polarizing angle relative to the unpolarized infrared beam of radiation. Figure 2.33 (Colthup *et al.*, 1975) describes the optical effects in schematic fashion.

The most efficient IR polarizer available is the wire grid polarizer. It has the advantage that it can readily be employed in any required position in the beam path. The Perkin–Elmer Corporation has a wire grid polarizer that uses silver bromide as substrate for the range 4000–250 cm^{-1}. [The substrate is the material on which a grating is formed, and then gold is vapor-deposited at an angle so that metal condenses only on the protruding portions, forming a network of fine parallel metal wires with the same interspacing as the grating (Fig. 2.34).] In addition to their high efficiency, wire grid polarizers are better than Brewster's polarizers (systems using parallel plates, e.g., AgCl as above) in several ways: (a) Being small, they can easily be placed before the entrance slit of the mono-

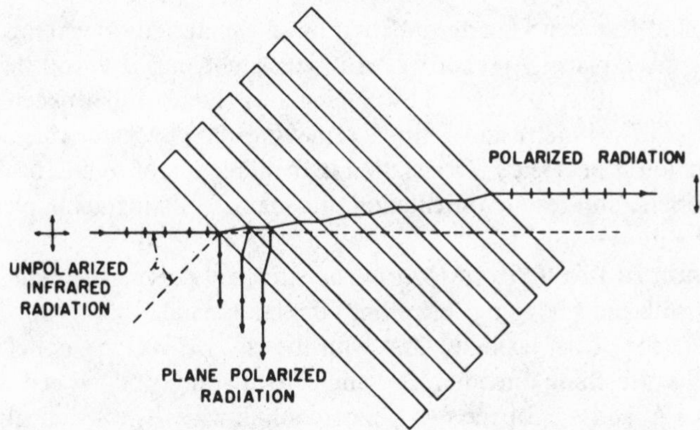

Fig. 2.33. Plane polarization of infrared radiation by several plates. If I is the incident parallel beam of radiation striking at the angle of the incidence, i, the reflected beam R is polarized. The incident plane is the plane of the paper, therefore beam R is vibrating normal to the paper. This reflection–polarization process occurs at each face of the stack of plates. After a sufficient number of reflections the transmitted beam T has been depleted of vibrations normal to the plane of incidence and consists almost completely of vibrations in this plane. Thus, two beams of polarized light are produced. (Colthup *et al.*, 1975.)

chromator, the most suitable position for measuring; this helps to eliminate the polarization caused by the instrument itself. (b) The beam is not displaced. (c) Beam convergence does not influence the polarization ratio ($I_{\|}/I_{\perp}$), which remains constant when the angle of incidence is altered by more than 50° (Young *et al.*, 1965).

Orientation of Films or Crystals. A single crystal or group of crystals oriented the same way may be used for crystalline compounds. Various conditions for producing crystallization exist: between salt plates from the melted compound (Halverson and Francel, 1949); out of the vapor phase (Zwerdling and Halford, 1955); or from solution (Hallman and Pope, 1958). Many spectra of crystalline compounds have been recorded but mostly as mulls or potassium halide pellets, techniques not very useful for polarization studies because of the finely ground state of the sample and the random distribution of the individual crystals. For work with crystals, a platelet of 5 to 20 μm thickness is usually needed and, unless an infrared microscope is used, the platelet should be large enough to cover the light beam of the spectrophotometer. The use of polarized single crystals has been described (Chance *et al.*, 1967) for obtaining infrared spectra of cytochrome *c* and myoglobin compounds.

Orientation of thin polymer films may be accomplished by stretching, which results in uniaxial orientation where a crystallographic axis, usually the long chain axis, tends

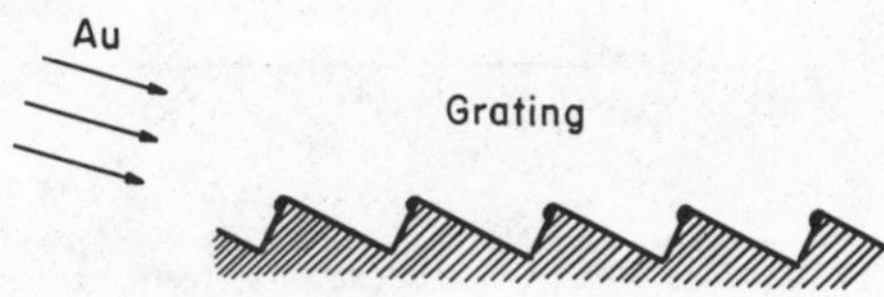

Fig. 2.34. Preparing a wire grid polarizer. (Courtesy of Perkin–Elmer Corp.)

to line up parallel to the stretching direction but no preferred orientation of crystallites about this axis takes place. Another way to orient a polymer is to roll the film between rollers. This can also be accomplished by placing the polymer film between silver chloride plates (Elliott *et al.*, 1948b) and rolling, removing the silver chloride with a sodium thiosulfate solution if necessary. Orientation in the direction of rolling occurs, but sometimes double orientation results (orientation of certain crystallographic planes parallel to the plane of the film).

Measurement of Dichroic Ratio. The stretched polymer is mounted in a spectrophotometer with the stretching (or rolling) direction parallel to the slit. The spectrum is then recorded for polarized light, first with the electric vector parallel and then perpendicular to the stretching direction. For any particular absorption band one determines the absorbances $A_{\parallel}$ and $A_{\perp}$ for the two polarization directions, respectively. The ratio R of these two values,

$$R = A_{\parallel}/A_{\perp}$$

is called the dichroic ratio for the particular absorption band. A sample which is not uniform in thickness and shape need not be moved. Instead, the polarizer is rotated, thus exposing the same part of the sample to the beam. Special care must be taken to avoid errors caused by this method as a result of partial polarization in the spectrometer itself.

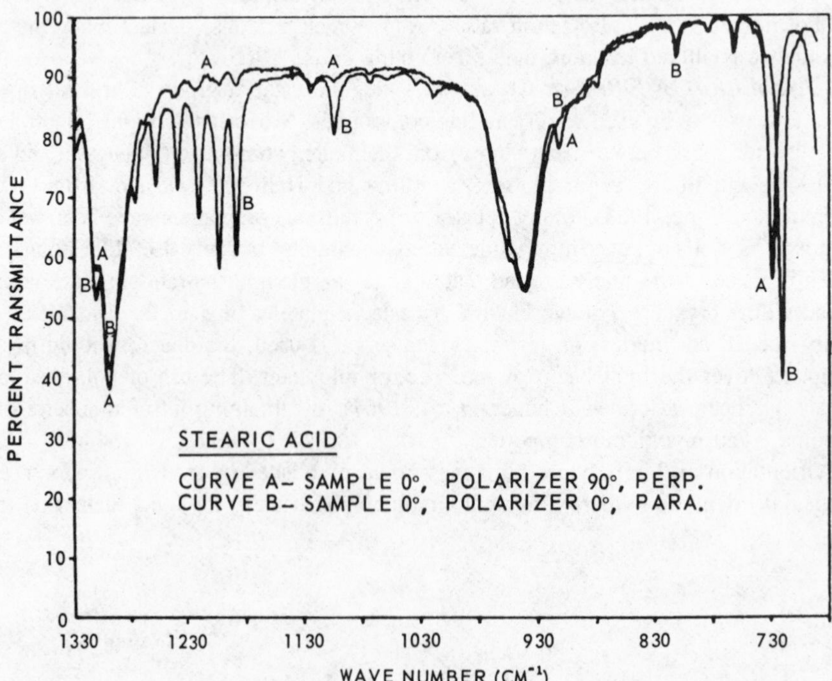

Fig. 2.35. Polarization spectra for stearic acid. (Hannah *et al.*, 1968, courtesy of Perkin–Elmer Corp.)

This polarization is caused by reflections at various mirrors and especially at the prism faces. To avoid these difficulties when rotating the polarizer, one can make measurements with the polarizer oriented 45° with respect to the slit, and then rotated 90° to a position 45° on the other side. If the sample has enough uniformity, one leaves the polarizer stationary in the position producing highest transmission while rotating the sample 90°. This is the most satisfactory way to measure polarization properties (Colthup *et al.*, 1964). Polarization effects due to the spectrometer itself have also been noted in grating instruments (George, 1966). Hannah *et al.* (1968) studied oriented stearic acid and polyvinyl alcohol using a wire-grid polarizer in an instrument equipped with an analytical data system composed of a digital data recorder, a paper tape punch, and a manual keyboard. With the polarizer placed between the sample and the monochromator, the monochromator polarization did not contribute to errors in the dichroic ratio. Figure 2.35 shows polarization spectra for stearic acid oriented by melting between two sodium chloride plates and then cooling one end, thus providing a temperature gradient during the crystallization. Koenig *et al.* (1967) have discussed the use of three-dimensional measurements for doing dichroic studies.

A review of the infrared dichroism technique as used for protein and peptide structure determination has been presented by de Lozé *et al.* (1978).

References

Anderson, D. H., and Miller, O. E., *J. Opt. Soc. Am.* **43,** 777 (1953a).

Anderson, D. H., and Woodall, N. B., *Anal. Chem.* **25,** 1906 (1953b).

Anderson, D. W. *Analyst* **84,** 50 (1959).

Anderson, J. L., and Kincaid, J. R., *Appl. Spectrosc.* **32,** 356 (1978).

Andrade, J. D., Lee, H. B., Jhon, M. S., Kim, S. W., and Hibbs, J. B., *Trans. Am. Soc. Artif. Intern. Organs* **19,** 1 (1973).

Antikainen, P. J., *Suomen Kemistilehti* **B31,** 223 (1958).

Baier, R. E., "Research Trends," Cornell Aeronautical Laboratory, Inc., Spring 1970.

Baier, R. E., Loeb, G. I., and Wallace, G. T., *Fed. Proc.* **30,** 1523 (1971).

Bailey, G. F., Kint, S., and Scherer, J. R., *Anal. Chem.* **39,** 1040 (1967).

Barer, R., Cole, A. R. H., and Thompson, H. W., *Nature* **163,** 198 (1949).

Barker, S. A., Bourne, E. J., Neely, W. B., and Whiffen, D. H., *Chem. Ind.* (*London:* 1954), 1418.

Barker, S. A., Bourne, E. J., and Whiffen, D. H., *Methods Biochem. Anal.* **3,** 213 (1956).

Barnes, R. B., Gore, R. C., Williams, E. F., Linsley, S. G., and Petersen, E. M., *Ind. Eng. Chem., Anal. Ed.*, **19,** 620 (1947).

Bentley, F. F., Wolfarth, E. F., Srp, N. E., and Powell, W. R., *Spectrochim. Acta* **13,** 1 (1958).

Bisset, F., Bluhm, A. L., and Long, L., Jr., *Anal. Chem.* **31,** 1927 (1959).

Blout, E. R., *Ann. N. Y. Acad. Sci.* 69, **Art** 1, 84 (1957).

Blout, E. R., and Lenormant, H., *J. Opt. Soc. Am.* **43,** 1093 (1953).

Bradbury, E. M., and Elliott, A., *Spectrochim. Acta* **19,** 995 (1963).

Bradley, K. B., and Potts, W. J., Jr., *Appl. Spectrosc.* **12,** 77 (1958).

Cameron, D. G., Casal, H. L., and Mantsch, H. H., *J. Biochem. Biophys. Methods* **1,** 21 (1979).

Caspary, R., *Spectrochim. Acta* **21,** 763 (1965).

Caspary, R., *Appl. Spectrosc.* **22,** 694 (1968a).

Caspary, R., *Appl. Spectrosc.* **22,** 689 (1968b).

Chance, B., Estabrook, R. W., and Yonetani, T., eds., *Hemes and Hemoproteins,* Academic Press, New York, 1967.

Chang, S. S., Ireland, C. E., and Tai, H., *Anal. Chem.* **33,** 479 (1961).

Clark, D. A., and Boer, A. P., *Spectrochim. Acta* **12,** 276 (1958).
Coblentz, W. W., *Investigations of Infrared Spectra,* Carnegie Institution of Washington, Washington, D.C. (1905).
Colthup, N. B., Daly, L. H., and Wiberley, S. E., *Introduction to Infrared and Raman Spectroscopy,* 2nd Ed., Academic Press, New York, 1975. (First Ed., 1964).
DeBari, V. A., and Needle, M. A., *J. Reticuloendothelial Soc.* **22,** 121 (1977).
de Lozé, C., Baron, M.-H., and Fillaux, F., *J. Chim. Phys.* **75,** 25 (1978).
Denton, S., *J. Forensic Sci. Soc.* **5,** 112 (1965).
Dinsmore, H. L., and Edmondson, P. R., *Spectrochim. Acta* **15,** 1032 (1959).
Dows, D. A., "Infrared Spectra of Molecular Crystals," in *Physics and Chemistry of the Organic Solid State,* Vol. I (D. Fox, M. M. Labes, and A. Weissberger, eds.), Interscience, New York, 1963.
Elliott, A., Ambrose, E. J., and Temple, R. B., *J. Opt. Soc. Am.* **38,** 212 (1948a).
Elliott, A., Ambrose, E. J., and Temple, R. B., *J. Chem. Phys.* **16,** 877 (1948b).
Fahrenfort, J., *Spectrochim. Acta* **17,** 698 (1961).
Farmer, V. C., *Spectrochim. Acta* **8,** 374 (1957).
Farmer, V. C., *Chem. Ind. (London)* **1959,** 1306.
Flett, M. St. C., and Hughes, J., *J. Chromatogr.* **11,** 434 (1963).
Fogelberg, B. C., and Kaila, E., *Paperi Ja Puu* **39,** 375 (1957).
Ford, T. A., Seto, P. F., and Falk, M., *Spectrochim. Acta* **25A,** 1650 (1969).
Fox, J. W., and Tu, A. T., *Appl. Spectrosc.* **33,** 647 (1979).
Freeman, S. K., and Landon, D. O., *The Spex Speaker,* Vol. VIII, No. 4, Dec., 1968 (Spex Industries, Metuchen, New Jersey 08840).
Fujiyama, T., Herrin, J., and Crawford, B. L., Jr., *Appl. Spectrosc.* **24,** 9 (1970).
Garland, C. W., and Schumaker, N. E., *J. Phys. Chem. Solids* **28,** 799 (1967).
Garner, H. R., and Packer, H., *Appl. Spectrosc.* **22,** 122 (1968).
George, R. S., *Appl. Spectrosc.* **20,** 101 (1966).
Goddu, R. F., and Delker, D. A., *Anal. Chem.* **32,** 140 (1960).
Gordon, J., in Jencks (1963).
Gore, R. C., Barnes, R. B., and Petersen, E., *Anal. Chem.* **21,** 382 (1949).
Goulden, J. D. S., *Spectrochim. Acta* **15,** 657 (1959).
Hacskaylo, M., *Anal. Chem.* **26,** 1410 (1954).
Hainer, R. M., and King, G. W., *Nature* **166,** 1029 (1950).
Hallam, H. E., *Unicam Spectrovision* No. 11, 1961.
Hallman, H., and Pope, M., "Technical Report on Preparation of Thin Anthracene Single Crystals," ONR Contract No. 285, (25), June 9, 1958, cited in Colthup *et al.,* 1964.
Halverson, F., and Francel, R. J., *J. Chem. Phys.* **17,** 694 (1949).
Hannah, R. W., Savitzky, A., and Kessler, H. B., Paper delivered at The Pittsburgh Conference, Cleveland, Ohio, March 3–8, 1968.
Hansen, W. N., Osteryoung, R. A., and Kuwana, T., *J. Am. Chem. Soc.* **88,** 1062 (1966).
Harrick, N. J., *Phys. Rev. Lett.* **4,** 224 (1960).
Harrick, N. J., *Internal Reflection Spectroscopy,* John Wiley and Sons, New York, 1967.
Hawes, R. C., Sloane, H. J., and Haber, H. S., *Eur. Congr. Mol. Spectrosc. 9th,* Madrid, 1967.
Hermann, T. S., *Appl. Spectrosc.* **19,** 10 (1965a).
Hermann, T. S., *Anal. Biochem.* **12,** 406 (1965b).
Hermann, T. S., *Appl. Spectrosc.* **23,** 461, 473 (1969).
Hermann, T. S., and Harvey, S. R., *Appl. Spectrosc.* **23,** 435 (1969).
Hermann, T. S., Harvey, S. R., and Honts, C. N., *Appl. Spectrosc. 23, 451 (1969).*
Hollingdale-Smith, P. A., *Can. Spectry.* **11,** 107 (1966).
Jencks, W. P., *Methods in Enzymology,* Vol. 6, Academic Press, New York, p. 914, 1963.
Johnson, R. D., *Anal. Chem.* **38,** 160 (1966).
Kapany, N. S., and Silbertrust, N., *Nature* **204,** 138 (1964).
Katlafsky, B., and Keller, R. E., *Anal. Chem.* **35,** 1665 (1963).
Katon, J. E., Miller, J. T., Jr., and Bentley, F. F., *Arch. Biochem. Biophys.* **121,** 798 (1967).

Katon, J. E., Miller, J. T., Jr., and Ferguson, R. R., Technical Report AFML-TR-68-169, Air Force Materials Laboratory, Wright-Patterson Air Force Base, 1968.
Katon, J. E., Miller, J. T., Jr., and Bentley, F. F., *Carbohyd. Res.* **10,** 505 (1969).
Kiefer, W., and Bernstein, H. J., *Appl. Spectrosc.* **25,** 500 (1971).
Koenig. J. L., Cornell, S. W., and Witenhafer, D. E., *Polymer Sci., Part A-2,* **5,** 301 (1967).
Koningstein, J. A., and Gächter, B. F., *J. Opt. Soc. Am.* **63,** 892 (1973).
Kuhn, L. P., *J. Am. Chem. Soc.* **74,** 2492 (1952).
Kuhn, L. P., *J. Am. Chem. Soc.* **76,** 4323 (1954).
Lecomte, J., *Anal. Chim. Acta* **2,** 727 (1948).
Lohr, L. J., and Kaier, R. J., *Anal. Chem.* **32,** 301 (1960).
Lord, R. C., and Yu, N.-T., *J. Mol. Biol.* **50,** 509 (1970).
Lutinski, C., *Anal. Chem.* **30,** 2071 (1958).
Lyman, D. J., and Kim, S. W., *Fed. Proc.* **30,** 1658 (1971).
Lyman, D. J., Metcalf, L. C., Albo, D., Jr., Richards, K. F., and Lamb, J., *Trans. Am. Soc. Artif. Int. Organs* **20B,** 474 (1974).
Malone, C. P., and Flournoy, P. A., *Spectrochim. Acta* **21,** 1361 (1965).
Mark, H. B., Jr., and Pons, B. S., *Anal. Chem.* **38,** 119 (1966).
Maxwell, J. C., and Caughey, W. S., *Biochemistry* **15,** 388 (1976).
McCall, E. R., Morris, N. M., Tripp, V. W., and O'Connor, R. T., *Appl. Spectrosc.* **25,** 196 (1971).
McNiven, N. L., and Court, R., *Appl. Spectrosc.* **24,** 296 (1970).
Michell, A. J., *Austral J. Chem.* **23,** 833 (1970).
Miyazawa, T., *J. Mol. Spectrosc.* **4,** 198 (1960a).
Miyazawa, T., *J. Chem. Phys.* **32,** 1647 (1960b).
Miyazawa, T., Shimanouchi, T., and Mizushima, S., *J. Chem. Phys.* **24,** 408 (1956).
Miyazawa, T., Shimanouchi, T., and Mizushima, S., *J. Chem. Phys.* **29,** 611 (1958).
Nachod, F. C., and Martini, C. M., *Appl. Spectrosc.* **13,** 45 (1959).
Newman, R., and Halford, R. S., *Rev. Sci. Instr.* **19,** 270 (1948).
Omori, T., and Kanda, Y., *Mem. Fac. Sci. Kyushu University, Ser. C. Chem.* **6**(1), 29 (1967).
Ondrias, M. R., Babcock, G. T., and Salmeen, I., *Fed. Proc.* **39,**(6), 2062 (1980).
Park, P. J. D., and Wyn-Jones, E., *Chem. Commun.* **1966,** 557.
Parker, F. S., *Appl. Spectrosc.* **12,** 163 (1958).
Parker, F. S., *Biochim. Biophys. Acta* **42,** 513 (1960).
Parker, F. S., *Perkin-Elmer Instrument News* **13,**(4), 1 (1962).
Parker, F. S., *Nature* **200,** 1093 (1963).
Parker, F. S., *Nature* **203,** 975 (1964).
Parker, F. S., in *Progress in Infrared Spectroscopy,* Vol. 3 (H. A. Szymanski, ed.), Plenum Press, New York, p. 75, 1967.
Parker, F. S., Mid-America Symposium on Spectroscopy, Chicago, Illinois, Paper 119, May, 1969.
Parker, F. S., and Kirschenbaum, D. M., *Spectrochim. Acta* **16,** 910 (1960).
Parker, F. S., and Ans, R., *Appl. Spectrosc.* **20,** 384 (1966).
Parker, F. S., and Ans, R., *Anal. Biochem.* **18,** 414 (1967).
Parker, F. S., and D'Agostino, M., *Canad. J. Spectrosc.* **21,** 111 (1976).
Parker, F. S., Kleinman, C. S., and Mittl, R., *Proceedings of the Eighth Meeting of the Career Scientists of the Health Research Council of N.Y. City,* New York Academy of Medicine, Dec. 1969.
Pawlak, J. A., Fricke, G., and Szymanski, H. A., in *Progress in Infrared Spectroscopy,* Vol. 3 (H. A. Szymanski, ed.), p. 39, Plenum Press, New York, 1967.
Pettit, D., and Carter, A. R., *J. Soc. Leather Trades' Chemists* **48,** 476 (1964).
Pez, G., *The Spex Speaker,* Vol. VIII, No. 4, December, 1968 (Spex Industries, Metuchen, New Jersey 08840).
Pierson, R. H., Fletcher, A. N., and Gantz, E. St. C., *Anal. Chem.* **28,** 1218 (1956).
Polchlopek, S. E., in *Applied Infrared Spectroscopy* (D. N. Kendall, ed.), p. 462, Reinhold, New York, 1966.
Potts, W. J., Jr., *Chemical Infrared Spectroscopy, Vol. I: Techniques,* John Wiley and Sons, New York, 1963.
Prager, M. J., and LaRosa, C. N., *Appl. Spectrosc.* **22,** 449 (1968).
Price, G. D., Sunas, E. C., and Williams, J. F., *Anal. Chem.* **39,** 138 (1967).

Randall, H. M., Fuson, N., Fowler, R. G., and Dangl, J. R., *Infrared Determination of Organic Structures,* Van Nostrand, Princeton, New Jersey, 1949.
Rao, C. N. R., *Chemical Applications of Infrared Spectroscopy,* Academic Press, New York, 1963.
Resnik, F. E., Harrow, L. S., Holmes. J. C., Bill, M. E., and Greene, F. L., *Anal. Chem.* **29,** 1874 (1957).
Richards, R. E., and Thompson, H. W., *Trans. Faraday Soc.* **41,** 183 (1945).
Robinson, T., *Nature* **184,** 448 (1959).
Robinson, F. P., and Vinogradov, S. N., *Appl. Spectrosc.* **18,** 62 (1964).
Rochkind, M. M., *Science* **160,** 196 (1968).
Schiele, C., *Appl. Spectrosc.* **20,** 253 (1966).
Schwing, K. J., and May, L., *Anal. Chem.* **38,** 523 (1966).
Senn, W. L., Jr., and Drushel, H. V., *Anal. Chim. Acta* **25,** 328 (1961).
Stewart, J. E., *Application Data Sheet IR-88-MI,* Beckman Instruments Division.
Stewart, J. E., *Anal. Chem.* **30,** 2073 (1958).
Stupp, S. I., Kauffman, J. W., and Carr, S. H., *J. Biomed. Mater. Res.* **11,** 237 (1977).
Szymanski, H. A., *IR—Theory and Practice of Infrared Spectroscopy,* Plenum Press, New York, 1964.
Tipson, R. S., *Natl. Bur. Std. Monograph 110,* 1968.
Tipson, R. S., and Isbell, H. S., *J. Res. Natl. Bur. Std.* **64A,** 239 (1960).
Tipson, R. S., and Isbell, H. S., *J. Res. Natl. Bur. Std.* **66A,** 31 (1962).
Tsuji, K., Yamada, H., Suzuki, K., and Nitta, I., *Spectrochim. Acta* **26A,** 475 (1970).
Van Oss, C. J., Good, R. J., and Neumann, A. W., *J. Electroanal. Chem.* **37,** 387 (1972).
Van Oss, C. J., and Gillman, C. F., *J. Reticuloendothel. Soc.* **12,** 283 (1972).
Vergoten, G., Fleury, G., and Moschetto, Y., Patent Anvar-Inserm (1976), quoted in Vergoten *et al., Advan. IR Raman Spectrosc.* **4,** 195 (1978).
Walsh, A., and Willis, J. B., *J. Chem. Phys.* **18,** 552 (1950).
White, J. U., and Alpert, N. L., *J. Opt. Soc. Am.* **48,** 460 (1958).
White, J. U., Weiner, S., and Alpert, N. L., *Anal. Chem.* **30,** 1694 (1958).
White, J. U., Alpert, N. L., Ward, W. M., and Gallaway, W. S., *Anal. Chem.* **31,** 1267 (1959).
Wilhite, R. N., and Ellis, R. F., *Appl. Spectrosc.* **17,** 168 (1963).
Wilks Scientific Corp., *Model 26 MIR Skin Analyzer* (1966).
Wilks, P. A., Jr., and Hirschfeld, T., *Appl. Spectrosc. Rev.* **1,** 99 (1967).
Woodruff, W. H., and Spiro, T. G., *Appl. Spectrosc.* **28,** 74 (1973).
Young, J. B., Graham, H. A., and Peterson, E. W., *Appl. Opt.* **4,** 1023 (1965).
Yu, N.-T. and Jo, B. H., *Arch. Biochem. Biophys.* **156,** 469 (1973).
Yu, N.-T., Liu, C. S., and O'Shea, D. C., *J. Mol. Biol.* **70,** 117 (1972).
Yu, N.-T., Jo, B. H., and Liu, C. S., *J. Am. Chem. Soc.* **94,** 7572 (1972a).
Zhbankov, R. G., Ivanova, N. V., and Komar, V. P., *Vysokomol. Soedin.* **8,** 1778 (1966).
Zwerdling, S., and Halford, R. S., *J. Chem. Phys.* **23,** 2221 (1955).

Chapter 3

POLYPEPTIDES AND PROTEINS

Polypeptides

Table 3.1 gives detailed information about the amide I, amide III, and skeletal vibrations of many polypeptides, some containing side chains of only one type of amino acid, others containing two, e.g., poly-(Ala-Gly) and poly-(Ser-Gly).

The amide I mode for the α-helix appears in the same region (1650–1657 cm^{-1}) for both Raman and IR spectra. For the antiparallel β-sheet conformation the amide I mode is divided: the Raman spectrum has a strong band at ~1670 cm^{-1}, while the IR spectrum shows two bands at ~1630 and 1685 cm^{-1}. The Raman amide I band for the disordered conformation is found at ~1665 cm^{-1}, whereas this conformation appears near 1655 cm^{-1} in the IR. Thus, the amide I in the Raman spectrum can be employed to distinguish between α-helix and β-sheet or disordered conformations but it cannot readily be used to distinguish between the latter two types.

The amide III band comes from C—N stretching and C—N—H in-plane bending modes but other motions appearing in this range may be coupled to it. This band is frequently used in describing conformations of proteins.

The region from 1230 to 1240 cm^{-1} in the Raman spectrum shows a readily distinguishable very strong amide III band for the antiparallel β-sheet conformation, whereas amide III for the disordered conformation lies at ~1245 cm^{-1}. α-Helical conformation in polypeptides does not show strong Raman bands between 1200 and 1300 cm^{-1}. This conformation has bands appearing between 1260 and 1295 cm^{-1}, and a strong band near 900 cm^{-1} (skeletal stretching) in the Raman spectrum (Koenig and Frushour, 1972; Yu *et al.*, 1973). The 900 cm^{-1} band is either missing or weak in the IR spectrum. β-Sheet polypeptides do not display any strong Raman bands in this spectral region.

Frushour and Koenig (1975) have also summarized the Raman bands that are sensitive to protein conformation. They have cautioned on the application of certain assignments for *quantitative* use until some problems are resolved. There must be a determination of what contribution the rather weak amide III band of the α-helix makes to the overall contour of the amide III band in the spectra of the protein. Also, significant variations

Table 3.1. Conformationally Sensitive Raman Lines of Polypeptides[a,b]

Polypeptide	Amide I	Amide III	Skeletal vibrations (850–950 cm^{-1})
		Alpha helix	
Poly-L-alanine	1655–1659s[b]	Triplet 1284vw 1274vw 1262vw	905s
Poly-γ-benzyl-L-glutamate (PBLG)	1650–1652s	1294w	931s
Poly-γ-methyl-L-glutamate	1656s	1250–1350w	924s
Poly-γ-benzyl-L-aspartate[c]	1663s	?	911m, 890s
Poly-γ-methyl-L-aspartate	1661s	1294w	
Poly-L-leucine	1653s	1261w, 1294w [d]	931s
Poly-L-methionine	1652s	1264w	907s
Poly-L-glutamic acid, solid	1656s	1246wbr	926s
Poly-L-glutamic acid, pH 4.8		1238m	926s
Poly-L-lysine, HCl, solid at 50% R.H.[e]	1655m	1295w	945m
Poly-L-lysine, pH 11.0	1645m (H_2O) 1632m (D_2O)	1200–1300w (H_2O)	945s (H_2O) 950s (D_2O)
Random copolymer of L-leucine and L-glutamic acid	1653s	1200–1300w	931s
Random copolymer of L- and D-leucine	1658s	1258w, 1294w [d]	931m
Racemic blend of PBLG and PBDG	1650s	1291w	931s
		Antiparallel pleated sheet	
Polyglycine I	1674s	1234s, 1221w	884w
Poly-L-alanine	1669s	1243s, 1231m	909s
Poly-L-valine	1666–1671s	1231s, 1277w[f]	959w, 942w
Poly-L-lysine gel	1670s	1240s	1002m
Poly-L-serine	1674s	1235s	894m
Poly-β-benzyl-L-aspartate	1679s	1236m	911m
Poly-β-methyl-L-aspartate	1668s	1230m	
Poly-S-methyl-L-cysteine	1675s	1238s	940m
Poly-(Ala-Gly)	1665s	1238s, 1271w	925m, 890m
Poly-(Ser-Gly)[g]	1668s	1236s	983m
		Random coil	
Poly-L-glutamic acid, pH 7.0	1665s	1248m	949s
Poly-L-lysine, pH 7.0	1665s (H_2O) 1660s (D_2O)	1243s	958m
Poly-L-ornithine, pH 7.0	1665s (H_2O)	1242s	960m
Copolymers of L-glutamic acid and L-tyrosine, pH 10.0		1251m	930s
Copolymers of L- and D-lysine at pH 7.0	1671s (H_2O)	1243s	959m
		3_1 Helix	
Polyglycine II	1654m	1238w, 1261m, 1244m	884s

Continued

Table 3.1. *(Continued)*

Polypeptide	Amide I	Amide III	Skeletal vibrations (850–950 cm^{-1})
	Poly imino acids		
	Carbonyl stretching		
Poly-L-proline I	1650s (solid)		
Poly-L-proline II	1650m (solid), 1630m (in H_2O)		
Poly-L-hydroxyproline	1628m (solid), 1630s (in H_2O)		

[a] Frushour and Koenig (1975).
[b] Abbreviations used here: s (strong), m (medium), w (weak), v (very), br (broad), sp (sharp), sh (shoulder), db (doublet), tr (triplet); all frequencies are in cm^{-1} and refer to the energy shift from the exciting radiation.
[c] This polypeptide is unusual because the chain of L-residues forms a left-handed alpha helix rather than a right-handed helix.
[d] The amide III mode is most likely one of these two lines.
[e] R.H. (relative humidity).
[f] Amide III assignments based on deuteration studies of Frushour and Koenig, *Biopolymers* **13,** 455 (1974).
[g] The sample of poly-(Ser-Gly) used contained a disordered chain fraction.

of frequency occur for the amide III mode of an individual conformation. These authors also stated that investigators tend to group together those regions of protein tertiary structures that are not like recognized conformations—e.g., α-helix, β-sheet, and 3_1 helix—as random and call them random coil conformations as with polypeptides. It must be understood that the frequencies for the disordered regions of all proteins are not identical.

Poly-L-alanine in the solid state is an α-helix, according to X-ray diffraction measurements (Elliott and Malcolm, 1959). Poly-γ-benzyl-L-glutamate is also α-helical, as determined by such measurements (Elliott *et al.*, 1965), and poly-L-valine has a β-pleated sheet conformation (Fraser *et al.*, 1965; Blout *et al.*, 1960; Bloom *et al.*, 1962). Chen and Lord (1974) have recorded the Raman spectra of poly-L-alanine, poly-γ-benzyl-L-glutamate, poly-L-lysine · HCl, and poly-L-valine, in order to study the correlations between various conformations and the frequencies of the peptide linkage. The α-helical conformation in the first two polymers displays amide III frequencies in the range 1265–1300 cm^{-1}. The amide III line is a triplet at 1265, 1275, and 1283 cm^{-1}, which correspond to A,E_1, and E_2 symmetry modes, respectively, of this kind of vibration in the $C_{18/5}$ group (18 residues in 5 turns) of which the α-helix is a member (Fanconi *et al.*, 1969; Fanconi *et al.*, 1971). The β-form in poly-L-valine shows a strong line at 1229 cm^{-1} and a weaker one at 1289 cm^{-1}. The random-coil conformation of poly-L-lysine shows an amide III centered at about 1245 cm^{-1}.

Chen and Lord have summarized the amide III frequencies of the three conformations as follows: random-coil, 1243–1253 cm^{-1} (medium strong); β-antiparallel-pleated sheet, 1229–1235 (strong) and 1289–1295 cm^{-1} (weak); and α-helix, 1265–1300 cm^{-1} (medium). Various workers have verified these frequencies (Lord, 1971; Yu *et al.*, 1972; Yu and Jo, 1973; Chen *et al.*, 1973; Chen *et al.*, 1974; Lord and Yu, 1970a,b). The amide III frequency is about three times as sensitive to conformational change as the amide I.

Rabolt *et al.* (1977) have calculated the optically active normal vibration frequencies of α-helical poly(L-alanine) and poly(L-alanine-*N-d*) (*N*-deuteration). They employed the 47/13 helical structure and included all atoms. The force-field analysis indicated that the amide II′ is in Fermi resonance with one component of the CH_3 asymmetric bend, thus producing a small change of C—N and C═O stretching force constants. These workers found very good agreement between calculated and observed IR and Raman bands. Their results allowed them to make calculations of the influence of small structural changes on the spectrum occurring with observed temperature changes. Their data included IR and Raman studies of the N—H and C—H stretching regions (2500–3500 cm^{-1}) at 300 K and 120 K; the region 1700–500 cm^{-1} (IR) and 1700–100 cm^{-1} (Raman); the far-IR 400–100 cm^{-1} region; the far-IR 100–25 cm^{-1} region; and the various N—D stretching regions.

Fasman *et al.* (1978) have presented a systematic Raman spectroscopic survey of the effect of the degree of polymerization (DP) on the conformation of poly(L-valine). They used materials ranging from two valine residues up to poly(L-valine) with a DP of 930, and investigated the gradual changes in conformation with increasing DP in order to find the DP number for nucleating a β-sheet in poly(L-valine) and the minimum DP necessary for some helical structure to appear. The hexapeptide spectrum (1750–40 cm^{-1}) was very similar to those of polymers with much larger DP, and the amide I (1650–1670 cm^{-1}) and III (1230–1290 cm^{-1}) frequencies showed the structures to be antiparallel β-pleated sheet. A small amount of helical structure exists in polymers with a DP above 500. Unstable α-helical structure was stabilized by the inclusion of L-alanine, which is a strong helix-former.

Itoh *et al.* (1976) have used IR spectroscopy to study the two β forms of poly(L-glutamic acid). In a gel state, α-helical poly(L-glutamic acid) was easily converted to the antiparallel β form by heating. Temperatures between 40 and 85°C yielded a β form with a certain spacing between pleated sheets, and the authors called this form β_1. Heating at a temperature greater than 85°C caused the β_1 form to undergo another conformational change, in which the spacing between pleated sheets was reduced (from 9.03 to 7.83 Å), and this conformation was called β_2. No prominent change was thereby produced in the fiber repeat distance (i.e., the polypeptide backbone conformation).

IR spectra of the samples were used to measure the time course of these two transitions. The $\alpha \rightarrow \beta_1$ transition in its initial stage has a pseudo-first-order rate process with activation enthalpy and entropy of 54 kcal/mol and 92 e.u., respectively. The IR spectra for these two β forms were measured and the data showed that the conformation of the side chains and the mode of the hydrogen bonding between the side-chain carboxyl groups undergo considerable change during the transition.

Heitz *et al.* (1975) have recorded IR spectra of films of the strictly alternating polymer poly (γ-benzyl-D-glutamate-γ-benzyl-L-glutamate). The spectra displayed the amide A band at 3290 cm^{-1}, amide I at 1664 cm^{-1}, and amide II at 1550 cm^{-1}. Polarized spectra showed an α-helical conformation for this polypeptide.

Painter and Koenig (1976) have investigated infrared and Raman spectra of poly (L-lysine) and poly (D,L-lysine) in solutions of low and high ionic strength. Figure 3.1 shows the Raman spectrum of poly (L-lysine), $(Lys)_n$, at pH 10.8 in both α-helical and β-sheet conformations. The β-structure is characterized here by a strong amide I line at 1672

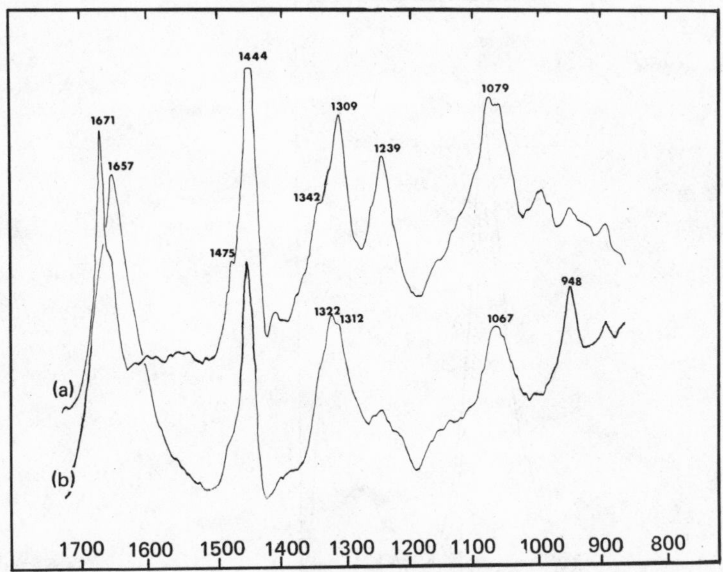

Fig. 3.1. Raman spectra of $(Lys)_n$: (a), $(Lys)_n$, pH 10.8, lyophilized, antiparallel β-sheet conformation; (b), $(Lys)_n$ solution, pH 10.8, α-helix conformation. (Painter and Koenig, 1976.)

cm^{-1} and an amide III line of medium to strong intensity at 1239 cm^{-1}. The spectrum of the α-helical form can be distinguished from that of the β-sheet by the amide I position, now at 1657 cm^{-1}, and a very weak amide III near 1260–1270 cm^{-1}. The α-helical form also displays a line at 948 cm^{-1} that is absent in the spectrum of the β-sheet. In 1 *M* $NaClO_4$, $(Lys)_n$ is subjected to an α-helix-forming effect. Salt solutions of 4 *M* $CaCl_2$ cause a disordering effect on the conformation of the polymer.

St. Pierre *et al.* (1978) have studied the conformations of three sequential copolypeptides, poly (L-tyrosyl-L-lysine), poly (L-tyrosyl-L-lysyl-L-lysine), and poly [L-tyrosyl-$(L\text{-lysyl})_2$-L-lysine] by a variety of techniques, including IR spectroscopy, circular dichroism, ultracentrifugation, and X-ray diffraction. IR dichroism experiments with poly (L-tyrosyl-L-lysine) in oriented polyoxyethylene (Fig. 3.2) showed parallel dichroism of the amide I band at 1623 cm^{-1} and perpendicular dichroism of the minor band at 1690 cm^{-1}, giving evidence of cross-β structure.

IR dichroism studies of poly (L-tyrosyl-L-lysyl-L-lysine) pointed to a helical structure in the solid state. This compound showed parallel dichroism of the amide I band (1651 cm^{-1}) and perpendicular dichroism of the two components of the amide II band (1544 and 1512 cm^{-1}), indicating α-helix structure.

The combined data obtained from the different types of measurements on poly-(L-tyrosyl-L-lysine) supported an antiparallel β-structure for this polymer in acidic or neutral aqueous solution in the solid state.

Oliveira *et al.* (1977) have studied the IR spectra (Fig. 3.3) of the synthetic renin-substrate tetradecapeptide, Asp-Arg-Val-Tyr-Ile-His-Pro-Phe-His-Leu-Leu-Val-Tyr-Ser. The spectrum for the solid state shows an intense amide I band at 1627 cm^{-1}, with a

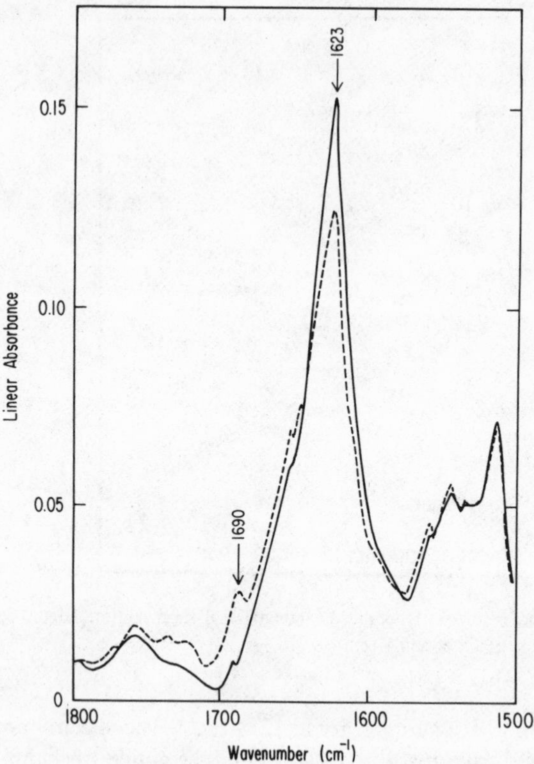

Fig. 3.2. Polarized IR spectrum of amide I and II bands of poly(L-tyrosyl-L-lysine) in a polyethylene oxide matrix: (——) parallel orientation, (- - -) perpendicular orientation. (St. Pierre *et al.*, 1978.)

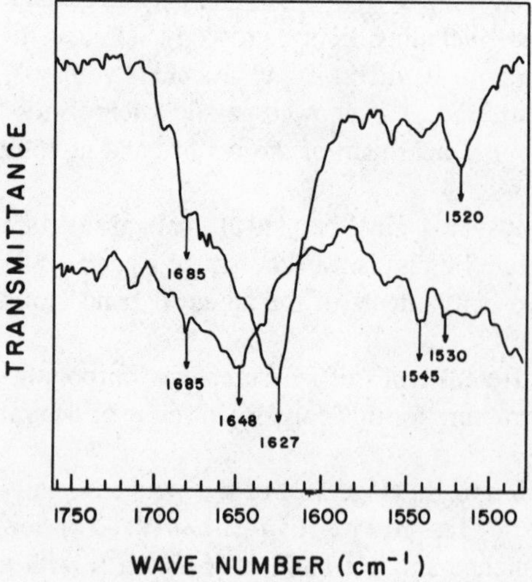

Fig. 3.3. Infrared spectra of the tetradecapeptide in solid film (upper tracing) and in D_2O solution (lower tracing), pD = 4.04. (Oliveira *et al.*, 1977.) Reprinted with the permission of the American Chemical Society. Copyright 1977.

weaker band at 1685 cm^{-1}, representative of antiparallel β structure, although the weak amide II band at 1520 cm^{-1} indicates unordered conformation (Krimm, 1962). At pD 4.04 in D_2O solution, a weak amide I band is displayed, showing antiparallel β structure (Krimm, 1962), but the main band is broadened, showing a peak at 1648 cm^{-1}. Oliveira *et al.* assigned this latter absorption to possible contributions from amide I bands from unordered structure (1658 cm^{-1}) and antiparallel β structure (1630 cm^{-1}). The amide II bands at 1530 and 1545 cm^{-1} also show the presence of antiparallel β structure.

A Polypeptide Antigen

Søruр *et al.* (1977) have synthesized a multichain branched polypeptide antigen and, as one means of characterizing this molecule, measured H–D exchange rates by IR spectroscopy (Parker, 1971). Figure 3.4 shows spectra of the antigen, poly(L-Tyr,L-Glu)-poly(D,L-Ala)-poly(L-lys), (T,G)-A–L, during the exchange progress. Rate curves of the exchange ($^1H \rightarrow {}^2H$) in the peptide groups are presented in Figure 3.5, where the dashed lines are calculated according to Eq. (3.1) (an approximate expression):

$$k_0(\text{per second}) \approx (10^{-pH} + 10^{pH-6})10^{0.05(T-298.16)} \qquad (3.1)$$

Figure 3.5 shows that the exchange rate curves measured for the polypeptide hydrogens deviate somewhat from the rate curves calculated for the exchange in solutions of randomly coiled, linear polypeptides, but the exchange rates observed for the polymer (T,G)-A-L

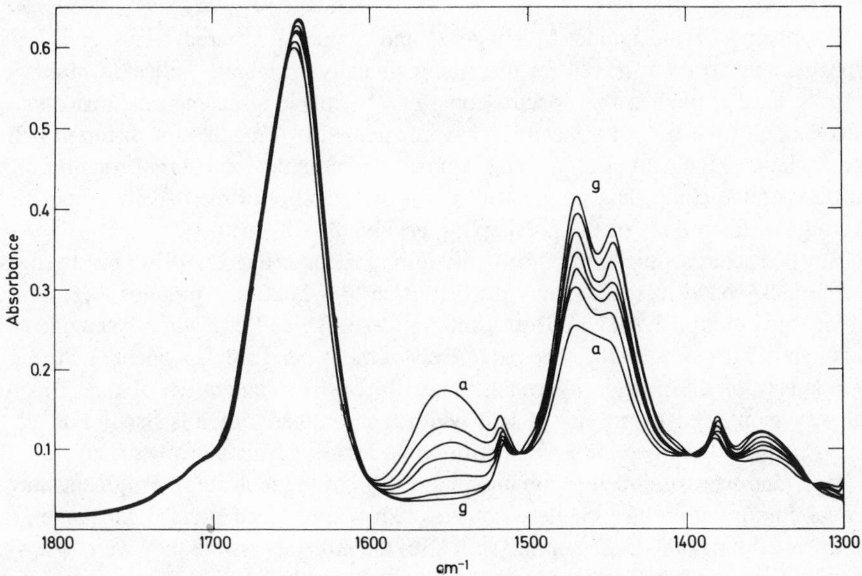

Fig. 3.4. Infrared spectra of (T,G)-A–L 2030 in ^2H_2O, pH 3.2, 10°C, recorded at: (a), 23 min; (b), 40 min; (c), 55 min; (d), 70 min; (e), 101 min; (f), 180 min; (g), 288 min after dissolution. (Scanning takes about 10 min, and the times given refer to the recording of the amide II band at 1553 cm^{-1}). (Sørup *et al.*, 1977.)

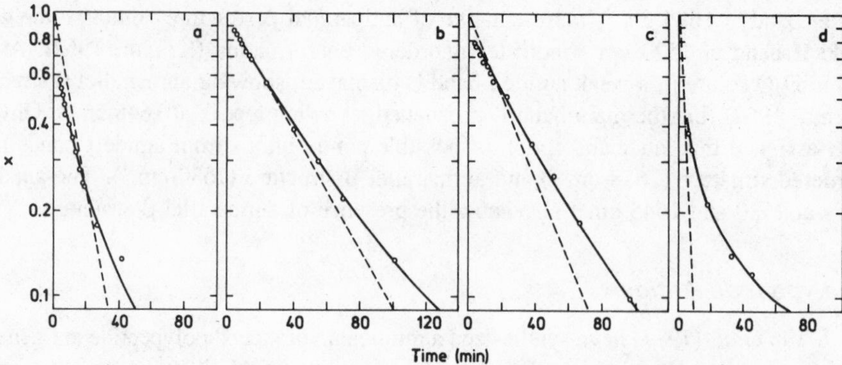

Fig. 3.5. Rate curves of the $^1H \rightarrow {}^2H$ exchange in the peptide groups of (T,G)-A–L 2030, dissolved in ^2H_2O at 10°C. (×) is the fraction of nondeuterated groups. (a), pH 2.2, salt-free solution; (b), pH 3.2, salt-free solution; (c), pH 3.2 and 0.13 *M* NaCl; (d), pH 4.4, salt-free solution. The dashed lines represent the exchange rates in solutions of randomly coiled, linear polypeptides, calculated according to Eq. (3.1) (Sørup *et al.*, 1977.)

are so fast compared with the exchange in solutions of globular proteins (Hvidt and Nielsen, 1966; Hvidt, 1973; Willumsen, 1971; and Hvidt and Pedersen, 1974), that they exclude the existence in this polymer of any proteinlike conformations.

α-Helix Distortion

Broadening of the amide A, amide I, and amide II infrared bands of α-helical polypeptides has been observed for thermodynamically unstable α-helices. Chirgadze *et al.* (1976) have explained this spectroscopic fact by invoking geometrical distortions of the backbone of the helical structure. They considered two models for distorted helices which included regular or irregular distortions of the angles of internal rotation of the main polypeptide chain, and pointed out that the instability of the α-helix is associated with irregular distortions of the polypeptide backbone.

The IR spectroscopic method of deuterium exchange (Parker, 1971) has been used in the amide-A-band region to study the distortion of α-helices in proteins, e.g., sperm-whale metmyoglobin, lysozyme from bacteriophage T4, and egg-white lysozyme (Brazhnikov and Chirgadze, 1978). The parameters of the A band are dependent on the length and geometry of the peptide hydrogen bond. Successive deuteration of the protein in ^2H_2O was monitored by means of half-width measurements of this band. For all the proteins examined the peptide groups with broad amide A bands were exchanged at the first stage (and were assigned to the unordered form of the molecule). Fragments having α-helical conformation had smaller values of half-widths (second-stage exchange). The more distorted the parts of the peptide structure, the more accessible they were to a water molecule. These workers assumed that the fragments most inaccessible to water are located in hydrophobic regions and that the less distorted fragments of the peptide structure have the least conformational flexibility.

Polyhistidine–Carboxylic Acid Hydrogen Bonding

In an IR investigation of the hydrogen bonding between polyhistidine and various carboxylic acids Lindemann and Zundel (1978) showed that $OH\cdots N \rightleftharpoons O^-\cdots H^+N$ bonds formed between carboxylic groups and histidine residues are easily polarizable proton-transfer hydrogen bonds when the protonated histidine residues have a pK_a about 2.8 units greater than that of the carboxylic groups ($\Delta pK_a = 2.8$). From their data they postulated that $OH\cdots N \rightleftharpoons O^-\cdots H^+N$ bonds between histidine and aspartic or glutamic acid side chains in proteins may be easily polarizable proton-transfer bonds. In the presence of water or polar surroundings $OH\cdots N \rightleftharpoons O^-\cdots H^+N$ bonds with smaller ΔpK_a values become easily polarizable proton-transfer hydrogen bonds. Such bonds between histidine and aspartic and glutamic acid residues are found in proteases, ribonucleases, and in hemoglobin, and are present in the active centers of the enzymes.

When equilibrium shifts toward the polar structure, a significant change occurs in the interaction of these bonds with their surroundings, since the dipole moments of the $OH\cdots N \rightleftharpoons O^-\cdots H^+N$ bonds are 8 to 10 D larger when the polar structure is established (Sobczyk, 1976; Ratajczak and Sobczyk, 1969; Nouwen and Huyskens, 1973; Sobczyk and Pawelka, 1973).

The results presented by Lindemann and Zundel show that pH-dependent maxima of the dielectric constant (Kirkwood and Shumaker, 1952) can be caused by $OH\cdots N \rightleftharpoons O^-\cdots H^+N$ hydrogen bonds. The interactions within protein molecules are profoundly affected by the presence or absence of such large dipoles.

Amino Acid Side Chains

Tsuboi (1977) has summarized the characteristic positions of Raman bands of certain amino acid side chains (Table 3.2).

Tyrosyl Residues

Siamwiza *et al.* (1975) have examined the possibility of determining the state of tyrosyl residues in proteins by means of the intensities of a Raman doublet at 850 and 830 cm^{-1}. The doublet was found to be due to Fermi resonance between the symmetric ring-breathing vibration ν_1, and the overtone $2\nu_{16a}$ of the nonplanar ring vibration of 413 cm^{-1} (Fig. 3.6). Examination of a series of model molecules allowed these authors to conclude that the intensity ratio of the doublet is related primarily to the state of the phenolic hydroxyl group, rather than to the environment of the benzene ring or to the conformation of the amino acid backbone.

The tyrosyl residues in proteins may be characterized as having a variety of properties, ranging from those in which the phenolic oxygen is the acceptor atom in a strong hydrogen bond for which the proton donor is very acidic hydrogen (I_{849}/I_{825} ratio 10:4 as in L-tyrosine·HCl solid) to those in which the phenolic hydroxyl is the proton donor in a strong hydrogen bond to a very negative acceptor such as a carboxylate ion (I_{845}/I_{830} ratio 3:10, as in L-tyrosine solid). Another tyrosyl state has the OH ionized to O^-, at pH $\geqslant$12, and the I_{850}/I_{830} ratio is 7:10. When a tyrosyl residue is on the surface of a protein in aqueous

Table 3.2. Characteristic Raman Bands (cm^{-1}) of Amino Acid Side Chains[a]

Phenylalanine	1004 (s), ⌈ weaker than 1004 ⌉ 1203, 1032 624
Tryptophan	1620 (w), 1553 (s), 1433 (w), 1358 (s), 1016 (s), 880 (m), 760 (s) (Certain of these are sensitive to environment of Trp, e.g., 1358 cm^{-1}).
Tyrosine	1621 (w), 1210 (s), 1180 (m), 850 (s), 830 (s), 650 (m) Doublet at 850 and 830 (see Siamwiza *et al.*, 1975). Ratio *I* (850) to *I* (830) depends on intramolecular environment. See also p. 93, this book).

Cys—S—S—Cys bridge; stretching of S—S bond (ν_{S-S}).
ν_{S-S} is sensitive to rotation angles around the two S—C single bonds. (see Sugeta *et al.*, 1972, 1973; Nakanishi *et al.*, 1974; and p. 98, this book): *gauche–gauche,* 510; *gauche–trans* or *trans–gauche,* 525; *trans–trans,* 540 cm^{-1}, Sugeta *et al.* (1971, 1972). (See also Van Wart and Scheraga, 1976a,b and Van Wart *et al.*, 1976a,b.)

[a] Adapted from Tsuboi (1977).

solution, the phenolic OH will exist simultaneously as acceptor and donor of moderate to weak hydrogen bonds, so it lies in the middle of the above range. A surface tyrosine is considered to be "normal" and the doublet intensity ratio is about 10:8, although the actual ratio may vary from 9:10 (L-tyrosine at pH 0.3 in water) to 10:7 (glycyl-L-tyrosine at pH 2.5 in water).

For a tyrosine "buried" in a hydrophobic region of a protein the phenolic oxygen will probably not be the acceptor of a strong hydrogen bond from a highly acidic group, that is, the properties of the buried tyrosines will usually vary from "normal" to those

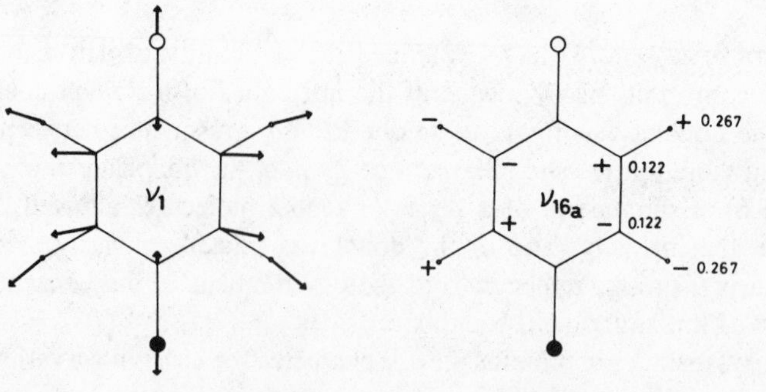

Fig. 3.6. Vibrational modes of ν_1 and ν_{16a} for *p*-cresol [numbering of Wilson (1934)]. The vectors in ν_1 show quantitatively the relative atomic displacements on the mass-reduced scale given by the vector for 0.1 amu$^{-1/2}$. For ν_{16a} the displacements are perpendicular to the molecular plane (+ for upward motion, – for downward) and the relative magnitudes are shown numerically. (Siamwiza *et al.*, 1975.) Reprinted with the permission of the American Chemical Society. Copyright 1975.

Table 3.3. Correlations of the Classification of Tyrosyl Residues in Proteins by Raman Spectra and by Other Methods[a]

Raman spectra		Other methods	
Intensity ratio $I(850)/I(830)$	State of phenolic hydroxyl group	"Buried" tyrosyl residue ($pK > 11$, decreased reactivity)	"Exposed" tyrosyl residue ($pK \sim 10$, normal reactivity)
10:4 [L-tyrosine · HCl (solid)]	Acceptor of strong H-bonds	(Acceptor of strong H-bonds)	
10:8 (e.g., glycyl-L-tyrosine aq soln, pH 3)[b]	Donor and acceptor of moderate H-bonds (N)	Moderate-to-weak H-bonds (A)	Moderate-to-weak H-bonds (N)
3:10 (L-tyrosine solid)	Donor of strong H-bonds to CO_2^- (H)	Donor of strong H-bonds (A)	
7:10 (L-tyrosine, aq soln pH 12)	Ionized (I)	Ionized (I) clearly distinguished by UV absorption	

[a] Siamwiza *et al.*, (1975).
[b] The range of values for the ratio in various tyrosyl model compounds in aqueous solution is 10:10–10:7.

forming strong hydrogen bonds with negative acceptors. Thus, the ratio I_{850}/I_{830} should vary from about 10:7 to about 3:10 in the extreme case of strong hydrogen bonding. Table 3.3 shows a correlation of the expected character of this ratio in the Raman effect of proteins with character suggested by UV spectra and chemical reactivity.

Siamwiza *et al.* (1975) also classified tyrosyl residues of five proteins—lysozyme, ribonuclease A, insulin, erabutoxin a, and cobramine B—according to the observed doublet ratio for each. The classification indicates that the number of moderately hydrogen-bonded tyrosines and of strongly hydrogen-bonded tyrosines obtained from the Raman spectrum of each protein is in good agreement with those of the "normal" tyrosines and "abnormal" tyrosines, respectively, obtained by other methods. The structural changes of proteins caused by lyophilization, heating, decreasing pH, and other denaturing agents commonly increase the intensity of the higher-frequency band of the doublet. These authors state that the intensity change is related to the disruption of the strong hydrogen bonds of the tyrosyl residues.

Disulfides. Cystine Linkages

Van Wart and Scheraga (1976a,b; 1977) and Van Wart *et al.* (1976a,b) have studied disulfides, and the data they collected pertain to the conformations of cystine linkages. They found that the range of S—S stretching frequencies in primary disulfides containing the C—C—S—S—C—C group (e.g., cystine) was from about 485 to 540 cm^{-1}, depending on the size of the dihedral angles of the SS—CC and CS—SC groups. They found, upon examining Raman spectra of model compounds with CS—SC dihedral angles near $\pm 90°$, that disulfides with the *trans* or either of two nonequivalent *gauche* conformations about their C—S bonds all have values of S—S stretching frequencies near 510 cm^{-1}. However, disulfides with A conformations, i.e., those with SS—CC dihedral angles near 30°, about one or both of their C—S bonds gave rise to disulfide stretching bands near 525 or 540 cm^{-1}, respectively. In contrast, S—S stretching frequencies at less than about 505 cm^{-1} arose from absolute values of the CS—SC dihedral angle less than about 60°. The S—S stretching frequencies at about 509, 495, and 486 cm^{-1} corresponded to CS—SC dihedral angles of ±60°, ±30°, and ±10°, respectively. It was predicted that an increase in the dihedral angle above 90° would probably decrease the S—S stretching frequency value. When the S—S stretching frequency ($\nu_{S—S}$) is much above 510 cm^{-1}, it indicates that A conformations exist about C—S bonds; and when $\nu_{S—S}$ is less than about 505 cm^{-1}, it indicates that the CS—SC dihedral angle is strained away from its regular value of ±90° by at least 30°. S—S stretching frequencies of 510 ± 5 cm^{-1} are shown by several conformations about the C—S bond and by CS—SC dihedral angles within about 30° of ±90°.

There are three disulfide bridges in porcine pepsin (Tang *et al.*, 1973). One bridge is part of a hexapeptide loop (-Cys-Ser-Ser-Leu-Ala-Cys-); another bridge is part of a pentapeptide loop, (-Cys-Ser-Gly-Gly-Cys-). The third bridge connects the Cys-233 and Cys-250 residues. To extend the information concerning the conformation of disulfide bridges in proteins Klis and Siemion (1978) synthesized two cyclic peptides (as *N*-acetylamides) very similar to the loop structures shown above. These were peptides I and II:

$$\text{Ac-Cys-Ser-Ser-Leu-Ala-Cys-NH}_2 \qquad \text{(I)}$$

$$\text{Ac-Cys-Ser-Gly-Gly-Cys-NH}_2 \qquad \text{(II)}$$

They recorded Raman spectra of these compounds and found $\nu_{S—S}$ for peptide I at 515 cm^{-1}. Peptide II displayed a band at 525 cm^{-1} and a weaker one at 512 cm^{-1}. Sugeta (1975) and Sugeta *et al.* (1972, 1973) had supported the concept that the frequency of $\nu_{S—S}$ was not dependent on the C—S—S—C dihedral angle ϕ, but rather on the torsional angles in the C—C—S—S—C—C chain. If ϕ has a value near 90°, a band for the $\nu_{S—S}$ frequency appears at 510 cm^{-1} for a *gauche–gauche–gauche* conformation; at 525 cm^{-1} for a *gauche–gauche–trans* conformation; and at 540 cm^{-1} for a *trans–gauche–trans* conformation. The data of other workers had suggested that for cystine peptides the C—C—S—S group most likely is in the synclinal *gauche–gauche–gauche* conformation. However, Klis and Siemion felt that the relationship found by Van Wart *et al.* (1973) was applicable to the analysis of the disulfide bridge conformation in peptides I and II. Van Wart *et al.* had found, for a series of model compounds with *gauche* orientation in the C—C—S—S group, that there is a linear relationship between the ϕ angle value and the location of the $\nu_{S—S}$ band. This relationship led to a value of approximately 85° for the dihedral ϕ angles of pepsin disulfide bridges. The 515 cm^{-1} band for $\nu_{S—S}$ found for peptide I by Klis and Siemion gave a value for ϕ of 75°. They concluded that for peptide II in aqueous solution (bands at 525 and 512 cm^{-1}) an equilibrium obtains between two conformers which probably differ in the chirality of the S—S bridge. The corresponding dihedral angles would have values of 75° and 90°, according to the Van Wart relationship. These authors also concluded that the most favorable conformation of S—S bridges in peptides having pepsin partial sequences is not the same as that in protein, the latter being determined by several extra factors.

Polypeptide Hormones

Lysine Vasopressin, Oxytocin, Mesotocin, Vasotocin, and Arginine Vasopressin. The conformations of the C—C—S—S—C—C chain in lysine vasopressin (III) and oxytocin (IV) were studied by Raman spectroscopy by Maxfield and Scheraga (1977). These two compounds showed an intense band at 508 cm^{-1} in water and in dimethyl sulfoxide (Figs. 3.7 and 3.8). The band was assigned to S—S stretching. Shoulders appearing on this band between 490 and 525 cm^{-1} gave evidence of an equilibrium existing among several conformations for the disulfide moiety of these hormones in solution. The CS—SC dihedral angles are mainly within 30° of ±90°; however, some of the molecules have angles that are strained away from this value by >30°. These authors re-examined earlier observed CD spectra and showed that these spectra gave

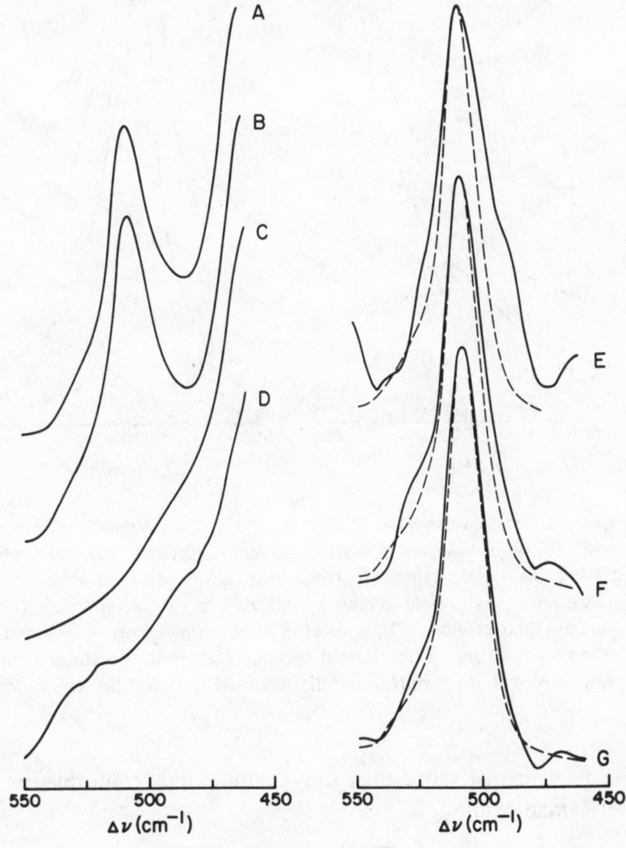

Fig. 3.7 Raman spectra of oxytocin in solution: (A), oxytocin in water at 5°C, adjusted to pH 4 with acetic acid; (B), oxytocin in water at room temperature, adjusted to pH 4 with acetic acid; (C), acetic acid in water, pH 4; (D), oxytocin (treated with performic acid) in water, adjusted to pH 4 with acetic acid; (E), oxytocin in dimethyl sulfoxide at room temperature; (F), same as in A, but with the solvent subtracted out; (G), same as in B, but with the solvent subtracted out. The dashed lines under curves E–G are cystine Raman spectra. The intensity scale is arbitrary. (Maxfield and Scheraga, 1977.) Reprinted with the permission of the American Chemical Society. Copyright 1977.

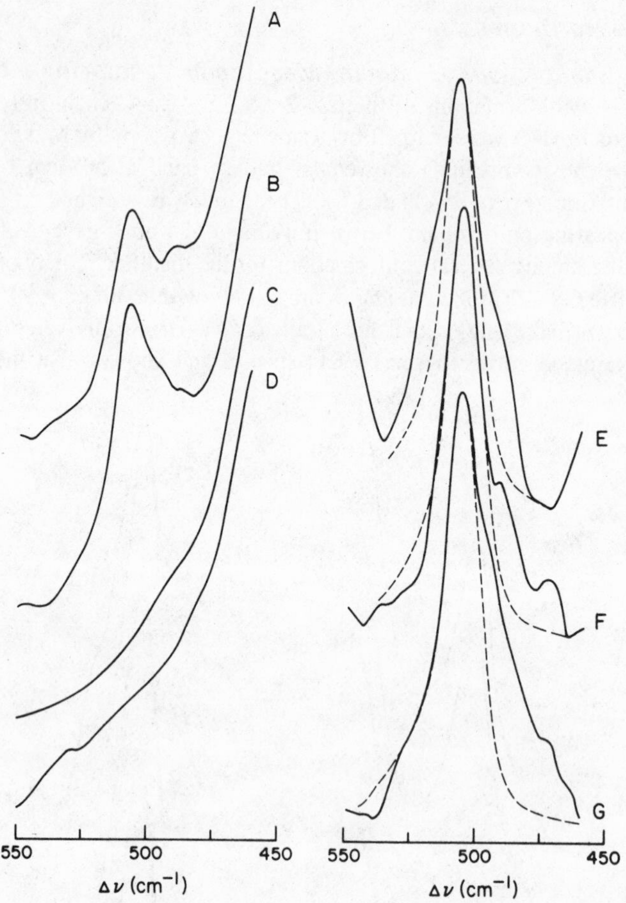

Fig. 3.8. Raman spectra of lysine vasopressin in solution: (A), lysine vasopressin in water at 5°C, adjusted to pH 4 with acetic acid; (B), lysine vasopressin in water at room temperature, adjusted to pH 4 with acetic acid; (C), acetic acid in water, pH 4; (D), lysine vasopressin (treated with performic acid) in water, adjusted to pH 4 with acetic acid; (E), lysine vasopressin in dimethyl sulfoxide at room temperature; (F), same as in A, but with the solvent spectrum subtracted out; (G), same as in B, but with the solvent spectrum subtracted out. The dashed lines under curves E–G are cystine Raman spectra. The intensity scale is arbitrary. (Maxfield and Scheraga, 1977.) Reprinted with the permission of the American Chemical Society. Copyright 1977.

evidence of the presence of more than one disulfide-unit conformation, a finding that agreed with the Raman results.

Lysine vasopressin: H-Cys-Tyr-Phe-Gln-Asn-Cys-Pro-Lys-Gly-NH_2 (III)

Oxytocin: H-Cys-Tyr-Ile-Gln-Asn-Cys-Pro-Leu-Gly-NH_2 (IV)

The conformation of the peptide chain of oxytocin was studied by Raman spectroscopy (Tu *et al.*, 1978). Figures 3.9 and 3.10 show spectra of the hormone in the solid

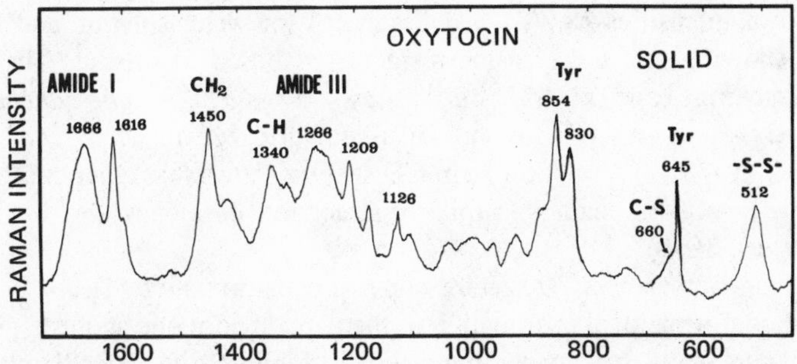

Fig. 3.9. Laser-Raman spectrum of oxytocin in solid phase in the region of 450 to 1740 cm^{-1}. Conditions: radiant power 150 mW, resolution at 5 cm^{-1}, integration time of 2 sec, scanning speed of 6 cm^{-1}/min. (Tu *et al.*, 1978.)

state and in aqueous solutions (H_2O and D_2O), respectively. In the solid phase the amide I band is located at 1666 cm^{-1} and in aqueous solutions at 1663 cm^{-1}, thus showing the absence of α-helical conformation (normally ca. 1650 cm^{-1}). The amide III region displays a clear band at 1260 cm^{-1} in H_2O, which shifts to 994 cm^{-1} in D_2O. Tu *et al.* interpret the relatively high frequency for amide III at 1260 cm^{-1} as ruling out α-helix, antiparallel β-sheet, or random coil structures; and they tentatively assign the band to "β-turn"-like conformations, primarily those which are not involved in intramolecular hydrogen bonding.

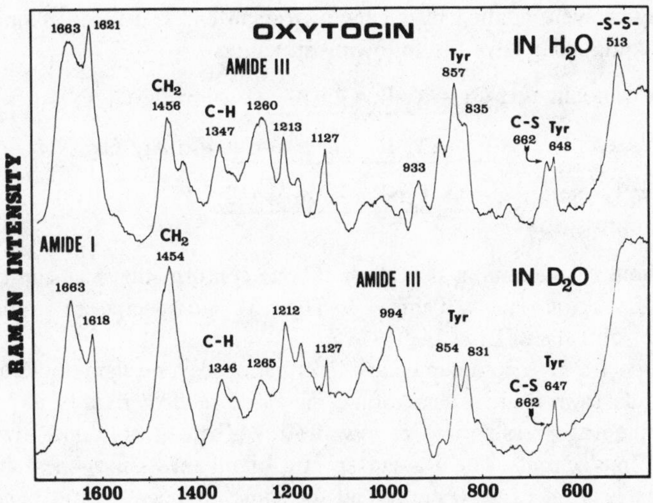

Fig. 3.10. Laser Raman spectra of oxytocin in aqueous solution (upper spectrum) and in ^2H_2O (lower spectrum). For deuteration, oxytocin was dissolved in ^2H_2O and the solution was allowed to stand at 25°C for 4 h, then lyophilized. The lyophilized oxytocin was then dissolved in ^2H_2O and the spectrum obtained. Conditions: radiant power, 500 mW, resolution at 5 cm^{-1}, integration time of 2 sec, scanning speed of 6 cm^{-1}/min. (Tu *et al.*, 1978.)

The ratio of intensities I_{850}/I_{830} is 1.4 and 1.3 for water solution and solid phase, respectively, showing that the tyrosine residue is exposed to solvent. The location of the S—S stretching band at 512 cm^{-1} shows a *gauche-gauche-gauche* conformation of C—C—S—S—C—C in the disulfide bridge, in accordance with data of Sugeta *et al.* (1973). Information from CD spectra shows that other conformations of the C—C—S—S—C—C chain in similarly structured hormones (see below) may be present (Tu *et al.*, 1979).

Studies using Raman and CD spectroscopy were carried out by Hruby *et al.* (1978), who compared conformational information of the peptide hormone agonists oxytocin and [4-glycine]oxytocin with that concerning oxytocin inhibitors [1-penicillamine]oxytocin and [1-penicillamine, 2-leucine]oxytocin. [Gly^4]oxytocin and oxytocin present very similar Raman spectra with an S—S stretching band at 510 cm^{-1} and amide III bands in the interval 1260–1272 cm^{-1}, whereas the oxytocin antagonists [Pen^1]oxytocin and [Pen^1, Leu^2]oxytocin present a singlet or doublet characteristic of S—S stretching and the amide III band at 1255 cm^{-1}.

From their Raman and CD data Hruby *et al.* drew several conclusions: (a) [Gly^4]oxytocin is similar conformationally to oxytocin about the disulfide grouping, the tyrosine residue, and the peptide chain backbone; (b) [Pen^1]oxytocin and [Pen^1, Leu^2]oxytocin are similar conformationally about the disulfide grouping and the peptide chain, but these differ from the ones in oxytocin and [Gly^4]oxytocin; (c) the C–S–S–C dihedral angle in [Pen^1, Leu^2]oxytocin and in [Pen^1]oxytocin is about 110°, with right-handed chirality; (d) the peptide chain in [Pen^1, Leu^2]- and [Pen^1]-oxytocins, and the tyrosine moiety in the latter one are apparently more rigid than in oxytocin, [Gly^4]oxytocin, and other oxytocin agonist analogs that have been studied with circular dichroism.

The neurohypophyseal hormones mesotocin, vasotocin, lysine vasopressin, and arginine vasopressin were studied by means of Raman and CD spectroscopy (Tu *et al.*, 1979). These compounds have the following structures:

Mesotocin: H-Cys-Tyr-Ile-Gln-Asn-Cys-Pro-Ile-Gly-NH_2

Vasotocin: H-Cys-Tyr-Ile-Gln-Asn-Cys-Pro-Arg-Gly-NH_2

Arginine (lysine): H-Cys-Tyr-Phe-Gln-Asn-Cys-Pro-Arg(Lys)-Gly-NH_2
vasopressin

All these hormones have similar peptide backbone conformations, as can be seen from amide I and III spectral data in Table 3.4. Their Raman spectra are similar to that of oxytocin in the solid state (Tu *et al.*, 1978).

The CD spectra give complementary information beyond the structural deductions that can be made from Raman data. Some increased relative rigidity is present in the backbone for the vasopressins and at low pH for all these hormones owing to local intramolecular interactions. The Raman spectra of all these hormones show a major S—S stretching band at ca. 510 cm^{-1}, and for some of them a lesser band at ca. 534 cm^{-1}. The chain structure C—C—S—S—C—C may have a *gauche–gauche–gauche* conformation, but other types are also present. CD spectra of these compounds show the presence of disulfide n-σ^* transition at wavelengths greater than 290 to 300 nm, indicating the presence of conformations of the C—S—S—C group having dihedral angles $\geqslant 120°$ (Tu *et al.*, 1979).

Table 3.4. Comparison of Characteristic Raman Lines of Neurohypophyseal Hormones[a,b]

Compounds	Amide I[c] (cm^{-1})	Amide III (cm^{-1})	$\nu_{s\text{-}s}$[c] (cm^{-1})
Oxytocin	1666	1266	512
Mesotocin	1662–1672	1278	516(m)[c] 534(s)[c]
Vasotocin	1668	1278	516
Arg-Vasopressin	1667(m), 1676	1274	510(m), 534(s)[c]
Lys-Vasopressin	1668–1674	1275	510
[Gly[4]]-Oxytocin	1670	1270	510

[a] Tu *et al.* (1979).
[b] Only solid sample data are assembled in this table.
[c] m, main peak; s, shoulder.

The Raman spectra of these four hormones display a greater intensity at 850 cm^{-1} than at 830 cm^{-1}, showing that their tyrosine side chain lies outside the ring structure of the hormone molecule, as in oxytocin (Tu *et al.*, 1978).

Somatostatin. Somatostatin, whose structure is

H-Ala-Gly-Cys-Lys-Asn-Phe-Phe-Trp-Lys-Thr-Phe-Thr-Ser-CysOH

has been isolated from sheep and pig hypothalami. It inhibits the spontaneous release of growth hormone from rat pituitary cells in culture (Brazeau *et al.*, 1972). Gilon *et al.* (1980) used IR dichroic and hydrogen–deuterium exchange measurements of somatostatin and certain of its analogs incorporated in uniaxially oriented polyoxethylene (POE), in order to study some aspects of their tertiary structure. Their aim was to obtain information that could aid in the design of new active analogs and in the understanding of the mechanism of action of this class of compounds. Band positions and dichroic ratios (obtained with a gold wire grid polarizer) in the N—H stretching and amide I and II regions of these compounds are similar to those of flexible and unordered peptides such as poly(sodium glutamate) and valinomycin. For somatostatin (~2 mg/ml) in oriented POE, ν(NH) is at 3360 cm^{-1} at pH values 2 and 10; amide I is at 1650 cm^{-1} at these pH's; and amide II is at 1535 cm^{-1}. These bands are not dichroic; the spectra are identical for the films at 50% relative humidity or when the films are dried over P_2O_5. Table 3.5 shows the dichroic properties of the acetamido-methylated derivative of dihydrosomatostatin, [$S^{3,14}$-Acm]-somatostatin, which cannot attain the covalent cyclic structure of native somatostatin assumed to be the active form on the receptor site (Gilon *et al.*, 1980). The information above, along with fast deuterium exchange rates, suggested to these authors that somatostatin exists in a flexible, unordered conformation in POE.

Angiotensinamide II. IR and Raman studies were done by Fermandjian *et al.* (1972) on angiotensinamide II conformations. These authors presented spectra of angiotensinamide II, some of its constitutive peptides, and an analog. Comparison of data for the solid hormone with those for the subunits yielded amide vibrations of the hormone

Table 3.5. IR Spectral Parameters of [$S^{3,14}$-Acm]-Somatostatin in Oriented Polyoxyethylene[a,b]

Band	pH 2[b,c] 1.4 mg/ml	pH 10[b,c] 22.0 mg/ml
νNH	3400 (1.0)	3400 (1.0)
	3280 (1.2)	3270 (1.7)
Amide I	1650 (1.0)	1650 (1.0)
	1621 (1.8)	1623 (2.4)
Amide II	1535 (1.0)	1540 (0.8)

[a] Gilon *et al.* (1970).
[b] Dichroic ratios ($D_{\parallel}$) are given in parentheses. Band positions are in cm^{-1}.
[c] Relative humidity, 50%.

in the 500 to 1700 cm^{-1} interval. Angiotensinamide II adopts a preferential antiparallel β conformation in both the solid state and concentrated aqueous solutions.

Glucagon. Characteristic Raman frequencies have been given for glucagon and several other polypeptides by Yu and Peticolas (1974). Among the polypeptides were polyglycine, poly(L-alanine), poly(L-glutamic acid) and poly(L-lysine). Glucagon displays some α-helix structure as shown by amide I and III bands at 1658 and 1266 cm^{-1}, respectively; some β structure as shown by amide I and III bands at 1672 and 1232 cm^{-1}, respectively; and some random coil structure as shown by an amide III band at 1248 cm^{-1}.

Proteins

Table 3.6 shows Raman modes that are applicable to the solving of protein structure.

Fibrous Proteins

Collagen, Gelatin, and Elastin. Frushour and Koenig (1975b) have investigated the Raman spectra of collagen, gelatin (Fig. 3.11), and elastin. In order to make assignments of the lines for the gelatin and elastin spectra, they deuterated the amide N—H groups in gelatin and studied the superposition spectra of the constituent amino acids, a technique used by Lord and Yu (1970a) with several globular proteins. The spectra of collagen (bovine Achilles tendon) and gelatin show lines at 1271 and 1248 cm^{-1} that were assigned to amide III (Table 3.7). Frushour and Koenig believed that the presence of two amide III lines is possibly related to the fact that the tropocollagen molecule has nonpolar (high proline content) and polar (low proline content) regions existing in the chain. They studied the melting, or collagen-to-gelatin transition, in water-soluble collagen of calf skin and assigned the 1248 cm^{-1} amide III line to the 3_1 helical regions of the tropocollagen molecule.

Table 3.6. Raman Modes Useful in the Interpretation of Protein Structure[a]

Origin and frequency (Δcm^{-1})	Assignment	Structural information
Backbone		
Skeletal acoustic		
25–30	Mode of large portion of protein; possibly intersubunit mode	Overall structure; subunit interactions
75	Torsion mode	α-Helix (tentative)
160	Bending mode	β-Structure (tentative)
Skeletal optical		
935–945 (950 D_2O)	C—C stretch (or Cα⁻ C—N stretch)	α-Helix
900	C—C stretch	Conformational change markers broaden and lose intensity with denaturation
963	—	
1002 (H_2O)	Cα—C′ or Cα—Cβ stretch	Suggestive of β-pleated sheet
1012 (D_2O)	—	—
1100–1110	C—N stretch	Conformational change marker broadens and loses intensity with denaturation
Amide I	Amide C═O stretch	A sharp band suggests homogenous structure and uniform hydrogen bonding. Strong hydrogen bonding lowers amide I frequencies
1655 ± 5	—	α-Helix
1632 (D_2O)	—	—
1670 ± 3	—	Antiparallel β-pleated sheet
1661 ± 3 (D_2O)	—	—
1665 ± 3	—	Disordered structure (solvated)
1658 ± 2 (D_2O)	—	—
1685	—	Disordered structure (non-hydrogen-bonded)
Amide III	N—H in-plane bend, C—N stretch	Strong hydrogen bonding raises amide III frequencies
>1275	—	α-helix, no structure below 1275 cm^{-1}
1235 ± 5 (sharp)	—	Antiparallel β-pleated sheet
983 ± 3 (D_2O)	—	—
1245 ± 4 broad	—	Disordered structure
1235	—	Disordered structure (non-hydrogen-bonded)
Amino acid sidegroups		
Tyrosine		
850/830	Fermi resonance between ring fundamental and overtone	State of phenolic-OH I_{850}/I_{830} = 9:10 to 10:7, "exposed" H-bond from acidic donor. 10:7 to 3:10 "buried," strong-OH H-bond to negative acceptor
Tryptophan		
1361	Indole ring	Ring environment. Sharp intense line for buried residue. Intensity diminished on exposure or environmental change. Band contour of doublet near 1360–1380 is sensitive to environment
1338	—	—
Phenylalanine		
1006	Ring breathe	Conformation-insensitive frequency/intensity reference

Continued

Table 3.6. ***(Continued)***

Origin and frequency (Δcm^{-1})	Assignment	Structural information
Amino acid sidegroups (cont.)		
Phenylalanine (cont.)		
624	Ring breathe	Ratio with tyrosine 664 cm^{-1} to estimate Phe/Tyr
Histidine		
1409 (D_2O)	*N*-deuteroimidazole	Possible probe of ionization state, metalloprotein structure, proton transfer
—S—S		
510	S—S stretch	*gauche–gauche–gauche*. Broadening and/or shifts may indicate conformational heterogeneity among disulfides
525	S—S stretch	*gauche–gauche–trans*
540	S—S stretch	*trans–gauche–trans*
C—S		
630–670	C—S stretch	*gauche*
700–745	—	*trans*
S—H		
2560–2580	S—H stretch	Environment, deuteration rate
Carboxylic acids		
1415	O C stretch O	State of ionization
1730	O C ,C═O stretch OH	
	O C ,C═O stretch OR	Metal complexation

[a] Spiro and Gaber (1977).

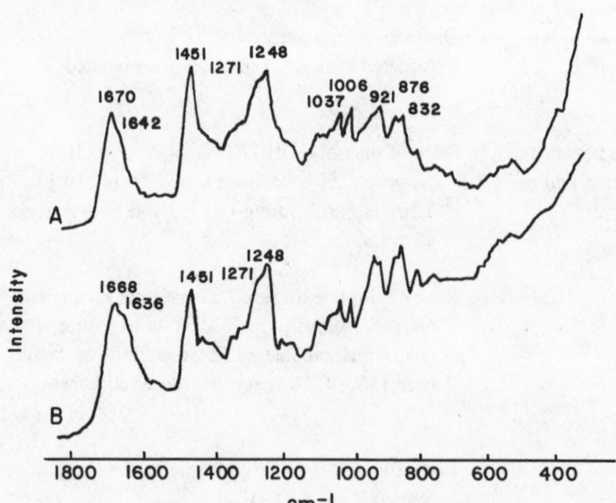

Fig. 3.11. Raman spectra of collagen from bovine Achilles tendon (A) and calf-skin gelatin (B): slit width 8 cm^{-1}, scan rate 10 cm^{-1} min^{-1}, laser power at sample 300 mW in (A) and 600 mW in (B), time constant 10 sec. (Frushour and Koenig, 1975b.)

Table 3.7. Raman Lines in Collagen and Related Spectra[a,b]

Collagen	Gelatin (10% aqueous solution)	Mixture of all amino acids pH 13	Mixture of all amino acids pH 2	Mixture of aromatic amino acids pH 13	Mixture of nonaromatic amino acids pH 13	Gelatin (10% in D_2O)	Assignment
			1746s				ν(C=O)
1670s	1668s					1664s	Amide I
1642s, sh	1636s, sh					1645s sh?	Amide I
					1601m		
	1608w	1611m	1611m	1604m		1611w	Phe, Tyr
		1589m	1589m	1585m			Pro., Hypro
	1566w				1576m		
				1483w			
1464s, sh	1464s, sh		1460s		1457s	1464s	$\delta(CH_3, CH_2)$
1451s	1451s	1450s		1451s			$\delta(CH_3, CH_2)$
			1438s				In-plane bend of carboxyl OH
1422m	1422m	1415vs			1408s	1415m	$\nu_s(COO^-)$
	1399m			1396s			
1392m	1389m						
1343m	1347m	1353s	1353m	1330m	1350m	1347m	$\gamma_w(CH_3, CH_2)$,
1314m	1320m	1323m	1324m	1320m	1320m	1330m	$\gamma_t(CH_3, CH_2)$, $\delta(C_\alpha—H)$
			1271w	1274w			
1271s	1271s					1274w	Amide III
		1245w	1235m	1238w			
1248s	1248s					1247w	Amide III
1211w	1211w	1211w	1211m	1211m		1211δ (D_2O)	Hypro, Tyr
	1198w						
1178w	1182w	1188w	1188w	1188m		1185w	Tyr
					1178w		
1161w	1165w				} 1145w	1135w	NH_3^+
1128w	1128w						
			1118m		1111m		ν(C—N)
1101w	1101w	1111m				1105	
1087w	1084w	1088m	1091w	1091m	1088m		ν(C—N)
1067w	1064w					1057w	
	1051w			1051w			
			1044s				Out-of-plane bend of carboxyl OH
1037m	1037m	1037w		1034m			Pro
					1024w		
1006m	1006m	1006m	1006m	1006s		1006s	Phe
		986w		982m		993m	Amide III′
966w	969sh		969m			966m	Amide III′
					942s		ν(C—C) of residues
938m	942s					942s	ν(C—C) of protein backbone

Continued

Table 3.7. *(Continued)*

Collagen	Gelatin (10% aqueous solution)	Mixture of all amino acids pH 13	Mixture of all amino acids pH 2	Mixture of aromatic amino acids pH 13	Mixture of nonaromatic amino acids pH 13	Gelatin (10% in D_2O)	Assignment
921m	925s					925s	ν(C—C) of Pro ring
918m			918m	918s			
		907vs			908s		
890w	890w						
876m	880s			873s		884s	ν(C—C) of Hypro ring
		856vs	870vs		866s		ν(C—C) of residues
856m	863s			859s		856s	ν(C—C) of Pro ring
			825vs				ν(C—C) of residues
821w	818m					814m	ν(C—C) of backbone
		783m					
769w	769w			786w	783w		
			741s		668w	759wbr	
			650m	646w			
622w	625w	629w	629m	629w			Phe
		593w	593w		593		
568w	572w		575w			575w	
533w	536w	543s	522m		540w	540w	
		518m	504s		522w		
		447w					
	425w			479m			Pro
			414m	447m			Hypro
396w						407w	
				345w			Pro

[a] Frushour and Koenig (1975).
[b] Key to abbreviations: s (strong), m (medium), w (weak), sh (shoulder), vw (very weak), ν (stretching coordinate), δ (deformation coordinate), γ_w (wagging coordinate), γ_t (twisting coordinate), Tyr (tyrosine), Phe (phenylalanine), Trp (tryptophan), Pro (proline), Hypro (hydroxyproline).

Strong lines in the spectrum of elastin at 1668 and 1254 cm^{-1} for amide I and amide III, respectively, and only weak scattering at 938 cm^{-1}, support the concept that elastin consists mostly of disordered structure.

The far-IR spectra of native, partially denatured, and fully denatured collagen were recorded in the range from 400 to 240 cm^{-1} by Gordon *et al.* (1974). Figure 3.12 (curve A) shows that native collagen has an absorption maximum at 345 cm^{-1}, which arises from the triple-helical (tertiary) structure of the molecule. This band is destroyed (see curve D) when the sample is prepared by casting at 60°C, a method that is used to prepare "hot-cast" gelatin (Bradbury and Martin, 1952), which retains the primary structure (amino

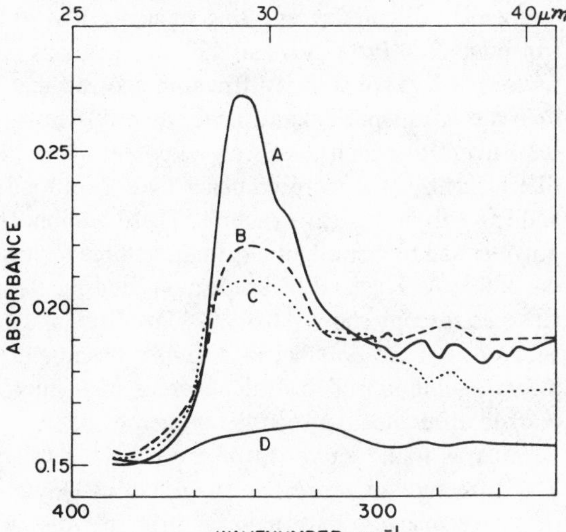

Fig. 3.12. The spectra of native, partly denatured, and fully denatured collagen films in the region 400–240 cm^{-1}: (A) native collagen, cast at 23°C; (B), denatured collagen (gelatin) cast at 4°C; (C), denatured collagen, cast at 23° C ("cold-cast" gelatin); (D), denatured collagen, cast at 60°C ("hot-cast" gelatin). All films cast from 0.05 *M* acetic acid solution. (Gordon *et al.*, 1974.) Reprinted with the permission of the American Chemical Society. Copyright 1974.

acid sequence) of collagen but contains no higher level of structural order. Curve C, recorded with a film cast at 23°C from a solution of denatured collagen, displays a band absorbance part way between that of curve A (native collagen) and that of the completely denatured film. Curve B was obtained with a sample of gelatin which was less denatured than the "cold-cast" gelatin of Bradbury and Martin (1952). This film was prepared by casting a gelatin solution at 4°C. Using a value of 0% helicity for gelatin cast at 60°C and 100% helicity for native collagen, these workers obtained values for the apparent percentages of helicity in gelatin cast at 4°C and 23°C. For the former, values of 76 ± 12% and 71 ± 9% helicity were obtained; for the latter, 46 ± 10% and 44 ± 10%. Values obtained from measurements by optical rotatory dispersion were in agreement.

Veis and Brownell (1977) have used hydrogen–deuterium exchange to study triple-helix formation of ribosome-bound nascent chains of procollagen. They obtained the polyribosomes containing the procollagen chains from chick embryo fibroblasts in culture, and the isolated chains were nearly completely hydroxylated. The IR spectrum of the polyribosomes in D_2O at 15°C was recorded with a reference cell that held collagen-depleted polyribosomes in D_2O with the concentration of RNA matched to that of the sample. When the polyribosomes were heated in D_2O at 44°C, the absorbance at 1480 cm^{-1} of the amide II N—D band increased greatly, showing that hydrogen–deuterium exchange had occurred. When polyribosomes had their collagen depleted, they displayed no such appearance of an N—D band in the 1480–1450 cm^{-1} region upon heating. Polyribosomes that were removed from the IR cells after treatment at 44°C and cooling still contained collagen. The results mentioned above showed that nascent collagen bound to the polyribosomes can have a hydrogen-bonded structure. These authors suggested that the collagen studied was in the conformation of a triple helix, and assumed that the formation of a triple helix can occur between nascent chains while they are bound to the surface of the endoplasmic reticulum.

Doyle *et al.* (1971) have carried out conformational studies on the polypeptide and oligopeptides with the repeating sequence L-alanyl-L-prolyl-glycine. They used the polytripeptide (Ala-Pro-Gly)$_n$ and the homogeneous oligopeptides Boc-(Ala-Pro-Gly)$_n$-OMe, where n was 1–6. X-ray diffraction showed that the polymer and the oligomers longer than the hexapeptide could have three different forms in the solid state, depending on the solvent to which they were exposed: (1) hydrogen-bonded sheets of polyproline-II-like helices; (2) a more compact hydrogen-bonded polyproline II type sheet structure; and (3) a triple helical structure. These authors studied the solid-state IR spectra of the polymer and higher oligomers in these three forms. They found clear correlations between the amide A, I, and II band locations and the structure for the higher oligomers. IR and CD spectroscopy showed that (Ala-Pro-Gly)$_n$ and the oligomers were unordered in aqueous solution, but the polymer was highly structured in ethylene glycol-hexafluoroisopropyl alcohol solution and had some order in trifluoroethanol solution. The data showed a marked difference between the properties of (Ala-Pro-Gly)$_n$ and those of (Pro-Ala-Gly)$_n$, which was found earlier to be triple helical in aqueous solution and in the solid state. The difference in properties suggested to Doyle *et al.*, that X–imino–acid–glycine tripeptide sequences may have a different role in collagen than imino-acid–X–glycine sequences.

Keratins. The IR spectra of α-keratins (e.g., silk, wool, feather, porcupine quill, human hair), which contain α-helices in the form of microfibrils, display an amide I band near 1650 cm^{-1} with parallel dichroism and an amide II band near 1550 cm^{-1} with perpendicular dichroism (Krimm, 1962). For the β-keratins, formed by stretching or LiBr contraction of the α-helices, amide bands in the IR are characteristic of β-sheet structure—a perpendicular dichroic amide I at 1630 cm^{-1} and a parallel dichroic amide II at 1535 cm^{-1}.

Figures 3.13 and 3.14 present Raman spectra of wool in the native state and immersed in ^2H_2O (D_2O), respectively. Table 3.8 gives the band assignments. The amide I band at 1658 cm^{-1} shows the high content of α-helix in native wool. Notice that the amide III region is wide and ill defined, typical of proteins that are mainly α-helical. After immersion in ^2H_2O the amide III region is weaker and the band center is at 1261 cm^{-1}. This band was ascribed to α-helices that were not deuterated completely (Lin and Koenig, quoted in Frushour and Koenig, 1975). The very strong band at 962 cm^{-1} for wool in ^2H_2O was assigned to the amide III mode of the deuterated disordered chains in the amorphous region.

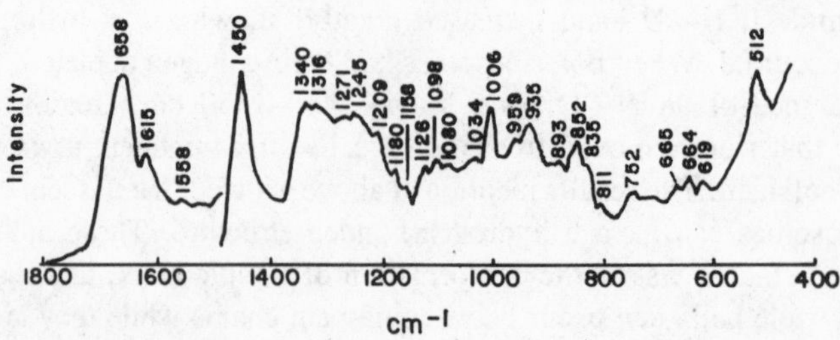

Fig. 3.13. Raman spectrum of wool in native state. (Lin and Koenig, quoted in Frushour and Koenig, 1975.)

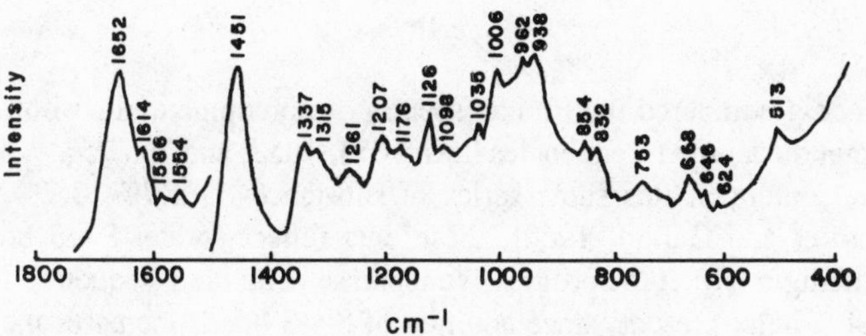

Fig. 3.14. Raman spectrum of wool in ^2H_2O. (Lin and Koenig, quoted in Frushour and Koenig, 1975.)

Table 3.8. Frequencies and Tentative Assignments of Raman Spectra of Wool in Native State and in Deuterium Oxide[a]

Wool	Wool in D_2O	Tentative assignments
1658vs		Amide I
	1652vs	Amide I
1615m	1614m	Tyr and Trp
	1586vw	Phe
1558vw	1554vw	Trp
1450vs	1451vs	CH_2 and CH_3 bending mode
1340sdb	1337sdb	CH bend, Trp
1316sdb	1315sdb	$C_\alpha H$ bend
1271mvbr		
	1261wvbr	Amide III (α-helical)
1245mvbr		Amide III (disordered)
1209m		Tyr and Phe
	1207m	D_2O
1180w	1176w	Tyr
1158vwsh		Skeletal stretch (β)
1226m	1126m	
1098wbr	1098wbr	C—N stretch
1080wbr		
1034w	1035w	Phe
1006s	1006s	Phe and Trp
	962sdb	Amide III′
959w		CH_2 rock
935m	938sdb	Skeletal stretch (α), residue C—C stretch
883wbr		Skeletal stretch (β)
852mdb	854mdb	Tyr
835mdb	832mdb	Tyr
811vw		
752wbr	753wbr	Trp
665vw	668w	Cys C—S stretch
644w	646w	Tyr
619vw	624vw	Phe
512m	513m	Cys S—S stretch

[a] Frushour and Koenig (1975).

Lin and Koenig compared the Raman spectra of porcupine quill, wool, and human hair and found that the amide I frequencies lie at 1653, 1658, and 1663 cm^{-1}, respectively. Since the sulfur content of this same series of substances is 2.7%, 3.7%, and 3.9%, respectively (Seifter and Gallop, 1966), these data reflect greater S—S bond content. Frushour and Koenig interpret the progressive increase in amide I frequency as a decrease in the amount of α-helix present, since addition of S—S bonds increases the amorphous content of the keratin.

When wool is stretched 30% and 60% beyond the unstretched state, the amide I band shifts from 1658 (unstretched) to 1670 and 1672 cm^{-1}, and the 1672 cm^{-1} band corresponds with both disordered and β-sheet conformations (Frushour and Koenig, 1975). The amide III band increases in intensity during the stretching process and its frequency shifts from 1245 to 1237 cm^{-1}, the latter being due to β-sheet conformation. Table 3.9 presents additional data on fibrous proteins.

Dimethylsulfoxide (Me_2SO), hexylmethylsulfoxide (HxMeSO), and decylmethylsulfoxide (DecMeSO)—all known to be enhancers of skin permeability—induced the formation of antiparallel-chain β-sheet conformation in *in vitro* human stratum corneum (the outermost layer of epidermis, ca. 10 μm) (Oertel, 1977). Me_2SO and HxMeSO appeared to act by displacing water molecules bound to polar side chains of protein, but DecMeSO probably interacted hydrophobically with the protein. The conformational transition did not result from the removal of lipid.

The β-sheet protein, most probably formed in normally α-helical portions of the intracellular keratin filaments, returned to α-helix form when the tissue was rehydrated. Oertel found that the order of ability to cause β-sheet formation is HxMeSO>DecMeSO>Me_2SO, for 1 molar solutions. He used a Digilab FTS-14 Fourier

Table 3.9. Conformationally Sensitive Raman Modes of Fibrous Proteins[a]

Predominant conformation	Amide I (cm^{-1})	Amide III (cm^{-1})	Skeletal modes (cm^{-1})
α-Helical			
Porcupine quill	1653s (sp)	~1260w	935s
Wool	1658s (bd)	1261w (α′) 1245m (disordered)	935s
β-Form			
β-Keratin	1672s	1237s	
Cross-β (steam supercontracted wool)	1671s	1239s	
Silk	1667s	1229s (β) 1259w (disordered)	880w (β) 1169m (β)
Feather	1669s	1240s (β) 1274m (disordered)	879w (β) 1162m (β)
Disordered			
LiBr supercontracted wool	1666s	1249s	

[a] Frushour and Koenig (1975).

transform spectrophotometer for the measurements of dependence of β-sheet formation on sulfoxide concentration, duration of treatment of tissue, pH, and degree of tissue hydration.

Muscle Proteins

Myosin. Carew *et al.* (1975) have used Raman spectroscopy to probe the substructure of myosin. Figure 3.15a shows the amide III region of myosin, which can be compared with aqueous solutions of amino acids [Fig. 3.15, (b)(c)(d)] at relative concentrations corresponding to their content in myosin. The amino acid spectra between 1230 and 1320 cm^{-1} lack the intense activity displayed by myosin near 1240 and 1310 cm^{-1}. The 1244-cm^{-1} band of myosin may indicate either β structure or random coil conformations, or a mixture of both; and the 1232 cm^{-1} shoulder may indicate β structure. By comparison, α-helical polypeptides do not show amide III frequencies below 1260 cm^{-1}.

Carew *et al.* made their interpretations (shown in Table 3.10 and Fig. 3.16) partly by using data already known about simpler molecules. Evidence of β structure would

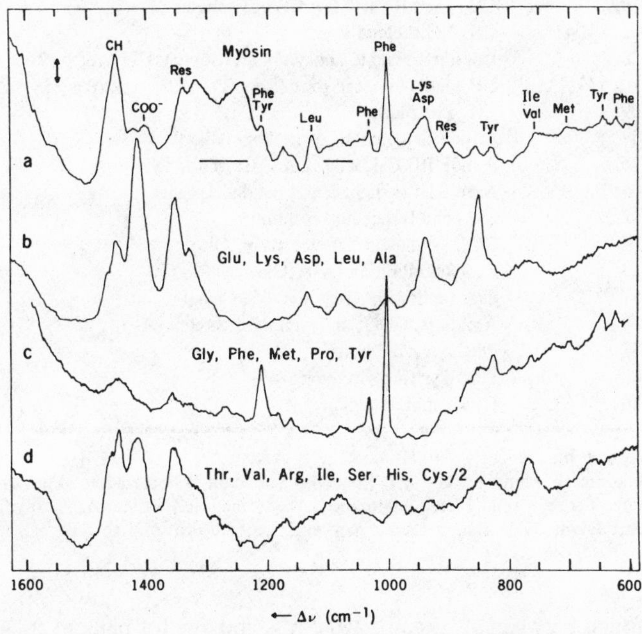

Fig. 3.15. Comparison of Raman spectra of aqueous solutions of myosin (a) and its constituent amino acids (b–d). The relative concentrations of the amino acids correspond to those in myosin. The five most prevalent amino acids appear in (b); those in (c) are most common in the globular head portions of the myosin molecule. Horizontal lines indicate regions containing contributions from vibrations near the peptide bond. Abbreviations: Res, residue vibration; Gly, glycine; Thr, threonine; Ser, serine; Ile, isoleucine; His, histidine; Cys/2, cystine; other abbreviations and more complete assignments appear in Table 3.10. Conditions were: (a) myosin, 35 mg/ml in 0.6 *M* KCl, pH 7.0; (b, c, and d) amino acids, pH 6.0, 3.0, and 11.0, respectively (to aid solution). Resolution, 5 cm^{-1}; scanning speed, 30 cm^{-1}/min; vertical arrow, 300 count/sec in (a) and 3000 count/sec in (b to d); laser power, 600 mw in (a) and 200 mw in (b to d); excitation, 5145 Å. (Carew *et al.*, 1975.)

Table 3.10. Raman Frequencies of Myosin[a,b]

Peak (cm^{-1})	Tentative assignment
622	CC ring twist (Phe)
645	CC ring twist (Tyr)
660–695	CS vibrations
704	CS stretch (Met)
755	Amide V; residue vibrations
780	
830	CC symmetric ring stretch (Tyr)
855	CH_2 residue rock (especially Tyr)
(882)	
901	CC residue stretch
940	CC residue stretch (Lys, Asp); CH_3 symmetric stretch (Leu, Val)
(962)sh	CH_3 symmetric stretch
1004	CC ring stretch (Phe); weak CC stretch (Lys, Glu)
1033	CC ring bend (Phe)
(1044)?	CH_2 twist (Glu, Arg, Pro)
(1058)?	CH_2 twist (Lys, Arg); CH, CC stretch (?)
1081, 1104	CN, CC skeletal stretch
1128	Isopropyl residue antisymmetric stretch; CN stretch (?)
1160, 1175	CH_3 antisymmetric rock (Leu, Val); CH rock (Phe, Tyr)
1209	Tyr, Phe modes
1244	Amide III (β chain + random coil)
(1265)sh	Amide III; CH bend (Leu, Asp, Glu, Tyr, Pro)
(1304)sh, 1320	Amide III (α-helix); CH bend CH_2 twist
1342	CH bend (especially residue)
1402	COO^- symmetric stretch (Asp, Glu)
1423	Residue vibration (Asp, Glu, Lys)
1451	CH_3 (antisymmetric), CH_2, CH bend
1553	Amide II; COO^- antisymmetric stretch (Asp)
1587	Phe, Arg vibrations
1607	Phe, Tyr ring vibrations
1650	H_2O, amide I regions

[a] Carew *et al.* (1975).
[b] Notation: sh, shoulder; and ( ), frequency uncertain. Abbreviations: Phe, phenylalanine; Tyr, tyrosine; Met, methionine; Lys, lysine; Val, valine; Asp, aspartic acid; Leu, leucine; Glu, glutamic acid; Arg, arginine; and Pro, proline.

generally be found between 1225 and 1240 cm^{-1}, and that for random coils between 1235 and 1250 cm^{-1}. Various polypeptides in the β conformation, e.g., poly-L-valine (Chen and Lord, 1974), and polyglycine I (Small *et al.*, 1970), and poly-L-lysine (Yu *et al.*, 1973) display amide III frequencies at 1229, 1234, and 1240 cm^{-1}, respectively; but the amide III frequencies shown by the random coil forms of insulin, glucagon, and poly-L-glutamic acid (Yu *et al.*, 1972) appear at 1239, 1235, and 1248 cm^{-1}, respectively.

The shoulders at 1265 cm^{-1} and 1304 cm^{-1} are at frequencies characteristic of α-helical structure (Chen and Lord, 1974; Yu *et al.*, 1973; Lord and Yu, 1970a; 1970b; Tobin, 1968; Yu *et al.*, 1972; Small *et al.*, 1970). The 1265-cm^{-1} shoulder is apparently

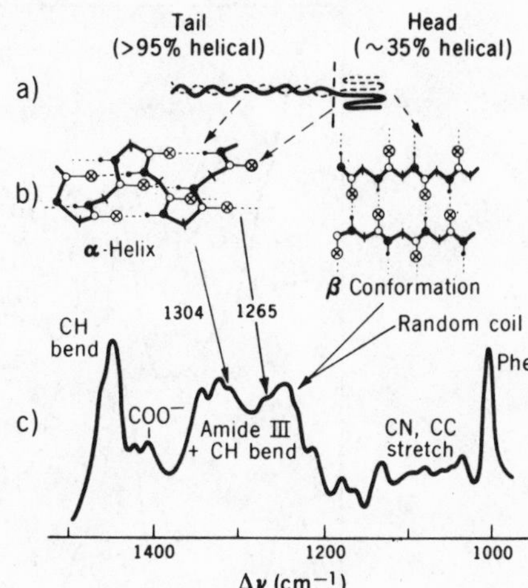

Fig. 3.16. One possible correlation of (a) the structurally distinct subregions of myosin, (b) their proposed conformational structures, and (c) the Raman spectrum of the amide III region of myosin. Arguments favoring this interpretation of the spectral results are presented in Carew *et al.*, 1975.

caused by the globular α-helical fraction, and the 1304 cm^{-1} region is characteristic of the fibrous α-helical conformation of myosin (Fig. 3.16). The region 1320 to 1360 cm^{-1} for myosin has many CH bending modes (Fig. 3.15, a–d).

Figure 3.17 compares myosin in H_2O and D_2O. The H_2O peak at ~ 1650 cm^{-1} moves to ~1200 cm^{-1} in D_2O, and thereby allows better examination of the amide I region (1620 to 1680 cm^{-1}). When myosin is dissolved in D_2O containing KCl, deuteration of the NH groups shifts amide III vibrations about 300 cm^{-1} downward. Figure 3.17b shows decreased intensities near 1244, 1265, and 1304 cm^{-1} in deuterated myosin. These decreases identify the missing sections of those bands as amide III vibrations. A similar reduction in the relative intensity of the 1304 cm^{-1} band happens in depolarized Raman spectra. Carew *et al.* believe this polarized portion of that band, which is shifted by deuteration, to be an α-helical amide III vibration.

Thermal denaturation of myosin increases the intensity of the 1244 cm^{-1} band (β structure or random coil), and also causes changes in the 1040–1120-cm^{-1} region. This region has conformationally sensitive skeletal vibrations (mainly CC and CN stretching).

Myosin displays CS stretching modes near 704 cm^{-1} (methionine moiety) and 669 and 693 cm^{-1} (disulfide linked). Ionized carboxyl groups of the aspartic and glutamic residues show peaks at 1402 cm^{-1} and 1533 cm^{-1} (symmetric and antisymmetric stretch, respectively). Some peaks not noted earlier represent CH and CC vibrations of the amino acid residues. These are shown in the spectra of Fig. 3.15, b–d.

Ring vibrations of phenylalanine (near 1004s, 1033s, and 1209s; and near 622w, 1587w, and 1607 cm^{-1}w), and of tyrosine (1209s, doublet near 830 and 850, and near 645 cm^{-1}w) are rather insensitive to thermal denaturation of the myosin molecule.

The conformation of myosin in films prepared in various ways has been studied by IR spectroscopy (Jacobson *et al.*, 1973). The precipitating conditions determine whether β structure as well as the α helix are present. In films which were prepared by precipitation from low ionic strength aqueous media (which approach physiological conditions) the myosin contains a mixture of α-helical and random forms. In films which were prepared

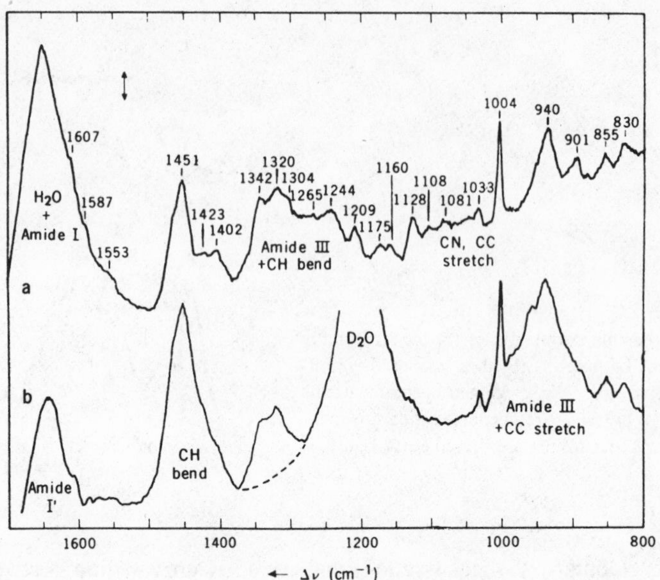

Fig. 3.17. Comparison of Raman spectra of myosin in (a) H_2O and (b) D_2O solution (0.6 *M* KCl, pH 7.0). Solvent peaks occur in different regions in these solvents (near 1650 and 1200 cm^{-1}, respectively). Deuteration of the NH groups of myosin shifts the amide III activity from the region 1220 to 1320 cm^{-1} to the region 890 to 990 cm^{-1}. The decrease in intensity from 1240 to 1310 cm^{-1} in D_2O solution verifies the presence of amide III contributions in that region in spectra of normal myosin. Spectral resolution, 5 cm^{-1}; scanning speed, (a) 30 cm^{-1}/min and (b) 60 cm^{-1}/min; vertical arrow, 300 count/sec; laser power, 500 mW; excitation, 5145 Å. (Carew *et al.*, 1975.)

by precipitation in aqueous media at low pH, the spectra showed that a significant amount of β structure was caused by the pH change. The observed conformations also depended on the precipitating media when organic solvents were used to prepare films. If pure methanol or a 50:50 mixture of methanol and chloroform were used in the precipitating media, β structure was observed. With pure chloroform no β structure was found.

Tropomyosin. Frushour and Koenig (1974) have studied the Raman spectra of tropomyosin denatured by pH changes. This protein in the native state contains about 90% α-helix, but this value decreases rapidly as the pH exceeds 9.5. The native state shows a Raman spectrum with a strong amide I band at 1655 cm^{-1}, very weak scattering in the amide III region at about 1250 cm^{-1}, and a band of medium intensity at 940 cm^{-1}. After denaturation, the tropomyosin displays a strong amide III band at 1254 cm^{-1}, and the 940-cm^{-1} band is weak. The intensities of these two bands are a sensitive indicator of the amount of α-helix and disordered chain present.

Troponin C. In vertebrate skeletal muscle, troponin and tropomyosin regulate the interaction between myosin and actin. Troponin consists of three polypeptide chains, one of which is called troponin C (TrC, molecular weight 18,000). TrC binds four Ca^{2+} ions, but with greater binding constants for two of the binding sites that also bind Mg^{2+}

(Ca^{2+}–Mg^{2+} sites), and lesser constants for the other two binding sites, which bind only Ca^{2+} (Ca^{2+}-specific sites) (Potter and Gergely, 1975). Carew *et al.* (1980) have used Raman spectroscopy as a means for probing Ca^{2+}-induced changes in TrC. Other spectroscopic measurements had been used earlier to show that Ca^{2+}-binding to the Ca^{2+}–Mg^{2+} sites induces large increases in α-helical and tertiary structure (van Eerd and Kawasaki, 1972; Murray and Kay, 1972; Potter *et al.*, 1976; Leavis *et al*, 1978; Seamon *et al.*, 1977; Levine *et al.*, 1977, 1978).

Figure 3.18 shows spectra of TrC in the absence and presence of Ca^{2+} (mole ratios 0–4 as indicated). The amide III region of TrC alone presents broad major bands at 1242 and 1268 cm^{-1} and two weaker bands (actually shoulders) at 1253 and 1260 cm^{-1}. The bands at 1268 and 1260 cm^{-1} show α-helical structure; those at 1242 and 1253 cm^{-1} are characteristic of disordered structure. β-Pleat structure is minimal, as judged by the absence of bands below 1240 cm^{-1}. With increasing addition of Ca^{2+} ions to TrC, the spectra point to an increase in α-helical content. The spectral changes are completed at a ratio of 2 mol Ca^{2+}/mol protein.

The conclusion drawn by Carew *et al.* from the amide III region was that major changes in the spectra are associated with Ca^{2+} binding to the Ca^{2+}–Mg^{2+} sites rather than the Ca^{2+}-specific sites. These changes include a sharpening of the 1260 cm^{-1} band and the appearance of two shoulders at 1272 and 1280 cm^{-1}. Other spectra (not shown) display changes corresponding to carboxylate, thiol, and phenol side chains.

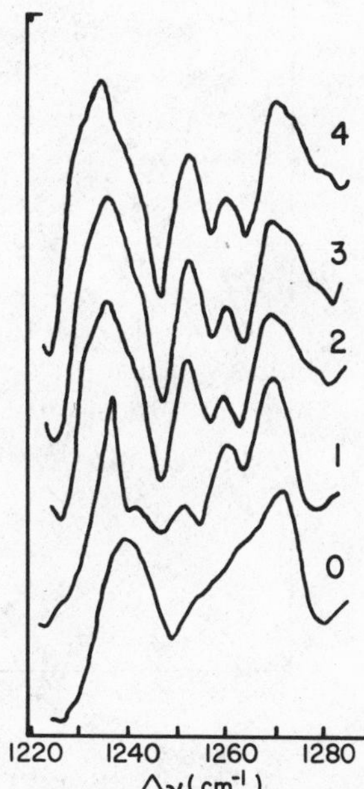

Fig. 3.18. Expanded Raman spectrum of amide III region. Conditions: 1 m*M* TrC in solution containing 0.1 *M* KCl, 25 m*M* Hepes buffer, pH 7.4, and Ca^{2+} at indicated mole ratios. Hepes is *N*-2-hydroxyethylpiperazine-*N'*-2-ethanesulfonate. (Carew *et al., 1980.)*

Muscle Fibers (Intact)

Intact single muscle fibers of the giant barnacle *(Balanus nubilus)* have been examined by Raman spectroscopy to determine the state of water in muscle tissue (Pézolet *et al.*, 1978). The spectra in the O—H (or O—^{2}H) stretching regions for water in unfrozen fibers showed that there is very little difference between the shape and relative intensity of the bands from water molecules situated inside the fiber and those of the corresponding bands in the pure water spectrum. These Raman measurements ruled out the presence of more than ~5% "structured" intracellular water.

The same type of muscle fibers have been studied by Pézolet *et al.* (1978a) to obtain conformational information. Figure 3.19 shows Raman spectra of an intact single muscle fiber in ^2H_2O and H_2O. Several bands of the myofibrillar proteins appear, several of which indicate α-helix as the predominant structure: amide I at 1648 cm^{-1}, relatively sharp and with no significant shoulder; very weak amide III scattering between 1225 and 1280 cm^{-1}; and a strong band in the skeletal C—C stretching region at 939 cm^{-1}. Such features occur in the spectra of α-helical poly-L-lysine (Yu *et al.*, 1973) and tropomyosin, which contains about 90% α-helix (Frushour and Koenig, 1974). Subtraction of the spectrum in ^2H_2O from that in H_2O gave a difference spectrum (not shown) which permitted a more detailed amide III band. In particular, a broad band at 1242 cm^{-1} was believed to be caused by flexible and weakly hydrogen-bonded disordered structure (Yu and East, 1975; Guillot *et al.*, 1977) rather than antiparallel β-structure, which displays

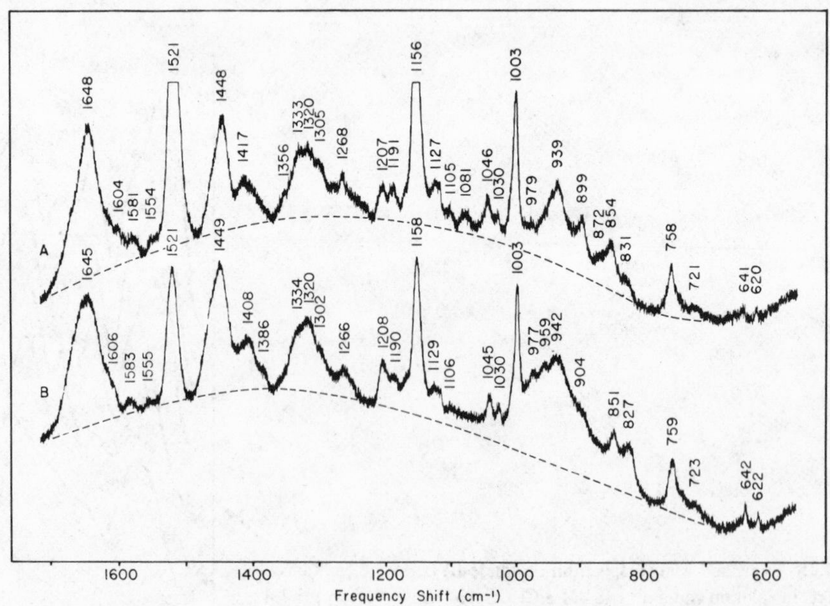

Fig. 3.19. Laser Raman spectra of intact single muscle fibers in ^2H_2O (A) and H_2O (B). Laser (514.5 nm) power at sample 120 mW; spectral slit width, 4 cm^{-1}; scan rate, 22 cm^{-1}/min; 10^4 counts full scale; time constant, 0.5 sec. (Pézolet *et al.*, 1978a.)

a sharp amide III band at 1240 cm^{-1} (Pézolet *et al.*, 1976). Other amide III bands in the difference spectrum (1282 and 1305 cm^{-1}) were assigned to α-helix. The low scattering intensity between 1230 and 1290 cm^{-1} was attributed to the presence of a protein, paramyosin, in the molluscan muscle fiber. Aromatic side chains (Phe, Tyr, Trp) were also identified from characteristic Raman bands. Resonance-enhanced Raman bands of membrane-bound β-carotene were observed at 1521 and 1156 cm^{-1}.

The predominantly α-helical structure of the muscle-fiber contractile proteins of *Balanus nubilus* is not affected when the fiber is converted from the relaxed form to the contracted one by the addition of ATP and Ca^{2+} (Pézolet *et al.*, 1980). Figure 3.20A shows the Raman spectrum of a fiber-filled glass capillary tube in a solution that causes relaxation of the fiber. The myoplasm of the fiber is exposed to the external solution at both ends of the capillary. The amide I and III regions (1620–1680 cm^{-1} and 1220–1320 cm^{-1}, respectively) of the muscle fiber spectrum show that the proteins are mainly in the α-helical conformation, since the α-helix gives a strong sharp amide I band at 1650 ± 5 cm^{-1} and weak scattering in the amide III region. Figure 3.20 shows the spectrum of the same fiber-filled capillary after the addition of Ca^{2+} and ATP, with conditions set to allow the complete diffusion of Ca^{2+} and ATP to the center of the capillary (a 15-h process), where the fiber is transversely illuminated.

Comparison of the two spectra suggested to Pézolet *et al.* that contraction does not significantly change the secondary structure of the contractile proteins. However, the addition of Ca^{2+} and ATP causes marked decreases in intensity of certain bands originating from amino acid side chains, especially from the acidic and tryptophan residues. [Decreases in the β-carotene (β-car) band intensities are not caused by contraction but rather by the bleaching of the photosensitive pigment after long exposure to the laser beam.] These authors have thus shown that these amino acids are involved during the generation of tension.

Caillé *et al.* (1982), in a continuation of Raman work on barnacle muscle fibers, have found that the conformational modification previously reported by Pézolet *et al.* (1980) was not induced by Mg^{2+} or ATP but was induced by Ca^{2+}

Globular Proteins

Insulin and Proinsulin. Yu *et al.* (1972) have recorded the Raman spectrum of native insulin in crystalline powder form (Fig. 3.21). The amide I mode of the native compound (dotted line) has a major band at 1662 cm^{-1} with a shoulder at 1680 cm^{-1}. The former was assigned to α-helix conformation and the latter to random coil. The amide III region, which is conformationally sensitive, is rather indistinct for insulin. This is characteristic of proteins containing much α-helix structure (Frushour and Koenig, 1975). The major band at 1270 cm^{-1} and the rather weak shoulder at 1288 cm^{-1} was assigned to α-helix. The 1239-cm^{-1} shoulder was probably caused by random coil and a small amount of β-sheet conformation. In a later paper Yu *et al.* (1974) used single-crystal native insulin and determined a corrected amide III region by difference spectroscopy, by subtracting the spectrum of the deuterated crystal from that of the nondeuterated one. Bands at 1303, 1269, and 1284 cm^{-1} were assigned to α-helix.

When the insulin is denatured (Fig. 3.21, solid line), the amide I band shifts to 1672

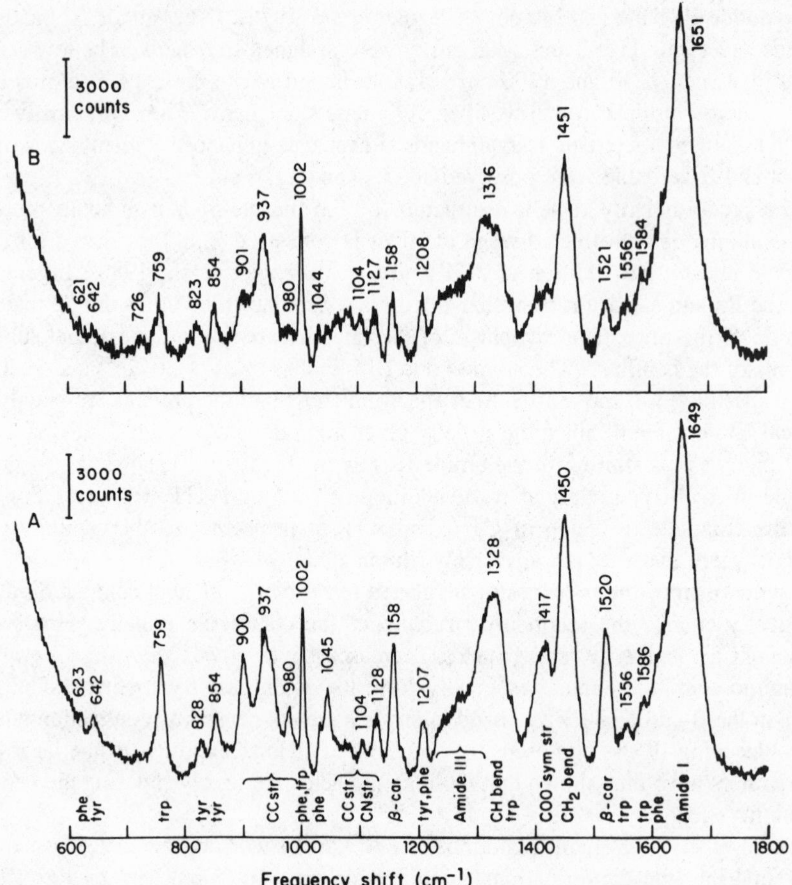

Fig. 3.20. Raman spectra of a fiber-filled capillary of 7 mm in length, illuminated at 3 mm from the end of the capillary and bathing in a solution containing 100 m*M* K, 25 m*M* Tris, pH 7.2, and 0.5 m*M* EGTA (curve A), and 100 m*M* K, 25 m*M* Tris, pH 7.2, 0.5 m*M* EGTA, 0.5 m*M* Ca, and 5.0 m*M* ATP (curve B), Excitation wavelength, 514.5 nm; laser power at the sample, 125 mW; spectral resolution, 5 cm^{-1}; integration period, 2 sec, frequency shift increment, 1 cm^{-1}. Both spectra are from single scans and are unsmoothed. However, each spectrum was corrected for a small fluorescent background by subtracting a quadratic polynomial function. Abbreviations: str, stretching; sym, symmetric; bend, bending; CC, carbon–carbon bond; CN, carbon–nitrogen bond; CH, carbon–hydrogen bond; β-car, β-carotene; phe, phenylalanine; tyr, tyrosine; trp, tryptophan. (Pézolet *et al.*, 1980.)

cm^{-1}. This may be caused by antiparallel β-sheet structure (Frushour and Koenig, 1975) rather than random coil formation as postulated for denatured lysozyme by other workers. Also, a band at 1230 cm^{-1} for denatured insulin agrees with the amide III frequency for certain model compounds having antiparallel β-sheet structure.

Koenig and Frushour (1972) and Frushour and Koenig (1974) have shown that strong bands appear in the region of 946 cm^{-1} and 934 cm^{-1} (C—C stretching of polypeptide backbone) for α-helical proteins and polypeptides with high helical content. Breaking of

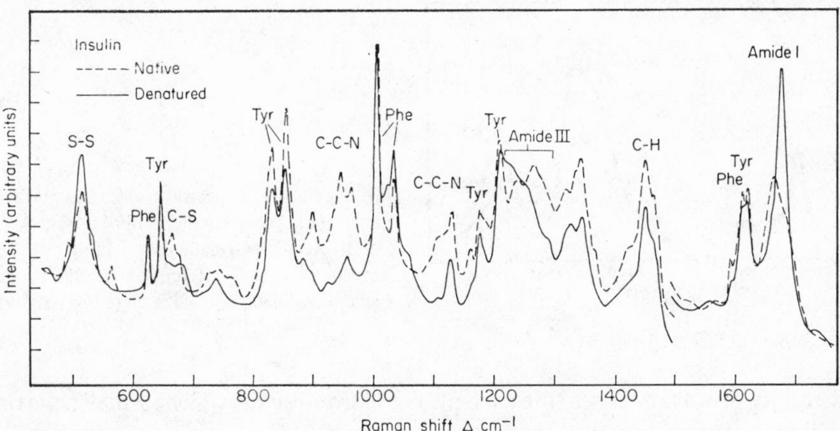

Fig. 3.21. Superimposed comparison between the spectra of native and denatured insulin. The line at 624 cm^{-1} due to the inplane ring vibration of phenylalanine residues is known to be conformation-independent and is used as an internal reference. This line has nearly equal intensity in both spectra. (Yu *et al.*, 1972.)

the helices, with the introduction of β-sheet structure, produces lowered intensities of these bands, such as those for denatured insulin (Fig. 3.21).

The denaturation of insulin causes the intensity of the S—S stretching vibration at 515 cm^{-1} to increase; the C—S stretching vibration at 668 cm^{-1} shifts to 657 cm^{-1} with an intensity decrease. Lord and Yu (1970a,b) believed that the spatial configurations of the two interchain S—S linkages change but the lone intrachain bonding does not.

When heated in acid, the globular protein insulin polymerizes to form submicroscopic fibrils. Primarily from their IR dichroism experiments, Ambrose and Elliott (1951) had decided that the fibrils contain polypeptide chains which are in the β conformation and which extend transverse to the fibril axis; that is, they have cross-β structure. Burke and Rougvie (1972) provided evidence supporting the essential correctness of the Ambrose–Elliott viewpoint. Low-angle X-ray diffraction data and electron microscopy suggested that the fibrils have a uniform cross section with dimensions of about 29 by 47 Å. Wide-angle X-ray diffraction and IR dichroism showed that the fibrils have cross-β structure, and circular dichroism and optical rotatory dispersion measurements supported the β conformation.

Figure 3.22 shows IR dichroism spectra of solid samples of oriented insulin fibrils. The amide I band at 1632 cm^{-1} indicates the presence of β structure. Although these spectra cannot be used to prove the absence of a weak band at 1690 cm^{-1} (antiparallel β structure) due to overlap by a band at 1740 cm^{-1}, they can be used to show that a 1690-cm^{-1} band of moderate strength is absent. Burke and Rougvie stated that if the 1690-cm^{-1} component of the amide I band of the fibrils were nearly as strong as it is in the antiparallel β structure of silk, the IR spectra of Fig. 3.22 would have exposed it readily. The dichroism of the amide I and II bands shows that the hydrogen bonds are parallel to the long axis of the fibrils, so that the peptide chains are perpendicular to the fibrillar long axis and, therefore, in a cross-β conformation. There is no β structure shown

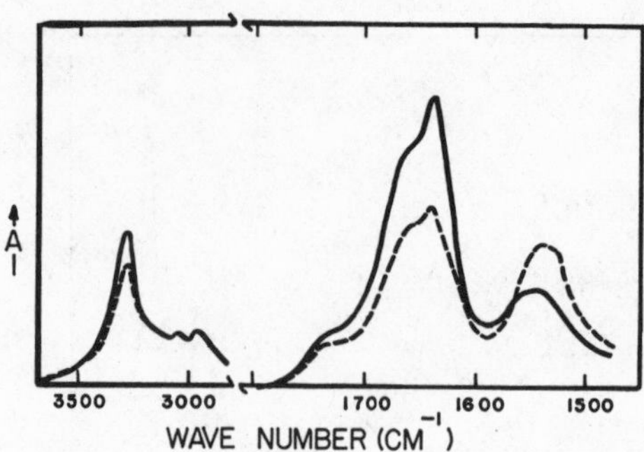

Fig. 3.22. Infrared dichroism of insulin fibrils. Solid line, electric vector parallel to fibril axis; broken line, electric vector perpendicular to fibril axis. (Burke and Rougvie, 1972.) Reprinted with the permission of the American Chemical Society. Copyright 1972.

in IR spectra of solid noncrystalline samples of native insulin, which display a maximum at 1660 cm^{-1}.

Yu *et al.* (1972) have studied porcine proinsulin by Raman spectroscopy for differences in conformation between the insulin moiety in proinsulin and that in free insulin; they found essentially no change in conformation. However, the spectra did differ in the amide I and amide III regions. These differences were ascribed to the extra peptide structure of proinsulin not present in insulin. This C-peptide is removed when proinsulin is activated. It contains 33 amino acid residues (Bhagavan, 1974). When Yu *et al.* (1972) found a band at 1663 cm^{-1} and a shoulder at 1685 cm^{-1} upon graphical subtraction of the proinsulin and insulin spectra in the amide I region, they attributed these difference bands to α-helix and random coil conformations, respectively, of the C-peptide. Frushour and Koenig (1975) question the former assignment and state that the 1663 and 1685 cm^{-1} bands may be associated with random coil structure.

Serum Albumin. Chen and Lord (1976) have studied the conformation of bovine serum albumin (BSA) by Raman spectroscopy. Figure 3.23 shows a spectrum of 4% BSA and Table 3.11 presents the frequencies and relative peak intensities of the Raman lines. The amide III vibration is a mixture of C—N bond stretching and in-plane N—H bending (Chen and Lord, 1974; Frushour and Koenig, 1974; T.J. Yu *et al.*, 1973a). It displays a frequency near 1230–1235 cm^{-1} for peptide links in a β-pleated sheet conformation, near 1240–1250 cm^{-1} for the random-coil form, and from 1260 to 1300 cm^{-1} for the α-helical form. Figure 3.23 has no peak intensity near 1230–1235 cm^{-1}, thus ruling out the presence of β form. The amide III intensity caused by α-helix in the range 1260 to 1300 cm^{-1} is much larger than that resulting from random coil (the line at 1248 cm^{-1}). Chen and Lord estimate the relative amounts of helical and random coil forms to be 55%–60% and 40%–45%.

In the spectrum of albumin a sharp, strong amide I line appears at 1652 cm^{-1}, which (in confirmation of the amide III line) results from a large amount of α-helix. The sharpness of this line is similar to that of the α-helix form of poly-L-alanine (Chen and Lord, 1974; Frushour and Koenig, 1974). Another line in support of helical conformation (Frushour and Koenig, 1974) is the strong one at 941 cm^{-1}.

The BSA molecule contains about 19 tyrosyl residues. The tyrosyl residue of proteins has been shown (Bellocq *et al.*, 1972; Yu *et al.*, 1973b; Siamwiza *et al.*, 1975) to display a doublet, the intensity ratio of which is a reflection of the degree of the hydrogen bonding

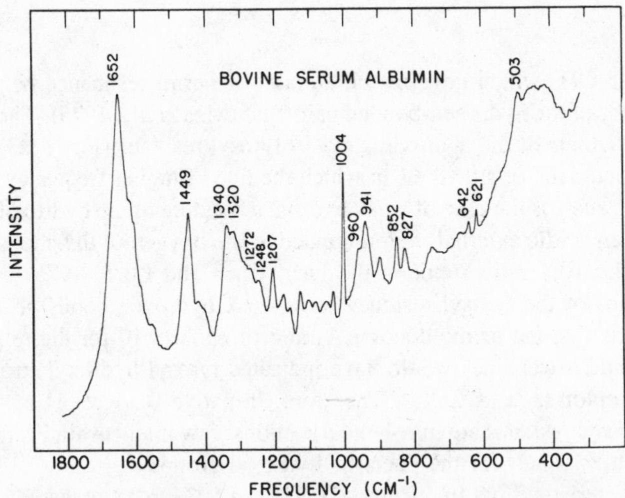

Fig. 3.23. Raman spectrum of 4% BSA as originally recorded: spectral slit width 5 cm^{-1}, time constant 2 sec, scan speed 0.5 cm^{-1}/sec, laser power at the sample approximately 150 mW at 4880 Å. (Chen and Lord, 1976.) Reprinted with the permission of the American Chemical Society. Copyright 1976.

Table 3.11. Raman Spectrum of BSA[a,b]

Frequency, (cm^{-1})	Tentative assignment	Frequency, (cm^{-1})	Tentative assignment
422 (0)		1055 (2)	
462 (0)		1083 (2)	
503 (2)	ν(SS)	1105 (2)	ν(CN)
573 (0)		1128 (3)	
621 (2)	Phe	1159 (2)	
642 (1)	Tyr	1177 (3)	Tyr + Phe
672 (1)	ν(CS)	1207 (5)	
697 (0)		1248 (5)	Amide III
750 (1)	Trp	1272 (5)	
804 (0)		1320 (8)	$\gamma(CH_2)$(?)
827 (3)	Tyr	1340 (9)	
852 (4)		1415 (3)	$\nu(CO_2^-)$
884 (0)		1449 (10)	$\delta(CH_2)$
900 (3)		1458 (5 sh)	
941 (6)	ν(CC)	1550 (2)	Amide II
960 (4 sh)		1587 (0)	Phe
1004 (10)	Phe	1605 (1)	Tyr + Phe
1032 (2)		1615 (0)	
		1652 (21)	Amide I + H_2O

[a] Chen and Lord (1976).

[b] Solution 4% by weight, pH 6, 0.1 *M* NaCl. Frequencies for sharp lines are accurate to ± 1 cm^{-1} and broad lines to ± 2 cm^{-1}. Numerical figures in parentheses are relative peak intensities with that of 1449 cm^{-1} taken as 10. sh denotes shoulder. ν means stretching vibration, δ deformation, and γ twisting.

of the phenolic OH, which governs the amount of Fermi resonance between the donor and the acceptor of the hydrogen-bonded pair (Siamwiza *et al.*, 1975). The mean intensity ratio for the doublets of the approximately 19 tyrosyl residues (Fig. 3.23) is 10:7, which is a ratio characteristic of situations in which the line of higher frequency is more intense than the lower, such as the case of weak hydrogen bonding of the hydroxyl to an acceptor, or the case of an acidic external proton bonded to the oxygen of the phenolic OH (typical ratio 10:4). The 10:7 ratio supports the idea (Chen and Lord, 1976) that most of the hydroxyl groups of the tyrosyl residues form weak hydrogen bonds or act as acceptors to other protons that are acidic donors. A ratio of about 3:10 for the relative intensities of the higher and lower lines would have indicated tyrosyl hydroxyl strongly bonded to a negative acceptor such as COO^-. Therefore, in native BSA at pH 6, where probably most of the tyrosyl residues are in α-helical sections, few if any available negative binding sites exist such as COO^- for the phenolic hydroxyl groups.

The characteristic S—S frequency of BSA is at 503 cm^{-1} and there is little indication of a shoulder on the high-frequency side of the S—S line caused by frequencies in the range 525–545 cm^{-1}. Several workers have assigned this range to the S—S stretching vibration of the —C—S—S—C groups in *gauche-gauche-trans* and *trans-gauche-trans* configurations (Sugeta *et al.*, 1972; Sugeta *et al.*, 1973; Miyazawa and Sugeta, 1974). If one assumes the validity of this assignment, most of the S—S groups in BSA are not in such conformations but rather must be in the *gauche-gauche-gauche* configuration, the S—S frequency of which ranges from 500 to 515 cm^{-1}.

Most of the intensity in the C—S stretching range (650–750 cm^{-1}) is caused by C—S vibrations of the cystine residues, since there are only three methionine residues per BSA molecule. The C—S bond vibrations do not display sharply defined bands in the spectrum. Lines appearing at about 672 and 697 cm^{-1} are due to conformations in which either the hydrogen atom on the α-carbon of cystine or the amino nitrogen is in the *trans* position across the C_α—C_β bond from the near sulfur atom (Miyazawa and Sugeta, 1974).

A similar study on bovine serum albumin has been carried out by Lin and Koenig (1976). These authors suggest a mechanism of denaturation, due to pH and temperature changes (pH range 1.7–10.9 and t°C 25–90), not previously proposed. The process involves the following: the native state; a reversible conformational change; irreversible unfolding of α-helices; aggregation, with disulfide exchanges, and further unfolding; gel formation and formation of intermolecular β-conformation, and further unfolding.

Hvidt and Wallevik (1972) have used the infrared hydrogen–deuterium exchange method to study conformational changes in human serum albumin. They presented their data as plots of the fraction of unexchanged peptide hydrogen atoms (X) at a given time (t) versus log (k_0t), where k_0 is the pH-dependent exchange rate constant of solvent-exposed peptide groups. The plots of X versus log (k_0t) are called "relaxation spectra." In an exchange experiment the protein is dissolved in water with an isotopic composition different from its own, and the irreversible approach (the relaxation) of the solution to isotopic equilibrium is followed.

For albumin at 25°C the plot of X versus log (k_0t) is a discontinuous function of pH, illustrating the changes in the stability of the conformation of the protein brought about by changes in pH.

At a pH value near 7 defatted albumin has maximum stability and at this pH less than 20% of the peptide groups are protected from solvent exposure by conformations that are stabilized by values of ΔG larger than 8 kcal mol^{-1}. For half of the peptide groups ΔG is smaller than 5 kcal mol^{-1}.

β-Lactoglobulin. Frushour and Koenig (1975a) have compared the Raman spectra of β-lactoglobulin in solution and in the lyophilized and crystalline forms. The spectra of the lyophilized and crystalline materials were essentially identical, except for an obvious difference of a weak line appearing at 1286 cm^{-1} in the latter. These authors concluded that the conformations of β-lactoglobulin in these two states are very similar. The intensity of the conformationally sensitive amide III line at 1242 cm^{-1} increased by 30% when the protein was dissolved in water. This change is interpreted as a conformational change in the disordered chains of the protein, and appears to be characteristic of globular proteins having a substantial amount of disordered chain structure.

When the pH of an aqueous solution of β-lactoglobulin was increased from 6.0 (native state) to 11.0 (denatured state), the amide III line moved from 1242 to 1246 cm^{-1}, decreased in intensity, and broadened. Other workers have proposed that such changes reflect the conversion of β-sheet structures to the disordered conformation, and the findings of Frushour and Koenig are consistent with their data. At pH 13.5, the amide III moved to 1257 cm^{-1}, a change characteristic of a completely disordered protein, i.e., that any remaining β-sheet structure in β-lactoglobulin is randomized.

With the denaturation, several changes occur in the intensities of the tryptophan and tyrosine vibrations. As the pH changes from 6.0 to 11.0, the intensity ratio of two ring vibrations of tyrosine, I_{855}/I_{830}, decreases from 1.0:0.9 to 1.0:1.3. At pH 7, a copolymer composed of 95% glutamic acid and 5% tyrosine (where the polymer is a random coil in which the tyrosines are available to the solvent) has the value of the ratio, I_{855}/I_{830}, at 1.0:0.62. This value of the ratio is much closer to that for β-lactoglobulin in the native state than in the denatured condition, and suggests that the average tyrosine in the denatured state (pH 11.0) may be in a more hydrophobic environment than in the native state (pH 6.0). As the pH is increased from 6.0 to 11.0, a tryptophan vibration displayed at 833 cm^{-1} for the protein in its native state becomes weak in the denatured state. A change in the intensity of this line may also indicate a change in the local environment of the tryptophan residue.

Ralston (1972) has studied the exchange of peptide hydrogens in β-lactoglobulin and its *N*-ethylmaleimide derivative (modification of the lone SH group of the protein) by measuring the amide II intensity decrease at 1547 cm^{-1}. The EX_2 mechanism of exchange (see Parker, 1971, Chap. 11) was observed for both proteins at pD values between 4 and 7. Between pD 7 and 9 both change conformation, resulting in increase in the observed rate of H–D exchange. Modification of the single SH group caused a more rapid exchange of peptide hydrogens at all pD values studied.

The alteration of the exchange behavior of α-lactoglobulin, brought about by alteration of the SH group, is similar to that caused in metmyoglobin by heme removal (Abrash, 1970). (See exchange models in Ralston, 1972).

Ralston (1972) used an empirical method in this study to correct for changes in the absorbance of the amide I band that accompanied the deuteration process.

Ceruloplasmin. Freeman and Daniel (1978) have studied human ceruloplasmin (a plasma protein) by several spectroscopic techniques—IR absorption, CD, and luminescence. The IR and CD methods indicated the presence of β-pleated sheet conformation in the native structure. Approximately half of the amino acids are in the β conformation and half in an unordered form. Absorption and CD spectra allowed these workers to conclude that the copper chromophores are involved in six electronic transitions between 300 and 900 nm. Their data provided evidence for an interaction between the copper chromophores responsible for the 330-nm absorption in ceruloplasmin. The number of copper atoms per molecule of ceruloplasmin may be six (Magdoff-Fairchild *et al.*, 1969), but the work of Freeman and Daniel supports the presence of as low as five atoms. The IR spectrum shows two strong bands at 1632 and 1530 cm^{-1}, amide I and amide II, respectively. Minor bands occur at ~1400 and ~1450 cm^{-1}.

Prothrombin and α-Casein. Frushour and Koenig (1974) have investigated the Raman spectra of prothrombin and α-casein, proteins having very little ordered secondary structure, if any. They found that the spectra of both proteins have strong bands at 1254 cm^{-1} in the amide III region but only weak scattering at 940 cm^{-1}, characteristics like those of denatured tropomyosin. The amide I band of α-casein is found at 1668 cm^{-1}, which value agrees with previously studied spectra of disordered polypeptides, poly-L-glutamic acid (Lord and Yu, 1970b) and poly-L-lysine at pH 7.0 and mechanically deformed poly-L-alanine studied in Koenig's laboratory.

Attachment of Water Molecules to Caseins. Rüegg and Häni (1975) studied the IR spectra of α_s-, β-, and micellar casein films at relative water vapor pressures (p/p_0) ranging from 0 to 0.98. The data that were gathered following the increases in intensities of the OH and O^2H absorption bands produced sigmoid curves (Fig. 3.24) similar to isotherms usually obtained by gravimetric sorption measurements. The considerable frequency and intensity changes in the amide I, II, and III bands in the p/p_0 range from 0 to about 0.10 (Fig. 3.25, for example) permitted the conclusion that water

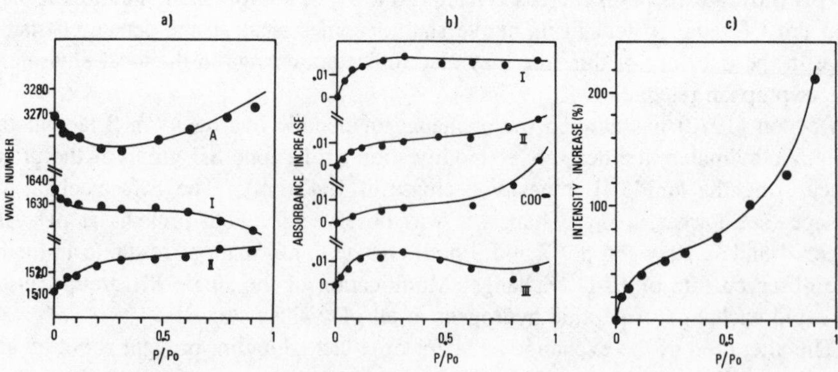

Fig. 3.24. Spectral changes of main infrared absorption bands of β-casein as a function of relative water vapor pressure. (a) Frequency shifts of amide A, amide I, and amide II bands. (b) Changes in absorbance of amide I, II, III, and COO^- bands. (c) Integral band intensity of amide A band region (amide A + amide B); increase calculated as percentage of area at p/p_0 = 0. The diameter of the circles represents the estimated standard deviation. (Rüegg and Häni, 1975.)

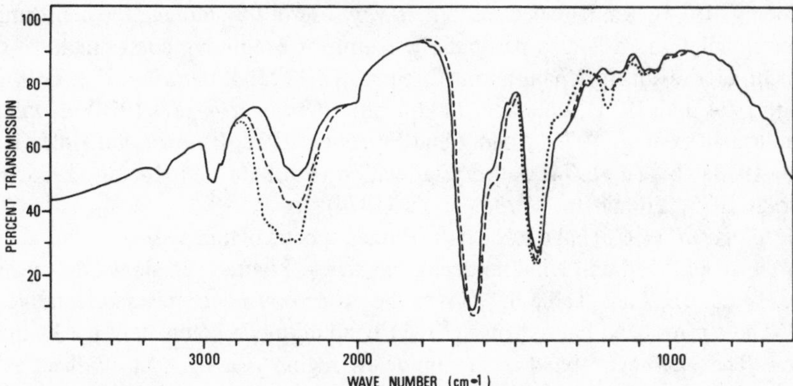

Fig. 3.25. Infrared spectrum of deuterated α_s-casein at different ^2H_2O vapor pressures. Curves of increasing intensities correspond to p/p_0 values of 0, 0.30, and 0.95. Feature at 2000 cm^{-1} is of instrumental origin. (Rüegg and Häni, 1975.)

molecules are already attached to the peptide repeat unit at very low humidities. The authors showed that much less than one water molecule per polar group of the casein molecule was needed to produce these significant spectral variations.

Lutropin, a Luteinizing Hormone. In order to test several possibilities which could explain the appearance of biological activity when the α and β subunits of a glycoprotein hormone bind to each other to form a dimer, Combarnous and Nabedryk-Viala (1978) have studied, as a model, the luteinizing hormone porcine lutropin. Among the approaches they used was that of hydrogen–deuterium exchange (Parker, 1971) of lutropin and its α and β subunits. They made the following plots of their data: (a) The percentage of unexchanged peptide hydrogens in lutropin and its α and β subunits at a variety of pH values, versus log (k_0t); (b) they compared the number of unexchanged peptide hydrogens in native lutropin with the sum of those in the isolated α and β subunits as a function of log (k_0t). The peptide hydrogens of the α subunit were much more easily exposed to solvent than those of the β subunit and those of native lutropin. Also, the difference between the sum of the numbers of unexchanged peptide hydrogens in the isolated subunits and those of lutropin was equal to about eight throughout the whole range of kinetics. The data suggested that there was little structural stabilization resulting from the binding of subunits, but rather shielding of eight peptide groups that were completely reached by solvent in the free subunits. The data showed that biological activity of the hormone, resulting from the binding of α and β subunits, is not caused by one subunit inducing a conformational change in the other but by certain amino acid residues from each subunit forming the active site of lutropin.

Immunoglobulins. Human IgG. Spectroscopic studies of immunoglobulins in solution have shown that there is little or no α-helix content in them, and they contain some β-sheet structure (Litman *et al.*, 1973; Dorrington and Tanford, 1970; Cathou *et al.*, 1968; Abaturov *et al.*, 1969; Termine *et al.*, 1972; Painter and Koenig, 1975). The proportion of β-structure is believed to be relatively high (Litman *et al.*, 1973; Painter and Koenig, 1975). Raman spectroscopy is very useful for studying the conformation of proteins in solution. It is also particularly useful for examining conformational changes that occur as a result of denaturation (Chen *et al.*, 1973; Yu *et al.*, 1972; Frushour and Koenig, 1974 and 1975), chemical modification (Brunner *et al.*, 1974), lyophilization (Frushour and Koenig, 1975; Koenig and Frushour, 1972; Brunner and Holz, 1975; Yu and Jo, 1973a, b; Yu, 1974) and crystallization (Frushour and Koenig, 1975; Brunner and Holz, 1975; Yu and Jo, 1973a, b; Yu, 1974).

Pézolet *et al.* (1976) have studied the Raman spectra of human IgG in neutral solution, and in the lyophilized and alkaline-denatured states. Figure 3.26 shows the spectrum of IgG in H_2O at pH 7 and Table 3.12 gives the observed frequencies and tentative assignments. The strong band caused by the H_2O bending mode completely masks the amide I region. The 1240-cm^{-1} band in the amide III region is assigned to β-sheet structure. From the intensity of this band, the β-structure content was calculated to be 37 ± 4%. This result was supported by the strong amide I′ at 1667 cm^{-1} and by the presence in the spectra of C—C and C—N skeletal stretching modes at 991 and 1078 cm^{-1}, respectively. Lyophilized immunoglobulin-G powder displays a spectrum different from that of the solution. The differences give evidence that lyophilization causes conformational

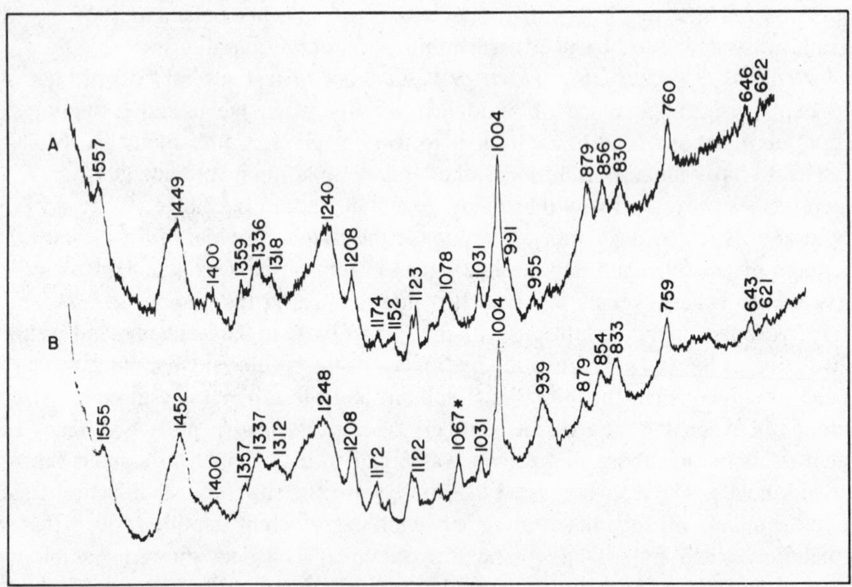

Fig. 3.26. Laser-Raman spectra (frequencies in cm^{-1}) of human IgG in 0.15 *M* NaCl aqueous solutions at 3% concentration and 25°C. (A) Native protein at pH 7. 150 mW (514.5 nm) laser power at sample; spectral slitwidth, 5 cm^{-1}; scan rate, 20 cm^{-1}/min; time constant, 2 sec. (B) Denatured protein at pH 12. Experimental conditions as above except laser power of 200 mW. The band at 1067 cm^{-1} in the spectrum of the denatured protein is not a Raman band of the sample. It was observed when basic solutions were stored in glass tubes for a period of time exceeding 24 h, and was not present in the spectra of freshly prepared solutions. (Pézolet *et al.*, 1976.)

Table 3.12. Tentative Assignment of the Observed Bands in the Raman Spectra of Human IgG in Solution[a,b]

Frequency (cm^{-1})		
H_2O	^2H_2O	Assignment
622 (1)	621 (1)	Phe
646 (1)	643 (1)	Tyr
760 (4)	758 (3)	Trp
830 (4)	830 (3)	Tyr
856 (4)	856 (4)	Tyr
879 (5)	877 (3)	Trp
955 (1) }	960 (3) }	C—C stretch
991 (4) }	991 (7) }	
1004 (10)	1004 (10)	Phe
1031 (4)	1032 (2)	Phe
1078 (4)	1084 (2)	C—N stretch
1123 (3)	1122 (1)	
1152 (2)		
1174 (2)		
1208 (5)		Tyr and Phe
1240 (7)		Amide III of β-structure
1318 (4)	1317 (1)	C—H_2 twist
1336 (4) }	1336 (3) }	Trp and C—H bend
1359 (3) }	1359 (1) }	
1400 (2)	1406 (3)	COO^- symmetric stretch
1449 (6)	1451 (10)	C—H_2 bend
1557 (3)	1554 (6)	Trp
	1584 (5)	Trp
	1613 (6)	Trp, Tyr, and Phe
	1667 (11)	Amide I′ of β-structure

[a] Pézolet *et al.* (1976).
[b] Relative peak height intensities are given in parentheses.

changes that perturb the local environment of some of the tryptophan residues and alter the secondary structure of the protein.

When the IgG is denatured at pH 11, the amide III and amide I′ lines become weaker and broader, and they shift, respectively, from 1240 to 1248 cm^{-1} and from 1667 to 1656 cm^{-1}. These alterations point to decreased β-structure content and a shift toward a much more disordered conformation. Many lines of the aromatic chromophores also change during the denaturation, particularly the tryptophan bands at 1573 cm^{-1} (increased intensity), 1359 cm^{-1} (decreased intensity), and 879 cm^{-1} (decreased intensity). With regard to the tyrosines, the intensity ratio of the 856-cm^{-1} band to that at 830 cm^{-1} decreases from 10:7 to 9:10, when the IgG is denatured. The 9:10 ratio probably indicates that some of the IgG tyrosines remain un-ionized even at pH 12, since a 7:10 ratio was measured for an ionized model molecule (Siamwiza *et al.*, 1975).

Human IgG1. Stability is maintained in the four-chain structure of the human immunoglobulin G1 (IgG1) by strong noncovalent interactions between the heavy and light chains, between the two heavy chains, and by interchain disulfide bonds (Fleischmann *et al.*, 1963; Edelman and Gally, 1964).

To discover the role of the interchain disulfide bonds in the dynamic behavior of IgG1 molecules, Venyaminov *et al.* (1976) have carried out comparative ^1H–^2H (H–D) exchange experiments (Parker, 1971) on native, nonalkylated-reassociated, and *S*-alkylated-reassociated IgG1 samples. Figure 3.27 gives a summary of hydrogen–deuterium exchange data for these three kinds of IgG1 molecules. The exchange rates in Fig. 3.27, examined in the pH range 6.17 to 8.45, were continuous for all three kinds of IgG1. This shows that no profound conformational changes occur in this pH range (Hvidt and Wallevik, 1972; Willumsen, 1971; Willumsen, 1966). Venyaminov *et al.* stress the fact that the most readily exchangeable peptide hydrogen atoms of the protein (which constitute about 50% of the total) were not covered by their experiments, done at 25°C and pH 6.17–8.45. Under those conditions exchange is too rapid to allow measurement. Con-

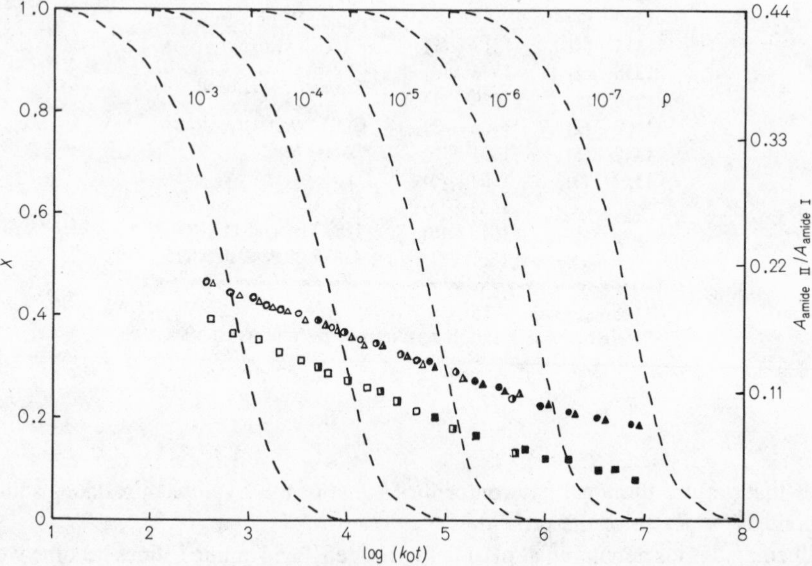

Fig. 3.27. Summarized hydrogen-deuterium exchange data for native IgGl (circles), *S*-alkylated-recombined IgGl (squares), and nonalkylated recombined IgGl (triangles), at 25°C in the pH range of 6.17–8.45. The left-hand ordinate X is the fraction of unexchanged peptide hydrogens. $A_{\text{amide II}}/A_{\text{amide I}}$, the right-hand scale, is the ratio of absorbances at the maxima of the amide II and amide I bands, after subtraction of the absorption. The value of k_0 was calculated by equation below, and t was time, measured from the start of ^1H–^2H exchange. The dashed curves represent hypothetical polypeptides characterized by ρ values indicated in the figure. ρ is the probability of solvent exposure of the peptide groups. Open, half-shaded, and full symbols represent measurements carried out at pH values 6.17, 7.28, and 8.45 respectively. $k_0 \approx (10^{-pH} + 10^{pH-6})\ 10^{0.05(\theta-25)}$ sec^{-1}, where θ = °C. (Venyaminov *et al.*, 1976.)

formational changes in this region of the molecules may occur, but are not observable by the method used. The lack of intersubunit disulfide linkages in *S*-alkylated-reassociated molecules resulted in a decreased conformational stability, as indicated by the displacement of the exchange curve in Fig. 3.27.

IgM-κ McE. The fragments Fabμ and $(Fc)_5\mu$ of IgM-κ McE have been studied by Raman spectroscopy (Thomas *et al.*, 1979) to allow the comparison of band intensities in the amide I and amide III spectral regions of both fragments, since Raman bands in these regions (1600–1700 cm^{-1} and 1200–1300 cm^{-1}, respectively) reflect protein chain conformation (Chen and Lord, 1974; Frushour and Koenig, 1975; Lippert *et al.*, 1976). Figure 3.28 shows that the above fragments are similar to each other and to IgM-κ McE in conformational structure, which is predominantly of the β-sheet type. The Raman spectra of several other human IgM and IgG proteins (not shown) indicate predominantly β-sheet type structures.

The tyrosine doublet is of interest in Fig. 3.28. For Fabμ, the intensity ratio I_{858}/I_{833} is 1.31 ± 0.10; for $(Fc)_5\mu$, this ratio is 1.00 ± 0.10 (see inset). The latter value is very low compared with all other immunoglobulins studied by Thomas *et al.* (1979) and others (Middaugh *et al.*, 1977; Painter and Koenig, 1975; Pezolet *et al.*, 1976), and shows that more tyrosine residues of $(Fc)_5\mu$ than of other immunoglobulins are involved in strong hydrogen bonds to COO^- groups, e.g., of aspartate or glutamate residues (Siamwiza *et al.*, 1975).

Mouse Myeloma Immunoglobulins. Kumaı *et al.* (1978) have used resonance Raman spectroscopy to study the binding of four dinitrophenyl (DNP) haptens to the mouse myeloma proteins MOPC 315 IgA (immunoglobulin A) and MOPC 460 IgA. Free haptens displayed spectra which showed certain features. When the haptens were bound to the proteins, features changed, and these changes were attributed to interactions of the binding sites of the proteins with either the *p*-nitro or the *o*-nitro/amine ("ortho structure") regions of the haptens. The interactions between each individual immunoglobulin and a given hapten were often quite different from one another. Also, for both antibodies the interactions between the 2,4-DNP chromophore and the protein were modified by the chemical nature of the side chain (R) of a dinitrophenyl-NHR compound. No unique set of contacts was found between the binding site and the DNP moiety.

Kumar and Carey (1975) have also recorded resonance Raman spectra of a series of nitroanilines and nitrophenols in connection with myeloma protein–hapten binding studies.

Cryoimmunoglobulins. Cryoimmunoglobulins are immunoglobulins that exhibit temperature-dependent, reversible insolubility when the temperature of certain sera is lowered from 37° to 0°C (Lerner *et al.*, 1947). Middaugh *et al.* (1977) used Raman spectra to determine if changes occurred in the secondary structure of certain IgM cryoimmunoglobulins upon precipitation from aqueous solution. They found no significant changes in the amide I and amide III lines which would suggest a significant structural difference between the dissolved and cryoprecipitated states of cryoimmunoglobulin McE. However, they found some induced changes in the intensities of the Raman lines of aromatic amino acid moieties, e.g., the 1320-cm^{-1} line of tryptophan was a bit sharper at −8°C than at 38.5°C. Such changes may suggest environmental differences in the two states. Other IgM and IgG molecules also showed little or no apparent changes in their Raman spectra.

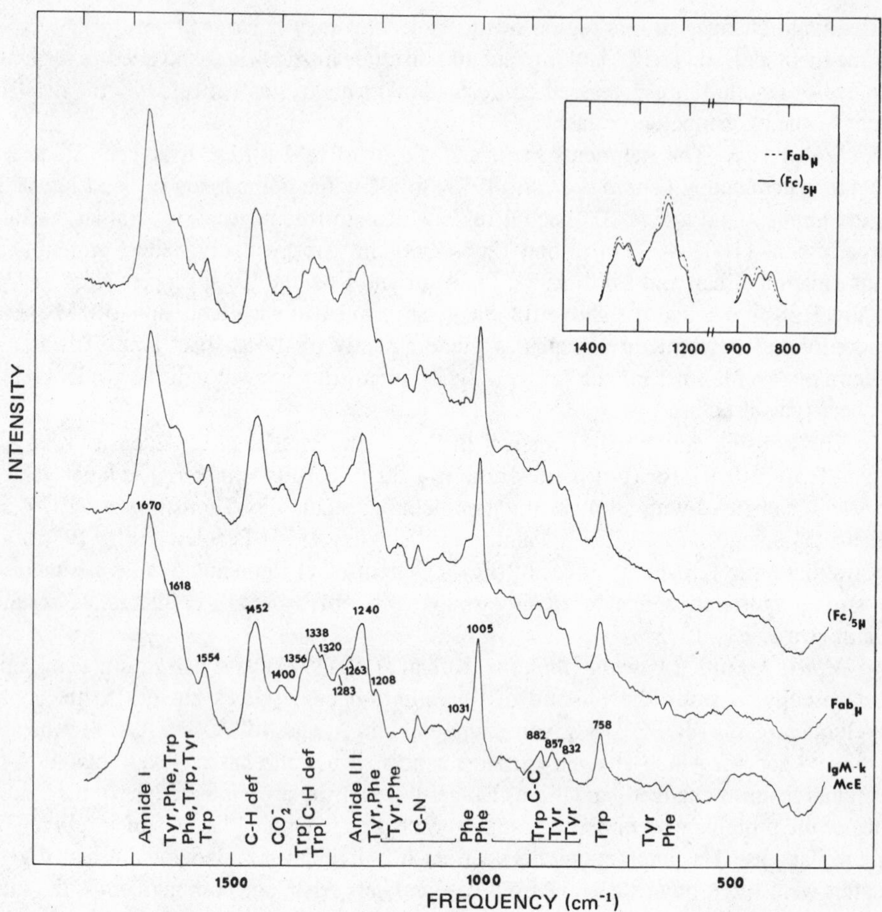

Fig. 3.28. Raman spectra at 39°C of the following in 0.15 *M* NaCl, pH 7.0: (top to bottom) IgM-κ McE $(Fc)_5\mu$ fragment, IgM-κ McE Fabμ fragment, and cryo IgM-κ McE. Inset: Amide III (1200–1350 cm^{-1}) and tyrosine doublet (800–900 cm^{-1}) regions of the Fabμ and $(Fc)_5\mu$ spectra, redrawn over a flat base line utilizing the 1452-cm^{-1} C—H deformation line for normalization. The base line used to redraw the tyrosine doublet was arbitrarily assumed to be a straight line extending from the minima in the spectra at 800 and 910 cm^{-1}. For the amide bands the straight base line extended from 1150 to 1375 cm^{-1}. (Thomas *et al.*, 1979.)

Antisera to Rabbit Serum Albumin and to Ovalbumin

Ockman (1978) has applied the method of ATR with polarized IR excitation to the study of the interactions of the antiserum to rabbit serum albumin with films of rabbit serum albumin, bovine serum albumin, and ovalbumin, as well as that of the antiserum to ovalbumin with films of ovalbumin and rabbit serum albumin. This investigator used a germanium–aqueous interface to obtain information about the conformation of the polypeptide chains and their orientation with respect to the surface. The amide I absorptions of the first three systems studied displayed intensities that indicated that the strengths of binding of these three albumin proteins with anti-rabbit serum albumin is, under appropriate conditions, in the order rabbit > bovine >> ovalbumin; with anti-ovalbumin, the order is ovalbumin >> rabbit.

The major contributions to their conformation comes from α-helical and random-coil structures (and possibly 20% β-conformations). The polypeptide chains are on the average approximately parallel to the surface for the interactions examined. Therefore, Ockman concluded that the orientation and conformation of the polypeptide chains appear to be little affected by specific antibody–antigen interaction.

Histones

Guillot *et al.* (1977) recorded Raman spectra of calf thymus histones H1, H2A, and H2B in aqueous solution. Table 3.13 gives assignments and intensities of the observed bands. Histone H1, which contains much lysine, shows an amide III band at 1245 cm^{-1}, very near that of the amide III of ionized poly (L-lysine). When the concentration of NaCl was increased from zero to 1 *M*, the amide III frequency shifted to 1250 cm^{-1}, owing to the formation of a more compact disordered structure of the *N*-terminal section of the histone. A change of pH from 3 to 5 caused the same shift in frequency.

Histone H2A (slightly rich in lysine) displayed two bands in the amide III region, 1247 and 1265 cm^{-1}. Histone H2B (also slightly high content of lysine) showed bands at 1254 and 1265 cm^{-1} in this region. Guillot *et al.* attributed these bands for the two histone molecules to vibrations involving the backbone of at least two structurally distinct sections. They assigned the 1247 and 1254 cm^{-1} frequencies to the random-coil regions of the histones, structural configurations appearing in the H1 and H2A proteins; and the bands at 1265 cm^{-1} to the α-helical and rigid disordered structures of H2A and H2B histones. The spectra of H2A and H2B also show C—C stretching vibrations around 935 and 960 cm^{-1}, which the authors assigned to α-helical and disordered structures, respectively.

Histones H2A and H2B have three and five tyrosyl residues, respectively; Hl has only one. The ratio of the intensities in the tyrosine doublet, I_{855}/I_{830}, for histones H2A and H2B was 10:7, giving evidence that the tyrosines are not strongly hydrogen bonded (Siamwiza *et al.*, 1975) in water or NaCl solutions, but are exposed to the solvent. Thus, the three tyrosines of H2A, and tyrosine-83 of H2B, which are situated in the globular regions, do not contribute directly to secondary structure formation. (Guillot *et al.*, 1977).

Infrared studies showed a lack of β-structure in the H2A histone of calf thymus (Bradbury *et al.*, 1975). The amino acid sequence of this histone, which has 129 amino acid residues, has been determined (Yeoman *et al.*, 1972; Sautiere *et al.*, 1974).

Fasman *et al.* (1977) discussed the application of various physical methods that have been used by investigators to determine the conformations of histones—NMR, CD, ORD, IR. The book edited by Ts'o (see reference to Fasman *et al.*, 1977) is a mine of information on the structure of histones, chromatin, nonhistone proteins, condensed DNA, and certain RNAs of low molecular weight.

Table 3.13. Assignments and Intensities of the Observed Raman Bands in the Spectra of Calf Thymus Histones H1, H2A, and H2B in Aqueous Solution[a,b]

Frequency (cm^{-1})			
H1	H2A	H2B	Assignment
654 (2)		647 (2)	
752 (2)	743 (0)	745 (2)	
	789 (4)		
825 (1sh)	826 (3)	830 (3)	Tyrosine
854 (2)	854 (4)	856 (5)	Tyrosine
902 (4)		902 (2)	CH_2 rock
938 (1sh)	934 (1sh)	930 (1sh)	C—C stretch
959 (1sh)	958 (1sh)	961 (1sh)	
979 (2)			SO_4^{2-}
1002 (2)	1003 (4)	1006 (7)	Phenylalanine
1033 (0sh)	1033 (0)	1034 (2)	Phenylalanine
1052 (4)	1054 (0)	1063 (2)	C—C and C—N stretch
		1087 (2)	
1100 (4)	1102 (2sh)	1105 (2)	
1128 (1)	1127 (5)	1130 (3)	
1165 (1)	1171 (3)	1178 (3)	
	1208 (2)	1210 (5)	Tyrosine + phenylalanine
1245 (7)	1247 (7)	1254 (9)	Amide III
	1265 (2sh)	1265 (2sh)	
1305 (1sh)	1305 (1sh)	1305 (1sh)	
1317 (6)	1316 (6)	1323 (7)	CH_2 twist
1340 (4sh)	1339 (5sh)	(1345) (5sh)	
1379 (2)			CH_2 wag
1389 (2)	1389 (0)	1394 (0)	
	1409[c] (2)	1414[c] (3)	N-deuterated imidazolium of His
	1420 (0)	1427 (0)	COO^- symmetric stretch
1449 (10)	1451 (10)	1454 (10)	CH_2 bend
	1612[c] (1)	1616[c] (3)	Tyrosine + phenylalanine
1654[c] (6)	1648[c] (7)	1661 (8)	Amide I′

[a] Guillot *et al.* (1977).
[b] Relative peak height intensities, with the 1450 cm^{-1} band taken as 10, are given in parentheses. sh denotes a shoulder.
[c] Frequency obtained from ^2H_2O solution spectra.

Ribosomal Proteins

The presence of little or no β-structure was indicated by IR examination of both solid films and solutions of the 16S RNA binding protein S4 from *E. coli* ribosomes (Morrison *et al.*, 1977). Figure 3.29 spectra show no absorption in the β-structure band at ca. 1620 cm^{-1}. CD measurements on protein S4 indicated an α-helix content of ca. 30% in buffers.

The pure protein L7 from *E. coli* ribosomes was studied in D_2O by IR spectroscopy

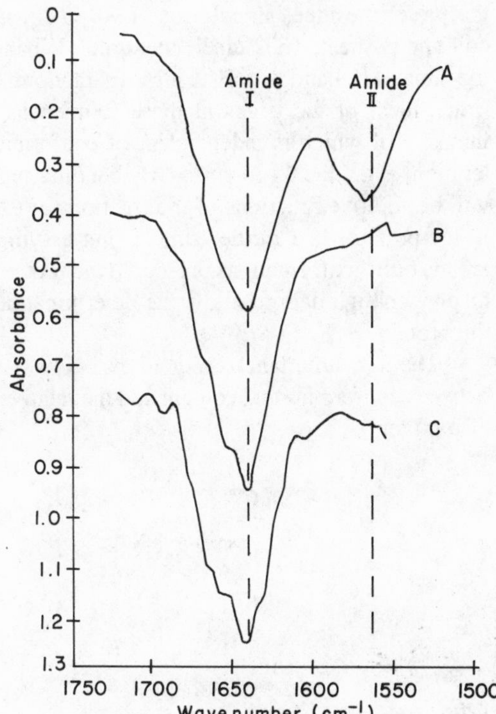

Fig. 3.29. Amide I and II region of infrared spectra of protein S4, 15 mg/ml in ^2H_2O at (A) p^2H 7.0, (B) p^2H 2.0, and (C) readjusted to p^2H 7.0. The vertical scale applies to spectrum (A) only and must be lowered for spectra (B) and (C). Amide I and II band positions are indicated. (Morrison *et al.*, 1977.)

(Gudkov *et al.*, 1978). The amide I band (1642 cm^{-1}) was resolved into two lesser bands at 1637 and 1648 cm^{-1}. The 1637 cm^{-1} band was indicative of helical conformation and the 1648 cm^{-1} band of random conformation. No evidence of β structure was found for this protein.

Estimation of Secondary Structure from Raman and Infrared Spectra

Lippert *et al.* (1976) have devised a model technique for the estimation of the secondary structural content of proteins from their Raman spectra in H_2O and ^2H_2O solutions. The method allows one to estimate the fractions of α-helix, β-sheet, and random coil conformations in proteins, and is based on an interpretation of their amide I′ and amide III Raman spectral intensities. The model amide intensities that are used are due to purely α-helical, β-sheet, and random conformations in proteins; and these intensities are those found for amide intensities of poly-L-lysine in its α-helical, β-sheet, and random conformations (Yu *et al.*, 1973; Chen and Lord, 1974) and of two proteins with proven secondary conformations, ribonuclease A and lysozyme. Lippert *et al.* used the model to estimate the secondary structures of α-chymotrypsin, chymotrypsinogen, bovine serum albumin, concanavalin A, pepsin, pepsinogen, and insulin. If poly-L-lysine is considered as a model, Lippert *et al.* state that the Raman spectra of proteins should have three frequencies, the intensities of which could be used for monitoring the conformation of the protein residues in solution: 1240 cm^{-1}, the amide III band in H_2O due to random coil and β-sheet; 1632 cm^{-1}, the amide I′ band in ^2H_2O due to α-helix; and 1660 cm^{-1}, the amide I′ band in ^2H_2O due to random coil and β-sheet. If the spectral heights (intensities) of the peaks at these frequencies are measured relative to some band the intensity of which is independent of conformational content and is not affected by deuteration, e.g., the 1448-cm^{-1} CH_2 bending mode, then these relative spectral intensities will be additive functions of the fractions of α-helix, β-sheet, and random coil contained in the protein. If a further simplifying assumption is made by these workers that those are the only conformations present in the protein, then the three relative spectral intensities display enough data to allow the determination of the fractions of each conformation in the protein.

The four simultaneous equations below (Lippert *et al.*, 1976) express the relationship between conformational content and the relative spectral intensities of the Raman spectrum of a protein:

$$C^{\text{protein}} I_{1240}^{\text{protein}} = f_\alpha I_{1240}^{\alpha} + f_\beta I_{1240}^{\beta} + f_R I_{1240}^{R} \tag{3.2}$$

$$C^{\text{protein}} I_{1632}^{\text{protein}} = f_\alpha I_{1632}^{\alpha} + f_\beta I_{1632}^{\beta} + f_R I_{1632}^{R} \tag{3.3}$$

$$C^{\text{protein}} I_{1660}^{\text{protein}} = f_\alpha I_{1660}^{\alpha} + f_\beta I_{1660}^{\beta} + f_R I_{1660}^{R} \tag{3.4}$$

$$1.0 = f_\alpha + f_\beta + f_R \tag{3.5}$$

$I_{1240}^{\text{protein}}$ is the spectral height of the spectrum of the protein observed at 1240 cm^{-1} in water relative to the spectral height of the methylene bending mode at 1448 cm^{-1}; I_{1240}^{α} is that for "pure" α-helical protein (poly-L-lysine); f_α, f_β, and f_R are the fractions of residues in the α-helix, β-sheet, and random coil conformations in the protein; and C^{protein} is a scaling constant (to be determined) for the methylene band intensity relative to the model compound, poly-L-lysine.

Figure 3.30 shows Raman spectra of 10% solutions of lysozyme in H_2O and ^2H_2O. Spectra of other proteins were recorded in similar fashion. The spectra of the other proteins (not shown here) and that shown in Fig. 3.30, display the spectral intensities which, measured relative to the methylene deformation intensity at 1448 cm^{-1}, are the quantities $I_{1240}^{\text{protein}}$, $I_{1632}^{\text{protein}}$, and $I_{1660}^{\text{protein}}$, applied for each protein in Eq. (3.2)–(3.5). Table 3.14 lists all these proteins and their relative spectral intensities and the relative spectral intensities of poly-L-lysine (α, β, and random coil) obtained from Yu *et al.* (1973). Table 3.15 lists revised spectral intensities, which take account of certain discrepancies occurring if poly-L-lysine serves as the model for the β component. (Proteins with large amounts of β structure have amide I′ maxima that are 5–7 cm^{-1} higher than in β-sheet poly-L-lysine.)

The relative spectral intensities in Table 3.15 were used in Eq. (3.2)–(3.5), along with the measured spectral intensities in Table 3.14, to predict the amounts of the various conformations in α-chymotrypsin, chymotrypsinogen, serum albumin, concanavalin A, pepsin, pepsinogen, and insulin. The predictions based on Raman methodology are in good agreement with those based on other techniques. Lippert *et al.* (1976) believe that their techniques may be applied with some surety (about ± 10%–15%) to other proteins,

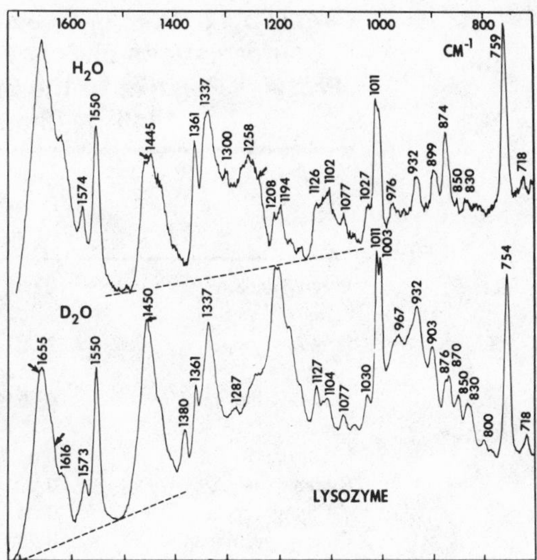

Fig. 3.30. Raman spectra of lysozyme solutions (10%) in H_2O and ^2H_2O. Arrows point to the frequencies whose spectral intensities are used to determine structural content. Conditions of spectra: spectral slit width ($\Delta\sigma$), 5 cm^{-1}; rate of scan (γ), 0.2 cm^{-1}/sec; time constant (τ), 10 sec; sensitivity (s) 3×10^{-7} A full scale; laser power (p) 400 mW. (Lippert *et al.*, 1976.) Reprinted with the permission of the American Chemical Society. Copyright 1976.

to the extent that they are composed only of α-helical, antiparallel β-sheet, and random coil peptide sections. The work in their paper quantitates a conformational difference between crystalline insulin and its aqueous solution, the first time this has been done for a protein. Yu *et al.* (1974) had suggested that insulin in solution contains much less α-helix than the crystal does, and the work by Lippert *et al.* confirms that, when insulin is dissolved, it loses about 10% to 17% of its helical content. The reader is advised to see a discussion of the method of Lippert *et al.* (1976) given in Bailey *et al.* (1979), in which suggestions are made for improving the method by including β-reverse turn structure components in the equations (page 147, this book).

The reader is also referred to page 147 for references to other methods for the estimation of secondary structure.

Eckert *et al.* (1977) have attempted to develop a procedure for curve-fitting analysis of IR spectra of *N*-deuterated proteins suitable for the determination of the conformation of globular proteins. They used ribonuclease, lysozyme, cytochrome *c*, insulin, papain, α-chymotrypsin, ox methemoglobin, concanavalin A, and they scanned from 1800 to 1600 cm^{-1} and 4000 to 2800 cm^{-1}. They used a computer program to estimate mean residue absorptions at each 2 cm^{-1} in the region 1720–1594.

In IR spectra of proteins, as much as 10% of the molar absorption may be caused by the side chains of various amino acids in the areas of their highest absorption (Chirgadze, 1973; Chirgadze *et al.*, 1975). Therefore, Eckert *et al.* ascertained whether the characteristic shapes of their "basis" spectra (calculated from the amide I′ bands of reference protein sets) and their standard deviations were affected by the corresponding correction of the reference protein spectra. Thus, they used the molar absorption coefficients of the side chains of asparagine, glutamine, arginine, and tyrosine, and of the carboxyl groups of the terminal aspartic and glutamic acids in the range 1594 to 1720 cm^{-1}, to calculate the side-chain absorption contributions for all reference proteins based

Table 3.14. Raman Spectral Intensities of Three Conformations of Poly-L-lysine and of Several Proteins Relative to the Spectral Intensity of the 1446–1448-cm^{-1} Band[a,b]

	H_2O soln,	^2H_2O soln	
	1240 cm^{-1}	1632 cm^{-1}	1660 cm^{-1}
Poly-L-lysine, α helix[c,d]	0.00	0.80	0.55
Poly-L-lysine, β sheet[c,e]	1.20	0.33	1.22
Poly-L-lysine, random coil[c,f]	0.60	0.21	0.70
Lysozyme	0.64	0.50	0.99
Ribonuclease A	0.92	0.52	0.99
α-Chymotrypsin	0.86	0.36	0.88
Chymotrypsinogen	0.78	0.44	0.83
Bovine serum albumin	0.30	0.69	0.86
Concanavalin A	1.39	0.58	1.28
Pepsin	1.20	0.48	1.28
Pepsinogen	1.10	0.56	1.13
Insulin, solution[g]	0.75	0.55	1.10
Insulin, crystalline[h]	0.60	0.65	0.91

[a] Lippert *et al.* (1976).
[b] Unless otherwise reported, all measurements are of Raman spectra of 5%–7% solutions in H_2O or ^2H_2O at pH 7. All values are ±0.05 and are the average of 2 or 3 measurements.
[c] Measured from the spectra of Yu *et al.* (1973) and repeated at least twice from spectra obtained for poly-L-lysine solutions, prepared as in Koenig (1972).
[d] pH or pD 11.8, T = 4°C.
[e] pH or pD 11.8, T = 52°C or 22°C, after heating to 52° C for 1 h.
[f] pH or pD 3.7, T = 22°C.
[g] pH or pD 2.1, T = 22°C.
[h] Crystallized from H_2O or ^2H_2O at pH 7.1. Spectra were obtained of crystals in contact with supernatant solution.

Table 3.15. Relative Raman Spectral Intensities, I_ν^α, I_ν^β, and I_ν^R, for Determination of Conformational Content in Proteins[a]

	H_2O soln,	^2H_2O soln	
	1240 cm^{-1}	1632 cm^{-1}	1660 cm^{-1}
Rel intensity for α helix	0.00	0.80	0.55
Rel intensity for β sheet	1.20	0.72	0.88
Rel intensity for random coil	0.60	0.08	0.78

[a] Lippert *et al.* (1976).

on their amino acid composition. Subtraction of these calculated values from the measured protein spectra allowed them to obtain the corrected spectra for side-chain absorption.

Comparison of the percentages of the β-conformation derived from X-ray structural analysis or calculated from IR spectra showed the suitability of the "basis" spectra for the rough estimate of the percentages of β-conformation content of proteins.

H–D Exchange of Protein Films

Hydrogen–deuterium exchange rates of protein films of myoglobin, lysozyme, β-lactoglobulin, cytochrome *c*, ribonuclease, and serum albumin have been studied by the FMIR method on germanium plates (Deutschmann and Ullrich, 1979). The exchange characteristics of the films were the same as those for corresponding solutions, provided that the films were maintained in a controlled atmosphere of at least 90% relative humidity. Table 3.16 shows a comparison of the number of relatively fast-exchanged peptide hydrogens (N_f) with the number of peptide hydrogens of the random coil conformation (N_p) as calculated from X-ray analysis. These data support the assumption that the fast ex-

Table 3.16. Comparison of Deuterium Exchange and X-Ray Analysis Data[a,b]

Protein	Method		Reference
	^1H–^2H exchange, percentage N_f[c]	X-ray analysis, percentage $N\rho$	
Myoglobin	29–31	23–32	*d*
Lysozyme	46–47	48–62	*e*
Ribonuclease	45–53	46–58	*f*
Cytochrome *c*	42–58	48–61	*g*
Cytochrome b_5	30–32	27	*h*

[a] Deutschmann and Ullrich (1979).
[b] The values listed for the ^1H–^2H exchange were obtained by either extrapolating the slow phase from a semilogarithmic plot and calculating the fast phase as the difference (first value) or by taking the first measurement as the amount of fast-exchanging hydrogens (second value). $N\rho$ represents the number of peptide hydrogens of the random coil conformation, whereby the lower value is calculated from an α-helical content involving the 3_{10}-helix and other distorted helices and the upper value is obtained from the regular α-helix content alone.
[c] At pH 7.4.
[d] Kendrew, J. G., Dickerson, R. E., Standberg, B. E., Hart, R. G., Davies, D. R., Phillips, D. G., and Shora, V. C., *Nature* **185,** 422 (1960).
[e] Blake, C. C. F., Koenig, D. F., Mair, G. A., North, A. C. T., Phillips, D. G., and Sharma, V. R., *Nature* **206,** 757 (1965).
[f] Kartha, G., Bello, J., and Harker, D., *Nature* **213,** 862 (1967).
[g] Dickerson, R. E., Takako, T., Eisenberg, D., Kallai, O. B., Samson, L., Cooper, A., and Margoliash, E., *J. Biol. Chem.* **246,** 1511 (1971).
[h] Mathews, F. S., Levine, M., and Argos, P., *J. Mol. Biol.* **64,** 449 (1972).

changing hydrogens at neutral pH values are present in peptide bonds of random coil structures without internal hydrogen bonding.

The validity of the film technique (first reported by Haggis, 1956, 1957) used here is further supported by comparison among three methods for measuring hydrogen isotope exchange—film H–D exchange by FMIR, tritium exchange (Englander and Staley, 1969), and Linderstrøm-Lang's (1955) gradient method in solution. A comparison is presented in Table 3.17. Although not all values were obtained under exactly the same conditions, the results of the three methods agree well except for one high value for unexchanged hydrogens for lysozyme by the gradient method. The higher temperature in the gradient experiments (38°C compared to 25°C for the other methods) may account for the discrepancy. The FMIR results for ^{1}H–^{2}H exchange measurements indicate that this technique produces essentially the same results as the tritium-exchange method but has the advantages of using smaller samples, being less costly, and avoiding radioactive hazards (Deutschmann and Ullrich, 1979).

Table 3.17. Comparison of the Isotope Exchange in Protein Films and Solutions[a,b]

		Method (% values)				
Protein	Compared data	Infrared[c]	^{3}H exchange[c]	Reference	Gradient technique[c]	Reference
Lysozyme	N_u	46–50	45	*d*	76	*g*
		(3.2–5.2; 25)	(4.7; 25)		(3.2; 38)	
Ribonuclease	N_u	45	40–50	*e*	49	*h*
		(3.8; 25)	(3.9; 25)		(3.9; 25)	
		71	70		—	
		(3.8; 25)	(4.7; 0)		—	
β-Lactoglobulin	N_u	105	—		97	*i*
		(4.2; 25)			(5.5; 25)	
Serum albumin	N_s	349	—		350	*j*
		(7.4; 25)	—		(7.0; 25)	
Myoglobin	N_s	118	117	*f*	—	
		(5.1; 25)	(4.7; 0)		—	
	N_{vs}	75	—		74	*k*
		(7.4; 25)	—		(7.0; 0)	

[a] Deutschmann and Ullrich (1979).
[b] *N* refers to the number of hydrogens; u, s, vs, are indexes for the unexchanged, slowly exchanged, and very slowly exchanged hydrogens, respectively.
[c] The pH and temperature (°C) are indicated in parentheses.
[d] Praissman, M., and Rupley, J. A., *Biochemistry* **7,** 2446 (1968).
[e] Englander, S. W., *Biochemistry* **2,** 798 (1963).
[f] Englander, S. W., and Staley, R., *J. Mol. Biol.* **45,** 277 (1969).
[g] Hvidt, A., and Kanarek, L., *Compt. Rend. Trav. Lab. Carlsberg* **33,** 463 (1963).
[h] Leach S. J., and Springell, P. H. *Australian J. Chem.* **15,** 350 (1962).
[i] Linderstrøm-Lang, K., Soc. Biol. Chemists India Silver Jubilee Souvenir, p. 191, 1955.
[j] Benson, E. S., Hallaway, B. E., and Lumry, R. W., *J. Biol. Chem.* **239,** 122 (1964).
[k] Benson, E. S., *Compt. Rend. Trav. Lab. Carlsberg* **31,** 235 (1959).

H–D Exchange by an Ultraviolet Method

A very simple and useful UV method for the measurement of peptide group H–D exchange has been described by Englander *et al.* (1979).

When the —CONH— group becomes —COND—, the absorbance envelope [consisting of three bands (Gratzer, 1967)] shifts to the blue to yield an absorbance change of ~5% in the 230–220-nm region. This change can be used to follow the kinetics of amide H–D exchange after dilution of an aqueous sample solution into D_2O in a quartz cuvette. Englander *et al.* stated that an advantage of the IR method is that it uses a larger fractional change in absorbance than the UV, which is partially offset by the higher quality of measurement in the UV. The major disadvantage of the UV method is its limited accuracy.

Protein Adsorption on Surfaces. Fourier Transform Studies

Complex biological systems may be examined advantageously with Fourier transform infrared techniques (Gendreau and Jakobsen, 1978). Parker (1963) had used the ATR technique to study peptides in aqueous solution, and Parker and Ans (1967) had studied human and animal tissues, which contain water, by this method. In the 1967 paper the authors had been able to distinguish normal from atherosclerotic human arterial tissue, and had discussed biochemical changes in the latter tissue. A disadvantage of the system, namely, that absorption bands of water obscure certain information, is correctable by the use of FTIR.

Gendreau and Jakobsen (1978) have used ATR in combination with FTIR to obtain spectra showing the proper intensity ratio of amide I/amide II in a specific system, the adsorption of proteins of rabbit blood plasma onto a germanium ATR plate, followed by a desorption process in saline solution. The sensitivity of the FTIR system allowed careful subtraction of the spectrum of water from that of the protein materials. Figure 3.31, A and C, shows the results of subtracting a water spectrum from the spectra of Fig. 3.32, A and B, respectively. The spectra of Figs. 3.31A and 3.31C are scale expanded by a factor of 10 compared to those in Fig. 3.32 and show the spectra of the proteins adsorbed on the ATR crystal. Spectrum 3.31A minus spectrum 3.31C, expanded, yields obvious differences, in Fig. 3.31B. Observe that, as the coated crystal is exposed longer to the saline solution, a loss of intensity of amide I at 1630 cm^{-1} and an increase of intensity of amide II at 1545 cm^{-1} occur. Also, an increase in amide III absorption at 1243 cm^{-1} and increases in absorption at 1399 and at 1463 cm^{-1} occur. These spectral changes characterize a change in protein conformation from that of a β-sheet to one of a random coil. The authors of this paper, attempting to study how plastic implants affect blood clotting, also showed spectra of amide I and amide II bands of protein adsorbed from plasma onto polyurethane.

Jakobsen and Gendreau (1978) have described a combined ATR and FTIR technique useful for studying the adsorption of blood plasma proteins onto polymer surfaces. When a surgeon implants an artificial organ which will be continually exposed to blood, the possibility of protein accumulation, tissue encapsulation or clot formation always exists.

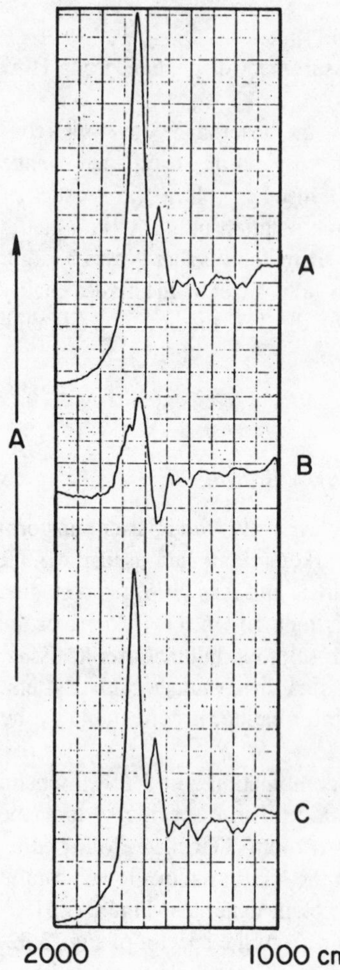

Fig. 3.31. ATR liquid cell infrared spectra. Desorption of components adsorbed from blood plasma. (A), 40 min after removal of blood plasma and introduction of fresh saline solution and after subtraction of H_2O spectrum; (B), result of subtraction of spectrum of Fig. 3.31C from the spectrum of Fig. 3.31A. (C), 85 min after removal of blood plasma and introduction of fresh saline solution and after subtraction of H_2O spectrum. (Gendreau and Jakobsen, 1978.)

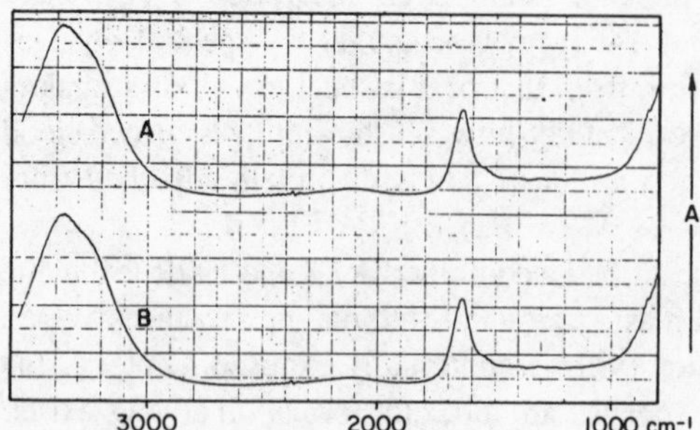

Fig. 3.32. ATR liquid cell infrared spectra. Desorption of components adsorbed from blood plasma. (A), 40 min after removal of blood plasma and introduction of fresh saline solution; (B), 85 min after removal of blood plasma and introduction of fresh saline solution. (Gendreau and Jakobsen, 1978.)

Relatively little is known about the composition and conformation of the protein layers that adsorb on the implant surfaces.

Investigators have postulated that a variety of factors are important in determining biocompatibility, one of which is the composition of the adsorbed protein layer (Layman *et al.*, 1975; Roohk *et al.*, 1976). Others are the relative amounts of fibrinogen and albumin adsorbed (Lee and Kim, 1974) and the conformation of the proteins adsorbed on the surface (Morrissey and Fenstermaker, 1976). Various workers have used ATR before the advent of FTIR to analyze adsorbed proteins (Lee and Kim, 1974; Baier *et al.*, 1971; Mattson *et al.*, 1975), but conventional dispersive IR had not been used with whole blood plasma.

Jakobsen and Gendreau prepared heparinized polyethylene strips (50 mm by 5 mm), which were then allowed to adsorb blood plasma components by exposure to freshly drawn rabbit plasma for one hour. The polyethylene strips were rinsed quickly in non-buffered isotonic saline solution, pH 7, and were dried in an environment of low humidity. Figure 3.33 shows ATR–FTIR spectra of heparinized and plain polyethylene strips, before and after their exposure to plasma. Figure 3.34 shows the spectral differences between untreated and treated surfaces observed after subtraction of the proper control spectrum from the spectrum of a sample exposed to plasma. Figure 3.34a shows that the polyethylene bands have disappeared as a result of subtraction, and the protein bands are seen more clearly. Figure 3.34b shows that the 1050- and 1250-cm^{-1} bands of heparin were not completely removed, even though the polyethylene bands were almost entirely removed by subtraction. When we compare Figs. 3.34a and 3.34b, we see basic differences between the layers adsorbed onto the heparinized and plain surfaces. The positions and intensities of the amide I and amide II bands are different. The amide I band of the protein layer adsorbed on an untreated surface always falls at a higher frequency (1660 cm^{-1}) than does the amide I band on a heparinized surface (1636 cm^{-1}). Jakobsen and Gendreau explain these differences in two ways: first, the interaction between heparin and adsorbed proteins may cause or prevent a change in conformation of the latter; second, the heparin, by smoothing the surface, may function to change the composition of the adsorbed protein layers.

Comparing Fig. 3.34b to Fig. 3.33d, we find evidence for a significant interaction between the surface proteins and the heparin. Figure 3.33d shows normal heparin bands in the region 1250–1000 cm^{-1}. Figure 3.34b (expanded 5× compared to Fig. 3.33d) shows that subtraction of the heparin control does not remove the heparin bands in the spectrum of heparin exposed to plasma. The polyethylene polymer bands, e.g., 1460 cm^{-1}, in Fig. 3.34b almost completely disappear upon subtraction. The 1050- and 1250-cm^{-1} bands of heparin, although decreased in intensity, are still present despite subtraction and display more splitting. The protein bands at 1636 and 1550 cm^{-1} appear to show little change; the increased peak height is caused by the 5× expansion of Fig. 3.34b as compared to Fig. 3.33d. If no interaction had occurred between heparin and protein, the heparin bands would have cancelled out as did the polyethylene bands. Thus, the fact that heparin did not cancel out upon subtraction shows that the heparin exposed to saline solution differs from the heparin in contact with plasma. These authors suggest that this difference is caused by protein binding to the heparin residues on the polymer surface.

Gendreau *et al.* (1981) have routed whole flowing blood of a beagle through an

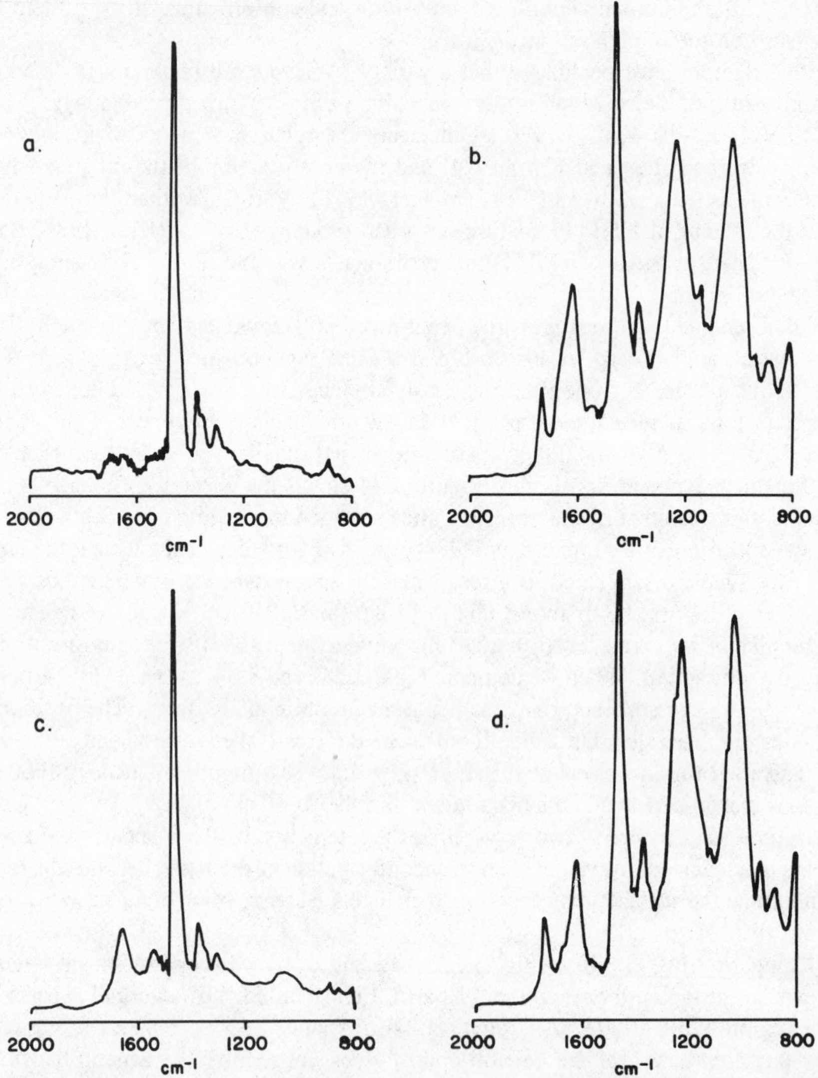

Fig. 3.33.(a) Spectrum of polyethylene. (b) Spectrum of heparin-treated polyethylene. (c) Spectrum of blood plasma-exposed polyethylene. (d) Spectrum of heparin-treated, blood plasma-exposed polyethylene. (Jakobsen and Gendreau, 1978.)

ATR germanium cell and have recorded at 5-sec intervals FTIR spectra of protein adsorption onto the Ge surface. They tentatively identified albumin and glycoproteins as the initially adsorbing material, followed by partial replacement of this protein layer by fibrinogen and other proteins. Although such studies are useful, it should be pointed out that the adsorption of protein followed by the blood-clotting process is much more complex

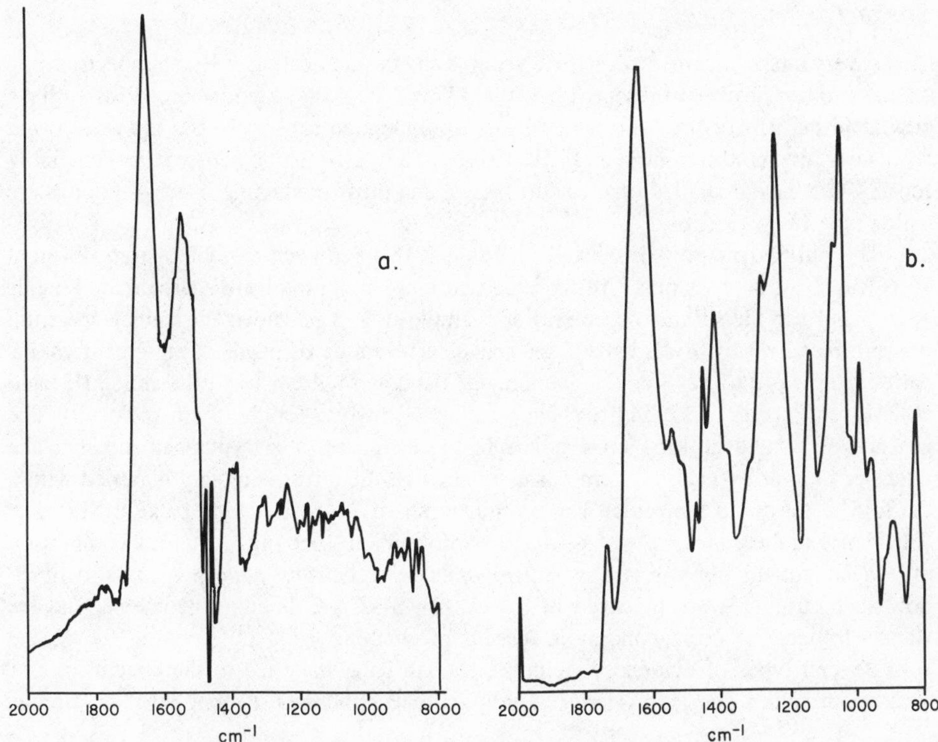

Fig. 3.34.(a) Blood-plasma-exposed polyethylene spectrum minus polyethylene control spectrum. (Jakobsen and Gendreau, 1978.) (b) Heparin-treated, blood plasma exposed polyethylene spectrum minus heparin-treated polyethylene control spectrum. (Jakobsen and Gendreau, 1978.)

than the authors indicate. It would be helpful if they instituted some parallel type of analysis to prove what the adsorbed materials are, e.g., immunochemical techniques.

Gendreau *et al.* (1982) have used FTIR coupled with attenuated total reflection (ATR) optics to study the first five seconds of the adsorption of proteins from unaltered, flowing, whole blood in live female dogs and live female sheep in an *ex vivo* manner with an implanted carotid-jugular shunt as the source of whole blood. These authors studied the adsorption of plasma proteins onto foreign surfaces (e.g., synthetic polymers); and obtained information about the kinetics of adsorption. *Ex vivo* derived blood gives adsorption patterns (FTIR spectra) which differ from those obtained when blood from the same animal source is drawn and bagged first. Adsorbed protein species can be differentiated when whole blood is compared to whole-blood fractions including platelet-rich plasma, platelet-poor plasma, or serum.

With a time resolution of better than one second, Gendreau *et al.* detected major differences in terms of protein adsorption patterns in the first few seconds after whole blood contacts various polymers. Glycoproteins were identified as components that display differential behavior, depending on the source of blood and the kind of foreign surface.

Snake Toxins

A very interesting review on toxic venoms and the application of Raman spectroscopy to snake toxins has been presented by Tu (1977a). The postsynaptic neurotoxins all have similar primary structures, i.e., they have a polypeptide backbone, but the amino acid sequences may be different (Tu, 1973; 1977b). The postsynaptic neurotoxins containing four S—S bonds are called type I neurotoxins, and those containing five S—S bonds are called type II.

The purified neurotoxins listed in Table 3.18 have been extracted from different species of sea snake venoms. All these neurotoxins show prominent sharp amide I bands at 1672 cm^{-1} in their Raman spectra. For example, Fig. 3.35 shows the Raman spectrum of *Lapemis hardwickii* (Hardwick's sea snake). The amide III bands of these neurotoxins varied from 1240 to 1248 cm^{-1}. The amide I band at 1672 cm^{-1} and the amide III band at 1240 cm^{-1} (Fig. 3.35) indicate antiparallel β-structure for the toxin molecule. The absence of a band at 1361 cm^{-1} points to the exposure of a tryptophan moiety. The intensity ratio at 846 and 834 cm^{-1} is evidence that the tyrosine moiety is buried within the fold of the toxin molecule. The prominent sharp S—S stretching band at 512 cm^{-1} is evidence of the *gauche–gauche–gauche* conformation. (See Fig. 3.36 for a comparison of *gauche–gauche–gauche*, *trans–gauche–gauche*, and *trans–gauche–trans* structures, how such structures are involved in C—C—S—S—C—C bonding geometry, and the Raman frequencies corresponding to these conformations.)

Several types of evidence, including Raman data, have led to the conclusion that snake neurotoxins consist primarily of antiparallel β-structures. Among the data, circular

Table 3.18. Characteristic Raman Lines of Snake Neurotoxins and Cobramine B[a]

Family	Hydrophilidae				Elapidae
Subfamily	Hydrophilinae			Laticaudinae	
Genus	*Pelamis*	*Lapemis*	*Enhydrina*	*Laticauda*	*Naja*
Species	*platurus*	*hardwickii*	*schistosa*	*semifasciata*	*Naja*
Origin	Costa Rica	Thailand	Malaya	Philippines	India
Name	Pelamis toxin a	Toxin	Toxin	Toxin b	Cobramine B
	Bands (cm^{-1})				
Amide I	1672	1672	1672	1672	1672
Heating	1664	—	—	—	—
Amide III	1245	1240	1242	1248	1235
Heating	1243 (broader)	—	—	—	—
—S—S	512	512	512	512	510
Heating	512 (major)	—	—	—	—
	546 (minor)	—	—	—	—

[a] Tu (1977a).

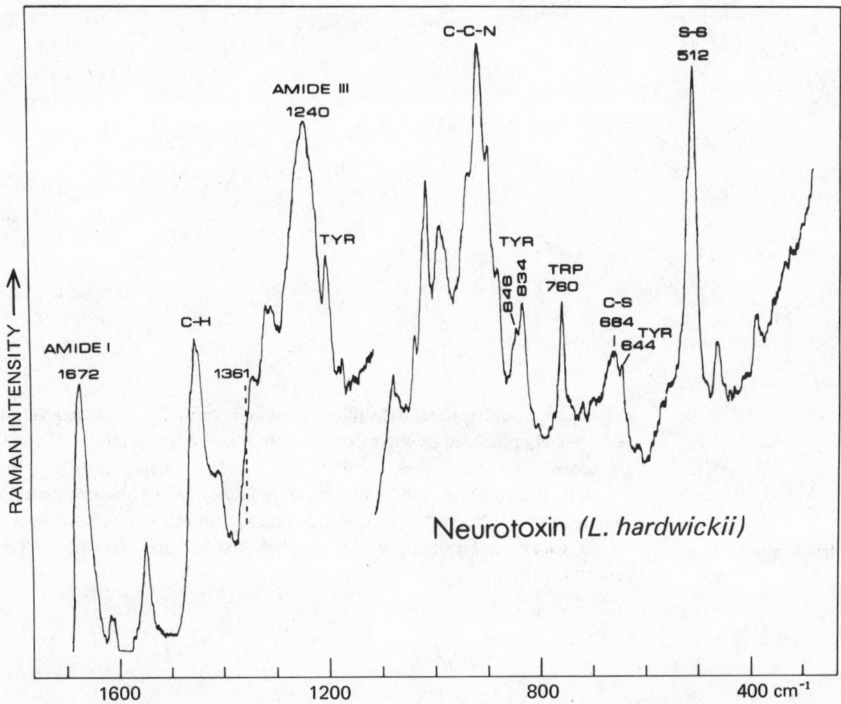

Fig. 3.35. Raman spectrum of purified sea snake neurotoxin. Most sea snake toxins show Raman spectra similar to the *Lapemis hardwickii* neurotoxin. The major toxin of *Lapemis hardwickii* (Hardwick's sea snake) is used as an example in this figure. This indicates that most sea snake neurotoxins have very similar protein conformation regardless of the species and geographical origins of the sea snakes themselves. (Yu *et al.*, 1975.)

dichroism had shown that two neurotoxins have β-sheet structure (Yang *et al.*, 1968; Hauert *et al.*, 1974). It is interesting to note that shortly after Raman spectra gave evidence of the antiparallel β-structure of sea snake neurotoxins, two laboratories published X-ray diffraction studies which verified this conclusion (Tsernoglou and Petsko, 1976; Low *et al.*, 1976; Tsernoglou *et al.*, 1977). This was a turning point for those using Raman spectroscopy. Previously, they had picked for examination only proteins whose structures had been worked out by means of X-ray crystallography.

The amide I and amide III bands of neurotoxins are the same in the solid state and aqueous solution (Yu *et al.*, 1975), showing thereby that the peptide backbone conformation remains the same when the solid form is put into solution.

Besides antiparallel β-structure, the snake neurotoxins contain β-turn structures, which are involved in a change in direction of the polypeptide backbone, like the change that takes place in a hairpin turn. Such structure was expected from theoretical considerations and X-ray diffraction subsequently proved its presence (Low *et al.*, 1976; Chen *et al.*, 1975).

Type I neurotoxins contain four disulfide linkages and type II, five. All the neurotoxin molecules have a very high number of such linkages. This fact explains the compactness and high stability of these molecules. We mentioned earlier the use of the position of

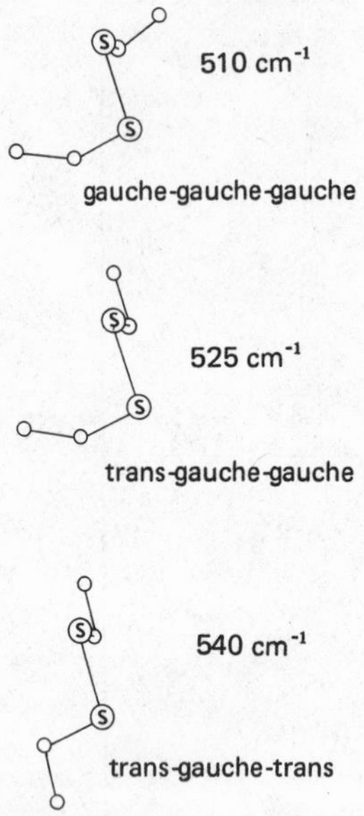

510 cm^{-1}

gauche-gauche-gauche

525 cm^{-1}

trans-gauche-gauche

540 cm^{-1}

trans-gauche-trans

Fig. 3.36. Conformation of disulfide bonds in proteins. The stretching vibration of S—S is influenced by the rotational conformation about the C—C and C—S bonds. The assignment numbers for C and S atoms are: C_2-C_1-S_1-S_2-$C_{1'}$-$C_{2'}$. The disulfide bond with *gauche–gauche–gauche* shows a strong and sharp band at 510 cm^{-1}. The disulfide bonds with *trans–gauche–gauche* and *trans–gauche–trans* conformation give bands at 525 and 540 cm^{-1} in Raman spectra. (Tu, 1977a.)

the S—S stretching band for characterizing the geometry of the C—C—S—S—C—C network. The sharp symmetrical band at 512 cm^{-1} found for all type I neurotoxins indicates that they have the *gauche–gauche–gauche* form. Takamatsu *et al.* (1976) obtained the Raman spectrum of toxin B, isolated from the Indian cobra, *Naja naja*. This is a type II neurotoxin. It displayed an S—S stretching vibration with its main peak at 510 cm^{-1} and a shoulder at 523 cm^{-1}. These characteristics show the presence of *trans–gauche–gauche* configuration besides the main *gauche–gauche–gauche* conformation.

The sea snake neurotoxins have only one tyrosine residue per molecule. This amino acid is necessary for poisonous activity, and is buried within the molecule (Raymond and Tu, 1972). Yu *et al.* (1973) showed that the environment of the tyrosine side chain is reflected by the intensity ratio, I_{830}/I_{853}. Figure 3.37 shows the spectrum of glycyltyrosine, in which the tyrosine is exposed. The intensity ratio, I_{853}/I_{828}, is 1.0/0.71. The ribonuclease molecule contains buried tyrosine. The Raman spectrum for this enzyme shows a ratio, I_{832}/I_{852}, of 1.0/0.8, but when the enzyme is heated to 85°C, it is denatured and the tyrosine side chain is exposed. This exposure results in a reversal of the ratio, i.e., I_{852}/I_{832} is 1.0/0.8. In Fig. 3.35, we see that the band at 846 cm^{-1} is less intense than that at 834 cm^{-1}. This shows that the one tyrosyl side chain does not have easy access to the solvent and may be bound somehow with other side-chain moieties.

Yu (1974) found that when the indole part of the tryptophan side chain is buried within a protein molecule and is possibly surrounded by a hydrophobic region, the band at 1361 cm^{-1} has a sharp peak. However, the intensity of this band is decreased when water is available to the tryptophan residue. Since no peak appears at 1361 cm^{-1} for neurotoxins (Fig. 3.35), the single tryptophanyl side chain is exposed. Chemical studies are in agreement with this conclusion.

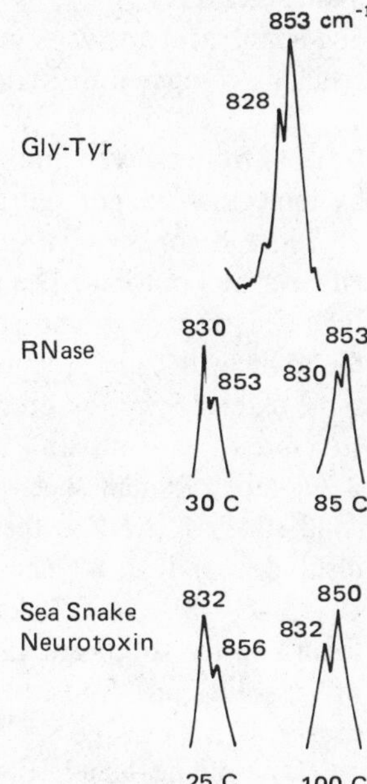

Fig. 3.37. Raman spectrum of tyrosine residue. Gly-Tyr is used as a model of exposed tyrosine residues. The intensity ratio at 853/830 is reversed when RNase is heat denatured. This indicates that a tyrosine residue originally buried is now exposed by heat denaturation. The same phenomenon can be observed for sea snake neurotoxin. (Tu, 1977a.)

The methionine residue can also be detected in neurotoxins by means of its C—S stretching band at 700 cm^{-1}. Tu *et al.* (1976) used this band to show the presence of methionine in one neurotoxin and the absence of it in another one.

A neurotoxin from the venom of *Lapemis hardwickii* has been studied by Fox *et al.* (1977), who determined its complete amino acid sequence, and detected the presence of a free SH group by means of a typical S—H stretching vibration at 2585 cm^{-1} (Fig. 3.38), in addition to the four S—S bonds. The previous report of eight half-cystines (Tu

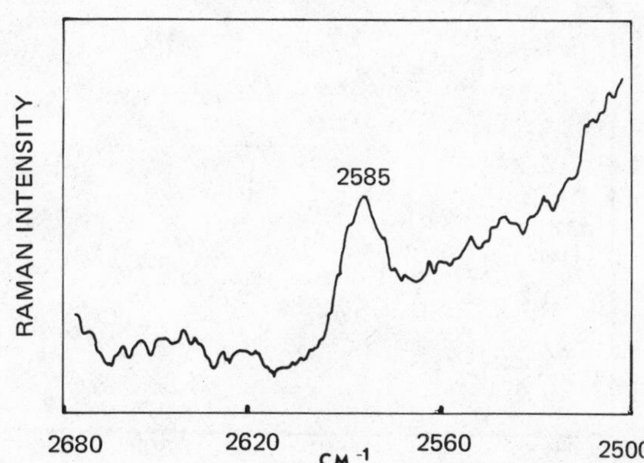

Fig. 3.38. Laser-Raman spectrum of *Lapemis hardwickii* major toxin (lyophilized powder) in the region of the S—H stretching vibration. (Fox *et al.*, 1977.)

and Hong, 1971; Yu *et al.*, 1975), determined by the methods of reduction, alkylation, and amino acid analysis, was incorrect "due to the lower sensitivity and accuracy of these methods compared to Raman spectroscopy and sequence determination."

Bjarnason and Tu (1978) have isolated five previously unknown hemorrhagic proteins (toxins) from the venom of the western diamondback rattlesnake and characterized them by molecular weight and amino acid composition. Toxin *e* has a molecular weight of 25,700 and contains 219 amino acids. All the toxins contain approximately one mol of zinc per mol of toxin. The proteolytic and hemorrhagic properties of toxin *e* were equally inhibited when the zinc atom was removed by 1,10-phenanthroline. Reconstitution of the toxin with zinc brought back the proteolytic and hemorrhagic characteristics to the same extent. Spectroscopic methods (CD, UV, and Raman) were used to investigate the structure of native toxin *e* and the structural changes induced by removal of zinc. Figure 3.39 shows Raman spectra of the zinc-containing native toxin and the apotoxin in the solid state, as well as the —S—S— region of native toxin in aqueous solution. The disulfide band at 509 cm^{-1} indicates a *gauche–gauche–gauche* configuration of the C—C—S—S—C—C linkage system. The upper spectrum represents hemorrhagic apotoxin *e* in the solid state and the lower two spectra represent native hemorrhagic toxin *e* in the solid state. The variance in spectra that were obtained is shown by the two lower curves.

The amide I and III regions display the most significant differences between the spectra of the holo- and apotoxins. The holotoxin shows an amide I doublet, composed of a band around 1655–1660 cm^{-1} (α-helix) and another one around 1665–1668 cm^{-1} (possibly "random coil," but could be β and/or β-turn structures). The meaning of "random

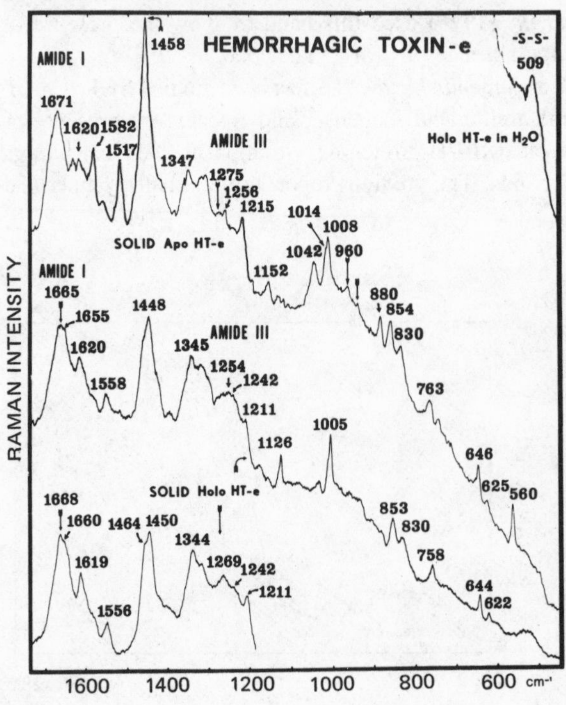

Fig. 3.39. Laser-Raman spectra of native hemorrhagic toxin e (two bottom curves) and hemorrhagic apotoxin e (top curve) in the solid state, and the disulfide region of aqueous native hemorrhagic toxin e (upper right-hand corner). (Bjarnason and Tu, 1978.)

coil" here is "different from normal α-helical or β-sheet structures." For the β-reverse turn conformation Tu *et al.* (1978) showed amide I at 1663–1666 cm^{-1} and amide III at 1260–1266 cm^{-1}. In the spectrum of apotoxin *e,* the amide I band is at 1671 cm^{-1} and no band for α-helical structure appears. Holotoxin *e* shows an amide III band at 1269 cm^{-1}, which may be due to β-reverse turn conformation (Bjarnason and Tu, 1978). This band is absent for apotoxin *e,* which may mean that when the zinc ion is missing, the β-turn is lost along with the α-helical conformation. CD spectral information was in fairly good agreement with the findings obtained from Raman spectra.

Sea Anemone Toxin II

Prescott *et al.* (1976) have studied the Raman spectra of sea anemone toxin II in both aqueous and solid states, and have found that the conformation-sensitive amide I and amide III frequencies are not much different after removal of the aqueous solvent. They concluded that the secondary structure of toxin II is not significantly affected in going from one state to the other. The spectral data indicate predominantly disordered structure for the peptidyl backbone.

In aqueous solution a line at 509 cm^{-1} (S—S stretching vibration) indicated a *gauche–gauche–gauche* configuration for each of the three C—C—S—S—C—C networks of the toxin. According to these authors, the "disordered" secondary structure does not imply a statistical random coil in which the chain is free to assume any orientation, but should be regarded as a structure in which the average configuration of peptidyl groups differs markedly from that normally occurring in either α-helical or β-sheet structures.

Many structural dissimilarities occur among toxins, when their conformational structures are examined. A remarkable diversity exists among them, ranging from a predominance of α-helix, to large amounts of β-structure, to a highly disordered structure.

Predictions of Conformation

Fox and Tu (1979) have used three methods to study the conformation of the major neurotoxin from *Lapemis hardwickii:* structure prediction from amino acid sequence (Chou and Fasman, 1974, 1977); circular dichroism (Chen *et al.*, 1972); and Raman spectroscopy (Lippert *et al.,* 1976; Pézolet *et al.*, 1976). The first method permitted the prediction of three regions of β-sheet structure (see Fig. 3.40), at residues 1–7, 13–18, and 41–45, but the group 13–18 was not considered as β-sheet because there is a β-turn near it and because of environmental factors (Fox and Tu, 1979). (The letter abbreviations for amino acids are as used in Dayhoff, 1972.) They predicted β-turns at residues 14–17, 20–23, 30–33, 37–40, and 46–49 and estimated the neurotoxin to have about one-third β-turn content and about one-fifth β-sheet. CD spectra show the presence of both types of structures. Raman spectra also show that both types are present. The method of Lippert *et al.* (using H_2O and D_2O protein solutions) yielded a value of 35% β-sheet structure, while that of Pézolet *et al.* gave a value of 36 ± 10% β-sheet structure for the neurotoxin in water. The last method as applied here used the amide III band intensity at 1243 cm^{-1} measured against a reference band (CH_2 bending modes) at 1452 cm^{-1} to perform the calculation. Fox and Tu discussed the advantages and disadvantages of the two Raman methods, and in addition, they felt that none of the spectroscopic methods they used shows adequately how much β-turn structure is present.

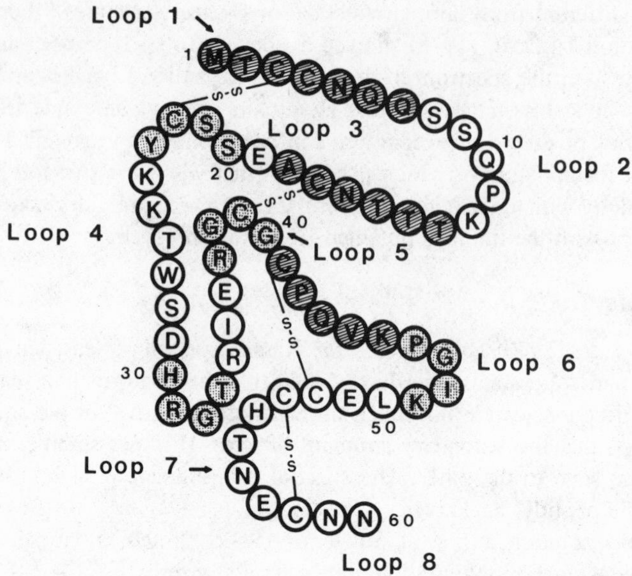

Fig. 3.40. Diagrammatic representation of *Lapemis hardwickii* major toxin depicting predicted β-sheet regions (▨), and β-turns (⁝⁝). (Fox and Tu, 1979.)

Bailey *et al.* (1979) have used the following three methods to study the secondary structure of myotoxin *a,* a polypeptide isolated (Cameron and Tu, 1977) from prairie rattlesnake *(C. viridis viridis)* venom: (1) prediction of structure from the amino acid sequence (Chou and Fasman, 1978; Dufton and Hider, 1977); (2) analysis by circular dichroism (CD) spectroscopy; and (3) analysis by Raman spectroscopy, applying the method of Lippert *et al.* (1976), which is described in this book on page 132. Prediction by the method of Chou and Fasman yielded a composition of up to 14% α helix, 57% β-sheet, and 26% reverse turn, when the analysis was based on parameters obtained from 15 proteins. These results agreed more closely with results from CD and Raman spectroscopy than the surprising results obtained with the Chou–Fasman method when the analysis was made with revised parameters based on almost twice as many proteins: up to 64% α-helix, 43% β-sheet, and 67% β-reverse turn, adding to more than 100%. This value of more than 100% is due to extensive overlapping of α helix and β-reverse turn regions, and one explanation offered by Bailey *et al.* is that myotoxin *a* is largely helical, the helical segments consisting of sequences of β-reverse turns as in a 3_{10} helix (page 29 in Dickerson and Geis, 1969), where 3 stands for the number of residues per turn and the subscript denotes the number of atoms in the rings formed by closing the hydrogen bond. (The α-helix is called the 3.6_{13} helix by this notation.)

Analysis by CD spectroscopy (method of Chen *et al.*, 1972) yielded a negative value for the content of helical structure, believed by Bailey *et al.* to be due to the fact that the equation given by them does not consider the presence of β-turn structures, and therefore is only applicable to relatively few cases. However, Bailey *et al.* did find that, qualitatively, the CD spectrum displayed properties of both β-turn and β-sheet structures.

The Raman spectroscopic method produced these values for myotoxin *a:* zero α helix, 72.6% β-sheet, and 27.4% random coil content. There is no term for β-reverse turn in the equations of Lippert *et al.* (1976), so that the β-turn content was not calculable. By combining their results from the three independent methods, Bailey *et al.* were able to decide that myotoxin *a* consists of significant amounts of β-sheet and β-turn structures, small amounts of random coil structures, and possibly α helix. These authors feel that the method of Lippert *et al.* for predicting protein conformation may improve if it is altered to encompass β-reverse turn structures.

References

Abaturov, L. V., Nezlin, R. S., Vengerova, T. I., and Varshavsky, Ja. M., *Biochim. Biophys. Acta* **194,** 386 (1969).

Ambrose, E. J., and Elliott, A., *Proc. R. Soc. London Ser. A* **208,** 75 (1951).

Baier, R. E., Loeb, G. I., and Wallace, G. T., *Fed. Proc.* **30,** 1523 (1971).

Bailey, G. S., Lee, J., and Tu, A. T., *J. Biol. Chem.* **254,** 8922 (1979).

Bellocq, A. M., Lord, R. C., and Mendelsohn, R., *Biochim. Biophys. Acta* **257,** 280 (1972).

Bhagavan, N. V., *Biochemistry,* J. B. Lippincott Co., Philadelphia, 1974.

Bjarnason, J. B., and Tu, A. T., *Biochemistry* **17,** 3395 (1978).

Bloom, S. M., Fasman, G. D., de Lozé, C., and Blout, E. R., *J. Am. Chem. Soc.* **84,** 458 (1962).

Bradbury, E. M., Cary, P. D., Crane-Robinson, C., Rattle, H. W. E., Boublik, M. and Sautière, P., *Biochemistry* **14,** 1876 (1975).

Bradbury, E. M., and Martin, C., *Proc. R. Soc. London Ser. A* **214,** 183 (1952).

Brazeau, P., Vale, W., Burgus, R., Ling, N., Butcher, M., Rivier, J., and Guillemin, R., *Science* **179,** 77 (1973).

Brazhnikov, E. V., and Chirgadze, Yu. N., *J. Mol. Biol.* **122,** 127 (1978).

Brunner, H., Holz, M., and Jering, H., *Eur. J. Biochem.* **50,** 129 (1974).

Burke, M. J., and Rougvie, M. A., *Biochemistry* **11,** 2435 (1972).

Caillé, J.-P., Pigeon-Gosselin, M., and Pézolet, M., Abstracts of the 26th Annual Meeting, *Biophys. J.* **37,** No. 2 Part 2, page 120a, February (1982).

Cameron, D. L., and Tu, A. T., *Biochemistry* **16,** 2546 (1977).

Carew, E. B., Leavis, P. C., Stanley, H. E., and Gergely, J., *Biophys. J.* **30,** 351 (1980).

Carew, E. B., Asher, I. M., and Stanley, H. E., *Science* **188,** 933 (1975).

Cathou, R. E., Kulczycki, A., Jr., and Haber, E., *Biochemistry* **7,** 3958 (1968).

Chen, M. C., and Lord, R. C., *J. Am. Chem. Soc.* **96,** 4750 (1974).

Chen, M. C., and Lord, R. C., *J. Am. Chem. Soc.* **98,** 990 (1976).

Chen, M. C., Lord, R. C., and Mendelsohn, R., *Biochim, Biophys. Acta* **328,** 252 (1973).

Chen, M. C., Lord, R. C., and Mendelsohn, R., *J. Am. Chem. Soc.* **96,** 3038 (1974).

Chen, Y.-H., Yang, J. T., and Martinez, H. M., *Biochemistry* **11,** 4120 (1972).

Chen, Y.-H., Lu, H. S., and Lo, T. B., *J. Chin. Biochem. Soc.* **4,** 69 (1975).

Chirgadze, Yu. N., *Molek. Biol. (USSR)* **1,** 9–60 (1973).

Chirgadze, Yu. N., Fedorov, O. V., and Trushina, N. P., *Biopolymers* **14,** 679 (1975).

Chirgadze, Yu. N., Brazhnikov, E. V., and Nevskaya, N. A., *J. Mol. Biol.* **102,** 781 (1976).

Chou, P. Y., and Fasman, G. D., *Annu. Rev. Biochem.* **47,** 251 (1978).

Chou, P. Y., and Fasman, G. D., *J. Mol. Biol.* **115,** 135 (1977).

Chou, P. Y., and Fasman, G. D., *Biochemistry* **13,** 222 (1974).

Combarnous, Y., and Nabedryk-Viala, E., *Biochem. Biophys. Res. Commun.* **84,** 1119 (1978).

Dayhoff, M. O., *Atlas of Protein Sequence and Structure,* Vol. 5, National Biomedical Research Foundation, Georgetown University Medical Center, Washington, D. C., 1972.

Deutschmann, G., and Ullrich, V., *Anal. Biochem.* **94,** 6 (1979).

Dickerson, R. E., and Geis, I., *The Structure and Action of Proteins,* W. A. Benjamin, Inc., Menlo Park, California, 1969.

Dorrington, K. J., and Tanford, C., *Advan. Immunol.* **12,** 333 (1970).

Doyle, B. B., Traub, W., Lorenzi, G. P., and Blout, E. R., *Biochemistry* **10,** 3052 (1971).
Dufton, M. J., and Hider, R. C., *J. Mol. Biol.* **115,** 177 (1977).
Eckert, K., Grosse, R., Malur, and Repke, K. R. H., *Biopolymers* **16,** 2549 (1977).
Edelman, G. M., and Gally, J. A., *Proc. Natl. Acad. Sci., USA* **51,** 846 (1964).
Elliott, A., and Malcolm, B. R., *Proc. R. Soc. London, Ser. A* **249,** 30 (1959).
Elliott, A., Fraser, R. D. B., and MacRae, T. P., *J. Mol. Biol.* **11,** 821 (1965), and references therein.
Englander, J. J., Calhoun, D. B., and Englander, S. W., *Anal. Biochem.* **92,** 517 (1979).
Englander, S. W., and Staley, R., *J. Mol. Biol.* **45,** 277 (1969).
Fanconi, B., Tomlinson, B., Nafie, L. A., Small, W., and Peticolas, W. L., *J. Chem. Phys.* **51,** 3993 (1969).
Fanconi, B., Small, E. W., and Peticolas, W. L., *Biopolymers* **10,** 1277 (1971).
Fasman, G. D., Chou, P. Y., and Adler, A. J., "Histone Conformation: Predictions and Experimental Studies," in *The Molecular Biology of the Mammalian Genetic Apparatus* Vol. 1, P.O.P. Ts'o ed., North-Holland Publ. Co., Amsterdam, pp. 1–52, 1977.
Fasman, G. D., Itoh, K., Liu, C. S., and Lord, R. C., *Biopolymers* **17,** 125 (1978).
Fermandjian, S., Fromageot, P., Tistchenko, A.-M., Leicknam, J.-P., and Lutz, M., *Eur. J. Biochem.* **28,** 174 (1972).
Fleischmann, J. B., Porter, R. R., and Press, E. M., *Biochem. J.* **88,** 220 (1963).
Fox, J., and Tu, A. T., *Arch. Biochem. Biophys.* **193,** 407 (1979).
Fox, J. W., Elzinga, M., and Tu, A. T., *FEBS Lett.* **80,** 217 (1977).
Fraser, R. D. B., Harrap, B. S., MacRae, T. P., Stewart, F. H. C., and Suzuki, E., *J. Mol. Biol.* **12,** 482 (1965).
Freeman, S., and Daniel, E., *Biochim. Biophys. Acta* **534,** 132 (1978).
Frushour, B. G., and Koenig, J. L., *Biopolymers* **13,** 455, 1809 (1974).
Frushour, B. G., and Koenig, J. L., *Biopolymers* **14,** 649 (1975a).
Frushour, B. G., and Koenig, J. L., *Biopolymers* **14,** 379 (1975b).
Frushour, B. G., and Koenig, J. L., in *Advances in Infrared and Raman Spectroscopy,* Vol. I (R. J. H. Clark and R. E. Hester, eds.), Heyden, London, pp. 35–97, 1975.
Gendreau, R. M., and Jakobsen, R. J., *Appl. Spectrosc.* **32,** 326 (1978).
Gendreau, R. M., Winters, S., Leininger, R. I., Fink, D., Hassler, C. R., and Jakobsen, R. J., *Appl. Spectrosc.* **35,** 353 (1981).
Gendreau, R. M., Winters, S., Brown, L. L., Burch, K., and Jakobsen, R. J., Abstracts of the Pittsburgh Conference, paper No. 655, 1982.
Gilon, C., Simmons, D., and Goodman, M., *Biopolymers* **19,** 341 (1980).
Gordon, P. L., Huang, C., Lord, R. C., and Yannas, I. V., *Macromolecules* **7,** 954 (1974).
Gratzer, W. B., in *Poly-α-Amino Acids* (G. D. Fasman, ed.), Marcel Dekker, New York, Chap. 5, pp. 177–238, 1967.
Gudkov, A. T., Khechinashvili, N. N., and Bushuev, V. N., *Eur. J. Biochem.* **90,** 313 (1978).
Guillot, J.-G., Pézolet, M., and Pallotta, D., *Biochim. Biophys. Acta* **491,** 423 (1977).
Haggis, G. H., *Biochim. Biophys. Acta* **19,** 545 (1956).
Haggis, G. H., *Biochim. Biophys. Acta* **23,** 494 (1957).
Hauert, J., Maire, M., Sussmann, A., and Bargetzi, J.P., *Int. J. Peptide Protein Res.* **6,** 201 (1974).
Heitz, F., Lotz, B., and Spach, G., *J. Mol. Biol.* **92,** (1975).
Hruby, V. J., Deb, K. K., Fox, J., Bjarnason, J., and Tu, A. T., *J. Biol. Chem.* **253,** 6060 (1978).
Hvidt, Aa. in *Dynamical Aspects of Conformation Changes in Biological Macromolecules* (C. Sandron, ed.), D. Reidel Publishing Co., Dordrecht, Holland, p. 103, 1973.
Hvidt, Aa., and Nielsen, S. O., *Advan. Protein Chem.* **21,** 287 (1966).
Hvidt, Aa., and Pedersen, E. J., *Eur. J. Biochem.* **48,** 333 (1974).
Hvidt, Aa., and Wallevik, K., *J. Biol. Chem.* **247,** 1530 (1972).
Itoh, K., Foxman, B. M., and Fasman, G. D., *Biopolymers* **15,** 419 (1976).
Jacobson, A. L., Vejdovsky, V., and Krueger, P. J., *Biochemistry* **12,** 2963 (1973).
Jakobsen, R. J., and Gendreau, R. M., *Artificial Organs* **2**(2), 183 (1978).
Kirkwood, J. G., and Shumaker, J. B., *Proc. Natl. Acad. Sci., USA* **38,** 855, 863 (1952).
Klis, W. A., and Siemion, I. Z., *Int. J. Peptide Protein Res.* **12,** 103 (1978).
Koenig, J. L., *J. Polymer Sci., Macromol. Rev.* **6,** 59 (1972).

Koenig, J. L., and Frushour, B. G., *Biopolymers* **11,** 1871 (1972).
Krimm, S., *J. Mol. Biol.* **4,** 528 (1962).
Kumar, K., and Carey, P. R., *J. Chem. Phys.* **63,** 3697 (1975).
Kumar, K., Phelps, D. J., Carey, P. R., and Young, N. M., *Biochem. J.* **175,** 727 (1978).
Layman, D. J., Knutson, K., McNeill, B., Shibatane, K., *Trans. Am. Soc. Artif. Intern. Organs* **21,** 49 (1975).
Leavis, P. C., Rosenfeld, S. S., Gergely, J., Grabarek, Z., and Drabikowski, W., *J. Biol. Chem.* **253,** 5452 (1978).
Lee, R. G., and Kim, S. W., *J. Biomed. Mater. Res.* **8,** 251 (1974).
Lerner, A. B., Barnum, C. P., and Watson, C. J., *Am. J. Med. Sci.* **214,** 416 (1947).
Levine, B. A., Thornton, J. M., Fernandes, R., Kelly, C. M., and Mercola, D., *Biochim. Biophys. Acta.* **535,** 11 (1978).
Levine, B. A., Mercola, D., Coffman, D., and Thornton, J. M., *J. Mol. Biol.* **115,** 743 (1977).
Lin, V. J. C., and Koenig, J. L., *Biopolymers* **15,** 203 (1976).
Lindemann, R., and Zundel, G., *Biopolymers* **17,** 1285 (1978).
Linderstrøm-Lang, K., Soc. Biol. Chemists India Silver Jubilee Souvenir, p. 191, 1955.
Lippert, J. L., Tyminski, D., and Desmeules, P. J., *J. Am. Chem. Soc.* **98,** 7075 (1976).
Litman, G. W., Litman, R. S., Good, R. A., and Rosenberg, A., *Biochemistry* **12,** 2004 (1973).
Lord, R. C., *Proc. Int. Congr. Pure Appl. Chem., Suppl., 23rd,* **7,** 179 (1971).
Lord, R. C., and Yu, N.-T., *J. Mol. Biol.* **50,** 509 (1970a).
Lord, R. C., and Yu, N.-T., *J. Mol. Biol.* **51,** 203 (1970b).
Low, B. W., Preston, H. S., Sato, A., Rosen, L. S., Searl, J. E., Rudko, A. D., and Richardson, J. S., *Proc. Natl. Acad. Sci. USA* **73,** 2991 (1976).
Magdoff-Fairchild, B., Lovell, F. M., and Low, B. W., *J. Biol. Chem.* **244,** 3497 (1969).
Mattson, J. S., Smith, C. A., and Paulsen, K. E., *Anal. Chem.* **47,** 736 (1975).
Maxfield, F. R., and Scheraga, H. A., *Biochemistry* **16,** 4443 (1977).
Middaugh, C. R., Thomas, G. J., Jr., Prsescott, B., Aberlin, M. E., and Litman, G. W., *Biochemistry* **16,** 2986 (1977).
Miyazawa, T., and Sugeta, H., U.S.-Japan Joint Seminar, "The Raman Spectroscopy of Biological Molecules," Cleveland, Ohio, 1974.
Morrison, C. A., Garrett, R. A., and Bradbury, E. M., *Eur. J. Biochem.* **78,** 153 (1977).
Morrissey, B. W., and Fenstermaker, C. A., *Trans. Am. Soc. Artif. Intern. Organs* **22,** 278 (1976).
Murray, A. C., and Kay, C. M., *Biochemistry* **11,** 2622 (1972).
Nakanishi, M., Takesada, H., and Tsuboi, M., *J. Mol. Biol.* **89,** 241 (1974).
Nouwen, R., and Huyskens, P., *J. Mol. Struct.* **16,** 459 (1973).
Ockman, N., *Biopolymers* **17,** 1273 (1978).
Oertel, R. P., *Biopolymers* **16,** 2329 (1977).
Oliveira, M. C. F., Juliano, L., and Paiva, A. C. M., *Biochemistry* **16,** 2606 (1977).
Painter, P. C., and Koenig, J. L., *Biopolymers* **14,** 457 (1975).
Painter, P. C., and Koenig, J. L., *Biopolymers* **15,** 229 (1976).
Parker, F. S., *Applications of Infrared Spectroscopy in Biochemistry, Biology, and Medicine,* Chap. 11, Plenum Press, New York, 1971.
Parker, F. S., and Ans, R., *Anal. Biochem.* **18,** 414 (1967).
Parker, F. S., *Nature* **200,** 1093 (1963).
Pézolet, M., Pigeon-Gosselin, M., and Coulombe, L., *Biochim. Biophys. Acta* **453,** 502 (1976).
Pézolet, M., Pigeon-Gosselin, M., Savoie, R., and Caillé, J.-P., *Biochim. Biophys. Acta* **544,** 394 (1978).
Pézolet, M., Pigeon-Gosselin, M., and Caillé, J.-P., *Biochim. Biophys. Acta.* **533,** 263 (1978a).
Pézolet, M., Pigeon-Gosselin, M., Nadeau, J., and Caillé, J.-P., *Biophys. J.* **31,** 1 (1980).
Potter, J. D., and Gergely, J., *J. Biol. Chem.* **250,** 4628 (1975).
Potter, J. D., Seidel, J. C., Leavis, P., Lehrer, S. S., and Gergely, J., *J. Biol. Chem.* **251,** 7551 (1976).
Prescott, B., Thomas, G. J., Jr., Béress, L., Wunderer, G., and Tu, A. T., *FEBS Lett.* **64,** 144 (1976).
Rabolt, J. F., Moore, W. H., and Krimm, S., *Macromolecules* **10,** 1065 (1977).
Ralston, G. B., *Compt. Rend. Trav. Lab. Carlsberg* **39,** 65 (1972).
Ratajczak, H., and Sobczyk, L., *J. Chem. Phys.* **50,** 556 (1969).
Raymond, M. L., and Tu, A. T., *Biochim. Biophys. Acta* **285,** 498 (1972).

Roohk, H. V., Pick, J., Hill, R., Hung, E., and Bartlett, R. H., *Trans. Am. Soc. Artif. Intern. Organs* **22,** 1 (1976).
Rüegg, M., and Häni, H., *Biochim. Biophys. Acta* **400,** 17 (1975).
Sautière, P., Tyrou, D., Laine, B., Mizon, J., Ruffin, P., and Biserte, G., *Eur. J. Biochem.* **41,** 563 (1974).
Seamon, K. B., Hartshorne, D. J., and Bothner-By, A. A., *Biochemistry* **16,** 4039 (1977).
Siamwiza, M. N., Lord, R. C., Chen, M. C., Takamatsu, T., Harada, I., Matsuura, H., and Shimanouchi, T., *Biochemistry* **14,** 4870 (1975).
Small, E. W., Fanconi, B., and Peticolas, W. L., *J. Chem. Phys.* **52,** 4369 (1970).
Sobczyk, L., in *The Hydrogen Bond, Recent Developments in Theory and Experiments,* Vol. 3 (P. Schuster, G. Zundel, and C. Sandorfy, eds.), Chap. 20, p. 937, North Holland, Amsterdam, 1976.
Sobczyk, L., and Pawelka, Z., *Roczn. Chem.* **47,** 1523 (1973).
Sørup, P., Junager, F., and Hvidt, Aa., *Biochim. Biophys. Acta* **494,** 9 (1977).
Spiro, T. G., and Gaber, B. P., *Annu. Rev. Biochem.* **46,** 553 (1977).
St. Pierre, S., Ingwall, R. T., Verlander, M. S., and Goodman, M., *Biopolymers* **17,** 1837 (1978).
Sugeta, H., Go, A., and Miyazawa, T., *Chem. Lett.,* 83–86 (1972); *Bull. Chem. Soc. Jpn* **46,** 2752, 3407 (1973).
Takamatsu, T., Harada, I., Shimanouchi, T., Ohta, M., and Hayashi, K., *FEBS Lett.* **72,** 291 (1976).
Tang, J., Sepulveda, P., Marciniszyn, J., Jr., Chen., K.C.S., Huang, W.-Y., Tao, N., Liu, D., and Lanier, J. P., *Proc. Natl. Acad. Sci. USA.* **70,** 3437 (1973).
Termine, J. D., Eanes, E. D., Ein, D., and Glenner, G. G., *Biopolymers* **11,** 1103 (1972).
Thomas, G. J., Jr., Prescott, B., Middaugh, C. R., and Litman, G. W., *Biochim. Biophys. Acta* **577,** 285 (1979).
Tobin, M. C., *Science* **161,** 68 (1968).
Tsernoglou, D., and Petsko, G. A., *FEBS Lett.* **68,** 1 (1976).
Tsernoglou, D., Petsko, G. A., and Tu, A. T., *Biochim. Biophys. Acta* **491,** 605 (1977).
Tsuboi, M., in *Vibrational Spectroscopy—Modern Trends* (A. J. Barnes and W. J. Orville-Thomas, eds.), Elsevier Publishing Co., New York, p. 405, 1977.
Tu, A. T., *Annu. Rev. Biochem.* **42,** 235 (1973).
Tu, A. T., *The Spex Speaker,* Spex Industries, Inc., 3880 Park Ave., Metuchen, New Jersey, Vol. XXII, No. 2, June, 1977a.
Tu, A. T., *Venoms: Chemistry and Molecular Biology,* John Wiley and Sons, New York, 1977b.
Tu, A. T., and Hong, B.-S., *J. Biol. Chem.* **246,** 2772 (1971).
Tu, A. T., Prescott, B., Chou, C. H., and Thomas, G. J., Jr., *Biochem. Biophys. Res. Commun.* **68,** 1139 (1976).
Tu, A. T., Bjarnason, J. B., and Hruby, V. J., *Biochim. Biophys. Acta* **533,** 530 (1978).
Tu, A. T., Jo, B. H., and Yu, N.-T., *Int. J. Peptide Prot. Res.* **8,** 337 (1976).
Tu, A. T., Lee, J., Deb, K. K., and Hruby, V. J., *J. Biol. Chem.* **254,** 3272 (1979).
van Eerd, J.-P., and Kawasaki, Y., *Biochem. Biophys. Res. Commun.* **47,** 859 (1972).
Van Wart, H. E., and Scheraga, H. A., *J. Phys. Chem.* **80,** 1812 (1976a).
Van Wart, H. E., and Scheraga, H. A., *J. Phys. Chem.* **80,** 1823 (1976b).
Van Wart, H. E., and Scheraga, H. A., *Proc. Natl. Acad. Sci. USA* **74,** 13 (1977).
Van Wart, H. E., Scheraga, H. A., and Martin, R. B., *J. Phys. Chem.* **80,** 1832 (1976b).
Van Wart, H. E., Cardinaux, F., and Scheraga, H. A., *J. Phys. Chem.* **80,** 625 (1976a).
Van Wart, H. E., Lewis, A., Scheraga, H. A., and Saeva, F. D., *Proc. Natl. Acad. Sci. USA* **70,** 2619 (1973).
Veis, A., and Brownell, A. G., *Proc. Natl. Acad. Sci. USA* **74,** 902 (1977).
Venyaminov, S. Yu, Rajnavölgyi, É., Medgyesi, G. A., Gergely, J., and Závodszky, P., *Eur. J. Biochem.* **67,** 81 (1976).
Willumsen, L., *Biochim. Biophys. Acta* **126,** 382 (1966).
Willumsen, L., *Compt. Rend. Trav. Lab. Carlsberg* **38,** 223 (1971).
Wilson, E. B., Jr., *Phys. Rev.* **45,** 706 (1934).
Yang, C. C., Chang, C. C., Hayashi, K., Suzuki, T., Ikeda, K., and Hamaguchi, K., *Biochim. Biophys. Acta* **168,** 373 (1968).
Yeoman, L. C., Olson, M. O. J., Sugano, N., Jordan, J. J., Taylor, C. W., Starbuck, W. C., and Busch, H., *J. Biol. Chem.* **247,** 6018 (1972).

Yu, N.-T., *J. Am. Chem. Soc.* **96,** 4664 (1974).
Yu, N.-T., and East, E. J., *J. Biol. Chem.* **250,** 2196 (1975).
Yu, N.-T., and Jo, B. H., *Arch. Biochem. Biophys.* **156,** 469 (1973a).
Yu, N.-T., and Jo, B. H., *J. Am. Chem. Soc.* **95,** 5033 (1973b).
Yu, N.-T., Jo, B. H., and O'Shea, D. C., *Arch. Biochem. Biophys.* **156,** 71 (1973).
Yu, N.-T., Jo, B. H., Chang, R. C. C., and Huber, J. D., *Arch. Biochem. Biophys.* **160,** 614 (1974).
Yu, N.-T., Lin, T.-S., and Tu, A. T., *J. Biol. Chem.* **250,** 1782 (1975).
Yu, N.-T., Liu, C. S., and O'Shea, D. C., *J. Mol. Biol.* **70,** 117 (1972).
Yu, T.-J., and Peticolas, W. L., in *Peptides, Polypeptides, and Proteins; Proceedings of the Rehovot Symposium on Poly (Amino Acids), Polypeptides, and Proteins and Their Biological Implications,* May 1974, John Wiley and Sons, New York.
Yu, T. J., Lippert, J. L., and Peticolas, W. L., *Biopolymers* **12,** 2161 (1973a).

Chapter 4

TWO SPECIALIZED TISSUES: THE EYE AND THE PURPLE MEMBRANE OF *HALOBACTERIUM HALOBIUM*

The Eye

Lens and Cornea

Harding and Dilley (1976) have written an extensive review of the structural proteins of the mammalian lens, with emphasis on changes in development, aging, and cataract. Included are discussions of lens anatomical structure; α, β, and γ-crystallins; changes in the soluble proteins of the lens during development and aging and in cataract; insoluble lens proteins; high-molecular-weight proteins, e.g., aggregates of α-crystallin and of —S—S— cross-linked protein; brown cataract non —S—S— covalently linked protein; and protein synthesis and degradation in the lens.

Raman spectroscopy has been applied to a study of the intact lens of the eye. With this technique, Schachar and Solin (1975) have obtained significant information on both structural and chemical properties of the bovine lens. The assignments shown in Table 4.1 are based upon a comparison of the Raman shifts of the bovine lens with those seen in the Raman spectra of amino acids and certain proteins (Lord and Yu, 1970; Koenig and Frushour, 1972). Schachar and Solin have assigned a strong line at 1675 cm^{-1} to amide I (C═O stretching vibration), which was evidence that the lens proteins have an antiparallel β-pleated sheet structure (Krimm and Abe, 1972). They observed no Raman lines between 1630 and 1654 cm^{-1}, the expected frequencies for the amide I band of the parallel β-pleated sheet and α-helix structures, respectively (Yu *et al.*, 1974; Miyazawa, 1960). Several of the frequencies of Table 4.1 display propagation-dependent depolarization ratios with $\rho_{\parallel} \neq \rho_{\perp}$. These ratios are evidence of microscopic anisotropy, which is responsible for the uniaxial quality of the lens observed by Brewster (1816).

On the basis of a mathematical analysis, the authors concluded that the CONH bonds of the antiparallel β-pleated sheet are arranged in an angular distribution that is not random but is skewed toward directions orthogonal to the lens optic axis. The work of

Table 4.1. Raman Shifts in cm^{-1} and Depolarization Ratios for Intact Bovine Lenses[a,b]

Raman shift $\Delta\tilde{\nu}$ (cm^{-1})	Depolarization ratios $\rho_\perp$	$\rho_\parallel$	Assignments
765	0	0	Trp
837	0	0	Tyr
861	0	0	Tyr
883	0	0	
940	0	0	
1009	0.14	0.06	Phe
1036	0.07	0.06	Phe
1132	0.28	0	Skeletal modes
1179	0.23	0	Tyr
1213	0.17	0.06	
1243	0.23	0.19	Amide III
1259	0.24	0.12	Amide III
1273	0.24	0.12	Amide III
1340	0.36	0.31	Trp
1450	0.50	0.64	CH bending
1552	0.25	0.15	Trp
1619	0.50	0.40	Trp, Tyr, and Phe
1675	0.26	0.16	Amide I
2881	0.09	0.08	Aliphatic CH stretching
2941	0.15	0.14	Aliphatic CH stretching
3068	0.13	0.14	Aromatic CH stretching
3308	0.13	0.14	Superposition of H_2O and H-bonded N—H stretching frequencies

[a] Schachar and Solin, (1975).
[b] The depolarization ratios $\rho_\parallel$ and $\rho_\perp$ were measured with 4,880 Å incident radiation propagating parallel and perpendicular, respectively, to the lens optic axis. The Raman shifts are accurate to ± 3 cm^{-1} and the depolarization ratios are accurate to within $\pm 15\%$.

Kuck *et al.* (1976), showing that bird lens protein is mainly α-helical, appears to contradict the suggestion of Schachar and Solin that the relationship between the fiber macrostructure of the lens proteins and the microstructure of the proteins is causal (Yu, 1977).

Schachar and Solin also discussed the relationship between lens microstructure and opacity; and presented polarized spectra of an intact bovine lens, bovine lens homogenate, and lysozyme in aqueous solution.

Raman spectra of isolated lens proteins have been recorded by Yu and East (1975). They separated the water-soluble proteins of the cow lens on Sephadex G-200 into the following fractions: α-, β_1-, β_2-, β_3-, and γ-crystallins.

Figure 4.1 shows the spectra of these crystallins and that of albuminoid (water insoluble). The amino acid side chains of phenylalanine, tyrosine, and tryptophan produce strong Raman lines at 624, 644, and 760 cm^{-1}, respectively. The ratio of Tyr to Phe is similar for β_1-, β_2-, and β_3-crystallins, but differs markedly from the Tyr/Phe ratios of α- and γ-crystallins. The α-crystallin has much phenylalanine and little tyrosine, while

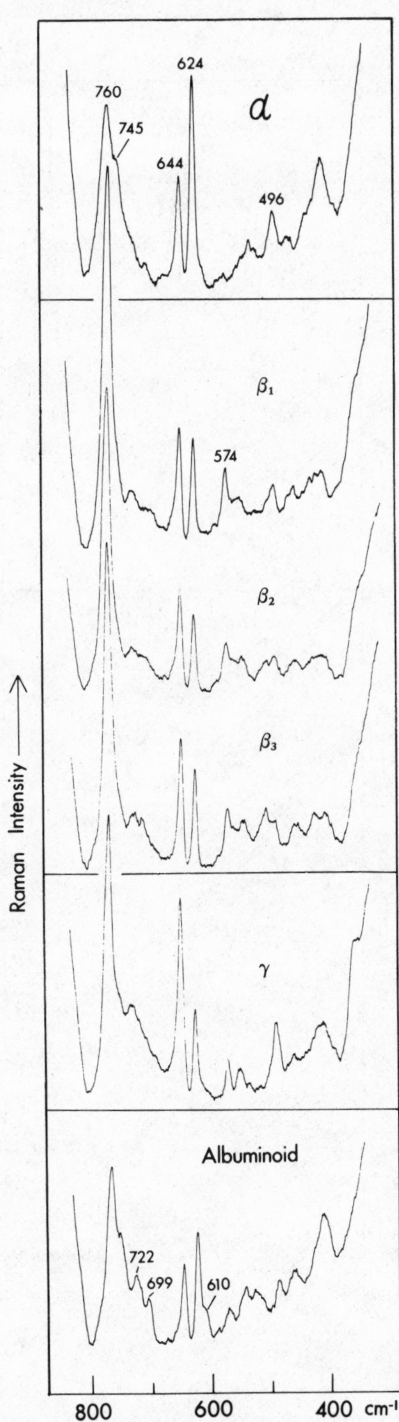

Fig. 4.1. Raman spectra of α_1, β_1, β_2, β_3, γ, and albuminoid (lyophilized) powder, 400–800-cm^{-1} region. Instrumental conditions: spectral slit width ($\Delta\sigma$), 4 cm^{-1}; sensitivity (s), 3×10^3 counts per sec full scale; rate of scan (γ), 0.5 cm^{-1}/sec; time constant (τ), 7 sec; laser power (p) at the sample approximately 80 mW. (Yu and East, 1975.)

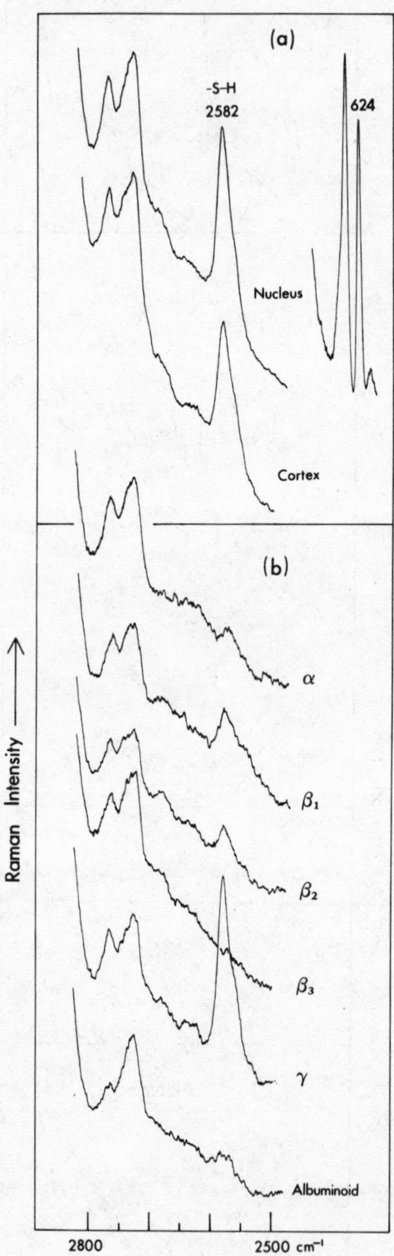

Fig. 4.2. Raman spectra in the 2500–2800-cm^{-1} region showing scattering from the —SH groups of intact lens and separated proteins. The phenylalanine line at 624 cm^{-1} from the nucleus is shown as a reference line. $\Delta\sigma$, 5.0 cm^{-1}; s, 3×10^3 cps, τ, 7 sec; γ, 0.5 cm^{-1}/sec; p = 180 mW for (a) and $\cong$ 80 mW for (b). (Yu and East, 1975.)

γ-crystallin contains little of the former and much of the latter. α-Crystallin and albuminoid display similar spectra. This suggested to these investigators that the albuminoid in bovine lens may be derived from α-crystallin. Chang (1976), however, studied rat lens proteins and found that rat albuminoid is very similar to γ-crystallin.

Figure 4.2 shows the interval of the spectra from 2500 to 2800 cm^{-1} for these fractions and for intact lens nucleus and cortex. Strong lines are displayed at 2582 cm^{-1} by the S—H stretching vibrations of the protein sulfhydryl groups. Note that the greatest intensity at 2582 cm^{-1} is displayed by the γ-crystallin, showing that almost all the —SH groups of this protein stay in the reduced form after chromatographic fractionation.

That γ-crystallin lacks disulfide formation is shown in Fig. 4.1 by the absence of a Raman line near 508 cm^{-1}. The spectra of β_1, β_2, and β_3 display increasing intensity at 512 cm^{-1} (tentatively assigned to the S—S bond), thus giving evidence of S—S bond content in the three β subfractions.

Table 4.2 shows the correlation between the chain conformation of proteins and amide I and III frequencies in Raman and IR spectra. Studies of synthetic polypeptides and proteins of established structure yielded the values in this table. The polypeptide chain conformations of α-, β-, and γ-crystallins are observed in the amide I (1650-1680 cm^{-1}) and amide III (1220–1300 cm^{-1}) regions, based on the following information: The α-helical conformation produces a strong amide I line at 1650 ± 5 cm^{-1}, and weak intensity in the amide III region; the antiparallel β-pleated sheet conformation displays an intense sharp line in the amide I region at 1670 ± 2 cm^{-1} and an intense amide III line at ca. 1240 cm^{-1}. The random coil or unstructured conformation displays a broad amide I line of medium intensity at ca. 1665 cm^{-1}, and a broad amide III line at ca. 1248 cm^{-1}. Figure 4.3, which shows α, β_1, γ, and albuminoid spectra, illustrates the fact that these isolated proteins have predominantly antiparallel β-pleated sheet conformations.

Table 4.2. Correlation between Protein Backbone Conformation and Amide I and III Frequencies[a,b]

	Amide I (cm^{-1})		Amide III (cm^{-1})	
Conformation	Raman	Infrared	Raman	Infrared
α-Helix	1660 ± 4	1657 S	>1264	~1262 S
β Structure: antiparallel	1673 ± 2 $\nu(0,0)$	1685 M $\nu(0,\pi)$ 1636 S $\nu(\pi,0)$	1227–1240	~1236 M
Parallel	~1655[c]	~1655[c]		
Random coil: solvated	~1655		1248 ± 4	
Non-hydrogen-bonded	1685		1235 ± 5	

[a] Yu and East (1975).
[b] Based on studies by T. J. Yu *et al.* (1973); Chen and Lord (1974); Frushour and Koenig (1974); N.-T. Yu *et al.* (1974); N.-T. Yu (1977); Yu and Liu (1972); and others.
[c] Theoretically predicted value (Krimm and Abe, 1972). At present, there is no experimental evidence for it.

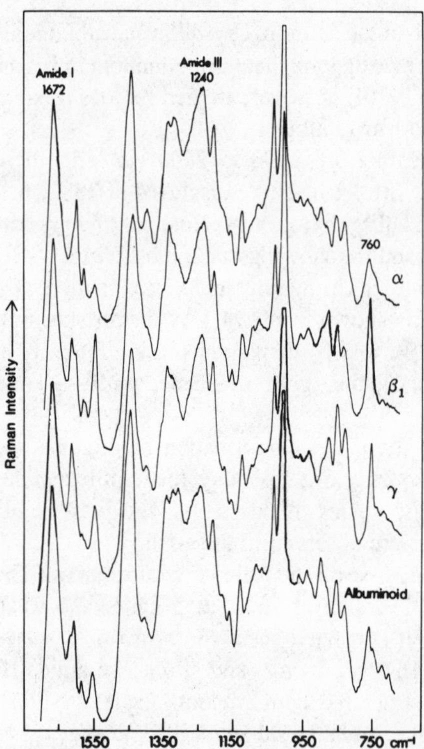

Fig. 4.3. Raman spectra of α, β_1, γ and albuminoid (lyophilized powder, 750 to 1700 cm^{-1} region). $\Delta\sigma$, 3.5 cm^{-1}, *s,* 3 × 10^3 cps, γ, 0.5 cm^{-1}/sec; τ, 7 sec; *p,* 80 mW. (Yu and East, 1975.)

Raman spectroscopy has been used by East *et al.* (1978) to study the relative concentrations of —SH and S—S in lenses surgically removed from rats and mice and maintained in a nutrient medium. These authors showed that —SH decreases while S—S increases correspondingly in the nucleus of the aging rat or mouse lens. These changes are caused only by postsynthetic modification of protein already present. The Raman lines used to identify —SH and S—S groups were as follows: 2580 cm^{-1}, —SH stretching vibration; 508 cm^{-1}, *gauche–gauche–gauche* form of disulfide cross links (S—S stretching vibration).

The ratio of tyrosine to phenylalanine was also measured by Raman spectra. The factor of 0.8 was used for converting I_{644}/I_{622} to this ratio (Yu and East, 1975), which was constant in the aging lens nucleus, an observation that supports the view that the amino acid composition of nuclear proteins remains as originally constituted. As the rat lens ages, the relative concentration of tyrosine–phenylalanine in the cortex decreases from 1.6 to 0.8, an observation correlating with the increasing proportion of α-crystallin (which contains relatively large amounts of Phe and small amounts of Tyr) and the decreasing proportion of γ-crystallin, which contains relatively small amounts of Phe and large amounts of Tyr.

These workers did not find that the lowered —SH concentration equaled the increase in S—S concentration; the decreased —SH was probably caused, they stated, by the faster synthesis of α- and β-crystallins, which have small amounts of —SH, and the slower synthesis of γ-crystallin, which contains large amounts of —SH.

When mice were continuously irradiated with UV light (300–400-nm region, with a maximum at ~353 nm), decreases in both —SH and S—S were produced in the lens nucleus and cortex.

To record the Raman spectra of an intact lens, the lens of the enucleated eye of a decapitated animal was immersed in culture medium in a glass tube (Yu *et al.*, 1977). A spectrum from a selected small zone (a column of 50 μm diameter × 0.6 mm length) of intact lens was obtained by focusing the laser beam (~50 μm) on that zone and by admitting Raman emission from only that zone to the spectrometer.

Askren *et al.* (1979) have used Raman spectroscopy to measure the variation of protein —SH and glutathione —SH levels along the visual axis of the lenses of rats and mice of different ages. Figure 4.4 shows an example of such visual-axis profiles for rat lenses of various ages. The ratio I_{2582}/I_{2731} is plotted against distance from the center of the lens (i.e., the distance in mm from the central axis to the position of laser-Raman sampling). Since lens anterior epithelium contains a higher concentration of glutathione than posterior capsule, the right side (with a higher end point) was labeled by these workers as "anterior" and the left side as "posterior." The intensity of the band at 2731 cm^{-1} was employed as an indicator of protein concentration, although its spectral origin is not known. [This band is present in isolated crystallins (Yu and East, 1975) and other proteins (unpublished data from Yu's laboratory)]. The intensity of the band at 2582 cm^{-1} indicates —SH concentration, and the ratio I_{2582}/I_{2731} allows comparisons to be made of the variation of total sulfhydryl concentration per unit protein along the visual axes of different lenses. Of general interest for analytical purposes is the fact that the laser Raman

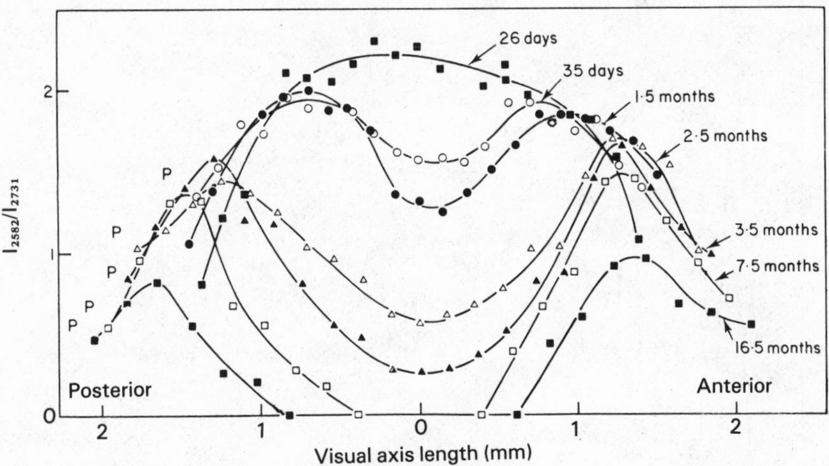

Fig. 4.4. Visual axis profiles for the rat lens. The intensity ratio I_{2582}/I_{2731} is plotted vs. position for seven rat lenses of varying age. The curves represent the distribution of sulfhydryl within the lens per unit protein. (Askren *et al.*, 1979.)

technique can be used to obtain structural information from a very small volume, ca. 1 × 10^{-3} μl, of an intact lens or some other piece of tissue.

The Raman spectra of the proteins of intact bird and reptile lenses have been recorded by Yu *et al.* (1977). δ-Crystallin, a protein which is unique to these classes, is the one present in the largest quantities, and exists mainly as α-helix, contrasting with the lens proteins of all the other species examined by these workers.

Figure 4.5 shows spectra of isolated chick δ-crystallin in aqueous solution and in the lyophilized state. Curve a is almost identical to the spectrum (not shown) of an intact 10-day-old chick lens. Raman lines at 1652, 940, and 525 cm^{-1} and the absence of an amide III line below 1264 cm^{-1} attest to the presence mostly of the α-helical conformation (estimated at about 75%, as corroborated by CD studies).

Parker and D'Agostino (1976) have examined human lenses and corneas by multiple internal reflection IR spectroscopy. They gave evidence for the presence of phospholipid (ester C═O, 1740 cm^{-1}) in normal lenses and much lower concentrations of phospholipid in cataractous lenses (see example in Figs. 4.6 and 4.7). Spectral evidence shows a loss of water in the opaque cornea compared to normal cornea. The amount of parallel-chain extended structure in the normal cornea is decreased in the opaque cornea. Also, the opaque cornea shows evidence of the destruction of mucopolysaccharide or glycoprotein when compared to normal cornea. Thus, an obvious disappearance of bands occurs (spectra not shown) in the region 1100–1000 cm^{-1} for opaque compared to normal cornea.

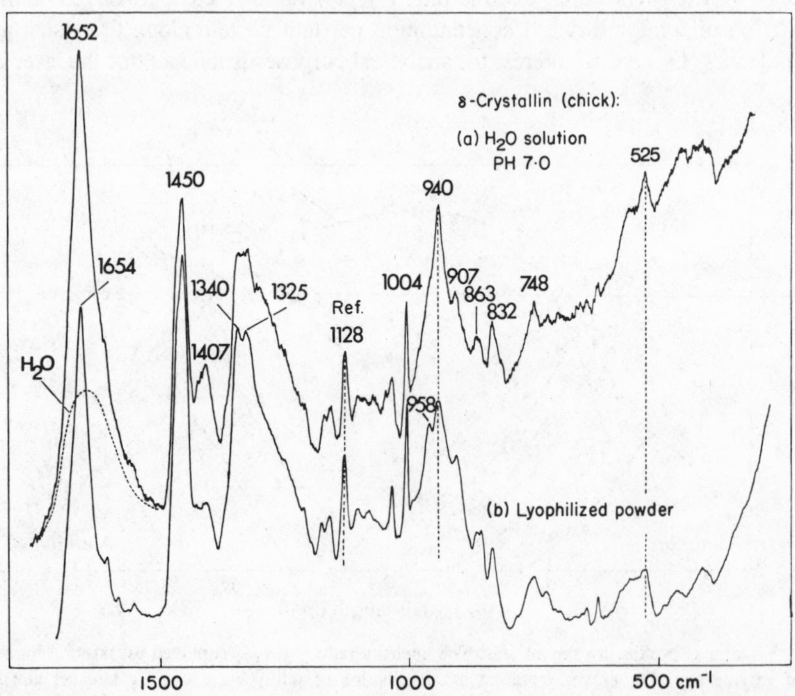

Fig. 4.5. Raman specura of isolated δ-crystallin in H_2O (pH 7.0) and lyophilized state. (Yu *et al.*, 1977.)

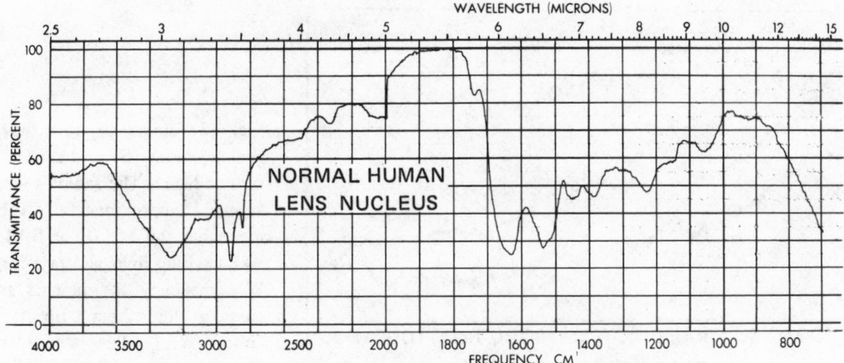

Fig. 4.6. Normal human lens nucleus. (Parker and D'Agostino, 1976.)

Raman spectra have been recorded for collagen from feline corneas in several states—fresh and intact, heat denatured, and incubated in ^2H_2O (Goheen *et al*, 1978). Bands at ca. 1630 and 1660 cm^{-1} and at ca. 1270 and 1247 cm^{-1} were displayed in the spectrum of the fresh and heat denatured collagen (Fig. 4.8). The latter two bands were assigned to amide III vibrations in the polar and nonpolar regions of the collagen. Corneas immersed in ^2H_2O for 30 h showed only one small amide I band (~1650 cm^{-1}), suggesting that some of the bands in this region for corneas in water are related to water vibrations.

Fluorphors in the Lens; Use of a Raman Spectrometer as a Fluorometer

Fluorescence in the human lens causes interference in measurements of Raman spectra (Kuck and Yu, 1978), as it does in other systems. These workers have used the fluorescence effect to advantage in the examination of a 23-year-old human lens. They discovered a new fluorphor which is excited maximally at 514.5 nm and has a broad

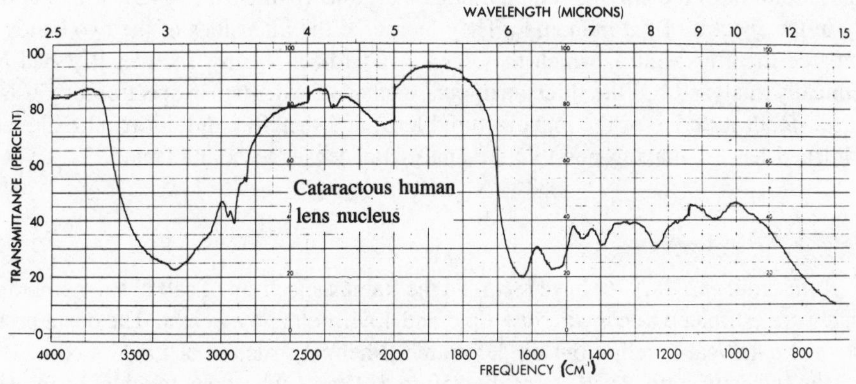

Fig. 4.7. Cataractous human lens nucleus. (Parker and D'Agostino, 1976.)

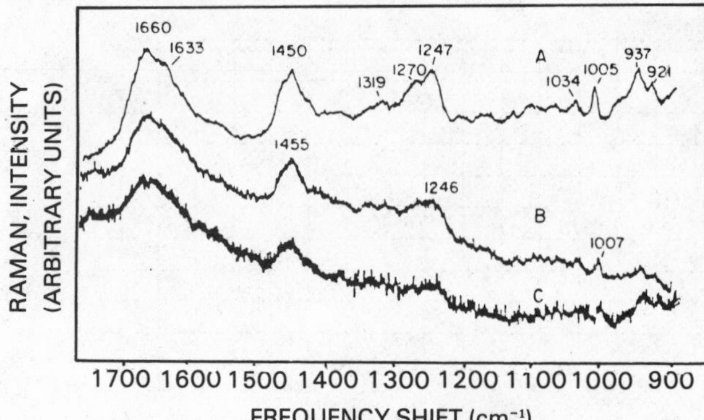

Fig. 4.8. Raman spectra of fresh (A), heat denatured (B), and aged (C) corneal collagen. The samples were examined at approximately 24°C by using either the 488.0- or 5.14.5-nm Ar^+ exciting lines at a power of 0.6 W. Frequency assignments are accurate to within 2 cm^{-1}. Spectra using either exciting line were identical. (Goheen *et al.*, 1978.)

emission peak with a maximum at 556.4 nm. Kuck and Yu attributed their success in finding the fluorphor to the use of a Raman spectrometer as a fluorometer, with a sensitivity ~100 times that of a conventional fluorometer having an uncooled photomultiplier tube. The study of fluorphors in the lens is of interest because they may be crucial compounds in the formation of protein aggregates of high molecular weight found in the nucleus of human nuclear cataracts. Tryptophan and bityrosine, as well as the reaction between two protein molecules (2 SH → S—S), and the reaction of glucose with the lysine of crystallins, have been considered in such involvement (see references in Kuck and Yu, 1978). The new fluorphor is not found in lenses seven years of age or younger; it is detectable at 23 years and is found in greater amounts progressively from 55 to 87 years of age.

Artificial Lens

Galin *et al.* (1977) have used IR and Raman spectroscopy and gas–liquid chromatography to ascertain the presence of methylmethacrylate in polymethylmethacrylate, which is used to manufacture intraocular lenses. They selected two spectral bands for analysis of implants—1730 cm^{-1} (ester carbonyl group), which appears in Raman and IR spectra of both the polymer and the monomer, and 1640 cm^{-1} (C═C), which appears only in the spectra of the monomer. They measured the intensities of the two bands and compared them as a ratio, which they then standardized against samples that had been chemically analyzed. Of the three analytical methods used, Raman spectroscopy was, of course, nondestructive of the manufactured lens, and therefore did not require alteration of finished lenses, making possible the analysis of lenses to be implanted.

Opsin Membranes

Rothschild *et al.* (1976) have recorded the Raman spectrum of calf opsin membranes, which were estimated to contain 50% lipid and 40% protein by weight. The opsin protein contained α-helical bonding but no detectable β-helical structure or random coil.

The intensity ratio, I_{850}/I_{831}, of the 850 to 831 cm^{-1} vibrations (doublet of tyrosine) is a sensitive indicator of hydrogen bonding to tyrosine (Siamwiza *et al.*, 1975). The ratio of these lines in opsin membranes is less than one, indicating that a significant fraction of tyrosine residues is strongly hydrogen bonded. The bonding may possibly be to aspartic and glutamic acid residues, which constitute 40% of the opsin polar residues (Daemen, 1973; De Grip, 1974). Rothschild *et al.* state that other work in their laboratories using Fourier transform IR absorption methodology indicates extensive α-helical structure and no detectable β-structure for both opsin and rhodopsin.

Hydrogen Isotope Exchange in Rhodopsin

In order to obtain unambiguous information about the solvent accessibility of rhodopsin's peptide hydrogens, Osborne and Nabedryk-Viala (1977) have studied by IR spectroscopy the degree of deuterium incorporation into membrane-bound rhodopsin (a transmembrane protein). This method offers an advantage over the tritium-exchange technique used by Downer and Englander (1975). The ^1H–^2H exchange method using IR spectroscopy measures directly and quantitatively the ^1H–^2H exchange of the peptide protons only, whereas the tritium (^3H) method measures peptide hydrogen exchange plus the exchange of all the labile protons, peptide and side chains.

The IR data showed that rhodopsin has a very low solvent accessibility and that a considerable number of its amide protons are buried in a core shielded from the solvent water.

The same authors (1978) have used the hydrogen–deuterium IR spectroscopic method and the tritium-exchange method to compare the conformation of bovine rhodopsin in its membrane-bound form with that in the detergent-solubilized form (nonionic detergents). The IR peptide exchange data showed that the highly hydrophobic character of rhodopsin is conserved in the presence of the two detergents used. Approximately half of the peptide hydrogens exchanged under conditions in which ca. 80% would exchange in most soluble proteins (Willumsen, 1971). When rhodopsin is bleached in the presence of these detergents, the exchange rates (measured by tritium-exchange kinetics) increase greatly, and a large conformational change occurs, as shown by shifts in the amide I band frequencies. Membrane-bound rhodopsin does not undergo such changes. These authors concluded that the conformation of rhodopsin is not changed by solubilization in the nonionic detergents.

The Visual Process. Rhodopsin and a Related Substance, Bacteriorhodopsin

Rhodopsin, the Visual Pigment of Vertebrates

Some of the interesting work concerned with the center of molecular activity in vision, the retinylidene chromophore, has led to a suggestion for the possible cause of the spectral perturbation of this chromophore by the opsin matrix, a confirmation that the chromophore and opsin are tied to each other by linkage through a Schiff base to a lysyl ϵ-amino group, and a finding that the Schiff base linkage is in the protonated form (Lewis, 1974; Lewis and Spoonhower, 1974; Lewis *et al.*, 1973).

Lewis and his co-workers obtained one of the first resonance Raman spectra of the retinylidene chromophore in an extract of bovine rhodopsin. Later, Lewis (1976a) obtained the spectrum with a live eye. By selectively observing the C—N vibration, he was able to demonstrate that the Schiff base is protonated, as shown in Fig. 4.9. Other workers have corroborated this finding of protonation of the Schiff base linkage (Oseroff and Callender, 1974; Mathies *et al.*, 1976; Callender *et al.*, 1976). It is a significant fact because, if the proton were not attached, the photochemical basis of visual transduction (conversion of light energy to chemical energy) would be altered dramatically.

When a photon interacts with rhodopsin, the energetics of the act can be formulated in two different ways, as shown in Fig. 4.10 (Lewis, 1976b). According to Lewis (1976b), if one assumes that the primary event involving the photon in vision is a simple *cis–trans* conversion, previous evidence from *cis–trans* isomerizations in other systems necessitates an assumption that bathorhodopsin is at a lower energy than the initial *cis*-rhodopsin (Fig. 4.10A). Lewis favors an alternative scheme of events (Fig. 4.10B) for what actually happens in vision. Here part of the energy of the 2-eV photon is absorbed by the *cis*-rhodopsin and is stored in bathorhodopsin. Then, the batho form is at a higher energy than the *cis*-retinal (Lewis *et al.*, 1974). Spectral experiments on bacteriorhodopsin (Busch *et al.*, 1972), which behaves similarly, further support this explanation. Rosenfeld *et al.* (1976) gave evidence supporting the idea that all bathorhodopsins are at higher energies than the *cis* forms.

In a later paper, Lewis (1978) does not consider the 11-*cis* to all-*trans* isomerization

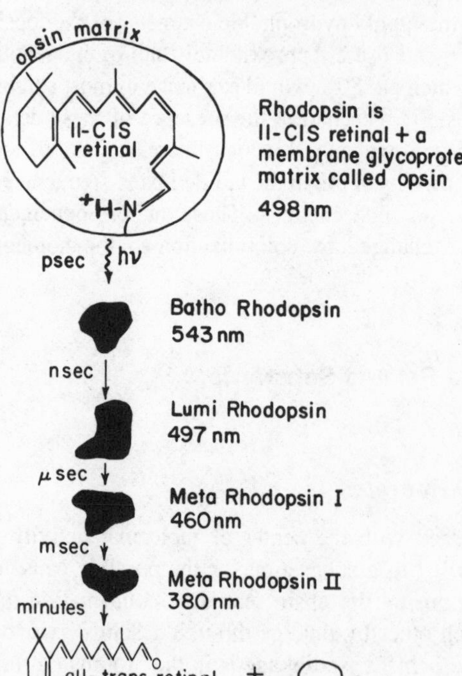

Fig. 4.9. The chemical composition and the light-induced conformational changes of rhodopsin, the principal light absorber in the eye. Rhodopsin is a large molecule consisting of retinal embedded in a matrix, opsin. The center of photochemical activity is the double-bonded segment of the retinal and those atoms electronically coupled to this region of the molecule. (Lewis, 1976b.)

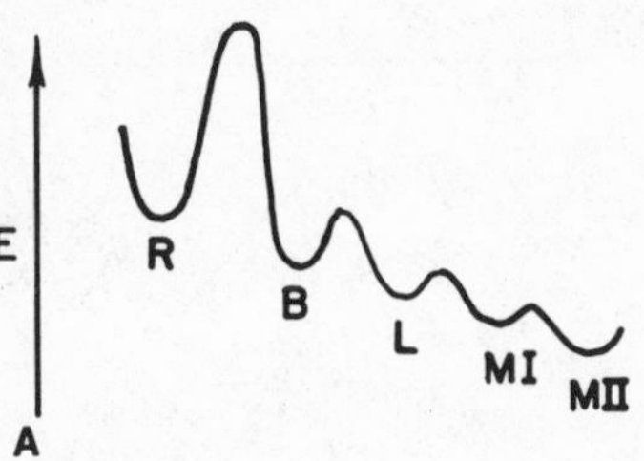

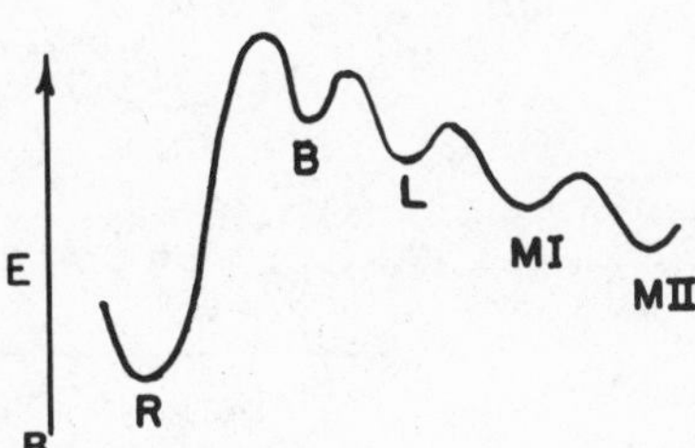

Fig. 4.10. Diagrams depicting two possible schemes for the relative energies of rhodopsin (R) and the intermediates (batho, B; lumi, L; Meta I, MI; and meta II, MII) generated as a function of time after light is absorbed. It can be shown that the lower diagram is correct. Here the part of the energy of the photon absorbed by rhodopsin in the first step is stored in B (the batho intermediate). This energizes the subsequent steps in the process. (Lewis, 1976b.)

of the retinylidene chromophore to be a primary mechanism of excitation. He proposes that an alternate biological role for this molecular process is to afford the irreversibility necessary for effective quantum detection on the time scale of a neural response. In the electronic theory of excitation put forth by him (1978) the primary action of light is to cause sufficient electron redistribution in the retinal to create thereby new interactions that vibrationally excite and perturb the conformation of the ground-state protein. A three-dimensional ground-state and excited-state energy surface (not shown here) is used by Lewis to define the molecular mechanism of excitation in rhodopsin and bacteriorhodopsin.

Lewis (1976a) has used resonance Raman spectroscopy to try to relate molecular changes in rhodopsin to the process of transduction (conversion of light energy into chemical energy) by performing a backscattering experiment on the live eye of an albino rabbit. He chose this kind of rabbit because it lacks a pigmented epithelium, which complicates the resonance Raman experiment. Figure 4.11 shows the spectrum of the retinylidene chromophore of rhodopsin in the live eye enhanced above the lipid and protein molecules in the eye, and the simultaneously recorded electroretinogram (ERG). In essence he observed no ERG, a finding which indicated, at least, that the reaction Meta I → Meta II was not being stimulated by the pulsed 614-nm laser experiments. If the laser power is increased from 5 to 10 mW without changing the pulsing frequency of the laser and the spot of illumination in the eye, then some of the rhodopsins can go through the bleaching sequence and these rhodopsins could then stimulate the primary ion movements and disk membrane changes that are basic to the process of visual transduction. If the laser power is increased under the above conditions, one observes (Fig. 4.11b) discrete changes in the spectrum of the retinylidene chromophore, and can detect an ERG. An increase of power to 15 mW (Fig. 4.11c) causes even greater changes in

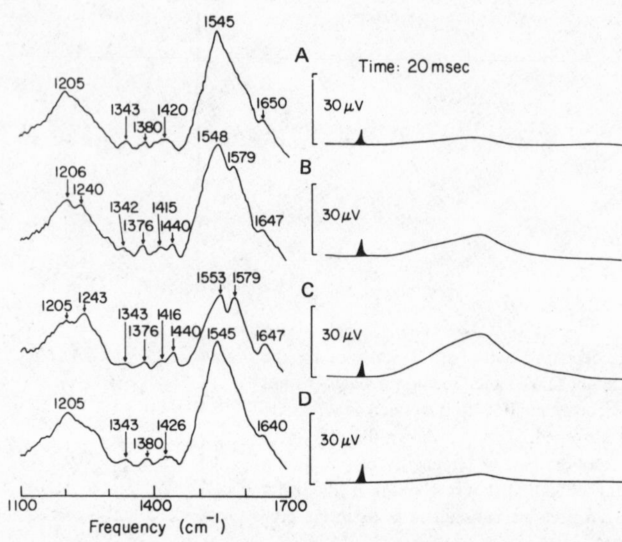

Fig. 4.11. Tunable laser resonance Raman spectra of the retinylidene chromophore in the live eye of an albino rabbit with a retrobulbar injection of 20 cm^3 Xylocaine and 1 cm^3 hyaluronidase to prevent eye movement. (a) The Raman spectra and the simultaneously detected electroretinogram obtained with a pulsed 614-nm emission of a tunable dye laser with 5 mW of continuous power. (b) Same as (a) except 10 mW of continuous power. (c) Same as (a) except 15 mW of continuous power. (d) Same as (a). (Lewis, 1976a.)

the spectrum to occur and an even larger ERG than before. A return to the power used for Fig. 4.11a produces (Fig. 4.11d) essentially the same spectrum as in Fig. 4.11a and the ERG is lost. The spectra in Figs. 4.11b and c cannot be duplicated by the spectra of the thermal intermediates of rhodopsin (Lewis, 1976a). The fact that the spectral changes seen in Fig. 4.11 are reversible in the retina within 40 sec after the laser power change argues against thermal intermediates being responsible for the changes in Figs. 4.11b and c.

An alternative explanation offered by Lewis is that the spectra of the retinylidene chromophores surrounding those rhodopsin molecules responsible for stimulating the visual response are indicative of the primary ion movements and the membrane changes accompanying the visual process.

Lewis mildly sonicated a suspension of rod outer segments, and when he increased the H^+ concentration in the suspension from pH 7 to pH 5, he began to induce changes in the retinylidene chromophore spectra similar to the ones seen in Figs. 4.11b and 4.11c. Before the pH was changed, the spectra of the sonicated rod outer segments looked similar to those seen in Figs. 4.11a and 4.11d. A comparison of the *in vivo* and *in vitro* data indicates that protons are released within the disk. According to Lewis, these proton movements are apparently the primary result of the energy transduction process.

The bleaching sequence of rhodopsin is given in Fig. 4.12, which shows transitions in the sequence rhodopsin, bathorhodopsin, lumirhodopsin, metarhodopsin I (meta I), metarhodopsin II (meta II), all-*trans* retinal plus opsin, and isorhodopsin (Doukas *et al.*,

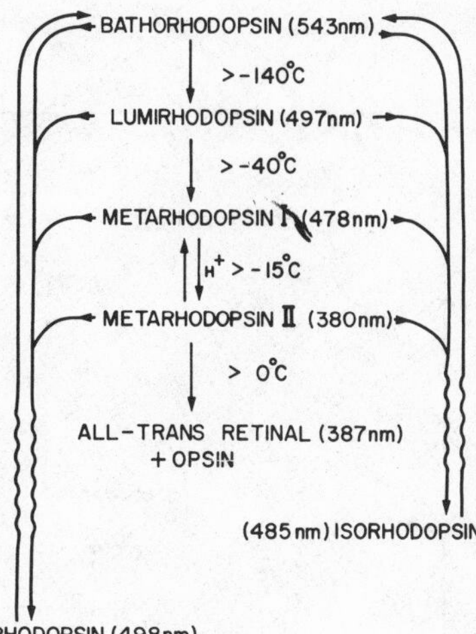

Fig. 4.12. The bleaching sequence of rhodopsin. Rhodopsin and isorhodopsin are placed lowest in the figure to indicate they have lower free energy than their common photoproducts. Wavy lines indicate photoreactions and straight lines thermal reactions. (Doukas *et al.*, 1978.) Reprinted with the permission of the American Chemical Society. Copyright 1978.

1978). These authors have recorded the resonance Raman spectra of bovine metarhodopsins I and II (Fig. 4.13) by using a flow technique, in which less than 5% of the Raman scattering arose from photoconverted material in the laser beam for the metarhodopsin I samples and possibly an even lower percentage for metarhodopsin II samples. By comparing these spectra with those of all-*trans*-retinal-*n*-butylamine hydrochloride and all-*trans*-retinal-*n*-butylamine, they found that metarhodopsin I is bonded to opsin by means of a protonated Schiff base linkage, while metarhodopsin II is linked to an unprotonated Schiff base. In the case of meta I, the HC═NH$^+$ bond stretching is located at 1656 cm^{-1} (Heyde *et al.*, 1971; Oseroff and Callender, 1974) and assignment was made on this basis. For meta II, assignment was made by comparison with a weak 1632 cm^{-1} band of unprotonated all-*trans*-retinal *n*-butylamine (not shown).

The Raman data of Doukas *et al.* (1978) have given direct evidence supporting the argument of Hubbard and Kropf (1958) that the chromophore in metarhodopsin I (Fig. 4.12) is in the all-*trans* conformation. The resonance Raman spectra of meta I and meta II (Fig. 4.13) are nearly identical with the spectra (not shown) of protonated and unprotonated all-*trans*-retinal-*n*-butylamine, respectively. The butylamine residue acts as a model for the ε-lysine group by which retinal is attached to opsin in the pigment.

Studies aimed at unraveling the complex resonance Raman (vibrational) spectrum of rhodopsin have been undertaken by Cookingham and Lewis (1978). They examined chemically modified retinals, a series of analogs of *n*-butyl retinals substituted at C9 and C13. The spectral data for the 11-*cis* isomer allowed them to assign the C—CH_3 vibrational frequencies seen in the spectrum of the retinylidene chromophore. They assigned the C9—CH_3 stretching vibration to the vibrational mode found in the 1017-cm^{-1} region; and the C13—CH_3 stretching vibration to the band at 997 cm^{-1}. The splitting in the C(*n*)—CH_3 vibration (*n* is 9 or 13) is characteristic of the 11-*cis* conformation (Fig. 4.14). With the structural changes involving *n*-butyl group substitutions at the C9 and/or C13 positions, these workers expected to see a striking effect between 1040 and 980 cm^{-1} (CH_3-stretching region). Figure 4.14, curve B shows this dramatic effect for the

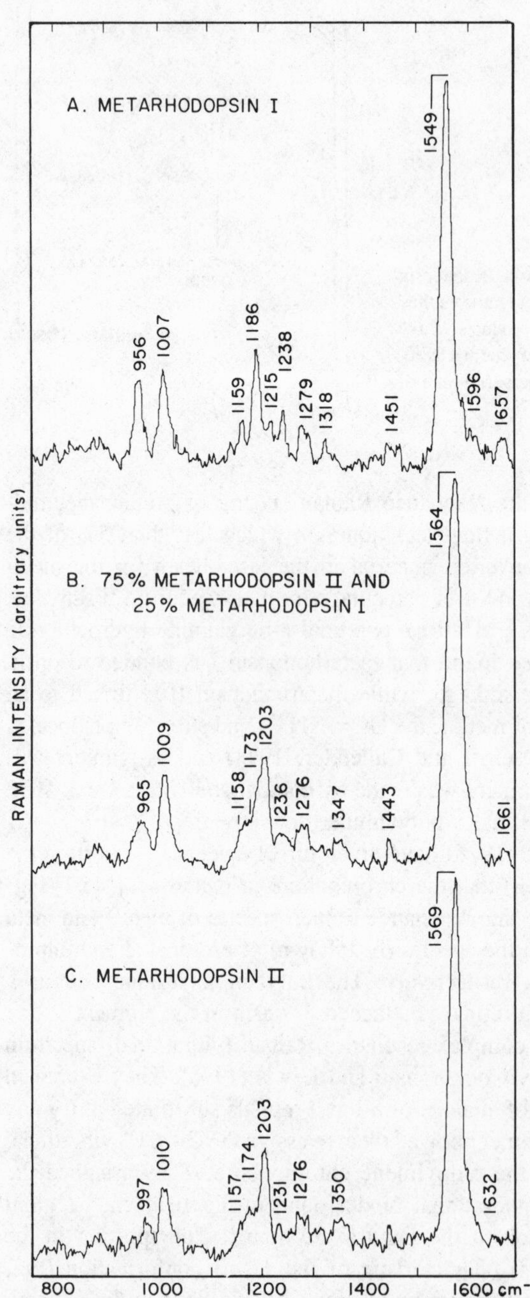

Fig. 4.13. Resonance Raman spectra of (A) metarhodopsin I, (B) mixture of 75% metarhodopsin II and 25% metarhodopsin I, and (C) essentially metarhodopsin II (spectra of A subtracted from spectra of B). The data were taken with a spectrometer resolution of 6 cm^{-1}. (Doukas *et al.*, 1978.) Reprinted with the permission of the American Chemical Society. Copyright 1978.

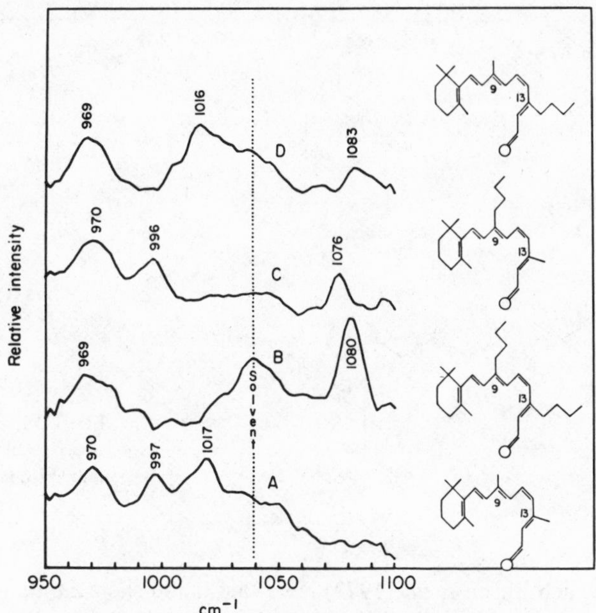

Fig. 4.14. The methyl stretching region of the Raman spectra of the 11-*cis* isomers of (curve A) retinal; (curve B) 9,13-dibutyl-retinal; (curve C) 9-butyl-retinal; (curve D) 13-butyl-retinal. (The solvent band of CH_3CN has not been removed from the spectrum shown. This band occurs at 1039 cm^{-1}.) Spectra were taken at room temperature in CH_3CN solvent with 2 cm^{-1} resolution. All labeled bands are accurate to $\pm$ 2 cm^{-1} and have been observed in multiple experiments. Probable schematic plane representations of the 11-*cis* isomers of each molecule are depicted. (Cookingham and Lewis, 1978.)

11-*cis*-dibutyl-retinal analog, where these two vibrational modes are completely missing from this region. Figure 4.14, curve C shows the spectrum of 11-*cis*-9-butyl retinal, which has a *n*-butyl group substituted at the C9 position. No band occurs near 1017 cm^{-1}, so the 996-cm^{-1} band was assigned to the C13—CH_3 stretching vibration. On the other hand, curve D shows the spectrum of 11-*cis*-13-butyl retinal, which displays no band near 997 cm^{-1} but does show one at 1016 cm^{-1}. The 1017-cm^{-1} band was assigned to the C9—CH_3 stretching vibration.

The results found for the modified retinals do not support the hypothesis of Callender *et al.* (1976), which predicts a degenerate C9—CH_3 and C13—CH_3 (12s-*cis*) vibration at 1018 cm^{-1} and a C13—CH_3 (12s-*trans*) vibration at 998 cm^{-1}, nor do the results support the idea that the splitting of the above-mentioned stretching mode (C9—CH_3 and C13—CH_3) originates from equilibrium mixtures of 11-*cis,* 12-s-*cis* and 11-*cis,* 12-s-*trans* in solution (Fig. 4.15).

Cookingham and Lewis (1978) used the presence of a strong band near 1271 cm^{-1} in bovine and squid rhodopsin spectra as evidence of 11-*cis,* 12-s-*trans* structure for the chromophore in these pigments.

Picosecond-Range Studies

Some picosecond-range investigations of the primary photochemical event that occurs when a visual pigment absorbs a photon have caused uncertainty about the validity of the original model of the event, which involves an isomerization of the chromophore from an 11-*cis* to an all-*trans* conformation (Hubbard and Kropf, 1958; Yoshizawa and Wald, 1963). The primary event takes less than 6 psec for completion at room temperature,

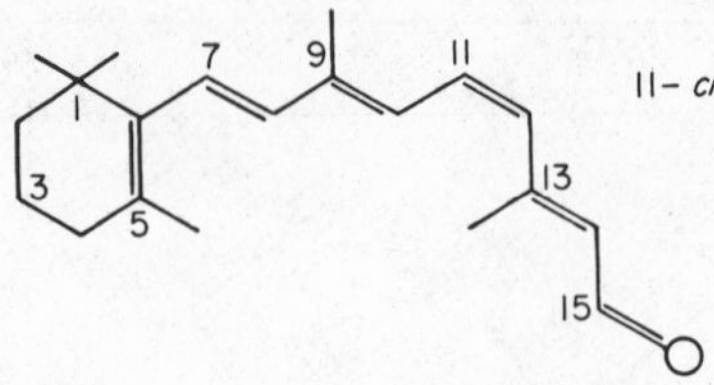

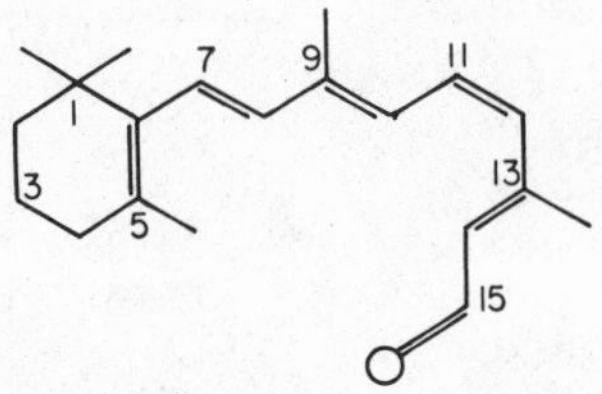

Fig. 4.15. Schematic plane representations of the two possible solution conformations of the 11-*cis*-retinal isomer. (Cookingham and Lewis, 1978.)

which Busch *et al.* (1972) have said is too short a time to change from the 11-*cis* to the all-*trans* form. However, Green *et al.* (1977) have concluded that photoisomerization times are greatly accelerated by chromophore–protein interactions, and *cis–trans* isomerization does occur in picoseconds. In 1977 Peters *et al.* observed that at low temperature, if the exchangeable proteins on the pigment are replaced by deuterium, the rate of the isomerization is inhibited significantly. Honig *et al.* (1979) have stated that, because only one proton on the chromophore is exchangeable, it is unlikely that this would have a measurable effect on the rate of isomerization.

Several studies in the picosecond range have led to models (van der Meer *et al.*, 1976; Peters *et al.*, 1977; Lewis, 1978; Warshel, 1978; Favrot *et al.*, 1979) that portray a photochemical proton transfer succeeded by a thermal *cis–trans* isomerization in a subsequent step. However, Honig *et al.* (1979) have pointed out that the original evidence supporting the idea of *cis–trans* isomerization has never been discredited, and therefore have presented a model that is consistent with all the evidence at hand.

The model (Fig. 4.16) proposed by Honig *et al.* (1979) for the early events in visual pigments and bacteriorhodopsin involves formation of a salt bridge between a negatively charged amino acid and the N^+ of the retinylic chromophore. In visual pigments the photochemical event is a *cis–trans* isomerization and in bacteriorhodopsin a *trans–cis* one. In each instance this results in cleavage of the salt bridge and charge separation within the protein. Evidence for the existence of interactions such as the one between an amino acid and the positively charged nitrogen has been obtained from resonance Raman spectroscopy, which has demonstrated protonation of the Schiff base of the chromophore (Oseroff and Callender, 1974; Sulkes *et al.*, 1978; Eyring and Mathies, 1979; Aton *et al.*, 1979).

The findings of Monger *et al.* (1979) have provided additional support for the *cis–trans* isomerization model of the primary event in vision. They excited bovine rho-

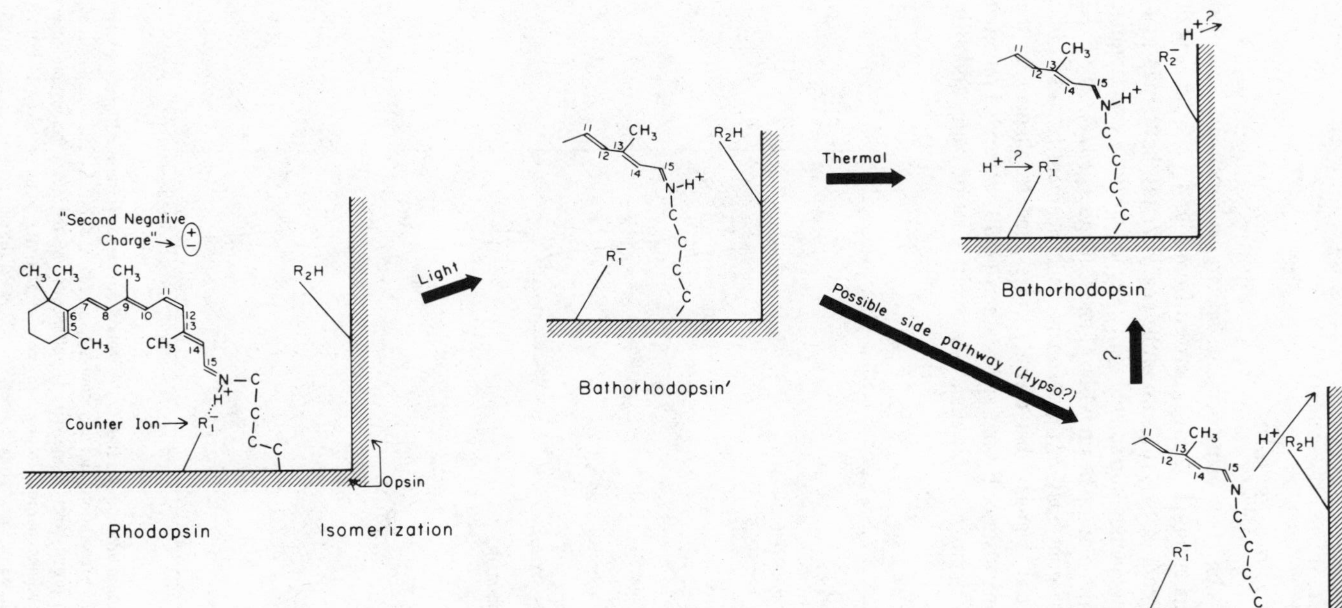

Fig. 4.16. Model for the early events in visual excitation. The 11-*cis* chromophore of rhodopsin is depicted with its Schiff base forming a salt bridge with a negative counter ion. The additional charge pair near the 11═12 double bond represents the group (or groups) that regulate the absorption maxima of different pigments. (Honig *et al.*, 1976.) The photochemical event is an isomerization about the 11═12 double bond in rhodopsin (probably about the 13═14 bond in bacteriorhodopsin), but any isomerization in any direction will produce charge separation as shown in the first step in the figure. The p*K* values of the Schiff base and those of other groups on the protein, such as R_1 and R_2, are strongly affected by photoisomerization because a salt bridge is broken, a positive charge has moved near R_2, and R_1 is now a bare negative charge. Possible proton transfer steps resulting from charge separation are depicted. For bacteriorhodopsin, the isomerization is *trans–cis* rather than *cis–trans,* but all other events are assumed to be equivalent. The proton transferred in the Batho′–Batho transition could be pumped in bacteriorhodopsin, as might the Schiff base proton if it were released at a later stage. The number of protons pumped in bacteriorhodopsin has not been firmly established. Batho′ is the transient observed by Peters *et al.* (1977). (Honig *et al.*, 1979.) Reprinted with the permission of the American Chemical Society. Copyright 1976.

dopsin and isorhodopsin with a single 530-nm, 7-psec light pulse and observed within 3 psec of excitation, absorbance changes caused by the formation of bathorhodopsin.

The Complete Bathorhodopsin Spectrum

Aton *et al.* (1980) have been able, using accurate spectral subtractions, to obtain the complete bathorhodopsin spectrum (Fig. 4.18, derived from Fig. 4.17) by means of a double-beam "pump-probe" method. A 476.2-nm probe beam excites the Raman scattering, while the sample is simultaneously irradiated with a second spatially overlapping, variable wavelength laser beam, the pump beam (Oseroff and Callender, 1974). The fixed probe wavelength maintains the constancy of the resonance enhancement factors, while the pump beam induces spectral changes that directly correspond to changes in the composition of the photostationary state. Eyring and Mathies (1979) have also reported a spectrum of bathorhodopsin, similar to this one. The Schiff base of bathorhodopsin

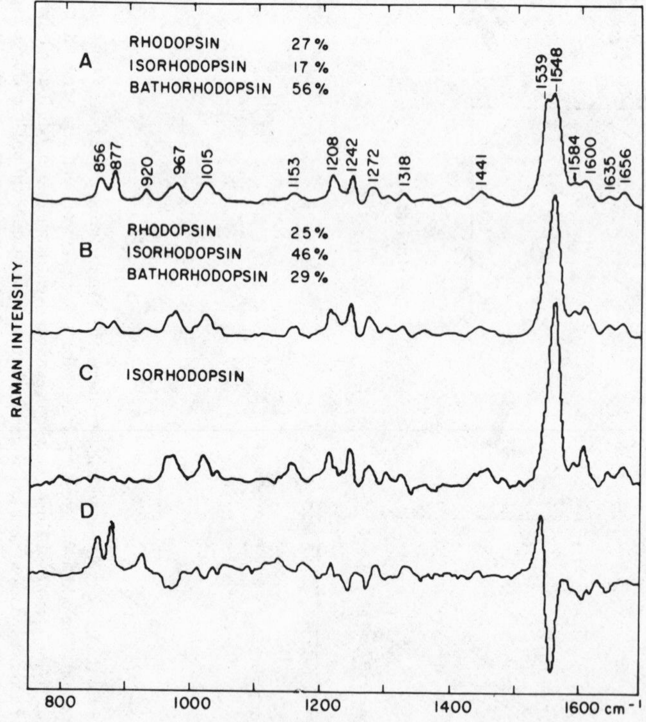

Fig. 4.17. The Raman data and one subtraction step leading to the bathorhodopsin spectrum. All spectra were taken with a 476.2-nm probe, at 7.3 cm^{-1} resolution, and at a sample temperature of 80° K. The compositions are given in the figure. (A) 476.2-nm probe beam alone. (B) Simultaneous 580-nm pump beam applied at seven times pump/probe ratio. (C) Simultaneous 568.2-nm pump beam at 25 times pump/probe ratio. (D) A − 1.08 B. A, B, and C are scaled for the same input probe power; since D is a different spectrum and relatively small, its scale has been increased by a factor of 3 for graphical purposes. (Aton *et al.*, 1980.)

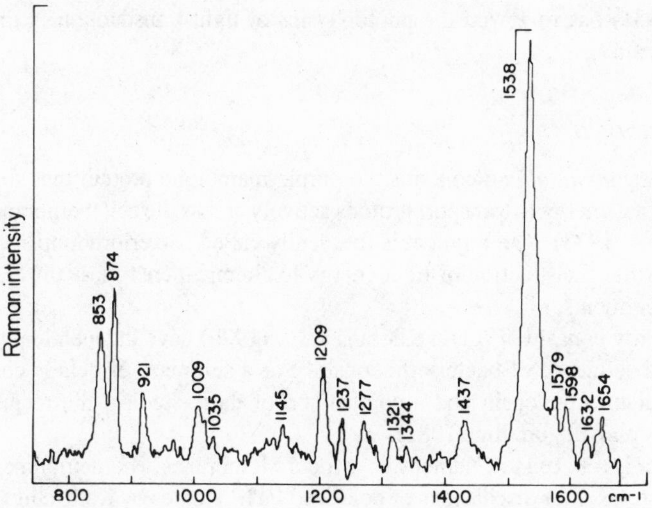

Fig. 4.18. The bathorhodopsin spectrum. Represents (Fig. 4.17D) + 0.25 (Fig. 4.17C). Sample temp. = 80 K; 476.2 nm probe. (Aton *et al.*, 1980.)

was shown by Aton *et al.* and Eyring and Mathies to be fully protonated and the extent of its protonation unchanged by its formation photochemically from either rhodopsin or isorhodopsin. The fact that the spectrum of bathorhodopsin differs from those of its precursor pigments lends support to the idea that a change in the geometry of the chromophore is significantly involved in the primary light-stimulated event in vision.

The results of Aton *et al.* and Eyring and Mathies show that the Schiff base nitrogen atom is *covalently* linked to the proton of rhodopsin and of bathorhodopsin, and disagree with models of the visual system that employ a changing state of protonation (H-bonding) of the nitrogen in the primary step. N—H bond strength does not change when bathorhodopsin is produced (Eyring and Mathies, 1979). The data of Aton *et al.* also suggest a *trans*-like conformation for bathorhodopsin, but not an all-*trans* one. The photochemical event in visual pigments, according to a model of Honig *et al.* (1979), is a *cis–trans* isomerization, and in bacteriorhodopsin it is *trans–cis*. In each case it is postulated that a salt bridge between the plus-charged N of the retinylic chromophore is cleaved, thus causing charge separation within the protein, with consequent increase in energy and a spectral red shift (see page 172).

Figure 4.17C shows a spectrum of essentially (95%) pure isorhodopsin. The procedure used here by Aton *et al.* (1980), by Oseroff and Callender (1974), and by Eyring and Mathies (1979) determines the relative pigment concentrations to no better than ± 5%, and therefore the spectrum of bathorhodopsin can have present some remnants of other pigments (about ± 10%–15%).

Mathies (1980) has been able to perform resonance Raman microscopy on photoreceptor cells and has obtained spectra of rhodopsin and bathorhodopsin in individual intact red rods of the toad. He claims 100-fold improvement in efficiency over conventional Raman spectrometers (Mathies and Yu, 1978).

Birge (1981) has reviewed the photophysics of light transduction in rhodopsin and bacteriorhodopsin.

Bacteriorhodopsin

Halobacterium halobium contains the purple membrane protein that absorbs visible light and uses that energy to transport protons actively across the cell membrane (Oesterhelt and Stoeckenius, 1973). The pigment is frequently called bacteriorhodopsin, although its physiological role, transduction of light energy to chemical energy, is different from that of the visual pigments.

Ovchinnikov *et al.* (1979) and Khorana *et al.* (1979) have independently determined the amino acid sequence of bacteriorhodopsin. For a summary of details concerning the structure of bacteriorhodopsin and a comparison of these studies, the reader is referred to Stoeckenius and Bogomolni (1982).

The reader is referred to Chapter 11, "Model Membranes, Biomembranes, and Lipid-containing Systems," for discussions of polarized FTIR studies by Rothschild *et al.* (1980) on the structure of bovine photoreceptor membrane and by Rothschild and Clark (1979) on the structure of the light-adapted form of the purple membrane from *Halobacterium halobium*.

When adapted to light, the purple membrane protein absorbs at about 568 nm. When this pigment (PM568) absorbs light, it changes to the batho form, which absorbs at longer wavelengths than the pigment. When the batho form is heated above a temperature where it is stable (77 K), it goes through a series of thermal (dark) reactions to form an intermediate at 412 nm, M412, which cycles back to PM568 ultimately (Lozier *et al.*, 1975). A proton is actively pumped across the cell membrane during the cycle (Oesterhelt and Stoeckenius, 1973).

Aton *et al.* (1977) have used resonance Raman spectroscopy to study the chromophores of the M412 and PM568 forms of light-adapted purple membrane. These authors used a flow technique (Callender (*et al.*, 1976; Mathies *et al.* 1976) to avoid the problem of serious perturbation of the composition of the sample by the laser light.

The main point made by several previous resonance Raman studies (Lewis *et al.*, 1974; Mendelsohn, 1973, 1976; Mendelsohn *et al.*, 1974) was that the PM568 chromophore is linked to the protein by a protonated Schiff base, but the M412 linkage is not protonated. Aton *et al.* used essentially pure PM568 and M412 forms of the pigment, and obtained Raman data on model chromophores, namely, all-*trans* and 13-*cis*-retinal *n*-butylamine, both as protonated and nonprotonated Schiff bases. They confirmed work by others by showing that lines at 1644 cm^{-1} for PM568, and 1623 cm^{-1} for M412 correspond to protonated and nonprotonated Schiff base (C═N) modes, respectively; and by showing that lines at 1532 cm^{-1} (PM568) and 1572 cm^{-1} (M412) correspond to C═C stretching vibrations of the retinal polyene chain.

As first shown by Rimai *et al.* (1971), the line position of the C═C stretching mode is a sensitive indication of electron delocalization (see discussion of electron delocalization in Honig *et al.*, 1976), since with increased delocalization the bond order is decreased, leading to a shift to lower frequency of the C═C band. Figure 4.19 shows that a good correlation exists between the C═C stretching frequencies of both protonated and non-

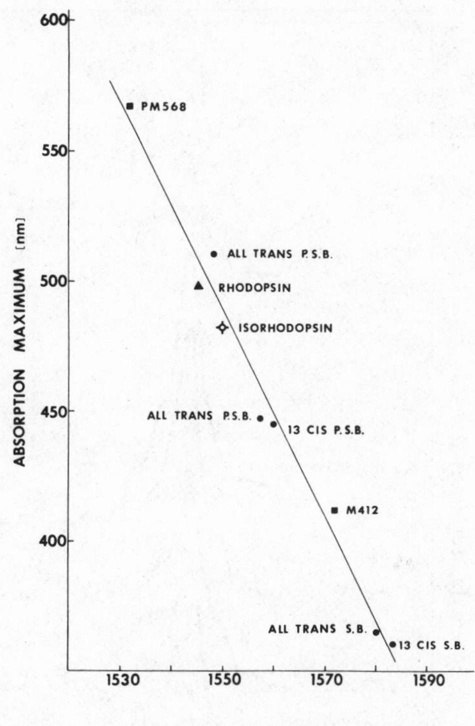

Fig. 4.19. Correlation of ethylenic (C═C) stretching frequency of retinal-based structures with their absorption maxima. All data are from figures in Aton *et al.* (1977), except the all-*trans*-protonated Schiff base absorbing at 512 nm (Heyde *et al.*, 1971) and rhodopsin and isorhodopsin (Oseroff and Callender, 1974; Callender *et al.*, 1976). p.s.b. = protonated Schiff base; s.b. = Schiff base. (Aton *et al.*, 1977.) Reprinted with the permission of the American Chemical Society. Copyright 1977.

protonated Schiff bases, as well as rhodopsin and isorhodopsin, and their λ_{max} values (see also Rimai *et al.*, 1971).

Kakitani *et al.* (1982) have pointed out that, although the frequency of the intense C═C stretching mode appearing in Raman spectra has been widely used and interpreted as a measure of π electron delocalization; and although an inverse correlation between absorption maxima and C═C stretching frequencies is expected and is experimentally observed, no similar relationship exists for the C═N stretching mode of the protonated Schiff base of retinal even though it too participates in the π electron system of the chromophore. Among the factors contributing to this anomalous behavior of the C═N stretch, these authors included coupling to bending modes and in particular to the adjacent C—C single bond. This bond is quite sensitive to changes in π electron delocalization and is involved in reversing the effects expected from the C═N stretch alone.

The kinetic resonance Raman technique of Marcus and Lewis (1977) combines resonance Raman spectroscopy and variable-speed continuous flow methods. These workers took kinetic spectra of bacteriorhodopsin with 30 mW of focused, 457.9-nm laser light for various transit times in the beam (beam diameter divided by bulk flow velocity). These spectra are presented in Fig. 4.20, (B) through (F). The 457.9-nm laser light provided selective enhancement of the M412 intermediate. Raman flow methods had been employed earlier by other investigators (Callender *et al.*, 1976; Mathies *et al.*, 1976) with the aim of observing the spectrum of rhodopsin with continuous instead of pulsed

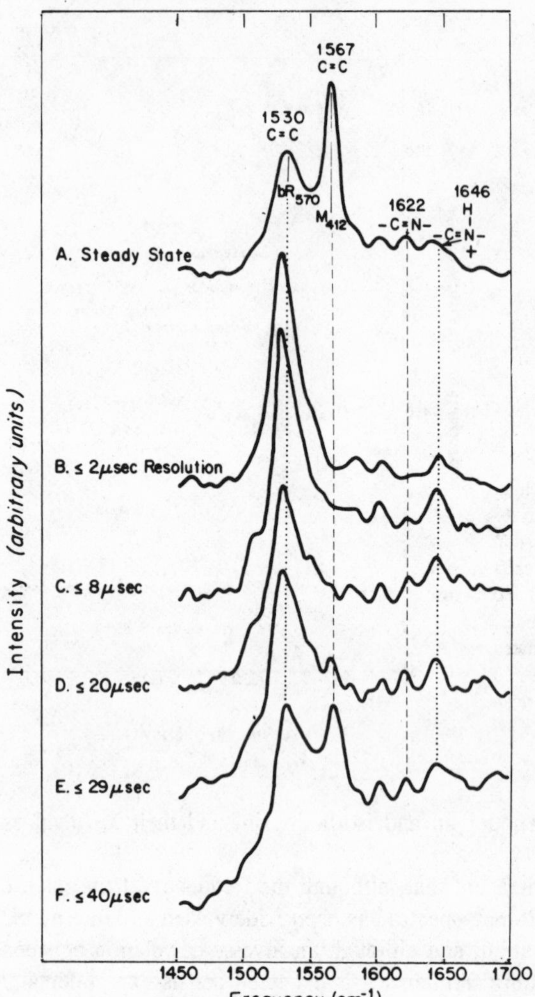

Fig. 4.20. Steady state and kinetic resonance Raman spectra of bacteriorhodopsin from *Halobacterium halobium* (S9) grown and isolated by the method of Kanner and Racker (1975). Bacteriorhodopsin (optical density 1.5) was flowed through stainless steel Yale syringe needles of varying diameter by a Cole Parmer masterflex tubing pump and a suitable reservoir to ensure that a bacteriorhodopsin molecule passed through the laser beam only once every few seconds. The vertically flowing bacteriorhodopsin was excited with 30 mW of 457.9-nm light from an argon ion laser focused to a 35 ± 2 μm beam diameter. The temperature of the sample was 29°C (thermocouple attached to the syringe needle). A JY Ramanor double monochromator with horizontal slits and an RCA C31034 photomultiplier tube were used to collect the 90° scattered light. Photon-counting techniques were employed, and a ModComp II computer was used for data reduction. All data presented were taken with two wavenumber steps, a 5-sec time constant, and are plotted with arbitrary units. Typically, the largest peak is 2.5 times the background. The spectra shown are only over a limited frequency region. The bands at 1646 cm^{-1} in steady state experiment are shifted to a slightly lower frequency. This is probably due to contributions from other protonated spectral intermediates. (Marcus and Lewis, 1977.)

laser beams, but they used low laser powers and frequencies of light for which the cross section of absorption was very low. Also, they did not vary the flow velocity enough to be able to observe the kinetics of formation and decay of rhodopsin and its spectral intermediates [see Fig. 4.21, which shows the proposed photochemical cycle of light-adapted bacteriorhodopsin (Stoeckenius and Lozier, 1974; Dencher and Wilms, 1975; Kung *et al.*, 1975; Chance *et al.*, 1975; and Lozier *et al.*, 1975)].

Marcus and Lewis (1977) excited the bacteriorhodopsin photosystem and observed the kinetics of the resulting changes by using greater laser powers, by exciting with laser wavelengths near the absorption maxima of the intermediates being studied, and by varying the flow velocities. The bacteriorhodopsin molecules enter the focused laser beam (35 ± 3 μm) in the 570-nm state. Many of these molecules absorb light and start to go

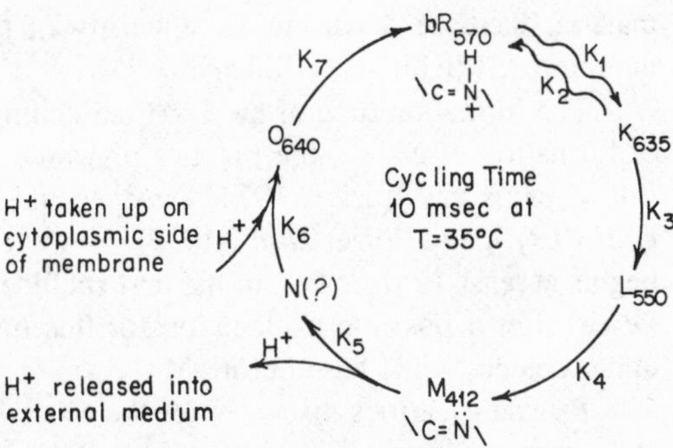

Fig. 4.21. The proposed photochemical cycle of light-adapted bacteriorhodopsin (Stoeckenius and Lozier, 1974; Dencher and Wilms, 1975; Kung *et al.*, 1975; Chance *et al.*, 1975; Lozier *et al.*, 1975.) The letters (K), (L), (M), (N), (O), and (bR) refer to distinct species, and the numbers to absorption spectra maxima. The absorption maximum of K varies in the literature from 590 to 640 nm. The first reaction is driven by light in the forward and reverse directions. All the other intermediates are the result of thermal change in the dark. K_{635} is batho-bacteriorhodopsin. It decays to L_{550} in the dark. M_{412} is the next intermediate. N(?) and O_{640} are two probable intermediates between the decay of M_{412} and the rise of bR_{570}, which takes place at physiological temperatures. (Marcus and Lewis, 1977.)

through the intermediate forms of the cycle induced photochemically. The concentrations of these intermediates are different throughout the beam. As a result, the recorded Raman scattering from any one species is related to the average concentration of that species in the beam. Thus, by changing the flow velocity these workers could obtain kinetic resonance Raman spectra having a time resolution of 500 nanoseconds.

Figure 4.20A shows the spectrum of the same bacteriorhodopsin sample recorded with steady-state conditions and the identical settings of the instrument that were used for recording spectra (B) through (F) in Fig. 4.20, and the same continuous laser frequency. Earlier work had shown that the 1530- and 1567-cm^{-1} bands in the steady-state spectrum (Fig. 4.20A) originate from the C=C stretching frequencies for bR570 and M412, respectively (Lewis *et al.*, 1974; Mendelsohn, 1973); the 1646-cm^{-1} band comes from the protonated Schiff-base linkage, and the 1622-cm^{-1} band represents an unprotonated linkage (Oseroff and Callender, 1974). However, when the bacteriorhodopsin spectrum was observed under rapid-flow conditions so that the sample remained in the beam for 2 μsec or less, the recorded spectra were identical, e.g., as in Fig. 4.20B. Note that bands assigned to M412 are absent at 1567 and 1622 cm^{-1}, thus confirming that the 1530- and 1646-cm^{-1} bands originate from bR570, a protonated Schiff base (Oseroff and Callender, 1974).

Figure 4.20C shows, however, that when the time resolution is 8 μsec or less, although no band is present at 1567 cm^{-1}, a small feature is apparent at 1622 cm^{-1}, and there are shoulders on both sides of the 1530-cm^{-1} band. These shoulders may be from the C=C stretching vibrations of K635 and L550 (Marcus and Lewis, 1977). When the time resolution is 20 μsec or less, the spectrum of Fig. 4.20D shows little contribution from the C=C stretching frequency of M412 but shows a greater intensity from the band near 1622 cm^{-1} when compared to the spectrum with a resolution of 8 μsec or less. The spectrum of Fig. 4.20E shows that when the resolution is 29 μsec or less, the C=C stretching frequency of M412 begins to appear more strongly. The 1622-cm^{-1} band makes an even stronger appearance. Figure 4.20F shows that when the resolution is 40 μsec or less, the C=C stretching frequency of M412 contributes more significantly, but no material increase occurs in the intensity of the 1622-cm^{-1} band, which is due to the unprotonated Schiff-base linkage.

The formation time of the C═C stretching frequency at 1567 cm^{-1} parallels the rate of formation of M412, the kinetics of which was observed by absorption spectroscopy (Stoeckenius and Lozier, 1974; Dencher and Wilms, 1975; Kung *et al.*, 1975; Chance *et al.*, 1975; and Lozier *et al.*, 1975). However, the development of the 1622-cm^{-1} band begins at least 12 μsec before the first shoulder is measurable at 1567 cm^{-1}. Marcus and Lewis offer a possible explanation for this finding: another intermediate exists with an unprotonated Schiff base before M412.

By using various distances for the space between where the two laser beams excite and probe the flowing sample, one can obtain resonance Raman spectra of practically pure kinetic intermediates. The authors point out that the technique is applicable not only to rhodopsin and its intermediates, but also to a wide variety of other biological systems, e.g., enzymatic reaction mechanisms, analyses of cooperativity in hemoglobin, and observations of molecular dynamics in other photobiological systems.

Lewis *et al.* (1982) have studied the kinetic resonance Raman spectroscopy of carotenoids as a sensitive kinetic monitor of bacteriorhodopsin-mediated membrane potential changes. Using the model system of β-carotene incorporated into reconstituted vesicles containing bacteriorhodopsin, they detected a rapid (14–22-μsec) lightinduced membrane potential. Their data demonstrated that the kinetic resonance Raman spectrum of β-carotene is an extremely sensitive monitor of kinetic alterations in membrane potential with μm spatial resolution in a highly scattering medium. These workers suggested that their techniques would be applicable to a wide variety of biological systems without the perturbing side effects that often attend the use of nonbiological, potential-sensitive dyes.

Campion *et al.* (1977a) have reported a time-resolved (pulse-excited) resonance Raman spectrum of an intermediate, bK_{590}, in the photochemical cycle of bacteriorhodopsin. The 1530-cm^{-1} band ($\nu_{C=C}$) for bacteriorhodopsin in a low resolution spectrum was asymmetric, so they recorded the band shape with higher resolution; this more highly resolved spectrum displayed a shoulder at a frequency ~ 10 cm^{-1} lower. The shoulder appeared only under conditions in which pulsed excitation was used and did not appear during the use of continuous excitation. These authors interpreted this feature as arising from Raman scattering by the intermediate bK_{590}.

Campion *et al.* (1977b) have developed apparatus which should, in principle, be capable of recording resonance Raman spectra with a time resolution of 100 nsec. These workers recorded spectra of the intermediates of the photosynthetic cycle of bacteriorhodopsin on the microsecond time scale. The kinetic results and the resonance enhancement profile suggested that deprotonation of the Schiff base results in an intermediate preceding bM_{412} that has an optical absorption maximum at a wavelength longer than that of bM_{412}.

Ehrenberg and Lewis (1978), using kinetic resonance Raman spectroscopy, have demonstrated that the rate of Schiff-base deprotonation is strongly affected by pH, and that light alters the pK of the group directly controlling this deprotonation from a pK greater than 12 before light absorption to a pK between 9.9 and 10.3, microseconds after light absorption. The method used for recording kinetic spectra was described by Marcus and Lewis (1977). It is based on varying the flow rate of a bacteriorhodopsin suspension and thus changing the time period that molecules spend in a laser beam of known diameter.

The kinetics of Schiff-base deprotonation was monitored by observing the time it took for specific vibrational modes in the spectra of bacteriorhodopsin to evolve when the bacteriorhodopsin was excited by an argon-ion laser emitting a wavelength of 457.9 nm. Among these developing modes were the growth of the 1619-cm^{-1} C═N stretching frequency of the unprotonated Schiff base and that of the 1566-cm^{-1} C═C stretching frequency, which originates from the chromophore in the M412 intermediate (Lewis *et al.*, 1974).

Figure 4.22 presents the region from 1450 to 1700 cm^{-1} as a function of time spent by the molecules in the laser beam and at two extreme pH values—Figs. 4.22A and 4.22B at pH 7 and 12, respectively. Figure 4.22 shows clearly that the C═C stretching frequency at 1566 cm^{-1} (for intermediate M412) develops much faster at pH 12. Application of the method of Sundius (1973) to the analysis of peak areas showed that the development of the C═N stretching band at 1619 cm^{-1} occurs similarly. Thus, such

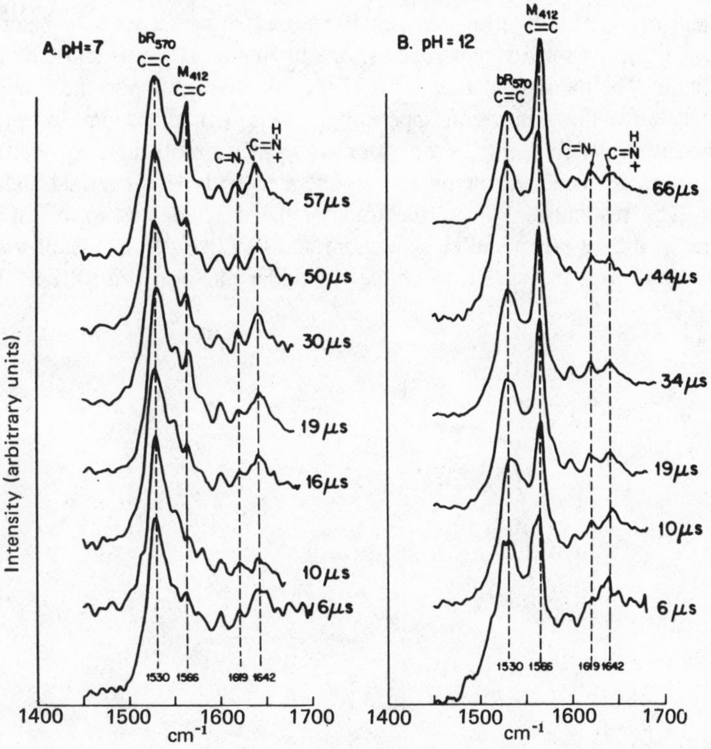

Fig. 4.22. Resonance Raman spectra of water suspensions of purple membrane fragments, ($\sim 5 \times 10^{-5}$ *M*) flowed through a capillary tube of 2 mm inner diameter and excited with 50 mW of 457.9-nm laser light at room temperature. Spectral resolution is 2 cm^{-1} in all cases. The pH of the suspensions are 7 in (A) and 12 in (B). Flow rates were varied to obtain the indicated average residence times of a molecule in the illuminated area. (Ehrenberg and Lewis, 1978.)

information can be used to determine the pK value of the pH dependence. The areas under the following bands were integrated as a function of time and pH: the C═N stretching band (1619 cm^{-1}), the $C{=}\underset{+}{N}{=}H$ stretching band (1642 cm^{-1}), the C═C stretching band (1566 cm^{-1}, M412), and the C═C stretching band (1530 cm^{-1}, ⊕ bR570). From these data Ehrenberg and Lewis calculated the initial rates of formation for the 1619- and 1566-cm^{-1} bands at each pH by comparing these bands with the integrated areas under the 1642- and 1530-cm^{-1} bands. The rates are plotted versus pH in Fig. 4.23. The values of the pK for deprotonation of the Schiff base and M412 development were 9.9 and 10.3, when the data were based on the C═N stretch and the M412 C═C stretch, respectively. It is possible that these are two distinct processes, with deprotonation preceding the formation of the M412 intermediate (Marcus and Lewis, 1977).

Aton *et al*. (1979) have compared the resonance Raman spectrum of the dark-adapted form of bacteriorhodopsin (purple membrane protein, with absorption maximum at 558 nm) to that of the light-adapted form (with absorption maximum at 568 nm) and those of model chromophores. The latter were all-*trans* and 13-*cis*-retinal-*n*-butylamine protonated Schiff bases. Earlier characterization of the dark-adapted form (Oesterhelt *et al.*, 1973; Pettei *et al.*, 1977; Sperling *et al.*, 1977) had shown it to be a heterogeneous mixture of equal proportions of two species, one having an all-*trans* and the other a 13-*cis* chromophore. The data of Aton *et al*. (1979) qualitatively support the conclusions of these workers, since the absorption spectrum (not shown) of the dark-adapted purple membrane is essentially the same as the absorption spectrum obtained by addition of the absorption spectra of bleached membrane regenerated with 13-*cis* retinal and with all-*trans* retinal. The resonance Raman spectrum of the dark-adapted form PM558 looks more like that of the 13-*cis* model chromophore and less like that of the all-*trans* model chromophore than does the spectrum of the light-adapted form (Fig. 4.24). The chro-

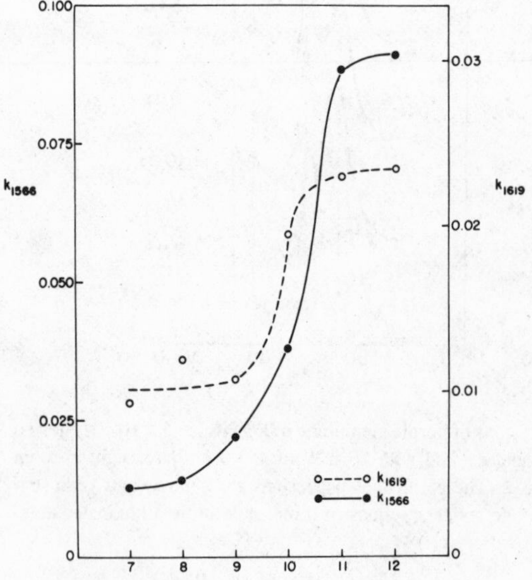

Fig. 4.23. pH titration curves of the formation rate constants of the unprotonated C═N stretching frequency, K_{1619} O-O-O and of the M412 C═C stretching frequency, K_{1566} ●—●. (Ehrenberg and Lewis, 1978.)

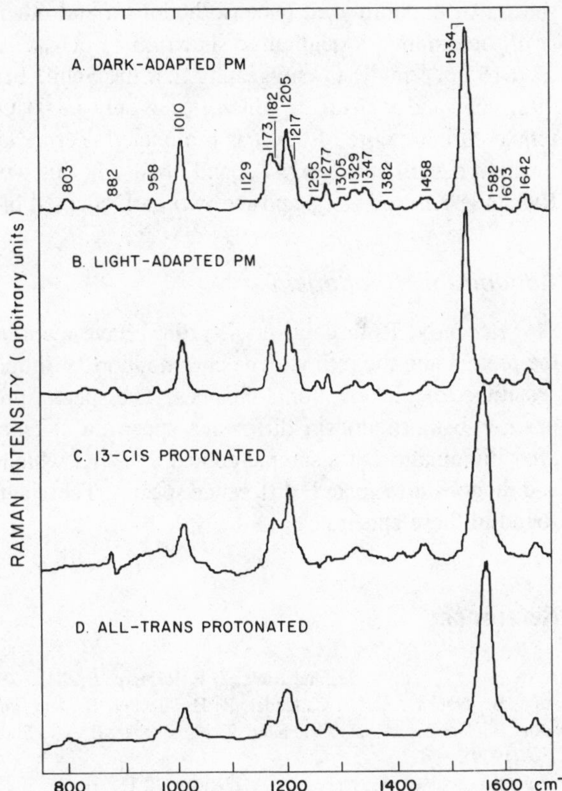

Fig. 4.24. Resonance Raman spectra of (A) PM558 dark-adapted purple membrane chromophore, (B) PM568 light-adapted purple membrane chromophore, (C) 13-*cis*-retinal *n*-butylamine·HCl, and (D) all-*trans*-retinal *n*-butylamine·HCl. The spectrometer resolution is (A) 7 cm^{-1}, (B) 7 cm^{-1}, (C) 6 cm^{-1}, and (D) 7 cm^{-1}. (Aton *et al.*, 1979.)

mophore–protein linkage in the dark-adapted form was found to be a protonated Schiff base as in the light-adapted form.

Terner *et al.* (1979) have used time-resolved resonance Raman spectra to characterize the bO_{640} intermediate (Fig. 4.25) that appears in the photoinduced cycle of bacteriorhodopsin (Lozier *et al.*, 1975). By using computer subtraction techniques and the Raman

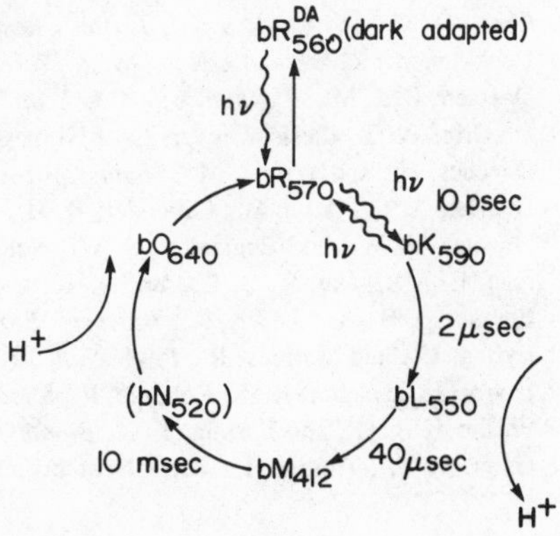

Fig. 4.25. The photoinduced cycle of bacteriorhodopsin according to Lozier *et al.* (1975). The subscripts are the absorption maxima in nanometers. The cycle is believed to be more complex than this simple scheme. (Terner *et al.*, 1979.) Reprinted with the permission of the American Chemical Society. Copyright 1979.

spectra of unphotolyzed bacteriorhodopsin and other intermediates in the cycle, Terner *et al.* determined a qualitative spectrum of bO_{640}. A band at 1630 cm^{-1} in H_2O moved to 1616 cm^{-1} in ^2H_2O, suggesting that the Schiff base of bO_{640} is protonated. Bands at 992, 965, and 947 cm^{-1}, which occur only in ^2H_2O suspensions, also showed that the retinal chromophore of bO_{640} is protonated. Terner *et al.* (1979a) determined that deprotonation occurs between bL_{550} and bM_{412}. In this work they also recorded the resonance Raman spectra of bL_{550} and the two dark-adapted bR_{560}^{DA} forms of bacteriorhodopsin.

Addendum: Rhodopsin

Recently, Rothschild *et al.* (1983) have recorded FTIR difference spectra for both the protein and the retinylidene chromophore of rhodopsin, showing that such spectra are sensititve to conformational changes. The spectra presented were the following: a rhodopsin-to-bathorhodopsin difference spectrum calculated from FTIR spectra before and after illumination of a sample cooled to 77 K, which stops the decay of bathorhodopsin; and rhodopsin-to-meta II difference spectra. These authors interpreted several of the peaks found in these spectra.

References

Askren, C. C., Yu, N.-T., and Kuck, J. F. R., *Exp. Eye Res.* **29,** 647 (1979).

Aton, B., Doukas, A. G., Callender, R. H., Becher, B., and Ebrey, T. G., *Biochemistry* **16,** 2995 (1977).

Aton, B., Doukas, A. G., Callender, R. H., Becher, B., and Ebrey, T. G., *Biochim. Biophys. Acta* **576,** 424 (1979).

Aton, B., Doukas, A., Narva, D., Callender, R. H., Honig, B., and Dinur, U., *Biophys. J.* **25,** 76a (1979).

Aton, B., Doukas, A. G., Narva, D., Callender, R. H., Dinur, U., and Honig, B., *Biophys. J.* **29,** 79 (1980).

Birge, R. R., *Ann. Rev. Biophys. Bioeng.* **10,** 315 (1981).

Brewster, D., *Phil. Trans. R. Soc. London* **106,** 311 (1816), quoted by Schachar and Solin, 1975.

Busch, G. E., Applebury, M. L., Lamola, A. A., Rentzepis, P. M., *Proc. Natl. Acad. Sci. USA* **69,** 2802 (1972).

Callender, R. H., Doukas, A., Crouch, R., and Nakanishi, K., *Biochemistry* **15,** 1621 (1976).

Campion A., Terner, J., and El-Sayed, M. A., *Nature* **265,** 659 (1977a).

Campion A., El-Sayed, M. A., and Terner, J., *Biophys. J.* **20,** 369 (1977b).

Chance, B., Porte, M., Hess, B., and Oesterhelt, D., *Biophys. J.* **15,** 913 (1975).

Chang, R. C. C., Ph.D. Thesis, Georgia Institute of Technology, Atlanta, Georgia, 1976, quoted by Yu, N.-T. and Kuck, J. F. R., Jr., in *The Spex Speaker,* Metuchen, New Jersey, September, 1978.

Chen, M. C., and Lord, R. C., *J. Am. Chem. Soc.* **96,** 4750 (1974).

Cookingham, R. E., and Lewis, A., *J. Mol. Biol.* **119,** 569 (1978).

Daemen, F. J. M., *Biochim. Biophys. Acta* **300,** 255 (1973).

De Grip, W. J., thesis, University of Nijmegen, Netherlands, 1974; quoted by Rothschild *et al.* (1976).

Dencher, N., and Wilms, M., *Biophys. Struct. Mech.* **1,** 259 (1975).

Doukas, A. G., Aton, B., Callender, R. H., and Ebrey, T. G., *Biochemistry* **17,** 2430 (1978).

Downer, N. W., and Englander, S. W., *Nature* **254,** 625 (1975).

East, E. J., Chang, R. C. C., Yu, N.-T., and Kuck, J. F. R., Jr., *J. Biol. Chem.* **253,** 1436 (1978).

Ehrenberg, B., and Lewis, A., *Biochem. Biophys. Res. Commun.* **82,** 1154 (1978).

Eyring, G., and Mathies, R., *Proc. Natl. Acad. Sci. USA* **76,** 33 (1979).

Favrot, J., Leclercq, J. M., Roberge, R., Sandorfy, C., and Vocelle, D., *Photochem. Photobiol.* **29,** 99 (1979).

Frushour, B. G., and Koenig, J. L., *Biopolymers* **13,** 455 (1974).

Galin, M. A., Turkish, L., and Chowchuvech, E., *Amer. J. Ophthalmol.* **84,** 153 (1977).

Goheen, S. C., Lis, L. J., and Kauffman, J. W., *Biochim. Biophys. Acta* **536,** 197 (1978).

Green, B. H., Monger, T. G., Alfano, R. R., Aton, B., and Callender, R. H., *Nature* **269,** 179 (1977).

Harding, J. J., and Dilley, K. J., *Exp. Eye Res.* **22,** 1–73 (1976).

Heyde, M., Gill, D., Kilponen, R., and Rimai, L., *J. Am. Chem. Soc.* **93,** 6776 (1971).

Honig, B., Greenberg, A. D., Dinur, U., and Ebrey, T. G., *Biochemistry* **15,** 4593 (1976).

Honig, B., Ebrey, T., Callender, R. H., Dinur, U., and Ottolenghi, M., *Proc. Natl. Acad. Sci. USA* **76,** 2503 (1979).

Hubbard, R., and Kropf, A., *Proc. Natl. Acad. Sci. USA* **44,** 130 (1958).

Kakitani, H., Kakitani, T., Honig, B., and Callender, R., Abstracts of the 26th Annual Meeting, *Biophys. J.* **37,** No. 2 Part 2, 228a, February (1982).

Kanner, B. I., and Racker, E., *Biochem. Biophys. Res. Commun.* **64,** 1054 (1975).

Kaufmann, K. J., Rentzepis, P. M., Stoeckenius, W., and Lewis, A., *Biochem. Biophys. Res. Commun.* **68,** 1109 (1976).

Khorana, H. G., Gerber, G. E., Herlihy, W. C., Gray, C. P., Anderegg, R. J., Nihei, K., and Biemann, K., *Proc. Natl. Acad. Sci. USA* **76,** 5046 (1979).

Koenig, J. L., and Frushour, B. G., *Biopolymers* **11,** 2505 (1972).

Krimm, S., and Abe, Y., *Proc. Natl. Acad. Sci. USA* **69,** 2788 (1972).

Kuck, J. F. R., Jr., and Yu, N.-T., *Exp. Eye Res.* **27,** 737 (1978).

Kuck, J. F. R., Jr., East, E. J., and Yu, N.-T., *Exp. Eye Res.* **23,** 9 (1976).

Kung, M. C., DeVault, D., Hess, B., and Oesterhelt, D., *Biophys. J.* **15,** 907 (1975).

Lewis, A., *Bull. Am. Phys. Soc.* II **19,** 375 (1974).

Lewis, A., *Fed. Proc.* **35,** 51 (1976a).

Lewis, A., *The Spex Speaker,* Spex Industries, Inc., Metuchen, New Jersey, Vol. 21, No. 2, June (1976b).

Lewis, A., *Proc. Natl. Acad. Sci. USA* **75,** 549 (1978).

Lewis, A., and Spoonhower, J., in *Spectroscopy in Chemistry and Biophysics* (S.-H. Chen and S. Yip, eds.), Academic Press, New York, p. 347, 1974.

Lewis, A., Fager, R., and Abrahamson, E. W., *J. Raman Spectrosc.* **1,** 465 (1973).

Lewis, A., Spoonhower, J., Bogomolni, R. A., Lozier, R. H., and Stoeckenius, W., *Proc. Natl. Acad. Sci. USA* **71,** 4462 (1974).

Lewis, A., Johnson, J. H., and Gogel, G., Abstracts of the 26th Annual Meeting, *Biophys. J.* **37,** No. 2 Part 2, 180a, February (1982).

Lord, R. C., and Yu, N.-T., *J. Mol. Biol.* **50,** 509 (1970).

Lozier, R. H., Bogomolni, R. A., and Stoeckenius, W., *Biophys. J.* **15,** 955 (1975).

Marcus, M. A., and Lewis, A., *Science* **195,** 1328 (1977).

Mathies, R., *Fed. Proc.* **39**(6), 2067 (1980).

Mathies, R., and Yu, N.-T., *J. Raman Spectrosc.* **7,** 349 (1978).

Mathies, R., Oseroff, A. R., and Stryer, L., *Proc. Natl. Acad. Sci. USA* **73,** 1 (1976).

Mendelsohn, R., *Biochim. Biophys. Acta* **427,** 295 (1976).

Mendelsohn, R., *Nature* **243,** 22 (1973).

Mendelsohn, R., Verma, A. L., Bernstein, H. J., and Kates, M., *Canad. J. Biochem.* **52,** 774 (1974).

Miyazawa, T., *J. Chem. Phys.* **32,** 1647 (1960).

Monger, T. G., Alfano, R. R., and Callender, R. H., *Biophys. J.* **27,** 105 (1979).

Oesterhelt, D., and Stoeckenius, W., *Proc. Natl. Acad. Sci. USA* **70,** 2853 (1973).

Oesterhelt, D., Meentzen, M., and Schuhmann, L., *Eur. J. Biochem.* **40,** 453 (1973).

Osborne, H. B., and Nabedryk-Viala, E., *Eur. J. Biochem.* **89,** 81 (1978).

Osborne, H. B., and Nabedryk-Viala, E., *FEBS Lett.* **84,** 217 (1977).

Oseroff, A. R., and Callender, R. H., *Biochemistry* **13,** 4243 (1974).

Ovchinnikov, Y. A., Abdulaev, N. G., Feigina, M. Y., Kiselev, A. V., and Lobanov, N. A., *FEBS Lett.* **100,** 219 (1979).

Parker, F. S., and D'Agostino, M., *Canad. J. Spectrosc.* **21,** 111 (1976).

Peters, K., Applebury, M. L., and Rentzepis, P. M., *Proc. Natl. Acad. Sci. USA* **74,** 3119 (1977).

Pettei, M. J., Yudd, A. P., Nakanishi, K., Henselman, R., and Stoeckenius, W., *Biochemistry* **16,** 1955 (1977).

Rimai, L., Gill, D., and Parsons, J. L., *J. Am. Chem. Soc.* **93,** 1353 (1971).

Rosenfeld, T., Honig, B., Ottolenghi, M., Hurley, J., Ebrey, T. G., Proceedings IUPAC Photochemistry Meeting, Aix en Provance, July, 1976.
Rothschild, K. J., and Clark, N. A., *Biophys. J.* **25,** 473 (1979).
Rothschild, K. J., Andrew, J. R., De Grip, W. J., and Stanley, H. E., *Science* **191,** 1176 (1976).
Rothschild, K. J., Cantore, W. A., and Marrero, H., *Science* **219,** 1333 (1983).
Rothschild, K. J., Sanches, R., Hsiao, T. L., and Clark, N. A., *Biophys. J.* **31,** 53 (1980).
Schachar, R. A., and Solin, S. A., *Invest. Ophthalmol.* **14,** 380 (1975).
Schachar, R. A., and Solin, S. A., *Invest. Ophthalmol.* **14,** 380 (1975).
Sperling, W., Carl, P., Rafferty, C. N., and Dencher, N. A., *Biophys. Struct. Mech.* **3,** 79 (1977).
Stoeckenius, W., and Bogomolni, R. A., *Annu. Rev. Biochem.* **51,** 587 (1982).
Stoeckenius, W., and Lozier, R. H., *J. Supramol. Struct.* **2,** 769 (1974).
Sulkes, M., Lewis, A., and Marcus, M. A., *Biochemistry* **17,** 4712 (1978).
Sundius, T., *J. Raman Spectrosc.* **1,** 471 (1973).
Terner, J., Hsieh, C.-L., Burns, A. R., and El-Sayed, M. A., *Biochemistry* **18,** 3629 (1979).
Terner, J., Hsieh, C.-L., and El-Sayed, M.A., *Biophys. J.* **26,** 527 (1979 a).
van der Meer, K., Mulder, J. J. C., and Lugtenberg, J., *Photochem. Photobiol.* **24,** 363 (1976).
Warshel, A., *Proc. Natl. Acad. Sci. USA* **75,** 2558 (1978).
Willumsen, L., *Compt. Rend. Trav. Lab. Carlsberg* **38,** 223 (1971).
Yoshizawa, T., and Wald, G., *Nature* **197,** 1279 (1963).
Yu, N.-T., *CRC Crit. Rev. Biochem.* **4,** 229 (1977).
Yu, N.-T., and East, E. J., *J. Biol. Chem.* **250,** 2196 (1975).
Yu, N.-T., and Liu, C. S., *J. Am. Chem. Soc.* **94,** 5127 (1972).
Yu, N.-T., Jo, B. H., Chang, R. C. C., and Huber, J. D., *Arch. Biochem. Biophys.* **160,** 614 (1974).
Yu, N.-T., East, E. J., Chang, R. C. C., and Kuck, J. F. R., Jr., *Exp. Eye Res.* **24,** 321 (1977).
Yu, T. J., Lippert, J. L., and Peticolas, W. L., *Biopolymers* **12,** 2161 (1973).

Chapter 5

ENZYMES

Raman and Infrared Studies

Lysozyme

Lord and Yu (1970b) have presented Raman spectra of lysozyme and its constituent amino acids. Figure 5.1 shows the spectrum of the enzyme (a) and the spectrum of the sum of the constituent amino acids (b). Table 5.1 gives the vibrational frequencies of lysozyme, with assignments stated for amino acid residues. The enzyme was dissolved in H_2O and ^2H_2O, pH (pD) = 5.2. Very intense Raman lines are found for the amino acids with aromatic side groups, such as tyrosine, tryptophan, and phenylalanine. Changes in environment and state of aggregation do not affect the characteristic frequencies of these rings. Several lines are assigned to the vibrations of the tryptophan indole ring: 544, 577, 761, 879, 1014, 1338, 1363, 1553, and 1582 cm^{-1}. A weak shoulder near 1432 cm^{-1} is due to the NH bending frequency of the indole ring. The monosubstituted phenyl ring of phenylalanine displays a "breathing" vibration (Wilmshurst and Bernstein, 1957) at 1006 cm^{-1}. The *p*-hydroxyphenyl ring of tyrosine shows two characteristic lines near 840 cm^{-1}. Lord and Yu provisionally assigned the lines at 836 and 858 cm^{-1} to tyrosine. Corresponding lines were found by Yu (1969) in the spectrum of ribonuclease.

The C—S stretching frequency of the C—S—S—C group is displayed rather strongly at 667 cm^{-1} in Fig. 5.1(b) but very weakly at 661 cm^{-1} in Fig. 5.1(a). The faintness of this C—S stretching line for native lysozyme when the S—S line at 509 cm^{-1} is strong suggested to these workers that the conformations of the C—S—S—C cross-linkages in the enzyme differ from those in the monomeric analogs in solution, e.g., cystine. They attempted to correlate the Raman frequencies of the S—S and C—S stretching vibrations with the C—S—S angles around the S—S cross-linkages in lysozyme and ribonuclease. In contrast, Martin (1974) and Bastian and Martin (1973) found no correlation between the Raman intensity ratio for C—S and S—S stretching bands and the C—S—S angle.

Lysozyme contains two methionine moieties. The structure of the methionine residue, $CH_3SCH_2CH_2$—, resembles that of 2-thiabutane. Hayashi *et al.* (1957) recorded three intense lines at 650, 675, and 723 cm^{-1} in the Raman spectrum of liquid 2-thiabutane,

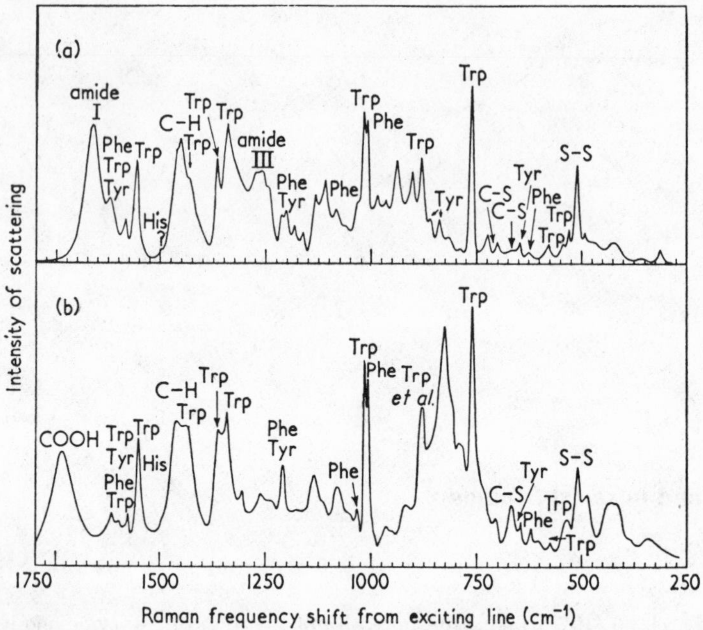

Fig. 5.1. (a) Raman spectrum of lysozyme in water, pH 5.2. (b) Sum of the constituent amino acids, pH 1.0. (Lord and Yu, 1970b.)

which they assigned to the C—S stretching frequencies: 723 cm^{-1}, *trans* form; 675 cm^{-1}, *gauche* form; and 650 cm^{-1}, both forms. Lord and Yu measured these frequencies as 654, 680, and 727 cm^{-1}. Aqueous L-methionine at pH 1.0 displayed three intense lines similar to those of 2-thiabutane at 655, 701, and 724 cm^{-1}.

The spectrum of aqueous glycyl-L-methionylglycine shows the analogous lines at 655, 700, and 724 cm^{-1}. In the crystal form, this substance yields a strong line at 705 cm^{-1} with a shoulder at 698 cm^{-1}. Lord and Yu assigned the weak line at 700 cm^{-1} in the spectrum of native lysozyme to this C—S frequency of methionine. It has been established from X-ray data on crystalline lysozyme that the two methionine side-chains are in the *gauche* form (Phillips, 1967; Blake *et al.*, 1967).

Lysozyme shows an amide I frequency at 1660 cm^{-1} and an amide III near 1260 cm^{-1}, which was resolved into a triplet at 1240, 1262, and 1274 cm^{-1}. The amide I and amide III frequencies have been assigned by Miyazawa *et al.* (1958) to C═O stretching and a coupled vibration of C—N stretching and N—H in-plane bending, respectively. Since Lord and Yu found that the amide III frequency of lysozyme shifted to 950 cm^{-1} on *N*-deuteration (ν_H/ν_D = 1.33), they deduced that its mode of vibration arises mainly from the N—H in-plane bending vibration.

All 20 amino acids examined by these authors show a line near 1730 cm^{-1} at acidic pH (pH <2), which is characteristic of the un-ionized carboxyl group. Another line, near 1415 cm^{-1}, is characteristic of the ionized carboxyl group, and it is displayed at alkaline and isoelectric pH's.

Table 5.1. Raman Spectra of Native Lysozyme in Water and Deuterium Oxide (250 to 1800 cm^{-1})[a,b]

Frequency (cm^{-1})			Frequency (cm^{-1})		
H_2O	D_2O	Tentative assignment	H_2O	D_2O	Tentative assignment
311(0)			1035(2s)	1035(2)	Phe
354(0)			1055(0)	1060(0)	
429(2B)			1078(1)	1080(0)	
	450(3B)		1109(4)	1110(1)	ν(C—N)
469(0)			1128(3)	1128(1)	
491(0)	489(0)		1160(1)	1161(0)	
509(5S)	509(5S)	ν(S—S)	1179(1)	1183(1)	Tyr: *O*-deuterated Tyr and N—D bending of the deuterated indole
529(1)	524(0)				
544(1)	544(1)	Trp; *N*-deuterated Trp	1198(3D)	1200(1s)	Tyr and Phe
577(1)	574(1)	Trp; *N*-deuterated Trp	1210(2D)	1211(2)	
603(0)?			1240(4)		Amide III
624(0)	625(0)	Phe		1253(1)	
646(0)	646(0)	Tyr; *O*-deuterated Tyr	1262(5)		Amide III
661(0)	662(0)	ν(C—S)(disulfide)	1274(2)		Amide III
700(1)	698(1)	ν(C—S)(Met)		1284(1)	
724(1)	723(1)		1290(0)?		
761(10S)	760(10S)	Trp; *N*-deuterated Trp	1304(0)		
800(0)?	804(1)		1338(8)	1337(6)	Trp; *N*-deuterated Trp
820(0)			1363(5S)	1362(2)	Trp
836(1)	835(2)	Tyr; *O*-deuterated Tyr		1385(2s)	*N*-deuterated Trp
	857(2)	*N*-deuterated Trp		1420(2s)	Symmetrical CO_2^- str.?
858(0)		Tyr	1432(4s)		N—H bending vibration of the indole ring
	873(2)		1448(9)	1447(8s)	
879(5)		Trp	1459(8s)	1459(10)	C—H deformation vibration
900(2)	907(2)	ν(C—C)		1482(1s)	
936(5)	934(7)	ν(C—C)	1494(0)?		His?
	950(6B)	Amide III′	1553(8S)	1553(8S)	Trp; *N*-deuterated Trp
964(0)	965(1)		1582(2)	1580(2)	Trp; *N*-deuterated Trp
984(1)			1622(4B)	1620(2s)	Trp; Tyr & Phe
1006(7SD)	1006(7SD)	Phe	1660(10)	1658(10)	Amide I; amide I′
1014(8SD)	1014(8SD)	Trp; *N*-deuterated Trp			

[a] Lord and Yu (1970b).

[b] Vibrational frequencies are expressed as the displacement in cm^{-1} of the Raman lines. The symbol B denotes broad, S sharp, D doublet, V very, s shoulder, ν(X—Y) a frequency assigned to an X—Y bond-stretching vibration and [m:] a partially resolved multiplet. Figures in parentheses are relative intensities based on a value of 10 for the strongest line in each spectrum. These intensities have been corrected for the frequency dependence of instrument response (determined with a standard lamp) by multiplying the observed intensity with the following factors: 250 to 500 cm^{-1}, 1.0; 500 to 750 cm^{-1}, 1.1; 750 to 1000 cm^{-1}, 1.2; 1000 to 1250 cm^{-1}, 1.3; 1250 to 1500 cm^{-1}, 1.5; 1500 to 1800 cm^{-1}, 1.8.

The lines at 2885, 2938, and the shoulder at 2985 cm^{-1} were assigned to aliphatic C—H stretching frequencies. The line at 3074 cm^{-1} arises mainly from aromatic C—H stretching.

In attempts to use Raman spectroscopy for the study of the denaturation of lysozyme by heating, Lord and Yu (1970b) found that optical inhomogeneity of heated concentrated solutions of the enzyme made the measurements difficult. The use of 6 *M* guanidine·HCl or 8 *M* urea, although producing optically satisfactory samples, causes the masking of the spectrum of the protein at much more dilute concentrations (<0.02 *M*). Denaturation was attained by use of 6 *M* LiBr without much effect on the spectrum of water; at the same time, the Raman spectrum of the enzyme was not masked (Lord and Mendelsohn, 1972). Figure 5.2 shows Raman spectra of lysozyme in solutions of LiBr of several concentrations at pH 4.2 and 25°C. Earlier work in Lord's laboratory had shown that α-

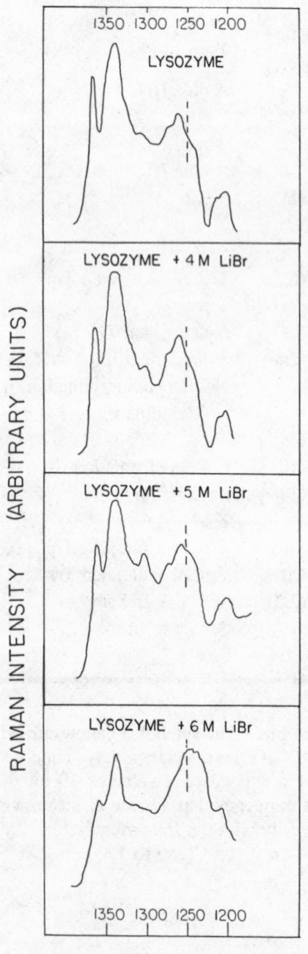

Fig. 5.2. Raman spectra of 5% aqueous lysozyme in LiBr solution (1200–1400 cm^{-1}). Excitation by 200 mW of 4880-Å radiation, pH 4.2, t = 25°C. The vertical dashed line indicates 1250 cm^{-1}. (Lord and Mendelsohn, 1972.) Reprinted with the permission of the American Chemical Society. Copyright 1972.

helical and β-pleated sheet conformations of proteins and model polypeptides display most of their Raman intensity from the amide III vibrational modes above 1260 cm^{-1}; the random-coil conformation has this frequency below 1250 cm^{-1}. Thus, Fig. 5.2 shows that the skeletal structure of lysozyme changes little in solutions of 0 to 4 *M* LiBr concentration. When the LiBr concentration is raised to 5 or 6 *M*, the center of gravity of the amide III band moves from above 1260 cm^{-1} to about 1245 cm^{-1}. This shift was interpreted as showing the removal of the ordered structure in the protein backbone; only a random-coil structure is left (bottom spectrum in Fig. 5.2).

Porubcan *et al.* (1978) have studied the denaturation of lysozyme by several reagents—dimethyl sulfoxide, guanidine hydrochloride, urea, sodium dodecyl sulfate, and LiBr. They used the intensity of the amide III Raman band at 1260 cm^{-1} relative to the intensity of the amide III band near 1240 cm^{-1} as a characteristic with which to compare other physical properties commonly measured to evaluate denaturation, e.g., viscosity and percentage of enzymic activity. The 1260-cm^{-1} band represents α-helical structure with strong hydrogen bonding; and the ~1240-cm^{-1} band characterizes more weakly hydrogen-bonded groups. The extents of denaturation with various denaturants as monitored by the ratio I_{1260}/I_{1240} paralleled those obtained from viscosity and activity measurements, thus indicating the value of amide III measurements for this purpose. The

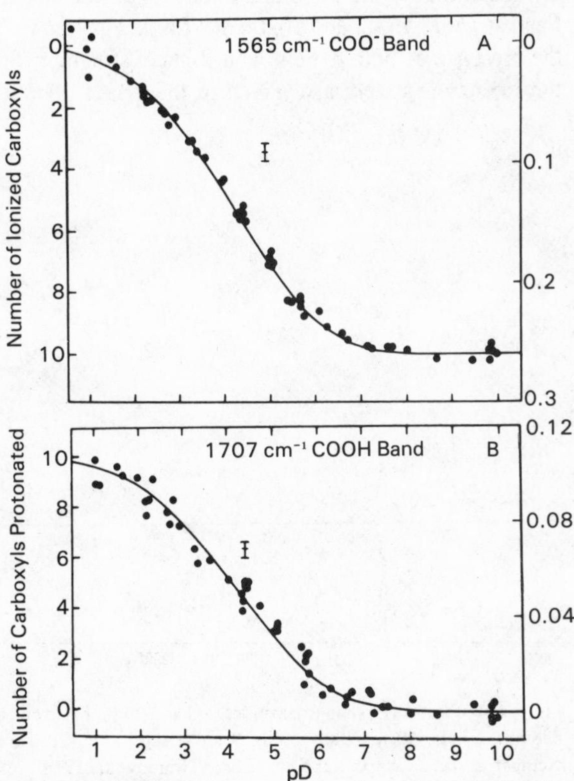

Fig. 5.3. Typical infrared titration curves as obtained on lysozyme. (A) 1565-cm^{-1} band; (B) 1707-cm^{-1} band. The dots are experimental points representing optical density differences from either the fully protonated (A) state or the fully ionized (B) state. The solid lines are calculated curves using best-fitting pK values. The error bars correspond to the experimental errors in reading differential band intensities. (Timasheff and Rupley, 1972.)

characteristics of the Raman spectra depend on the type of denaturant used, and Porubcan *et al.* interpreted this to mean that lysozyme does not have one unique denatured state.

Use has been made of difference IR spectroscopy of proteins to follow carboxylic acid ionization processes (Timasheff *et al.*, 1973; Koeppe and Stroud, 1976) and hydrogen bonding of sulfhydryl groups (Bare *et al.*, 1975). Figure 5.3 shows typical IR titration curves obtained in studies with lysozyme (Timasheff and Rupley, 1972). The difference spectra of lysozyme are characterized by certain bands which are pD dependent: 1565 cm^{-1}, the COO^- band, and 1707 cm^{-1}, the COOD band.

The data of Figs. 5.3A and 5.3B fitted well with the following values: one ionizable carboxyl group with pK_D (pK in D_2O medium) 2.5, three groups with pK_D 5.0, two groups with pK_D 6.0, and one group with pK_D 7.0. These values correspond to intrinsic pK values in H_2O of 2.0, 3.5, 4.5, 5.5, and 6.5.

Acid Phosphatase

Twardowski (1978) has recorded Raman spectra of two isoenzymes of acid phosphatase from rat liver (Fig. 5.4), and assigned S—S bands at 515, 525, 545, and 555 cm^{-1} to *gauche–gauche–gauche, gauche–gauche–trans, trans–gauche–trans,* and *trans–gauche–trans* forms, respectively. In acid phosphatase I and II a strong band for the tyrosine residue appeared at 850 and 845 cm^{-1}, respectively. Amide III bands occurred at 1235 and 1275 cm^{-1}. Amide I bands were located at 1670 and 1620 cm^{-1} for isoenzyme I and at 1668, 1632, and 1620 cm^{-1} for isoenzyme II. These amide I and III bands showed the presence of both α-helix and β-sheet structure. Spectra of H_2O and D_2O solutions of the isoenzymes were also given in this paper. The tertiary structures varied, depending

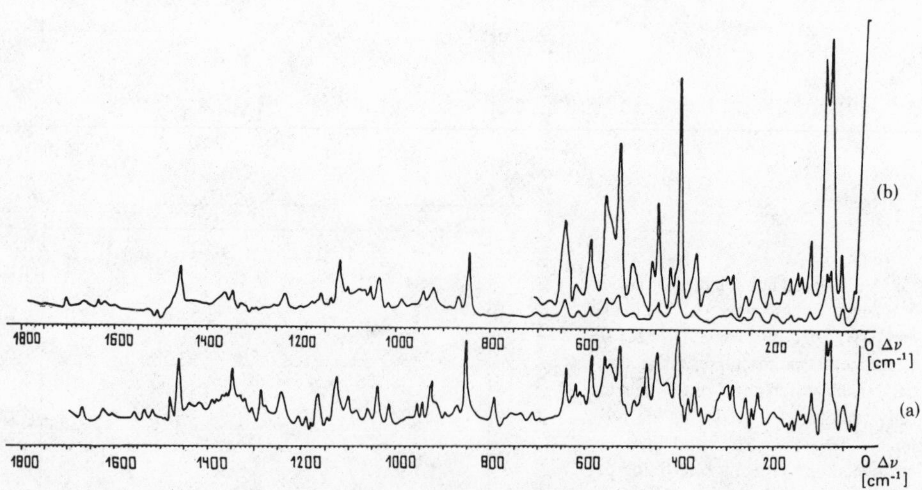

Fig. 5.4. The spectra of solid isoenzymes (a) I and (b) II of acid phosphatase. Instrumental conditions: excitation, 488 nm; power, 50 mW; slit width, 3 cm^{-1}; sensitivity, 20,000 cps (strong "b" spectra: 10,000 cps); time constant, 5 sec; scan speed 0.5 cm^{-1}/sec. (Twardowski, 1978.)

on whether measurements were made on the solid state or in H_2O or D_2O. The Raman spectra showed greater differences for the isoenzymes than did the IR spectra. Also, the IR spectra of these isoenzymes displayed no significant differences between them in the range between 700 and 2000 cm^{-1}.

Alcohol Dehydrogenase

Investigating the binding of inhibitors on the enzymic activity of various alcohol dehydrogenases, Young and Wang (1971) have studied the binding of azide by yeast alcohol dehydrogenase in the presence of NAD^+. Besides the absorption band of free azide (peak at 2048 cm^{-1}), a band was also observed at 2070 cm^{-1} in the presence of the enzyme and NAD^+ (Fig. 5.5 and Table 5.2). The absorption peak of free azide only was observed with solutions of coenzyme and azide, coenzyme and azide when an inert protein was present, or the enzyme and azide in the absence of coenzyme. Saturation of

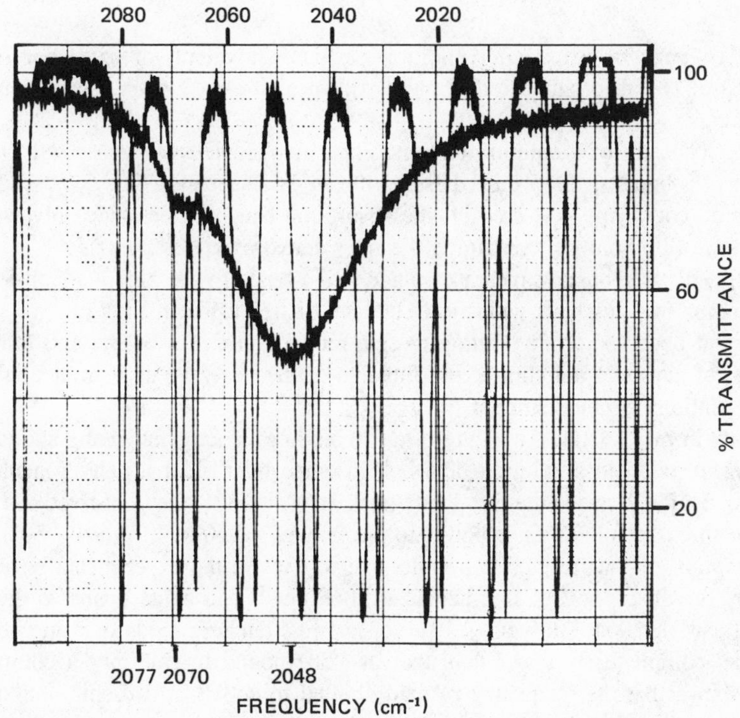

Fig. 5.5. Difference infrared spectrum of azide binding to yeast alcohol dehydrogenase in the presence of coenzyme. 5.6×10^{-4} *M* enzyme, 5.7×10^{-2} *M* NAD^+, and 2.0×10^{-3} *M* azide. The peak at 2070 cm^{-1} represents azide bound to the enzyme–coenzyme complex, and the peak at 2048 cm^{-1} represents free azide in solution. The gas phase infrared spectrum of deuterium chloride was superimposed on that of the aqueous enzyme solution for calibration (see Hardy *et al.*, 1932). (Young and Wang, 1971.)

Table 5.2. Frequencies of Infrared Absorption Maxima of Azide and Its Derivatives[a,c]

Compound	Absorption maximum (cm^{-1})
N_3^-(aqueous)[b]	2049
N_3–Zn(II)-yeast alcohol dehydrogenase-NAD^+	2070
N_3–Zn(II)-liver alcohol dehydrogenase-NAD^+	2065
N_3–Co(II)-Zn(II)-liver alcohol dehydrogenase-NAD^+	2064–2065
N_3–Zn(II)-diethylenetriamine[b]	2085
N_3–Co(II)-diethylenetriamine[b]	2067
N_3–Zn(II)-carbonic anhydrase[b]	2094
N_3–Co(II)-carbonic anhydrase[b]	2085
CH_3N_3[c]	2143

[a] Young and Wang (1971).
[b] Riepe and Wang (1968).
[c] Eyster and Gillette (1940).

the 2070-cm^{-1} peak occurred after addition of four equivalents of azide per mole of dehydrogenase. The absorbance of the bound azide peak lessened when ethanol was added to the mixture.

The IR spectrum of a solution of horse liver alcohol dehydrogenase, NAD^+, and azide displays a band at 2065 cm^{-1} besides the free-azide band (Fig. 5.6A and Table 5.2). When no coenzyme was present, the absorption band of free azide only was seen in the spectrum of a solution containing the liver dehydrogenase and azide.

The stretching frequency increases when azide binds to the zinc alcohol dehydrogenases. Comparing this shift with the frequency of free azide in solution, we see that it parallels that observed when binding to zinc model complexes occurs and in binding to the metal of carbonic anhydrase, but the size of the observed shift in alcohol dehydrogenase solutions is much smaller.

To study in more detail the nature of azide binding, Young and Wang analyzed the IR spectrum of azide in a solution of a hybrid horse liver alcohol dehydrogenase and NAD^+ (Fig. 5.6B). Besides the free-azide peak at 2048 cm^{-1}, they observed a shoulder in the range characteristic of azide bound to the enzyme, 2060 to 2070 cm^{-1}. Subtraction of the absorption of the free azide permits one to explain the experimentally recorded spectrum by the absorption of free azide (at 2048 cm^{-1}) and azide bound to the dehydrogenase (peak at 2064–2065 cm^{-1}). The observed stretching mode of azide bound to cobalt model complexes is lower than that of azide bound to zinc model compounds. Table 5.2 shows that the frequency of azide bound to cobalt in carbonic anhydrase is lower than that of azide bound to zinc in that enzyme. An absorption band at 2064–2065 cm^{-1} emanating from the azide binding to hybrid enzyme, when compared to the band at 2065 cm^{-1} coming from azide bound to the zinc enzyme, does not favor the idea that azide binds to cobalt in the hybrid enzyme.

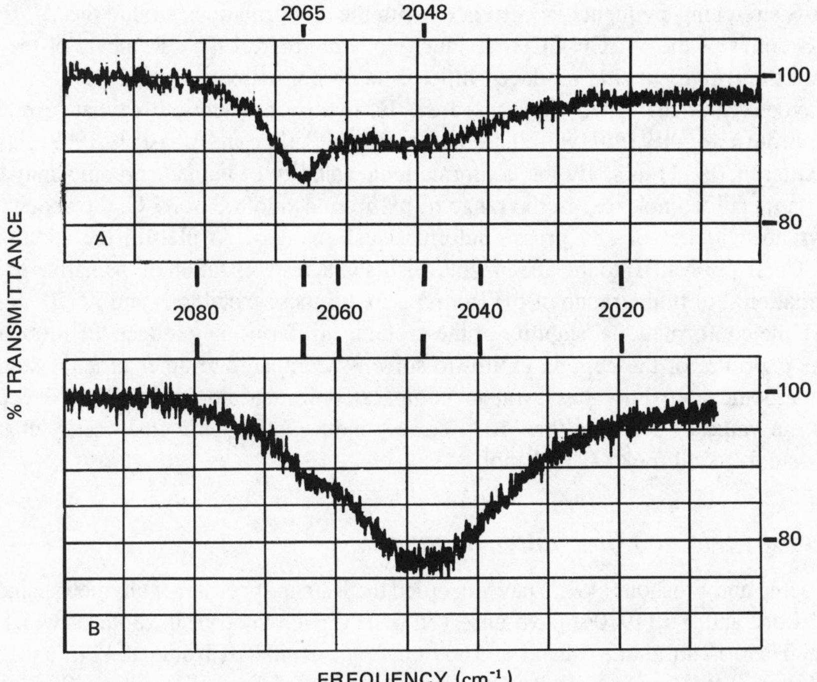

Fig. 5.6. Difference infrared spectra of azide binding to horse liver alcohol dehydrogenases. A, 3.1×10^{-3} *M* (w/w) native enzyme, 5.9×10^{-3} *M* NAD^+, and 6.4×10^{-3} *M* azide. The peak at 2065 cm^{-1} represents azide bound to the enzyme–coenzyme complex, while the free azide has a broad absorption with a maximum at 2048 cm^{-1}. B, 7.7×10^{-3} *M* NAD^+, 7.4×10^{-3} *M* azide, and hybrid enzyme. The hybrid concentration was approximately 1 m*M*, which is much lower than that of the native enzyme in solution A. This enzyme contained 1.3 g atoms of zinc per 80,000 g of protein after metal ion exchange with cobalt. The observed spectrum can be fitted by the absorption peak due to free azide (2048 cm^{-1}) and a peak with a maximum at 2064 to 2065 cm^{-1}. (Young and Wang, 1971.)

Carbonic Anhydrase B

Craig and Gaber (1977) have recorded Raman spectra of human carbonic anhydrase B and compared the structure deduced from those data with that determined by X-ray diffraction. Raman data yielded 19% helix, 39% β structure, and 42% disordered structure, compared with X-ray values of 17%, 40%, and 43%, respectively. Quantitative calculations using the Raman data yielded 3.7 "buried" tyrosine residues and 4.3 "exposed" to solvent, compared with four buried and four exposed tyrosines derived from X-ray data. The Raman data showed that the two methionine residues of the enzyme have conformations in which C_α is *trans* to sulfur and C_β is *gauche* to the methionyl methyl group. These conformations were based on data obtained for simple alkyl sulfides, where the C—S stretching frequency is correlated with the conformations around the C—C and CH_2—S bonds of the —CCH_2SCH_3 residue (Nogami *et al.*, 1975). Removal of the zinc ion from the native enzyme produced little or no conformational change.

Zavodszky *et al.* (1975) have used the IR hydrogen-deuterium exchange method (Linderstrøm-Lang, 1958; Hvidt and Nielsen, 1966; Parker and Bhaskar, 1969; Parker, 1971; Ottesen, 1971) to study the conformational stability of human carbonic anhydrase B and its metallocomplexes. In the range of pH from 4.6 to 8.8 at 25°C, the apoenzyme displays no evidence of any gross conformational changes. At pH 4.6 the addition of Co(II), Cd(II), or Mn(II) to the apoenzyme results in a destabilization of the native protein conformation, but in the range of pH from 5.5 to 8.8 these metal ions and Zn(II) slightly increase the conformational stability of the protein, insofar as they reduce the probability ρ of the exposure of the peptide groups to solvent. Compared to other proteins studied, native carbonic anhydrase has a rather compact conformation; for half of the peptide groups the value of ρ is less than 10^{-4}, corresponding to standard free energy changes larger than 5.5 kcal mol^{-1} (23 kJ mol^{-1}).

α-Chymotrypsin and Chymotrypsinogen

Koenig and Frushour (1972) have recorded the Raman spectrum of chymotrypsinogen A, and Lord and Yu (1970a) have observed that of α-chymotrypsin, both of which are similar. There is an amide I band at 1669 cm^{-1} and amide III bands at 1245 and 1260 cm^{-1}. Koenig and Frushour studied the reversible heat denaturation of the chymotrypsinogen molecule. They observed a 3-cm^{-1} decrease in the amide I frequency when the enzyme precursor was dissolved in ^2H_2O and was heated through the reversible transition temperature of the denaturation process; they ascribed this shift to a disturbance of the native structure such that ^2H_2O was permitted to permeate into the protein. There was no shift in frequency in the amide III region when the chymotrypsinogen was dissolved in H_2O, an observation that suggested to the authors that little or no change in the secondary structure occurs during the denaturation process.

Trypsin and Trypsin Inhibitor

Pershina and Hvidt (1974) have used the hydrogen–deuterium exchange method to study the complex formed between the basic pancreatic trypsin inhibitor and trypsin. They studied the kinetics of the ^1H–^2H exchange of the peptide groups in ^2H_2O solutions of the inhibitor, of trypsin, and of the inhibitor–trypsin complex. On the average the exchange rates in solutions of the complex were slower than the exchange rates in solutions of the constituent proteins. These authors expressed their experimental results in terms of the probability of exposure to the solvent of the peptide groups of the proteins, and in the changes in ΔG^0 (standard free energy) of the conformational transitions by which the exposure is brought about. These changes in free energy were found to be larger for the inhibitor–trypsin complex than for trypsin, indicating that the interior regions of the complex are less accessible to the solvent. Figure 5.7 shows the types of experiments carried out to study the kinetics.

Under identical conditions of pH the hydrogen–deuterium exchange of the peptide

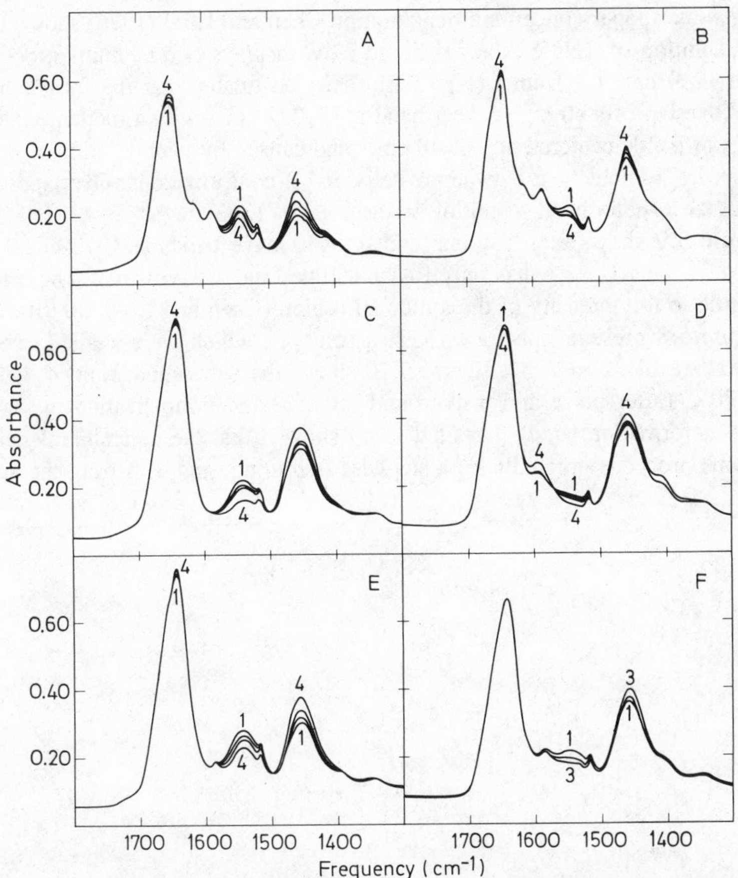

Fig. 5.7. The time course of the infrared absorption of the basic pancreatic trypsin inhibitor, trypsin and of the inhibitor–trypsin complex dissolved in ^2H_2O, 25°C. (A) 1.8% inhibitor, pH 3.4, (1) 35 min, (2) 97 min, (3) 340 min, (4) 25 h. (B) 2.0% inhibitor, pH 6.2, (1) 27 min, (2) 64 min, (3) 136 min, (4) 187 min, (C) 2.2% trypsin, pH 3.0, (1) 32 min, (2) 60 min, (3) 102 min, (4) 18 h. (D) 2.0% trypsin, pH 6.0, (1) 33 min, (2) 82 min, (3) 153 min, (4) 29 h. (E) 0.6% inhibitor + 2.0% trypsin, pH 3.1, (1) 27 min, (2) 63 min, (3) 180 min, (4) 21 h. (F) 0.5% inhibitor + 2.0% trypsin, pH 6.2, (1) 27 min, (2) 118 min, (3) 23 h. The times given are those of recording the peak of the amide II band. (Pershina and Hvidt, 1974.)

groups of the bovine basic pancreatic trypsin inhibitor is considerably slower than the exchange of the peptide groups of porcine insulin (Hvidt and Pedersen, 1974). This fact indicated that the interior regions of the inhibitor molecules are less accessible to the solvent than are those of the insulin molecules.

Ribonuclease A

Lord and Yu (1970a) had recorded the Raman spectrum of ribonuclease A in aqueous solution, but experimental difficulties had prohibited study of thermal denaturation at that time. In a later paper concerning denaturation Chen and Lord (1976) showed that a 7% aqueous solution of RNase A at pH 5 and 32°C displays two medium strong amide III bands at 1239 and 1263 cm^{-1} (Fig. 5.8). They postulated that the former band is the result of overlap of a strong β-sheet band at ~1232 cm^{-1} with a moderately strong and broad set of bands centered at ~1248 cm^{-1} and caused by peptide units of intermediate geometry, i.e., somewhere between α-helix and β-conformations. Overlapping also occurs with the α-helix band, resulting in the band at 1263 cm^{-1}.

Figure 5.9 shows that histidine and tyrosine have bands at 1272 and 1266 cm^{-1}, respectively; but RNase A has only four histidines and six tyrosines. These bands contribute little to the intensity of the amide III region (Chen and Lord, 1976).

Figure 5.8 presents spectra of RNase solutions which were studied over the temperature range of 32 to 70°C. Figure 5.10 shows the superimposition of the spectra at 32 and 70°C (after noise and background removal and normalization to the 147-cm^{-1} —CH_2— deformation band). Detailed study showed that the reversible denaturation of the enzyme proceeds gradually by a stepwise unfolding, and data from the amide I and

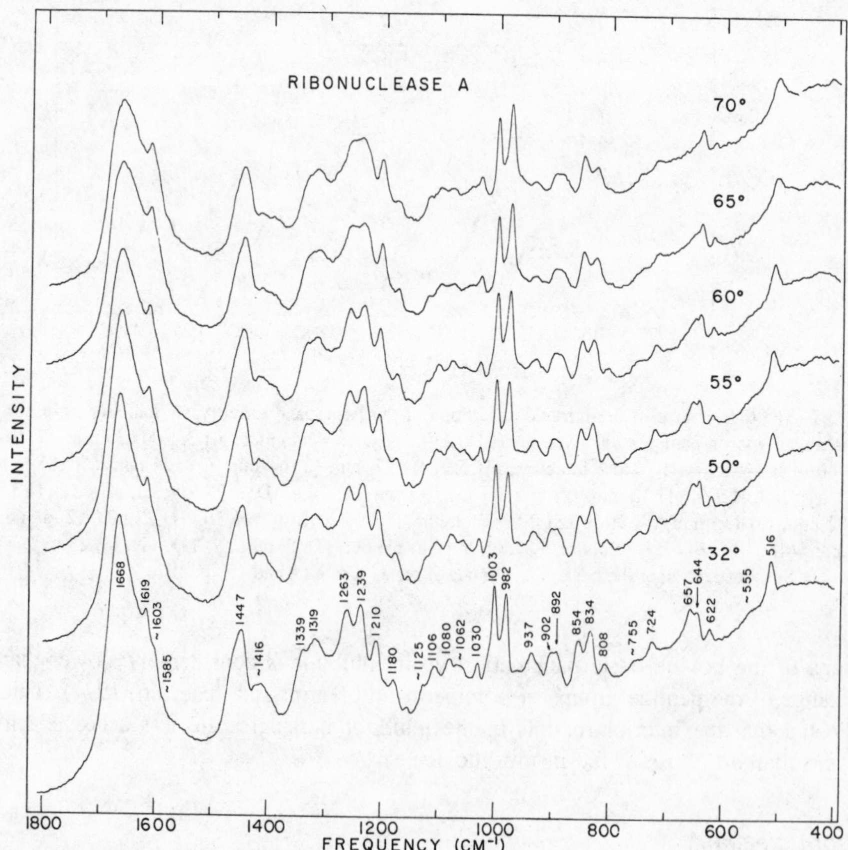

Fig. 5.8. Original Raman spectra of 7% aqueous RNase A at 32–70°C, ionic strength 0.1 *M* NaCl, pH 5. (Chen and Lord, 1976, with spectra redrawn by Dr. Lord's laboratory.)

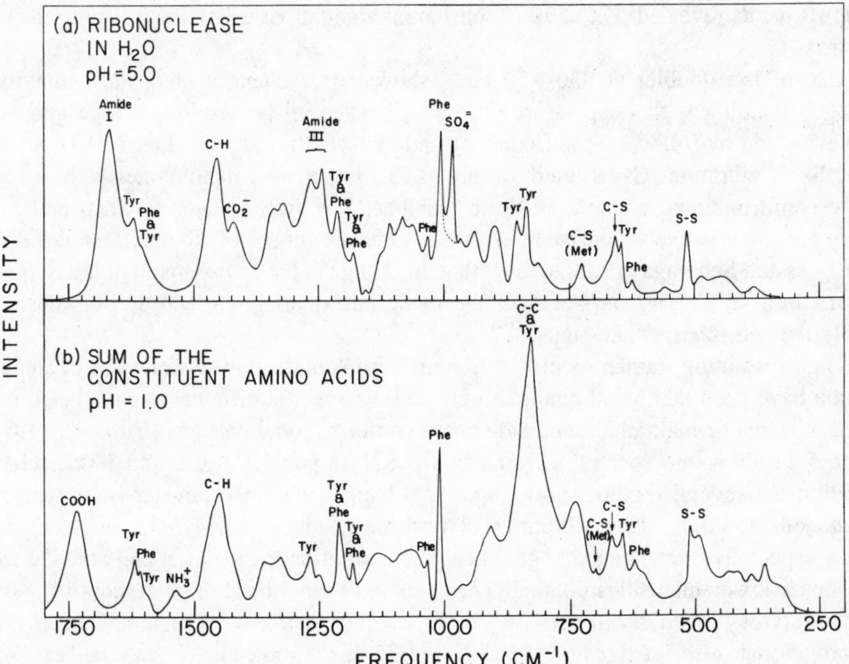

Fig. 5.9. Comparison of Raman spectrum of native RNase A with the spectrum of the sum of the constituent amino acids at pH 1 (from Lord and Yu, 1970a). Note that tyrosine has a weak line at 1265 cm^{-1}. (Chen and Lord, 1976.)

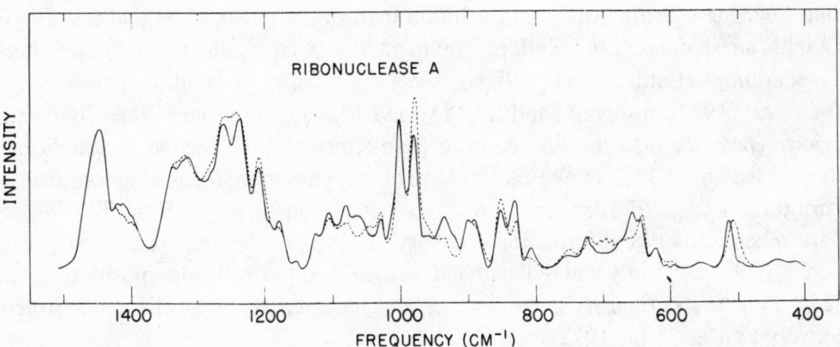

Fig. 5.10. Raman spectra (400–1500 cm^{-1}) of native and denatured RNase A, at 32 and 70°C, respectively, after correction of the water background and being normalized to the intensity of the methylene deformation mode at 1447 cm^{-1}. - - -, 70°C; ——, 32°C. (Chen and Lord, 1976.)

amide III bands revealed that at 70°C substantial amounts of α-helix and β-pleated sheet remain.

Use of the doublet at 830–850 cm^{-1} showed that changes take place in tyrosyl hydrogen bonding with temperature change. Changes in the disulfide-bridge geometry were evaluated by following the frequency and half-width of the band near 510 cm^{-1} due to the S—S vibration. C—S bond vibrations in cystines and methionines were used to follow conformational changes in these residues. The Raman data of Chen and Lord (1976) are quantitatively consistent with the six-stage model of ribonuclease unfolding of Burgess and Scheraga (1975), except that no change in the environment of the tyrosines appears until 45°C. The six stages of the model are overlapping and do not consist of sharply defined intermediate steps.

High-resolution Raman spectra of ribonuclease A in the powder form and in aqueous solution have been compared quantitatively and the spectral differences have been interpreted in terms of main chain and side chain conformational changes (Yu *et al.*, 1972). Figure 5.11 shows no spectral differences for RNase powder at 0% and 100% relative humidity; but several regions show important changes when the powder is compared to the aqueous solution. Ring vibrations of tyrosine residues at 644, 832, and 852 cm^{-1} display especially interesting changes. In aqueous solution the bands at 644 and 852 cm^{-1} have much lower intensities compared with the 832 cm^{-1} band. The authors suggested that these tyrosyl band differences may be ascribed to local environmental changes, i.e., conformational ones, of tyrosines 25, 92, and 97 of the molecule. Such changes would involve not only the breaking of tyrosyl-COO^- ion interactions but also unfolding of the main chain of the enzyme.

Lord and Yu (1970a) had assigned bands at 514 and 655 cm^{-1} to the S—S and C—S stretching of the four disulfide bridges (residues 26–84, 40–95, 58–110, and 65–72). In the powder spectra (Fig. 5.11a and b), the half-width of the 514-cm^{-1} band is 20 cm^{-1} and the intensity ratio $I_{C—S}/I_{S—S}$ is about 0.5. For the aqueous solution (Fig. 5.11c) the S—S band has a half-width of only 15 cm^{-1}, and the ratio of the intensities is about 0.9. Because the band width is an indication of the uniformity of the C—S—S—C geometry, and the ratio $I_{C—S}/I_{S—S}$ a measure of the C—S—S bond angles (Lord and Yu, 1970a,b), Yu *et al.* (1972) interpreted these spectral changes to mean that the geometry of the disulfide linkages is more uniform in solution than in the powder and that the C—S—S bond angles are decreased in solution. The band at 724 cm^{-1}, due to C—S stretching of *trans*-methionines (Lord and Yu, 1970a), sharpens in aqueous solution.

Yu *et al.* (1972) observed bands at 1239 and 1260 cm^{-1} for the powder in the amide III region. They assigned the first band to β-structure and the second to α-helices; the latter comprise about 15% of the molecule. The enzyme in aqueous solution displays a shift from 1260 cm^{-1} to 1265 cm^{-1}. At ~810 cm^{-1} and in the interval 870–980 cm^{-1} there are other signs of conformational change. In the powder spectra a band occurs at 1669 cm^{-1} (1667 cm^{-1} in water). This band was believed to be a superposition of amide I frequencies near 1660 and 1672 cm^{-1}, which are characteristic of α-helix and β-structure, respectively (Yu and Liu, 1972).

Matthies (1977) has used IR spectroscopy to study the interaction (by hydrogen bonding) of ribonuclease A with the phosphate groups of the inhibitors cytidine-2′-phosphate and cytidine-5′-phosphate, respectively, and compared the spectra of the nucleotides either free in solution or bound to RNase at pH values of 3.0, 5.5, and 7.5.

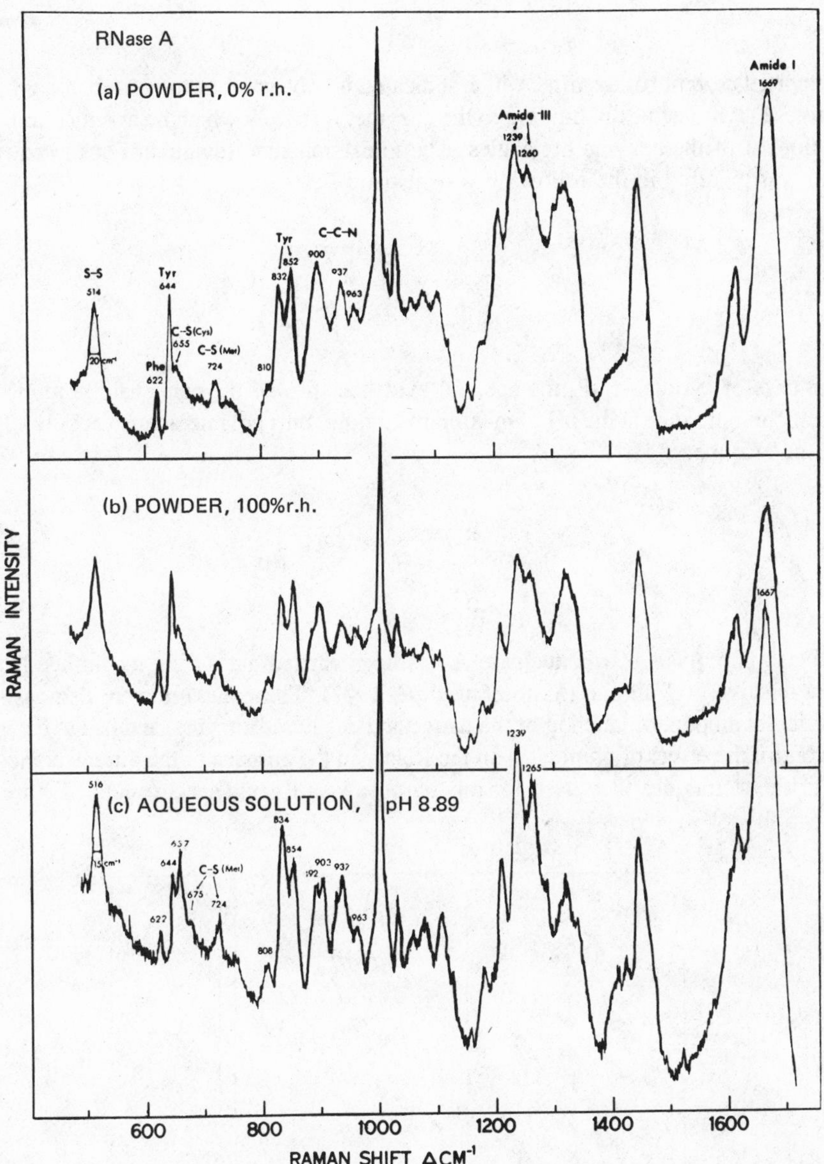

Fig. 5.11. Raman spectra of ribonuclease A in the solid and aqueous solution. (a) Spectrum of the lyophilized powder of RNase A in 0% relative humidity (rh): slit width ($\Delta\sigma$), 4 cm^{-1} (200 μ); sensitivity (*s*), 5000 counts per second (cps) full scale; rate of scan (γ), 10 cm^{-1}/min; standard deviation (sd), 0.7%; laser power (*p*) at the sample, 153 mW at 514.5 nm. The sample in powder form was packed into a conical depression at the end of an 1/8-in. gold-plated copper rod, which was fastened in a Thermovac flask equipped with an "O" ring and a vacuum-tight stopcock. The powder was dried *in vacuo* at 25°C over phosphorus pentoxide for 2 h before the experiment. The flask containing the sample and phosphorus pentoxide was then filled with dry nitrogen at 1 atm and used for laser Raman scattering work. (b) Spectrum of the lyophilized powder of RNase A in 100% rh : $\Delta\sigma$, 4 cm^{-1} (200 μ); *s*, 5000 cps; γ, 10 cm^{-1}/min; sd, 0.5%; *p*, 160 mW. The powder was equilibrated at room temperature with saturated H_2O vapor pressure (about 24.00 mm) for 5 h. (c) Spectrum of RNase A in aqueous solution at pH 8.89: $\Delta\sigma$, 4 cm^{-1} (200 μ); *s*, 5000 cps; γ, 25 cm^{-1}/min; sd, 1%; *p*, 200 mW. The solution sample at 200 mg/ml concentration and pH 8.89 was obtained by dissolving the required amount of the powder used in (a) and (b) without adding HCl or NaOH solution. (Yu *et al.*, 1972.) Reprinted with the permission of the American Chemical Society. Copyright, 1972.

He reported certain band shifts, which indicated involvement of the phosphate groups of cytidine-2′-P molecules in binding to the enzyme, a process which he assumed to involve histidine 12 of the enzyme molecules. The ionization state (dianionic) of the phosphate group was pictured in the following way at pH 5.5:

R—O—P⋯O · · · H—N N—H (structure: dianionic phosphate, charge $\overline{2}$, hydrogen-bonded to the protonated (+) imidazole ring)

In the case of cytidine-5′-P, the spectral evidence showed the phosphate group to be in the monoanionic state at the pH of maximum binding, thus not interacting (or only slightly) with the enzyme:

R—O—P—O—H · · · N NH (structure: monoanionic phosphate, charge ⊖, hydrogen-bonded to the imidazole ring)

Flash photolysis of ribonuclease A has been carried out during its surface labeling with a reactive aryl nitrene (Matheson *et al.*, 1977). These authors were demonstrating the different amount of labeling of the native and denatured enzyme, and had to determine as a control the effect of photolysis on the RNase in the absence of the nitrene compound. One of the techniques they used for this purpose was Raman spectroscopy. Figure 5.12

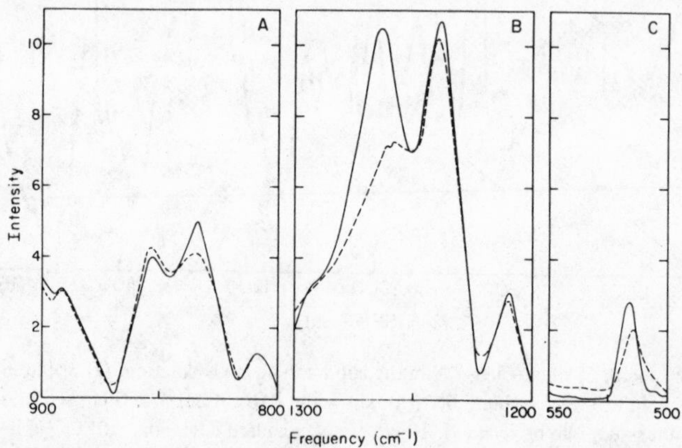

Fig. 5.12. (A) Raman spectrum of RNase A in the region 800–900 cm^{-1} showing the tyrosine doublet, before (solid curve) and after (broken curve) flashing. The curves have been redrawn to eliminate noise and baseline differences (peak intensities in any spectrum are always relative to the intensity of the δ_{CH_2} vibration which is taken to be 10). (B) Raman spectrum of RNase A in the region 1200–1300 cm^{-1}, showing the amide III region before (solid curve) and after (broken curve) photolysis. (C) Raman spectrum of RNase A in the region 500–550 cm^{-1} before (solid curve) and after (broken curve) photolysis, showing the S—S stretch vibration. In this figure, all peak assignments were taken from Chen and Lord (1976). (Matheson *et al.*, 1977.) Reprinted with the permission of the American Chemical Society. Copyright 1977.

shows three sections of the spectra of RNase A before and after flash photolysis. The tyrosine doublet, which Siamwiza *et al.* (1975) had shown to be indicative of the hydrogen-bonding condition of the phenolic moiety, appears in Fig. 5.12A. The size of the differences in the doublet for the unflashed and flashed enzyme indicates the exposure of only one buried tyrosine (Chen and Lord, 1976). The location and intensity of the band at ~1239 cm^{-1} in the amide III region show that practically no β structure is broken. The large change in the band at ~1260 cm^{-1} is partly explained by changes in tyrosine moieties (Frushour and Koenig, 1975). The S—S bond stretching region displays a change in Fig. 5.12C, showing that the photolysis affects such a bond.

Creatine Kinase

Creatine kinase (adenosine 5′-triphosphate-creatine phosphotransferase) is inhibited by a specific group of anions which stabilize the dead-end complex, kinase-divalent cation-ADP-creatine (Watts, 1973). These anions inhibit the reaction by simulating the equatorial PO_3 plane formed by the migrating phosphoryl group in the transition state of the reaction. Reed *et al.* (1978) have used IR spectroscopy to study the mode of binding of nitrate, thiocyanate, and azide to the dead-end complex of creatine kinase. An example of the type of experiment carried out is presented in Fig. 5.13, which shows the IR spectrum for solutions of KSCN and difference spectra for SCN^- recorded in the presence of enzyme, metal ions (Mn^{2+} and Co^{2+}), and substrates. A solution containing enzyme, KSCN, $MnCl_2$, ADP, and creatine shows a spectral band at 2093 cm^{-1}, the same position SCN^- displays when bound to Mn(II) (Fronaeus and Larsson, 1962). When $CoCl_2$ replaces

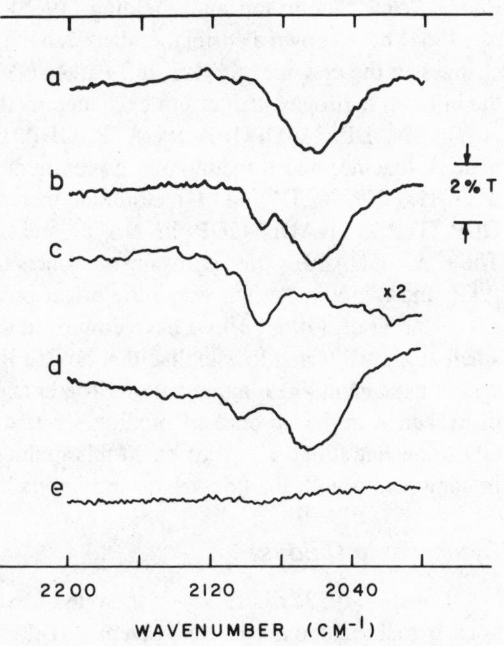

Fig. 5.13. Infrared difference spectra for KSCN. All solutions were buffered at pH 7.9 with 10 m*M* Tris/Cl^-. The cells were cooled by circulating water at 20°C through the jacketed cell holders. (a) KSCN (6 m*M*) + enzyme (6 m*M* sites) + ADP (6 m*M*) + creatine (50 m*M*) versus enzyme (6 m*M* sites) + ADP (6 m*M*) + creatine (50 m*M*) + KNO_3 (6 m*M*). (b) same compositions as in *a* with the addition of $MnCl_2$ (6 m*M*) to the sample cell. (c) KSCN (6 m*M*) + enzyme (6 m*M* sites) + ADP (6 m*M*) + $MnCl_2$ (6 m*M*) + creatine (50 m*M*) versus identical solution without $MnCl_2$. (d) same composition of sample and reference solutions as (b) except $CoCl_2$ (6 m*M*) replaces the $MnCl_2$. (e) enzyme was not present in sample or reference. KSCN (6 m*M*) + ADP (6 m*M*) + creatine (50 m*M*) + $MnCl_2$ (6 m*M*) versus similar solutions without metal ion. (Reed *et al.*, 1978.)

$MnCl_2$ in the above solution, the new absorption band is at 2109 cm^{-1}. The absorption band for a simple complex of SCN^- and Co(II) is ca. 3 cm^{-1} higher; the small downward shift for the more complicated solution is probably due to the other ligands to the Co(II) (Nakamoto, 1970). The close similarities in spectral position of the bands for the enzymic complexes and those for the corresponding SCN^--metal complexes, and the determination of the band location by the type of metallic ion present, are evidence that the shift in wavenumber for the enzyme-bound SCN^- originates in the direct binding of thiocyanate to the metal ion at the active site of the creatine kinase (Reed *et al.*, 1978).

IR spectra for complexes of nitrate and azide with kinase-divalent cation-ADP-creatine species also coincide with the spectra found for the simpler anion–metal ion complexes, and are evidence for the binding of these anions to the activating divalent cation.

None of the anions studied shows a high affinity for Mn(II), Mg(II), and Co(II) in free solution. Thus, the producing of coordination of anion and metal ion with the dead-end complex of the enzyme signifies coordination of the metal ion to an oxygen atom of the equatorial PO_3 plane in the transition state of the reaction (Reed *et al.*, 1978).

L-Glutamate Dehydrogenase

L-Glutamate dehydrogenase (GDH) has been extensively studied as a model allosteric enzyme (Tomkins *et al.*, 1963; Yielding *et al.*, 1964; Frieden, 1965; Henderson *et al.*, 1969). The theoretical basis of allosteric control is that small molecules (effectors) can bind to regulatory sites on an enzyme, triggering a conformational change which affects the activity of the enzyme. Many small molecules have been shown to influence the activity of bovine liver GDH. Some studies with fluorescence techniques (Dodd and Radda, 1969; Thompson and Yielding, 1968) and immunochemical methods (Talal *et al.*, 1964) have shown a correlation between the presence of effectors and conformational changes in the enzyme. Stryker and Parker (1970) have studied bovine liver GDH with the infrared hydrogen–deuterium exchange method. They measured the effects of NAD, NADH, NADP, NADPH, ADP, ATP, GDP, GTP, diethylstilbestrol, ethanol, L-glutamate, L-leucine, and L-methionine on the hydrogen–deuterium exchange characteristics of GDH. GTP, NADP, and L-methionine increased the extent of exchange, while NAD, GDP, GTP plus NADH, GDP plus NADP, and L-leucine decreased the extent of exchange. The effect of NADH varied with enzyme concentration. Diethylstilbestrol, NADPH, ADP, ATP, and L-glutamate had very little effect on the exchange.

Sund *et al.* (1968, 1969) have employed sedimentation, light scattering, and X-ray small-angle scattering to establish that bovine liver GDH undergoes a reversible concentration-dependent linear aggregation. Stryker and Parker (1970) have found that the extent of exchange in the absence of modifiers was markedly decreased by an increase of the GDH concentration. They attributed this finding to a change in conformation or motility in conjunction with the known linear aggregation of GDH.

Cytochrome Oxidase

Capaldi (1973) has followed the rate of exchange of peptide hydrogens of membranous cytochrome oxidase for solvent (^2H_2O) deuterium by IR spectroscopy. The IR spectrum showed that the enzyme contains α helix and possibly β conformation. The exchange data showed that a maximum of 59% of the polypeptide chains are in these conformations and also indicated that at least 60% of the cytochrome oxidase is exposed to the aqueous milieu surrounding the membrane.

Na^+, K^+-Dependent ATPase

Brazhnikov *et al.* (1978) have studied the secondary structure of the protein component of a highly active Na^+, K^+-dependent adenosine triphosphatase system. They examined the amide region of the protein–lipid complex by IR spectroscopy and showed that in the protein component, which was comprised of two polypeptides with molecular weights 50,000 and 100,000, the α-helical content was ca. 20% and the antiparallel β-pleated sheet content was ca. 25%. Regions with regular structure were mainly situated in protein sections to which water was not admitted; these regions appeared to be situated in the hydrophobic core. About 55% of the protein structure appeared to be unordered and readily admitted water molecules.

ATPase from Bacterium PS3

Ohta *et al.* (1978) have reported the effects of various adenine nucleotides on the kinetics of the H–D exchange reaction of peptide groups of an ATPase (TF_1) from a thermophilic bacterium PS3. Some examples of the kinetic plots obtained from their IR measurements are shown in Fig. 5.14. The exchange rates of peptide hydrogens were slower in the presence of ADP and adenylylimidodiphosphate [AMP-P(NH)P]. A series of ρ values was calculated for the peptide hydrogens of TF_1 under various conditions. ρ is the probability of solvent exposure of the peptide hydrogens (Hvidt and Wallevik, 1972). The authors concluded that TF_1 does not break up into subunits even when nucleotides are absent, so that differences in rates of exchange appeared to be the result of differences in the conformations of subunits.

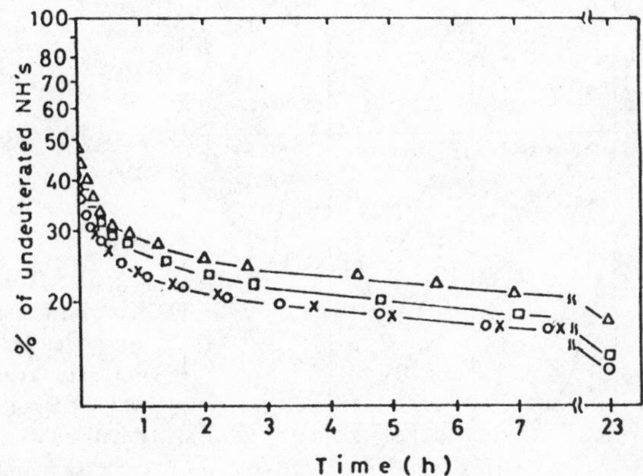

Fig. 5.14. Semilogarithmic plots of the hydrogen–deuterium exchange of the peptide hydrogen in TF_1 at 40°C and pH 7.52 in 200 m*M* Hepes–Na buffer. (○), TF_1 with 5 m*M* $MgCl_2$, (×), with 20 m*M* AMP and 20 m*M* $MgCl_2$, (□), with 20 m*M* ADP and 20 m*M* $MgCl_2$, (△), with 20 m*M* AMP-P(NH)P and 20 m*M* $MgCl_2$. Hepes is *N*-2-hydroxyethylpiperazine-*N'*-2-ethanesulfonic acid. (Ohta *et al.*, 1978.)

Spinach Chloroplast Coupling Factor

Coupling factor 1 (CF_1) from spinach chloroplasts is an enzymatic protein containing five subunits (Racker *et al.*, 1971) which catalyzes the formation of ATP from ADP and inorganic phosphate during photosynthesis. The IR method of studying hydrogen–deuterium exchange (Parker, 1971) has shown that when different adenine nucleotides (ATP, AMP, ADP, and 1,N^6-ethenoadenosine triphosphate) interact with CF_1 they all increase the conformational stability of the enzyme except AMP (Nabedryk-Viala *et al.*, 1977). The interaction corresponds to a nucleotide binding on three sites of CF_1.

Triose Phosphate Isomerase and β-Lactamase

Belasco and Knowles (1980) have used FTIR spectroscopy to study covalent and noncovalent enzyme-substrate complexes. When dihydroxyacetone phosphate binds to triose phosphate isomerase, a C═O absorption band shifts 19 cm^{-1} to lower frequency. The new band originates from the bound substrate. Also, in the β-lactamase-catalyzed hydrolysis of the 7α-methoxycephalosporin, cefoxitin, an acyl enzyme intermediate was shown to be on the reaction pathway.

Effect of SDS Treatment on IR Spectra of Elastase and Other Enzymes

Films of native elastase have a major infrared amide I band at 1635 cm^{-1}, which is greatly shifted to 1650 cm^{-1} when sodium dodecyl sulfate (SDS) is present (Visser and Blout, 1971). A shift in the opposite direction to 1622–1625 cm^{-1} occurs after the removal of SDS by dialysis. The amide II (NH) frequencies are respectively 1515, 1525, and 1512 cm^{-1}. The amide I frequency shifts suggested some α-helix and/or unordered struc-

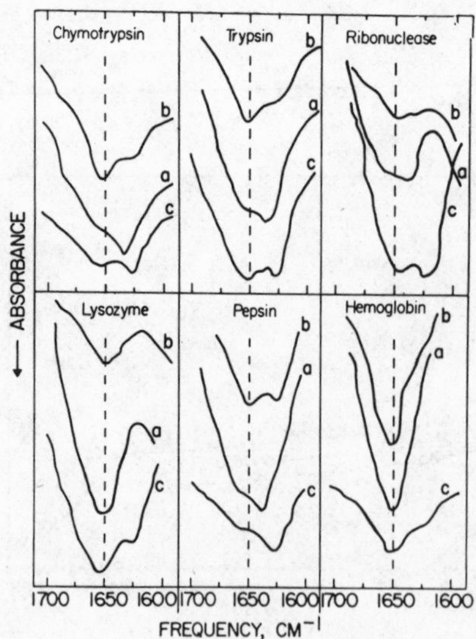

Fig. 5.15. The effect of sodium dodecyl sulfate treatment and subsequent dialysis on the infrared spectra of several proteins. The protein solutions were evaporated on silver chloride disks *in vacuo*. (a) native enzyme; (b) treated with SDS; (c) treated with SDS followed by dialysis. (Visser and Blout, 1971.) Reprinted with the permission of the American Chemical Society. Copyright 1971.

ture in elastase in the presence of SDS (1650 cm^{-1}); and a type of β-structure (1622–1632 cm^{-1}) was found after the removal of SDS.

The IR data for chymotrypsin, trypsin, pepsin, and lysozyme (Fig. 5.15) were qualitatively similar, and each involved the appearance of a new band or intensification of a shoulder near 1650 cm^{-1} when SDS was present. After dialysis a new band or an enhanced existing band or shoulder were observed at 1625 ± 3 cm^{-1}, with the appearance sometimes of a diffuse shoulder around 1680 cm^{-1}. Note in Fig. 5.15 that the presence of SDS did not alter the location of the amide I band of lysozyme and hemoglobin, substances containing large amounts of α-helical structure. The IR data on amide I frequencies of films given in this paper supported conclusions based on CD spectra. An important point is that great caution must be used for the interpretation of protein conformations when detergents are present or after they have been used during extraction processes.

Resonance Raman Studies

Carbonic Anhydrase

Kumar *et al.* (1976) have recorded resonance Raman spectra of several sulfonamide derivatives bound to various isoenzymes of carbonic anhydrase (CA) by exciting into the sulfonamide absorption bands in the 400 to 500 nm region. (A variety of aromatic sulfonamides are CA inhibitors.) In this way they were able to obtain vibrational spectra of the sulfonamides in the active site unobscured by bands from the vibrational frequencies of the protein and the water solvent. The sulfonamide derivatives were 4-sulfonamido-4′-dimethylaminoazobenzene (I), 4-sulfonamido-4′-hydroxyazobenzene (II), and 4-sulfonamido-4′-aminoazobenzene (III) (Fig. 5.16). The group —SO_2NH^- was present in the complexes (see also Kumar *et al.*, 1974). The spectra of the following complexes displayed no differences from each other: 4-sulfonamido-4′-dimethylaminoazobenzene bound to

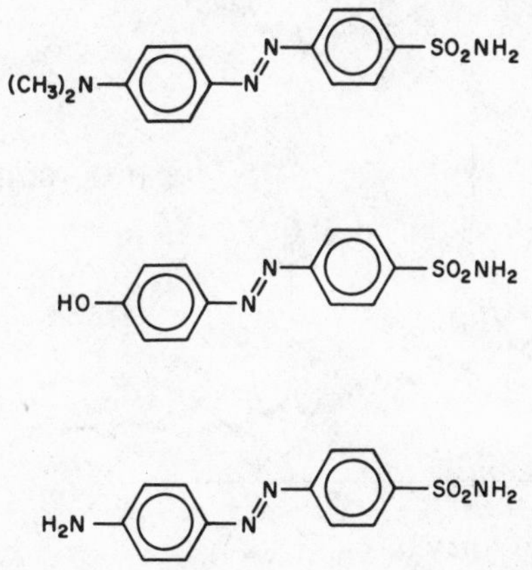

$(CH_3)_2N-C_6H_4-N{=}N-C_6H_4-SO_2NH_2$ I

$HO-C_6H_4-N{=}N-C_6H_4-SO_2NH_2$ II

$H_2N-C_6H_4-N{=}N-C_6H_4-SO_2NH_2$ III

Fig. 5.16. The structures of sulfonamides I, II, and III.

human carbonic anhydrase B and C, or between these and the sulfonamide bound to bovine carbonic anhydrase (BCA) or cobalt(II) human carbonic anhydrase B (HCA B). The bound sulfonamides display a new band (e.g., Fig. 5.17), which was interpreted as reflecting a spatial change in the vicinity of the sulfonamido sulfur atom, by virtue of the strong similarity between the bound sulfonamide group and the transition state of the reactants in the reversible hydration of carbon dioxide (Fig. 5.18).

This study and some of the others that follow use resonance Raman labeling. [See P. R. Carey, H. Schneider, and H. J. Bernstein, *Biochem. Biophys. Res. Commun.* **47,** 588 (1972).] A resonance Raman label furnishes a detailed vibrational spectrum from a

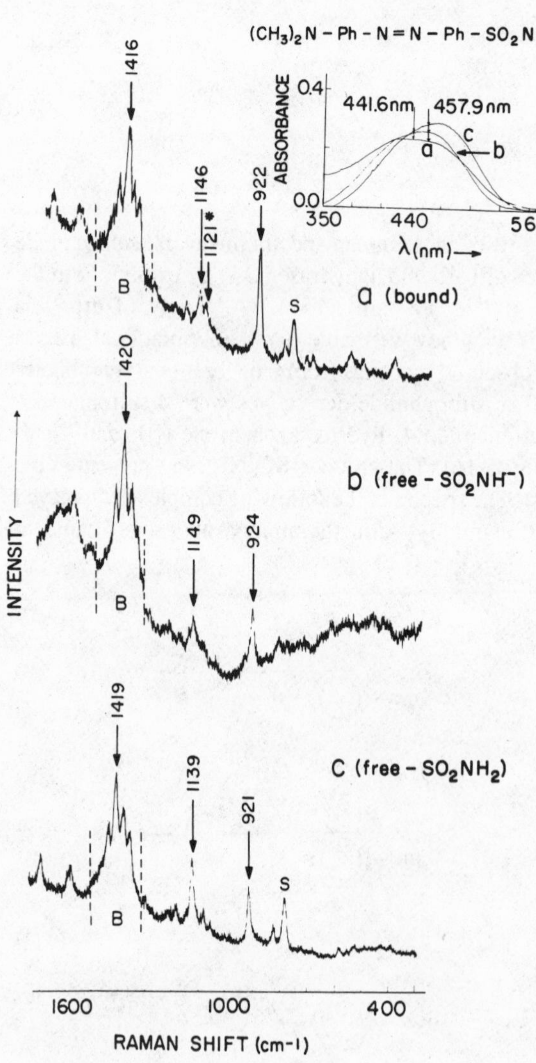

Fig. 5.17. The resonance Raman spectra (441.6-nm excitation) of sulfonamide (I): (a) bound to HCA(C), acetone-Tris-SO_4 buffer (1:10 v/v), pH 9.0, sulfonamide ~6 × 10^{-5} *M*, enzyme ~10^{-4} *M;* (b) free anionic sulfonamide (SO_2NH^-) ~2 × 10^{-4} *M*, 0.1 *N* KOH, pH 13.0; (c) free —SO_2NH_2 form, ~1.0 × 10^{-2} *M* in acetone. Spectral slit width ~9 cm^{-1} in all cases and power ~20 mW for the bound and ~60 mW for the unbound sulfonamide spectra. The asterisk indicates a new band. S denotes acetone solvent peaks. The experimental conditions for the inset (A) absorption spectra: (c) free (—SO_2NH_2) form, 6.6 × 10^{-5} *M*, pH 8.0; (b) free (SO_2NH^-) form, 6.6 × 10^{-5} *M*, pH 13.6; (a) bound to BCA, sulfonamide (I) 6 × 10^{-5} *M*, enzyme ~1 × 10^{-4} *M*, pH 8.0. All the spectra were recorded in 1-cm cells and the compound was dissolved in acetonitrile-Tris-SO_4 buffer (32:68 v/v). Experimental conditions for inset B resonance Raman spectra (457.9-nm excitation): (a) bound to BCA, CD_3CN:-Tris-SO_4 buffer (5:95 v/v), pH 8.8, sulfonamide ~4.5 × 10^{-4} *M*, enzyme ~1 × 10^{-3} *M;* (b) free anionic (—SO_2NH^-) form, ~1 × 10^{-3} *M*, CD_3CN:1 *N* KOH (50:50 v/v); (c) free (—SO_2NH_2) form, ~1 × 10^{-3} in CD_3CN. Slit width ~9 cm^{-1} in all cases. Power ~15 mW for unbound (—SO_2NH_2) and bound sulfonamide spectra and ~30 mW for unbound (—SO_2NH^-) spectra. Due to the overlapping with an acetone peak ~1430 cm^{-1}, the 1400-cm^{-1} region (inset B) is presented with CD_3CN as a solvent. (Kumar *et al.*, 1976.)

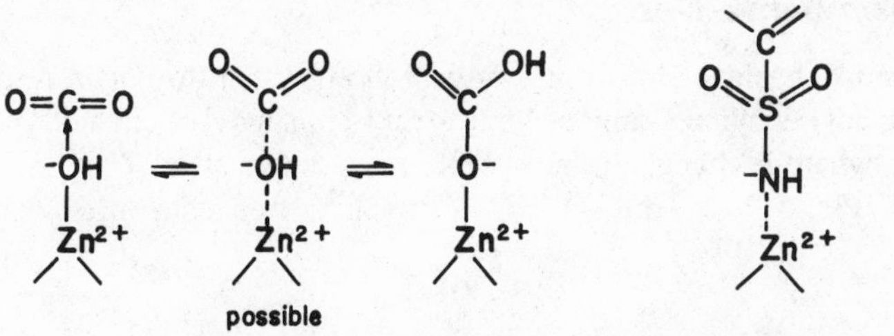

Fig. 5.18. Transition state of the reactants and its similarity to the structure of the bound sulfonamide group. (Adapted from Kumar *et al.*, 1976.) Reprinted with the permission of the American Chemical Society. Copyright 1976.

specially designed chromophore when it is bound to a biochemically active site [Carey and Schneider, *Accounts Chem. Res.* **11,** 122 (1978)].

In contrast to the study by Kumar *et al.* (1976) is the determination by Petersen *et al.* (1977) of the ionization state of another sulfonamide bound to the enzyme carbonic anhydrase. They used resonance Raman spectroscopy to study the interaction between the enzyme and 4′-sulfamylphenyl-2-azo-7-acetamido-1-hydroxynaphthalene-3,6-disulfonate (Neoprontosil, a CA inhibitor) shown here (Fig. 5.19). On the basis of spectral bands in the regions 900 to 1000 cm^{-1} and 1100 to 1200 cm^{-1}, Petersen *et al.* determined that the sulfonamide was bound to the carbonic anhydrase as SO_2NH_2 rather than as SO_2NH^-. The pH dependence of the Neoprontosil spectrum showed the pK_a of sulfonamide ionization to be 10.5. The enzyme–sulfonamide complex has a spectrum (not shown) identical to that of the low-pH acid form even at pH values for which the free sulfonamide is ionized, thus permitting these workers to conclude that the protonated form of the sulfonamide is the one that binds to the enzyme.

A later study (Carey and King, 1979) on the sulfonamide Neoprontosil used alkalimetric, spectrophotometric, NMR, and resonance Raman titrations to determine the ionization state of this drug when it is bound to carbonic anhydrase. It was not possible to draw a definitive conclusion on the ionization state of this sulfonamide when it is bound to the enzyme. The fact that the microscopic p*K*s for the groups —OH and —SO_2NH_2 are very close to each other in the pH range 10.5 to 11.5 precluded spectroscopic characterization of the separate —O^-, —SO_2NH_2 or —OH, —SO_2NH^- species.

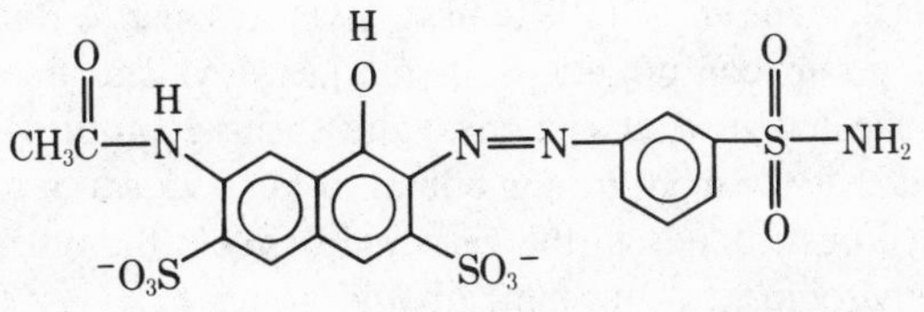

Fig. 5.19. Neoprontosil.

Chymotrypsin Derivatives

Carey and Schneider (1976) have used a flow system to obtain resonance Raman spectra of the acyl enzyme 4-amino-3-nitro-*trans*-cinnamoyl-α-chymotrypsin throughout the pH range where it changes from a stable to an active state. They assigned the band at 1625 cm^{-1} (Fig. 5.20) to the —C═C— stretching frequency in the acryloyl moiety

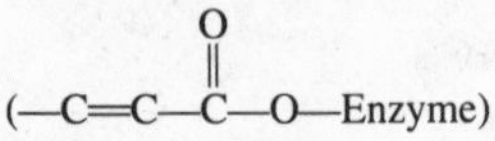

$$(\text{—C=C—}\overset{\overset{\text{O}}{\|}}{\text{C}}\text{—O—Enzyme})$$

and showed that bands in the range from 1200 to 1450 cm^{-1} were associated with the aromatic residue. Using analogs of the substrate, they established that use of the aromatic and the —C═C— bands could independently gauge chemical activity in the aromatic moiety and the

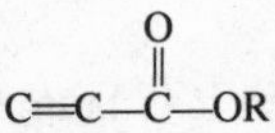

$$\text{C=C—}\overset{\overset{\text{O}}{\|}}{\text{C}}\text{—OR}$$

moiety, respectively. Thus, activity in the acyl enzyme which yielded spectral changes involving the catalytic site could be differentiated from those involving the aromatic site.

At pH 3, the structure of the acylating group in the stable acyl enzyme was similar to that of the substrate in aqueous solution, the conformation about the —C═C— bonds being essentially planar, *trans,* and probably s-*trans* about the single bond of the C═C—C═O group. It was found, however, that in the pH range from 5.7 to 7.0, the acyl enzyme produces a large change in the 1625-cm^{-1} band unaccompanied by any changes spectrally in the aromatic bands. This means that the enzyme causes a change (within about 10 sec) in a submolecular grouping of which the bond being hydrolyzed is part. Since the change in the 1625-cm^{-1} band was invariant at the pH of 5.7 (enzyme inactive) and at pH 7.0 (enzyme active), the spectral change must precede and be independent of the rate-controlling step in deacylation. These authors believe that the change in the 1625-cm^{-1} band probably indicates an ionization in the acyl enzyme between pH 3.0 and 5.7, and that above pH 5.7 deacylation occurs from a structure which is different from the one at the pH where it is stable, i.e., pH 3.0.

The resonance Raman spectra of the intermediate form give evidence supporting either the stabilization of resonance structures of the type,

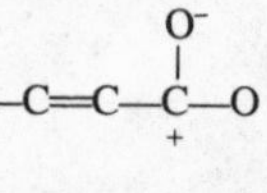

$$\text{—C=C—}\underset{+}{\overset{\overset{\text{O}^-}{|}}{\text{C}}}\text{—O}$$

or deformation to reduce planarity within the acryloyl moiety.

MacClement *et al.* (1981) have given resonance Raman spectroscopic evidence for two acyl group conformations in furylacryloyl and thienylacryloylchymotrypsins. For most of the native acyl enzymes, the spectral profiles in the C═O stretching region suggested that the acyl groups bound to Ser-195 adopt two conformations: one, having strong hydrogen bonds to the carbonyl oxygen; the other, having a nonbonding hydrophobic environment about the carbonyl group.

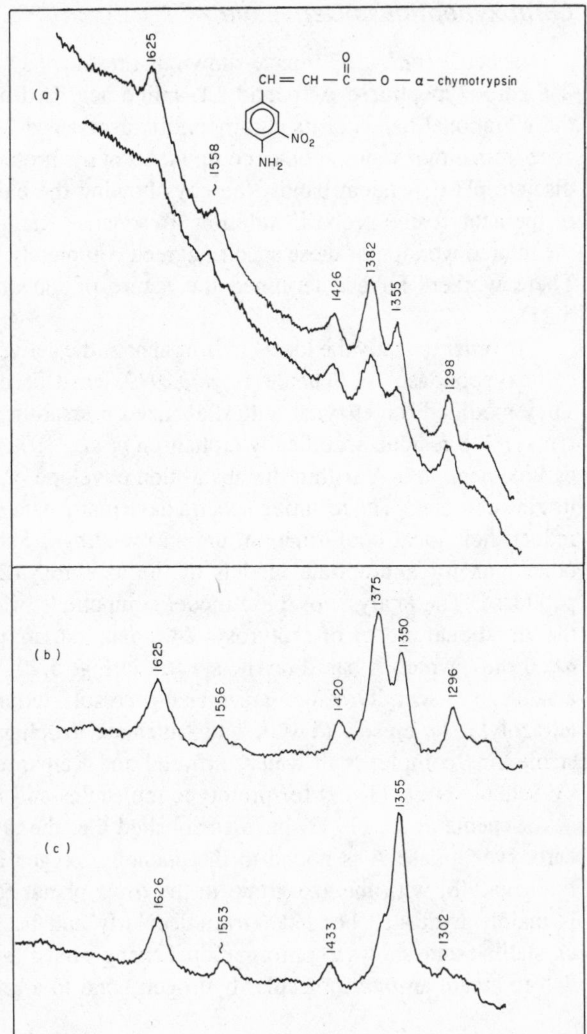

Fig. 5.20. Resonance Raman spectra of (a) 4-amino-3-nitro-*trans*-cinnamoyl-α-chymotrypsin, in a stationary quartz cell at pH 3.0 (top), in a flow system 15 sec after raising pH to 5.9 (identical spectra were obtained from pH 5.7 to 7.0) (bottom). (b) 4-amino-3-nitro-*trans*-cinnamic acid methyl ester in 60% H_2O, 40% C^2H_3CN. (c) 4-amino-3-nitro-*trans*-cinnamic acid methyl ester in 60% ^2H_2O, 40% C^2H_3CN. Concentrations: (a) 2×10^{-4} *M* in protein, approximately 30% of active sites are acylated. (b) and (c) 2.5×10^{-3} *M*. Typical spectral conditions 80 mW He–Cd laser power at 4416 Å, 9 cm^{-1} spectral slit width, 20,000 cts/sec full scale, time constant 2 sec. Under these conditions spectra taken in H_2O have a very weak contribution from an H_2O feature near 1650 cm^{-1}. (Carey and Schneider, 1976.)

Phelps *et al.* (1981) have used resonance Raman spectroscopy to study correlations between reactivity (rates of deacylation) and structure of some chromophoric acylchymotrypsins. They found that the intensity patterns in the spectra of the 4-amino-3-nitrocinnamoyl intermediate obtained by using 350.7-nm excitation are very different from those seen earlier (Carey and Schneider, 1976) as a result of 441.6-nm excitation. The C═O profiles or C═C stretching profiles in the resonance Raman spectra of 3-methylthienylacryloyl-, α-methyl-5-methyl-thienylacryloyl-, and α-methyl-5-methyl-furylacryloyl chymotrypsin, along with their anomalous correlations with respect to $\nu_{C=C}$ vs. λ_{max}, showed that enzyme–substrate interactions perturb the geometry and π-electron distribution of the chymotrypsin-bound chromophore compared to the free moiety in solution.

Carboxypeptidase Derivative

Scheule *et al.* (1977) have shown that resonance Raman spectra of arsanilazotyrosyl-248 carboxypeptidase A (peptidyl-L-amino acid hydrolase, E.C. 3.4.12.2) display only the vibrational bands of its chromophoric azotyrosyl-248 residue uncluttered by interference from either water or other components of the protein. The spectra contained multiple, discrete pH-dependent bands, thereby showing the existence of interconvertible species of the azotyrosine probe in solution. Resonance Raman data concerning pK' values for the interconversion of these species agreed completely with data from absorption spectra. These workers have determined the nature of the azotyrosyl-248–zinc complex (Fig. 5.21).

In order to study the local environment of the active center of the zinc metalloenzyme, carboxypeptidase A, Scheule *et al.* (1979) have used resonance Raman spectroscopy. They modified the enzyme with diazotized *p*-arsanilic acid, which labels the active-site tyrosyl-248 residue specifically (Johansen *et al.*, 1972), and then excited the arsanilazo-carboxypeptidase A within the absorption envelope of the azotyrosine chromophore and its zinc complex. The resulting spectra have many bands whose frequencies and intensities reflect their local conformation and surroundings. Scheule *et al.* reported spectral and other data for appropriate models of the azotyrosyl-248 residue of arsanilazocarboxypeptidase. The analysis of these model compounds offered a basis for the assignment of the vibrational modes of azotyrosyl-248, data that are presented in Table 5.3 (Scheule *et al.*, 1980), which is based on the spectra in Fig. 5.22. The model compounds used were arsanilazo-*N*-acetyltyrosine, arsanilazo-*p*-cresol, tetrazolylazo-*N*-acetyltyrosine (TAT), tetrazolylazo-*p*-cresol, (TAC), and tridentate azophenols TAT and TAC, which form stable zinc complexes in water. Infrared and Raman data were also given in the paper by Scheule *et al.* (1979) for prototype molecules and model azophenols.

Scheule *et al.* (1980) have established that the active-site zinc atom of arsanilazo-carboxypeptidase A is bound to the phenolic oxygen and an azo nitrogen atom of azo-tyrosine-248, with the azo group in the *trans* planar form. In solution, azotyrosine-248 is mainly hydrogen bonded *intra*molecularly and has water-like surroundings; but the crystalline state shows a conformational change that allows the phenolic proton of tyrosine-248 to create an *inter*molecular hydrogen bond to a group of the protein.

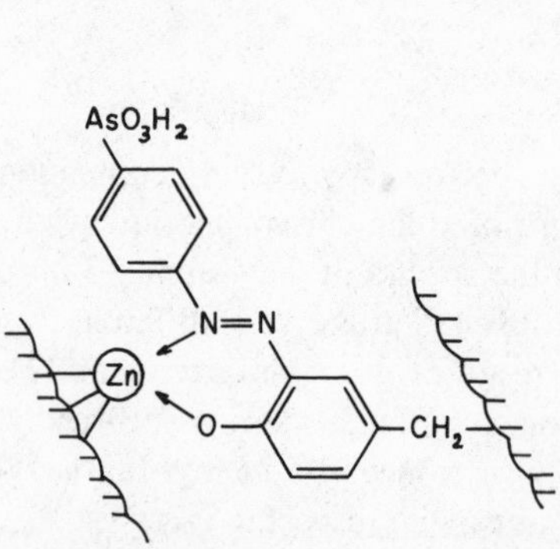

AsO_3H_2 N=N Zn O CH_2

Fig. 5.21. Structure of the azotyrosyl-248-zinc complex of arsanilazotyrosyl-248 carboxypeptidase A deduced from its resonance Raman spectra. (Scheule *et al.*, 1977.)

Table 5.3. Resonance Raman Spectra of Arsanilazotyrosine-248 Carboxypeptidase[a,b]

Azophenol, pH 6.2	Complex, pH 8.5	Azophenolate, pH 11.0	Assignment[b]
—	1538m	1539w	$\nu^{\phi O} + {}^2\nu_{8b}{}^{CC}$
?	1486w	1481w	${}^2\nu_{19a}{}^{CC}$
1456*m 1427*s	1436s	1440s	$\nu^{NN} + {}^2\nu_{19b}{}^{CC}$
?	1408sh	1404w	${}^1\nu_{19b}{}^{CC}$
1394m	1389s	1389s	${}^2\nu_{19b}{}^{CC} + \nu^{NN}$
1359w	1356w	~1355sh	${}^2\beta_3{}^{CH}$
—	1338s	1340m, b	${}^2\nu_{14}{}^{CC} + \beta^{\phi O}$
1319w	1314w	1310sh	
1247w, b	1240w	1244w	${}^2\nu_{7a}{}^{CX}$
1205w, b	1216m	1215w	$\nu^{\phi N} + {}^2\nu_{13}{}^{CX}$
	1187w	1203w	${}^2\nu_{13}{}^{CX} + \nu^{\phi N}$
1162m	1165s	1161m	${}^1\nu_{13}{}^{\phi X}$
1150*m 1123*w	1150m	1145m	${}^2\beta_{15}{}^{CH} + {}^2\nu_1{}^{CC} + \nu^{\phi N}$
—	1120m	1114w	${}^2\nu_{12}{}^{CC} + {}^2\beta_{18b}{}^{CH}$
1092w	1089w	1089w	${}^1\nu_1{}^{CC}$

[a] Scheule *et al.* (1980).

[b] Only bands in the 1100–1550-cm^{-1} region are listed. Bands with asterisks arise from rotamers (Fig. 5.23). All spectra are of 0.05 m*M* aqueous samples in 0.5 *M* NaCl and 2 m*M* Tris-HCl at 22°C. A dash denotes the absence of a band, and ? denotes uncertainty as to its presence. Abbreviations: s, strong; m, medium; w, weak; b, broad; sh, shoulder; v, very.

[c] Abbreviations: ν, stretch; β, in-plane bend; X, ring substituent. The subscripts (lower right) identify the normal mode according to the nomenclature of Wilson (1934). The superscripts (upper left) refer to modes arising from the aromatic ring on the left (Ar_1) or right (Ar_2) side of the azo group when the molecule is viewed as shown in Fig. 5.24. Where more than one mode designation is used, the order given reflects their relative contributions to the total potential energy change occurring during the vibration, the largest contribution coming from the mode listed first.

Papain Derivatives

Carey *et al.* (1976) have obtained resonance Raman spectra for the acyl enzyme 4-dimethylamino-3-nitro(α-benzamido)cinnamoyl-papain, which they had prepared by using the chromophoric substrate methyl 4-dimethylamino-3-nitro(α-benzamido) cinnamate. The aim of this study was to see if changes occurred in specific submolecular groupings of the acylating group while the enzymic reaction (deacylation) progressed, and then to establish the characteristics of such changes. The resonance Raman spectra gave positive evidence for a radical change in the geometry of the acyl residue upon binding to papain.

The spectra of the product from the enzymatic hydrolysis were obtained after concentration via dialysis at pH 7.0, followed by freeze-drying. Bands were displayed at

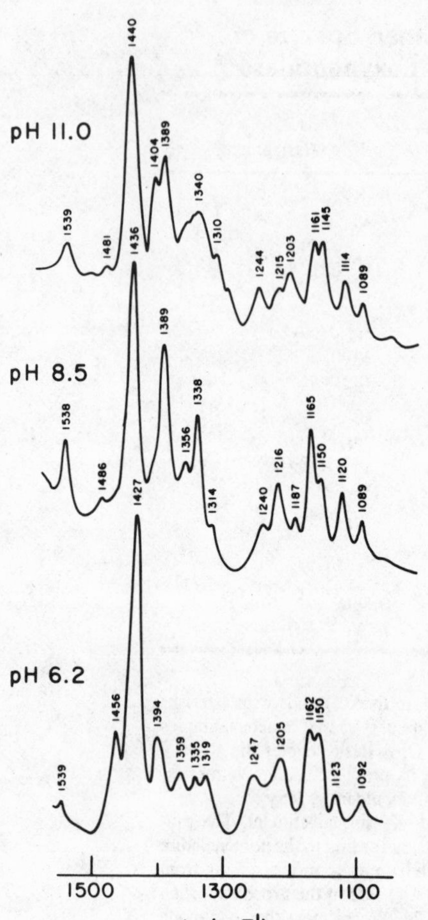

Fig. 5.22. Resonance Raman spectra of the three species of arsanilazotyrosine-248 carboxypeptidase at the pH values indicated. (Scheule *et al.*, 1980.) Reprinted with the permission of the American Chemical Society. Copyright 1980.

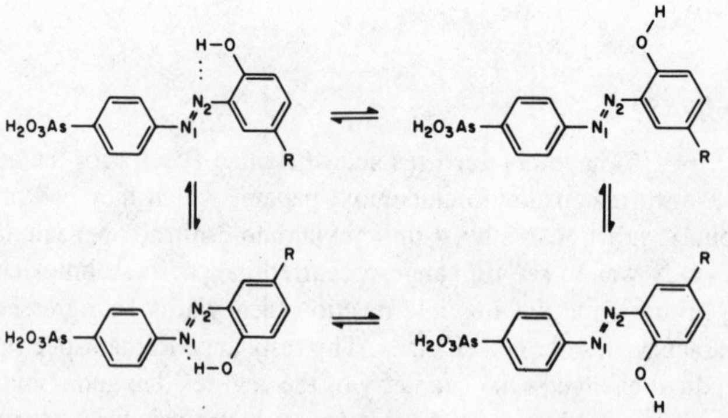

Fig. 5.23. Possible rotational isomers for *o*-hydroxyazobenzenes. (Scheule *et al.*, 1980.) Reprinted with the permission of the American Chemical Society. Copyright 1980.

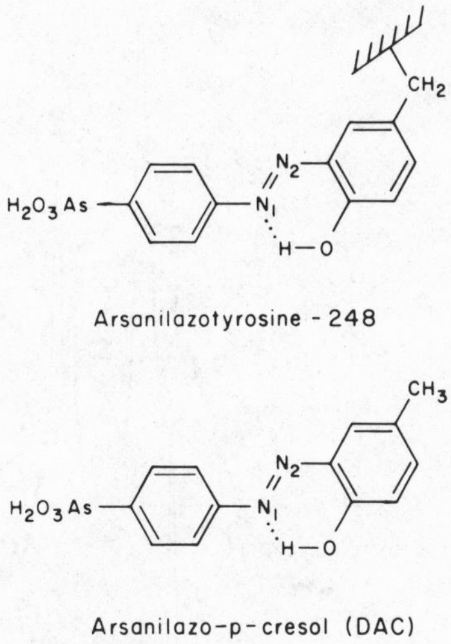

H_2O_3As — N_1 = N_2 — CH_2 — H—O

Arsanilazotyrosine - 248

H_2O_3As — N_1 = N_2 — CH_3 — H—O

Arsanilazo-p-cresol (DAC)

Fig. 5.24. Azophenol forms of the azoTyr-248 residue of arsanilazocarboxypeptidase and the model azophenol, arsanilazo-*p*-cresol (DAC) (Scheule *et al.*, 1980.) Reprinted with the permission of the American Chemical Society. Copyright 1980.

1355 and 1618 cm^{-1}, the same locations as those of the strong peaks produced by the cinnamic acid derivative (Fig. 5.25). The band located at 1649 cm^{-1} in Fig. 5.25 was hidden by a water peak in the enzymatically produced product. The spectrum of the product is very similar to that of the substrate (methyl 4-dimethylamino-3-nitro(α-benzamido)cinnamate, or αBA) (Figs. 5.26, 5.27) and was completely different from that of the acyl-enzyme (Fig. 5.26). The spectrum for the substrate (Figs. 5.26 and 5.27) is characteristic of a cinnamic acid ester (Carey and Schneider, 1974, 1976). The strong band at 1353 cm^{-1} was assigned to the symmetric stretching frequency of the nitro group.

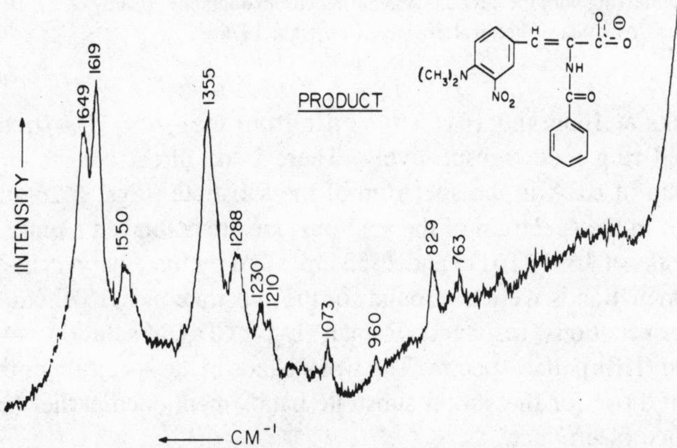

Fig. 5.25. Resonance Raman spectrum of 4-dimethylamino-3-nitro(α-benzamido)cinnamic acid, pH 13.0, concentration ~3 × 10^{-4} *M;* 441.6-nm excitation; typical conditions, 50 mW power, 4000 counts full scale, 9 cm^{-1} spectral slit width. (Carey *et al.*, 1976.) Reprinted with the permission of the American Chemical Society. Copyright 1976.

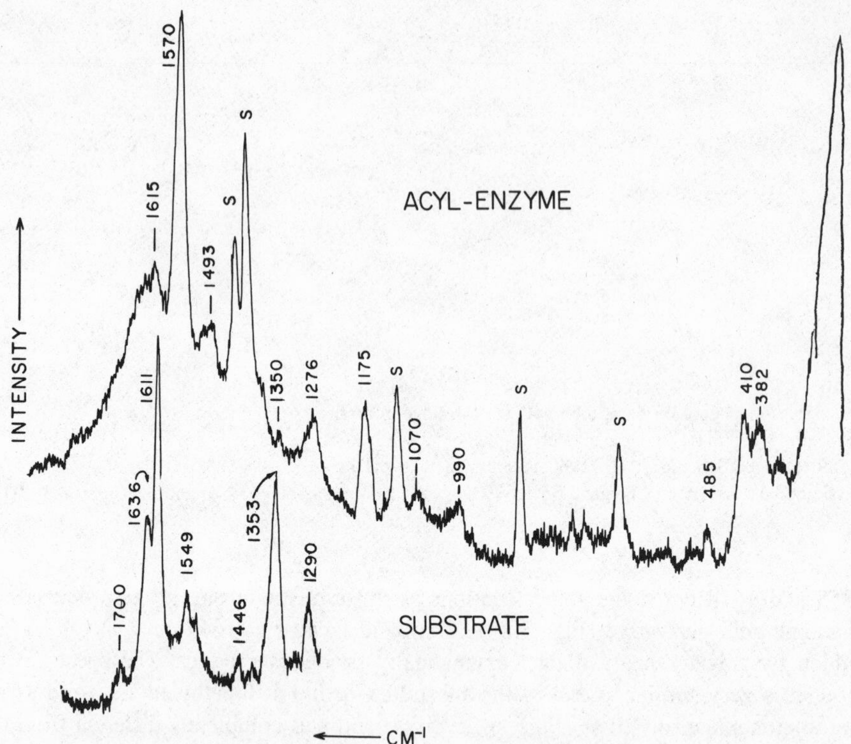

Fig. 5.26. Resonance Raman spectra of 4-dimethylamino-3-nitro(α-benzamido)cinnamoyl-papain (commercial crystallized): intermediate (top) and the substrate (αBA) alone (bottom). Concentrations: acyl-enzyme; enzyme ~6.5×10^{-5} *M;* bound substrate ~1.4×10^{-5} *M;* substrate 10^{-4} *M* in 80% D_2O, 20% CD_3CN, 441.6-nm excitation, 50-mW power, ~9-cm^{-1} spectral slit width. The substrate spectrum below 1290 cm^{-1} contains only weak features (see Fig. 5.27) which are obscured by solvent bands, S = solvent bands, resulting from dimethylformamide moving with the acyl-enzyme during chromatography. (Carey *et al.*, 1976.) Reprinted with the permission of the American Chemical Society. Copyright 1976.

The strong peaks at 1636 and 1611 cm^{-1} arise from the —C═C— stretching vibration and a benzenoid ring mode, respectively. There is no direct contribution from the α-benzamido group of αBA in the spectrum of the substrate (Figs. 5.26 and 5.27).

Differences in the spectrum of the acyl-enzyme are displayed mainly by the absence of the strong peaks at 1631, 1611, and 1353 cm^{-1} and by the emergence of a new intense peak at 1570 cm^{-1}. Bands were not found for the substrate near 1570 cm^{-1} in the Raman (solid, 647-nm excitation), resonance Raman (D_2O, CD_3CN solution, 441.6 nm excitation), or infrared (KBr pellet) spectra. The weak bands in the spectrum of the acyl-enzyme that are close to those for the strong substrate bands mentioned earlier (Fig. 5.26) arise from the presence of product.

The great spectral differences indicate the occurrence of a big change in the structure of the acyl residue. The characteristics of this change are indicated by the similarity of the resonance Raman spectrum of the intermediate to those of azlactones and of polyenes, and by the fact that the Raman intensity of the intermediate is greater than that of the substrate and product.

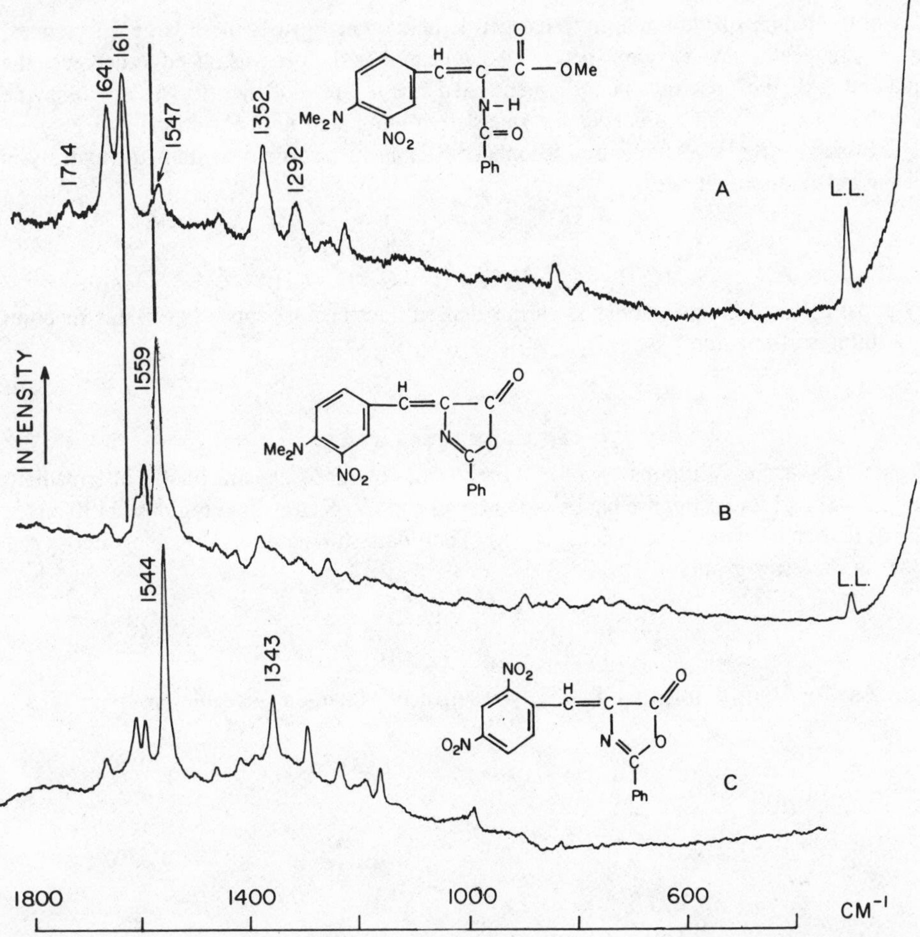

Fig. 5.27. Resonance Raman or Raman spectra of the substrate and two azlactones. (A) Substrate, methyl 4-dimethylamino-3-nitro(α-benzamido)cinnamate; (B) 4-(4-dimethylamino-3-nitro)benzylidene-2-phenyloxazolin-5-one; (C) 4-(2,4-dinitro)benzylidene-2-phenyloxazolin-5-one. All spectra are of solid samples in a KBr matrix and were taken with a rotating sample holder. (A) and (B), resonance Raman, 441.6-nm excitation. (C), normal Raman, 647.1-nm excitation. (Carey *et al.*, 1976.) Reprinted with the permission of the American Chemical Society. Copyright 1976.

The resonance Raman and normal Raman spectra of azlactones and of the complex all display a prominent band in the vicinity of 1560 cm^{-1}. Fig. 5.27 shows this band for two azlactones and the substrate. Such a band for azlactones has much —C=N— stretching character (Carey, P. R., Kumar, K., and Phelps, D. J., 1976, unpublished work). This fact plus the spectral characteristics of Fig. 5.27 show that a strong peak in the vicinity of 1560 cm^{-1} is a good indication of the PhC=C—N=CPh structure. Therefore, the strong intermediate band at ca. 1570 cm^{-1} arises from a stretching mode of a —C=C—N=C(—OX)— grouping formed from the α-benzamido side chain —C=C—NH—C(=O)—. The spectra are consistent with rearrangement of the α-benzamido group in the bound substrate, —NH—C(=O)Ph yielding —N=C(—OX)Ph, where the nature of the bonding to oxygen is unknown. Besides these large differences, small changes in acyl-enzyme spectra also appeared as the pH was raised to decrease the half-life. All the spectral changes mentioned above are evidence for an acyl-enzyme structural change which precedes the rate-determining step in deacylation.

Storer *et al.* (1979) have used resonance Raman spectroscopy to study the hydrolysis of methyl thionohippurate

$$(C_6H_5—\overset{\overset{\displaystyle O}{\|}}{C}—NH—CH_2—\overset{\overset{\displaystyle S}{\|}}{C}—O—CH_3)$$

by papain. It had been established earlier that an intermediate appearing in this reaction is a dithioacyl-enzyme,

$$C_6H_5—\overset{\overset{\displaystyle O}{\|}}{C}—NH—CH_2—\overset{\overset{\displaystyle S}{\|}}{C}—S—$$

papain (Lowe and Williams, 1965). Storer *et al.* obtained spectra of the intermediate (Fig. 5.28) and found intense bands in the C═S and C—S stretching regions (1130 cm^{-1}, C═S; 560 and 598 cm^{-1}, transient, C—S). Their data show that, by replacing an oxygen atom in the ester grouping

$$—\overset{\overset{\displaystyle O}{\|}}{C}—S—$$

with a sulfur atom to form the intensely resonance-Raman-enhanceable grouping

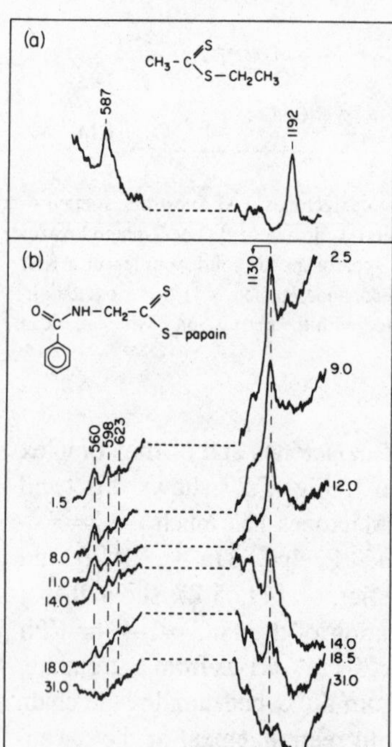

Fig. 5.28. The C—S and C═S stretching vibration regions in the resonance Raman spectra of (*a*) a buffer solution of 166 μM ethyl dithioacetate and (*b*) buffer solutions of dithioacylpapain obtained at various times after mixing the substrate and enzyme. The numbers at the side of each acylpapain spectrum refer to the times at which the 560 cm^{-1} peak (C—S region) and the 1130 cm^{-1} peak (C═S region) were scanned through. Each section took approximately 4 min to scan. The concentration of dithioacylpapain at 2 min after mixing was approximately 92 μM (A_{315} = 1.46, log ϵ_{315} = 4.2) (Lowe and Williams, 1965). (Storer *et al.*, 1979.)

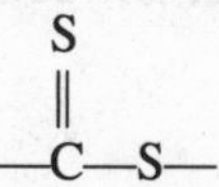

the catalytically crucial bonds in the ester moiety can be observed during enzymolysis. The authors stated that their method can also be used to monitor other enzymic reactions in which a thiol–acyl intermediate is formed.

Trypsin

Another enzyme that has been investigated by the resonance Raman method is trypsin. Dupaix *et al.* (1975) have studied the interactions between trypsin and a competitive inhibitor, 4-amidino-4′-dimethylamino azobenzene, which inhibits the trypsin-catalyzed hydrolysis of *N*-α-benzyloxycarbonyl-L-lysine-*p*-nitrophenyl ester. They recorded spectra of free or enzyme-bound inhibitor in aqueous solution.

When the complex formed between enzyme and inhibitor, the principal changes occurred in the relative intensities of bands at 1608, 1315, 1206, and 1171 cm^{-1}, and no large frequency shifts appeared. For a *trans* coplanar system of benzene rings N═N stretching is expected in the interval from 1440 to 1380 cm^{-1}. Twisting of the benzene rings out of this conformation should produce a shift of N═N stretching to higher frequency and a change in the relative intensity of the band (Carey *et al.*, 1973). Since, when the inhibitor was bound to the enzyme, no increase in frequency occurred for the N═N linkage nor any change in intensity, the conclusion may be drawn that during the binding process no twisting out of the *trans* coplanar conformation occurred for the benzene rings.

Lysozyme

Simplification of the lysozyme Raman spectrum (excitation, 488.0 nm) to the spectrum due only to tryptophan (Trp) residues in the enzyme molecule (excitation by near-UV argon ion laser line at 363.8 nm) has been achieved by Brown *et al.* (1977) with the preresonance Raman effect, just as the Raman spectrum of formyl methionine tRNA has been simplified to that of one organic base (4-thiouridine) by use of the preresonance effect (Nishimura *et al.*, 1976). Curve b of Fig. 5.29 (363.8-nm excitation) is in marked contrast to curve a (488.0-nm excitation) for solutions of egg-white lysozyme. This enzyme contains an active site involving Trp in the binding of the substrate, and thus lends itself to a preresonance Raman study of changes in the Trp spectrum due to complex formation involving Trp and the substrate. Note that this study approaches the problem of what happens in the *enzyme* molecule. Other workers have observed spectra characteristic of the *substrate* molecule (Carey *et al.*, 1973; Carey and Schneider, 1974; Kumar *et al.*, 1974, 1976). The procedure using UV excitation at 363.8 nm to interpret the Trp spectral bands and their intensity changes was applied by Brown *et al.* (1977) to a study of the weak complex of lysozyme with glucose. Chicken egg-white lysozyme contains six Trp residues, three of which are in active-site locations. The results of Brown *et al.* are consistent with the concept that the N—H groups of Trp residues 62 and 63 of lysozyme form hydrogen bonds to oxygen atoms of the hydroxyl groups of the carbohydrate substrates [see Dickerson and Geis (1969); Stryer (1981)].

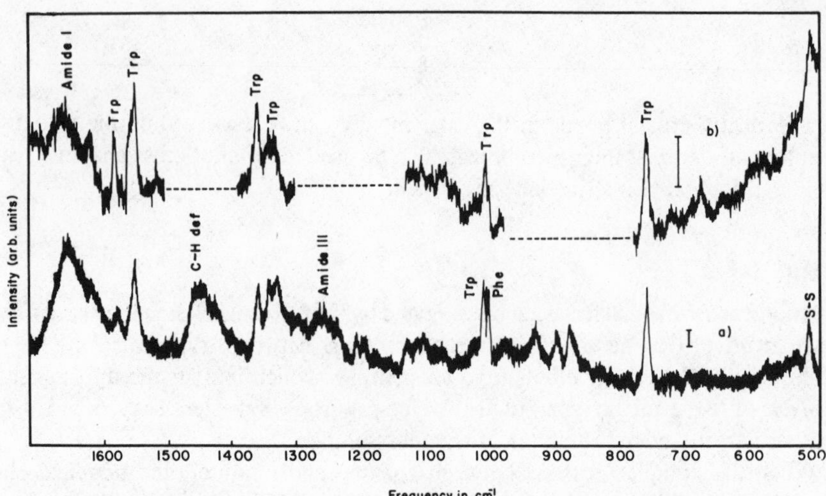

Fig. 5.29. Raman spectra of aqueous solutions of 29% (by weight) egg-white lysozyme (Sigma grade I): (a) 488.0-nm excitation with 200 mW of power at the sample; the spectral slit width was 4 cm^{-1} and the vertical line represents 1000 counts/sec; (b) 363.8-nm excitation with 20 mW of power at the sample; the spectral slit width was approximately 4 cm^{-1} and the vertical bar represents 100 counts/sec. Only the portions of the spectrum that correspond to significant bands observed with visible excitation are shown here since the sample was observed to decompose with the longer exposures required for a complete scan. (Brown *et al.*, 1977.) Reprinted with the permission of the American Chemical Society. Copyright 1977.

Protocatechuate 3,4-Dioxygenase

Protocatechuate 3,4-dioxygenase is an enzyme which acts on 3,4-dihydroxybenzoate (protocatechuate) to form β-carboxy-*cis, cis* muconate:

$$\text{(COO}^-\text{, OH, OH-substituted benzene ring)} + O_2 \longrightarrow \text{(COO}^-\text{, COO}^-\text{, COO}^-\text{ muconate)} + 2H^+$$

Keyes *et al.* (1978) have recorded the resonance Raman spectrum of this enzyme, which displayed four principal resonance-enhanced bands at 1602, 1503, 1263, and 1171 cm^{-1} with laser excitation of 514.5 nm. The authors attributed these bands to ring mode vibrations of one or more Fe(III)-coordinated tyrosinate residues as permanent iron ligands. Similar resonance Raman spectra have been obtained for Fe(III)-transferrins (Tomimatsu *et al.*, 1976; Gaber *et al.*, 1974; Carey and Young, 1974) and for human lactoferrin (Loehr *et al.*, unpublished). These Fe(III)-tyrosinate proteins appear to belong to a distinct class of non-heme iron proteins, which is easily identified as such by resonance Raman spectroscopy.

Pyrocatechase

Que and Heistand (1979) have recorded the resonance Raman spectra of pyrocatechase from *Pseudomonas arvilla* C-1 and its complexes with catechol and 4-nitrocatechol. This enzyme and protocatechuate 3,4-dioxygenase catalyze the cleavage of catechols to produce *cis, cis*-muconic acids. These spectra and those of several synthesized iron–catechol complexes permitted the authors to suggest that catechol binds to the active-site iron of the enzyme and that catechol loses both its protons when binding occurs.

Dihydrofolate Reductase

The binding of methotrexate and folate to three dihydrofolate reductases has been studied by Ozaki *et al.* (1981). They used resonance Raman spectroscopy to characterize the pteridine and *p*-aminobenzoyl binding sites by using, respectively, 350.6- and 324-nm excitation. The data supported the conclusions of other workers that methotrexate binds as the protonated pteridine ring whereas folate binds in the neutral form. The resonance Raman spectra (not shown) demonstrated that slight differences exist between the pteridine sites for methotrexate in the three enzymes studied whereas no differences were found among the *p*-aminobenzoyl sites; but in each methotrexate–protein complex, there is apparently a marked change in the geometry of the amide group in the benzoyl linkage of methotrexate compared to the geometry existing in the free ligand.

β-(2-Furyl)acryloylglyceraldehyde-3-phosphate Dehydrogenases

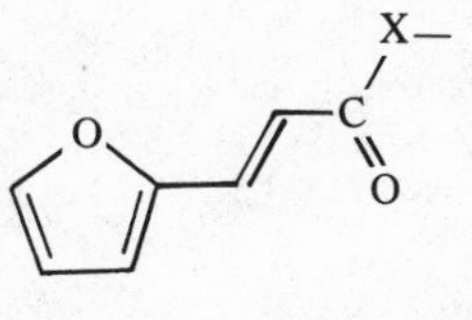

Some β-(2-furyl)acryloylglyceraldehyde-3-phosphate dehydrogenases have been studied by means of resonance Raman and electronic absorption spectroscopy (Storer *et al.*, 1981). Data from these studies and from rates of arsenolysis were used to show that the acylated subunits of the rabbit enzyme, but not the sturgeon enzyme, exist as a mixed population of at least two forms. One form of the rabbit enzyme exhibits a high rate of arsenolysis and its near-UV absorption maximum (λ_{max}) is red shifted when it binds to NAD$^+$, whereas the other has a lower rate of arsenolysis, and its absorption spectrum is not affected by excess NAD$^+$. The sturgeon enzyme exists as a single population with a high rate of arsenolysis, and its λ_{max} is red shifted when it binds to NAD$^+$. This shift to the red and the accompanying location of $\nu_{C=C}$, the band originating from the ethylenic double bond stretching vibration in the resonance Raman spectrum (not shown), show that with NAD$^+$ present the π electrons of the furylacryloyl chromophore are polarized, i.e., a permanent dipole has been established along the long axis of this group. Storer *et al.* believe that this does not result in activation of the C=O group by establishing polarization (i.e., $^{\delta+}$C—O$^{\delta-}$) within that group alone. A comparison of $\nu_{C=C}$'s and λ_{max}'s for a series of furylacryloyl derivatives of the type where X = H,N, or O, indicates a clear correlation; but when X = S (e.g., as in thiol esters), the correlation does not hold.

Flavoenzymes, Flavoproteins, and Flavin Derivatives

Riboflavin Derivatives. The resonance Raman spectra of riboflavin and several of its derivatives have been measured in the range 700 to 1700 cm^{-1} by Nishina *et al.* (1978). In order to avoid fluorescence of riboflavin they complexed it with riboflavin-binding proteins from egg white and yolk. Among the derivatives of riboflavin were the 3-deuterated, 3-methyl, 3-carboxymethyl, and 7,8-dichloro compounds. Figure 5.30 shows spectra of riboflavin bound to the egg-yolk protein in H_2O and in D_2O. Figure 5.31 shows a stick bar diagram with a comparison of the resonance Raman lines of several riboflavin derivatives bound to egg-white riboflavin-binding protein. In this figure the most intense line (1355 cm^{-1}) of each molecule is normalized to unity and the length of a stick bar is proportional to the Raman line intensity. Shifts in the Raman line positions due to substitutions at positions 7,8, and 3 are shown by broken lines. The authors made assignments of observed Raman lines to individual ring modes of isoalloxazine. The notation of the three rings is shown in Fig. 5.32.

Kitagawa *et al.* (1979) have used ^{13}C- and ^{15}N-labeled riboflavin which was bound to egg-white riboflavin-binding protein (apo-RBP) in order to make assignments from resonance Raman spectra. Binding to apo-RBP quenches the strong fluorescence of riboflavin. Observation of isotopic frequency shifts allowed calculations to be made for the in-plane displacements of the C2, C4a, N1, N3, and N5 atoms during each Raman active vibration (see Fig. 5.33 for the numbering system of riboflavin and for low-frequency spectra). Figure 5.34 shows the higher-frequency region.

There is a band at 1252 cm^{-1} for the N3—H form of riboflavin which is associated

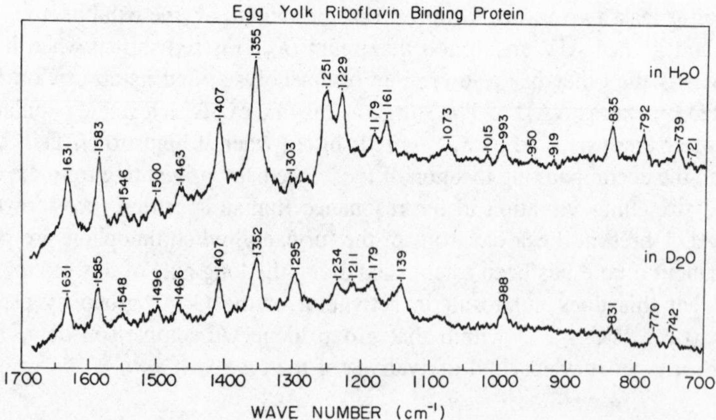

Fig. 5.30. Resonance Raman spectra of riboflavin bound to egg-yolk riboflavin binding protein in H_2O (upper) and in D_2O (lower). The concentration of riboflavin was 4.3×10^{-5} *M* for H_2O solution and 2.6×10^{-5} *M* for D_2O solution. The concentration of protein was 8.5×10^{-5} *M* for H_2O solution and 1.96×10^{-4} *M* for D_2O solution. (Nishina *et al.*, 1978.)

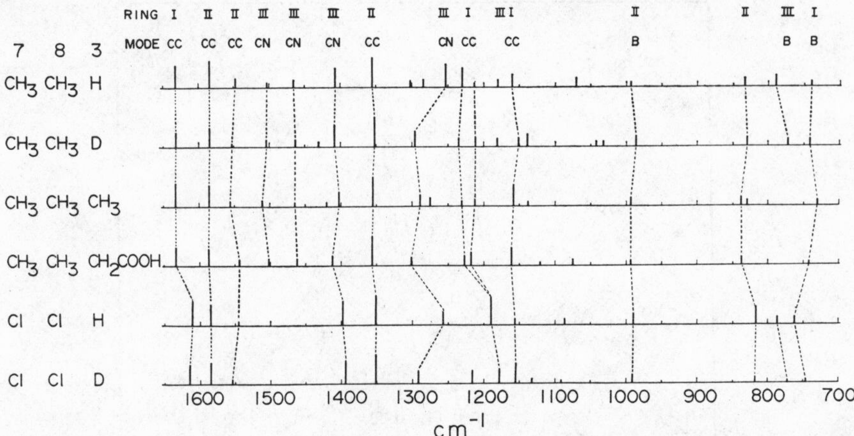

Fig. 5.31. Stick bar representation of the resonance Raman lines of several riboflavin derivatives bound to egg-white riboflavin binding protein. The most intense Raman line of each molecule is normalized to unit length and relative peak intensities are shown by the lengths of the stick bars. The atoms (or groups) attached to C(7), C(8), and N(3) are shown on the left and assignments of Raman lines to the three rings of isoalloxazine are designated at the top. The vibrations of C(4a)—C(10a) and C(5a)—C(9a) bonds are represented as Ring II modes. B denotes a breathinglike mode. The notation of the three rings is shown in Fig. 5.32. (Nishina *et al.*, 1978.)

with large vibrational displacements of the C2 and N3 atoms and is strongly coupled with the N3—H bending vibration. This band can be used to characterize the state of N3—H···protein bonding. The 1584-cm^{-1} band reflects, in particular, displacement of the N5 atom. The 1355-cm^{-1} band is associated with displacements of all ring III carbon atoms. Kitagawa *et al.* (1979) showed that these two bands are useful for studying the structure of isoalloxazine in various electronically excited forms.

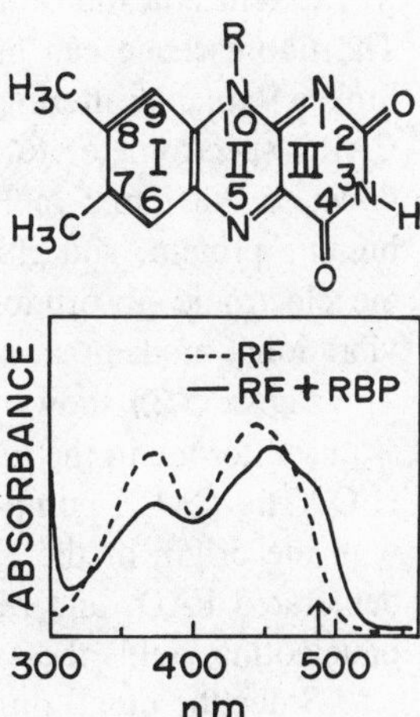

Fig. 5.32. Notation of atoms and rings of isoalloxazine (upper), and absorption spectra of riboflavin (broken line) and riboflavin bound to egg-white riboflavin binding protein (solid line). The arrow shows the excitation wavelength of Raman scattering. (Nishina *et al.*, 1978.)

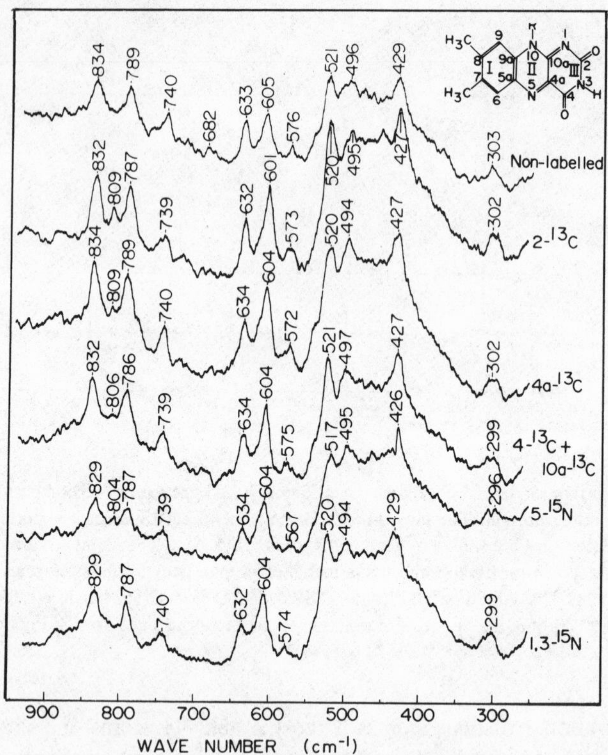

Fig. 5.33. Resonance Raman spectra in the lower-frequency region of the N(3)–H form of the labeled RF in the bound state to RBP. The inset indicates the numbering of atoms and rings of the isoalloxazine. (Kitagawa *et al.*, 1979.) Reprinted with the permission of the American Chemical Society. Copyright 1979.

CARS (Resonance Coherent Anti-Stokes Raman Scattering) Flavin Spectra. Flavin is a molecule whose fluorescence overwhelms its regular Raman spectrum. The fluorescence can be practically eliminated by the use of resonance coherent anti-Stokes Raman scattering (CARS) spectroscopy. Dutta *et al.* (1978) have used a modified CARS spectrometer to record spectra of flavin-containing molecules from 1654 cm^{-1} down to 300 cm^{-1} in H_2O and ^2H_2O. Flavin adenine dinucleotide (FAD), riboflavin binding protein, and glucose oxidase all contain the isoalloxazine moiety. Excitation in the electronic absorption bands of such flavin compounds yields resonance-enhanced vibrational modes that arise from the isoalloxazine rings.

Figure 5.35 shows the absorption spectrum of FAD and the locations of ω_1, ω_2, and ω_{as} used to obtain the CARS spectra, and the low-frequency CARS spectrum of FAD in H_2O. The CARS pump laser was tuned to 480.0 nm, since this wavelength coincides with the origin of the electronic transition. Figure 5.36 shows CARS spectra of FAD, deuterated FAD, and riboflavin binding protein in H_2O and ^2H_2O (D_2O). Flavin mononucleotide yields the same resonance CARS spectra as FAD.

Since the uracil ring is present in the structure of isoalloxazine, one can analyze the

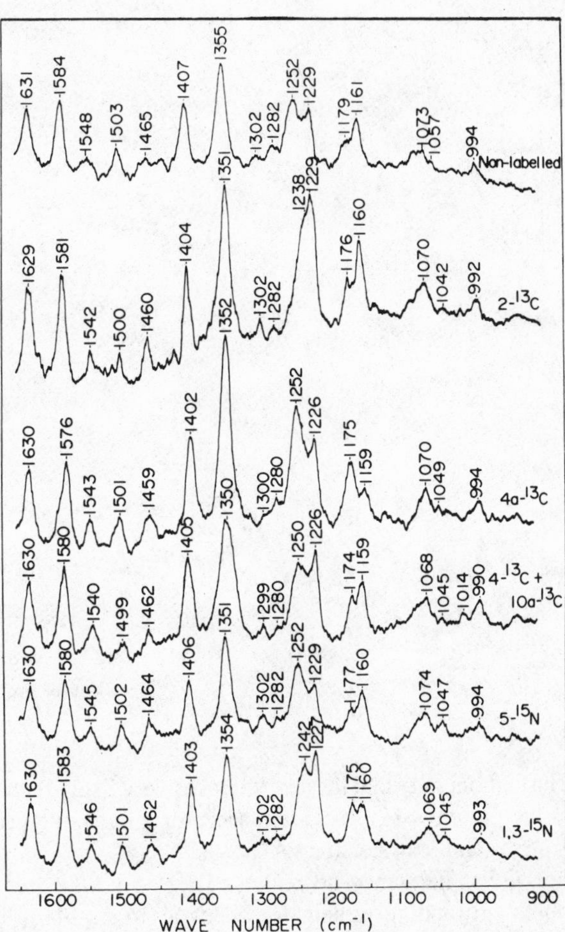

Fig. 5.34. Resonance Raman spectra in the higher-frequency region of the N(3)-H form of the labeled RF in the bound state to RBP. (Kitagawa *et al.*, 1979.) Reprinted with the permission of the American Chemical Society. Copyright 1979.

spectrum of the latter by analogy with that of the former and other simple ring systems. The uracil part of isoalloxazine contains an easily exchangeable proton at N3. The spectrum of FAD in H_2O (Fig. 5.35b) shows a strong peak at ca. 600 cm^{-1}, which does not shift in ^2H_2O. Dutta *et al.* (1978) attribute this band to an in-plane vibration due to breathing of the ring; the bands at 1635 and 1584 cm^{-1} (Fig. 5.36) may be C=O stretching vibrations.

The spectra in H_2O and ^2H_2O are quite different in the interval from 1100 to 1300 cm^{-1}, but corresponding band correlations were difficult to make. The spectrum of FAD shows a weak band at ca. 1160 cm^{-1} which moves to ca. 1140 cm^{-1} in ^2H_2O. A strong band (named H-band by Dutta *et al.*) at ca. 1255 cm^{-1} in H_2O moves to ca. 1295 cm^{-1} (named D-band) in ^2H_2O.

The H-band frequency of FAD decreases by 10 cm^{-1} in riboflavin binding protein; the D-band frequency decreases by 4 cm^{-1}. No H-band at all is found for glucose oxidase (not shown) but a D-band occurs at 1287 cm^{-1}, 10 cm^{-1} lower than in the FAD spectrum.

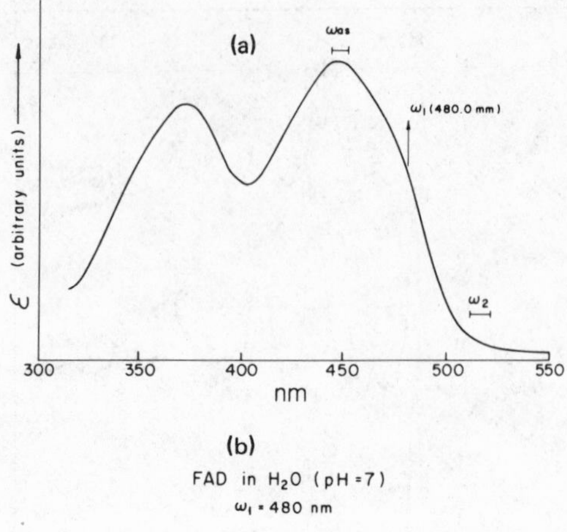

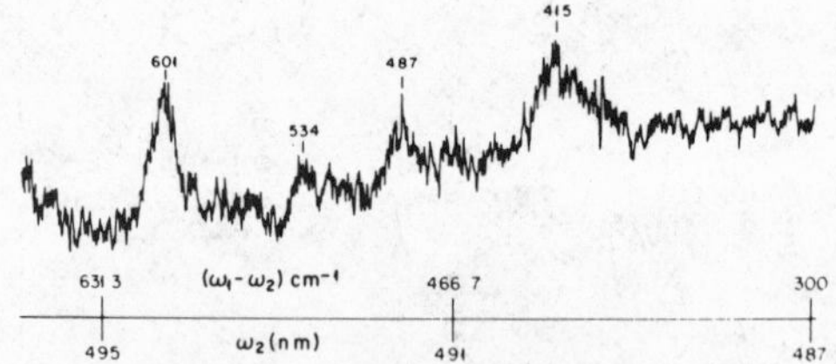

Fig. 5.35. (a) Absorption spectrum of FAD, showing the positions of ω_1, ω_2, and ω_{as} used to record CARS spectra. (b) Low-frequency CARS spectrum of FAD. (Dutta *et al.*, 1978.)

The authors state that these differences very likely reflect changes in the hydrogen bonding of the proton at N3, to proton acceptor groups in glucose oxidase and riboflavin binding protein, and to water for FAD.

Other frequency decreases of 3 to 5 cm^{-1} occur between the spectra of FAD and riboflavin binding protein for bands at 1635, 1416, 1185, and 1164 cm^{-1}, and similar shifts occur for glucose oxidase. These shifts may be displaying the differences between the surroundings in aqueous solution and in the areas where protein binding occurs (Dutta *et al.*).

Flavoenzymes. When various flavoenzymes react with their substrates, intermediates with nonsemiquinone structures often occur and display a broad absorption band at the longer wavelengths. In the case of NADPH oxidoreductase, Old Yellow Enzyme (OYE), which contains one flavin mononucleotide (FMN) per monomer, many phenolic compounds are strongly bound to the enzyme when the flavin is in the oxidized form, resulting in similar long-wavelength absorption (Massey and Ghisla, 1974). Abramovitz and Massey (1976) have shown a systematic correlation between the maximum absorption wavelength of the complexes and Hammett's σ_{para} constant for phenolic para substituents.

Kitagawa *et al.* (1979a) have used resonance Raman spectra as evidence for charge-transfer interactions of phenols with the FMN of the NADPH oxidoreductase—the phenol being the donor and the oxidized flavin the acceptor of the charge transfer. These authors studied the OYE-pentafluorophenol complex by exciting Raman scattering at 568.2 nm (Fig. 5.37). The Raman bands at 1475, 1310, 561, and 454 cm^{-1} were assigned to

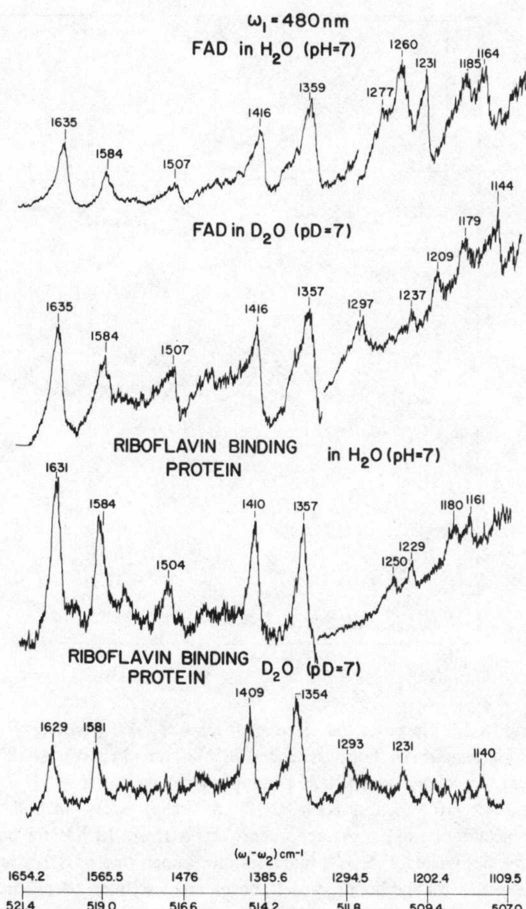

Fig. 5.36. CARS spectra of FAD, deuterated FAD, riboflavin binding protein in H_2O and D_2O; ω_1 = 480 nm, ω_2 scan speed = 0.6 nm/min; laser repetition rate = 10 pulses/sec; 30 pulses averaged. Laser pulse energies of ω_1 and ω_2 were ~10–25 μJ at the sample. (Dutta *et al.*, 1978.)

pentafluorophenol (F_5Ph) because they were observed for free F_5Ph but not for other flavoproteins (Dutta *et al.*, 1978a; Dutta *et al.*, 1978b; Nishina *et al.*, 1978; Kitagawa *et al.*, 1979). Also, these bands were absent in the spectra of the charge-transfer complexes of other phenol derivatives with the Old Yellow Enzyme. The isoalloxazine structure accounted for all other vibrational bands. Kitagawa *et al.* (1979a) showed that the bands of F_5Ph at 1475 and 454 cm^{-1} were greatly enhanced for the OYE-pentafluorophenol complex over that for F_5Ph alone and over that for OYE alone. The 1588-cm^{-1} band of the flavin group was also greatly enhanced. This band is known to be associated with vibrational movements of the C4a and N5 atoms of isoalloxazine (Kitagawa *et al.*, 1979). The 561- and 1470-cm^{-1} bands of F_5Ph were assigned to the totally symmetric C—F stretching and the benzene breathing mode, respectively (Green and Herrison, 1976; Long and Steele, 1961).

The charge-transfer interaction was postulated to involve the C4a—N5 region but not the N3—H bond. It is interesting to note that the group of Raman bands for the

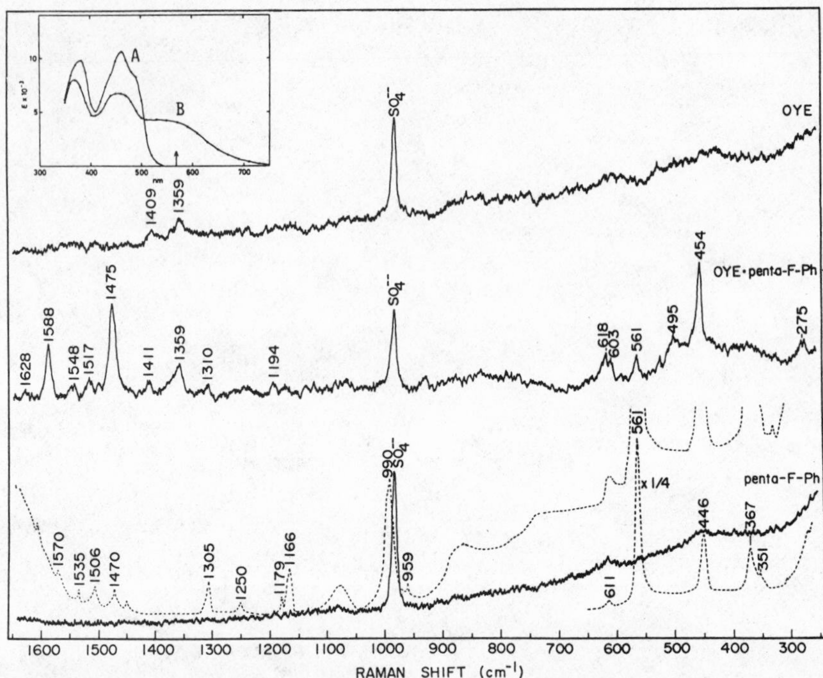

Fig. 5.37. The resonance Raman spectra of OYE (top), OYE-F_5Ph complex (OYE-penta-F-Ph, middle), and F_5Ph (penta-F-Ph, bottom) excited at 568.2 nm (35 mW). All the samples commonly contained 3.3% $(NH_4)_2SO_4$ and the solvent was 0.1 *M* sodium phosphate buffer at pH 7.0. The concentrations of OYE and F_5Ph were 8.77×10^{-4} *M* and 1.84×10^{-3} *M*, respectively, in both their pure solutions and mixed solution. The concentration of excess apoprotein was 6.15×10^{-5} *M* for both OYE and OYE-F_5Ph solutions. The Raman line designated as SO_4^{--} indicates the Raman line of SO_4^{2-} at 981 cm^{-1}. The broken line at the bottom was observed for the hundred-fold concentrated solution (0.112 m*M*) of F_5Ph in 0.1 *M* sodium phosphate buffer without $(NH_4)_2SO_4$. The inset figure displays the absorption spectra of OYE (A) and OYE-F_5Ph (B) in 0.1 *M* sodium phosphate buffer at pH 7.0. Concentrations of OYE and F_5Ph were 2.3×10^{-5} *M* and 2.5×10^{-4} *M*, respectively. An arrow indicates the excitation wavelength of Raman scattering. (Kitagawa *et al.*, 1979a.) Reprinted with the permission of the American Chemical Society. Copyright 1979.

phenolate ion observed in this work is totally different from those found for several Fe^{3+}–phenolate complexes when they were excited at their charge-transfer transition; e.g., protocatechuate-3,4-dioxygenase showed phenolate ion bands at ~ 1600, 1500, 1260, and 1170 cm^{-1} (Tatsuno *et al.*, 1978). The 1260-cm^{-1} band (C—O stretching of phenolate) was absent in the case of OYE-penta-F-Ph but was very strong for the Fe^{3+}–phenolate complexes, thus suggesting different interaction modes of phenolate anion for the two varieties of complex.

Nishina *et al.* (1980) have extended this work to include various OYE–phenol complexes and a study of the dependence of the Raman intensities of the OYE–F_5Ph complex on the wavelength of excitation (568.2 nm). The phenolic compounds complexed with OYE included phenol, *p*-chlorophenol, *p*-nitrophenol, salicylic acid, *o*-hydroxybenzaldehyde, *p*-cresol, and *p*-fluorophenol. The pentahalogenated and *para*-substituted phenols in the complexed form commonly display an intense Raman band of a ring-deformation mode (ν_{6a}) when excited at 568.2 nm. The pentahalogenated phenols exhibited the Raman line of a ring-breathing mode (ν_1) but the *para*-substituted phenols did not.

Certain intensity-enhanced in-plane mode bands of bound phenols appear, including ν_3, ν_{8a}, ν_{12}, ν_{14}, and ν_{19a}. Several Raman bands of flavin mononucleotide were also resonance-enhanced by the charge transfer band. Raman bands at 1585 and 1550 cm^{-1},

Table 5.4. Raman Frequencies of Flavins and Flavoproteins[a,b]

FMN	FAD	FAD in D_2O	N-1 protonated FMN	1,10-bridged flavinium ion, N-3 protonated
		1144m		
1162m	1161m	1161m		
1182sh	1182sh			
1232m	1228m		1207m	
1256m	1253m		1239s	1236m
1281sh		1297s	1290m	1272m
1354s	1354s	1354s	1348s	1368s
1410s	1411s	1411s		1412m
1467w			1454w	
1502w	1508w	1508w	1515s	1513s
1548w			1549m	
1585s	1585s	1584s		
1629s	1630s	1631s	1614s	1626s

N-3 ionized	FMN cationic semiquinone	Uncomplexed enzyme[c]	Dehydrogenase + AcAc-CoA	Dehydrogenase + crotonyl-CoA
	1135s			
			1162w	
1187m	1174s			
	1204w			
1241m	1239w	1234	1237m	1237m
1288s	1271s	1259	1262m	1257m
1365s	1329m	1350	1352s	1350s
1413s		1407	1412s	1407s
1516s				
			1587m	1587m
1629s			1626m	1627m

[a] Benecky *et al.* (1979).
[b] Relative intensity: s, strong; w, weak; m, medium; sh, shoulder.
[c] Signal-to-noise ratio was not large enough to determine the relative intensities.

involving vibrational displacements of the N5 and C4a atoms of isoalloxazine, showed remarkable resonance enhancement.

The above enhancement characteristics and certain others suggested to these authors that π electrons of the benzene ring interact as a whole with ring III of isoalloxazine so that their molecular planes are parallel with phenolate oxygen above the N5—C4a bond. They did not believe that there was localized interaction between the phenolate oxygen atom and a particular atom of FMN.

In order to evaluate the utility of resonance Raman spectroscopy for the examination of flavoprotein interactions, Benecky *et al.* (1979) have studied the spectra of several flavin model compounds and of the flavoprotein enzyme fatty acyl-CoA dehydrogenase. Table 5.4 shows Raman frequencies of flavin mononucleotide (FMN), flavin adenine dinucleotide (FAD), N1 protonated FMN, the uncomplexed dehydrogenase, the dehydrogenase plus acetoacetyl-CoA, the dehydrogenase plus crotonyl-CoA, and several other related compounds. Potassium iodide was used to reduce fluorescence.

The spectra (not shown) and Table 5.4 indicate that the spectrum of FAD is very similar to that of FMN, owing to the fact that the resonance enhancement effect comes only from the isoalloxazine moiety. Six of the bands found in the FMN spectrum (1629, 1585, 1410, 1354, 1256, and 1232 cm^{-1}) are also observed in the spectra of the dehydrogenase and its complexes. The spectrum of the dehydrogenase complex (Table 5.4) is almost identical with that of free FAD except that the protein FAD complex displays shifts of 1253 to 1259 cm^{-1}, and 1228 to 1234 cm^{-1}. In the case of the FAD enzyme–acetoacetyl–CoA complex, shifts also occur: from 1253 to 1262 cm^{-1}, and from 1228 to 1237 cm^{-1}. Otherwise, the spectra of FAD and of enzyme complex show few differences.

The shifts observed by Benecky *et al.* (1979) during the complexing of FAD to enzyme or enzyme complexes lie in the same region as the largest shifts seen by Dutta *et al.* (1978) in CARS spectra—the 1250-cm^{-1} region—arising from an N—H deformation vibration at N3 of the isoalloxazine ring. N3 was shown to be involved in hydrogen bonding in flavin systems. Benecky *et al.* attributed the shifts observed when protein binds FAD to stronger hydrogen bonding in the case of FAD bound at the active site of the enzyme than in the case where FAD is in aqueous solvent.

References

Abramovitz, A. S., and Massey, V., *J. Biol. Chem.* **251,** 5327 (1976).

Bare, G. H., Alben, J. O., and Bromberg, P. A., *Biochemistry* **14,** 1578 (1975).

Bastian, E. J., Jr., and Martin, R. B., *J. Phys. Chem.* **77,** 1129 (1973).

Belasco, J. G., and Knowles, J. R., *Fed. Proc.* **39,** (6), 1667 (1980).

Benecky, M., Li, T. Y., Schmidt, J., Frerman, F., Watters, K. L., and McFarland, J., *Biochemistry* **18,** 3471 (1979).

Blake, C. C. F., Mair, G. A., North, A. C. T., Phillips, D. C., and Sarma, V. R., *Proc. R. Soc. London Ser. B* **167,** 365 (1967).

Brazhnikov, E. V., Chetverin, A. B., and Chirgadze, Yu. N. *FEBS Lett.* **93,** 125 (1978).

Brown, K. G., Brown, E. B., and Person, W. B., *J. Am. Chem. Soc.* **99,** 3128 (1977).

Burgess, A. W., and Scheraga, H. A., *J. Theor. Biol.* **53,** 403 (1975).

Capaldi, R. A., *Biochim. Biophys. Acta* **303,** 237 (1973).

Carey, P. R., and King, R. W., *Biochemistry* **18,** 2834 (1979).

Carey, P. R., and Schneider, H., *Biochem. Biophys. Res. Commun.* **57,** 831 (1974).
Carey, P. R., and Schneider, H., *J. Mol. Biol.* **102,** 679 (1976).
Carey, P. R., and Young, N. M., *Can. J. Biochem.* **52,** 273 (1974).
Carey, P. R., Froese, A., and Schneider, H., *Biochemistry* **12,** 2198 (1973).
Carey, P. R., Carriere, R. G., Lynn, K. R., and Schneider, H., *Biochemistry* **15,** 2387 (1976).
Chen, M. C., and Lord, R. C., *Biochemistry* **15,** 1889 (1976).
Craig, W. S., and Gaber, B. P., *J. Am. Chem. Soc.* **99,** 4130 (1977).
Dickerson, R. E., and Geis, I., *The Structure and Action of Proteins,* p. 76. W. A. Benjamin, Menlo Park, California, 1979.
Dodd, G. H., and Radda, G. K. *Biochem. J.* **114,** 407 (1969).
Dupaix, A., Bechet, J.-J., Yon, J., Merlin, J.-C., Delhaye, M., and Hill, M., *Proc. Natl. Acad. Sci. USA* **72,** 4223 (1975).
Dutta, P. K., Nestor, J., and Spiro, T. G., *Biochem. Biophys. Res. Commun.* **83,** 209 (1978).
Dutta, P. K., Nestor, J. R., and Spiro, T. G., *Proc. Natl. Acad. Sci., USA* **74,** 4146 (1977).
Eyster, E. H., and Gillette, R. H., *J. Chem. Phys.* **8,** 369 (1940).
Frieden, C. in *Developmental and Metabolic Control Mechanisms and Neoplasia,* Williams and Wilkins, Baltimore, p. 392, 1965.
Fronaeus, S., and Larsson, R., *Acta Chem. Scand.* **16,** 1433, 1447 (1962).
Frushour, B. G., and Koenig, J. L., in *Advances in Infrared and Raman Spectroscopy,* Vol. 1 (R. J. H. Clark and R. E. Hester, eds.), Heyden and Son, Ltd., p. 58, 1975.
Gaber, B. P., Miskowski, V., and Spiro, T. G., *J. Am. Chem. Soc.* **96,** 6868 (1974).
Green, J. H. S., and Herrison, D. J., *J. Chem. Thermodyn.* **8,** 529 (1976).
Hardy, J. D., Barker, E. F., and Dennison, D. M., *Phys. Rev.* **42,** 279 (1932).
Henderson, T. R., Henderson, R. F., and Johnson, G. E. *Arch. Biochem. Biophys.* **132,** 242 (1969).
Hvidt, Aa., and Nielsen, S. O., *Adv. Protein Chem.* **21,** 287 (1966).
Hvidt, Aa., and Pedersen, E. J., *Eur. J. Biochem.* **48,** 333 (1974).
Hvidt, Aa., and Wallevik, K., *J. Biol. Chem.* **247,** 1530 (1972).
Johansen, J. T., Livingston, D. M., and Vallee, B. L., *Biochemistry* **11,** 2584 (1972).
Keyes, W. E., Loehr, T. M., and Taylor, M. L., *Biochem. Biophys. Res. Commun.* **83,** 941 (1978).
Kitagawa, T., Nishina, Y., Kyogoku, Y., Yamano, T., Ohishi, N., Takai-Suzuki, A., and Yagi, K., *Biochemistry* **18,** 1804 (1979).
Kitagawa, T., Nishina, Y., Shiga, K., Watari, H., Matsumura, Y., and Yamano, T., *J. Am. Chem. Soc.* **101,** 3376 (1979a).
Koenig, J. L., and Frushour, B. G., *Biopolymers* **11,** 2505 (1972).
Koeppe, R. E., II, and Stroud, R. M., *Biochemistry* **15,** 3450 (1976).
Kumar, K., King, R. W., and Carey, P. R., *Biochemistry* **15,** 2195 (1976).
Kumar, K., King, R. W., and Carey, P. R., *FEBS Lett.* **48,** 283 (1974).
Linderstrøm-Lang, K., *Symposium on Protein Structure* (A. Neuberger, ed.), Methuen, London, pp. 1–34, 1958.
Loehr, T. M., Ainscough, E., and Brodie, A. M., unpublished, quoted in Keyes *et al.,* (1978).
Long, D. A., and Steele, D., *Spectrochim. Acta* **19,** 1955 (1961).
Lord, R. C., and Mendelsohn, R., *J. Am. Chem. Soc.* **94,** 2133 (1972).
Lord, R. C., and Yu, N.-T., *J. Mol. Biol.* **51,** 203 (1970a).
Lord, R. C., and Yu, N.-T., *J. Mol. Biol.* **50,** 509 (1970b).
Lowe, G., and Williams, A., *Biochem. J.* **96,** 189 (1965).
MacClement, B. A. E., Carriere, R. G., Phelps, D. J., and Carey, P. R., *Biochemistry* **20,** 3438 (1981).
Martin, R. B., *J. Phys. Chem.* **78,** 855 (1974).
Massey, V., and Ghisla, S., *Ann. N. Y. Acad. Sci.* **227,** 446 (1974).
Matheson, R. R., Jr., Van Wart, H. E., Burgess, A. W., Weinstein, L. I., and Scheraga, H. A., *Biochemistry* **16,** 396 (1977).
Matthies, M., *FEBS Lett.* **81,** 183 (1977).
Miyazawa, T., Shimanouchi, T., and Mizushima, S., *J. Chem. Phys.* **29,** 611 (1958).
Nabedryk-Viala, E., Calvet, P., Thiéry, J. M., Galmiche, J. M., and Girault, G., *FEBS Lett.* **79,** 139 (1977).
Nakamoto, K., *Infrared Spectra of Inorganic and Coordination Compounds,* 2nd Ed., Wiley-Interscience, New York, pp. 150–294, 1970.

Nishimura, Y., Hirakawa, A. Y., Tsuboi, M., and Nishimura, S., *Nature (London)* **260,** 173 (1976).
Nishina, Y., Kitagawa, T., Shiga, K., Horiike, K., Matsumura, Y., Watari, H., and Yamano, T., *J. Biochem. (Tokyo)* **84,** 925 (1978).
Nishina, Y., Kitagawa, T., Shiga, K., Watari, H., and Yamano, T., *J. Biochem. (Tokyo)* **87,** 831 (1980).
Nogami, N., Sugeta, H., and Miyazawa, T., *Chem. Lett.* **147,** (1975).
Ohta, S., Nakanishi, M., Tsuboi, M., Yoshida, M., and Kagawa, Y., *Biochem. Biophys. Res. Commun.* **80,** 929 (1978).
Ottesen, M., *Methods Biochem. Anal.* **20,** 135 (1971).
Ozaki, Y., King, R. W., and Carey, P. R., *Biochemistry* **20,** 3219 (1981).
Parker, F. S., *Applications of Infrared Spectroscopy in Biochemistry, Biology, and Medicine,* Chap. 11, Plenum Press, New York, 1971.
Parker, F. S., and Bhaskar, K. R., *Appl. Spectrosc. Rev.* **3,** 91 (1969).
Pershina, L., and Hvidt, Aa., *Eur. J. Biochem.* **48,** 339 (1974).
Petersen, R. L., Li, T.-Y., McFarland, J. T., and Watters, K. L., *Biochemistry* **16,** 726 (1977).
Phelps, D. J., Schneider, H., and Carey, P. R., *Biochemistry* **20,** 3447 (1981).
Phillips, D. C., *Proc. Natl. Acad. Sci. USA* **57,** 484 (1967).
Porubcan, R. S., Watters, K. L., and McFarland, J. T., *Arch. Biochem. Biophys.* **186,** 255 (1978).
Que, L., Jr., and Heistand, R. H., *J. Am. Chem. Soc.* **101,** 2219 (1979).
Racker, E., Hauska, G. A., Lien, S., Berzborn, R. J., and Nelson, N., in *Proc. 2d Int. Congr. Photosynth., Stresa* (Forti, G., ed.), Vol. 2, p. 1097, 1971.
Reed, G. H., Barlow, C. H., and Burns, R. A., Jr., *J. Biol. Chem.* **253,** 4153 (1978).
Riepe, M. E., and Wang, J. H., *J. Biol. Chem.* **243,** 2779 (1968).
Scheule, R. K., Van Wart, H. E., Vallee, B. L., and Scheraga, H. A., *Biochemistry* **19,** 759 (1980).
Scheule, R. K., Van Wart, H. E., Zweifel, B. O., Vallee, B. L., and Scheraga, H. A., *J. Inorg. Biochem.* **11,** 283 (1979).
Scheule, R. K., Van Wart, H. E., Vallee, B. L., and Scheraga, H. A., *Proc. Natl. Acad. Sci. USA* **74,** 3273 (1977).
Siamwiza, M. N., Lord, R. C., Chen, M. C., Takamatsu, T., Harada, I., Matsuura, H., and Shimanouchi, T., *Biochemistry* **14,** 4870 (1975).
Storer, A. C., Phelps, D. J., and Carey, P. R., *Biochemistry* **20,** 3454 (1981).
Storer, A. C., Murphy, W. F., and Carey, P. R., *J. Biol. Chem.* **254,** 3163 (1979).
Stryer, L., *Biochemistry,* 2nd Ed., W. H. Freeman, San Francisco, California, 1981.
Stryker, M. H., and Parker, F. S., *Arch. Biochem. Biophys.* **141,** 313 (1970).
Sund, H., and Burchard, W., *Eur. J. Biochem.* **6,** 202 (1968).
Sund, H., Pilz, I., and Herbst, M., *Eur. J. Biochem.* **7,** 517 (1969).
Talal, N., Tomkins, G. M., Mushinski, J. F., and Yielding, K. L., *J. Mol. Biol.* **8,** 46 (1964).
Tatsuno, Y., Saeki, Y., Iwaki, M., Yagi, T., Nozaki, M., Kitagawa, T., and Otsuka, S., *J. Am. Chem. Soc.* **100,** 4614 (1978).
Thompson, W., and Yielding, K. L., *Arch. Biochem. Biophys.* **126,** 399 (1968).
Timasheff, S. N., and Rupley, J. A., *Arch. Biochem. Biophys.* **150,** 318 (1972).
Timasheff, S. N., Susi, H., and Rupley, J. A., *Methods Enzymol.* **27,** Part D, 548 (1973).
Tomimatsu, Y., Kint, S., and Scherer, J. R., *Biochemistry* **15,** 4918 (1976).
Tomkins, G. M., Yielding, K. L., Talal, N., and Curran, J. F., *Cold Spring Harbor Symp. Quant. Biol.* **28,** 461 (1963).
Twardowski, J., *Biopolymers* **17,** 181 (1978).
Visser, L., and Blout, E. R., *Biochemistry* **10,** 743 (1971).
Watts, D. C., in *The Enzymes,* 3rd Ed., Vol. 8 (P. D. Boyer, ed.), Academic Press, New York, pp. 383–455, 1973.
Wilson, E. B., Jr., *Phys. Rev.* **45,** 706 (1934).
Yielding, K. L., Tomkins, G. M., Bitensky, M. W., and Talal, N., *Can. J. Biochem.* **42,** 727 (1964).
Young, J. M., and Wang, J. H., *J. Biol. Chem.* **246,** 2815 (1971).
Yu, N.-T., Ph.D. thesis, Massachusetts Institute of Technology, quoted in Lord and Yu (1970).
Yu, N.-T., and Liu, C. S., *J. Am. Chem. Soc.* **94,** 5127 (1972).
Yu, N.-T., Jo, B. H., and Liu, C. S., *J. Am. Chem. Soc.* **94,** 7572 (1972).
Zavodszky, P., Johansen, J. T., and Hvidt, Aa., *Eur. J. Biochem.* **56,** 67 (1975).

Chapter 6

PORPHYRINS AND HEMEPROTEINS

Infrared Studies

Parker (1971) has discussed many infrared spectroscopic applications in the study of porphyrins and hemeproteins. Most of the topics in this chapter refer to more recent investigations.

Alben *et al.* (1973) have presented an FTIR spectroscopic study of porphyrins. They studied two general series of compounds: one, with hydrogens on the meso or methine carbons of the porphyrin ring, and either CH_3 groups or a number of electron-withdrawing groups on the pyrrole β-carbons, e.g., the etio- and deuteroporphyrin series; the other, with substituents on the mesocarbons and hydrogens on the β-carbons of the pyrrole rings.

Figure 6.1 shows schematic spectra of porphin, etioporphyrin, tetra-(*n*-propyl)porphyrin, and tetraphenylporphyrin. The authors stated that the spectra of the four etioporphyrin isomers (Etio-II) are essentially the same. The spectrum displays bands typically found in the whole deuteroporphyrin series of compounds, and the tetraphenylporphyrin (TPP) spectrum is representative of spectra of the whole series of substituted TPP derivatives. The band from 772 to 805 cm^{-1} is present in spectra of the whole series except etioporphyrin, and is caused by out-of-plane bending of the β-hydrogens on the pyrrole rings. The band at 835 to 840 cm^{-1}, which is present in etioporphyrin II and porphin, is caused by the out-of-plane bending vibration of the meso-CH (methine) groups; it is not present in the spectra of the meso-substituted derivatives. A band at about 1060 cm^{-1} may be the in-plane CH deformation. The in-plane NH-bending vibration appears to lie at 960 to 990 cm^{-1}; and the out-of-plane NH-bending vibration is present from 690 to 710 cm^{-1}. From 950 to 970 cm^{-1} a strong band, caused by a ring vibration of the porphyrin skeleton, is displayed; it is not involved with hydrogens. This band occurs in all the highly symmetrical derivatives presented by Alben *et al.* (1973). The range 1100 to 1250 cm^{-1} contains several bands from ring vibrations; and the range 1350 to 1600 cm^{-1} has some rather constant aromatic ring vibrational bands, and some rather variable ones.

Figure 6.2 shows schematic IR spectra of a series of metal complexes of deuterium-substituted tetraphenylporphyrin (TPP). The strong vibration of the porphyrin ring near

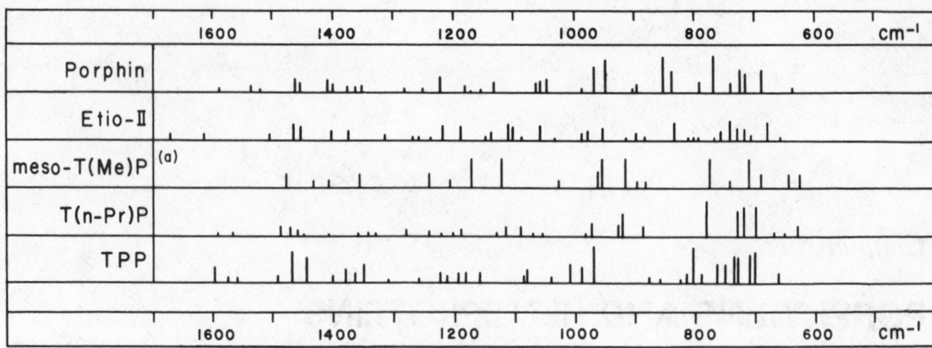

Fig. 6.1. Schematic representation of infrared spectra of porphyrins dispersed in KBr pellets. All spectra were obtained at 2-cm^{-1} resolution with a Digilab Model FTS-14 infrared interferometer, except that of meso-tetramethylporphin [Meso-T(Me)P], which was plotted from data of Mason (1958). Other abbreviations are as follows: Etio-II, etioporphyrin II; T(*n*-Pr)P, meso-tetra-*n*-propyl-porphin; and TPP, meso-tetraphenylporphin. (Alben *et al.*, 1973.)

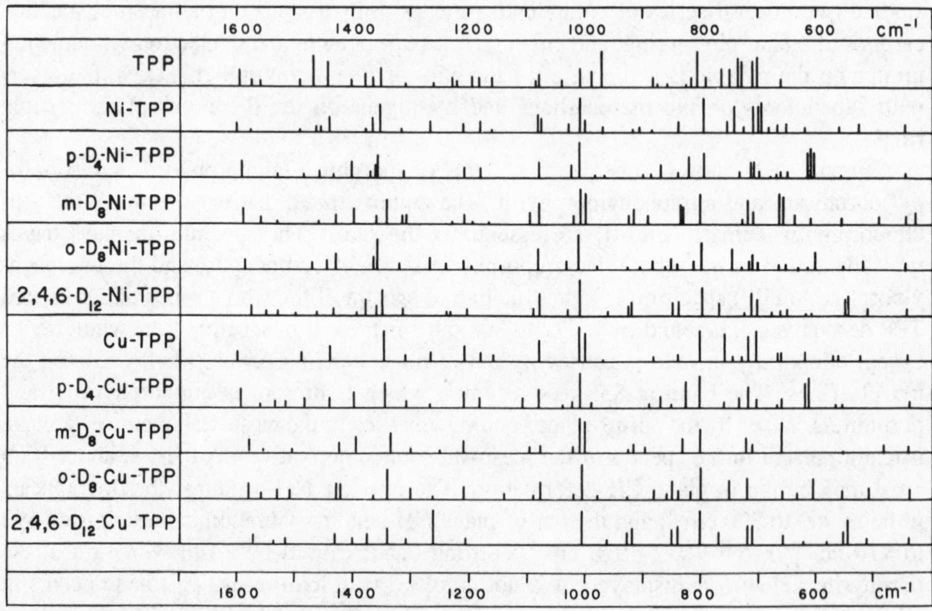

Fig. 6.2. Representation of infrared spectra of meso-tetraphenylporphin, its nickel (II) and copper (II) complexes, and deuterated derivatives, dispersed in KBr pellets. Spectra were obtained with a Perkin–Elmer Model 337 grating spectrometer with a "normal" resolution program. The following abbreviations were used for the nickel (II) and copper (II) complexes: *p*-D_4-TPP, tetra(4-deuterophenyl)porphin; *m*-D_8-TPP, tetra(3,5-dideuterophenyl)porphin; *o*-D_8-TPP, tetra(2,6-dideuterophenyl)porphin; 2,4,6,-D_{12}-TPP, tetra (2,4,6,-tri-deuterophenyl)porphin. (Alben *et al.*, 1973.)

970 cm^{-1} in TPP is moved to 1001–1010 cm^{-1} for the metal porphyrin compounds; deuterium substitution has no effect. Other bands do overlap this region and they are affected slightly by deuterium substitution.

Figure 6.3 shows schematic IR spectra of a group of TPP derivatives and their Fe(III) mu-oxo complexes. (See also Fig. 6.7). The 800-cm^{-1} band is quite constant. It is caused

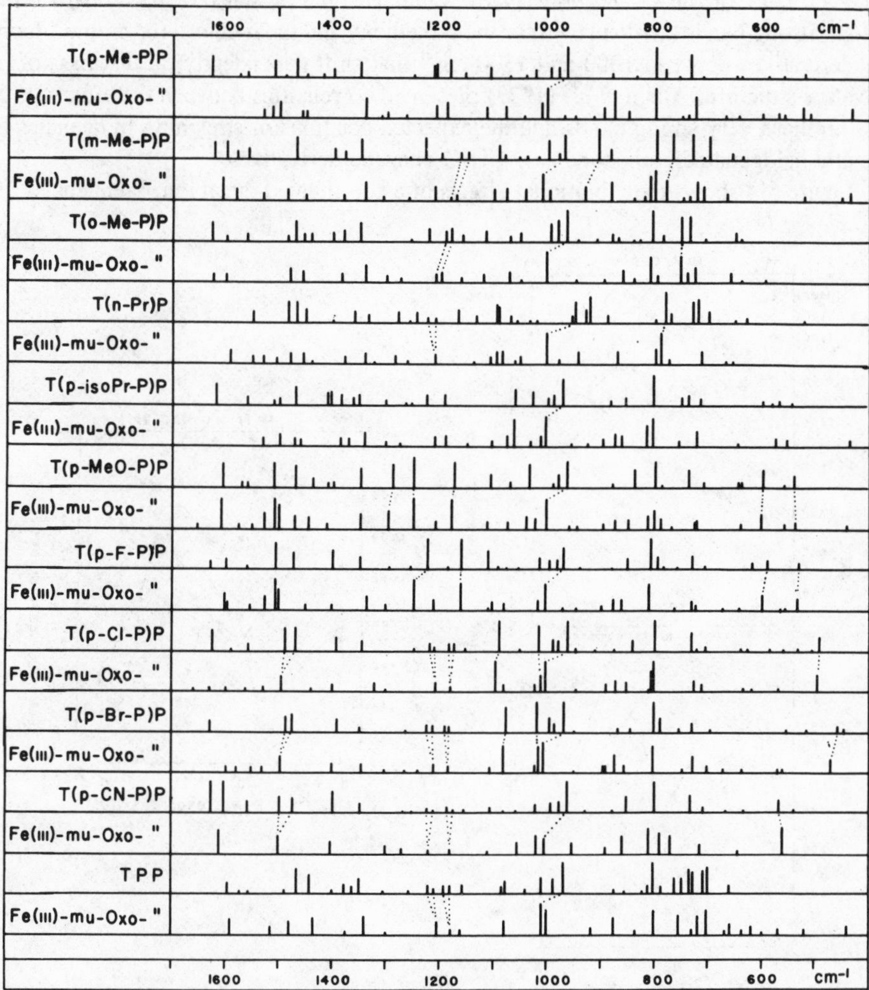

Fig. 6.3. Representation of infrared spectra of meso-substituted porphin derivatives and their iron (III) mu-oxo complexes, dispersed in KBr pellets. Spectra of all metal-free porphin derivatives except tetra(*p*-isopropylphenyl)porphin were measured with a Digilab model FTS-14 infrared interferometer at 2 cm^{-1} resolution, and as in Fig. 6.2. Other spectra were measured as in Fig. 6.2. The following abbreviations are used: T(*p*-Me-P)P, tetra(*p*-methylphenyl)-porphin; T(*m*-Me-P)P, tetra(3-(or 5-)monomethylphenyl)porphin; T(*o*-Me-P)P, tetra(2- or 6-)monomethylphenyl) porphin; T(*n*-Pr)P, meso-tetra-*n*-propyl-porphin; T(*p*-isoPr-P)P, tetra(*p*-isopropylphenyl)porphin; T(*p*-MeO-P)P, tetra(*p*-methoxyphenyl)-porphin; T(*p*-F-P)P, tetra(*p*-fluorophenyl)porphin; T(*p*-Cl-P)P, tetra(*p*-chlorophenyl)porphin; T(*p*-Br-P)P, tetra(*p*-bromophenyl)porphin; T(*p*-CN-P)P, tetra(*p*-cyanophenyl)porphin; TPP, tetraphenylporphin. (Alben *et al.*, 1973.)

by the out-of-plane bending vibration of the pyrrole CH groups. There is also a constant strong band from 960 to 970 cm^{-1} for the nonmetallic porphyrins that is moved to the range 1002–1010 cm^{-1} in the metal complexes. It is caused by a vibrational frequency of the porphyrin ring. The first, third, and fifth spectra are ortho-, meta-, and para-methyl-substituted TPP, and the second, fourth, and sixth spectra are for their respective metal complexes. The compounds contain one CH_3 group on each phenyl ring. A split band near 1200 cm^{-1} or the *o*- and *m*-derivatives indicates the presence of rotational isomerization. Strong bands are displayed for the *p*-methoxy and the *p*-chloro (or bromo, fluoro) TPP derivatives between 1000 and 1300 cm^{-1} that shift dependently on the mass of the derivative structure. Alben *et al.* (1973) presented correlations between the effects of the mass on these very strong bands and the expected positions of stretching frequencies for aromatic halogenated compounds, or CH_3O-benzene derivatives.

Figure 6.4 shows the vibrational effects of a coordinated metal ion. Frequencies and

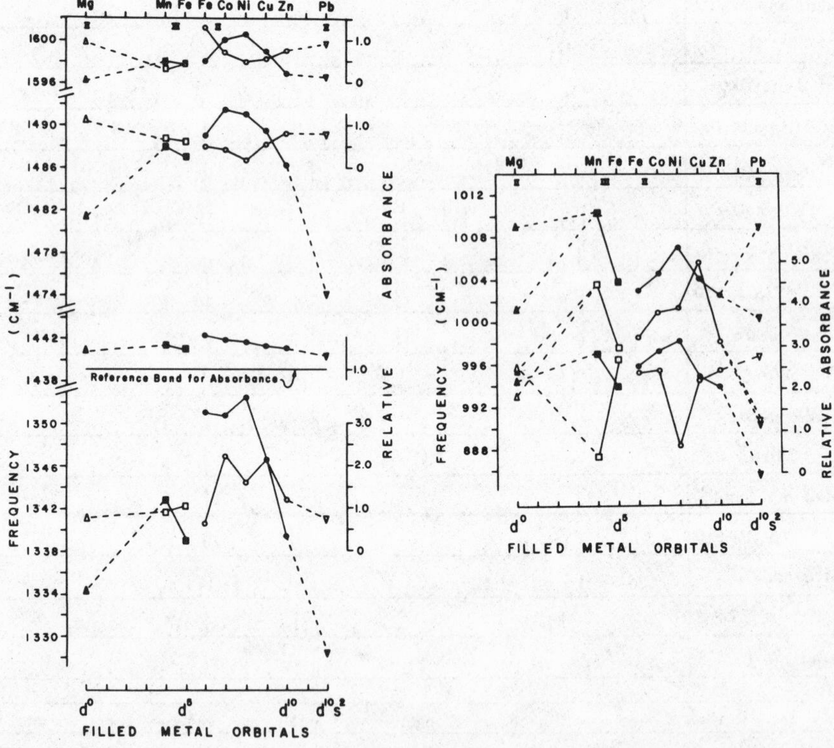

Fig. 6.4. Dependence of frequencies and relative absorbances of porphyrin vibrations on the nature of the coordinated metal in metal complexes of tetraphenylporphin dispersed in KBr pellets. Frequencies (closed symbols) were measured to ± 0.5 cm^{-1} at 2 cm^{-1} resolution, and absorbances to ± 0.01, with the Perkin–Elmer Model 102 infrared spectrometer with a grating-prism double monochromator. Relative absorbances (± 0.1, open symbols) were obtained as the ratio of absorbance of a band to that of the band centered at 1440–1442 cm^{-1}. Two bands are included between 888 and 1012 cm^{-1}. The higher-frequency band is the more intense, with its frequency and relative absorbance being affected in the same direction by ligand field stabilization energy, while the frequency and absorbance of the less intense lower-frequency band are affected in opposite directions by ligand field stabilization energy. (Alben *et al.*, 1973.)

relative absorbances of metal-sensitive bands versus the metal are plotted. According to the particular metal complex, characteristic shifts of these bands occur with respect to both frequency and absorbance. Alben *et al*. suggested that these shifts are at least partially caused by ligand field stabilization of the square, planar metal complexes.

Figure 6.5 shows similar effects of metal substitution on 2,4-diacetyl-deuteroporphyrin. The IR spectra of the 2,4-disubstituted deuteroporphyrin derivatives and their Fe(II) and Fe(III) complexes had been presented earlier (Caughey *et al*., 1966; Alben *et al*., 1968) and bands were numbered as seen in Figs. 6.5 and 6.6. Alben *et al*. (1973) also showed the extent of the effects of the coordinated metal on the porphyrin vibrational modes. They gave nine plots (not shown here) of wavenumbers of the vibrational modes versus number of *d* electrons for bands numbered A_2, A_3, 5_A, 7, 8, M, 9, and 16, as in Fig. 6.6. Bands 5_A and 7 represent ring vibrations quite sensitive to the metal. Band M represents the methyl ester C—O—C vibrations of the propionic acid esters at positions 6 and 7 on the porphyrin. A_2 and A_3 are caused by the acetyl groups at positions 2 and 4 on the porphyrin.

Table 6.1 gives the N—H stretching frequencies and relative absorbances of a series of tetraphenylporphyrin derivatives. These frequencies lie in a higher range than those for the deuteroporphyrin series. The N—H stretching frequencies of the compounds prepared as KBr pellets are about 5 to 6 cm^{-1} lower than when prepared as bromoform solutions.

The spectra of tetraphenylporphyrin derivatives (not shown here) have a few very

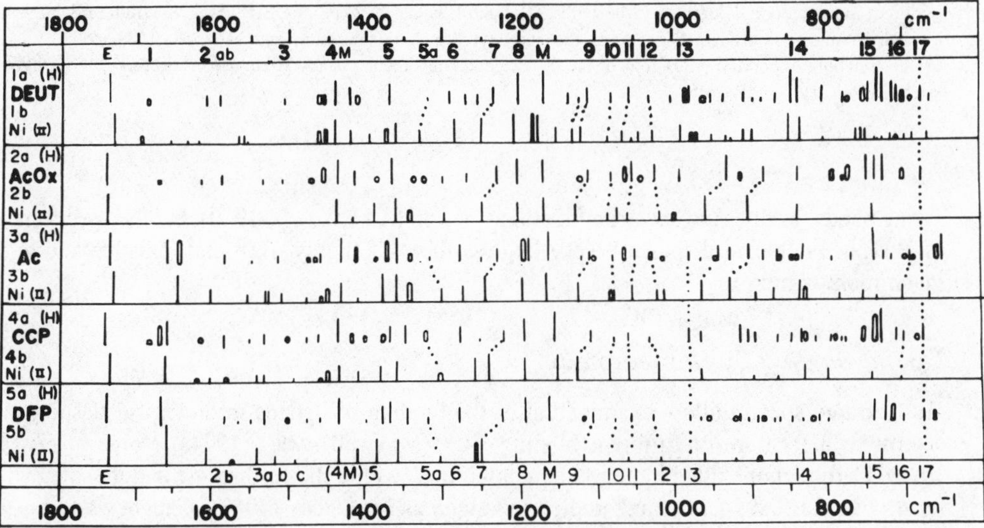

Fig. 6.5. Representation of infrared spectra of deuteroporphyrin derivatives and their nickel (II) complexes dispersed in KBr pellets. Spectra were obtained with a Perkin–Elmer Model 21 spectrometer with a sodium chloride prism, in the laboratory of Dr. Winslow Caughey at The Johns Hopkins University School of Medicine. Bands are numbered as indicated previously (Caughey *et al*., 1966). The following abbreviations are used: DEUT, deuteroporphyrin IX dimethyl ester; AcOx, 2,4-diacetyldeuteroporphyrin IX dimethyl ester dioxime; Ac, 2,4-diacetyldeuteroporphyrin IX dimethyl ester; CCP, 2-(and 4-)formyl-4-(and 2-)vinyldeuteroporphyrin IX dimethyl ester; DFP, 2,4-diformyldeuteroporphyrin IX dimethyl ester. (Alben *et al*., 1973.)

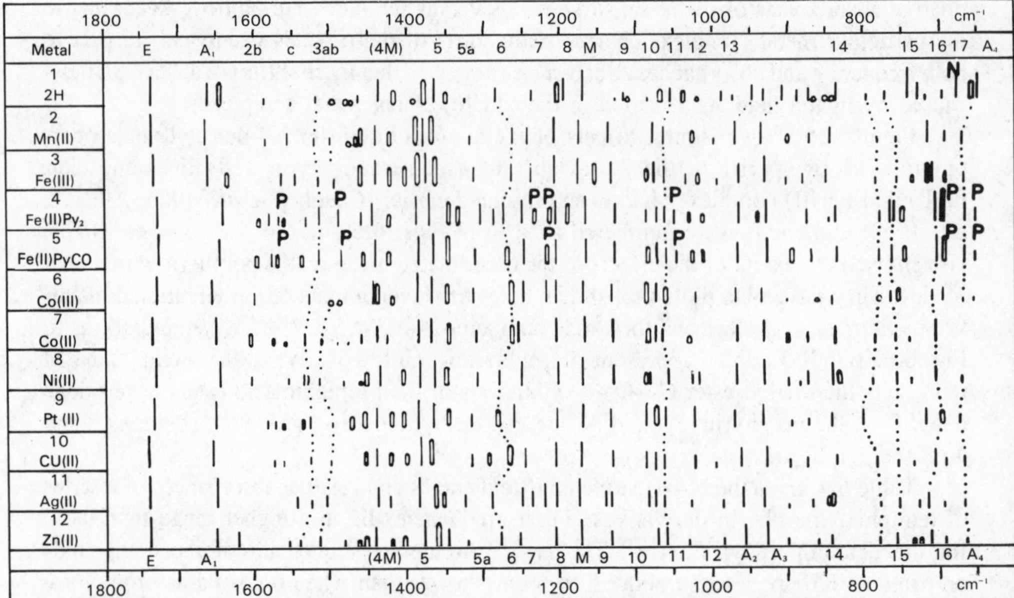

Fig. 6.6. Representation of infrared spectra of 2,4-diacetyldeuteroporphyrin IX dimethyl ester (1-2H), and its metal complexes dispersed in KBr pellets. Spectra were obtained as in Fig. 6.4. Bands marked P are due to pyridine axially coordinated to an Fe(II) porphyrin in the bis-pyridino complex [4-Fe(II)Py_2] and in the mono-pyridino, mono-carbonyl complex [5-Fe(II)PyCO]; band E is the ester carbonyl-stretching vibration of the 6,7-di(methylpropionate) groups; band A_1 is the carbonyl-stretching vibration, and bands A_2, A_3, and A_4 are C—C stretching or bending vibrations of the 2,4 diacetyl substituents; bands M are deformations of the methyl ester. (Alben *et al.*, 1973.)

strong bands, as does the spectrum of etioporphyrin (Alben *et al.*, 1973); but the spectrum of tetraphenylchlorin displays a very large number of highly overlapping bands and is much more complex.

Ligand Binding by Hemeproteins

Infrared spectra allow examination of the binding of ligands in biological tissue to determine the nature of different binding sites (Maxwell *et al.*, 1974a; Volpe *et al.*, 1975). Carbon monoxide binding to hemoglobin A (within the erythrocyte and in isolated form) was the first such ligand studied by Alben and Caughey (1968). Caughey and his colleagues then studied carbonyl complexes of abnormal hemoglobins (Caughey *et al.*, 1969), myoglobins (McCoy and Caughey, 1971; Maxwell *et al.*, 1974b), hemoglobins and myoglobins reconstituted with abnormal hemes (McCoy and Caughey, 1971; O'Toole, 1972), acid-denatured hemoglobins (O'Toole, 1972), cytochrome *c* oxidase (Caughey *et al.*, 1970; Volpe *et al.*, 1975), and certain tissues, including heart (Maxwell *et al.*, 1974a).

Table 6.1. Frequencies and Relative Absorbances of Pyrrole NH-Stretching Vibrations of Porphyrin Derivatives in Bromo Solution or in KBR Pellets[a]

	in Bromoform		in KBr
Porphyrin[b]	ν_{NH} (cm^{-1})	A_{NH}[c] / A_α	ν_{NH} (cm^{-1})
Porphin			3311.1
T(*p*-F-P)P			3316.3
Etio II			3317.0
T(*p*-CN-P)P	3322.5	0.057	
T(*o*-Me-P)P	3324.8	0.035	
T*p*-Cl-P)P	3322.7	0.013	
TPP	3323.0	0.022	3318
T(*m*-Me-P)P	3323.7	0.025	
T(*p*-phenyl-P)P	3323.8	0.013	
T(*p*-Me-P)P	3323.9	0.013	
T(*p*-MeO-P)P	3323.7	0.014	3317.1
T(*n*-Pr)P	3325.2	0.036	
TPC			3343.9

[a] Alben *et al.* (1973).
[b] Abbreviations: Etio-II, etioporphyrin II; T(*p*-F-P)P, tetra(*p*-fluorophenyl)porphin; T(*o*-Me-P)P, tetra((2- or 6-)monomethylphenyl)porphin; T(*p*-Cl-P)P, tetra (*p*-chlorophenyl)porphin; TPP, tetraphenylporphin; T(*m*-Me-P)P, tetra ((3- or 5-)monomethylphenyl)porphin; T(*p*-Me-P)P, tetra(*p*-methylphenyl)porphin; T(*p*-MeO-P)P, tetra(*p*-methoxyphenyl)porphin; T(*n*-Pr)P, meso-tetra-*n*-propylporphin; TPC, tetraphenylchlorin.
[c] Ratio of absorbance of the NH-stretching vibration to that of the α band in the visible spectrum.

Binding of CO and Other Ligands. Satterlee *et al.* (1978) have used IR and ^{13}C-NMR spectroscopy to study the nature of carbon monoxide binding to hemoglobins from human, horse, and rabbit. Human and horse carbonmonoxyhemoglobins display a single IR absorption band (1951 cm^{-1}) for CO bound to heme. IR bands for CO bound either to the α or β subunits of most hemoglobins are superimposed, as is the case with normal human hemoglobin. However, rabbit carbonmonoxyhemoglobin displays absorption bands at 1928 cm^{-1} and 1951 cm^{-1} (Barlow *et al.*, 1973) which were assigned (by use of ^{13}C-NMR and IR spectra of Hb-^{13}CO) to CO bound to the α and β subunits, respectively, within the intact four-unit structure.

Although the rabbit β-CO subunit displays similarities of frequency and band intensity with human and horse COHb's, the α-CO subunit has a much lower frequency

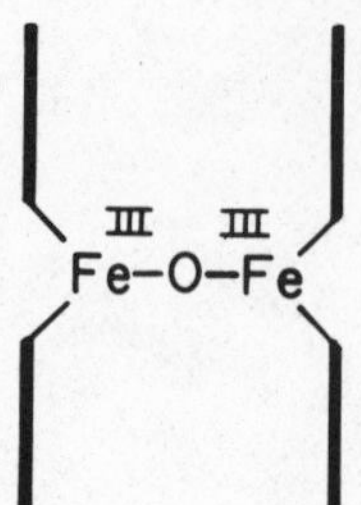

Fig. 6.7. A representation of the μ-oxo linkage between iron(III) porphyrins. Heavy vertical lines indicate the plane of the porphyrin ring. (O'Keeffe *et al.*, 1975.)

(ca. 23 cm^{-1}) and much smaller intensity (by about a factor of 2) compared with the other COHb's. Satterlee *et al.* interpreted these results to support the specific role of the E7 distal histidine (Dickerson and Geis, 1969) as a nucleophilic donor to coordinated CO (Fig. 6.8), first proposed by Caughey (1970).

Hemoglobin from carp, *Cyprinus carpio,* is a useful model for examining conformational changes in hemoglobins because it can be changed from an *R*-state (high affinity) to a *T*-state (low affinity) while fully saturated with ligand by the addition of inositol hexaphosphate (IHP) at low pH values (Tan and Noble, 1973; Tan *et al.*, 1973). Onwubiko *et al.* (1982) have used IR and ^{13}C NMR spectroscopy to demonstrate IHP-induced structural changes at ligand binding sites in carp hemoglobin carbonyl. IR and ^{13}C NMR spectra for carbonyl ligands bound to carp Hb at pH 6.0 and 25°C display marked shifts when IHP is added to the solution.

The relative intensities of the two C—O stretch bands at 1951 and 1968 cm^{-1} arising from two ligand-site conformers are changed by IHP in the same way as is observed when the pH is lowered in the absence of IHP. The lowering of pH and the binding of IHP both enhance the conformer corresponding to the 1968-cm^{-1} band, which is normally of much lower stability. In addition, ^{13}C NMR spectra show that the CO sites for only one kind of subunit, probably the β-subunit, are changed by IHP (Onwubiko *et al.*, 1982). These results show directly that the observed IHP-induced switch from the *R*- to the *T*-form has a concomitant significant change in the ligand binding site structure at two of the four subunits.

IR spectra have demonstrated the existence of multiple species of hemoglobin carbonyls and myoglobin carbonyls and azides that have not been found by crystallographic methods (Caughey *et al.*, 1978; McCoy and Caughey, 1971; Tucker *et al.*, 1978).

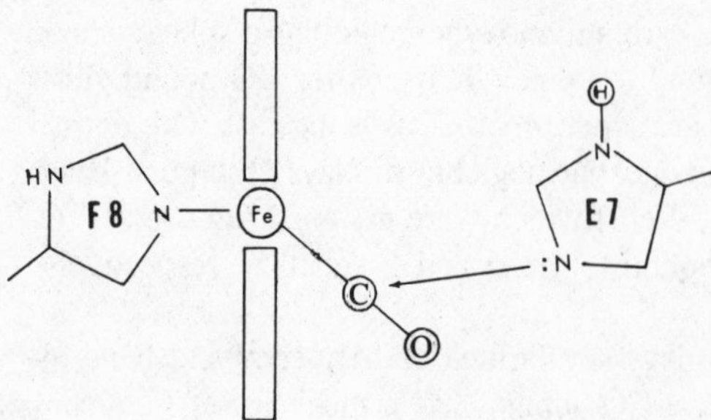

Fig. 6.8. Suggested mode of interaction between the distal histidine (E7) and heme bound CO; nucleophilic interaction. (Satterlee *et al.*, 1978.) Reprinted with the permission of the American Chemical Society. Copyright 1978.

Caughey *et al.* (1982) have also found that multiple conformers can exist in hemeproteins. Their evidence was from IR spectra for carbonyl, oxygenyl, and cyano ligands. For example, four CO stretch bands found near 1950 cm^{-1} can be ascribed to four discrete protein conformers in dynamic equilibrium for myoglobin carbonyls; and hemoglobin carbonyls also exhibit multiple bands but only two or three have been detected for a given subunit. Also, oxymyoglobins and oxyhemoglobins display multiple (5 or 6) IR bands for bound oxygen near 1130 cm^{-1}; thus, multiple conformers for myoglobin and hemoglobin oxygenyls are indicated, although some band splitting owing to Fermi resonance may occur. These authors also discussed conformational changes in cytochrome *c* oxidase upon sequential addition of electrons to the fully oxidized oxidase when the oxidase binds cyanide.

The ΔH and ΔS values have been calculated for the rapid interconversions of four discrete conformers of the carbon monoxide complex with myoglobin that had been isolated as the oxygenyl species from bovine heart muscle (Caughey *et al.*, 1981). The ΔH and ΔS values were estimated to range from -8 to 34 kJ per mol and -27 to 87 J mol^{-1} K^{-1}, respectively. The structures of the conformers were, therefore, expected to vary significantly.

Caughey (1980) has discussed the use of IR spectroscopy for the determination of environments of metal ions in proteins. Among the topics treated were CO binding to heme iron of hemoglobin, measurement of IR spectra in aqueous media (H_2O and D_2O), IR band frequencies for ligands bound to metal ions in proteins—(e.g., CO bound to Cu (I) in hemocyanin (not a porphyrin), CO bound to Fe (II) in horseradish peroxidase, CO bound to Fe (III) in metmyoglobin, N_3^- bound to Fe (III) in metmyoglobin, CN^- bound to Fe (III) in cytochrome *c* oxidase—and others I have mentioned in this chapter.

It is interesting to show how the use of isotopes allowed confirming evidence for the assignment of the C—O stretch in hemoglobin A—CO to be obtained by Alben and Caughey (1968). $^{12}C^{16}O$ absorbed at 1951 cm^{-1} but shifted to 1907 cm^{-1} for $^{13}C^{16}O$ and also to 1907 cm^{-1} for $^{12}C^{18}O$. The frequency shift in each case was 44 cm^{-1}, which is nearly the shift predicted for a simple diatomic molecule as calculated from the reduced mass relationship.

Caughey (1980) has given several examples of the use of isotopes for the study of ligand binding to metalloproteins.

An interesting group of hemeproteins are the leghemoglobins, which are involved in nitrogen fixation by rhizobium bacteroids in legume root nodules. Fuchsman and Appleby (1979) have studied the effects of pH upon IR spectra in the carbonyl stretching region of the CO complexes of soybean leghemoglobins (Lb) *a*, c_1, and c_2, sperm whale myoglobin, and human hemoglobin A. At neutral pH the carbonyl stretching frequency for Lb—CO complexes was 1947.5 cm^{-1} (Fig. 6.9). At acidic pH myoglobin—CO and hemoglobin—CO complexes displayed CO stretching bands at 1966 to 1968 cm^{-1}, and Lb—CO complexes formed such bands at ca. 1957 cm^{-1} (Fig. 6.9). Apparent p*K* values of 4.0 to 4.6 were determined for all the CO-complexes from values of ν_{CO} as affected by pH change (plots of absorbance vs. pH). The pK_{app}^{CO} values in the 4.0 to 4.6 region appeared to be p*K* values of the distal histidines (see Fuchsman and Appleby, 1979, for an exception based on visible spectra—a $pK_{app}^{CO} = 5.8$ for myoglobin—CO, and attributed by them to a group other than distal and proximal histidines).

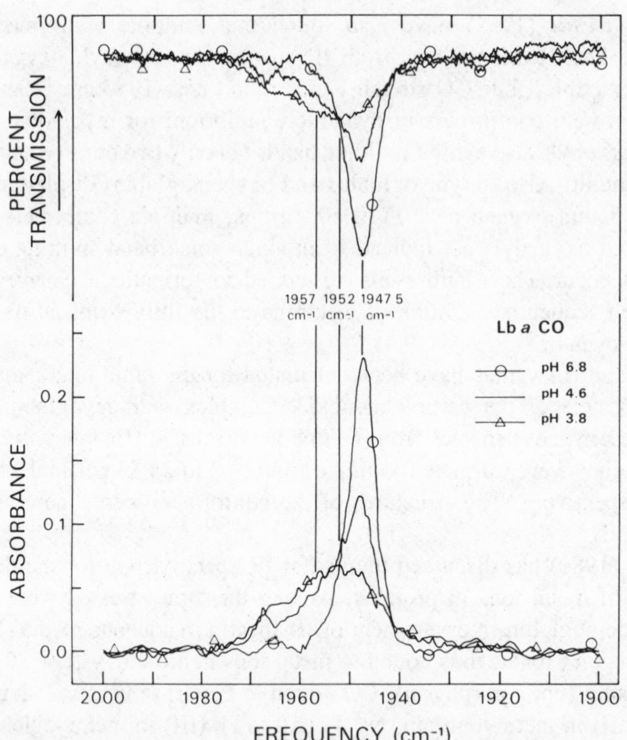

Fig. 6.9. Percent transmission scans and corrected absorbance plots of ν_{CO} region of IR spectra of soybean leghemoglobin *a*-CO at pH 6.8 (○), pH 4.6 (——), and pH 3.8 (△). Base lines of percent transmission scans have been brought into coincidence by slight vertical movement of entire scans. Conditions: ~2°C, 9.5 m*M* leghemoglobin *a*-CO, 1.0 m*M* EDTA, 0.10 *M* citrate, 0.10 *M* Mes, 0.10 *M* Mops. EDTA is ethylenediaminetetraacetate; Mes is 2-(*N*-morpholino)ethanesulfonate; Mops is 3-(*N*-morpholino)propanesulfonate. (Fuchsman and Appleby, 1979.) Reprinted with the permission of the American Chemical Society. Copyright 1979.

Exchange of CO between Cytochrome c Oxidase and Hemoglobin A. Yoshikawa *et al.* (1977) have recorded the infrared carbon monoxide stretch bands for CO liganded to hemoglobin A (1951 cm^{-1}, Alben and Caughey, 1968) and completely reduced cytochrome *c* oxidase (1963.5 cm^{-1}, Caughey *et al.*, 1970; Caughey, 1971; Volpe *et al.*, 1975), and have used these bands to determine the amount of the gas bound, to monitor the exchange of the gas from the oxidase to hemoglobin, and to study the nature of the CO and O_2 binding sites. They found an ϵ_{mM} for the 1963.5 cm^{-1} band of 4.9 ± 0.3 for the oxidase carbonyl and an ϵ_{mM} for the 1951-cm^{-1} band of HbCO of 3.7 ± 0.1.

The oxidase carbonyl band is exceptionally narrow, narrower than any other carbonyl of a hemeprotein, indicating that the CO ligand is not exposed to the aqueous medium, but is in a highly ordered environment. [See p. 5506 of Yoshikawa *et al.* for a discussion of the effects of solvents on band frequency and intensity.] The oxidase CO site was shown to be similar to hemoglobin sites rather than to those in cytochrome P-450, cytochrome P-420, and peroxidases. The binding of CO to hemoglobin is ca. 50 times as great as that for the oxidase under specific conditions; the CO:Fe ratio is 0.5, showing that only one metal atom (the classical a_3 heme) of the four possible sites ($2Fe^{2+}$, $2Cu^{1+}$) is attached to CO under these conditions (Yoshikawa *et al.*, 1977).

Yoshikawa and Caughey (1982) have more recently studied the effects of the oxidation state of cytochrome *c* oxidase (bovine heart muscle) on the binding of CO. Carbon monoxide stretch band intensities show that one CO molecule is bound when one to four electron equivalents are added (varying amount of NADH added to the fully oxidized enzyme in the presence of CO). Carbonyl stretch band frequencies are consistent with CO binding to Fe^{2+} but not to Cu^+. The CO stretch band was deconvolutable into two separate bands; these bands, separated by 4 cm^{-1} in an intensity ratio of five to one, suggested to these workers two different carbonyl conformations, one more stable than the other. Thus, IR parameters for the bound CO show that CO always binds to Fe^{2+} of heme A and that changes in the oxidation state of the other metals do not greatly affect either the electron donation from Fe^{2+} to bound CO or the environment about the CO. (See detailed discussion of these points in Yoshikawa and Caughey, 1982.)

CO Binding by Hemocyanin, a Nonporphyrin. Alben *et al.* (1970) have tested for the possibility of a bridging copper carbonyl structure in the hemocyanin–CO complex. They determined the IR absorption maximum of the complex (ν_{CO}) and compared it with those of simple copper carbonyls in solution. The spectrum of carboxyhemocyanin has a band at 2063 ± 1 cm^{-1} caused by bound CO (Fig. 6.10). This frequency lies in the range where mononuclear, nonbridging metal carbonyls should absorb and 100 to 200 cm^{-1} higher than that for a bridging carbonyl (Nakamoto, 1963). From their own spectral data and the ruling out of certain types of structures by other investigators, Alben *et al.* were able to give probable structures for CO-hemocyanin as I or II (Fig. 6.11). In structure I the second copper does not bind CO but may stabilize the protein. In structure V the second copper may affect the binding of CO through a bridging ligand or metal–metal bond. The IR data show that CO is coordinated to only one copper atom per binding unit in hemocyanin.

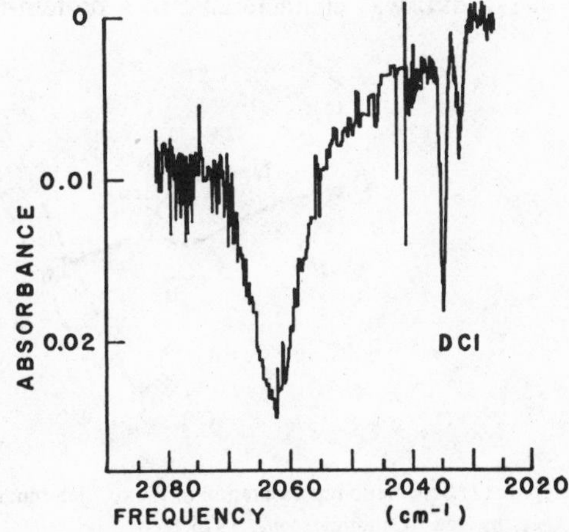

Fig. 6.10. Infrared spectrum of carboxyhemocyanin (ν_{CO}) recorded at 4 cm^{-1} spectral resolution and 50× absorbance, with a 0.054-mm sample cell path. Frequency calibration curves constructed with DCl and CO were compared with DCl peaks recorded with the carboxyhemocyanin spectrum. (Alben *et al.*, 1970.) Reprinted with the permission of the American Chemical Society. Copyright 1970.

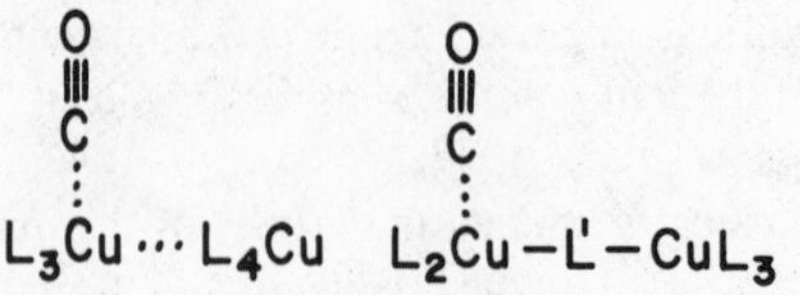

$$\begin{array}{cc} \overset{\displaystyle O}{\underset{\vdots}{\overset{|||}{C}}} & \overset{\displaystyle O}{\underset{\vdots}{\overset{|||}{C}}} \\ L_3Cu \cdots L_4Cu & L_2Cu - L' - CuL_3 \\ \text{I} & \text{II} \end{array}$$

Fig. 6.11. Probable structures for CO-hemocyanin. (Alben *et al.*, 1970.) Reprinted with the permission of the American Chemical Society. Copyright 1970.

Binding of Oxygen. Caughey *et al.* (1975) have discussed reactions of oxygen with hemoglobin, cytochrome *c* oxidase, and other hemeproteins. The reactions of these hemeproteins with oxygen have been discussed widely and with much dispute (Antonini and Brunori, 1971; Caughey, 1967). The general characteristics of the O_2-binding site in myoglobins and hemoglobins have been shown by X-ray crystallography. These sites are similar but not exactly the same. Protoheme is the iron porphyrin in the usual myoglobins and hemoglobins; histidine is a proximal, or *trans,* ligand; and a pocket exists, which is lined by certain amino acid residues from the O_2-binding site at the heme iron to the protein outer surface. Often, but not always, a distal histidine is placed spatially so that when oxygen binds to iron, it is found between the iron atom and the histidine residue closest to it. Figure 6.12 shows these characteristics, in which the space where bound oxygen is found is indicated by the hatched area. The characteristics shown are based on X-ray crystallographic data (Stryer *et al.,* 1964).

Caughey *et al.* (1973) have used IR and equilibrium studies of carbonyl myoglobins, hemoglobins, and hemes to observe *trans* effects (due to substitution in axial ligand), *cis* effects (due to different substituents on the porphyrin ring), and the effects of proteins or solvent environment near the bound CO molecule. Such data led them to predict bent end-on binding for CO in carboxyhemoglobin A, seen in Fig. 6.13 (Caughey, 1971). Huber *et al.* (1970), in an X-ray study, reported bent FeCO bonds for the CO complexes of erythrocruorin, the hemoglobin of insects. Barlow *et al.* (1973) have confirmed that oxygen is bound with a similar spatial arrangement. By means of IR difference spectra of erythrocytes treated with $^{16}O_2$, $^{18}O_2$, or CO, they found one isotope-sensitive band for $^{16}O_2$ at 1107 cm^{-1} and one for $^{18}O_2$ at 1065 cm^{-1}. Oxyhemoglobin A also displayed a band at 1107 cm^{-1} (^{16}O—^{16}O stretch) in either H_2O or ^2H_2O. The band width (8–9 cm^{-1} half-width) was characteristic of a protein-bound ligand (e.g., in COHb), and the band

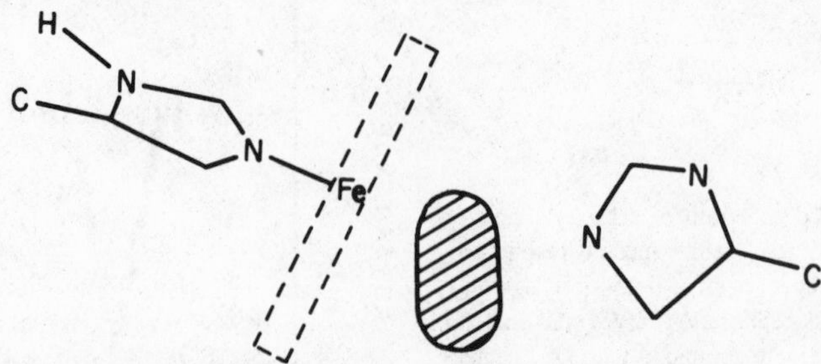

Fig. 6.12. Schematic representation of the oxygen binding site in hemoglobins. The hatched area shows where O_2 is located. (Caughey *et al.,* 1975.)

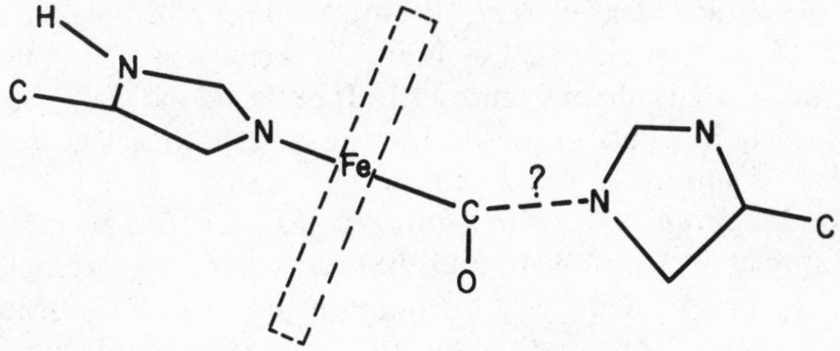

Fig. 6.13. Schematic representation of CO binding to hemoglobin. (Caughey *et al.*, 1975.)

intensity was very high for a symmetrical O_2, thus suggesting that O_2 is bound asymmetrically. Data gathered by various workers on the frequency of the O—O stretch for certain dioxygen compounds (see Tables in Caughey *et al.*, 1975) led to the conclusion that oxygen is bound in the bent end-on configuration, metal $—O\diagdown_{O}$. Oxyhemerythrin, on the other hand, has its $\nu_{O—O}$ band at 840 cm^{-1} (res. Raman data), characteristic of $M\langle\big|^{O}_{O}$ peroxidic binding of oxygen (where M stands for metal).

Oxygen bound to hemoglobin has two absorption bands (1107 cm^{-1} and 1156 cm^{-1}) (Alben, 1978) rather than only one at 1107 cm^{-1} as previously reported. These arise from splitting by Fermi resonance with the FeO_2 band at 567 cm^{-1}. This large resonant splitting is evidence for strong coupling between the Fe—O and O—O vibrations, and accordingly, strongly covalent bonds.

Hemoglobins that are reconstituted from globin and porphyrins modified by changing the substituents on the porphyrin ring or by replacing iron (II) with cobalt (II) can retain the ability to bind oxygen reversibly (Rossi Fanelli and Antonini, 1959; Sugita and Yoneyama, 1971; Hoffman and Petering, 1970; Yonetani *et al.*, 1974). Maxwell and Caughey (1974) have used IR methods to study the effects of a change in metal ion on the O_2-to-metal bonding. These authors examined IR data for O_2 complexes of cobalt(II) deuteroporphyrin IX and iron(II) deuteroporphyrin IX combined with globin from human hemoglobin A. IR bands for bound O_2 occurred at 1106 and 1105 cm^{-1} for the iron (II) and cobalt (II) complexes, respectively, thus providing direct evidence of very similar bent end-on metal-O_2 bonding, i.e., $M{=}{=}O{\diagdown}_{O}$, for the two metals. This type of structure for the cobalt attachment is supported by the reasons used earlier in the case of the iron atom (Barlow *et al.*, 1973; Maxwell *et al.*, 1974b; Caughey *et al.*, 1975).

Resonance Raman data showed that, in contrast, oxyhemerythrin has its $\nu_{O—O}$ band at 840 cm^{-1}, thus proving that the peroxide type of bonding ($M\langle\big|^{O}_{O}$) exists for the oxygen (Dunn *et al.*, 1973; Caughey *et al.*, 1975).

IR spectra of hemoglobins of the insect *Chironomus thummi thummi* have been observed by Wollmer *et al.* (1977). Using difference spectroscopic methods employed by Barlow *et al.* (1973) and Maxwell *et al.* (1974b), Wollmer *et al.* obtained the stretching band of O—O for O_2—hemoglobin III by using O_2—hemoglobin III in the sample cell and CO—hemoglobin III in the reference cell. They found the O—O stretch at 1107 cm^{-1}, the same location as that of human hemoglobin, showing that the geometry of oxygen binding is asymmetric in the insect hemoglobin also.

Binding of Oxygen and Carbon Monoxide to Cytochrome o (Vitreoscilla). The oxygenated species of cytochrome *o* from the filamentous myxobacterium *Vitreoscilla* has been studied by IR spectroscopy to determine the nature of FeO_2 bonding (Choc *et al.*, 1982). These authors have presented the first direct evidence of oxygen bound to a terminal oxidase (ν_{O-O} at 1134 cm^{-1}). Under turnover conditions in which NADH, O_2, and NADH-cytochrome *o* reductase were present at 4°C, they observed a characteristic "oxy" visible spectrum (λ_{max}, 415, 543, and 577 nm) and an IR band at 1134 cm^{-1} arising from bound $^{16}O_2$. This band has a half-bandwidth of 17 cm^{-1} and an integrated intensity of about 10 mM^{-1} cm^{-2}. When the available oxygen was used up, the IR band disappeared and the characteristic "deoxy" visible spectrum (λ_{max}, 423 and 553 nm) appeared. The IR band shifted to 1078 cm^{-1} with $^{18}O_2$. Choc *et al.* observed only a single band for either $^{16}O_2$ or $^{18}O_2$, in contrast with oxymyoglobins and oxyhemoglobins, for which multiple bands are seen. Since the frequencies for O_2 vibrations are in similar regions, these authors say that bent-end-on oxygenyl-type bonding between ferrous iron and dioxygen must occur in oxycytochrome *o* as well as in oxymyoglobins and oxyhemoglobins.

Of notable interest here are characteristic differences in IR spectra that indicate significant bonding differences among the three hemeproteins. Carbon monoxide bound to the fully reduced enzyme at 26°C displays a stretch band with ν_{CO} at 1964 cm^{-1}, a half-bandwidth of 9 cm^{-1}, and an integrated intensity of 14 mM^{-1} cm^{-2}. The intensity of the CO band is unusually sensitive to temperature changes; the integrated intensity went to 21 mM^{-1} cm^{-2} with the temperature of the sample going from 26° to 4°C. This C—O stretch frequency is much higher than the major bands of HbACO (1951 cm^{-1}) or bovine MbCO (1944 cm^{-1}); but this ν_{CO} is nearly identical to that observed for the carbonyl of another oxidase, cytochrome *c* oxidase. The bandwidth for cytochrome *c* oxidase (half-bandwidth of 4 cm^{-1}) is much less, being consistent with a more stable and less exposed ligand-binding site (Choc *et al.*, 1982).

Binding of Carbon Monoxide to Cytochrome P-450$_{cam}$. O'Keeffe *et al.* (1978) have used IR spectroscopy to study directly the dioxygen-binding site of cytochrome P-450$_{cam}$ (cytochrome P-450 isolated from *Pseudomonas putida* grown on *d*-camphor), with carbon monoxide as the spectrally observable ligand. These workers studied ferrous–carbonyl complexes of the P-450 and a denatured form, P-420, of this enzyme, in addition to cytochromes P-450 and P-448 induced in liver microsomes in rats by pretreatment with 3-methylcholanthrene and phenobarbital. The *d*-camphor (K^+)-bound P-450$_{cam}$ produced an absorbance band for the heme-bound carbonyl (ν_{CO}) at 1940 cm^{-1} with a half-bandwidth (width at half band height, $\Delta\nu_{1/2}$) of 13 cm^{-1}; but the camphor-free enzyme displayed two stretching frequencies of equal area at 1963 and 1942 cm^{-1}, with half-bandwidths of 11–12 and 19–21 cm^{-1}, respectively. When *d*-camphor and potassium ion were added to the camphor-free Fe^{2+}-carbonyl-enzyme, the spectrum reverted to that of the original enzyme–camphor complex. The combined areas of the 1963- and 1942-cm^{-1} bands equaled that of the 1940-cm^{-1} band.

When the native P-450_{cam} enzyme was converted to the P-420 form, the product displayed a CO stretching band at 1966 cm^{-1} with $\Delta\nu_{1/2}$ of 23 cm^{-1}.

Rat liver microsomal cytochromes (P-450 and P-448) displayed ν_{CO} and $\Delta\nu_{1/2}$ different from each other and P-450_{cam}. Cytochrome P-450 showed a ν_{CO} at 1948 cm^{-1} with $\Delta\nu_{1/2}$ of ca. 25 cm^{-1}. Cytochrome P-448 showed a ν_{CO} at 1954 cm^{-1} with $\Delta\nu_{1/2}$ of ca. 30 cm^{-1}

The data mentioned above allowed several conclusions to be made: (a) the iron–carbonyl bonding in each of the P-450 cytochromes is similar to that seen for the FeCO complexes of hemoglobin and myoglobin; (b) when *d*-camphor is present, the heme-carbonyl of the P-450_{cam} displays nonlinear bonding, but when it is absent, linear and nonlinear bonding occur—this is evidence that there is a close spatial relationship between the heme-carbonyl (oxygenyl) and *d*-camphor binding site; (c) in P-420 the FeCO bonding is almost the same as that of denatured ferrous–carbonyl complexes of hemoglobin and myoglobin, a fact that suggests linear heme–carbonyl bonding and a nitrogenous *trans* axial ligand.

Near-IR Observation of Oxygen Supply to Tissues. A near-IR (700–865 nm) *noninvasive* method has been developed by Jöbsis (1977) to monitor cerebral and myocardial oxygen sufficiency and circulatory parameters. Observations by IR transillumination in the exposed cat (or dog) heart and in the brain in cephalo without surgical intervention show that oxygen sufficiency for cytochrome a,a_3 function (cytochrome *c* oxidase), changes in tissue blood volume, and the average hemoglobin–oxyhemoglobin equilibrium can be recorded effectively and in a continuous manner for research and clinical purposes. These studies were aimed at application of this technique to the monitoring of the adequacy of intracellular oxygenation in clinical ischemia (local diminution in the blood supply) and hypoxemia.

Binding of CO and O_2 by Hemoglobin Zurich. The combined use of IR spectroscopy and reactivity probes can aid toward better understanding of the origins of hemoglobin diseases due to abnormal hemoglobin structure. Wallace *et al.* (1976) have provided a correlation of structure with the pathological manifestations of hemoglobin Zurich (HbZ, in which β_{63} distal histidine of HbA becomes arginine) by using such techniques.

When the β_{63} histidine is replaced by arginine, HbZ has enhanced susceptibility to autoxidation in the presence of "oxidant drugs" (Frick *et al.*, 1962; Bachmann and Marti, 1962). Thus, the effect of the substitution by arginine on the bonding of CO and O_2 (Caughey *et al.*, 1969; Caughey, 1971; Caughey *et al.*, 1973; Alben and Caughey, 1968; and Geraci *et al.*, 1969) are of interest in explanations of the pathology of this and other abnormal hemoglobins

HbZ displays abnormal chemical properties. For example, Table 6.ν_{CO} is displayed at 1951 cm^{-1} for HbA—CO, whereas ν_{CO} for HbZ—CO is composed of bands at 1950 cm^{-1} and 1958 cm^{-1} for CO bound in α and β chains (Fig. 6.14), respectively. Also, the β CO's are displaced less readily by O_2. The one-electron donor hydroquinone yields methemoglobin and peroxide more rapidly from HbZ—O_2 than from HbA—O_2. The stereochemistry of the interaction between CO or O_2 and HbA is disturbed when HbA becomes HbZ by virtue of the change from β_{63} histidine to arginine. In the β-subunits of HbZ the interaction is no longer possible.

Binding of Nitric Oxide. Maxwell and Caughey (1976) have studied the bonding of NO to heme B and hemoglobin A (HbA). They prepared five-coordinate and six-

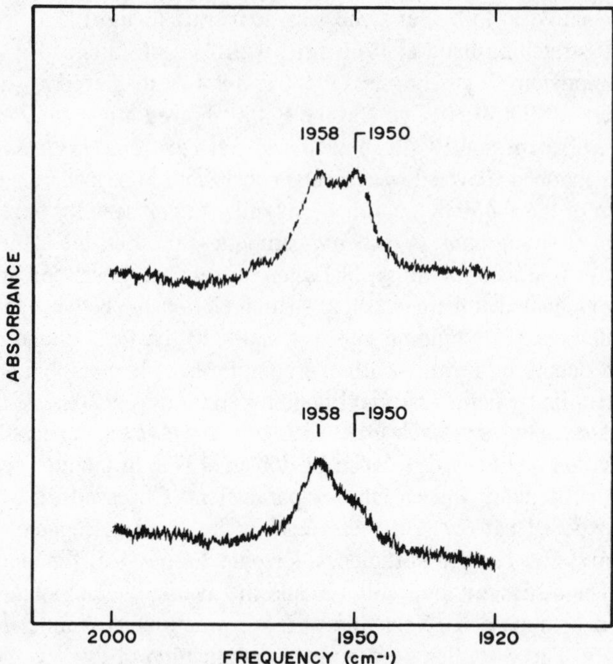

Fig. 6.14. The infrared spectrum of HbZCO. Upper spectrum: after saturation with CO, the 1950-cm^{-1} band is assigned to the normal α-subunit and the 1958-cm^{-1} band to the β-subunit. Lower spectrum: after exposure of a CO-saturated solution to oxygen to partially displace bound CO by O_2. (Wallace *et al.*, 1976.)

coordinate nitrosyl hemes, and compared their spectra (IR, electron paramagnetic resonance and visible-Soret) with the corresponding ones for nitrosyl HbA (HbA-NO). By carefully matching CaF_2 cells, these workers were able to use difference spectroscopy in ^2H_2O solutions (i.e., HbA-NO vs. HbA-CO) to demontrate that $^{15}N^{16}O$ bound to HbA absorbs at 1587 cm^{-1}, which corresponds to a six-coordinate nitrosyl heme; $^{14}N^{16}O$ absorbed at about 1615 cm^{-1}.

A bent-end-on bonding of the type (Fe—N$\searrow_{O}$) fits these data, whereby the Fe—N bond has appreciable double-bond character, the Fe^{II} acts as a π donor and the NO as a σ donor with a net donation of electron density from Fe^{II} to NO. This bonding is similar to that in carboxyhemoglobin A and oxyhemoglobin A as displayed by ν_{co} at 1951 cm^{-1} and ν_{O_2} at 1106 cm^{-1}.

When inositol hexaphosphate (IHP) is present, the NO stretching vibration in HbA—NO produces two approximately equal bands, the first at the wavenumber value in the absence of IHP and the second at a wavenumber about 50 cm^{-1} higher (1668 cm^{-1} for $^{14}N^{16}O$ and 1635 cm^{-1} for $^{15}N^{16}O$). The wave number shifts induced by IHP are very similar to changes in $\nu_{^{14}N^{16}O}$ from l-methylimidazole-Fe^{II}-NO in 1-methylimidazole (1618 cm^{-1}) to five-coordinate Fe^{II}-NO in 1,2-dichloroethane (1668 cm^{-1}). The spectral data support the concept that the binding of IHP to protein (globin) causes a break of the bonds between the proximal histidines and iron in two of the four subunits (Fig. 6.15).

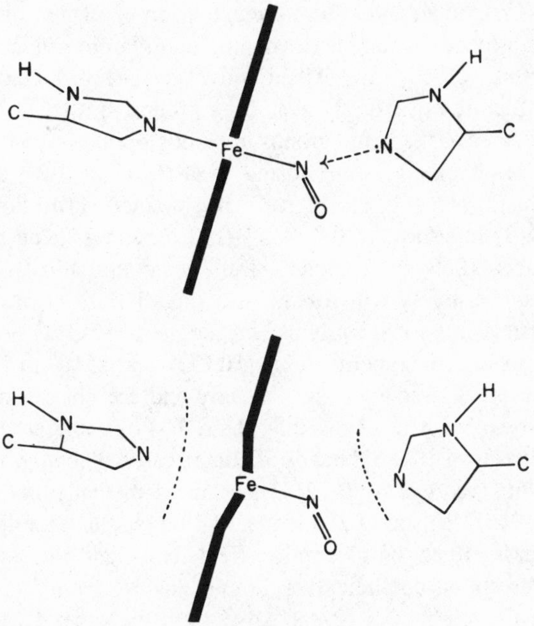

Fig. 6.15. Schematic representations of NO binding to hemoglobin. The upper drawing represents the six-coordinate NO heme present at all heme sites in stripped or native HbA—NO. The lower drawing represents the five-coordinate heme NO present in half the subunits of HbA as a result of IHP binding where infrared data provide convincing evidence for complete cleavage of the proximal histidine–Fe bond but changes in the distal histidine–NO interaction are less clear. (Maxwell and Caughey, 1976.) Reprinted with permission of the American Chemical Society. Copyright 1976.

Maxwell and Caughey offered the idea that the proximal histidines "pull away" from the heme iron, as a plausible explanation for the transmission of the effect from the IHP-protein binding site to the heme site. It was not possible with the evidence at hand to evaluate the relative importance of a number of possible IHP-induced changes of interactions between amino acid residues with the porphyrin ring and the NO ligand.

Binding of CO by Sulfmyoglobin. Clinical cyanosis is frequently caused by sulfhemoglobin (sulf-Hb), which is a nonfunctional pigment found abnormally in erythrocytes (Wintrobe, 1967; Cartwright, 1970; Jaffé, 1971; Finch, 1948). Berzofsky *et al.* (1972), in order to understand the chemistry of sulf-Hb, have studied the reactions of a very similar substance, sulfmyoglobin (sulf-Mb). They measured the stretching frequency of CO and correlated their observations with the decrease (by at least 1500) in sulf-Mb binding of this gas compared to the binding of myoglobin. The C—O stretching frequency of CO attached to the iron of ferrous sulfmyoglobin was 1953.5 ± 0.5 cm^{-1}, which is about 10 cm^{-1} higher than that of CO bound to myoglobin. These authors found a linear correlation of the logarithms of the affinity constants of O_2 and of CO with the C—O stretching frequency, which they used as a measure of the electron density at the iron of the *chlorin* prosthetic group of sulf-Mb. The data were collected with an IR interferometer, which was interfaced with a digital data system that computed a fast Fourier transform to produce the spectra.

Studies of —SH Groups in Hemeproteins

Carboxyhemoglobins and Other Derivatives. Fourier transform IR (FTIR) spectroscopy provides a significant improvement in the signal-to-noise ratio over previous

IR methods, and allows the extension of IR investigations of small molecules to include substances as complex as hemoglobin. For example, Alben and Caughey (1968), Caughey *et al.*, (1969), and Alben and Fager (1972) had studied the effects of local molecular structure on strongly absorbing groups, e.g., CO (1968, 1969) and azide (1972) coordinated to the heme groups in hemoglobin or myoglobin. Later, Bare *et al.* (1975), using FTIR methods, investigated the SH groups of cysteine residues in hemoglobins from man, cow, pig, and horse. They defined absorption bands caused by SH groups at the $\alpha_1\beta_1$ interface (α-104 and β-112 cysteine). [The reader should consult Dickerson and Geis (1969) for a good diagram depicting the $\alpha_1\beta_1$ and $\alpha_1\beta_2$ interchain contacts and the numbering system for the structures.] Bare *et al.* used FTIR spectroscopy to observe bands due to SH groups (ν_{SH}) of α-104 (G11) and β-112 (G14) cysteine residues of human carboxyhemoglobin (HbCO) near 2560 cm^{-1}. The SH groups of the β-93 cysteine absorb IR radiation very weakly and are not distinguishable from the background, but single SH bands caused by the α-104 cysteine groups of horse and pig hemoglobins are displayed. The SH bands of human carboxyhemoglobin show the same isotopic frequency shift (ν_{SD}/ν_{SH} = 0.7267) as that of methanethiol. The reader is referred to Herzberg (1945), Wilson *et al.* (1955), and Pinchas and Laulicht (1971) for discussions of isotopic mass effects on absorption frequencies and intensities. The value of the integrated IR absorption coefficient, ϵ_{mM}(area) in mM^{-1} cm^{-2}, for 0.1 *M* ethanethiol increased with hydrogen-bond acceptor solvents in the order CCl_4(0.07), H_2O(0.21), acetone (0.43), and *N,N*-dimethylacetamide (1.35 mM^{-1} cm^{-2}).

$$\epsilon_{mM}(\text{area}) = \frac{1}{C1}\int A\, d\nu$$

where A is absorbance.

When Bare *et al.* (1975) compared the integrated absorption coefficients for the α-104 cysteine SH (2.43), and the β-112 SH (0.80), of human carboxyhemoglobin, with those of ethanethiol solutions, they concluded that specific hydrogen-bonded structures exist with peptide carbonyl groups four (or three) residues back in the G helices.

This study is an elegant application of FTIR for the observation of SH groups in a native protein. Tbe legend to Fig. 6.16 explains how spectra were recorded by this method. The figure shows spectra of carboxyhemoglobin of man, pig, and horse.

Alben and Bare (1978) have used FTIR spectroscopy to study molecular interactions in hemoglobin (Hb). They showed how FTIR can be applied to measure differences in protein structure that are related to biological control mechanisms. Many details were discussed about obtaining difference spectra experimentally. The original paper should be consulted for these details.

These workers employed a Digilab-14D interferometer to record IR spectra of the α-104 (G11) cysteine SH group for aqueous solutions of hemoglobin derivatives from human beings, pigs, and horses. The interferometer was equipped with a digital data system and a liquid-nitrogen-cooled InSb detector for measurements above 1800 cm^{-1}.

The α-104 cysteine SH vibrational absorptions appear between 2552.6 and 2556.3 cm^{-1} in aqueous solutions of human Hb. The SH bands are much narrower ($\Delta\nu_{1/2}$ = 13.5 cm^{-1}) and much more intense (a_{mM} = 0.17 mM^{-1} cm^{-1} for α-104 SH and 0.055

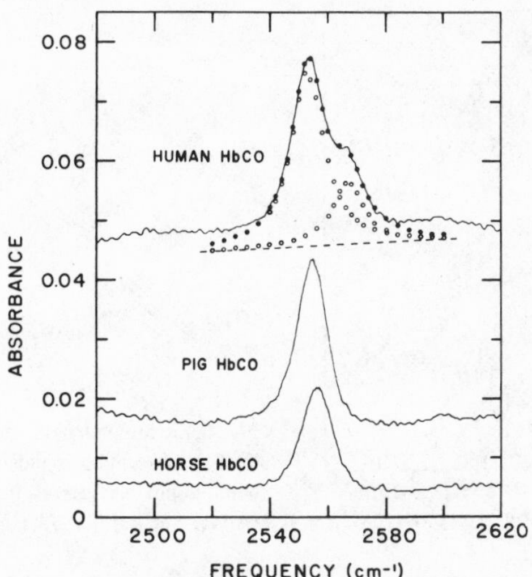

Fig. 6.16. Infrared absorbance difference spectra of human (17 m*M* heme), pig (16 m*M*), and horse (11 m*M*) carboxyhemoglobin, each vs. cow carboxyhemoglobin as reference. Single-beam spectra were collected in the following manner: the optical path of the CaF_2 cell was 0.2 mm, and the spectra of the above-mentioned solutions were recorded on a Digilab Model FTS-14D interferometer equipped with a liquid-N_2 cooled InSb detector. Four sets of 64 coherently coadded interferograms were independently subjected to Fourier transformation and were, in turn, coadded to yield a digital single-beam spectrum at 2 cm^{-1} resolution. This was ratioed against a similarly obtained spectrum of air derived from four sets of 256 interferograms and plotted with absorbance as ordinate and frequency as abscissa. A similarly obtained digital absorbance spectrum of water was then subtracted from each sample spectrum. The water content of the samples was determined from a near-IR band of water at 1.92 μm. The spectrum of cow hemoglobin (Hb) was treated similarly, and after normalization to the appropriate Hb concentration was subtracted from each of the human, pig, and horse absorbance spectra. This produced spectra containing only the Hb SH absorptions under study. Hb solutions were buffered to pH 7.1 with 0.05 *M* bis-tris and contained 0.1 *M* chloride. A sum of two Lorentzian (Cauchy) functions (●) (Ramsey, 1952), computed to fit the observed spectrum of human HbCO, was used to estimate individual SH absorption bands (○) contributed by the α-104 and β-112 cysteines. (Bare *et al.*, 1975.) Reprinted with the permission of the American Chemical Society. Copyright 1975.

mM^{-1} cm^{-1} for β-112 SH in HbCO) than for thiols in water ($\Delta\nu_{1/2}$ = 39.5 cm^{-1}, a_{mM} = 0.0073 mM^{-1} cm^{-1}), thereby giving evidence of strongly hydrogen-bonded SH groups in an environment of nonpolar and unexchanging nearest-neighbor van der Waals contacts that exist at the $\alpha_1\beta_1$ interface (Bare *et al.*, 1975). The SH group donates a H-bond to the peptide C═O oxygen of α-100 leucine, four residues back in the G-helix. Cow Hb, which contains cysteine at the β-93 position only, was used as a reference to permit subtraction of absorptions originating from the β-93 cysteine SH and other protein absorptions, such as the wings of CH-stretching vibrations centered at higher frequencies.

The centers of the SH bands display ligand-sensitive patterns that are similar for man, pig, and horse. The relative positions for the centers of various SH bands for human and pig hemoglobins are as follows: νSH(HbCO) < νSH(HbO_2 ~ HbCN) < νSH(Hb^+) << νSH(deoxy Hb). The α-104 SH group is most strongly hydrogen bonded (smallest νSH) and has the widest range of νSH(Hb→HbCO) in human hemoglobin; it is least

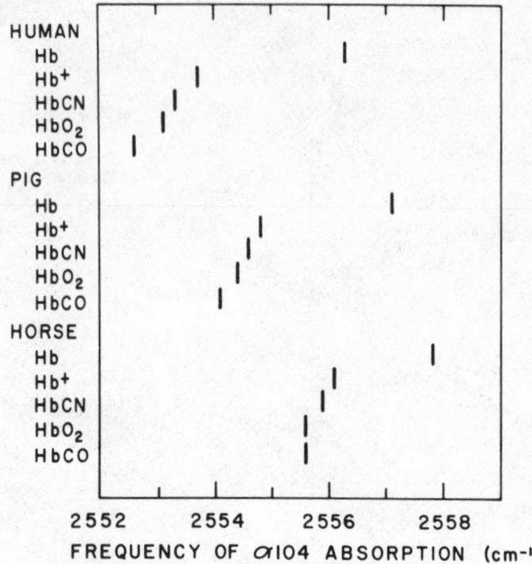

Fig. 6.17. Summary of center frequencies (± 0.3 cm^{-1}) of SH absorption bands of the α-104 cysteine in hemoglobin derivatives from human, pig, and horse. (Alben and Bare, 1978.)

strongly hydrogen bonded and has the narrowest range of νSH (HB→HbCO) in horse hemoglobin (Fig. 6.17). The β-112 cysteine SH in human Hb is more weakly hydrogen bonded than is the α-104 SH.

Distinguishing of Cysteines in α and β Chains. An interesting and important paper (Alben and Bare, 1980) describes the use of the —SH vibrational absorption of the α-104 (G11) cysteine residue as a probe that is specific to α-chain tertiary structure of the G helix of human hemoglobin, while also reflecting differences in ligation at the α-heme and the broader effects of hemoglobin quaternary structure. FTIR spectroscopy allows one to distinguish the —SH vibration of cysteine at the 104 position of the α chain from the —SH vibrations of cysteines in the β chain at positions 112 (G14) and 93 (F9). The α-104 —SH band furnishes a test at the α-chain of Perutz's allosteric mechanism (1970, 1976) for biological control of oxygen transport by hemoglobin.

There is a strong hydrogen bond between the —SH group of the α-104 cysteine and the peptide >C=O of the α-100 (G7) leucine, four residues back in the G-helix; and the position of the absorption band is, as mentioned above, sensitive to alteration of the tertiary structure of the α-chain G helix at the $\alpha_1\beta_1$ interface.

Ligated hemoglobin derivatives (e.g., HbCO, HbNO, CN-metHb, HbO_2, aquomet Hb) display a lower value of ν_{SH} than α chains; this indicates a stronger O···HS hydrogen bond to the peptide >C=O of α-100 leucine. Deoxyhemoglobin, on the other hand, displays a greater value of ν_{SH}, reflecting a weaker hydrogen bond (Alben and Bare, 1980). The differences in —SH frequencies for the α-104 cysteine of α_2 dimer derivatives are much less than those for corresponding derivatives of the tetramer (Fig. 6.18). The $\alpha_2\beta_2$ structure of hemoglobin thus emphasizes ligand-dependent changes in tertiary structure at the α-chain G helix.

Inositol hexaphosphate (IHP) has been shown to alter the $R \rightleftharpoons T$ equilibrium. It binds very strongly to deoxyHb between the β chains (Arnone and Perutz, 1974), to stabilize the *T* quaternary structure, and more weakly to other hemoglobin derivatives.

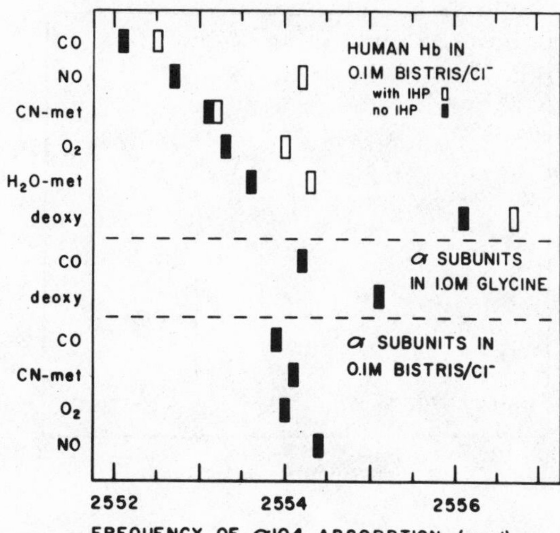

Fig. 6.18. Summary of sulfhydryl group center frequencies from α-104 cysteine —SH in human hemoglobin derivatives and isolated α chains. ■, data obtained in the absence of IHP; □, data obtained with IHP. IHP is inositol hexaphosphate. (Alben and Bare, 1980.)

When IHP is present (Fig. 6.18), a shift occurs toward higher ν_{SH} for the α-104 cysteine —SH. This figure and a table (not shown here, Alben and Bare, 1980) summarize the effects of quaternary structure and of IHP binding. IHP lowers the oxygen affinity of hemoglobin. It is closely related to the inositol pentaphosphate which replaces D-2,3-diphosphoglycerate (of human erythrocytes) in regulating the oxygen affinity of bird erythrocytes (Rapoport and Guest, 1941; Johnson and Tate, 1969).

Spectra of aqueous protein solutions were recorded (Alben and Bare, 1980) on a Fourier transform interferometer (Digilab FTS-14D) at 2 cm^{-1} resolution and were analyzed as described by Alben and Bare (1978), who also gave a summary (Fig. 6.17) of the wavenumber values of —SH absorption bands of the α-104 cysteine in hemoglobin derivatives from man, pig, and horse. The H-bond strength and the range of differences in H-bond strength with ligation increase proportionately in the series, horse < pig < human. Alben and Bare (1978) say that these data suggest that comparisons of ligated hemoglobin derivatives from the horse with human deoxyhemoglobin should be interpreted cautiously, and that differences among ligated human hemoglobin derivatives are greater than among similar derivatives from the horse.

The combined use of very good signal-to-noise ratio and accurate frequency determination in this FTIR work has made possible the quantitative measurement of the —SH absorption bands and the analysis of frequency differences among *R*-state derivatives. Alben and Bare (1978) were able to simplify small shifts of absorption band center frequencies in absorbance difference spectra, and determined the amount of shift ($\Delta\nu_0$) from the measured absorbance difference by comparison with the difference of computed Lorentzian functions.

Cytochrome Conformation. Rieske *et al.* (1975) have studied the infrared perturbations of sulfhydryl groups as a possible probe of conformational changes in Complex III (cytochrome b–c_1 segment) of the respiratory chain. The presence of 19 to 23 sulfhydryl

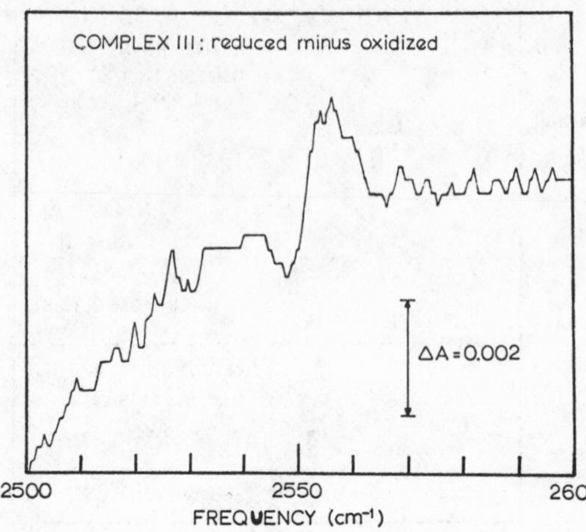

Fig. 6.19. Signal-averaged FTIR difference spectrum (reduced minus oxidized) of Complex III. Reduced and oxidized samples of Complex III were prepared by treating identical solutions of the complex with dithionite and ferricyanide, respectively, after which the complex was sedimented (3 h at $144{,}000 \times g$) into optically clear pellets. These pellets were pressed between CaF_2 windows of the spectroscopic cell with a 0.4-mm path length. (Rieske *et al.*, 1975.)

groups per molecule of cytochome c_1 prompted these authors to determine whether SH perturbations, linked to a conformational transition on the complex, are observable in this complex.

Figure 6.19 shows a Fourier transform difference spectrum between reduced and oxidized Complex III, and displays a band shift in the SH-group region (~2560 cm^{-1}), indicative of such a redox mediated conformational change. This band shift is similar to the shift in frequencies of the SH absorption bands of hemoglobin upon change of ligation of the heme group.

Resonance Raman Studies

Introduction

The visible and near-UV spectra of metalloporphyrins display a very intense absorption band near 400 nm, called the Soret, *B*, or γ band, and a much weaker one near 550 nm, called the α, or Q_0, band. There is also a β or Q_v band, which lies about 1300 cm^{-1} higher than the α-band center. The spectrum of ferrocytochrome *c* in Fig. 6.20 is such an example.

As shown in Fig. 6.21, the laser excitation wavelength determines the characteristics of the resonance Raman spectrum, e.g., that of oxy- and deoxyhemoglobin in the α–β and Soret regions.

Vibrational Modes of Heme. Spiro (1975a,b) has reviewed the subject of resonance Raman spectroscopy of heme proteins. Figure 6.22 shows the structure of heme. Porphyrin ring in-plane vibrational modes are prominent in the region 1650–750 cm^{-1}. When heme compounds are excited in the α–β region, there are essentially no bands below 750 cm^{-1} (Strekas and Spiro, 1972a,b), where iron atom vibrations should be displayed. When heme is excited in the Soret region, moderately intense low-frequency Raman bands appear (Salmeen *et al.*, 1973; Brunner *et al.*, 1972; Brunner and Sussner,

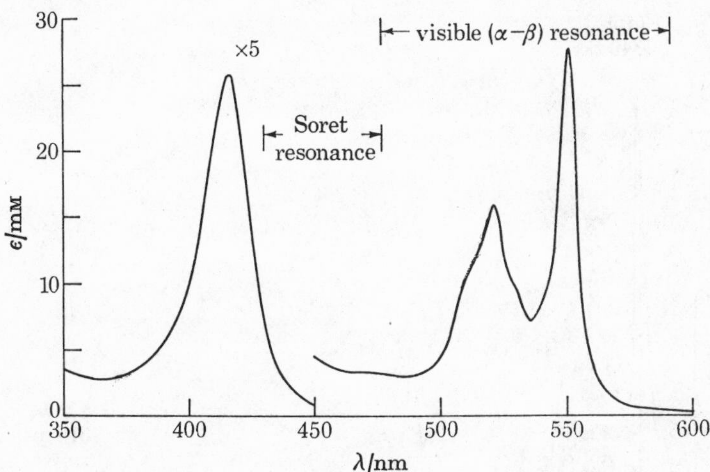

Fig. 6.20. The near-UV (Soret) and visible (α-β) absorption spectrum of ferrocytochrome *c*. Arrows span the approximate regions in which resonance with each of the two kinds of optical transitions dominates the Raman spectrum. (Spiro and Strekas, 1974.) Reprinted with the permission of the American Chemical society. Copyright 1974.

1973; Brunner, 1973). Figure 6.23 shows spectra of cytochrome *c* and cytochrome *c* oxidase. Brunner and Sussner (1973) made tentative assignments to Fe—N stretching vibrations for the following bands in the spectra of oxy- and deoxyhemoglobin: oxy-, 351, 380, and 424 cm^{-1}; deoxy-, 339, 364, and 412 cm^{-1}. Brunner (1974) found an Fe—O_2 out-of-plane stretching vibration at 567 cm^{-1} for oxyhemoglobin.

Although heme displays many porphyrin vibrational modes of high frequency, one can classify them with certainty by means of their polarizations. Also, by varying the wavelength of excitation one can resolve overlapping bands with different symmetries (Spiro and Strekas, 1974). One finds extra bands for the whole range of high frequencies for hemoglobin (Spiro and Strekas, 1974) and cytochromes *b* (Adar and Erecinska, 1974), produced by the two vinyl substituents of heme, which are conjugated to the π system of porphyrin. The —CH═CH_2 groups linked to the porphyrin ring should yield two modes near 1620 cm^{-1}, which might be resonance enhanced. Also, the asymmetric arrangement of the —CH═CH_2 groups removes the effective center of symmetry of the chromophore, and can cause infrared-active vibrational frequencies (E_u) to have Raman activity. One finds that the appearance of all the classified bands of cytochrome *c* can be explained by reference to the expected porphyrin skeletal modes. It is unnecessary to look at the peripheral substituents, because they are linked to the pyrrole rings through saturated carbon atoms and therefore maintain electronic symmetry close to the D_{4h} (Herzberg, 1945) type.

Figure 6.24 shows a correlation diagram for resonance Raman bands of hemoglobin and cytochrome *c*. When the iron atom in heme changes from a low spin to a high spin, certain bands undergo frequency decreases, e.g., at ~1640 cm^{-1} (depolarized) (~1620 cm^{-1} for Fe II), 1590 cm^{-1} (anomalous polarization), and 1500 cm^{-1} (polarized), shift to

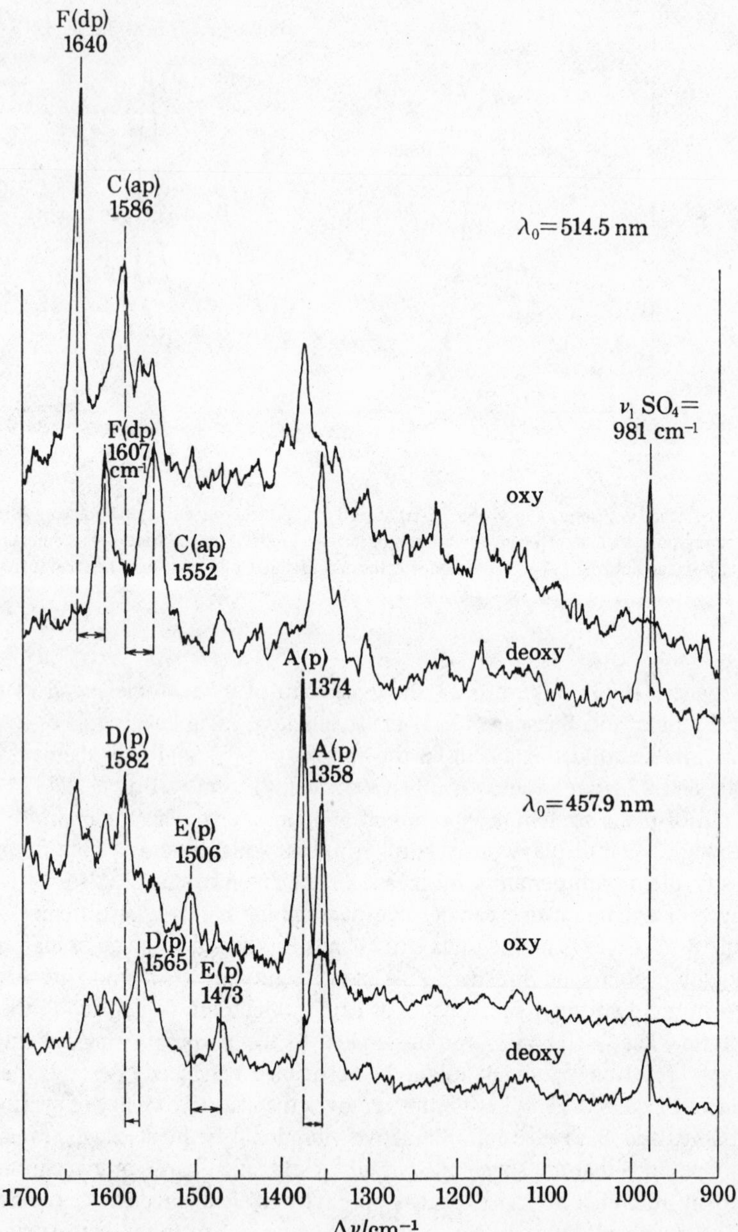

Fig. 6.21. Resonance Raman spectra of oxy- and deoxyhemoglobin in the α-β (λ_0 = 514.5 nm) and Soret (λ_0 = 451.9 nm) scattering regions. The solutions were 0.68 and 0.34 m*M* in heme for oxy- and deoxyhemoglobin, respectively, and the latter contained 0.4 *M* $(NH_4)_2SO_4$, the $\nu_1(SO_4)^{2-}$ band (981 cm^{-1}) of which is indicated. Frequency shifts for corresponding bands are marked by the arrows. p = polarized, dp = depolarized, ap = anomalous polarization. (Spiro and Strekas, 1974.) Reprinted with the permission of the American Chemical Society. Copyright 1974.

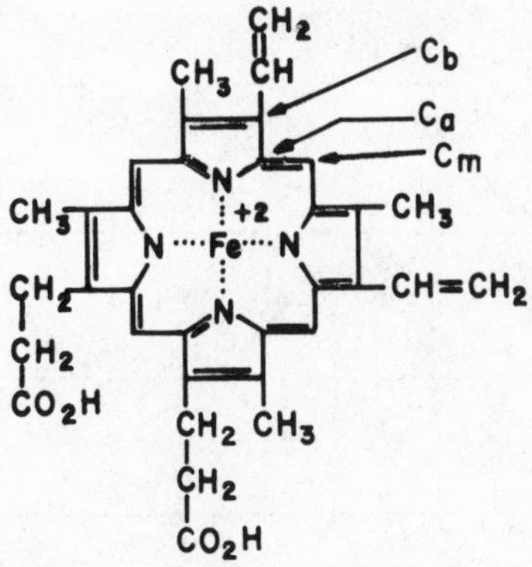

Fig. 6.22. Heme (ferrous iron and protoporphyrin IX).

~1607, 1555, and 1478 cm^{-1}, respectively. Changes in heme structure can account for these shifts (Hoard, 1971): Hemes of low spin are six-coordinate, the iron atom remaining planar with the heme; hemes of high spin are five-coordinate, or if six-coordinate, one axially attached group is less strongly linked than the other. The iron atom is no longer planar with the porphyrin ring, which adopts a domed structure, in the same direction (Fig. 6.25), a conformational change which is probably responsible for the changes in vibrational frequencies.

Figure 6.26 shows groupings of porphyrin Raman frequencies which are characteristic of spin state (F, C, and E) or oxidation state (F and A) of the heme. Ferricytochrome *c′* (from *Rhodopseudomonas palustris*) at pH values 10.3 and 6.9 (Strekas and Spiro,

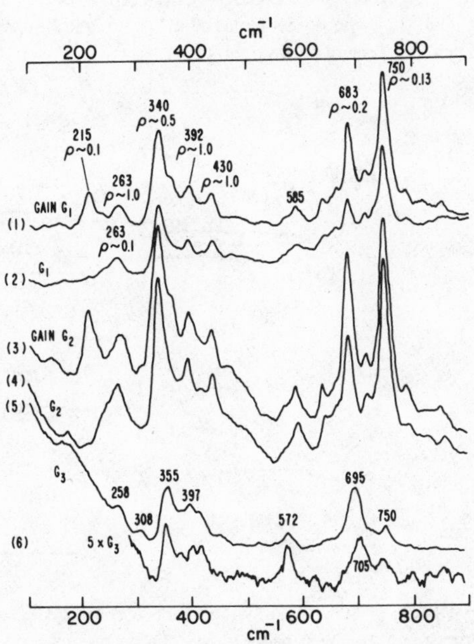

Fig. 6.23. Raman spectra in the 100–900-cm^{-1} region. Relative system gains indicated adjacent to each spectrum. Spectra are unpolarized, reflection mode, single scans using standard 5 mm × 10 mm rectangular cuvettes, slit width ca. 5 cm^{-1}; scan rate of 1 cm^{-1}/sec; sample volumes ca. 0.3 ml. (1) Cytochrome *c* oxidase ca. 70 μ*M* reduced with 0.5 μ*M* cytochrome *c* and a few crystals of sodium dithionite. (2) Oxidized cytochrome *c* oxidase, same sample as in (1) prior to reduction. (3) ETP, ca. 70 mg/ml protein, reduced with a few crystals of sodium dithionite. (4) ETP sample as in (3) oxidized by addition of ferricyanide. (5) Cytochrome *c* ca. 40 μ*M* reduced with dithionite. (6) Oxidized cytochrome *c* same sample as in (5) prior to reduction. ETP = electron transport particles. (Salmeen *et al.*, 1973.)

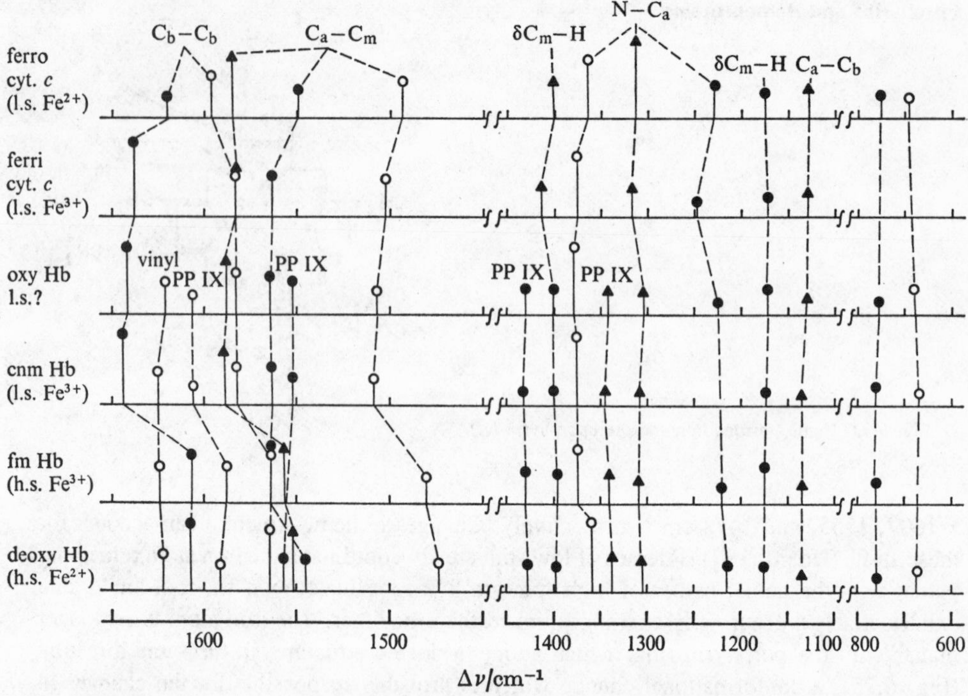

Fig. 6.24. Correlation diagram for resonance Raman bands of hemoglobin and cytochrome *c*. Spin and oxidation states of the various species are indicated. The lengths of the solid lines are roughly proportional to the observed relative intensities at 514.5 nm (5145 Å) excitation (496.5 nm for fmHb) for anomalously polarized (▲) and depolarized (○) bands, and at 457.9 nm (4579 Å) for polarized (●) bands. Suggested assignments (approximate) to various heme internal coordinates are indicated at the top. The bands marked PP IX (protoporphyrin) and vinyl are observed for hemoglobin derivatives only. l.s. = low spin; h.s. = high spin; cnm = cyanomethemoglobin; fm = fluoromethemoglobin. (Spiro and Strekas, 1974.) Reprinted with the permission of the American Chemical Society. Copyright 1974.

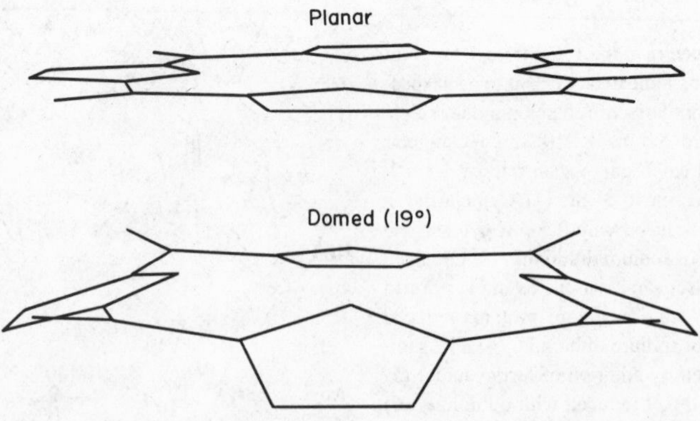

Fig. 6.25. Perspective drawing of the porphyrin skeleton in planar and domed (19° angle between the pyrrole groups and the mean beam plane) conformations. (Spiro, 1975a.)

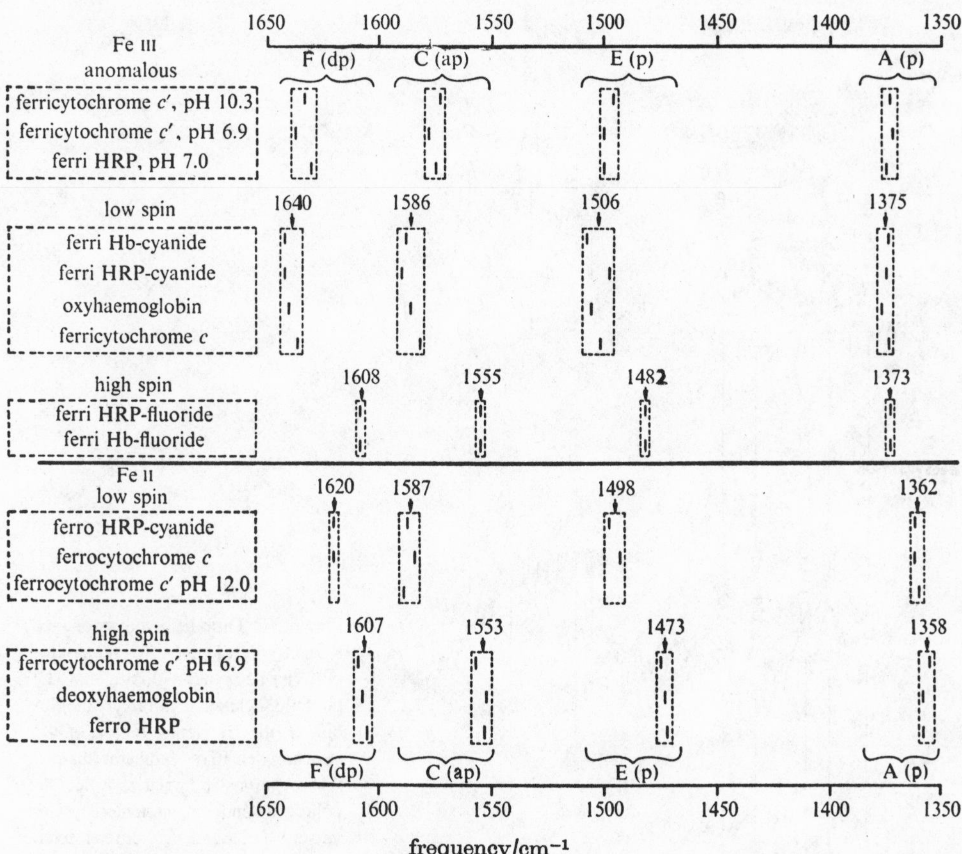

Fig. 6.26. Groupings of porphyrin Raman frequencies (cm^{-1}) which are sensitive to spin (*F*, *C*, and *E*) and oxidation (*F* and *A*) state of the heme. HRP means horseradish peroxidase. (Rakshit and Spiro, 1974.) Reprinted with the permission of the American Chemical Society. Copyright 1974.

1974) and Fe III horseradish peroxidase at pH 7 (Rakshit and Spiro, 1974) are anomalous, because they do not fit the correlation. Their frequencies lie near the low-spin Fe III frequencies but they are not of the low-spin type (Keilin and Hartree, 1951; Moss *et al.*, 1969; and Blumberg *et al.*, 1968).

Hemoglobin and Myoglobin Derivatives

Time-Resolved Spectra. Woodruff and Farquharson (1978) have recorded time-resolved resonance Raman spectra of oxyhemoglobin, deoxyhemoglobin, carboxyhemoglobin (COHb), and the corresponding myoglobin derivatives. They used 7-nsec excitation pulses at a wavelength of 531.8 nm. Figure 6.27 shows time-resolved spectra of the structure-sensitive spectral intervals, from 1300 to 1700 cm^{-1}, of the first two substances and of photodissociated COHb. The spectrum of the COHb represents approxi-

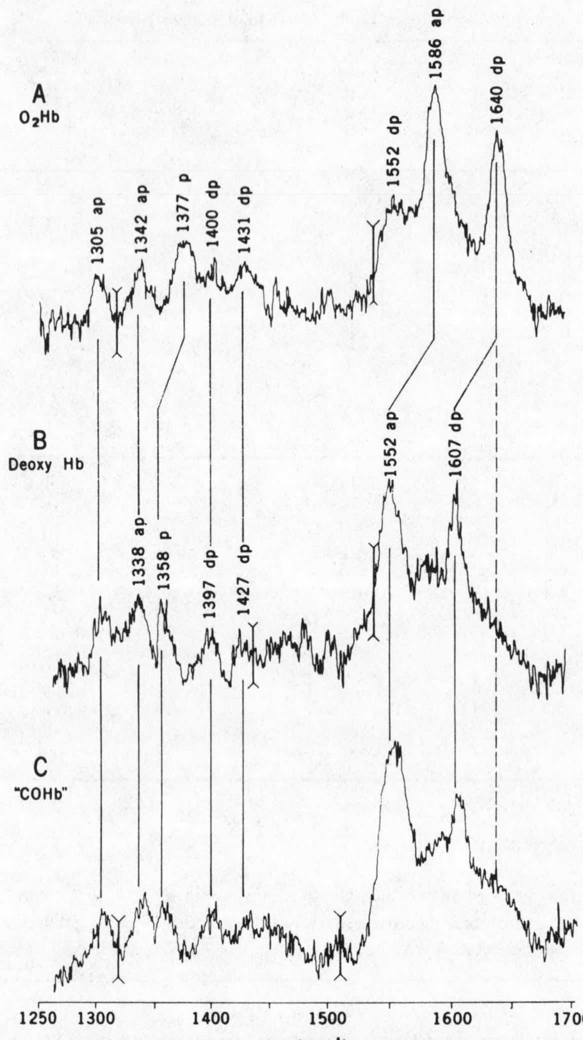

Fig. 6.27. Time-resolved resonance Raman spectra of (A) oxyhemoglobin, (B) deoxyhemoglobin, and (C) photodissociated carboxyhemoglobin, in the frequency region of the structure-sensitive indicator bands. Abbreviations: p, polarized; dp, depolarized; and ap, anomalously (inversely) polarized. Conditions: excitation, 531.8 nm; pulses, 7 nsec; pulse repetition frequency, 10 Hz; pulse energy, 10 mJ; and accumulation time, 165 sec. The double-ended arrows in each spectrum denote the points where adjacent vidicon frames were joined to make up complete spectra. (Woodruff and Farquharson, 1978.)

mately 86% $Hb_{4°R}^{3°oxy}$, 10% $Hb_{4°R}^{3°deoxy}$, and 4% COHb, where 3° and 4° indicate the tertiary and quaternary structures of the protein. The three marker frequencies that indicate the heme spatial or electronic structure shift from 1377, 1586, and 1640 cm^{-1} in oxyhemoglobin to 1358, 1552, and 1607 cm^{-1} in deoxyhemoglobin, in accordance with Spiro and Loehr (1975) and Spiro and Strekas (1974). The two bands of highest frequency are believed to reflect mainly the movement of the iron atom in or out of the heme plane, especially when both are not coplanar, as in deoxyhemoglobin (Spiro and Strekas, 1974; Scholler and Hoffman, 1978; Spiro and Burke, 1976).

The spectra of deoxyhemoglobin and photolyzed COHb are practically the same (Fig. 6.27). Myoglobin gave the same result (not shown). These data suggested to the authors that the structural relaxations of the heme moiety after photolysis of COHb are complete in much less than 7 nsec (Fe in plane becomes Fe out of plane), even though there is evidence that the protein structural relaxations occur during a longer time period. Their conclusions are twofold:

(1) The change in the heme structure, acting as a trigger for cooperativity in hemoglobin, does not take place in the same time period as do the protein reorganizations that it is thought to trigger;

(2) The nonequilibrium globin structure in photolyzed $Hb^{3°}_{4°R}{}^{oxy}$ does not seem to exert any significant transient constraint on the heme structure.

Lyons *et al.* (1978) have obtained transient resonance Raman spectra of photolyzed carboxyhemoglobin. The use of nanosecond time-resolved spectra permitted the conclusion that, after the removal of the CO molecule within picoseconds, the main reorganization of the porphinato core occurs within 5 nsec of photolysis.

Spatial Arrangement of Dioxygen in Oxyhemoglobin. The spatial arrangement of dioxygen with respect to the iron atom in oxyhemoglobin has been shown to be Fe—O—O (bent, end-on) rather than Fe<(O|O) (side-on) by means of resonance Raman spectroscopy (Duff *et al.*, 1979). A spectrum in the ν_{Fe-O_2} region of oxyhemoglobin prepared with $^{16}O^{18}O$ gas showed two bands for ν_{Fe-O} at 567 and 540 cm^{-1}, which are comparable with 567 and 540 cm^{-1} for $^{16}O_2$ and $^{18}O_2$ oxyhemoglobin, respectively. No band appeared at 555 cm^{-1}, which was calculated for Fe<(O|O) with $^{16}O^{18}O$. Thus, the spectrum indicates that the two oxygen atoms are in different environments, thereby supporting the linear structure above (see also, Barlow *et al.*, 1973 and Maxwell *et al.*, 1974).

Carbon Monoxide Hemoglobin and Myoglobin. Dallinger *et al.* (1978) have used resonance CARS spectra of carbon monoxide hemoglobin (HbCO) to show that the iron atom is displaced from the heme plane to its position in deoxyhemoglobin, a distance of 0.60 Å (Fermi, 1975), within 6 nsec after photolysis of the carbon monoxide ligand.

Figure 6.28 presents the CARS spectrum of HbCO observed with ω_1 (the frequency of one of the laser beams) in resonance with the Q_0 transition at 569 nm. By causing the polarizations of the two incident beams to be parallel or perpendicular to each other, one can do CARS polarization measurements (Nestor, 1978). The depolarization ratio, $\rho = I_\perp/I_\parallel$, has a value about equal to that of the square of the regular Raman depolarization ratio (Yuratich and Hanna, 1977). Therefore, polarized, depolarized, and anomalously polarized modes should have ρ_{CARS} approximately equal to ρ^2_{Raman}, with values of $<9/16$, $9/16$, and $>9/16$, respectively (Nestor, 1978).

We observe in the polarized spectra of Fig. 6.28 two depolarized bands at 1609 and 1549 cm^{-1} and an anomalously polarized band at 1559 cm^{-1}, characteristics of deoxyhemoglobin as recorded separately in Spiro's laboratory.

Since certain heme marker bands are absent from the spectra of Fig. 6.28—namely, an anomalously polarized band at 1583–1589 cm^{-1} and a depolarized band at 1617–1642

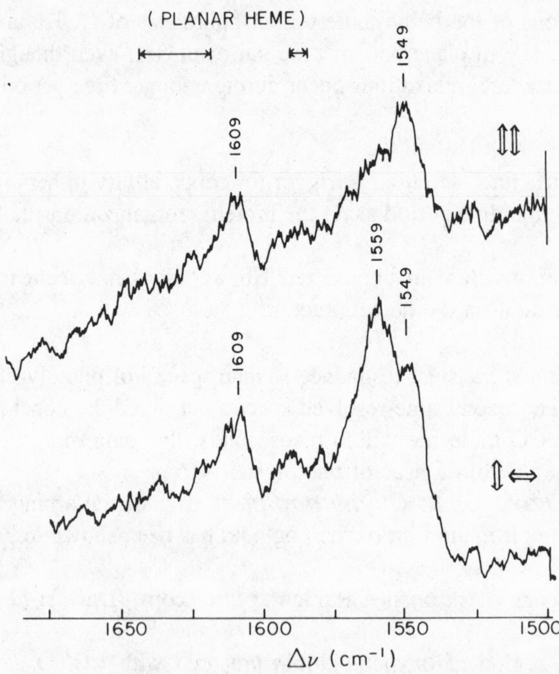

Fig. 6.28. Polarized CARS spectrum of COHb: ⇕⇕, dye lasers polarized in the same direction; ⇔ dye lasers polarized orthogonal to each other; ω_1 = 569 nm (pulse energy ~ 6 μJ); ω_2 = 622-629 nm (pulse energy ~ 20 μJ; [COHb] = 0.69 m*M*; laser repetition rate = 10 pulses/sec; 100-pulse average; ω_2 scan rate = 0.75 nm/min. Sample in Teflon-stoppered 1-mm path length visible cell. (Dallinger *et al.*, 1978.) Reprinted with the permission of the American Chemical Society. Copyright 1978.

cm^{-1} (the latter from in-plane iron)—Dallinger *et al.* (1978) concluded that the iron atom does not remain in the heme plane for an appreciable fraction of 6 nsec.

Yu *et al.* (1982) have used resonance Raman spectroscopy to investigate the carbon monoxide bonding in carbon monoxide hemoglobin and myoglobin. They detected the Fe—CO stretching, Fe—C—O bending, and bound C—O stretching vibrations at 507 (512), 578 (577), and 1951 (1944) cm^{-1}, respectively, in carbon monoxy HbA (or sperm whale Mb) when they excited at 406.7 nm within the Soret band. Examination of the Fe—CO stretching mode at 507 cm^{-1} in carbon monoxy HbA and Hb Kansas both with and without inositol hexaphosphate showed no changes in either frequency or intensity. According to Yu *et al.*, these facts imply that no significant change in the Fe—C bond energy is induced by switching the quaternary structure from the *R*- to the *T*-form in ligated HbCO Kansas.

Yu *et al.* (1982a) have used resonance Raman spectroscopy to study CO and O_2 binding to elephant myoglobin, which has a distal glutamine at E7 instead of the usual histidine in the heme binding site.

Kinetic measurements showed that elephant Mb has an affinity for CO ~6 times higher than that for human or sperm whale Mb. If this increased affinity were solely the result of the removal of some of the steric hindrance that normally tilts the CO off the heme axis, these workers would have expected the ν(Fe—CO) frequency to decrease and the ν(C—O) frequency to increase relative to the corresponding values in sperm whale Mb; however, they found the opposite effect. Yu *et al.* (1982a) believe that these results may imply that factors such as interaction between the amide oxygen of glutamine E7 and the carbon of the CO ligand, as well as a hydrogen bonding between the amide hydrogen and the oxygen of the ligand, may cause the enhanced binding of CO by elephant Mb.

Hemoglobin Complexes with NO, CH_3NO, and C_6H_5NO. Chottard and Mansuy (1977) have used resonance Raman spectra to characterize the iron-ligand bonding in hemoglobin complexes with NO, CH_3NO, and C_6H_5NO, although they gave only tentative assignment to the bonding in the latter two complexes. Low-frequency bands were recorded at 549, 631, and 450 cm^{-1} in Hb—NO, Hb—CH_3NO, and Hb—C_6H_5NO, respectively. The band at 549 cm^{-1} was assigned to the Fe—NO stretching vibration by isotopic substitution of ^{15}N for ^{14}N. Its excitation profile suggested that the Raman intensity of the Fe—NO vibration originated in delocalization of the π electrons of the porphyrin ring toward the Fe—NO grouping.

T and R Conformations of Carp Hemoglobin Derivatives. Resonance Raman and electron paramagnetic resonance (EPR) spectroscopies have been employed to study the properties of liganded and of unliganded hemes in both low- and high-affinity quaternary conformations (T and R, respectively) of carp hemoglobin (Scholler and Hoffman, 1979). This approach was possible because a simple change in pH and inositol hexaphosphate concentration can bring about a change between the T and R conformations. A change from T to R structure did not cause any shifts in the resonance Raman frequencies of the heme spectrum of the unliganded Hb, or of either low- or high-spin liganded derivatives (cyanomet-Hb, deoxy-Hb, fluoromet-Hb); but the induced structural alteration did change the high-spin/low-spin equilibrium of azidomet-Hb from 10%/90% to 45%/55%, with changes showing up in the EPR parameters of the low-spin liganded derivatives. These workers stated that resonance Raman spectroscopy should be insensitive to small out-of-plane motions of the metal ion in the case of a low-spin six-coordinate heme, whereas EPR responds sensitively to deformations in the geometry of coordination. They believe that the former spectroscopic tool might be sensitive to strains in a high-spin heme, whereas EPR is not.

Methemoglobin Derivatives. Asher *et al.* (1977) have obtained resonance Raman spectra for the OH^-, N_3^-, and F^- derivatives of methemoglobin (metHb) by excitation in the 550–650 nm region. They observed a selective enhancement with excitation in the charge-transfer bands for peaks at 413 and 497 cm^{-1} and a doublet at 471 and 443 cm^{-1} in the N_3^-, OH^-, and F^- (Fig. 6.29) complexes, respectively. They assigned these peaks to Fe-axial ligand stretches. The doublet recorded at 471 and 443 cm^{-1} may have reflected a heterogeneity in the heme cavity caused by hydrogen bonding of H_2O to the F^- ligand in both the α and β subunits. These workers suggested that the frequency of the Fe—F^- vibration reflected the out-of-plane distortion of the iron from the heme plane. The absence of a shift in the frequency of the Fe—F^- vibration indicated that there is little or no movement of the iron relative to the heme plane when inositol hexaphosphate (IHP) is added. IHP is thought to alter the allosteric equilibrium between the R and T forms of metHb. An X-ray crystallographic study (Fermi and Perutz, 1977) of a complex between IHP and metHb-F^- was consistent with this finding. The lack of movement of the iron when IHP is added to metHb-F^-, is consistent with a second interpretation of Fermi and Perutz: that the positions of the iron atoms in the R and T forms of metHb-F^- are nearly identical and the iron is close to in-plane.

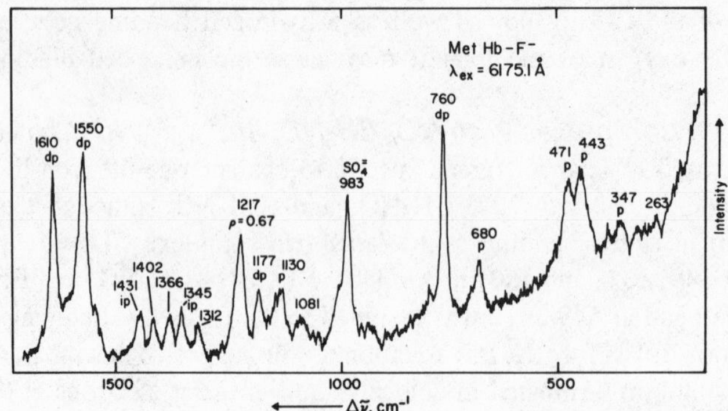

Fig. 6.29. Resonance Raman spectrum of MetHb-F^- (0.76 m*M* heme) containing 0.25 *M* Na_2SO_4 as an internal standard: λ_{ex} 6175.1 Å; energy 2 × 10^{-3} J/pulse; pulse repetition rate 30 Hz; scan speed 23 Å/min; slit width 3.7 Å; p, polarized; dp, depolarized; ip, inversely polarized. (Asher *et al.*, 1977.) Reprinted with the permission of the American Chemical Society. Copyright 1977.

Methemoglobin and Myoglobin Spin-Marker Frequencies. Spiro *et al.* (1979) have obtained resonance Raman spectra for high-spin bis(dimethyl sulfoxide) complexes of protoporphyrin IX dimethyl ester, octaethylporphyrin, and tetraphenylporphyrin, and for aquo- and fluoromethemoglobin and myoglobin. In particular, they have examined the II, IV, and V marker bands, which are sensitive to the spin state of the iron atom. Their data supported porphyrin core expansion as the determining factor of the metHb and Mb spin-marker frequencies and negated the hypothesis that doming (in which the metal ion lies above the porphyrin plane and the tilting of the pyrrole rings also involves their swiveling about the methine bridge linkages, causing the nitrogen plane to lie above the mean plane of the porphyrin, Fig. 6.30) is a protein-induced process. It is known from X-ray crystallographic studies of metmyoglobin that the iron atom is 0.40 Å from its mean heme plane (Takano, 1977) and that the iron lies 0.23 and 0.07 Å from the mean heme plane (Ladner *et al.*, 1977) in the β and α chains of methemoglobin. Since

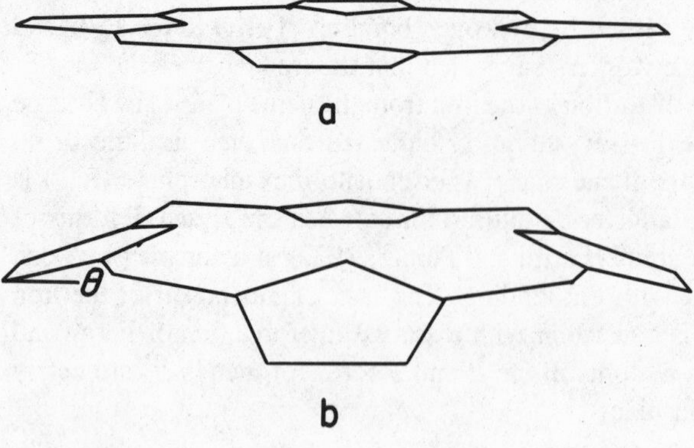

Fig. 6.30. Perspective drawings for (a) planar and (b) domed porphyrin skeletons from about 10° above. θ is the angle by which the pyrroles tilt out of the mean porphyrin plane. (Spiro *et al.*, 1979.) Reprinted with the permission of the American Chemical Society. Copyright 1979.

no associated variation in the spin-marker frequencies occurred, it follows that the core size (measured by the C_t–N distance, or the distance from the center of the porphyrin ring to the pyrrole nitrogen atoms) is not governed by the disposition of the iron atom, but probably by steric interactions of the axial ligands with the pyrrole N atoms (Spiro *et al.*, 1979).

Native horseradish peroxidase (Lanir and Schejter, 1975) and cytochrome *c′* (Kitagawa *et al.*, 1977) display spin-marker frequencies which are similar to those in several previously examined five-coordinate Fe(III) hemes. These similarities showed that these proteins have five-coordinate instead of six-coordinate Fe(III) hemes. For the horseradish peroxidase exogenous ligands can produce six-coordination. Spiro *et al.* (1979) suggested that the intermediate-spin state found for these proteins is caused by weakening of the bond between Fe(III) and the single axial ligand.

These authors again studied the correlation of core size with the spin-marker frequencies. Nonplanar hemes have frequencies that are considerably lowered, as a result of the loss of π conjugation at the methine bridge. Spiro *et al.* developed a model that gives good estimates of the σ and π contributions to the methine bond-stretching force constant. They concluded that both core expansion and pyrrole tilting (Fig. 6.30) are factors in the lowering of frequencies. For moderate angles of tilt the effect of core expansion is more important. In all the heme proteins that have been studied so far, no evidence has been found for the induction of extra doming by the protein moiety.

Myoglobin Complexes. Twelve different complexes of oxidized and reduced myoglobin of horse heart have been studied by resonance Raman spectroscopy (441.6

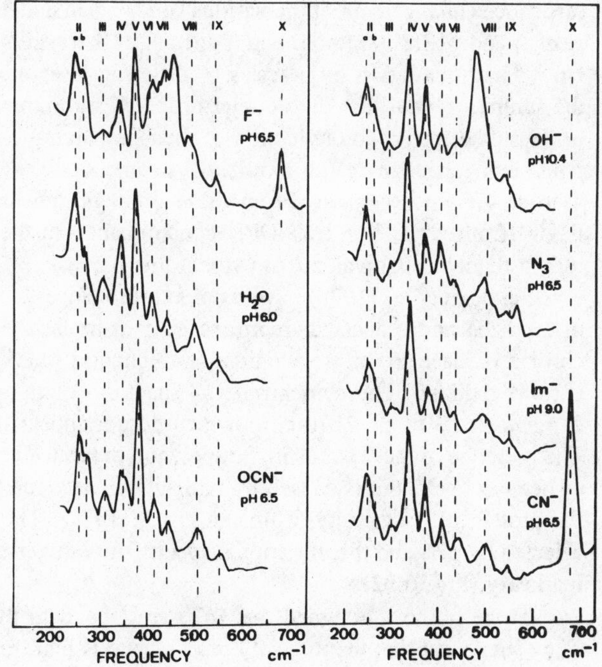

Fig. 6.31. Resonance Raman spectra of horse heart metmyoglobin with various ligands. Conditions: excitation wavelength is 441.6 nm; scanning is 25 cm^{-1}/min; slit width is 8 cm^{-1}; laser power is 50 mW; summation of six scans; heme concentration is 1 m*M*; nature of the ligand and pH conditions are indicated on the right side of the spectra. Im^- = imidazole. (Desbois *et al.*, 1979.) Reprinted with the permission of the American Chemical Society. Copyright 1979.

nm excitation) in the region from 150 to 700 cm^{-1} (Desbois *et al.*, 1979). Examples of the spectra are given in Fig. 6.31. In this region, spectra display ten bands common to all myoglobin derivatives. These are numbered I–X (bands I lie in the range from 194 to 109 cm^{-1}, and are not present in Fig. 6.31). The positions of bands II_a and II_b vary with spin state, coordination number (c.n.), and oxidation number (o.n.) of myoglobin derivatives, as follows: high-spin ferric derivative, c.n. 6, o.n. 3, II_a at 249–252 cm^{-1}, II_b at 270–272; low-spin ferric, c.n. 6, o.n. 3, II_a at 254–255; high-spin ferrous, c.n. 5, o.n. 2, II_a at 222, II_b at 243; and low-spin ferrous, c.n. 6, o.n. 2, II_a at 218 cm^{-1}.

The relative intensities of bands IV, V, and X characterize the doming state of the heme and the spin state of the iron atom. The authors made the following assignments of bands: II_a, stretching mode of Fe—N(pyrrole ring); IV and V, heme in-plane deformation; VI, stretching mode of Fe—N_ε(histidine, proximal); VIII, possibly heme deformation plus Fe—N(pyrrole); X, heme deformation; ligand-specific bands, stretching of Fe-sixth ligand.

Cytochromes

Cytochrome Oxidase. Adar and Yonetani (1978) have recorded resonance Raman spectra of cytochrome oxidase that they had solubilized in Tween 20 and in sodium cholate, and then excited at 413.1 nm. They found minimal differences in the spectra of the two preparations, thus indicating that the local environment of the hemes is similar in the two preparations. The strongest band was displayed at 1358 cm^{-1} as in the work of Salmeen *et al.* (1973). Some of the other bands had shapes and frequencies slightly different from those of Salmeen *et al.*, and the reasons for these were the differences in resonance enhancement of the various bands when under excitation of 441.6 nm (Salmeen *et al.*,) and 413.1 nm (Adar and Yonetani). Observation of the range from 1350 to 1380 cm^{-1} under excitation by a series of differing laser intensities (from 10 to 130 mW on the sample) reveal that the doublet observed by Salmeen *et al.* at 1358 and 1372 cm^{-1} had resulted from photoreduction of their preparations. When Adar and Yonetani added potassium ferricyanide (an oxidizer) to some samples, broad luminescence bands were seen at 476 and 641 nm. From these findings they inferred that catalytic amounts of flavin (from NADH and NADPH contaminants) in the samples are photoreduced, providing reducing equivalents to cytochrome oxidase.

Salmeen *et al.* (1978) have recorded resonance Raman spectra (441.6 nm excitation) of oxidized and reduced monomeric heme *a*-imidazole complex, of ligand-bound cytochrome oxidase in different oxidation–reduction states, and of alkaline denatured cytochrome oxidase. The cytochrome a_3 spectrum of the reduced oxidase shows bands at 215, 364, 1230, and 1670 cm^{-1} which do not appear in the spectrum of cytochrome *a*. The presence of these bands in the spectrum of reduced cytochrome a_3 is not characteristic of heme *a* itself, but is caused by interactions between the heme *a* of cytochrome a_3 and the protein in its vicinity (Salmeen *et al.*, 1978). The protein–heme *a* interactions are affected by pH, and the vibrational spectra of both heme *a* groups reflect these influences markedly (Fig. 6.32).

These authors assigned the 1670-cm^{-1} band to the formyl substituent of heme *a*. They suggested that the intensity of this band is high for reduced cytochrome a_3 because

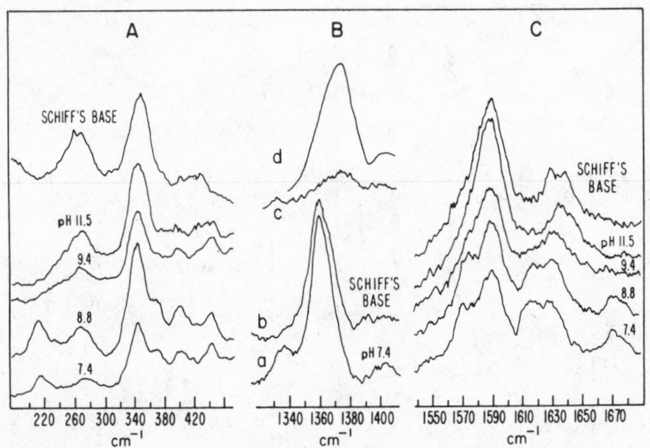

Fig. 6.32. Major Raman bands of reduced cytochrome oxidase as a function of pH. Spectra under (A) and (C) labeled according to pH. Traces a and b under (B) are reduced protein at two values of pH. Trace c is experimental data obtained after adding 100 μM potassium ferricyanide to reduced sample at pH 11.4. Trace d is a manual replot of trace c normalized to peak intensity of trace a. (Salmeen *et al.*, 1978.) Reprinted with the permission of the American Chemical Society. Copyright 1978.

the carbonyl group is situated *in* the plane of the porphyrin and is very weak for reduced and oxidized cytochrome *a*, oxidized cytochrome a_3, and reduced and oxidized heme *a*-imidazole because the carbonyl group is positioned *out* of the plane.

To aid their interpretation of resonance Raman spectra (excitation λ, 441.6 nm) of oxidized cytochrome oxidase and its inhibitor complexes with cyanide and formate (Fig. 6.33), Babcock and Salmeen (1979) have also recorded spectra of low-spin and high-spin ferric heme *a* model complexes. The spectra of the model compounds (not given here) showed that the characteristics of heme *a* vibrational bands between 1540 and 1660 cm^{-1} depend on the spin state of the iron atom; and showed that the Raman spectrum of the oxidized cyctochrome oxidase is produced mainly from vibrations of low-spin cytochrome a^{3+}. Data from the spectra of the enzyme-inhibitor complexes fall in line with this concept (Babcock and Salmeen, 1979).

By using excitation at 441.6 nm, which preferentially enhances the Raman lines of reduced cytochrome oxidase, Adar and Erecinska (1979) have carried out a photoreductive titration of the resonance Raman spectra of cytochrome *c* oxidase in whole mitochondria at about −10°C (Fig. 6.34). The spectrum of the most oxidized sample (bottom curve) does not display certain bands that appear in the spectrum of the sample reduced by dithionite—namely, 216, 363, 560, and 1665 cm^{-1}. During the photoreduction process, the bands at 216 and 560 cm^{-1} show up before those at 363 and 1665 cm^{-1}. The former bands develop complete intensity at photoreduction, while the latter bands are reduced to half the intensity of the dithionite-reduced system. The band at 1609 cm^{-1} is weaker near the beginning of the titration than that at 1623 cm^{-1}, but as the photoreduction continues, both bands of this doublet reach equal intensities.

Adar and Erecinska interpreted their data as showing that there are three states in

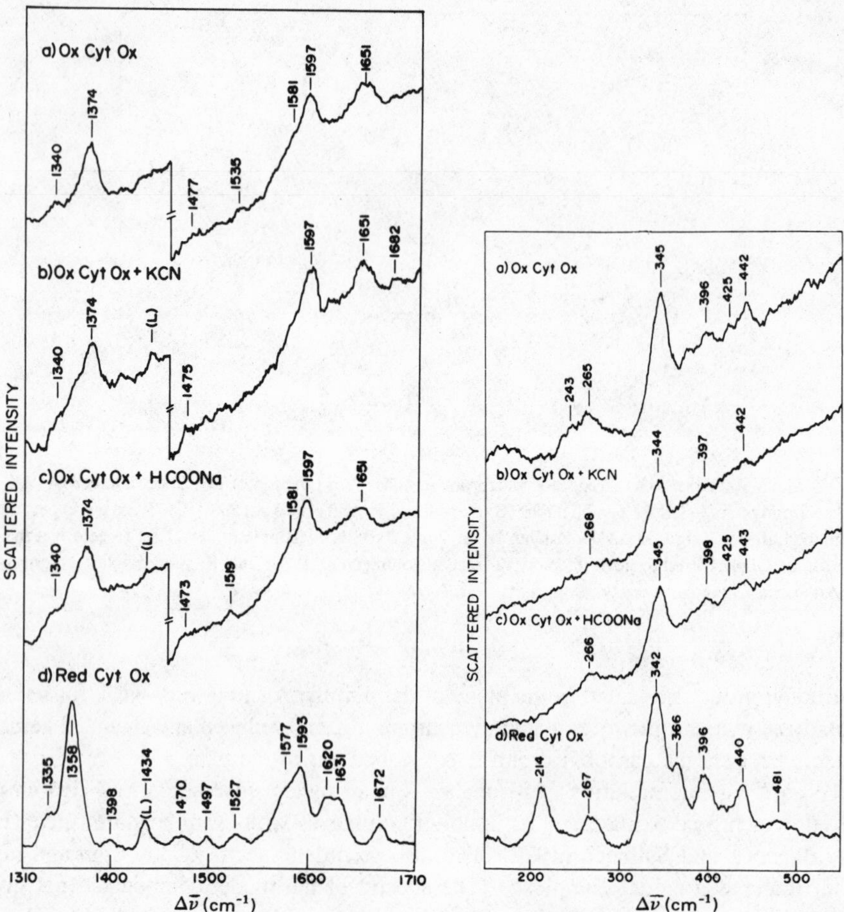

Fig. 6.33. Raman spectra of (a) oxidized cytochrome oxidase, its inhibitor complexes with (b) cyanide and (c) formate, and (d) reduced cytochrome oxidase in the low- (right panel) and high-(left panel) frequency regions. In (a), (b), and (c), the heme *a* concentration was 45 μ*M*; in (d) the sample was 25 μ*M* in heme *a*. The cyanide concentration was 5 m*M* and the formate concentration was 60 m*M*. Stock solutions of each inhibitor were neutralized prior to addition to the protein solution. In (d), the line at 1434 cm^{-1} corresponds to laser fluorescence. The overall instrument sensitivity was the same for all four samples. (Babcock and Salmeen, 1979.) Reprinted with the permission of the American Chemical Society. Copyright 1979.

the reduction represented by the resonance Raman spectra of cytochrome *c* oxidase. The changes in intensity of the 1609- and 1623-cm^{-1} bands indicate the non-equivalent but interacting condition of the high- and low-electropotential hemes (Wilson and Leigh, 1972; Leigh *et al.*, 1974; Wilson *et al.*, 1976).

The resonance Raman spectra of cytochrome *c* oxidase (oxidized, reduced, and oxidized cyanide-bound) have been recorded by Bocian *et al.* (1979) with excitation at 610, 600, and 590 nm. The spectra of these three forms of the oxidase look alike. Thus,

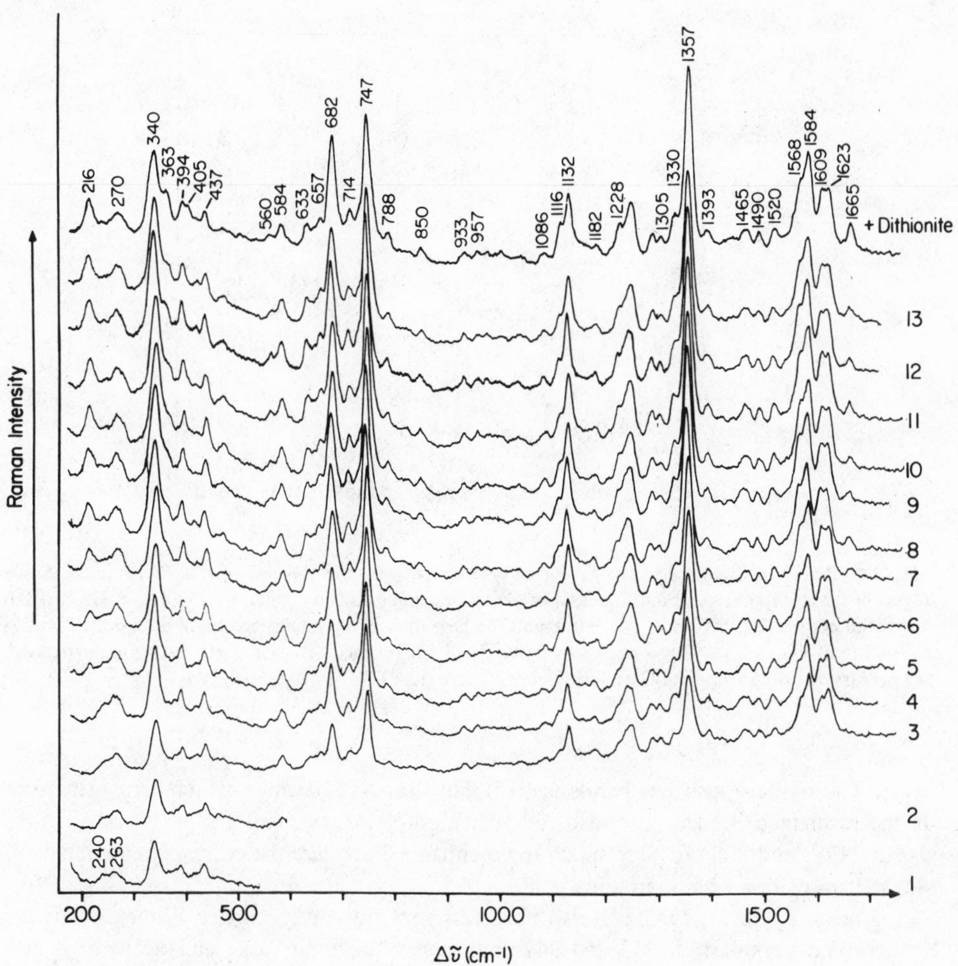

Fig. 6.34. Photoreductive titration of the resonance Raman spectrum of mitochondria excited at 441.6 nm. Heme *a* concentration was 25 μ*M*; mitochondria were suspended in 100 m*M* phosphate buffer, pH 7.4. Samples were saturated with oxygen before transferring to capillaries and cooled with nitrogen gas to about −10°C. The spectrometer slits were set at 600 μm which corresponds to a bandpass of about 8 cm^{-1}. Scans were recorded at the rate of 100 cm^{-1}/min with a 0.1-sec time constant. Time lapse between the beginning of the scans was about 18 min. (Adar and Erecinska, 1979.) Reprinted with the permission of the American Chemical Society. Copyright 1979.

Fig. 6.35 shows spectra quite representative of them all. No Raman bands are present that can readily be assigned to either cytochrome *a* or cytochrome a_3, exclusively. Some bands do appear, however, which are not seen in the resonance Raman spectra of other hemoproteins (and metalloporphyrins, generally) when the excitation is in their absorption bands in the visible region. These other hemoproteins show a single Raman band at ca. 750 cm^{-1}, while the spectra of cytochrome *c* oxidase display a doublet at 732 and 744

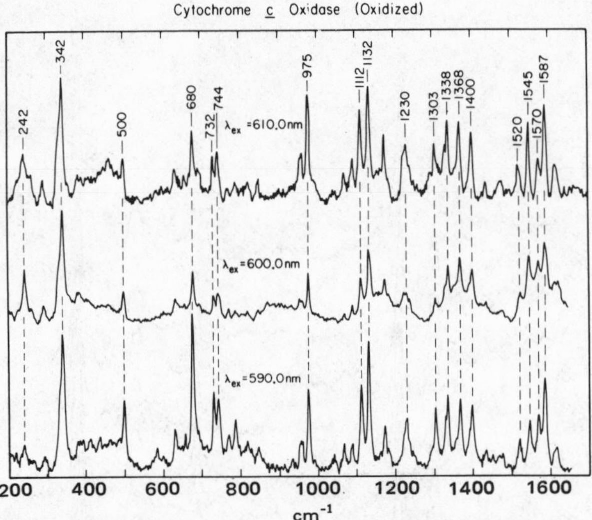

Fig. 6.35. Resonance Raman spectra of oxidized cytochrome *c* oxidase obtained with excitation in the visible region of the absorption spectrum. The laser power at the sample was approximately 150 mW at λ_{ex} = 610.0 and 600.0 nm and 220 mW at λ_{ex} = 590.0 nm. The base lines of the Raman spectra of the oxidized protein shown in this figure have been corrected for background fluorescence. (Bocian *et al.*, 1979.) Reprinted with the permission of the American Chemical Society. Copyright 1979.

cm^{-1}. The oxidase also has bands at 975, 1112, and 1520 cm^{-1}, which are not present in the resonance Raman spectra of other hemoproteins, whereas bands at 1570, 1545, 1400, 1172, and 960 cm^{-1} are much more enhanced than are the corresponding bands in the spectra of other hemoproteins.

Bocian *et al.* (1979) have also observed very intense resonance Raman bands for cytochrome *c* oxidase at 242 and 342 cm^{-1} and a relatively large enhancement at 500 cm^{-1}. The 242 and 342 cm^{-1} band positions are similar to those of the copper-ligand vibrations found for type 1 copper proteins when they are excited in their 600-nm absorption band. Type 1 copper proteins, when excited within their characteristic 600-nm absorption band, display intense resonance Raman bands between 200 and 500 cm^{-1} (Miskowski *et al.*, 1975; Siiman *et al.*, 1976). When type 1 copper proteins are reduced, the 600-nm absorption band disappears along with the low-frequency Raman bands. When cytochrome *c* oxidase is reduced, these bands remain, and thus they are not caused by copper-ligand vibrations. The 500-cm^{-1} band cited above has no corresponding band in the resonance Raman spectra of other hemoproteins or metalloporphyrins.

All the bands seen in the resonance Raman spectra of Bocian *et al.* (1979) were attributed to the two heme *a* moieties in the enzyme.

Cytochrome c and Cyanocobalamin. In the technique called coherent anti-Stokes Raman scattering (CARS) (Begley *et al.*, 1974), the Raman signal is generated as a coherent beam of light, separated readily from the nondirectional fluorescence. The system still lends itself to studies of resonance enhancement; e.g., Nestor *et al.* (1976) have observed resonance CARS spectra of dilute aqueous cytochrome *c* and vitamin B_{12}.

The CARS spectrum of ferrocytochrome *c* is shown in Fig. 6.36. For comparison, the conventional resonance Raman spectrum, obtained with 5145 Å excitation, is also shown. The inset absorption spectrum shows the resonance conditions for the CARS experiment. The resonance Raman bands correspond to vibrations of the porphyrin ring. All these bands are seen in the resonance CARS spectrum, including the inverse polarized bands which originate from A_{2g} porphyrin vibrations and which become allowed only in resonance scattering (Spiro and Strekas, 1972). The slightly different relative intensities displayed in the two spectra were attributed to the use of different resonance conditions.

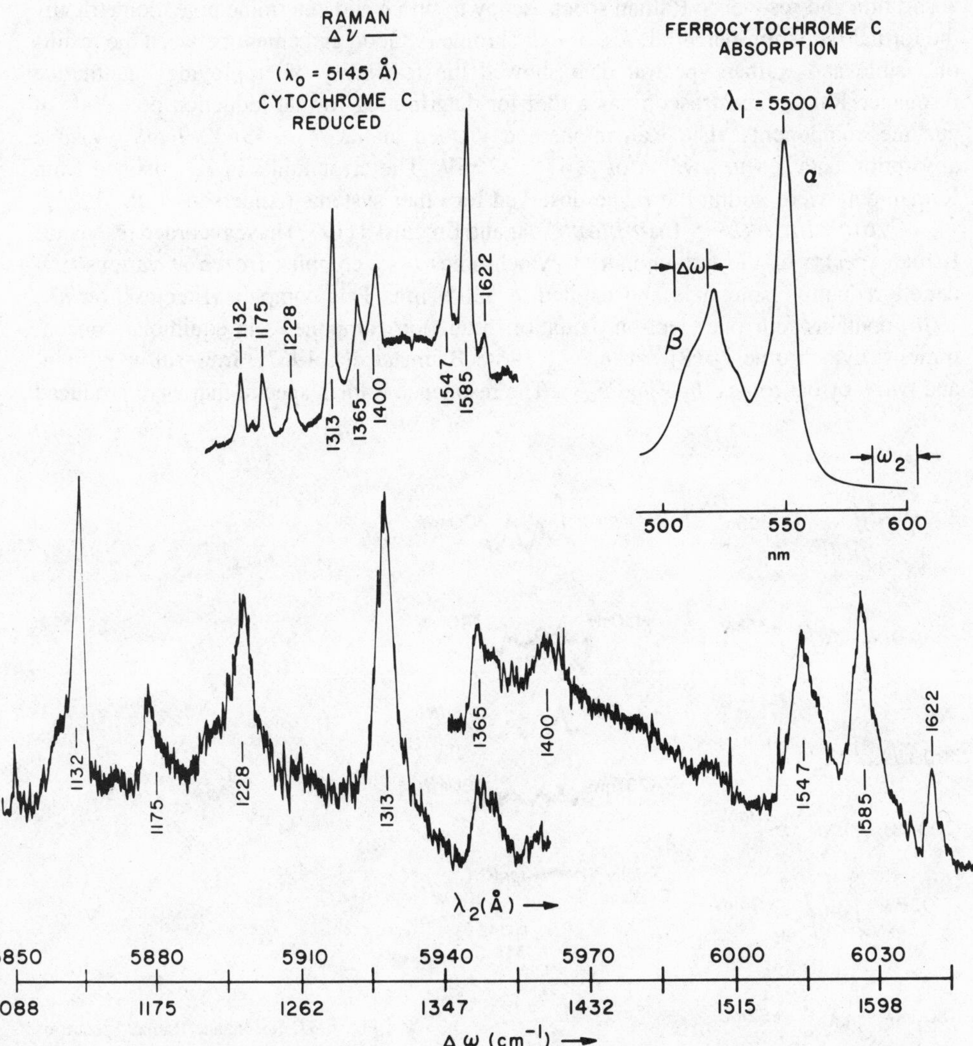

Fig. 6.36. Resonance CARS spectrum of ferrocytochrome *c*, 1 m*M* in H_2O, λ_1 = 5500 Å, λ_2 scan speed = 2 Å/min, laser repetition rate = 10 pulses per sec, 30 pulses averaged. The discontinuity in the spectrum is due to the limited tuning range of the apparatus (Chabay *et al.*, 1976). Insets: resonance Raman and visible absorption spectra of the same solution (Spiro and Strekas, 1972). (Nestor *et al.*, 1976.)

Although ferrocytochrome *c* is usually accepted as not being fluorescent, one can see a broad fluorescence band underlying the Raman spectrum. The CARS spectrum does not show such fluorescence. Its slightly sloping base line is caused by changing laser power levels with wavelength.

Nestor *et al.* have also presented CARS spectra of cyanocobalamin, and also showed inverse Raman effects of this compound, i.e., a decrease in intensity of a band (1500 cm^{-1}), rather than an enhancement. The results of Nestor *et al.* (1976) show that CARS dominates at resonance, but the inverse Raman effect dominates off-resonance.

Anderson and Kincaid (1978) have coupled electrochemical methods with visible absorption and resonance Raman spectroscopy to titrate and determine potentiometrically the formal reduction potential, E_0', of cytochrome *c*. Good agreement between the results of visible and Raman spectral data showed the feasibility of employing quantitative resonance Raman spectroscopy as a tool for determining formal reduction potentials of enzyme components. The Raman method yielded an E_0' of -53 ± 9 mV. Visible absorption data led to a value of -47 ± 2 mV. The error limits in E_0' obtained from Raman data were within the range observed for other systems (Anderson *et al.*, 1976).

Cytochromes b–c₁ Complex. Adar and Erecinska (1977) have recorded resonance Raman spectra of the mitochondrial cytochromes b–c_1 complex frozen at various oxidation–reduction potentials and excited at 530.9 nm. This complex (Erecinska *et al.*, 1976) contains four one-electron oxidation–reduction components in equimolar concentrations: cytochrome c_1 (Rieske *et al.*, 1964; Baum *et al.*, 1967), iron–sulfur protein, and two *b* cytochromes (b_{561} and b_{566}). The resonance Raman spectra that were produced

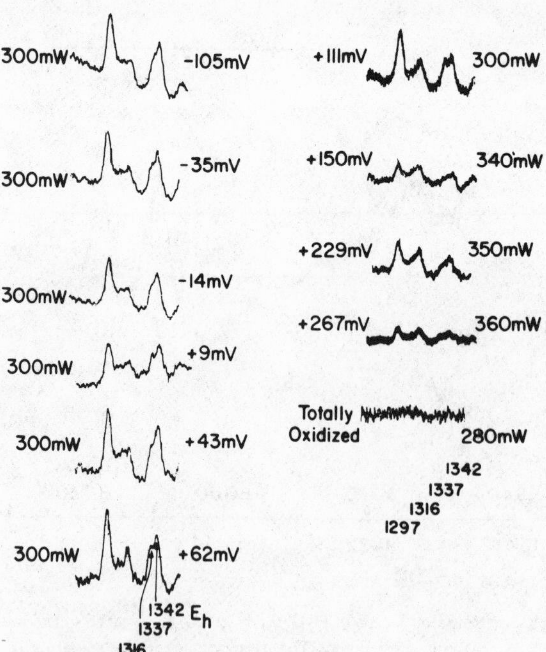

Fig. 6.37. Resonance Raman spectra of the bands between 1290 and 1350 cm^{-1} as a function of redox potential. Concentration of cytochrome c_1 was 66 μM. (Adar and Ericinska, 1977.)

in the region 1290–1350 cm^{-1} as a function of oxidation–reduction potential between −105 and +276 mV are shown in Fig. 6.37. The intensities of the bands decrease as oxidation of the sample proceeds, and the spectra given here represent reduced cytochromes. These spectra and a plot of intensity at 1298 cm^{-1} versus oxidation–reduction potential (based on this figure) allowed the conclusion to be made that there are interactions between membrane-bound *b* and *c* type hemes in the membranous preparation of the cytochromes b–c_1 complex. The physical mechanism of these interactions (i.e., geometry between hemes) was not clear at the time of the study.

Cytochrome b_{562}. Resonance Raman spectra of ferric and ferrous cytochrome b_{562} of *E. coli* have been recorded in the region 1300–1700 cm^{-1} (Bullock and Myer, 1978). Both forms display features (Figs. 6.38 and 6.39) characteristic of heme iron in the low-spin electronic configuration (Spiro and Strekas, 1974; Spiro and Loehr, 1975). Similarities and several differences occur among the spectra of cytochrome b_{562} and other *b*-type cytochromes (Adar and Erecinska, 1974; Kitagawa *et al.*, 1975). The spectra of b_{562} show polarized bands at 1084, 1362, and 1493 cm^{-1}, depolarized bands at 1174 and 1621 cm^{-1}, strongly inversely polarized bands at 1128 and 1586 cm^{-1}, and a band at 1340 cm^{-1}—characteristics present in all cytochromes of the *b* variety. However, the inversely polarized band at 1300 cm^{-1} is 5 cm^{-1} lower than that for cytochrome b_5 and 15 cm^{-1} lower than that for cytochrome *b*. The spectrum of b_{562} displays a depolarized band at 1223 cm^{-1}, lower by 12 and 3 cm^{-1} than for b_5 or *b*, respectively; the depolarized band at 1548 cm^{-1} is higher by 13 and 9 cm^{-1}, respectively. Ferrous cytochrome b_{562} does not

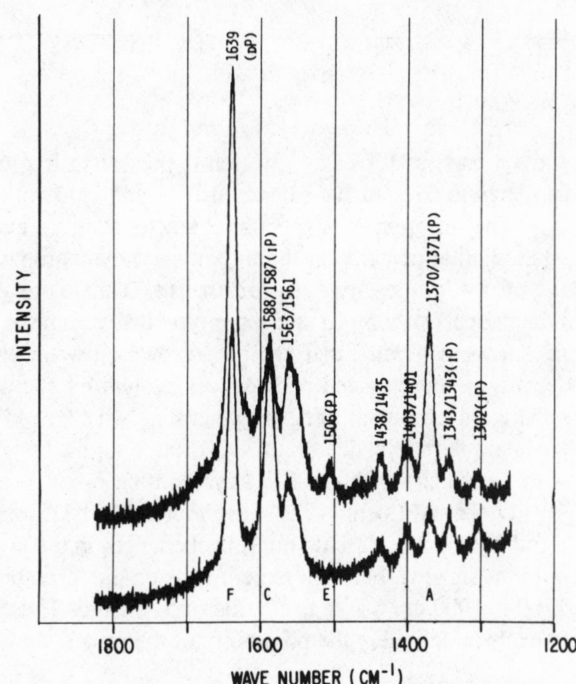

Fig. 6.38. Resonance Raman spectra of ferric cytochrome b_{562} from *E. coli* in 0.1 *M* phosphate, pH 6.8. Conditions: excitation wavelength 5145 Å; flux, 20 mW; slit width, 300 μm; wavelength steps, 0.1 Å; integration time, 1 sec; p, polarized, dp, depolarized; ip, inversely polarized: classified by Spiro and Strekas (1974). Protein concentration, 0.23 m*M*. (Bullock and Myer, 1978.) Reprinted with the permission of the American Chemical Society. Copyright 1978.

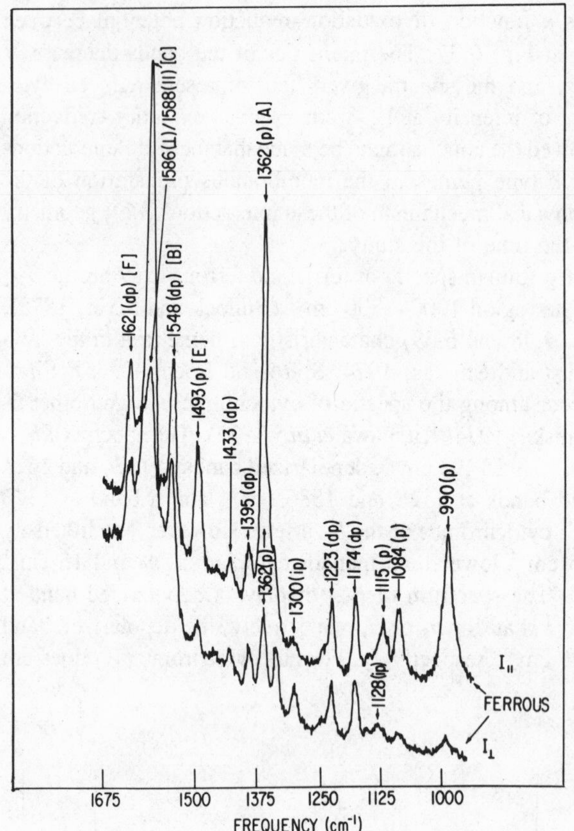

Fig. 6.39. Resonance Raman spectra of ferrous cytochrome b_{562} from *E. coli* in 0.1 *M* phosphate, pH 6.8. Conditions and other details are the same as in Fig. 6.38. (Bullock and Myer, 1978.) Reprinted with the permission of the American Chemical Society. Copyright 1978.

show a band at 1562 cm^{-1}, which is present in ferrous cytochrome b_5. The cytochrome b_5 spectrum does not have bands at 1433 and 1115 cm^{-1}, which are present for cytochrome b_{562}. The differences in spectral characteristics are evidence of differences in the spatial configuration of heme in the three type-*b* cytochromes (Bullock and Myer, 1978).

An Algal c-Type Cytochrome. Yamamoto *et al.* (1976) have used resonance Raman spectroscopy to study a *c*-type cytochrome isolated from a blue-green alga, *Synechococcus lividus,* that had been grown on water and ^2H_2O media. They recorded the spectra on the oxidized and reduced protein by exciting the molecules within the Soret band at 441.6 nm, in order to determine whether individual resonance Raman bands of the heme shift upon deuterium substitution for the hydrogens, and to provide a comparison with the spectra of horse heart cytochrome *c*.

Loehr and Loehr (1973) have found that the resonance Raman spectra of cytochrome *c* and microperoxidase do not reflect changes in the protein environment around the heme or in the ligands, but their experiments on the ferric proteins only covered the range from 1700 to 1000 cm^{-1} with excitation of 514.5 nm. These conditions were not broad enough to exclude definitely the possibility of the protein and ligand effects in the spectra. Other laboratories have produced spectra of *c*-type and *b*-type cytochromes, which allowed comparison of the Raman spectra of heme in different protein environments. Kitagawa and Iizuka (1974) and Adar and Erecinska (1974) have found that differences between the derivatives is rather small (e.g., a maximum of 7 cm^{-1} difference for four derivatives of *c*-type cytochromes, and other small differences for the *b* and b_5 cytochromes). The latter workers attributed the *b*-type cytochrome spectral differences to the protein environment. By comparing the spectra of cobalt-protoporphyrin in hemoglobin, myoglobin, and piperidine, Woodruff *et al.* (1975) were able to conclude that the spectra of the hemes in different proteins and in solution are not different.

Yamamoto *et al.* (1976) have found that the algal cytochrome yields resonance Raman spectra roughly similar to those of cytochrome *c* of horse heart, but there are differences between the spectra of the two proteins, which may reflect the effect of nonmethionine ligands and protein environment on the vibrations of the *c*-type heme in the cytochrome of *S. lividus*.

Other c-Type Cytochromes. Kitagawa *et al.* (1977) have investigated the pH dependence of the resonance Raman spectra and structural alterations at heme moieties of several *c*-type cytochromes (*c*, c_2, c_3, *c*-551, and *c*-555). The frequencies of the 1539-cm^{-1} (Fe^{2+}) and 1565-cm^{-1} bands (Fe^{3+}) changed with replacement of the sixth ligand. The titration curve for the 1565-cm^{-1} band of cytochrome *c* paralleled that for the 695-nm band. The pH dependence of the 1539-cm^{-1} band of cytochrome c_3 (Fe^{2+}) gave evidence of a stepwise replacement of the sixth ligand of its four hemes, although such pH dependence was not displayed for the Raman spectra of other ferrous cytochromes examined. Detergents had the effect of changing the relative intensities at 1639, 1587, and 1561 cm^{-1} in the spectra of a ferric protoporphyrin bis-imidazole complex. When the corresponding Raman bands of cytochromes b_5 and *c* were examined in the presence and absence of detergents, their relative intensities were located close to those of the ferric porphyrin complex, thus hinting at a considerable difference in their heme environments.

Kihara *et al.* (1978) have investigated the structure of the thermoresistant cytochrome *c*-552 extracted from *Thermus thermophilus* by using absorption and resonance Raman spectroscopy. They compared this structure with that of horse heart cytochrome *c*. At neutral pH histidine and methionine were considered to be the ligands of the ferricytochrome *c*-552; and the ligands of ferrocytochrome *c* were histidine and histidine or histidine and lysine. The thermodynamic parameters for an alkaline isomerization of ferric cytochrome *c*-552 with a p*K* of 12.3 (25°C), and three isomerization states of alkaline pH (p*K* 9.3, 12.9, and >13.5 at 25°C) were estimated. The entropy change for the *c*-552 cytochrome was found to be negative, whereas that for horse heart cytochrome *c* was positive.

Adar (1977) has recorded resonance Raman spectra of protozoan cytochromes c_{557} and c_{558}, which supported the chemical evidence (Pettigrew *et al.*, 1975) for one vinyl group on the heme moiety and one thioether linkage to the protein.

Ikeda-Saito *et al.* (1975) have studied the resonance Raman spectra of ferrous alkylated cytochrome *c* (dicarboxymethylmethionyl-cytochrome *c*). Brunori *et al.* (1972) had attributed the pH-dependent change of absorption spectra of this compound to the

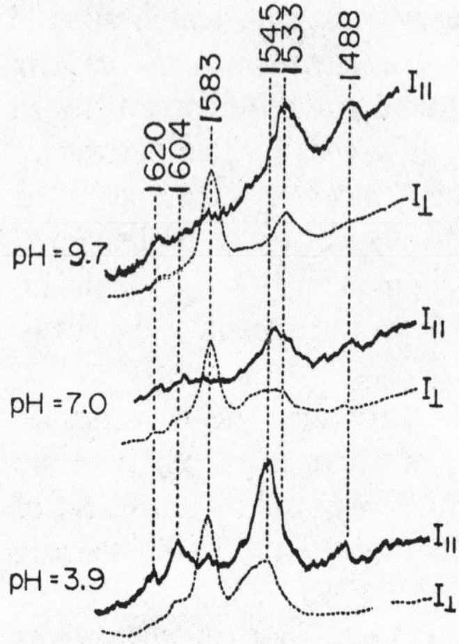

Fig. 6.40. Polarized Raman spectra of ferrous alkylated cytochrome *c* near 1540 cm^{-1} band at pH 3.9, 7.0, and 9.7. Instrumental conditions: excitation, 514.5 nm line of Ar^+/Kr^+ laser (Spectra Physics model 164-02); power, 200 mW; time constant, 0.5 sec; sensitivity, 4000 counts full scale; rate, 0.25 sec; slit width, 6 cm^{-1}. Solid line: electric vector of the scattered radiation is parallel to that of the incident radiation. Broken line: electric vector of the scattered radiation is perpendicular to that of incident radiation. (Ikeda-Saito *et al.*, 1975.)

transition between low-spin and high-spin states. A pH-sensitive Raman band was observed (Ikeda-Saito *et al.*) near 1540 cm^{-1} and it was concluded that the pH-dependent change results from the transition between two kinds of low-spin state with a p*K* of 7.9.

Parallel and perpendicular components of Raman scattering near 1540 cm^{-1} of ferrous alkylated cytochrome *c* in 0.1 *M* phosphate buffer at pH 3.9, 7.0, and 9.7 are shown in Fig. 6.40. As the pH rises from 3.9 to 9.7, the inverse polarized band at 1583 cm^{-1} and the polarized band at 1488 cm^{-1} do not display pH dependence; the depolarized band at 1545 cm^{-1} disappears and a new depolarized band appears at 1533 cm^{-1}.

Figure 6.41 shows the apparent observed frequency of the pH-sensitive Raman band plotted versus pH, along with the change in absorbance of the α-band at 550 nm of ferrous alkylated cytochrome *c*. The p*K* value for this transition is 7.9.

Resonance Raman spectroscopy indicates that the compound is in the low-spin state between pH values 3.0 and 11.0, since the Raman band characteristic of high-spin state (Yamamoto *et al.*, 1973) is never displayed. The Raman spectra of alkylated cytochrome

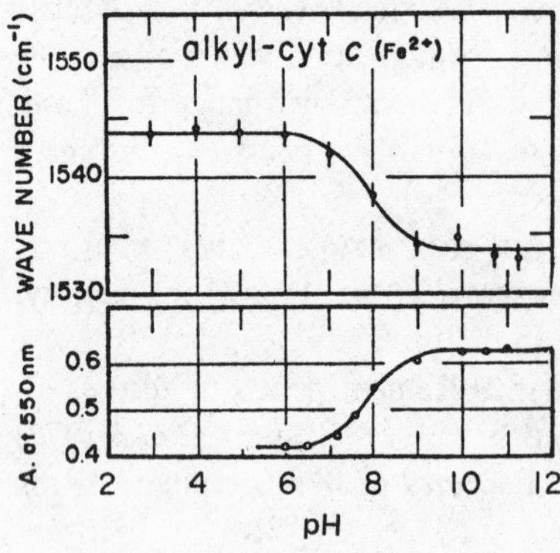

Fig. 6.41. pH-Dependence of the apparently observed wavenumber of the ligand sensitive Raman band (top) and that of optical absorbance at 550 nm (bottom) of ferrous alkylated cytochrome *c*. (Ikeda-Saito *et al.*, 1975.)

c between pH 3.0 and 11.0 and of native cytochrome c, which has a low-spin state, are alike.

Lysine-79 was believed to be coordinated to the heme iron of the modified cytochrome in the low-spin state, in the pH range cited. The origin of the pH-sensitive band was not yet clear, but the band reflects a "delicate change of the interaction between ligand and heme iron" (Ikeda-Saito *et al.*, 1975).

Kitagawa *et al.* (1975) have recorded resonance Raman spectra of mammalian cytochrome c, bacterial cytochrome c_3, algal photosynthetic cytochrome f (another c-type), alkylated cytochrome c, and cytochrome b_5 in their reduced and oxidized states. For ferrous cytochrome f, a Raman line sensitive to the replacement of an axial ligand of the heme iron was observed at 1545 cm^{-1}, the same frequency as for cytochrome c. A similar Raman line was observed at ca. 1540 cm^{-1} for ferrous alkylated cytochrome c (see page 276). Thus, histidine and methionine were shown to be likely axial ligands as in cytochrome c. For ferrous cytochrome c_3, the ligand-sensitive Raman line occurs between those of cytochrome c and cytochrome b_5. With histidines accounting for two axial ligands in cytochrome b_5, it is possible that histidines account for most of the axial ligands of the four hemes in cytochrome c_3. On the other hand, Kitagawa *et al.* (1975) could not rule out completely a combination of histidine and methionine as a pair of axial ligands for one or two of the four hemes.

The spectra of ferrous cytochrome b_5 displayed two weak Raman lines at 1302 and 1338 cm^{-1} instead of the strongest line at 1313 cm^{-1} of ferrous cytochromes of the c-type. The authors suggested the use of these bands to identify types of cytochromes.

Cytochrome c' is the generic name for a group of proteins with covalently bound heme groups, which have been obtained from a denitrifying species of bacteria and from purple photosynthetic bacteria. At pH >11.5, the reduced Fe(II) protein displays a standard low-spin hemochrome absorption spectrum (Horio and Kamen, 1961). At pH <11, the spectrum is similar to that of deoxyhemoglobin, except for a split in the Soret band. The reduced protein is high-spin at neutral pH. The oxidized form is also low-spin in alkaline solution and shows a hemichrome spectrum. Below pH 11, the spectrum returns to a high-spin heme-Fe(III) form, also with a split Soret band, but it changes with lowering pH values. The measured magnetic susceptibilities lie somewhere between values expected for low-spin and high-spin Fe(III).

Strekas and Spiro (1974) have recorded resonance Raman spectra for cytochrome c' of *Rhodopseudomonas palustris* by using laser excitation frequencies in the region of the Soret band and the α-β bands. They compared the vibrational frequencies of the porphyrin ring with those of other heme proteins known to have definite oxidation and spin states. At pH >11.5, the oxidized and reduced forms of this cytochrome c' show normal low-spin heme spectra. At lower pH, the reduced form, which is high-spin, displays a spectrum like that of deoxyhemoglobin, i.e., high-spin five-coordinate ferroheme. At lower pH, the oxidized protein spectra are unusual, displaying frequencies more like low-spin than high-spin values; but their magnetic moments, although intermediate, are nearer to high-spin values. Hydroxymethemoglobin, a proven mixed-spin heme protein, has two overlapping sets of vibrational frequencies, and two such sets are absent for the cytochrome c'. Strekas and Spiro (1974) suggested that the heme groups are more planar than are normal hemes of high spin, and that the spin state may be intermediate,

$s = 3/2$. At pH 6.9, the tentative structure is more planar, a fact which parallels the spectral changes. They measured the depolarization ratios of the Raman band at ca. 1370 cm^{-1} and concluded that there was a lowering of 4-fold symmetry of the heme chromophore in the lower pH forms of both oxidized and reduced cytochrome c'. The splittings of the Soret band are consistent with this.

Cytochrome P450$_{cam}$. Cytochromes P450 are *b*-type hemeproteins, which catalyze the monoxygenase or "mixed-function" oxidase reactions that utilize a two-electron reduction of dioxygen coupled with hydrocarbon oxygenation. Gunsalus *et al*. (1973), Gunsalus and Wagner (1978), and Katagiri *et al*. (1968) identified and purified a soluble three-component P450 system, the camphor 5-monoxygenase [EC.1.14.15.1] from *Pseudomonas putida*. This system, P450$_{cam}$ (or cytochrome *m*), catalyzes the hydroxylation of camphor to the 5-exo alcohol as displayed in the reaction cycle of Fig. 6.42.

Champion *et al*. (1978) have obtained resonance Raman spectra of cytochrome P450$_{cam}$ in several stable reaction states with laser excitations in the Soret region (350–450 nm). Resonance in this region produces vibrational modes of greatly enhanced intensity resulting from the large transition moments associated with the absorption band. Table 6.2 shows resonance Raman peak locations of various P450 complexes. In the table, the abbreviations of Fig. 6.42 are used. The table shows data for ferric P450$_{cam}$-substrate complex (m^{os}), ferric cytochrome (m^{o}), ferrous P450$_{cam}$-substrate complex (m^{rs}), and P420 (a catalytically inactive form of cytochrome *m*). The oxidized cytochrome, in the substrate-free ferric form, shows a selective enhancement of vibrational modes at 676 and 1372 cm^{-1}, which were assigned to normal coordinates of expansion for the porphyrin in the excited state. The substrate-bound cytochrome in the oxidized and reduced states shows unusual band positions, which are evidence of a weakening in bond strengths of

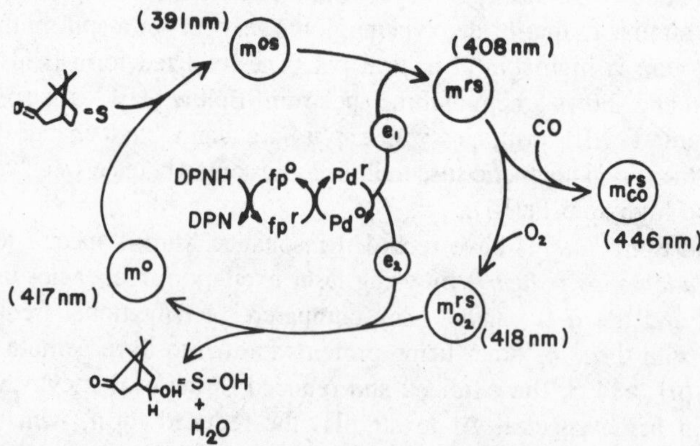

Fig. 6.42. Cytochrome P450$_{CAM}$ reaction cycle. Cytochrome P450 is abbreviated *m* for monoxygenase in order to simplify labeling of ligand and redox states; m^{o}, ferric cytochrome; m^{os}, ferric P450$_{CAM}$-substrate complex; m^{rs}, ferrous P450$_{CAM}$-substrate complex; $m^{rs}_{O_2}$, oxygenated ferrous P450$_{CAM}$-substrate complex; m^{rs}_{CO}, carbon monoxide ferrous P450$_{CAM}$-substrate complex; DPNH, diphosphopyridine nucleotide; fp, flavoprotein putidaredoxin reductase; Pd, putidaredoxin. The reaction states are also labeled by the wavelength of their Soret maxima. (Champion *et al*., 1978.) Reprinted with the permission of the American Chemical Society. Copyright 1978.

Table 6.2. Resonance Raman Peak Positions of P450 Complexes[a]

m^{os} $\Delta\nu$(I, pol)[b]	m^{o} $\Delta\nu$(I, pol)	m^{rs} $\Delta\nu$(I, pol)	P420 $\Delta\nu$(I, pol)	m^{os} $\Delta\nu$(I, pol)	m^{o} $\Delta\nu$(I, pol)	m^{rs} $\Delta\nu$(I, pol)	P420 $\Delta\nu$(I, pol)
311 (w,u)				1342 (w,u)	1343 (w,u)		
318 (w,u)				1368 (s,p)		1344 (s,p)	
		322 (w,p)		1372 (s,p)[c]	1372 (s,p)	1361 (v,p)	1372 (s,p)
342 (m,p)	345 (m,p)		345 (m,u)	1380 (sh,u)			1382 (sh,u)
351 (m,p)				1396 (sh,dp)	1395 (w,dp)	1391 (s,dp)	1400 (w,u)
		362 (w,p)		1422 (sh,u)			
377 (m,p)	378 (m,p)	375 (w,p)		1429 (m,u)	1432 (w,u)	1425 (s,dp)	1430 (m,u)
		403 (w,p)			1464 (w,u)	1445 (sh,u)	1468 (w,u)
421 (w,p)	425 (w,p)	416 (w,p)		1488 (m,p)		1466 (s,p)	1490 (m,u)
675 (m,p)	676 (s,p)	673 (m,p)	676 (s,p)	1502 (sh,p)	1502 (m,p)		1502 (w,u)
691 (w,u)		629 (sh,u)		1526 (w,p)			1525 (m,u)
716 (w,u)		713 (w,dp)	715 (w,u)			1534 (s,dp)	
721 (w,u)			736 (w,u)	1550 (m,u)	1550 (sh,u)		1548 (sh,u)
754 (m,dp)	751 (m,dp)	745 (m,dp)	752 (w,dp)	1570 (s,p)	1560 (m,u)	1563 (s,p)	1570 (s,u)
784 (w,u)		783 (w,dp)		1581 (sh,p)	1581 (m,u)	1584 (s,p)	1585 (m,u)
796 (w,u)	797 (w,u)				1601 (m,p)		
821 (w,u)		815 (w,p)			1619 (m,p)		
	930 (w,p)	925 (w,u)		1623 (s,dp)	1635 (s,dp)	1601 (s,dp)	1627 (s,dp)
970 (sh,u)						1612 (sh,u)	
984 (w,u)	987 (w,u)	974 (m,u)			2049 (w,u)		
1004 (w,p)	1006 (w,u)	1000 (m,p)					
1086 (w,u)	1087 (w,u)	1080 (w,p)	1070 (w,u)				
		1115 (w,p)					
1125 (m,u)	1128 (m,p)	1125 (sh,p)	1125 (m,u)				
		1140 (w,u)					
1170 (m,u)	1170 (m,u)	1176 (m,dp)	1170 (w,u)				
1214 (sh,u)							
1225 (m,dp)	1223 (w,u)	1213 (m,dp)	1220 (sh,u)				
1232 (sh,u)	1245 (w,u)	1235 (sh,u)	1238 (w,u)				

[a] Champion *et al.* (1978).
[b] $\Delta\nu$ is Stokes shift in cm^{-1} (± 2 cm^{-1}) relative to the excitation wavelength. I corresponds to relative intensity of peaks in the spectrum where they appear the strongest: s = strong, sh = shoulder, m = medium, w = weak, v = variable, pol refers to the polarization of peaks: p = polarized ($I_{\parallel}/I_{\perp} \lesssim 1/8$), dp = depolarized ($I_{\parallel}/I_{\perp} \sim 3/4$), u = undetermined.
[c] Possibly due to a strongly enhanced low-spin fraction.

the porphyrin ring. Champion *et al.* (1978) interpreted this effect as an increase in the population of the π^* antibonding orbitals of the porphyrin ring, which are produced by the interaction of a strongly electron-donating axial ligand (e.g., mercaptide S of cysteine). The spectra are sensitive to spin-state equilibria, and show predominantly low-spin ferric heme in the substrate-free cytochrome and a mixture of high- and low-spin ferric heme in the substrate complex.

Table 6.3 shows marker band positions of certain heme proteins. The high-frequency porphyrin ring modes in cytochrome *m* differ greatly from the resonance Raman spectra of other hemeproteins (Spiro and Burke, 1976). The oxidation state marker band I(p), at 1368 cm^{-1} in m^{os} and 1344 cm^{-1} in m^{rs}, is shifted to lower frequency than one finds in most other high-spin ferric and ferrous heme complexes (Spiro and Burke, 1976). The location of band I in the ferrous cytochrome gives the first evidence of mercaptide sulfur coordination to the heme in the m^{rs} complex (Ozaki *et al.*, 1976). The shifts of band II,

Table 6.3. Marker Band Positions of Some Heme Proteins[a,b]

Hemeprotein	I (p)	II (p)	V (dp)
High-spin ferric			
Horseradish peroxidase	1375	1500	1630
Hemoglobin fluoride	1373	1482	1608
Cytochrome *c′*	1372	1500	1637
Chloroperoxidase	1369	1490	1627
Cytochrome m^{os}	1368	1488	1623
Low-spin ferric			
Horseradish peroxidase (CN^-)	1375	1497	1642
Hemoglobin cyanide (CN^-)	1374	1508	1642
Cytochrome *c* (Fe(III))	1374	1502	1636
Chloroperoxidase (CN^-)	1373	1503	1633
Cytochrome m^{o}	1372	1502	1635
High-spin ferrous			
Horseradish peroxidase (Fe(II))	1358	1472	1605
Deoxyhemoglobin	1358	1473	1607
Cytochrome *c′* (Fe(II))	1355	1475	1609
Chloroperoxidase (Fe(II))	1348	1470	1612
Cytochrome m^{rs}	1344	1466	1601 (?)[b]

[a] Champion *et al.* (1978).
[b] 1601-cm^{-1} band assignment to band V not decided.

from 1502 to 1488 cm^{-1}, and of band V, from 1635 to 1623 cm^{-1}, support a transition from a low- to a high-spin ferric state on substrate association. A shoulder still present at 1502 cm^{-1} in the m^{os} spectra signifies a residual low-spin species in the camphor bound complex. Also, the formation of a ternary complex of cytochrome, substrate, plus iron-sulfur effector protein (putidaredoxin) causes a decrease in the intensity of the m^{os} band II at 1488 cm^{-1} (high-spin) compared to the 1502-cm^{-1} band (low-spin), and thus demonstrates a shift from high to low spin in the spin-state equilibrium.

Raman spectra were recorded *in vivo* by means of resonance excitation in the Soret region of intact *P. putida* cells, and spectra of other hemeproteins were compared. Differential excitation will be required to assign the spectral components of the individual hemeproteins *in vivo,* a method of clear potential.

Chloroperoxidase

Chloroperoxidase, a heme protein, is an enzyme that catalyzes peroxidative oxidation of classical peroxide substrates, such as pyrogallol and guaiacol; it also catalyzes the chlorination reactions involved in the biosynthesis of caldariomycen (2,2-dichloro-1,3-cyclopentenedione). Although similar to other protoheme peroxidases, such as horseradish peroxidase, Japanese radish peroxidase, and cytochrome *c* peroxidase—they are all isolated as monomeric proteins with molecular weights ranging from 35,000 to 50,000, they have similar optical spectra for the native (high-spin Fe III) and reduced (high-spin Fe

II) forms—chloroperoxidase displays a basic catalytic difference in its ability to catalyze chlorination reactions at low pH values. Therefore, it was believed that chloroperoxidase differs significantly from the others near the active site.

Champion *et al.* (1976) have used resonance Raman spectroscopy to study the active site of chloroperoxidase. Figure 6.43 shows the resonance Raman spectrum of native chloroperoxidase, containing the characteristic marker bands discussed by Spiro (1975) and Rakshit and Spiro (1974). Marker bands have been used successfully to correlate frequency and polarization of the Raman bands for a given heme protein with high or low spin and ferrous or ferric condition of the central iron atom; but some problems seem to occur when the system is applied to high-spin ferric heme proteins. Figures 6.43a and 6.43b show the marker bands, A,E,F (Spiro's notation, 1975), but note that they occur at frequencies atypical for a high-spin ferric heme protein. The A band, which indicates the oxidation state, lies at 1368 cm^{-1}, which is shifted 5 cm^{-1} lower than the normal value $\approx$1373 cm^{-1}. The E band at 1488 cm^{-1} is situated above the typical high-spin position of 1482 cm^{-1} and below the low-spin area near 1500 cm^{-1}. There are several

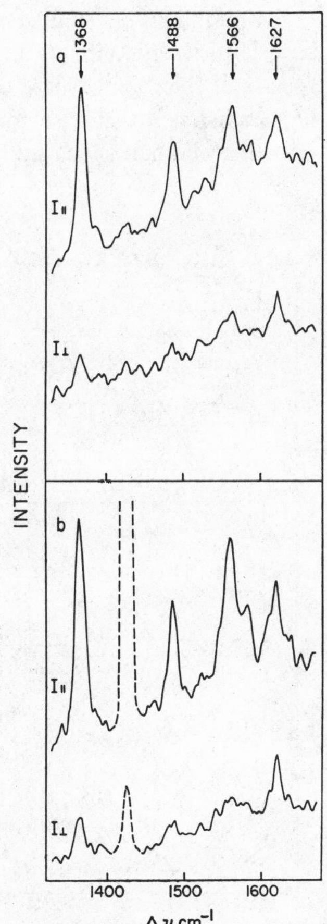

Fig. 6.43. The resonance Raman spectrum of native chloroperoxidase taken with (a) 4579 Å (10 mW incident) and (b) 4416 Å (1 mW incident) laser excitation. The dashed peak in (b) is the He (I) emission at 4713 Å from the He–Cd laser. Instrumental conditions: slit width, 5 cm^{-1}; step size, 0.1 Å; time per step, 3 sec. (a), 20 sec. (b); vertical scale, 300–1600 counts (a), 400–1200 counts (b). (Champion *et al.*, 1976.)

peaks for the c band region, which mainly denotes spin state. Polarized peaks lying at 1566 cm^{-1} and 1588 cm^{-1} (near marker band D) are located quite similarly to polarized peaks in fluoromethemoglobin (Spiro and Strekas, 1974), but are very different from the single 1575-cm^{-1} polarized peak of horseradish peroxidase (Rakshit and Spiro, 1974). The F marker band at 1627 cm^{-1} is depolarized. This location is unusual for a high-spin ferric heme protein and lies between the ordinary high-spin band (1608 cm^{-1}) and the irregular high-spin bands (1630–1640 cm^{-1}) discussed by Rakshit and Spiro (1974).

Champion *et al.* (1976) have concluded that the correlation proposed for the marker-band system and spin or oxidation state does not hold for the spectra of native and reduced chloroperoxidase, and that high-spin ferric heme proteins cannot be classified as easily as was first thought. In the case of reduced chloroperoxidase, the marker band A at 1348 cm^{-1} lies about 10 cm^{-1} lower than the position commonly found (Spiro, 1975) for high-spin ferrous heme proteins; and the A band marker for the native enzyme is 5 cm^{-1} lower than for high-spin ferrous heme proteins. The iron atom, in the native and reduced enzyme, exists in a very low symmetry environment (Champion *et al.*, 1973; Champion *et al.*, 1975) and this fact could reasonably be explained by an axial ligand that is unusually bonded to the iron atom (Champion *et al.*, 1976).

Continuing the studies of Champion *et al.* (1976), Remba and his colleagues (1979) have used laser excitation in the Soret region (350–450 nm) to obtain resonance Raman spectra of chloroperoxidase (native and reduced, and complexed with various small ions), horseradish peroxidase, cytochrome $P450_{cam}$, and cytochrome *c*. Figure 6.44a shows the resonance Raman spectrum of native (oxidized) chloroperoxidase observed when 457.9

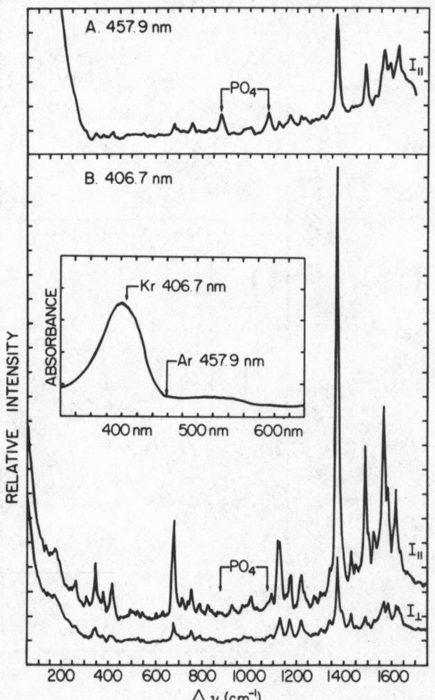

Fig. 6.44. Comparison of visibly excited and Soret-excited resonance Raman spectra of native chloroperoxidase. (A) Parallel polarized Raman spectrum taken with 457.9 nm laser excitation. The two peaks at 877 and 1075 cm^{-1} are due to nonresonant scattering from the phosphate ion in solution. (B) Parallel and perpendicular polarized resonance Raman spectra of chloroperoxidase taken with 406.7 nm laser excitation. The insert shows the absorption spectrum of native chloroperoxidase. Arrows indicate the relative positions of the krypton 406.7-nm and argon 457.9-nm lines. Typical experimental conditions; protein concentration, 15–50 μM for 90° scattering geometry; laser power 10–20 mW; effective slit resolution after averaging data in adjacent channels, 6–8 cm^{-1}; step size, 0.5–2.0 cm^{-1}; counting time, 3–10 sec/channel. (Remba *et al.*, 1979.) Reprinted with the permission of the American Chemical Society. Copyright 1979.

nm excitation is used, and Fig. 6.44b shows the result with 406.7 nm excitation (near the peak of the Soret band). The degree of enhancement of bands in the latter spectrum (approximately 20 times that in the former) permitted the observation of many more bands of chloroperoxidase than had been previously recorded. Remba *et al.* have presented tables of these Raman frequencies and marker band frequencies of various derivatives of chloroperoxidase (not shown).

The active-site environments of cytochrome $P450_{cam}$ and chloroperoxidase were shown to be quite similar to those found in earlier work. Also, native and reduced chloroperoxidase display unusual band locations (band I) which indicate a weakening of bond strengths of the porphyrin ring. These anomalous locations are probably caused by the presence of a strongly electron-donating axial ligand (Champion *et al.*, 1976), increasing the population of the π^* antibonding orbitals. Remba *et al.* (1979) believe that the mercaptide sulfur of cysteine may be involved in this way.

Certain low-spin ferrous forms of chloroperoxidase display very large selective enhancements of bands at 1360 and 674 cm^{-1}, which were assigned to primary normal vibrational modes of expansion of the porphyrin ring upon electronic excitation.

Chlorophylls a and b

Chlorophylls *a* and *b* (see Fig. 6.45) display many resonance Raman spectral characteristics like those of the porphyrins (Lutz and Breton, 1973; Lutz, 1974). Lutz (1974) has obtained resonance Raman spectra of chlorophylls *a* and *b* and of pheophytins *a* and

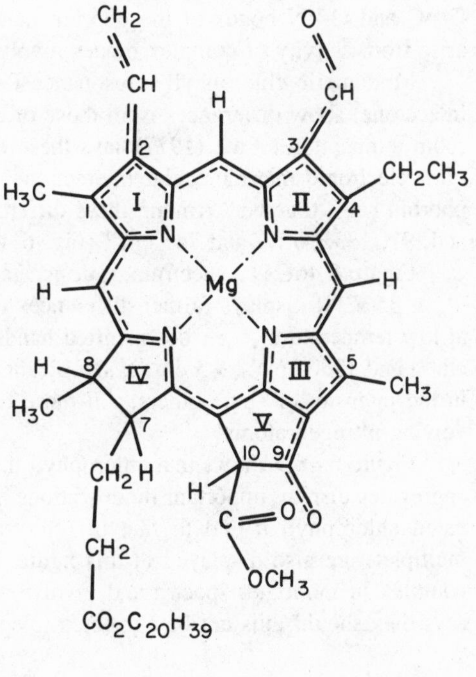

Fig. 6.45. Molecular structure of chlorophyll *b*. Interrupted lines indicate the extension of the conjugated π electron system. Chlorophyll *a* differs from chlorophyll *b* by substitution of a methyl group in place of the formyl radical in position 3, and chlorophyll *d* by further substitution of a formyl radical in place of the vinyl function in position 2. Pheophytins *a, b,* and *d* are the corresponding magnesium free, dihydro compounds. (Lutz, 1974.)

b in various media by excitation with ten different wavelengths in the region from 441.6 to 514.5 nm. The characteristics of the spectra of chlorophylls a and b are dependent on the excitation frequency. In polar solvents chlorophyll a displays a non-bonded C9═O stretching vibration (1700–1695 cm^{-1}), which shifts when aggregates form, as already known from IR investigations (Ballschmiter and Katz, 1969).

A strong band at 1664 cm^{-1} arises from the aldehyde C═O stretching at position 3. The 1175-cm^{-1} band of chlorophyll b was assigned to C—C stretching of the same group (Lutz, 1974). Chlorophylls a and b and pheophytins a and b display a medium band between 1705 and 1695 cm^{-1}, arising from carbonyl stretching of a ketone.

More recently, Lutz (1977) has obtained spectra of much higher quality from chloroplasts of various organisms. Using resonance enhancement in the Soret bands of chloroplast antenna of green plants and of monocellular algae, he has selectively recorded Raman spectra of their chlorophylls a and b at 35 K. Figure 6.46 shows spectra of chlorophyll a in various forms.

According to Lutz, the carbonyl stretching region (1750–1550 cm^{-1}) displays, besides bands arising from the stretching motion of the C9═O ketone groups of chlorophyll a, three bands near 1560, 1585, and 1617 cm^{-1} arising from C┈C stretching modes of the phorbin ring, without noticeable involvement of nitrogen motions; and the 1617-cm^{-1} band probably involves stretching motions of the methine bridges.

At 35 K the nonbonded C9═O stretching frequency mentioned above is found at 1690 cm^{-1} in dry, solid samples and at 1682 cm^{-1} for 0.01 M solutions in acetone (Fig. 6.46B). Dry (chlorophyll $a)_n$ oligomers at 35 K have a broad band at 1650 cm^{-1}, characteristic of Mg-bound C9═O carbonyls.

In the 1550–700-cm^{-1} region, resonance Raman bands of chlorophyll most likely originate from in-plane stretching (1600–1000 cm^{-1}) and angular modes of the conjugated C┈C and C┈N bonds of the phorbin macro ring (Lutz, 1974). Most of these bands arise from activity of complex modes involving simultaneous motions of several atoms.

Monomeric chlorophyll a resonance Raman spectra (frozen concentrated solutions in acetone) show differences from those of chlorophyll a aggregates in polar solvents at room temperature. Lutz (1977) says these differences probably arise as a consequence of π electronic intermolecular interactions, which occur when aggregation crowds the phorbin rings together. Among these differences are the relative enhancement of bands at 1391, 1331, 757, and 745 cm^{-1} (Fig. 6.46B).

Figure 6.46C is a spectrum of desiccated, self-associated chlorophyll a (chlorophyll $a)_n$ at 35 K. It displays further differences when compared to monomeric chlorophyll a at low temperature, e.g., downshifted bands at 1527 cm^{-1} (-8 cm^{-1}), 1289 cm^{-1} (-4 cm^{-1}) and 1266 cm^{-1} (-5 cm^{-1}); significant enhancement of some bands; and a decrease in the intensity of some others. All these bands originate from modes significantly involving nitrogen atoms.

Figure 6.46D shows that chlorophyll molecules that exist as (chlorophyll $a \cdot nH_2O)_m$ aggregates display important modifications in their spectra compared to that of unaggregated chlorophyll a and to that of (chlorophyll $a)_n$ oligomers. Several doublets and multiplets are also displayed in this figure. Most of them correlate with bands that are complex in monomer spectra and involve accidentally degenerate modes. Lutz (1977) says they should thus not be attributed to splittings by excitonic coupling.

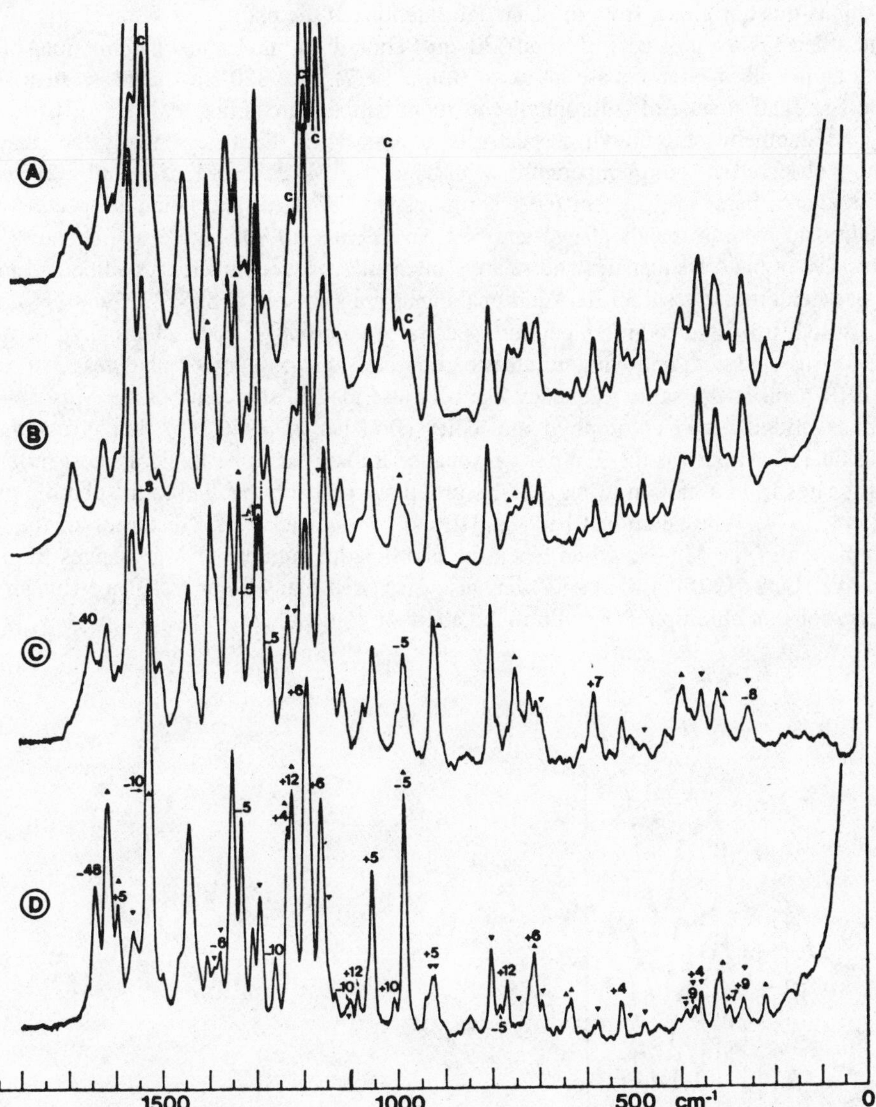

Fig. 6.46. Resonance Raman spectra of chlorophyll *a* at 35 K. (A), antenna chlorophyll *a* of whole chloroplasts of mutant barley lacking chlorophyll *b*. C labels bands from carotenoids. (B), monomeric chlorophyll *a*, 10^{-2} *M* in acetone. Arrows indicate bands enhanced in relative intensity with respect to spectra of monomer at room temperature. (C), (chlorophyll $a)_n$ oligomers, desiccated, solid deposit. Numbers and arrows: frequency shifts (cm^{-1}) and variations in relative intensity with respect to monomers at 35 K. (D), (chlorophyll *a*, $nH_2O)_m$ polymers in cyclohexane. Numbers and arrows as in (C). In case of ambiguous correlations with monomer spectra, the lowest possible frequency shifts have been indicated. Excitation wavelengths: (A), (B), 441.6 nm; (C), (D), 457.9 nm. Resolution at 1000 cm^{-1} is 8 cm^{-1}. (Lutz, 1977.)

In the region 700–50 cm^{-1}, many of the resonance Raman bands that chlorophyll *a* exhibits must originate from in-plane deformations of the phorbin skeleton (Lutz, 1974); but at least two bands near 355 and 320-cm^{-1} come from modes in which motions of the Mg atom and of N atoms are involved (Lutz, 1977). The 320-cm^{-1} band is sensitive to the aggregation state of chlorophyll *a* at room temperature (Lutz, 1974).

Monomeric chlorophyll *a* spectra in acetone at 35 K are essentially the same as those observed at room temperature, with bands at 354, 321, 295, 262, and ~210 cm^{-1} (not shown here). Self-associated chlorophyll *a* at 35 K displays similar spectra, with additional medium bands at 392 and 582 cm^{-1} (Lutz, 1977). Figure 6.46D shows differences in band frequencies and relative intensities between microcrystalline and monomeric chlorophyll *a* at 35 K. Additional bands are also seen at 635, 372, and 218 cm^{-1}.

Both low- and room-temperature spectra of chlorophyll *a* and chlorophyll *b*, when self-associated with and without intrusion of water, display another band near 310 cm^{-1}. An IR band at the same frequency has been ascribed to stretching of the Mg···O═C9 bonds bridging two chlorophyll molecules (Boucher *et al.*, 1966), but is not thusly attributed by Lutz for the 310-cm^{-1} resonance Raman band; he thinks it most probably originates from a motion of the $Mg–N_4$ grouping, probably parallel to the phorbin plane (Lutz, 1974). The additional band at 310 cm^{-1} is indicative of a distortion of the conformation of the $Mg–N_4$ group when the coordination number of Mg changes from six to five (Lutz, 1975). Lutz (1977) has also discussed many other resonance Raman observations on chlorophyll *b*, with much attention to detail, as with chlorophyll *a*.

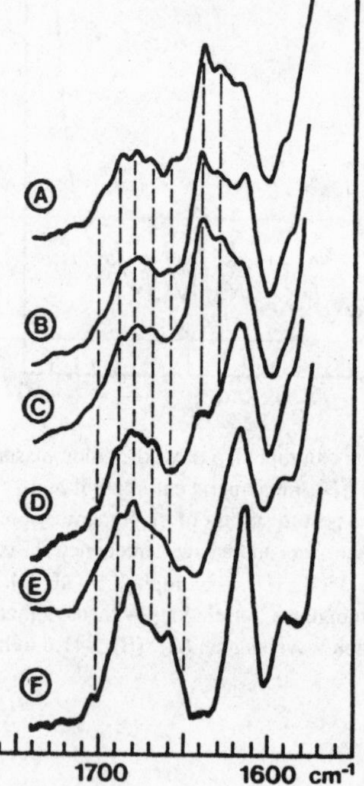

Fig. 6.47. Resonance Raman spectra of antenna chlorophyll *a*, region of carbonyl stretching modes, averaged by summation, 35 K, in intact chloroplasts of: (A), spinach; (B), *C. vulgaris* (whole cells); (C), normal barley; (D), greening maize; (E), mutant barley lacking chlorophyll *b*; (F), *B. alpina* (whole cells). Excitation wavelength: 441.6 nm. Resolution: (A), (B), (D), 5 cm^{-1}; (C), (E), (F), 7 cm^{-1}. Long dashed lines refer to stretching modes of 3—C═O groups of chlorophyll *b*. (Lutz, 1977.)

Figure 6.47 presents the carbonyl stretching regions of resonance Raman spectra of antenna chlorophyll *a* in intact chloroplasts of six different organisms. Lutz (1977) has discussed these in detail, also. Antenna chlorophyll *a* molecules occur in at least five discrete categories, characterized by different extramolecular bonding of their C9═O groups. Antenna molecules are bound to foreign molecules, probably proteins, through H-bonding on their formyl and/or keto carbonyl groups and through bonding of their Mg atoms.

Chlorophyll *b* molecules occur in at least two different categories that differ in the strength of the interactions of their 3-formyl C═O groups. Most chlorophyll *a* and *b* molecules have their Mg atoms bound to a single foreign ligand, whose nature may depend on the population observed.

Bacteriochlorophyll a and Bacteriopheophytin

Lutz and Breton (1973) and Lutz (1975, 1977) have shown that resonance Raman scattering of the chlorophylls gives information about their interactions with their environment in chloroplasts of green plants and algae. In order to extend their studies to photosynthetic bacteria, they observed resonance Raman spectra of bacteriochlorophyll *a* (BChl *a*) and bacteriopheophytin (BPheo *a*) *in vitro* and in chromatophores and reaction centers of *Rhodopseudomonas spheroides,* strains Y and R26.

Figure 6.48A shows a spectrum of BPheo *a* excited in the Q_x band, at 35 K. Stretching modes of ketone and acetyl C═O bonds are displayed as very weak bands at 1698 and 1665 cm^{-1}, respectively. These correspond to even weaker bands at about 1700 and 1670 cm^{-1} for BChl at 35 K. Excitation in the Soret region shows these modes at 1670 and 1695 cm^{-1} to be more active in room-temperature spectra of BChl in acetone excited at 363.8 nm. However, Q_x resonance strongly enhances low-frequency modes, which are much weaker in Soret resonance. These workers observed the same results for pheophytins *a* and *b*. BChl shows under Q_x resonance a complex band at 295 cm^{-1}, which is absent from the corresponding BPheo spectra. The chlorophylls have a similar band near 300 cm^{-1}, so the 295-cm^{-1} band appears to come partly from a vibration of the Mg–N_4 group (Lutz, 1974; Lutz *et al.*, 1975). BChl *a* should also have an Mg–N_4 contribution in the 352 cm^{-1} band (Lutz *et al.*, 1975).

There are four molecules of BChl *a* and two of BPheo *a* bound in the reaction centers of *R. spheroides* to a protein (Reed and Ke, 1973). Resonance Raman spectra of either BChl or BPheo, respectively, were produced by excitation with 600- and 529-nm light, at 35 K, in reaction centers of *R. spheroides* Y and R26. This confirmed earlier attribution of the 535- and 600-nm absorption bands of these reaction centers to Q_x electronic transitions of BPheo and BChl, respectively (Reed and Ke, 1973; Parson and Cogdell, 1975).

Reaction centers of *R. spheroides* R26 display a resonance Raman spectrum which arises mainly from BPheo *a* scattering (Fig. 6.48B). Several supplementary bands are also produced, most of which appear to couple with BPheo bands slightly shifted from their *in vitro* positions. These pairs are located at 1457–1467 cm^{-1}, 1055–1068 cm^{-1}, 1032–1050 cm^{-1}, 962–973 cm^{-1}, 925–932 cm^{-1}, 777–790 cm^{-1}, and 676–687 cm^{-1}. In addition, *Y* reaction centers display two weak bands at 1705 and 1685 cm^{-1}, both arising from the ketone C═O stretching mode of BPheo.

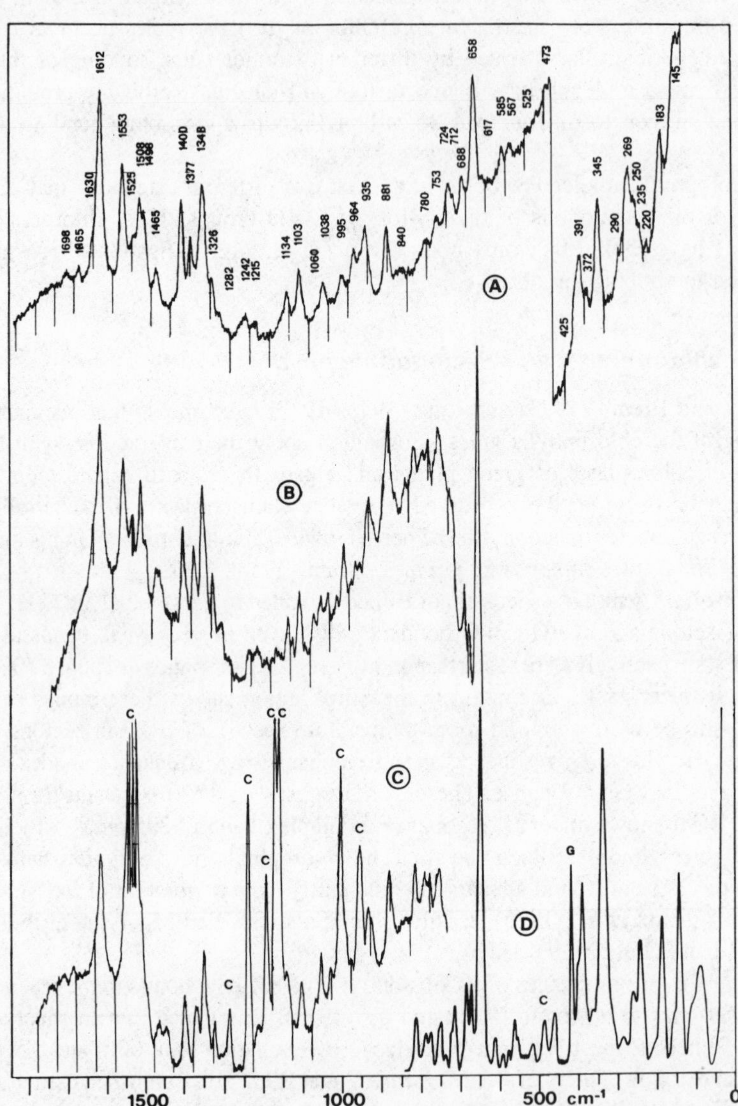

Fig. 6.48. Resonance Raman spectra obtained at 35 K, 528.7 nm excitation, and 5.5 cm^{-1} resolution at 1000 cm^{-1} from: Ⓐ, solid bacteriopheophytin cast from dry CCl_4; Ⓑ, reaction centers of *R. spheroides* R 26, 600–1750 cm^{-1} region; Ⓒ, reaction centers of *R. spheroides* Y, 700–1750 cm^{-1} region. C: RR bands of spheroidene; Ⓓ, same sample as in Ⓒ, 0–850 cm^{-1} region after subtraction of a 539 nm fluorescence band. Relative intensities and half bandwidths of RR bands may have been slightly altered by this operation. G: spurious line. (Lutz *et al.*, 1976.)

Although resonance Raman spectroscopy allows one to observe vibrational spectra of chromophores, e.g., of chlorophylls in complex media, such as chlorophyll–protein complexes, with a high selectivity (Lutz *et al.*, 1979; Lutz and Breton, 1973; Lutz *et al.*, 1976; Lutz, 1981 and 1982)—it is apparent that with such selectivity one obtains only indirect information about the binding sites of chlorophyll. For this reason Lutz *et al.* (1982) have studied the resonance Raman spectra of bacteriochlorophyll *a* in the soluble protein complex $(BChl_7—P)_3$ from *Prosthecochloris aestuarii* at 20–35 K (resonance effect on the Q_x or Soret electronic bands). The spectra show that the acetyl carbonyl groups of at least four of the seven molecules bound to the monomer subunit of the complex and the ketone carbonyls of at least five of them are oriented close to the mean plane of the conjugated part of the dihydrophorbin macrocycle. Up to three bacteriochlorophyll molecules may have their ketone carbonyls free from H-bonding and up to two may have their acetyl carbonyls similarly free.

The spectra of Lutz *et al.* (1982) (not shown here) confirm that the Mg atoms of most of the seven bacteriochlorophylls have pentacoordination. They also demonstrate that polarization effects from their local environments induce changes in the ground-state structures of the dihydrophorbin skeletons of the complexed molecules with respect to those of isolated, monomeric bacteriochlorophyll. These environmental effects (essentially identical for the seven molecules) predominate over any structural change caused by intermolecular bonding of the conjugated carbonyls or of the Mg atoms.

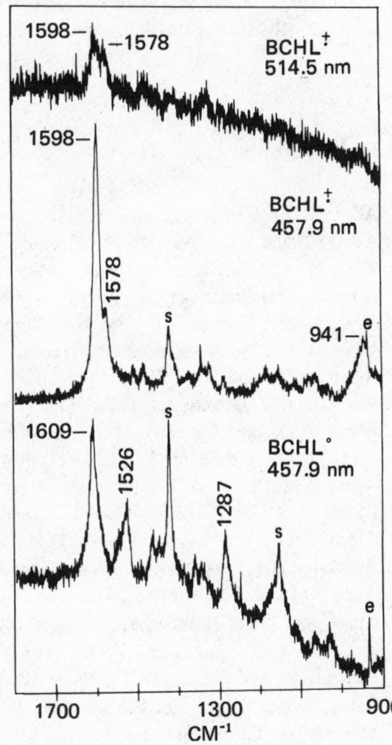

Fig. 6.49. Resonance Raman spectra of a 1.3×10^{-3} *M* solution of BChl cation radical in CH_2Cl_2 + 0.1 *M* TBAP obtained by excitation at 514.5 nm (upper spectrum) and 457.9 nm (middle spectrum). The lower spectrum is for an identical concentration of BChl neutral in CH_2Cl_2 obtained by excitation at 457.9 nm. $BChl^{\ddagger}$ was electrogenerated at +0.35 volts vs PtQRE. Spectra were recorded at 30 mW power, 3 cm^{-1} bandpass, scan rate = 0.167 cm^{-1} sec^{-1} and a 1.0-sec counting interval. S refers to solvent bands; e is an electrolyte band (ClO_4^-). TBAP is tetrabutyl ammonium perchlorate (a supporting electrolyte). PtQRE means platinum quasireference electrode. (Cotton and Van Duyne, 1978.)

Resonance Raman Spectroelectrochemistry of Bacteriochlorophyll a

The technique of resonance Raman spectroelectrochemistry (RRSE), which combines resonance Raman spectroscopy with various electrochemical procedures for radical ion generation, has already proven to be a sensitive technique with high molecular specificity for the identification and electronic structure characterization of the anion radicals for tetracyanoethylene (Jeanmaire *et al.*, 1975), tetracyanoquinodimethane (Jeanmaire and Van Duyne, 1976a,b), and the cation radicals of tetramethyl-*p*-phenylenediamine (Jeanmaire and Van Duyne, 1975) and tetrathiafulvalene (Suchanski, 1977; Van Duyne, 1977).

Cotton and Van Duyne (1978) have applied RRSE to the study of the radical ion species involved in the primary photochemistry of photosynthesis. Using controlled potential coulometry combined with resonance Raman spectroscopic detection, they have obtained the vibrational spectrum of the bacteriochlorophyll *a* (BChl) cation radical $BCHl^{\ddagger}$. They compared this spectrum with the corresponding one of the parent molecule (Fig. 6.49). We observe that the cation radical spectrum is significantly different from the neutral spectrum both in band frequencies and intensities. The 1609-cm^{-1} band of the neutral molecule is intensified by a factor of approximately 3 and shifted to 1598 cm^{-1}, or 11 cm^{-1} to lower frequency. The vibrations at 1528 and 1286 cm^{-1} are reduced in intensity and cannot be assigned at this concentration. A new broad band is observed near 941 cm^{-1}. Thus, RRSE is capable of differentiating between neutral BChl and $BChl^{\ddagger}$.

These results show that resonance Raman vibrational spectra may be applicable to the identification of radical ion formation and the monitoring of the kinetics of this process both in photosynthetic model systems and *in vivo*.

References

Adar, F., *Arch. Biochem. Biophys.* **181,** 5 (1977).
Adar, F., and Erecinska, M., *Arch. Biochem. Biophys.* **165,** 570 (1974).
Adar, F., and Erecinska, M., *Biochemistry* **18,** 1825 (1979).
Adar, F., and Erecinska, M., *FEBS Lett.* **80,** 195 (1977).
Adar, F., and Yonetani, T., *Biochim. Biophys. Acta* **502,** 80 (1978).
Alben, J. O., 29th Pittsburgh Conference, Cleveland, Ohio, Abstract No. 178, February 27–March 3, 1978.
Alben, J. O., and Bare, G. H., *Appl. Opt.* **17,** 2985 (1978).
Alben, J. O., and Bare, G. H., *J. Biol. Chem.* **255,** 3892 (1980).
Alben, J. O., and Caughey, W. S., *Biochemistry* **7,** 175 (1968).
Alben, J. O., and Fager, L. Y., *Biochemistry* **11,** 842 (1972).
Alben, J. O., Choi, S. S., Adler, A. D., and Caughey, W. S., *Ann. N.Y. Acad. Sci.* **206,** 278 (1973).
Alben, J. O., Fuchsman, W. H., Beaudreau, C. A., and Caughey, W. S., *Biochemistry* **7,** 624 (1968).
Alben, J. O., Yen, L., and Farrier, N. J., *J. Am. Chem. Soc.* **92,** 4475 (1970).
Anderson, J. L., and Kincaid, J. R., *Appl. Spectrosc.* **32,** 356 (1978).
Anderson, J. L., Kuwana, T., and Hartzell, C. R., *Biochemistry* **15,** 3847 (1976).
Antonini, E., and Brunori, M., *Hemoglobin and Myoglobin in Their Reactions with Ligands*, North-Holland Publ. Co., Amsterdam, 1971.
Arnone, A., and Perutz, M. F., *Nature* **249,** 34 (1974).
Asher, S. A., Vickery, L. E., Schuster, T. M., and Sauer, K., *Biochemistry* **16,** 5849 (1977).
Babcock, G. T., and Salmeen, I., *Biochemistry* **18,** 2493 (1979).
Bachmann, F., and Marti, H. R., *Blood* **20,** 272 (1962).

Ballschmiter, K., and Katz, J. J., *J. Am. Chem. Soc.* **91,** 2661 (1969).
Bare, G. H., Alben, J. O., and Bromberg, P. A., *Biochemistry* **14,** 1578 (1975).
Barlow, C. H., Maxwell, J. C., Wallace, W. J., and Caughey, W. S., *Biochem. Biophys. Res. Commun.* **55,** 91 (1973).
Barlow, C. H., Maxwell, J. C., Harris, J., and Caughey, W. S., *Fed. Proc.* **32,** 552 (1973).
Begley, R. F., Harvey, A. B., Byer, R. L., and Hudson, B. S., *J. Chem. Phys.* **61,** 2466 (1974).
Berzofsky, J. A., Peisach, J., and Alben, J. O., *J. Biol. Chem.* **247,** 3774 (1972).
Blumberg, W. E., Peisach, J., Wittenberg, B. A., and Wittenberg, J. B., *J. Biol. Chem.* **243,** 1854 (1968).
Bocian, D. F., Lemley, A. T., Petersen, N. O., Brudvig, G. W., and Chan, S. I., *Biochemistry* **18,** 4396 (1979).
Boucher, L. J., Strain, H. H., and Katz, J. J., *J. Am. Chem. Soc.* **88,** 1341 (1966).
Brunner, H., *Biochem. Biophys. Res. Commun.* **51,** 888 (1973).
Brunner, H., *Naturwissenschaften* **61,** 129 (1974).
Brunner, H., and Sussner, H., *Biochim. Biophys. Acta* **310,** 20 (1973).
Brunner, H., Mayer, A., and Sussner, H., *J. Mol. Biol.* **70,** 153 (1972).
Brunori, M., Wilson, M. T., and Antonini, E., *J. Biol. Chem.* **247,** 6076 (1972).
Bullock, P. A., and Myer, Y. P., *Biochemistry* **17,** 3084 (1978).
Cartwright, G. E., in *Harrison's Principles of Internal Medicine,* 6th ed. p. 1644 (M. M. Wintrobe, G. W. Thorn, R. D. Adams, I. L. Bennett, Jr., E. Braunwald, K. V. Isselbacher, and R. G. Petersdorf, eds.), McGraw-Hill, New York, 1970.
Caughey, W. S., *Adv. Chem. Ser.* **100,** 248 (1971).
Caughey, W. S., *Ann. N. Y. Acad. Sci.* **174,** 148 (1970).
Caughey, W. S., *Annu. Rev. Biochem.* **36,** 611 (1967).
Caughey, W. S., in *Methods for Determining Metal Ion Environments in Proteins* (R. Wilkins and D. Darnall, eds.), p. 75 Elsevier, New York, 1980.
Caughey, W. S., in *Proceedings of the First Inter-American Symposium on Hemoglobins. Genetical, Functional, and Physical Studies of Hemoglobins,* p. 180 (T. Arends, G. Bemski, and R. L. Nagel, eds.), S. Karger, Basel, Switzerland, 1971.
Caughey, W. S., Barlow, C. H., O'Keeffe, D. H., and O'Toole, M. C., *Ann. N.Y. Acad. Sci.* **206,** 296 (1973).
Caughey, W. S., Barlow, C. H., Maxwell, J. C., Volpe, J. A., and Wallace, W. J., *Ann. N.Y. Acad. Sci.* **244,** 1 (1975).
Caughey, W. S., Houtchens, R. A., Lanir, A., Maxwell, J. C., and Charache, S., in *Biochemical and Clinical Aspects of Hemoglobin Abnormalities* (W.S. Caughey, ed.), p. 29, Academic Press, New York, 1978.
Caughey, W. S., Alben, J. O., McCoy, S., Boyer, S. H., Charache, S., and Hathaway, P., *Biochemistry* **8,** 59 (1969).
Caughey, W. S., Bayne, R. A., and McCoy, S., *J. Chem. Soc. (D),* 950 (1970).
Caughey, W. S., Alben, J. O., Fujimoto, W. Y., and York, J. L., *J. Org. Chem.* **31,** 2631 (1966).
Caughey, W. S., Shimada, H., Choc, M. G., and Tucker, M. P., *Proc. Natl. Acad. Sci. USA* **78,** 2903 (1981).
Caughey, W. S., Shimada, H., Tucker, M. P., and Yoshikawa, S., Abstracts of the 26th Annual Meeting, *Biophys. J.* **37,** 371a, No. 2 Part 2, February (1982).
Chabay, I., Klauminzer, G., and Hudson, B., *Appl. Phys. Lett.* **28,** 27 (1976).
Champion, P. M., Münck, E., Debrunner, P. G., Hollenberg, P. F., and Hager, L. P., *Biochemistry* **12,** 426 (1973).
Champion, P. M., Chiang, R., Münck, E., Debrunner, P. and Hager, L. P., *Biochemistry* **14,** 4159 (1975).
Champion, P. M., Remba, R. D., Chiang, R., Fitchen, D. B., and Hager, L. P., *Biochim. Biophys. Acta* **446,** 486 (1976).
Champion, P. M., Gunsalus, I. C., and Wagner, G. C., *J. Am. Chem. Soc.* **100,** 3743 (1978).
Choc, M. G., Webster, D. A., and Caughey, W. S., *J. Biol. Chem.* **257,** 865 (1982).
Chottard, G. and Mansuy, D., *Biochem. Biophys. Res. Commun.* **77,** 1333 (1977).
Cotton, T. M., and Van Duyne, R. P., *Biochem. Biophys. Res. Commun.* **82,** 424 (1978).
Dallinger, R. F., Nestor, J. R., and Spiro, T. G., *J. Am. Chem. Soc.* **100,** 6251 (1978).
Desbois, A., Lutz, M., and Banerjee, R., *Biochemistry* **18,** 1510 (1979).
Dickerson, R. E., and Geis, I., *The Structure and Action of Proteins,* W. A. Benjamin, Inc., Menlo Park, California, 1969.

Duff, L. L., Appelman, E. H., Shriver, D. F., and Klotz, I. M., *Biochem. Biophys. Res. Commun.* **90,** 1098 (1979).

Dunn, J. B. R., Shriver, D. F., and Klotz, I. M., *Proc. Natl. Acad. Sci. USA* **70,** 2582 (1973).

Erecinska, M., Wilson, D. F., and Miyata, Y., *Arch. Biochem. Biophys.* **177,** 133 (1976).

Fermi, G., *J. Mol. Biol.* **97,** 237 (1975).

Fermi, G., and Perutz, M. F., *J. Mol. Biol. 114, 421 (1977).*

Finch, C. A., *New Eng. J. Med.* **239,** 470 (1948).

Frick, P. G., Hitzig, W. H., and Betke, K., *Blood* **20,** 261 (1962).

Fuchsman, W. H., and Appleby, C. A., *Biochemistry* **18,** 1309 (1979).

Geraci, G., Parkhurst, L. J., and Gibson, Q. H., *J. Biol. Chem.* **244,** 4664 (1969).

Gunsalus, I. C., and Wagner, G. C., in *Methods in Enzymology,* Vol. LII, Part C, page 166, (S. Fleischer and L. Packer, eds.) Academic Press, New York, 1978.

Gunsalus, I. C., Tyson, C. A., Tsai, R., and Lipscomb, J., in *Oxidases and Related Redox Systems* (T. King, H. Mason, and M. Morrison, eds.), p. 583 University Park Press, Baltimore, 1973.

Herzberg, G., *Molecular Spectra and Molecular Structure. II. Infrared and Raman Spectra of Polyatomic Molecules,* Van Nostrand, Princeton, New Jersey, 1945.

Hoard, J. L., *Science* **175,** 1295 (1971).

Hoffman, B. M., and Petering, D. H., *Proc. Natl. Acad. Sci. USA* **67,** 637 (1970).

Horio, T., and Kamen, M. D., *Biochim. Biophys. Acta* **48,** 266 (1961).

Huber, R., Epp, O., and Formanek, H., *J. Mol. Biol.* **52,** 349 (1970).

Ikeda-Saito, M., Kitagawa, T., Iizuka, T., and Kyogoku, Y., *FEBS Lett.* **50,** 233 (1975).

Jaffé, E. R., in *Cecil-Loeb Textbook of Medicine,* 13th ed. p. 1699, (P. B. Beeson and W. McDermott, eds.), W. B. Saunders Co., Philadelphia, 1971.

Jeanmaire, D. L., and Van Duyne, R. P., *J. Am. Chem. Soc.* **98,** 4029 (1976a); **98,** 4034 (1976b).

Jeanmaire, D. L., and Van Duyne, R. P., *J. Electroanal. Chem.* **66,** 235 (1975).

Jeanmaire, D. L., Suchanski, M. R., and Van Duyne, R. P., *J. Am. Chem. Soc.* **97,** 1699 (1975).

Jöbsis, F. F., *Science* **198,** 1264 (1977).

Johnson, L. F., and Tate, M. E., *Can. J. Chem.* **47,** 63 (1969).

Katagiri, M., Ganguli, B. N., and Gunsalus, I. C., *J. Biol. Chem.* **243,** 3543 (1968).

Keilin, D., and Hartree, E. F., *Biochem. J.* **49,** 88 (1951).

Kihara, H., Hon-Nami, K., and Kitagawa, T., *Biochim. Biophys. Acta* **532,** 337 (1978).

Kitagawa, T., and Iizuka, T., *Seibutsu Butsuri (Biophysics)* **14,** 272 (1974).

Kitagawa, T., Ozaki, Y., Teraoka, J., Kyogoku, Y., and Yamanaka, T., *Biochim. Biophys. Acta* **494,** 100 (1977).

Kitagawa, T., Ozaki, Y., Kyogoku, Y., and Horio, T., *Biochim. Biophys. Acta* **495,** 1 (1977).

Kitagawa, T., Kyogoku, Y., Iizuka, T., Ikeda-Saito, M., and Yamanaka, T., *J. Biochem. (Tokyo)* **78,** 719 (1975).

Ladner, R. C., Heidner, E. J., and Perutz, M. F., *J. Mol. Biol.* **114,** 385 (1977).

Lanir, A., and Schejter, A., *Biochem. Biophys. Res. Commun.* **62,** 199 (1975).

Leigh, J. S., Jr., Wilson, D. F., Owen, C. S., and King, T. E., *Arch. Biochem. Biophys.* **160,** 476 (1974).

Loehr, T. M., and Loehr, J. S., *Biochem. Biophys. Res. Commun.* **55,** 218 (1973).

Lutz, M., *Biochim. Biophys. Acta* **460,** 408 (1977).

Lutz, M., *J. Raman Spectrosc.* **2,** 497 (1974).

Lutz, M., in *Lasers in Physical Chemistry and Biophysics* (J. Joussot-Dubien, ed.), pp. 451–463, Elsevier, Amsterdam, 1975.

Lutz, M., in *Proceedings of the Seventh International Conference on Raman Spectroscopy* (W. F. Murphy, ed.), pp. 520–523, North Holland, Amsterdam, 1980.

Lutz, M., in *Photosynthesis III. Structure and Molecular Organization of the Photosynthetic Apparatus* (G. Akoyunoglou, ed.), pp. 461–476, Balaban Internatl. Sci. Services, Philadelphia, 1981.

Lutz, M., and Breton, J., *Biochem. Biophys. Res. Commun.* **53,** 413 (1973).

Lutz, M., Kléo, J., and Reiss-Husson, F., *Biochem. Biophys. Res. Commun.* **69,** 711 (1976).

Lutz, M., Kléo, J., Gilet, R., Henry, M., Plus, R., and Leicknam, J. P., in *Proc. 2nd Int. Conf. Stable Isotopes* (E. R. Klein and P. D. Klein, eds.), p. 462, Oak Brook, Illinois, 1975. U.S. Department of Commerce, Springfield, Virginia.

Lutz, M., Brown, J. S., and Rémy, R., in *Chlorophyll Organization and Energy Transfer in Photosynthesis* (Ciba Foundation Symp. 61), (G. Wolstenholme and D. W. Fitzsimons, eds.), pp. 105–125, Excerpta Medica, Amsterdam, 1979.

Lutz, M., Hoff, A. J., and Brehamet, L., *Biochim. Biophys. Acta* **679,** 331 (1982).

Lyons, K. B., Friedman, J. M., and Fleury, P. A., *Nature* **275,** 565 (1978).

Mason, S. F., *J. Chem. Soc.* **1958,** 976 (1958).

Maxwell, J. C., and Caughey, W. S., *Biochemistry* **15,** 388 (1976).

Maxwell, J. C., Barlow, C. H., Spallholz, J. E., and Caughey, W. S., *Biochem. Biophys. Res. Commun.* **61,** 230 (1974a).

Maxwell, J. C., Volpe, J. A., Barlow, C. H., and Caughey, W. S., *Biochem. Biophys. Res. Commun.* **58,** 166 (1974b).

McCoy, S., and Caughey, W. S., in *Probes of Structure and Function of Macromolecules and Membranes,* Vol. II. *Probes of Enzymes and Hemoproteins* (B. Chance, T. Yonetani, and A. S. Mildvan, eds.) p. 289, Academic Press, New York, 1971.

Miskowski, V., Tang, S.-P. W., Spiro, T. G., Shapiro, E., and Moss, T. H., *Biochemistry* **14,** 1244 (1975).

Moss, T. H., Ehrenberg, A., and Bearden, A. J., *Biochemistry* **8,** 4159 (1969).

Nakamoto, K., *Infrared Spectra of Inorganic and Coordination Compounds,* p. 180, John Wiley and Sons, New York, 1963.

Nestor, J. R., *J. Raman Spectrosc.* **70,** 90 (1978).

Nestor, J., Spiro, T. G., and Klauminzer, G., *Proc. Natl. Acad. Sci. USA* **73,** 3329 (1976).

O'Keeffe, D. H., Ebel, R. E., Peterson, J. A., Maxwell, J. C., and Caughey, W. S., *Biochemistry* **17,** 5845 (1978).

O'Keeffe, D. H., Barlow, C. H., Smythe, G. A., Fuchsman, W. H., Moss, T. H., Lilienthal, H. R., and Caughey, W. S., *Bioinorg. Chem.* **5,** 125 (1975).

Onwubiko, H. A., Hazzard, J. H., Noble, R. W., and Caughey, W. S., *Biochem Biophys. Res. Commun.* **106,** 223 (1982).

O'Toole, M. C., Ph.D. thesis, Arizona State University, 1972, quoted in Maxwell and Caughey (1976).

Ozaki, Y., Kitagawa, T., Kyogoku, Y., Shimada, H., Iizuka, T., and Ishimura, Y., *J. Biochem. (Tokyo)* **80,** 1447 (1976).

Parker, F. S., *Applications of Infrared Spectroscopy in Biochemistry, Biology and Medicine,* Plenum Press, New York, 1971.

Parson, W. W., and Cogdell, R. J., *Biochim. Biophys. Acta* **416,** 105 (1975).

Perutz, M. F., *Brit. Med. Bull.* **32,** 195 (1976).

Perutz, M. F., *Nature* **228,** 726 (1970).

Perutz, M. F., Muirhead, H., Cox, J. M., Goaman, L. C. G., Mathews, F. S., McGandy, E. L., and Webb, L. E., *Nature* **219,** 29 (1968).

Pettigrew, G. W., Leaver, J. L., Meyer, T. E., and Ryle, A. P., *Biochem. J.* **147,** 291 (1975).

Pinchas, S., and Laulicht, I., *Infrared Spectra of Labeled Compounds,* p. 297, Academic Press, New York, 1971.

Rakshit, G., and Spiro, T. G., *Biochemistry* **13,** 5317 (1974).

Ramsey, D. A., *J. Am. Chem. Soc.* **74,** 72 (1952).

Rapoport, S., and Guest, G. M., *J. Biol. Chem.* **138,** 269 (1941).

Reed, D. W., and Ke, B., *J. Biol. Chem.* **248,** 3041 (1973).

Remba, R. D., Champion, P. M., Fitchen, D. B., Chiang, R., and Hager, L. P., *Biochemistry* **18,** 2280 (1979).

Rieske, J. S., Alben, J. O., and Liao, H. T., in *Electron Transfer Chains and Oxidative Phosphorylation,* p. 119 (E. Quagliariello, S. Papa, F. Palmieri, E. C. Slater, and N. Siliprandi, eds.) North-Holland Publishing Co., Amsterdam, 1976.

Rieske, J. S., Zaugg, W. S., and Hansen, R. E., *J. Biol. Chem.* **239,** 3023 (1964); Baum, H., Silman, H. I., Rieske, J. S., and Lipton, S. H., *J. Biol. Chem.* **242,** 4876 (1967).

Rossi Fanelli, A., and Antonini, E., *Arch. Biochem. Biophys.* **80,** 299 (1959).

Salmeen, I., Rimai, L., Gill, D., Yamamoto, T., Palmer, G., Hartzell, C. R., and Beinert, H., *Biochem. Biophys. Res. Commun.* **52,** 1100 (1973).

Salmeen, I., Rimai, L., and Babcock, G., *Biochemistry* **17,** 800 (1978).

Satterlee, J. D., Teintze, M., and Richards, J. H., *Biochemistry* **17,** 1456 (1978).
Scholler, D. M., and Hoffman, B. M., *J. Am. Chem. Soc.* **101,** 1655 (1979).
Scholler, D. M., and Hoffman, B. M., in *Porphyrin Chemistry* (F. R. Longo, ed.), Ann Arbor Sciences Publishers, Michigan, 1978.
Siiman, O., Young, N. M., and Cary, P. R., *J. Am. Chem. Soc.* **98,** 744 (1976).
Smith, K. M., ed., *Porphyrins and Metalloporphyrins,* Elsevier Scientific Publ. Co., New York, 1975.
Spiro, T. G., *Proc. R. Soc. London Ser. A* **345,** 89 (1975a).
Spiro, T. G., *Biochim. Biophys. Acta* **416,** 169 (1975b).
Spiro, T. G., and Burke, J. M., *J. Am. Chem. Soc.* **98,** 5482 (1976).
Spiro, T. G., and Loehr, T. M., in *Advances in Infrared and Raman Spectroscopy* (R. J. H. Clark and R. E. Hester, eds.), Vol. 1, pp. 98–142, Heyden, London, 1975.
Spiro, T. G., and Strekas, T. C., *J. Am. Chem. Soc.* **96,** 338 (1974).
Spiro, T. G., and Strekas, T. C., *Proc. Natl. Acad. Sci. USA* **69,** 2622 (1972).
Spiro, T. G., Stong, J. D., and Stein, P., *J. Am. Chem. Soc.* **101,** 2648 (1979).
Strekas, T. C., and Spiro, T. G., *Biochim. Biophys. Acta* **263,** 830 (1972a).
Strekas, T. C., and Spiro, T. G., *Biochim. Biophys. Acta* **278,** 188 (1972b).
Strekas, T. C., and Spiro, T. G., *Biochim. Biophys. Acta* **351,** 237 (1974).
Stryer, L., Kendrew, J. C., and Watson, H. C., *J. Mol. Biol.* **8,** 96 (1964).
Suchanski, M. R., Ph.D. thesis (1977) Northwestern University, Evanston, Illinois, quoted in Cotton and Van Duyne (1978).
Sugita, Y., and Yoneyama, Y., *J. Biol. Chem.* **246,** 389 (1971).
Takano, T., *J. Mol. Biol.* **110,** 537 (1977).
Tan, A. L., and Noble, R. W., *J. Biol. Chem.* **248,** 7412 (1973).
Tan, A. L., Noble, R. W., and Gibson, Q. H., *J. Biol. Chem.* **248,** 2880 (1973).
Tucker, P. W., Phillips, S. E. V., Perutz, M. F., Houtchens, R. A., and Caughey, W. S., *Proc. Natl. Acad. Sci. USA* **75,** 1076 (1978).
Van Duyne, R. P., *J. Phys. (Paris)* **38,** quoted in Cotton and Van Duyne (1978).
Volpe, J. A., O'Toole, M. C., and Caughey, W. S., *Biochem. Biophys. Res. Commun.* **62,** 48 (1975).
Wallace, W. J., Volpe, J. A., Maxwell, J. C., Caughey, W. S., and Charache, S., *Biochem. Biophys. Res. Commun.* **68,** 1379 (1976).
Wilson, D. F., and Leigh, J. S., Jr., *Arch. Biochem. Biophys.* **150,** 154 (1972).
Wilson, D. F., Erecinska, M., and Owen, C. S., *Arch. Biochem. Biophys.* **175,** 160 (1976).
Wilson, E. B., Decius, J. C., and Cross, P. C., *Molecular Vibrations. The Theory of Infrared and Raman Vibrational Spectra,* p. 182, McGraw-Hill, New York, 1955.
Wintrobe, M. M., *Clinical Hematology,* 6th ed., p. 195, Lea and Febiger, Philadelphia, 1967.
Wollmer, A., Steffens, G., and Buse, G., *Eur. J. Biochem.* **72,** 207 (1977).
Woodruff, W. H., and Farquharson, S., *Science* **201,** 831 (1978).
Woodruff, W. H., Adams, D. H., Spiro, T. G., and Yonetani, T., *J. Am. Chem. Soc.* **97,** 1695 (1975).
Yamamoto, T., Palmer, G., and Crespi, H., *Biochim. Biophys. Acta* **439,** 232 (1976).
Yamamoto, T., Palmer, G., Gill, D., Salmeen, I. T., and Rimai, L., *J. Biol. Chem.* **248,** 5211 (1973).
Yonetani, T., Yamamoto, H., and Woodrow, G. V., III, *J. Biol. Chem.* **249,** 682 (1974).
Yoshikawa, S., and Caughey, W. S., *J. Biol. Chem.* **257,** 412 (1982).
Yoshikawa, S., Choc, M. G., O'Toole, M. C., and Caughey, W. S., *J. Biol. Chem.* **252,** 5498 (1977).
Yu, N.-T., Tsubaki, M., and Srivastav, R. B., Abstracts of the 26th Annual Meeting, *Biophys. J.* **37,** No. 2 Part 2, 173a, February (1982).
Yu, N.-T. Kerr, E. A., Bartnicki, D. E., Mizukami, H., and Romero-Herrera, A. E., Abstracts of the 26th Annual Meeting, *Biophys. J.* **37,** No. 2 Part 2, 174a, February (1982a).
Yuratich, M. A., and Hanna, D. C., *Mol. Phys.* **33,** 671 (1977).

Additional References

Cary, P. R., Resonance Raman Spectroscopy in Biochemistry and Biology, *Quart. Rev. Biophys.* **11,** 309–370 (1978).

Dolphin, D., ed., The Porphyrins, Vols. I–VII, Academic Press, New York, 1978 and 1979. These volumes contain much detailed information on structure and synthesis, physical chemistry, and biochemistry. The work is divided as follows: structure and synthesis, Vol. I, Part A and Vol. II, Part B; physical chemistry, Vol. III, Part A, with a section on IR by J. O. Alben, and one on resonance Raman by R. H. Felton and N.-T. Yu, Vol. IV, Part B and Vol. V, Part C; biochemistry, Vol. VI, Part A and Vol. VII, Part B.

Smith, K. M., ed., *Porphyrins and Metalloporphyrins,* Elsevier Scientific Publ. Co., New York, 1975. This is based on the original volume by J. E. Falk. Section C of Chapter 3 in Smith's book discusses the coordination chemistry of metalloporphyrins. There are chapters on vibrational spectroscopy, NMR, mass spectrometry, Mössbauer spectroscopy, EPR [including copper, silver, vanadyl, and molybdyl porphyrins, Co(II), Fe(II), Fe(III) and Mn(II) porphyrins], and stereochemistry, among others. The book also contains a chapter on the photochemistry of porphyrins in membranes and in photosynthesis.

Spiro, T. G., Resonance Raman Spectra of Hemoproteins, in *Methods in Enzymology,* Vol. LIV, *Biomembranes,* Part E (S. Fleischer and L. Packer, eds.), pp. 233–249, Academic Press, New York, 1978.

Chapter 7

RESONANCE RAMAN STUDIES OF COPPER PROTEINS AND OTHER METALLOPROTEINS

Copper Proteins

Spiro and Loehr (1975) have reviewed resonance Raman spectroscopy of heme proteins and other biological systems, e.g., copper proteins, among which are hemocyanin and the "blue" copper proteins—stellacyanin, laccase, ceruloplasmin, azurin, and plastocyanin; and non-heme iron-proteins, e.g., adrenodoxin, which is an iron–sulfur protein, hemerythrin, and transferrin.

Figure 7.1 shows the resonance Raman spectra of some "blue" copper proteins. A model of the "blue" copper site was offered by Miskowski *et al.* (1975): it has an approximately trigonal bipyramidal structure with two nitrogen atoms (not imidazole) and one sulfur atom (cysteine) in the equatorial plane.

Siiman *et al.* (1974) have used the resonance Raman spectra of model copper–peptide complexes with known structure to make vibrational assignments (Table 7.1) for atoms at the type-1 (or "blue") copper site of copper proteins. (Type-1 or blue copper absorbs intensely near 600 nm and has extinction coefficients from 3,000 to 11,000 M^{-1} cm^{-1}.)

Siiman *et al.* (1976) have measured the resonance Raman spectra of *Japanese lacquer-tree* stellacyanin and laccase, spinach plastocyanin, ascorbate oxidase of green zucchini squash, and human ceruloplasmin in the region 150 to 1700 cm^{-1}, by using laser excitation into their ca. 600-nm electronic absorption bands (region of maximum absorbance in the visible region).

The spectra (Fig. 7.2) of ceruloplasmin, stellacyanin, laccase, plastocyanin, and ascorbate oxidase display at least two intense bands in the region from 330 to 470 cm^{-1}. These workers (1974) previously assigned these bands to preferably Cu—N (of a peptide bond or an amide side chain) or possible Cu—O stretching modes. Evidence in favor of the first choice was obtained from the location and intensities of $\nu_{Cu—N}$, peptide in Cu(II)-peptide complexes and from the finding of three weak bands, in the proteins, in the region 600 to 1700 cm^{-1}. These were assigned to vibrations of the amide group. Ascorbate oxidase spectra and those of plastocyanin also display these characteristic weak bands. Also, the medium band ca. 260 cm^{-1} was assigned to $\nu_{Cu—S}$ of cysteine or $\nu_{Cu—N}$ of

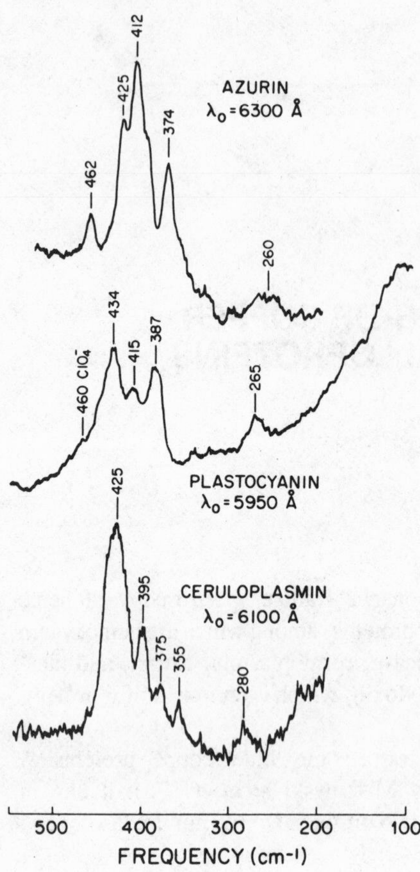

Fig. 7.1. Resonance Raman spectra of "blue" copper proteins, obtained with tunable dye (Rhodamine 6G) excitation at the indicated wavelengths; spectrometer parameters: slit width, 10 cm^{-1}; scan rate, 50 cm^{-1}/min; sensitivity, 10^{-9} A; laser power, ~60 mW. (Miskowski *et al.*, 1975.) Reprinted with the permission of the American Chemical Society. Copyright 1975.

Table 7.1. Vibrational Assignments in "Blue" Copper Proteins[a]

cm^{-1}	Assignment
350–400 (strong)	Cu—N (peptide N) and/or Cu—O (peptide CO oxygen)
260 (weak)	Cu—S (cys) and/or Cu—N (imidazole)
330,450 (weak)	Bending modes (CNC or CCN)
500,750 (weak)	Bending modes (C=O)
1240 (weak)	C—N (peptide)
1650 (weak)	C=O (peptide CO with bond to Cu)

[a] Siiman *et al.* (1974).

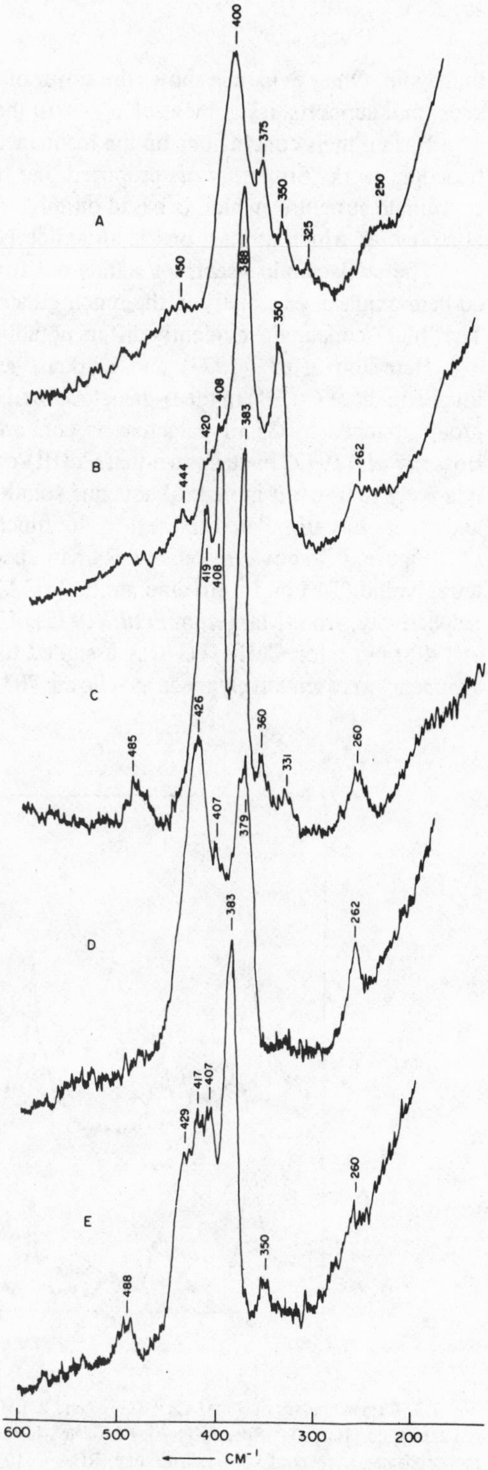

Fig. 7.2. Resonance Raman spectra of ceruloplasmin, 11.6 mg/ml, in 0.05 *M* acetate buffer, pH 5.5 (A); stellacyanin, 8.4 mg/ml, in 0.05 *M* phosphate buffer, pH 5.5 (B); laccase, 42.5 mg/ml, in 0.05 *M* phosphate buffer, pH 5.5 (C); plastocyanin, 1.5 mg/ml, in 0.05 *M* phosphate buffer, pH 6.9 (D); ascorbate oxidase, 42.4 mg/ml, in 0.05 *M* phosphate buffer, pH 7.0 (E). Experimental condition: time constant, 5 sec; scan rate, 30 cm^{-1}/min; excitation, 647.1 nm Kr^+. (Siiman *et al.*, 1976.) Reprinted with the permission of the American Chemical Society. Copyright 1976.

	Power (mW)	Slit width (cm^{-1})	Sensitivity (cps)
(A)	80	10.0	2000
(B)	80	7.0	1000
(C)	60	7.0	2000
(D)	40	7.0	2000
(E)	30	7.0	1000

imidazole. Other evidence shows the sulfur of cysteine to be a ligand in the "blue" copper sites, and supports assignment of ν_{Cu-S} to the 260-cm^{-1} band.

Basing their conclusions on the resonance Raman spectra, chemical and other spectroscopic work, Siiman *et al.* proposed that the "blue" copper site has a distorted four-coordinate structure, which is based on the binding of copper to one cysteine S and three N atoms, of which at least one is an amide N.

The stellacyanin spectrum stands out from those of plastocyanin, laccase, and ascorbate oxidase, especially in the much greater intensity of the ν_{Cu-N} band at 350 cm^{-1}. The "blue" copper site evidently differs notably from those in the other substances studied.

Hamilton *et al.* (1973) and Dyrkacz *et al.* (1976) have given evidence for the involvement of Cu(III) in the D-galactose oxidase catalyzed reaction, whereby the CH_2OH group attached to C5 of galactose is converted to CHO. Margerum *et al.* (1975) and Bossu *et al.* (1977) have shown that Cu(III)-complexes of deprotonated oligopeptides are relatively long-lived in neutral aqueous solution. Such experimentation by these workers and others has stimulated interest in the function of Cu(III) in biochemical processes.

Figure 7.3 shows resonance Raman spectra recorded by Kincaid *et al.* (1978) for tetraglycine (TG) and triglycine amide (TGA) complexes of Cu(III) (structures I and II, respectively, from Margerum *et al.*, 1975). The strong band at 417 cm^{-1} for Cu(III)TGA and 420 cm^{-1} for Cu(III)TG was assigned to amide stretching [Cu(III)—N]. Other assignments were tentatively made as follows: 703 cm^{-1}, Cu(III)TG and 711 cm^{-1}, Cu(III)TGA,

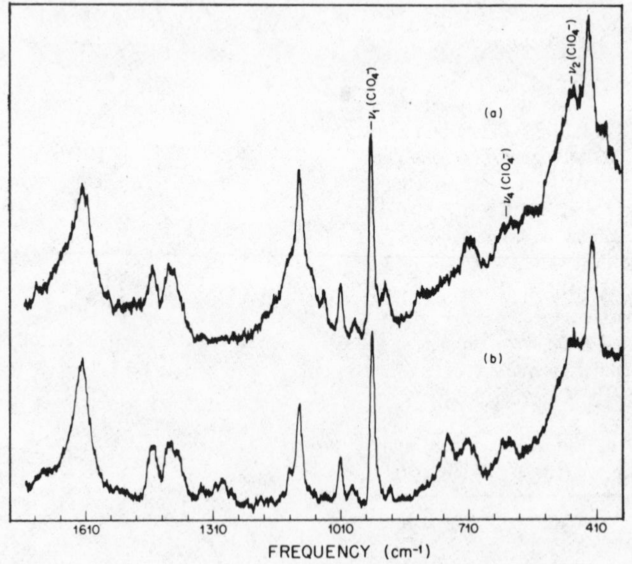

Fig. 7.3. Raman spectra of 3 m*M* Cu^{III} TG (a) and 2 m*M* Cu^{III}TGA (b) in 0.2 *M* $NaClO_4$ recorded with 363.8-nm excitation [100 mW, Spectra Physics 170 Ar^+ laser with UV optics and grating filter; Spex 1401 double monochromator, spectral slit width 8 cm^{-1}; RCA C31034A photomultiplier with photon counting detection, sensitivity (counts/sec) 250 (a) and 330 (b); rise time 10 sec; scanning speed 12 cm^{-1}/min]. The broad feature between 360 and 540 cm^{-1} is due to glass scattering. (Kincaid *et al.*, 1978.) Reprinted with the permission of the American Chemical Society. Copyright 1978.

$\delta_{C=O}$; 759 cm^{-1}, Cu(III)TGA, π(N—H) or out-of-plane primary amide deformation; 1286 cm^{-1}, Cu(III)TGA, δ(N—H) of primary amide; 1406 cm^{-1}, Cu(III)TG, and 1392, 1410 cm^{-1} doublet, Cu(III)TGA, ν(C—N); 1449 cm^{-1}, Cu(III)TG and 1449 cm^{-1}, Cu(III)TGA, $\delta(CH_2)$; and 1614 cm^{-1}, Cu(III)TG and 1616 cm^{-1}, Cu(III)TGA, $\nu_{C=O}$ (amide I). Thus, the resonance Raman technique furnishes a sensitive probe for the copper (III)-peptide linkage.

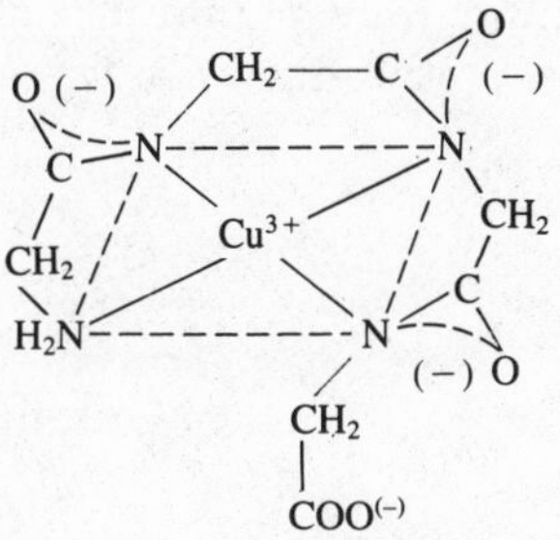

I

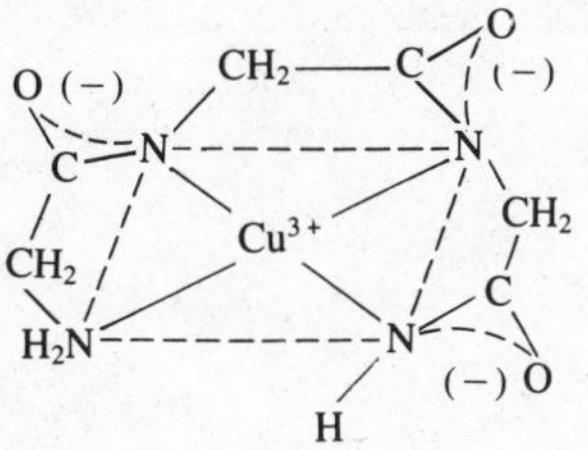

II

The structure of the "blue" or "type 1" active site in copper-containing proteins has been studied by many physical and chemical techniques (discussed in Thompson *et al.*, 1979). These studies have resulted in two proposed basic structural models, A and B, for the blue, type 1 active site (McLendon and Martell, 1977; Siiman *et al.*, 1976; Miskowski *et al.*, 1975; Tosi *et al.*, 1975; Gray, 1971):

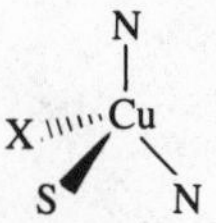

A

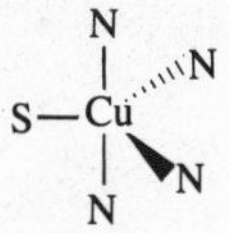

B

Thompson *et al.* (1979) have studied the synthesis, chemistry, spectroscopy, and structures of Cu(I) and Cu(II) coordination complexes with geometry as in A, where X is a heterocyclic nitrogenous base, the two coordinated nitrogens are in histidine imidazoles, and S is present in a cysteinate mercaptide. They synthesized Cu(I) and Cu(II) complexes of structure C, in which B is boron and SR is *p*-nitrobenzenethiolate (or *O*-

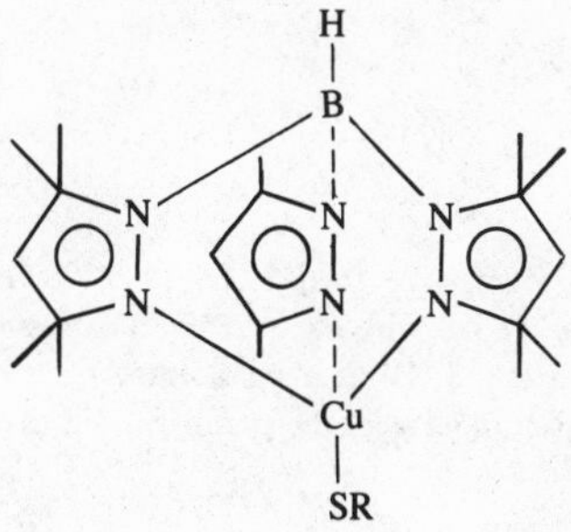

C

ethylcysteinate). These complexes were the only well-defined, mononuclear oxidation-reduction pairs (up to 1978) that were similar in structure to any of the proposed type-1 active sites (Thompson *et al.*, 1979) and that also display many of the unusual spectra of the blue copper proteins (e.g., intense absorption in the 600-nm region, ϵ 3500–5000 M^{-1} cm^{-1} for the oxidized cupric forms; and small copper hyperfine coupling (Ingram, 1969) in the EPR spectra, $A_{\|} = 3.3\text{–}9.0 \times 10^{-3}$ cm^{-1}). The results of Thompson *et al.* with these complexes (1979) and with similar cobalt(II)-substituted complexes (1979a) provide further understanding of the type-1 core with respect to structure and function.

Resonance Raman studies of the blue copper proteins show that they display char-

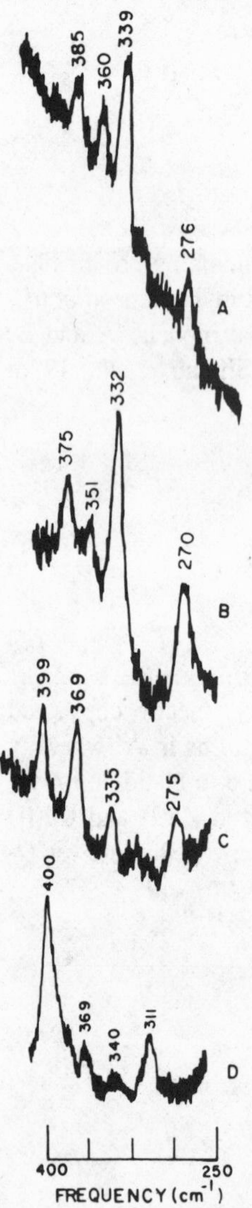

Fig. 7.4. Raman spectra of **2a** (A) at −80°C in THF, with Kr^+ (6471 Å) excitation; **1a** (B) as a solid with Ar^+ (5145 Å) excitation; **1b** (C) as a solid with Ar^+ (5145 Å) excitation; **3** (D) as a solid with Ar^+ (5145 Å) excitation. (Thompson *et al.*, 1979.) Reprinted with the permission of the American Chemical Society. Copyright 1979.

acteristic enhanced vibrational scattering (Spiro and Stein, 1977; Johnson and Peticolas, 1976) when they are laser-excited at their 600-nm electronic transition (Siiman *et al.*, 1974 and 1976; Miskowski *et al.*, 1975; Tosi *et al.*, 1975; Ferris *et al.*, 1978). The oxidized species of the proteins usually display several transitions in the region from 350 to 450 cm^{-1} and a lone one at ca. 270 cm^{-1}. The 350–450-cm^{-1} bands have been attributed mainly to Cu—N or Cu—O stretching and the 270-cm^{-1} band to Cu—S stretching. Thompson *et al.* (1979) assessed the former group as reasonable but not absolute, and the Cu—S assignment as possibly ambiguous.

Figure 7.4 shows Raman spectra for the following compounds:

(i) $K[Cu(HB(3,5\text{-}Me_2\text{–}1\text{-pyrazolyl})_3(SC_6H_4NO_2)] \cdot 2C_3H_6O$, **1a;**
(ii) $K[Cu(HB(3,5\text{-}Me_2\text{–}1\text{-pyrazolyl})_3)(SCH_2CH(NH_2)(COOC_2H_5)]$, **1b;**
(iii) $Cu(HB(3,5\text{-}Me_2\text{–pyrazolyl})_3)(SC_6H_4NO_2)$, **2a** in tetrahydrofuran; and
(vi) $Cu(HB(3,5\text{-}Me_2\text{–pyrazolyl})_3)(OC_6H_4NO_2)$, **3.**

The bands in the 270 cm^{-1} region were assigned to Cu—S stretching for these compounds. The bands at 332, 335, 339, 311, 351, 369, 360, 340, 375, 339, and 385 cm^{-1} were assigned to Cu—N stretching. The band at 400 cm^{-1} was attributed to Cu—O stretching. Thompson *et al.* used these results plus UV–visible and EPR data to show that the Cu(II) complexes have a very similar structure to that of the Cu(I) complex they had characterized by X-ray diffraction methods. Blue copper proteins have spectral properties similar to those of the $Cu(II)N_3(SR)$ complexes studied by Thompson *et al.* (1979). The absorption at 600 nm displayed by all the blue copper proteins was shown to originate from mercaptide (cysteinyl) sulfur-to-copper charge transfer. A rather unusual coordination geometry for Cu(II) appears to be required—tetrahedral or distorted tetrahedral instead of square planar or octahedral.

Hemerythrins and Hemocyanins

Hemerythrin

Hemerythrin contains eight subunits of molecular weight of about 7000. Two iron atoms are present at the O_2-binding site in each subunit, and tyrosine and histidine are potential iron ligands. Dunn *et al.* (1973) have shown that oxygen is bound to hemerythrin as peroxide. When they excited the molecule within the 500-nm oxygen → iron charge-transfer band, they produced resonance-enhanced Raman bands at 844 and 500 cm^{-1}. Substitution of $^{18}O_2$ for $^{16}O_2$ caused these bands to shift to 798 and 478 cm^{-1}, respectively, demonstrating that both vibrational frequencies involve the bound O_2 molecule. The band at 798 cm^{-1} was assigned to the O—O stretching of peroxide. The 500-cm^{-1} band of hemerythrin was assigned to an iron–peroxide stretching mode.

Oxyhemerythrins

The resonance Raman spectra of oxyhemerythrin from four species of marine worms—*Phascolopsis gouldii, Phascolosoma agassizii, Themiste dyscritum,* and *Themiste pyroides*—have been recorded by Dunn *et al.* (1977), and these spectra all consist of two major peaks at 503 and 844 cm^{-1}. These peaks had been identified in oxyhemerythrin from *P. gouldii* as resulting from Fe—O and O—O stretching modes, respectively, for a bound peroxide ion (Dunn *et al.*, 1973). Figure 7.5 presents a resonance Raman

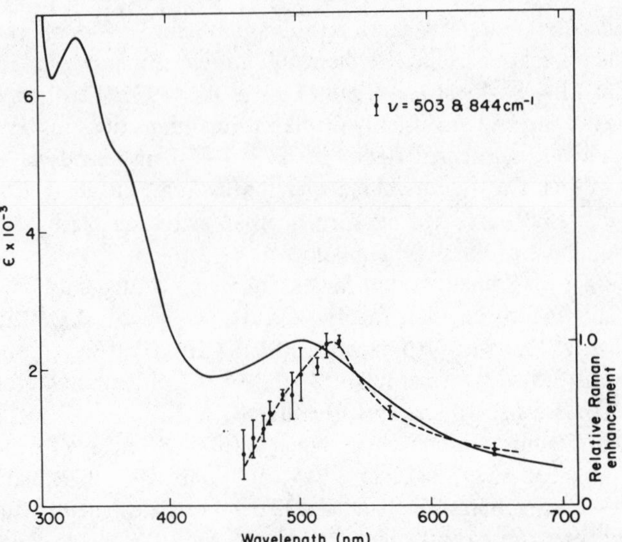

Fig. 7.5. Absorption spectrum (——) and resonance Raman enhancement profile (- - -) of vibrational modes associated with dioxygen bound at the active site in oxyhemerythrin from *P. gouldii*. Each error bar on the enhancement profile indicates the standard deviation for two sets of measurements of both the 503- and 844-cm^{-1} resonance Raman peaks. (Dunn *et al.*, 1977.) Reprinted with the permission of the American Chemical Society. Copyright 1977.

enhancement profile of these modes in oxyhemerythrin from *P. gouldii*. Maximum enhancement of the 503- and 844-cm^{-1} peaks in this species and in *T. dyscritum* occurred at ca. 525 nm excitation.

Metazidohemerythrin

The resonance Raman spectra of metazidohemerythrin, an inactive oxidized form of a model compound, consist of the following peaks for all four species: 295 (weak), 375 (strong), 508 (strong), 750–850 (broad and poorly defined), and 2049 cm^{-1} (moderate). The positions of the 375- and 2049-cm^{-1} bands are known to be dependent on the nitrogen isotope in the bound azide (Dunn *et al.*, 1975). Dunn *et al.* (1977) assigned these peaks to ν_sFe—N_3 (375 cm^{-1}) and ν_{as}N—N—N (2049 cm^{-1}), which are associated with the bound azide. Those vibrations and the 295-cm^{-1} band display maximum enhancement of their Raman intensities with excitation of ca. 505 nm. The ~510-cm^{-1} peak possibly originates from an iron–tyrosine mode in resonance with a tyrosine → iron charge-transfer transition (Dunn *et al.*, 1977).

Sulfidomethemerythrin

Freier *et al.* (1979) have used resonance Raman spectroscopy to study the structure of a sulfide complex of methemerythrin. Figure 7.6 presents spectra of oxyhemerythrin

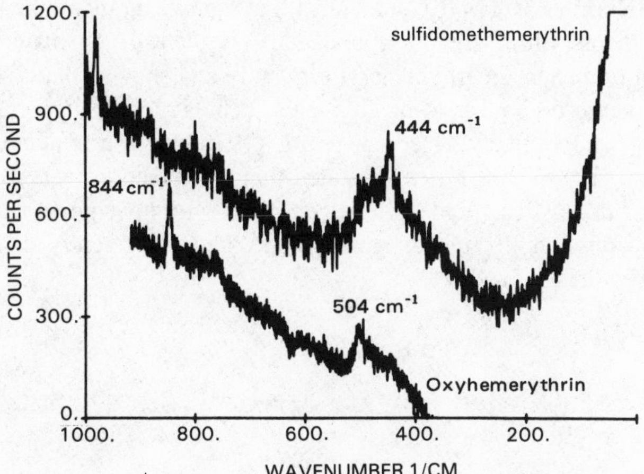

Fig. 7.6. Resonance Raman spectra of 4×10^{-3} *M* sulfidomethemerythrin and the product generated after addition of oxygen to sulfidomethemerythrin. Laser excitation, 514.5 nm, 50 mW; 4-cm^{-1} spectral slit. The 981-cm^{-1} band is due to sulfate added as an internal standard. In both spectra, the very broad band centered at 500 cm^{-1} is background Raman scattering of the Pyrex sample tube. (Freier *et al.*, 1979.) Reprinted with the permission of the American Chemical Society. Copyright 1979.

and sulfidomethemerythrin. The latter shows only one vibration, at 444 cm^{-1}, which the authors assigned to an iron-sulfide stretch. They proposed that the complex has the S^{2-} (per monomer unit) and two ferric iron atoms at each active site, with the S^{2-} attaching to a single iron or acting as a bridge between the two irons as depicted here:

Fe · · · Fe—S^{2-} or Fe—S^{2-}—Fe

The authors felt that it is also possible that sulfidomethemerythrin contains both a μ-sulfido and a μ-oxo bridge,

Fe(O^{2-})(S^{2-})Fe

which could explain the fact that the 444-cm^{-1} band for the Fe—S vibration is unusually high, from 50 to 100 cm^{-1} higher than a series of Fe—S frequencies of iron–sulfur compounds listed by them.

Hemocyanins

Thamann *et al.* (1977) have used resonance Raman spectroscopy to study the reaction between hemocyanin, a copper-containing respiratory protein, and a mixed isotope molecular oxygen. Kurtz *et al.* (1976) had shown that the two atoms in the bound peroxide grouping in oxyhemerythrin are not equivalent. The data of Thamann *et al.* differed significantly from those in the oxyhemerythrin study and showed that the oxygen atoms bound to hemocyanin do appear to be equivalent. The structure supported by their data is that previously proposed by Freedman *et al.* (1976), with a nonplanar, μ-dioxygen bridged geometry (III). Thamann *et al.* stated that resonance Raman data for oxyhemerythrin (Kurtz *et al.*, 1976) could only be consistent with nonequivalent oxygens in the Fe_2O_2 site, and consequently the oxygen is coordinated differently in this protein, as suggested by structures IV and V.

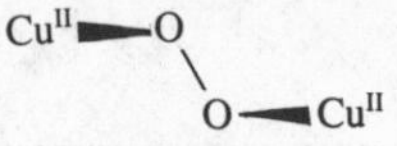

Cu^{II}—O

O—Cu^{II}

III

O

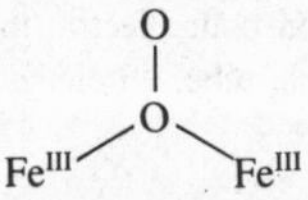

O

Fe^{III} Fe^{III}

IV

O

Fe^{III} ········ Fe^{III}—O

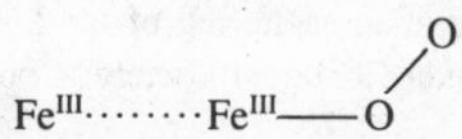

V

Loehr *et al.* (1974) have presented a partial assignment of the resonance Raman spectrum of the hemocyanin of *Cancer magister,* an arthropod. Because data from various investigations seemed to suggest dissimilar oxygen-binding site structures for arthropod and molluscan hemocyanins (Van Holde and Van Bruggen, 1971; Nickerson and Van Holde, 1971; Ke and Schubert, 1972; Bonaventura *et al.*, 1974), Freedman *et al.* (1976), in Loehr's laboratory, extended their study to compare the resonance Raman spectrum of hemocyanin from the mollusc, *Busycon canaliculatum,* to that of the Pacific crab, *Cancer magister*.

Resonance Raman spectroscopy identified bound oxygen as O_2^{2-} and the oxidation states of copper as Cu(II). For *C. magister,* the O_2^{2-} vibration occurred at 744 cm^{-1}; and for *B. canaliculatum,* at 749 cm^{-1}. Excitation profiles for this vibration showed that the absorption bands or circular dichroism bands at ca. 490 and ca. 570 cm^{-1} arise from $O_2^{2-} \rightarrow$ Cu(II) charge transfer. Resonance Raman frequencies for the oxyhemocyanins of the two phyla at 282 and 267 cm^{-1}, respectively, were assigned to Cu—N(imidazole) vibrations of histidine ligands.

This resonance Raman study showed that the mode of oxygen binding, oxidative addition of oxygen, is the same for the two phyla. The closeness of the values, 749 cm^{-1} and 744 cm^{-1} (O_2^{2-} vibration), suggested to Freedman *et al.* that the oxygens are bound in rather similar environments.

The oxyhemocyanin of *Achatina fulica,* Taiwan snails, has been studied by Chen *et al.* (1979). The resonance Raman spectrum of this compound (excitation λ, 514.5 nm) shows the O_2^{2-} vibration at 752 cm^{-1}. For the O_2^{2-} vibration in the hemocyanins of the arthropod *Cancer magister* and the mollusc *Busycon canaliculatum* the bands were located at 744 and 749 cm^{-1}, respectively, by other workers. Chen *et al.* concluded from NMR, other spectrophotometric data, and the position of the O_2^{2-} band, that the hemocyanins of the three organisms have similar structures and properties.

Larrabee *et al.* (1977) have obtained resonance Raman spectra (400–100 cm^{-1}) of the hemocyanins of a mollusc and an arthropod with excitation of 351.1 nm. In this range the spectra showed appreciably different characteristics. For the arthropod *(Limulus polyphemus)* the band at 287 cm^{-1} had the greatest intensity; for the mollusc *(Busycon canaliculatum)* the band at 266 cm^{-1} was greatest. The probable cause of these differences is a difference in the coordination geometry around the Cu^{2+} atoms of the hemocyanins.

Other Metalloproteins

Ovo- and Serum Transferrins, and Lactoferrin

Tomimatsu *et al.* (1973) have recorded the resonance Raman spectra of Fe(III)-ovotransferrin of chicken and Fe(III)-transferrin of human serum in aqueous solution. Figure 7.7 shows spectra of these substances and the corresponding substances in which iron is absent. This was one of the first observations of resonance Raman scattering attributed to amino acid ligand vibrational modes of an iron protein not containing heme. Figure 7.7 shows the spectra of the transferrins to be similar except that the resonance band near 1270 cm^{-1} is shifted to a higher frequency for human Fe(III)-transferrin than is that of Fe(III)-ovotransferrin. Tomimatsu *et al.* stated that the bands seen near 1170, 1270, 1500, and 1600 cm^{-1} may be due to resonance enhancement of *p*-hydroxyphenyl frequencies of tyrosine residues and/or imidazolium frequencies of histidine moieties. Ovotransferrin and serum transferrin have also been studied by other investigators (Carey and Young, 1974; Gaber *et al.*, 1974).

Besides the resonance Raman spectra of Fe(III)-transferrins, Tomimatsu *et al.* (1976) have studied those of copper(II)-, cobalt(III)-, and manganese(III)-transferrins, and that of a model for Cu(II) binding to transferrins. Their data indicate that the resonance Raman bands arise from enhancement of phenolic vibrational modes. Recordings of spectra of the model compound, bis(2,4,6-trichlorophenolato)diimidazolecopper(II)monohydrate, allowed a normal coordinate analysis to be made and used for assigning observed resonance bands at 1562, 1463, 1311, and 1122 cm^{-1} to A_1 vibrational modes of the 2,4,6-trichlorophenolate moiety. These assignments were consistent with those made for Cu(II)-transferrins.

The resonance Raman spectra of the several metal chelates of ovo- and serum transferrins are quite similar (Fig. 7.8) and they indicate that Mn(III), Co(III), Cu(II), and Fe(III) all occupy the same binding sites. Overall, the studies with resonance Raman spectra reflect the ligation of phenolate (i.e., tyrosine) to Fe(III) in transferrins. Tyrosine is also a ligand in the binding of copper to transferrins.

Lactoferrin, a non-heme iron-binding protein is present in high concentration in human milk and it occurs in a variety of other bodily secretions and in intracellular components (Aisen, 1973; Bezkorovainy, 1977; and Feeney and Komatsu, 1966). Ains-

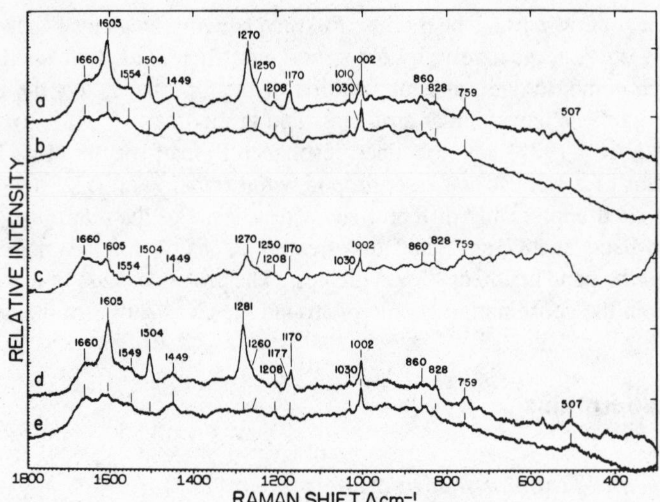

Fig. 7.7. Solution Raman spectra of Fe(III)-ovotransferrin (curves a and c), ovotransferrin (curve b), Fe(III)-serum transferrin (curve d), and serum transferrin (curve e) near 0°C, pH 7. Experimental conditions:

Curve	Protein conc. $\times 10^4 (M)$	Excitation (nm)	Power (mW)	Slit width (cm^{-1})	Scan rate (cm^{-1}/min)	Time constant (sec)
(a)	1.01	488.0	150	4	5	10
(b)	1.27	488.0	250	4	5	10
(c)	1.01	647.1	200	4	5	3.3
(d)	1.23	488.0	200	4	5	10
(e)	1.03	488.0	200	4	5	10

(Tomimatsu *et al.*, 1973.)

cough *et al.* (1980) have given visible absorption spectral data for several metal-complexed lactoferrins, e.g., Fe(III)-, Cr(III)-, Mn (III)-, Co(III)-, and Cu(II)-lactoferrins. Figure 7.9 presents resonance Raman spectra for two of these, the Fe(III)- and Co(III)-lactoferrins. The spectra of these and the others (not shown) have prominent strong resonance-enhanced vibrational modes of metal ion coordinated tyrosinates at ca. 1600, 1500, 1260–1290, and 1170 cm^{-1}. The 1600- and 1500-cm^{-1} frequencies have been assigned principally to C—C stretching, the 1260-cm^{-1} band to C—O stretching, and the 1170-cm^{-1} frequency to C—O bending of the phenolate ligand (Tomimatsu *et al.*, 1976). The use of resonance Raman, fluorescence, and electron paramagnetic resonance spectroscopy confirmed the close similarity between lactoferrin and serum transferrin.

Corrin Ring Vibrations

Several studies of vitamin B_{12} derivatives have observed very similar resonance Raman spectra for cobalamins, the Co(III) vitamin B_{12} derivatives, no matter what the

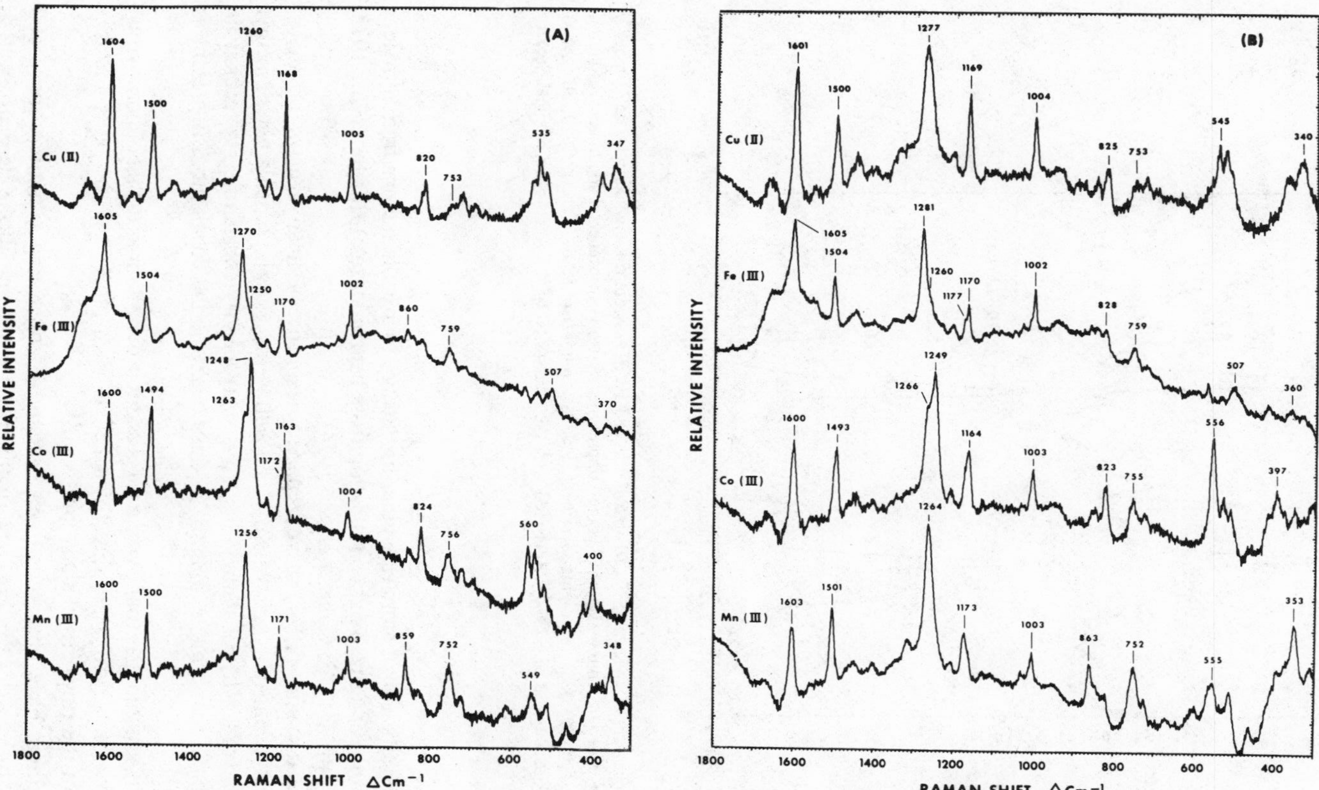

Fig. 7.8. (A) Raman spectra of aqueous solutions of ovotransferrin-metal chelates obtained by using 488.0-nm laser excitation and 4.0-cm^{-1} slit width. Protein concentration and laser power were as follows: Fe(III), 1.01×10^{-4} *M*, 150 mW; Cu(II), 0.91×10^{-4} *M*, 200 mW; Co(III), 0.88×10^{-4} *M*, 100 mW; and Mn(III), 1.16×10^{-4} *M*, 100 mW. (B) Raman spectra of aqueous solutions of human serum transferrin–metal chelates obtained by using 488.0-nm laser excitation and 4.0-cm^{-1} slit width. Protein concentrations and laser power were as follows: Fe(III), 1.23×10^{-4} *M*, 200 mW; Cu(II), 0.86×10^{-4} *M*, 170 mW; Co(III), 0.87×10^{-4} *M*, 230 mW; and Mn(III), 1.11×10^{-4} *M*, 160 mW. (Tomimatsu *et al.*, 1976.) Reprinted with the permission of the American Chemical Society. Copyright 1976.

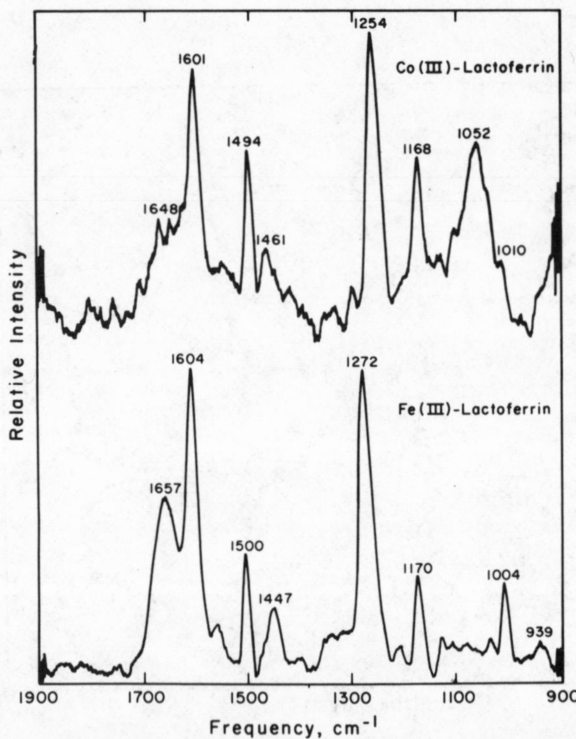

Fig. 7.9. Resonance Raman spectra of Co(III)-lactoferrin and Fe(III)-lactoferrin obtained with 457.9 nm (25 mW at the sample) and 514.5 nm (~100 mW at the sample), respectively. Instrumental conditions: resolution, 8 cm^{-1}; scan rate, 5.0 cm^{-1} sec^{-1}; digitizing increment, 1.0 cm^{-1}; number of scans, for Co(III)-Lf, 50, and for Fe(III)-Lf, 10. (Ainscough *et al.*, 1980.) Reprinted with the permission of the American Chemical Society. Copyright 1980.

sixth ligand is to the cobalt atom (the other ligands are four coordinations from nitrogens of the corrin ring, and one axial coordination from benzimidazole) (Mayer *et al.*, 1973; Wozniak and Spiro, 1973; Galuzzi *et al.*, 1974).

Salama and Spiro (1977) have reported resonance Raman spectra of aquo-, methyl, 5′-deoxyadenosyl-(coenzyme B_{12}), thiosulfate and cysteinyl cobalamin, of methyl cobinamide, and of Co(II) cobalamin (B_{12r}). 5′-Deoxyadenosylcobalamin is the coenzyme of methylmalonyl coenzyme A mutase. In this coenzyme the cobalt atom is present as Co(I).

The corrin structure has four pyrrole rings and it is quite similar to that of porphyrins, except that two of the rings are directly linked to each other, and the other linkages joining three of the rings are methene bridges.

These authors have observed a great variety of corrin ring vibrations above 600 cm^{-1}, but the frequencies are constant (within ± 3 cm^{-1}) for all the derivatives. B_{12r}

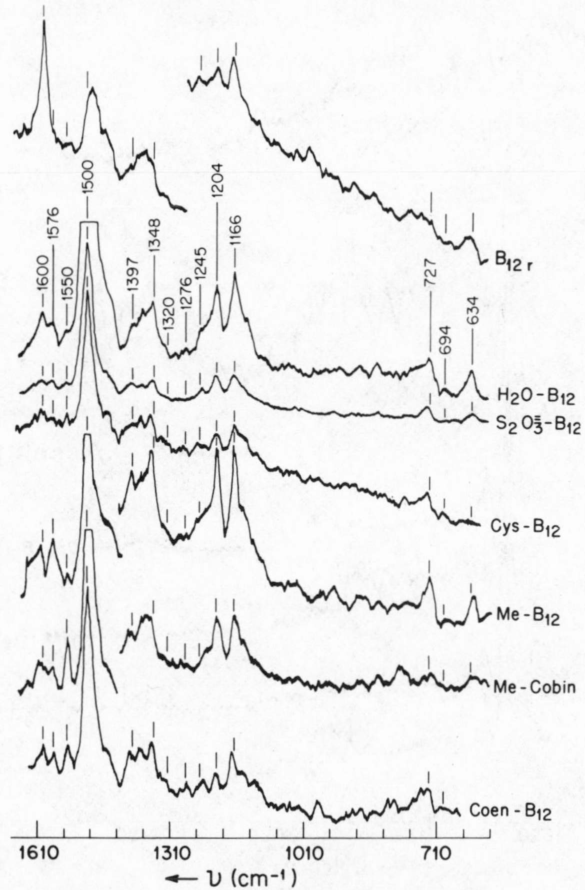

Fig. 7.10. Raman spectra of the indicated B_{12} derivatives with 5145 Å (H_2O-B_{12}, Coen-B_{12}, $S_2O_3^{2-}$-B_{12}, Me-B_{12}) or 4880 Å (Cys-B_{12}, Me-Cobin, B_{12r}) excitation. Conditions used: sensitivity (counts sec^{-1}), spectral slit (cm^{-1}), time constant (sec), scanning speed (cm^{-1} sec^{-1}): B_{12r} (1000, 11, 5, 0.5), H_2O-B_{12} (1000, 10, 5, 0.5); $S_2O_3^{2-}$-B_{12} (2000, 11, 5, 0.5); Cys-B_{12} (500, 9, 3, 1); Me-B_{12} (200, 8, 2, 1); Me-Cobin (500, 7, 2, 1); and Coen-B_{12} (1000, 8, 3, 1). (Salama and Spiro, 1977.)

shows large relative intensity changes in bands at 1500 and 1600 cm^{-1} relative to the Co(III) derivatives. Although methyl cobinamide has an absorption spectrum that looks like that of B_{12r}, the former displays Raman intensity behavior typical of the other Co(III) derivatives.

Figures 7.10 and 7.11 show spectra of the B_{12} derivatives obtained with visible and near-UV excitation, respectively. Abbreviations in the figures have the following meanings: Cys-B_{12}, cysteinyl cobalamin; Me-Cobin, methyl cobinamide; Coen-B_{12}, 5′-deoxyadenosyl-B_{12}. A rapid-flow technique held laser-induced photodecomposition to acceptable levels.

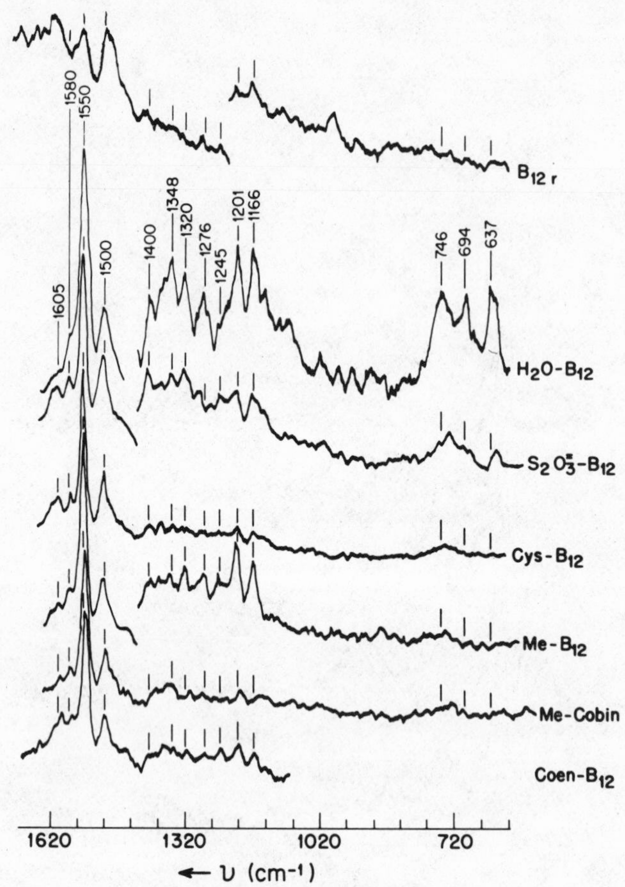

Fig. 7.11. Raman spectra of the indicated B_{12} derivatives with 3638 Å excitation. Conditions used: sensitivity (counts sec^{-1}), spectral slit (cm^{-1}), time constant (sec), scanning speed (cm^{-1} sec^{-1}): B_{12r} (500, 16, 5, 1); H_2O-B_{12} from 600–1400 cm^{-1} (200, 16, 5, 1) and from 1400–1650 cm^{-1} (1000, 16, 5, 1); $S_2O_3^{2-}$-B_{12} from 600–1400 cm^{-1} (200, 16, 10, 0.5) and from 1400–1650 cm^{-1} (500, 16, 10, 0.5); Cys-B_{12} (500, 16, 10, 0.5); Me-B_{12} from 600–1400 cm^{-1} (200, 16, 10, 0.5) and from 1400–1650 cm^{-1} (500, 16, 10, 0.5); Me-Cobin (200, 16, 10, 0.5) and Coen-B_{12} (200, 16, 10, 0.5). (Salama and Spiro, 1977.)

References

Ainscough, E. W., Brodie, A. M., Plowman, J. E., Bloor, S. J., Loehr, J. S., and Loehr, T. M., *Biochemistry* **19,** 4072 (1980).

Aisen, P., in *Inorganic Biochemistry* (G. L. Eichhorn, ed.), Vol. 1, Chap. 9, Elsevier Publishing Co., Amsterdam, 1973.

Bezkorovainy, A., *J. Dairy Sci.* **60,** 1023 (1977).

Bonaventura, C., Sullivan, B., Bonaventura, J., and Bourne, S., *Biochemistry* **13,** 4784 (1974).

Bossu, F. P., Chellappa, K. L., and Margerum, D. W., *J. Am. Chem. Soc.* **99,** 2195 (1977).

Carey, P. R., and Young, N. M., *Can. J. Biochem.* **52,** 273 (1974).
Chen, J. T., Shen, S. T., Chung, C. S., Chang, H., Wang, S. M., and Li, N. C., *Biochemistry* **18,** 3097 (1979).
Dunn, J. B. R., Shriver, D. F., and Klotz, I. M., *Biochemistry* **14,** 2689 (1975).
Dunn, J. B. R., Addison, A. W., Bruce, R. E., Loehr, J. S., and Loehr, T. M., *Biochemistry* **16,** 1743 (1977).
Dunn, J. B. R., Shriver, D. F., and Klotz, I. M., *Proc. Natl. Acad. Sci. USA* **70,** 2582 (1973).
Dyrkacz, G. R., Libby, R. D., and Hamilton, G. A., *J. Am. Chem. Soc.* **98,** 626 (1976).
Feeney, R. E., and Komatsu, S. S., *Struct. Bonding (Berlin)* **1,** 149 (1966).
Ferris, N. S., Woodruff, W. H., Rorabacher, D. B., Jones, T. E., and Ochrymowycz, L. A., *J. Am. Chem. Soc.* **100,** 5939 (1978).
Freedman, T. B., Loehr, J. S., and Loehr, T. M., *J. Am. Chem. Soc.* **98,** 2809 (1976).
Freier, S. M., Duff, L. L., Van Duyne, R. P., and Klotz, I. M., *Biochemistry* **18,** 5372 (1979).
Gaber, B. P., Miskowski, V., and Spiro, T. G., *J. Am. Chem. Soc.* **96,** 6868 (1974).
Galuzzi, F., Garozzo, M., and Ricci, F. F., *J. Raman Spectrosc.* **2,** 351 (1974).
Gray, H. B., *Bioinorganic Chemistry* (R. F. Gould, ed.), pp. 365–389, American Chemical Society, Washington, D.C., 1971.
Hamilton, G. A., Libby, R. D., and Hartzell, C. R., *Biochem. Biophys. Res. Commun.* **55,** 333 (1973).
Ingram, D. J. E., *Biological and Biochemical Applications of Electron Spin Resonance,* p. 23, Plenum Press, New York, 1969.
Johnson, B. D., and Peticolas, W. L., *Annu. Rev. Phys. Chem.* **27,** 465 (1976).
Ke, C. H., and Schubert, J., *Radiat. Res.* **49,** 507 (1972).
Kincaid, J. R., Larrabee, J. A., and Spiro, T. G., *J. Am. Chem. Soc.* **100,** 334 (1978).
Kurtz, D. M., Jr., Shriver, D. F., and Klotz, I. M., *J. Am. Chem. Soc.* **98,** 5033 (1976).
Larrabee, J. A., Spiro, T. G., Ferris, N. S., Woodruff, W. H., Maltese, W. A., and Kerr, M. S., *J. Am. Chem. Soc.* **99,** 1979 (1977).
Loehr, J. S., Freedman, T. B., and Loehr, T. M., *Biochem. Biophys. Res. Commun.* **56,** 510 (1974).
Margerum, D. W., Chellappa, K. L., Bossu, F. P., and Burce, G. L., *J. Am. Chem. Soc.* **97,** 6894 (1975).
Mayer, E., Gardiner, D. J., and Hester, R. E., *Biochim. Biophys. Acta* **297,** 568 (1973); *Molec. Phys.* **26,** 783 (1973); *J. Chem. Soc. Faraday Trans.* II **69,** 1350 (1973).
McLendon, G., and Martell, A., *J. Inorg. Nucl. Chem.* **39,** 191 (1977).
Miskowski, V., Tang, S.-P. W., Spiro, T. G., Shapiro, E., and Moss, T. H., *Biochemistry* **14,** 1244 (1975).
Nickerson, K. W., and Van Holde, K. E., *Comp. Biochem. Physiol. B* **39,** 855 (1971).
Salama, S., and Spiro, T. G., *J. Raman Spectrosc.* **6,** 57 (1977).
Siiman, O., Young, N. M., and Carey, P. R., *J. Am. Chem. Soc.* **96,** 5583 (1974).
Siiman, O., Young, N. M., and Carey, P. R., *J. Am. Chem. Soc.* **98,** 744 (1976).
Spiro, T. G., and Loehr, T. M., in *Advances in Infrared and Raman Spectroscopy,* Vol. 1 (R. J. H. Clark and R. E. Hester, eds.), Chap. 3, p. 98, Heyden, London, 1975.
Spiro, T. G., and Stein, P., *Annu. Rev. Phys. Chem.* **28,** 501 (1977).
Thamann, T. J., Loehr, J. S., and Loehr, T. M., *J. Am. Chem. Soc.* **99,** 4187 (1977).
Thompson, J. S., Marks, T. J., and Ibers, J. A., *J. Am. Chem. Soc.* **101,** 4180 (1979).
Thompson, J. S., Sorrell, T., Marks, T. J., and Ibers, J. A., *J. Am. Chem. Soc.* **101,** 4193 (1979a).
Tomimatsu, Y., Kint, S., and Scherer, J. R., *Biochem. Biophys. Res. Commun.* **54,** 1067 (1973).
Tomimatsu, Y., Kint, S., and Scherer, J. R., *Biochemistry* **15,** 4918 (1976).
Tosi, L., Garnier, A., Herve, M., and Steinbuch, M., *Biochem. Biophys. Res. Commun.* **65,** 100 (1975).
Van Holde, K. E., and Van Bruggen, E. F. J., in *Subunits in Biological Systems,* Part A (S. N. Timasheff and G. D. Fasman, eds.), pp. 1–53, Marcel Dekker, New York, 1971.
Wozniak, W. T., and Spiro, T. G., *J. Am. Chem. Soc.* **95,** 3402 (1973).

Chapter 8

CARBOHYDRATES

Infrared Spectroscopic Studies

Parker (1971) and Tipson and Parker (1980)* have reviewed the subject of the infrared spectroscopy of carbohydrates. They have discussed in detail functional groups that occur in these compounds and their derivatives, with sections on C—H, N—H, and O—H bands; C≡C and C=C bands; C=O bands (aldehydes and ketones, un-ionized carboxylic acids, lactones, acetates, and other esters, primary amides, and *N*-acetyl and *S*-acetyl groups).

These authors have also discussed C—O bands (esters and carboxylate ions); N≡N, N=N, and NO_2 bands; and S=O, —SO_2—, and C=S bands, as they occur in carbohydrates.

El Khadem and Parker (1980) have discussed electronic (UV) spectra of carbohydrates, including those of simple carbohydrates and their derivatives; and those of glycoproteins, nucleic acids, and related substances.

IR Correlations for Certain Aldopyranose Derivatives

In the range 960 to 730 cm^{-1}, Barker *et al.* (1954a,b,c) have found bands characteristic of several aldopyranoses and their derivatives. They identified (1954a), for D-glucopyranose derivatives, three principal sets of bands, which are presented in Table 8.1. These were as follows: for α anomers, type 1a at 917 ± 13 cm^{-1}, type 2a at 844 ± 8 cm^{-1}, and type 3a at 766 ± 10 cm^{-1}; and for β anomers, type 1b at 920 ± 5 cm^{-1}, type 2b at 891 ± 7 cm^{-1}, and type 3b at 774 ± 9 cm^{-1}.

If the bands are to be useful for distinguishing (D or L)-glucopyranose derivatives, α anomers should not display type 2b absorption, and β anomers should not display type 2a absorption. However, Barker *et al.* (1954a) found that (a) some α anomers show type-1 absorption in the region of type 2b bands, and (b) some β anomers exhibit "weak peaks

*The present chapter is based in part on this work.

Table 8.1. Positions (Mean Values, cm^{-1}) of Various Types of IR Bands for D-Glucopyranose Derivatives[a]

Linkage	Type 1	Type 2a	Type 2b	Type 3
α Anomeric				
Monosaccharides	915	847		767
	900	842		751
Higher saccharides	930	843		761
	917	839		768
β Anomeric				
Monosaccharides	914		896	
	918[b]		891	772[c]
Higher saccharides	921		890	774[d]

[a] Barker *et al.* (1954a).
[b] Six of ten compounds did not show this band.
[c] Eleven of sixteen compounds did not show this band.
[d] Five of sixteen compounds did not show this band.

of type 2a," which they thought were caused by traces of the α anomers. The type 2a band was found to be applicable with confidence to diagnosis of the α-anomeric form, particularly in polymers of glucopyranose. The type 2b band was not found useful to diagnose the β-anomeric form, but the *absence* of the type 2a band was applicable for the recognition of the β-anomeric form. Types 1 and 3 were useful only for determination of linkage points in α-glucopyranose polymers.

When they reported (Barker *et al.*, 1954b) the spectra of additional derivatives of glucopyranose, they found slightly different positions for type 2a and 3 bands (see Table 8.2). As before, some of the α anomers displayed type 1 bands in the range of type 2b bands. Also, some derivatives having a phenyl ring may absorb in the region of the type 2a band, and the acetates may absorb in the region of the type 2b band. The findings (Barker *et al.*, 1954b) for bands characteristic of four other aldopyranoses and their derivatives are also summarized in Table 8.2. The type 2a band can be used for diagnosing α anomers in manno- and galactopyranose derivatives; absence of the type 2a band can be used to diagnose the β-anomeric form. A type 2c band at 876 ± 9 cm^{-1} was characteristic of mannopyranose derivatives, and a type 2c band at 871 ± 7 cm^{-1} was characteristic of galactopyranose derivatives. The mean frequency for a particular type of band may vary with the configuration of the group; for example, the mean for type 3 absorption lies at 791 cm^{-1} for the *manno* configuration, but at 752 cm^{-1} for the *gluco* and *galacto* configurations.

Barker *et al.* (1954c) also observed that 2- and 3-deoxy derivatives of gluco-, manno-, and galactopyranose absorb at 869–865 cm^{-1}; seven 6-deoxy derivatives of mannopyranose or galactopyranose absorb near 967 cm^{-1}.

Application of these correlations (Barker *et al.*, 1954a,b,c) has proved useful (Barker *et al.*, 1956) in the study of many related compounds, including oligo- and polysaccharides. Assignments suggested (Barker *et al.*, 1956; 1954a,b,c) for the bands are presented in Table 8.3. In the region 1000–667 cm^{-1}, methyl β-D-xylopyranoside has only three

Table 8.2. Infrared Bands (Mean Values, cm^{-1}) Shown by Five (D or L)-Aldopyranoses and Their Derivatives[a]

Band type	Xylose	Arabinose	Glucose	Mannose	Galactose
			Both anomers		
1	?	?	917	?	?
2c	—	—	—	876	871
3	—	763	770[b] 753[b]	791[b]	752[b]
			α anomers only		
2a	—	—	844[b] 843[b]	833[b]	825[b]
3	749	—	—	—	—
			β anomers only		
2a	—	843 845	—	—	—
2b	—	—	891[c] 890[c]	893[d]	895[d]

[a] Barker *et al.* (1954a,b).
[b] Many derivatives containing a benzene ring absorb here.
[c] Must be confirmed by absence of absorption at ~844 cm^{-1}.
[d] Bands for other types of vibration also occur here.

Table 8.3. IR Bands Possibly Characteristic of Various Features of Some Aldopyranose Derivatives[a]

Band type	Structural feature	Bands,[b] (cm^{-1})	References
	Terminal *C*-methyl-group rocking[c]	967	*e*
1	Antisymmetrical ring-vibration[d]	917	*f*
2b	Anomeric C—H axial bond	891	*f*
2c	Equatorial C—H deformation (other than anomeric C—H)	880	*g*
	Ring-methylene rocking vibration (if not adjacent to the ring-oxygen atom)	867	*e*
2a	Anomeric C—H equatorial bond	844	*f*
3	Symmetrical ring-breathing vibration	770	*f*

[a] Barker *et al.* (1956; 1954a,b,c).
[b] Mean value.
[c] This band may not have diagnostic value.
[d] For glucopyranose derivatives.
[e] Barker *et al.* (1954*c*).
[f] Barker *et al.* (1954*a*).
[g] Barker *et al.* (1954*b*).

bands. (Tipson and Isbell, 1960b), namely, at 976, 963, and 898 cm^{-1}; indeed, β-D- or β-L-xylopyranose derivatives cannot be characterized by any of the bands listed in Table 8.2. Thus, none of the bands listed in Tables 8.2 and 8.3 can be considered as characteristic of the pyranoid ring, *per se,* of aldopyranoid derivatives.

IR spectra of β-D-glucopyranose have been recorded in the regions 4000–400 cm^{-1} and 500–50 cm^{-1} by Hineno (1977). Absorption bands were assigned as follows: 3250 and 2980–2820 cm^{-1}, due to O—H and C—H stretching modes, respectively; 1465 cm^{-1}, CH_2 scissoring; bands in the 1474–1199 cm^{-1} region, due to O—C—H, C—C—H, and C—O—H bending vibrations; 1153–904 cm^{-1} region bands, C—O and C—C stretching; bands in the 800–519 cm^{-1} region, internal rotation modes about C—OH (side) bonds; bands in the 463–250 cm^{-1} region, mainly C—C—O bending vibrations; 127 cm^{-1}, due to C—C—O—, C—C—C, and C—O—C bending modes, coupled with C—O ring internal-rotation modes; 104, 88.6, and 60.1 cm^{-1}, C—C internal-rotation modes and possibly lattice vibrations.

IR Correlations for Pyranoid and Furanoid Forms of Aldose and Ketose Derivatives

For aldo- and ketofuranose derivatives, Barker and Stephens (1954) found absorption frequencies at the following mean values: type A, 924 cm^{-1}; and type D, 799 cm^{-1}. Type-A absorption was not distinguishable from types 1 or 2b of aldopyranoses and therefore has no diagnostic value for differentiating between furanoid and pyranoid aldoses. Most of the furanoid compounds also displayed type B absorption at 879 cm^{-1} and type C absorption at 858 cm^{-1}. It has been observed (Tipson *et al.*, 1967, 1959; Tipson and Isbell, 1961a) that these correlations are, for the most part, restricted to the compounds they studied, and cannot be extended to have a wider diagnostic applicability to related compounds.

Tipson and Isbell (1962) recorded the spectra of most of the readily available, unsubstituted aldo- and ketopentoses and aldo- and ketohexoses. Shortly thereafter, Verstraeten (1964, 1966) studied these spectra, included some additional 2-ketoses, and observed that most of the common sugars having a cyclic structure, and their derivatives, show type 1 absorption at a mean value of 929 cm^{-1}. Hence, the type 1 (type A) band is not useful for distinguishing between aldoses and ketoses, or between glycofuranoses and glycopyranoses. Besides, since the type 1 band appears (Tipson *et al.*, 1967) for acyclic 1-acylamido derivatives of sugars, it cannot be applied to distinguish between cyclic and acyclic forms of such compounds.

Verstraeten (1964) noted that some ketopyranoses, as well as aldopyranoses, display a type 3 band at 781 cm^{-1}. Thus, this band also has limited diagnostic value. He concluded that type 3 absorption is displayed provided that two conditions are met. First, the sugar must have a pyranoid ring, and second, this pyranoid form must assume a conformation having at least one axial hydroxyl group. If the number of axial hydroxyl groups is increased (resulting in decreased conformational stability), type 3 absorption appears. For example, β-D- or β-L-xylopyranose, which presents no type 3 absorption, has no axial hydroxyl groups, whereas the α anomer in the favored conformation, which has an axial hydroxyl group at C-1, displays a band at 760 cm^{-1}.

It was observed (Verstraeten, 1964; 1966) that 2-ketoses show "type I" bands at 875 cm^{-1} and "type IIA" bands at 817 cm^{-1}, regardless of whether the 2-ketoses are pyranose or furanose. These bands were ascribed to the presence of structural feature I and were tentatively assigned to a skeletal vibration. However, six aldoses also display these bands.

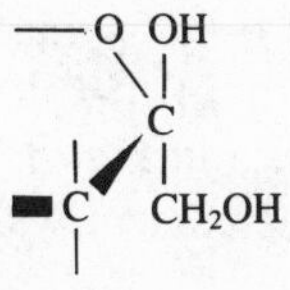

—O OH
C
C CH_2OH

I

The type I band, which appears to be the same as Barker's type B band for aldo- and ketofuranoses at 879 cm^{-1}, has no diagnostic value for 60 aldofuranoid, aldopyranoid, and acyclic 1-acylamido derivatives (Tipson *et al.*, 1967). The type IIA band is found in about the same region as Barker's type D band for aldo- and ketofuranose derivatives, which is at 799 cm^{-1}. If the hydroxyl groups of a 2-ketofuranose are substituted, or if C-2 of the 2-ketofuranose is attached to a pyranoid or furanoid structure, a type IIB band is presented at 834 cm^{-1}, in addition to the type IIA band or instead of it.

Verstraeten found that only furanoses have "type 2" absorption at 850 cm^{-1}. He stated that his type 2 absorption is the same as the type C absorption of Barker and Stephens (1954), and to avoid confusion, it should be referred to as the latter. The type C band is displayed by both aldo- and ketofuranoses, and therefore cannot be used for distinguishing between them.

Tipson *et al.* (1967) found that, if an *N*-acetyl group (but no ester group) is present, the bands of types C, 3, IIA, and IIB may have diagnostic value; if an *N*-benzoyl group (but no ester group) is present, the bands of types 3, IIA, and IIB may have diagnostic value. For *N*-acetyl-*O*-acetyl derivatives of sugars, the bands of types IIA and IIB may differentiate between ketoses and nonketoses, but not between cyclic and acyclic compounds.

Conformational Studies by IR Spectroscopy

In determining the conformations of sugars and their derivatives, the most direct information is obtained by NMR spectroscopy (Eliel *et al.*, 1965). However, the empirical correlation of infrared spectra has been used to give conformational information (Tipson and Isbell, 1960b). Suppose that the spectra of the α and β anomers of the methyl pyranosides of the four aldopentoses and eight aldohexoses were to be determined. There would be recorded 24 spectra of closely related compounds. Each compound has C—H, C—OH, C—OCH_3, and a pyranoid ring, and yet the spectrum of each is unique because the precise positions of the various bands change from one compound to another, depending on interactions arising from configuration and conformation and on whether or not there is a CH_2OH group present.

As an example, the spectra of the α and β anomers of methyl D-xylopyranoside and methyl L-arabinopyranoside were studied (Tipson and Isbell, 1960b). All the common bands shared by the four glycosides were ignored. All bands then shown in common by the two xylosides were regarded as characteristic of the *xylo* configuration and were ignored; similarly, all bands appearing in common for the two arabinosides were disregarded. This left a set of bands distinguishing between the anomers of the xylosides, on the one hand, and between the arabinoside anomers, on the other (see Table 8.4). This indicated a similarity between the β-D-xylopyranoside and the α-L-arabinopyranoside. Inasmuch as the conformation of methyl β-D-xylopyranoside has been shown by X-ray studies (Brown, 1960) to be that shown in Fig. 8.1, the conformational correlations are as indicated. These formulas are in agreement with the conformation predicted by considerations of interaction energies.

This correlation is purely empirical, but similar comparisons have been made for other pairs of anomers of methyl aldopyranosides (Tipson and Isbell, 1960b), acetylated methyl aldopyranosides (Tipson and Isbell, 1960a), and fully acetylated aldopyranoses (Tipson and Isbell, 1961b). In each case, the empirical correlations made from the spectra agreed with the predicted conformations. Those sugar derivatives for which one chair conformation is not predicted to be strongly favored over the other produced data that did not fit in the correlations. Examples are methyl α- and β-D-lyxopyranoside and their triacetates, methyl α-D-gulopyranoside and its tetraacetate, and penta-*O*-acetyl-α-D-gulopyranose.

In a group of fully acetylated monosaccharides, those containing an axial OAc at C-1 exhibited a band (Isbell *et al.*, 1957), possibly for a C—O stretching vibration, at 1174–1153 cm^{-1}; if the group was equatorial, a band was displayed at 1127 cm^{-1}. For each region, the other anomer showed the band only weakly or not at all. The data for

Table 8.4. Anomer-Differentiating IR Bands, in cm^{-1}, Shown by Four Methyl Pyranosides[a]

D-Xylo		L-Arabino	
β	α	α	β
3448		3460	
1385		1395	
1295		1295	
1218		1227	
1060		1058	
976		973	
645		646	
473		487	
	3333		3322
	2710		2695
	741		744
	437		433

[a] Tipson and Isbell (1960b).

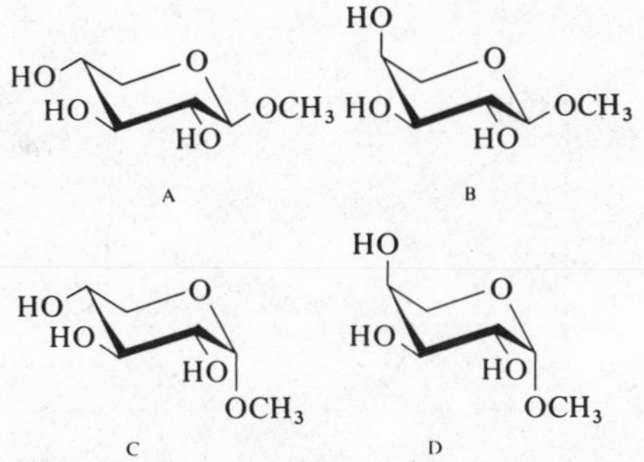

Fig. 8.1. The structures of (A) methyl-β-D-xylopyranoside-CA and (B) methyl α-L-arabinopyranoside-CE compared with those of (C) methyl α-D-xylopyranoside-CA and (D) methyl β-L-arabinopyranoside-CE. (Parker, 1971.)

compounds having the *gulo, ido,* and *talo* configurations indicated that they exist in the 1C(D) or C1(L) conformation, or as a mixture of the chair conformations.

Acetylated methyl glycosides (Isbell *et al.*, 1957) possessing an axial OMe at C-1 had bands at 1203–1198 and 1143–1130 cm^{-1}, whereas those with an equatorial OMe did not absorb in either range.

Glycosaminoglycans

Orr (1954) has made some assignments for the chondroitin sulfates. [Note that Turvey *et al.* (1967), Peat *et al.* (1968), and Harris and Turvey (1970) have commented on the dangers inherent in using only IR data to assign positions to sulfuric ester groups on sugar rings.] Since polysulfated hyaluronic acid (Fig. 8.2), which has equatorial sulfate groups only, absorbs at 820 cm^{-1}, he concluded that the sulfate group of chondroitin sulfate C showing a band at 825 cm^{-1} is equatorially attached, and that the sulfate group of chondroitin sulfate A showing a band at 855 cm^{-1} is axially attached. He ascribed the bands to the C—O—S vibration. Lloyd and Dodgson (1961) later found that the equatorial sulfate group in D-glucose 3-sulfate has a band at 832 cm^{-1}, and that a band at 820 cm^{-1} is displayed by the 6-sulfates of D-galactose, D-glucose, and 2-acetamido-2-deoxy-D-glucose (*N*-acetyl-D-glucosamine), in which the ester group is on the equatorial primary hydroxyl group. Hence, chondroitin sulfate C (and D) has an equatorial sulfate group on C-6, and chondroitin sulfate A (and B) has an axial sulfate on C-4 of the 2-acetamido-2-deoxy-D-galactose residues. The chondroitin sulfates have also been studied by Meyer *et al.* (1956); Mathews (1958); and Hoffman *et al.* (1958).

Mathews (1958) has shown that chondroitin sulfates A and C can be distinguished by their infrared spectra (Fig. 8.3) in the 1000–700 cm^{-1} region. The C form has unique bands at 1000, 820, and 775 cm^{-1}; the A form has bands at 928, 852, and 725 cm^{-1}. Chondroitin sulfate B resembles A, with bands at 928, 855, 840, and 712 cm^{-1}.

The mucopolysaccharide from nuclei pulposi, called "chondroitin sulfuric acid B" (Orr, 1954), and the shark cartilage chondroitin sulfate (Nakanishi *et al.*, 1956), may also be regarded as the C sulfate.

Suzuki and Strominger (1960) have indicated that the spectra of acetylgalactosamine 4-sulfate obtained from chondroitin sulfate A and B, and of acetylgalactosamine 6-sulfate from chondroitin sulfate of shark cartilage, corresponded, respectively, to those of chondroitin 4-sulfate and the 6-sulfate. Therefore, the spectral differences would be attributed

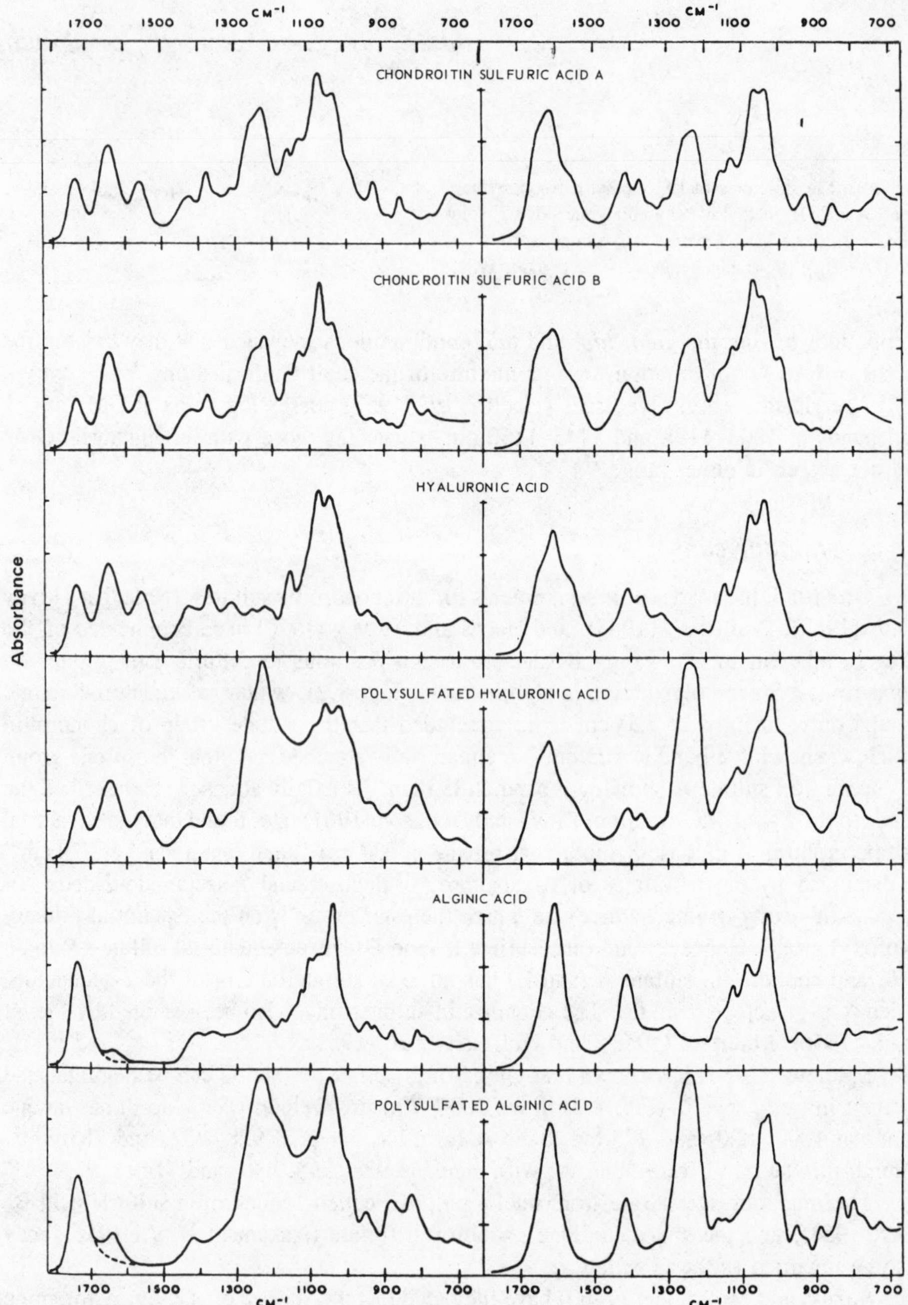

Fig. 8.2. The infrared spectra of the polysaccharides, on the left as free carboxylic acids, and on the right as salts. (Orr, 1954.)

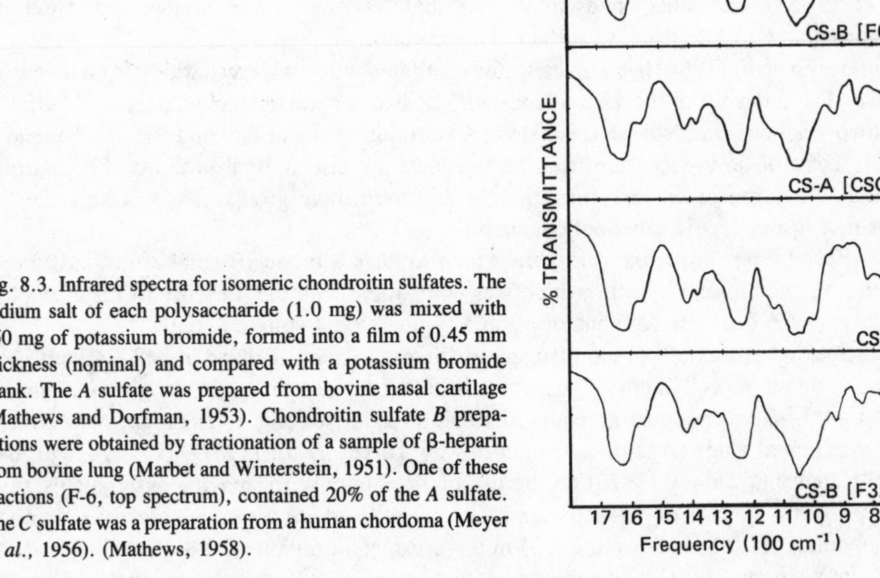

Fig. 8.3. Infrared spectra for isomeric chondroitin sulfates. The sodium salt of each polysaccharide (1.0 mg) was mixed with 150 mg of potassium bromide, formed into a film of 0.45 mm thickness (nominal) and compared with a potassium bromide blank. The *A* sulfate was prepared from bovine nasal cartilage (Mathews and Dorfman, 1953). Chondroitin sulfate *B* preparations were obtained by fractionation of a sample of β-heparin from bovine lung (Marbet and Winterstein, 1951). One of these fractions (F-6, top spectrum), contained 20% of the *A* sulfate. The *C* sulfate was a preparation from a human chordoma (Meyer *et al.*, 1956). (Mathews, 1958).

to the presence of the C-4 and C-6 sulfate in the galactosamine moieties of chondroitin sulfates. Although chondroitin sulfate from shark cartilage has a high sulfur content of 7.6% (1.3 residues of sulfate per acetylgalactosamine residue), the spectra are identical to that of chondroitin C from chordoma (Mathews, 1958; Nakanishi *et al.*, 1956).

In a report by Murata (1962) infrared spectroscopic data showing a high degree of sulfation for chondroitin polysulfate have been related to the anticoagulant activity of the polysulfate. (The chondroitin polysulfate had been prepared by sulfation of shark cartilage chondroitin sulfate with chlorosulfonic acid and pyridine.) Chondroitin polysulfate has sulfate groups at both C-4 and C-6.

Chondroitin sulfates B and C, a small amount of a glucosamine-containing sulfated mucopolysaccharide, and hyaluronic acid were found in human umbilical cord (Danishefsky and Bella, 1966). Characterization of the chondroitin sulfate B was based on characteristic bands at 928, 860, and 840 cm^{-1}; chondroitin sulfate C was identified by bands at 825 and 1000 cm^{-1}.

Scheinthal and Schubert (1963) have obtained fractionated materials from alkaline or trypsin degradation of the proteinpolysaccharide light fraction (PP-L) from cartilage and from alkaline degradation of proteinpolysaccharide heavy fraction (PP-H), and found IR absorption bands at 1648 and 1410 cm^{-1} (carboxylate) and 1230 to 1255 cm^{-1} (sulfate) and at 928, 852, and 725 cm^{-1} (axial 4-sulfate). Certain other fractions from alkaline degradation of PP-L and PP-H showed bands at 940, 890, 817, and 775 cm^{-1}, which were assigned to an equatorial 6-sulfate configuration.

Hunt and Jevons (1966) have isolated a polysaccharide (glucan) sulfate from the hypobranchial mucin of the whelk *Buccinum undatum*. The infrared spectra of the glucan sulfate contained intense absorptions at 1240 cm^{-1} and over the range 840–800 cm^{-1}. These two bands disappeared entirely on desulfation, indicating ester sulfate (Hoffman *et al.*, 1958), and other bands (940, 920, and 1000 cm^{-1}) also disappeared. Hunt and Jevons also attribute these to sulfate disappearance.

Terho *et al.* (1966) have investigated a sulfated mucopolysaccharide of human gastric juice. The intensity of the 1260–1230-cm^{-1} IR band characterized the degree of sulfation of two fractions that had been isolated. Absorption bands at 820 and 775 cm^{-1} found in the spectra of polysaccharide from gastric juice are also present in those of chondroitin sulfate C. Terho *et al.* therefore thought that the sulfate groups might occupy the C-6 position in the gastric mucopolysaccharide.

Rees (1963) have found λ-carrageenan to have a broad IR band at 860–810 cm^{-1}, with a maximum at 827 cm^{-1} which was compatible with the presence of residues of D-galactose 2,6-disulfate (di-equatorial) and D-galactose 4-sulfate (axial).

Sulfonic esters of pyranoid sugars exhibited a strong IR band at 848–840 cm^{-1} and a weak one at 893–877 cm^{-1}. These bands have been ascribed (Onodera *et al.*, 1965) to the C—O—S vibration of an equatorial and an axial sulfonic ester group, respectively, on a pyranoid ring. [Again, note cautions by Turvey *et al.* (1967), Peat *et al.* (1968), and Harris and Turvey (1970) on the use of IR data only for making assignments.]

The IR spectra of aqueous solutions of sulfated glycosaminoglycans and model compounds (Fig. 8.4 and Table 8.5) in the transparent region of water (1400–950 cm^{-1}) present a strong, complex absorption centered at ca. 1230 cm^{-1} and originating from the antisymmetric stretching vibrations of the S═O groups (Cabassi *et al.*, 1978). The sulfated compounds examined were: D-glucose 6-sulfate, dextran sulfate, methyl α-D-glucopyranoside 2-sulfate, 2-deoxy-2-sulfoamino-D-glucose, heparin, *N*-desulfated heparin, heparan sulfate, and chondroitin sulfates A, B, and C.

Primary and secondary *O*-sulfate groups absorb at somewhat higher frequencies (1260–1200 cm^{-1}) than *N*-sulfates (ca. 1185 cm^{-1}) (Cabassi *et al.*, 1978). Quantitative analysis can be done with each sulfate band, particularly within a given class of sulfated polysaccharide (e.g., Fig. 8.5). The ratio A(1280)/A(1225) is employed to evaluate the total sulfate/*N*-acetyl ratio in samples of heparan sulfate (Casu, Cifonelli, and Perlin, unpublished, quoted in Cabassi *et al.*, 1978).

The data in Table 8.5 suggested to these authors that the overall shape of the ν_{as}S═O band (as evidenced by the main absorption maxima and the approximate frequency of the center of gravity of the band) can be used to distinguish the type of sulfate group producing the absorption. Although the difference between primary and secondary *O*-sulfates is not large enough to differentiate between these groups in sulfated polysaccharides, the *N*-sulfate group should be detectable from the band or inflection at ca. 1180 cm^{-1}.

Cabassi *et al.* have presented Raman spectra of dextran, dextran sulfate, and B-type heparin, as well as the IR spectrum of heparin. The most prominent Raman band (at ca. 1060 cm^{-1}) was assigned to the symmetrical vibration of the S═O groups. *N*-sulfates scatter at frequencies a little lower (ca. 1040 cm^{-1}) than *O*-sulfates.

This paper also presented an essentially linear plot of the IR absorbance of the

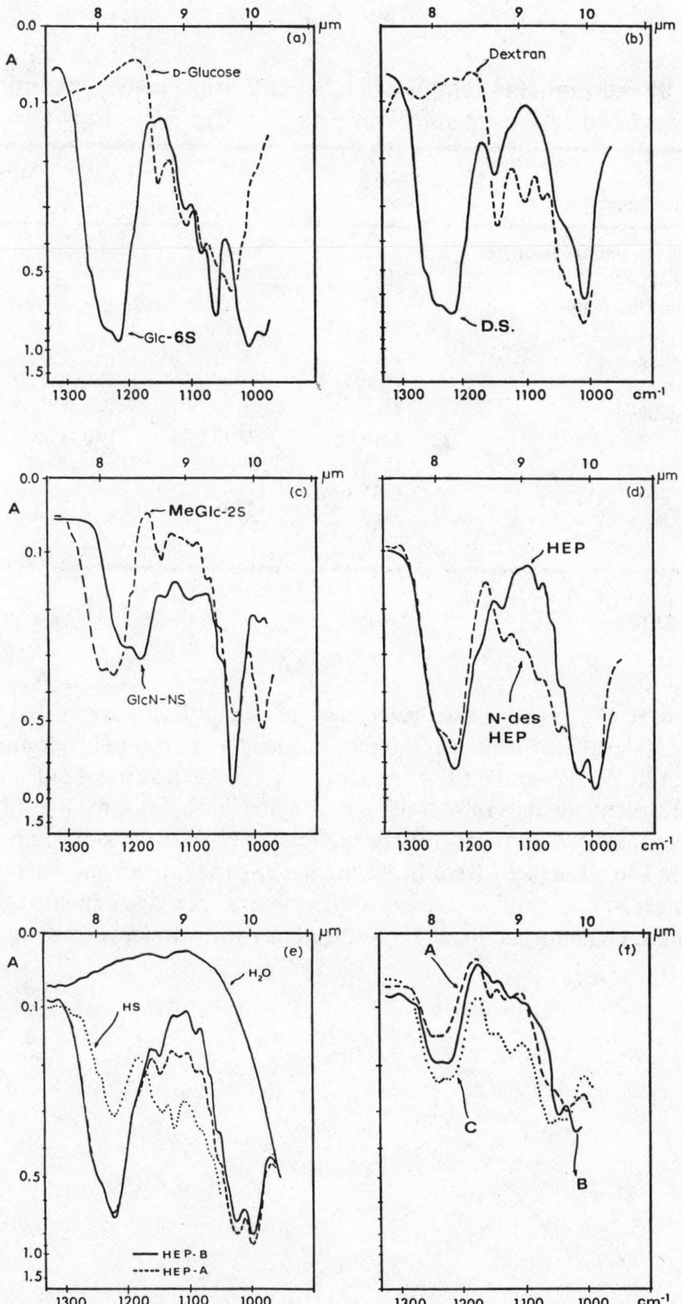

Fig. 8.4. IR spectra, in aqueous solution, of (a) D-glucose, D-glucose 6-sulfate (Glc-6S); (b) dextran, dextran sulfate (D.S.); (c) methyl α-D-glucopyranoside 2-sulfate (MeGlc-2S), 2-deoxy-2-sulfoamino-D-glucose (GlcN-NS); (d) B-type heparin (HEP), *N*-desulfated heparin (*N*-des HEP); (e) B-type and A-type heparin (HEP-B, HEP-A), heparan sulfate (HS); and (f) chondroitin sulfates A, B (dermatan sulfate), and C (A, B, and C); c = 10% w/v, except for D-glucose (6.6%), dextran and D.S. (6%), MeGlc-2S and GlcN-NS (7%), and A, B, and C (4%); l = 0.050 mm. (Cabassi *et al.*, 1978.)

Table 8.5. IR Frequencies (cm^{-1}) of $\nu_{as}S{=}O$ Bands[a] of Glycosaminoglycans and Model Compounds in Solution (Na Salts in H_2O)[b]

Compound	Frequency (cm^{-1})			Center of gravity of band
Methyl α-D-glucopyranoside 2-sulfate	1245	*1226*		(1240)
D-Glucose 6-sulfate	1245	*1216*		(1230)
2-Deoxy-2-sulfoamino-D-glucose		1210	*1182*	(1190)
Dextran sulfate	*1240*			(1240)
Chondroitin 4-sulfate	*1240*	1230		(1240)
Dermatan sulfate	1240	*1225*		(1230)
Chondroitin 6-sulfate	*1240*	1220		(1230)
Heparin	1245	*1222*	~1115 (sh)	(1230)
Heparan sulfate	1245	*1220*		(1220)
N-Desulfated heparin	1245	*1225*		(1235)
COOH-reduced heparin	1245	*1220*		(1230)
Sodium sulfate			1095	

[a] The strongest peaks are shown in *italics*.
[b] Cabassi *et al*. (1978).

$\nu_{as}S{=}O$ band at 1225 cm^{-1} versus percentage of sulfate in the compound for a set of heparins and *N*-desulfated heparin. The plot included a point for chondroitin sulfate A, which did not lie on the straight line; its absorbance value was higher than expected.

Infrared spectra in the 1800–1400-cm^{-1} region have been recorded for aqueous solutions (^2H_2O and ^2HCl) of several glycosaminoglycans and model compounds (Casu *et al*., 1978). The substances were D-glucuronic, D-galacturonic, and methyl glycopyranosiduronic acids; 2-acetamido-2-deoxy-D-galactose; *N*-acetylneuraminic acid; gluconic and acetic acids; chondrosine; methyl 6,6′-dicarboxy-β-maltoside; 6-carboxy-cyclohexa-

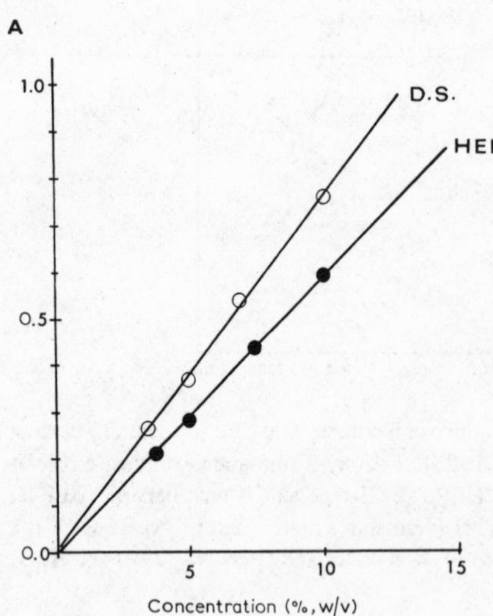

Fig. 8.5. Absorbance–concentration relationship for the IR $\nu_{as}S{=}O$ band of dextran sulfate (D.S.) and heparin (HEP) in aqueous solution. (Cabassi *et al*., 1978.)

(and hepta) amyloses; alginic and pectic acids; hyaluronic acid; chondroitin sulfates A and C; dermatan, heparan, and keratan sulfates; and heparin.

In ^2H_2O solution, the uronic COO^- and —$NHCOCH_3$ groups show characteristic absorption bands ($\nu_{as}COO^-$, 1605 ± 5; ν_sCOO^-, 1420 ± 5; amide I, 1623 ± 3; and amide II, 1480 ± 2 cm^{-1}). The amide bands change very little in 1 *M* ^2HCl, but the bands of COO^- ions expectedly disappear, being supplanted by a band of the COOH group at 1723 ± 3 cm^{-1}. These bands are useful for the quantitative determination of the uronic and 2-acetamido-2-deoxyhexose groups of glycosaminoglycans, with either standard polysaccharides or the corresponding monosaccharides as reference compounds. Figure 8.6 shows plots of absorbance versus concentration for the COO^- and COOH bands of D-glucuronic and/or D-galacturonic acid, and for the acetamido amide I band (carbonyl) of 2-acetamido-2-deoxy-D-glucose and/or -D-galactose. The plots follow Beer's law.

Glycosaminoglycans have also been examined by polarized IR spectroscopy. Samples of oriented, crystalline samples of hyaluronates, chondroitin 4-sulfate, chondroitin 6-sulfate, dermatan sulfate, and a cartilage proteoglycan, having different known chain conformations as found by X-ray diffraction, have been studied by Cael *et al.* (1976). These authors used a Fourier transform IR spectrophotometer with multiple scanning and signal-averaging capabilities.

Sodium hyaluronate samples had the fourfold conformation (in both the contracted and extended states). The chondroitin 6-sulfate specimen had the eightfold conformation, chondroitin 4-sulfate the threefold. Dermatan sulfate had the eightfold conformation. Eighty-five percent of the proteoglycan–hyaluronic acid complex was chondroitin 4-sulfate, which had crystallized in the twofold conformation.

All the glycosaminoglycans displayed perpendicular dichroism for both the amide I and II absorption bands (1660 to 1650 cm^{-1} and 1560 to 1550 cm^{-1}, respectively). Thus, the plane of the amide group is approximately perpendicular to the helix axis.

Cael *et al.* have assigned the absorption bands at ca. 1620 and ca. 1410 cm^{-1} to the antisymmetric and symmetric stretching vibrations of the planar carboxyl group; the dichroic properties of the former stretching mode depend on the rotational conformation of the CO_2^- group. All the glycosaminoglycan samples exhibited perpendicular dichroism

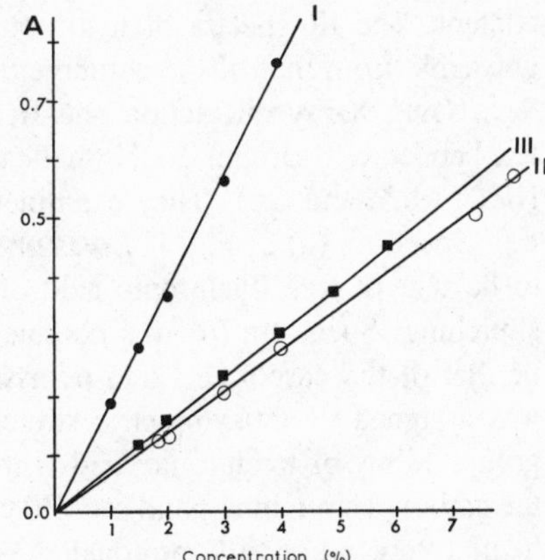

Fig. 8.6. A plot of absorbance (A) *vs*. concentration (%) for the carboxylate (I) and carboxyl (II) absorption bands of D-glucuronic and/or D-galacturonic acid, and for the acetamido Amide I band (III) of 2-acetamido-2-deoxy-D-glucose and/or -D-galactose. (Casu *et al.*, 1978.)

for the antisymmetric CO_2^- stretching band. For the symmetric stretching mode, chondroitin 6-sulfate, dermatan sulfate, and the proteoglycan-hyaluronate complex displayed perpendicular dichroism at ca. 1410 cm^{-1}. On the other hand, chondroitin 4-sulfate and the hyaluronate specimens exhibited parallel dichroism at this frequency. Cael *et al.* indicated, from a consideration of the greater extension of the types of helix for chondroitin 6-sulfate and the proteoglycan-hyaluronate complex, compared to the lesser degree of extension for chondroitin 4-sulfate and hyaluronate, that there is a transition from perpendicular to parallel dichroism for the symmetric carboxyl-stretching band with helix contraction, corresponding to a decrease in the angular projection of the C5–C6 bond on the axis of the molecule.

Dermatan sulfate, in which the C5–C6 bond lies axially to the carbohydrate ring and not equatorially as, for example, in chondroitin 4-sulfate, presents a different situation. When the α-L-idopyranuronic acid moieties of dermatan sulfate have the $^4C_1$ conformation, the C5–C6 bond lies axially to the pyranose ring. Comparison of molecular models with the disposition of the C5–C6 bond in the α-L-idopyranuronic acid–$^4C_1$ residue showed that the C5–C6 bond in this structure would lie approximately perpendicular to the helix axis. Thus, the perpendicular dichroism displayed by the carboxyl symmetric stretch supports the α-L-idopyranuronic acid $^4C_1$-chair conformation (Cael *et al.*, 1976).

For the sulfated glycosaminoglycans, strong absorption bands at ca. 1255 and ca. 1230 cm^{-1} were assigned to sulfate-related modes, probably due to antisymmetric and symmetric stretching of the sulfate group, respectively.

Absorption bands in the range 1200–1000 cm^{-1} are associated with C—C—C, C—C—O, and C—O—C stretchings of the pyranose ring structures. In general, four major bands are found at ca. 1150, 1130, 1065, and 1040 cm^{-1} in each spectrum. In most cases they display parallel dichroism. These workers stated that these vibrational modes can probably be related to those observed for the extended, twofold helix of cellulose, which also exhibit strong parallel dichroism (Liang and Marchessault, 1959; Blackwell *et al.*, 1970).

The chondroitin sulfate produced by B16 melanoma cells has been characterized by several analytical procedures (Bhavanandan and Davidson, 1977). IR and ORD spectra characterized the compound as similar to chondroitin 4-sulfate from cartilage, except for the lack of absorption in the IR spectrum between 1230 and 1250 cm^{-1}, due to low sulfate content. The IR spectra of hyaluronic acid and of cartilage proteoglycan were distinguishable from that of the compound from the mouse melanoma.

Using X-ray diffraction and IR methods, Sheehan *et al.* (1977) have proposed a left-handed antiparallel double-helical structure for hyaluronate in the presence of K^+, NH_4^+, Rb^+, and Cs^+. They examined films cast from solutions between pH 2.0 and pH 8.0. Spectrum (a) of Fig. 8.7 was obtained from a film cast at pH 2.2, and was assumed to be that of free hyaluronic acid. The band at 1730 cm^{-1} was attributed to carboxyl stretching. Spectrum (d) was obtained from a film cast at pH 7.0, and was assumed to be that of the carboxylic acid potassium salt. The band in the region 1600–1620 cm^{-1} was assigned to antisymmetric stretching in the resonance-stabilized carboxylate anion group. Films of hyaluronic acid with counterions of Rb^+, Cs^+, and NH_4^+ all showed the carboxyl stretching band at 1730 cm^{-1}. The double-helical structure is probably either hemiprotonated or fully protonated.

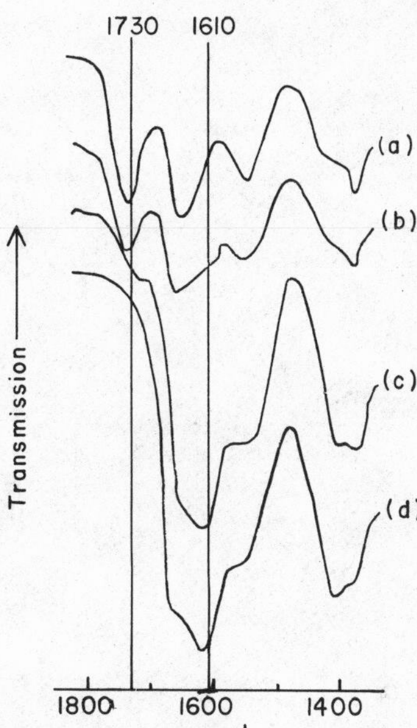

Fig. 8.7. Infrared spectra from potassium hyaluronate films cast from solutions at various pH values. Below pH 4.0 a band at 1730 cm^{-1} was observed. (a) pH 2.2; (b) pH 3.1; (c) pH 3.9; (d) pH 7.0. (Sheehan *et al.*, 1977.)

Raman Spectroscopic Studies

Mono- and Disaccharides. Some Polysaccharides.

She *et al.* (1974) have obtained Raman spectra of α-D-glucose, β-D-glucosamine hydrochloride, *N*-acetyl-α-D-glucosamine, and α- and β-D-glucuronic acid. Figure 8.8 presents spectra of deuterated D-glucose, D-glucose, and D-glucuronic acid. Figure 8.9 shows spectra of D-glucosamine hydrochloride and its deuterated form; Fig. 8.10 displays those of the *N*-acetylated forms. These authors have made assignments of many of the bands in these spectra (Table 8.6).

We have referred earlier in this chapter to the fact that several absorption bands in IR spectra are characteristic of α or β anomers. These were grouped as types 1, 2a, 2b, and 3. Type 1 bands have been assigned to ring asymmetrical vibration, similar to that of 1,4-dioxane (Parker, 1971; Avram and Mateescu, 1972; Lee *et al.*, 1976). Type 2 bands come from the anomeric C—H bending mode, the equatorial C—H vibration of α-D anomers in the $^4C_1$(D) conformation. Type 3 vibrational bands arise from the symmetrical ring vibration of the hexopyranose ring. There is much overlap of these types of absorption bands for the α and β conformations.

Raman studies of carbohydrates (Nakanishi and Solomon, 1977) have also shown that the type 2 band can be applied to the determination of anomeric linkages.

She *et al.* (1974) and Tu *et al.* (1977) have studied glycosidic linkages extensively by Raman spectroscopy. They found that α-D anomers display a band at 840 cm^{-1}, β-D anomers at 890 cm^{-1}, both types originating from the anomeric C—H deformation. [The

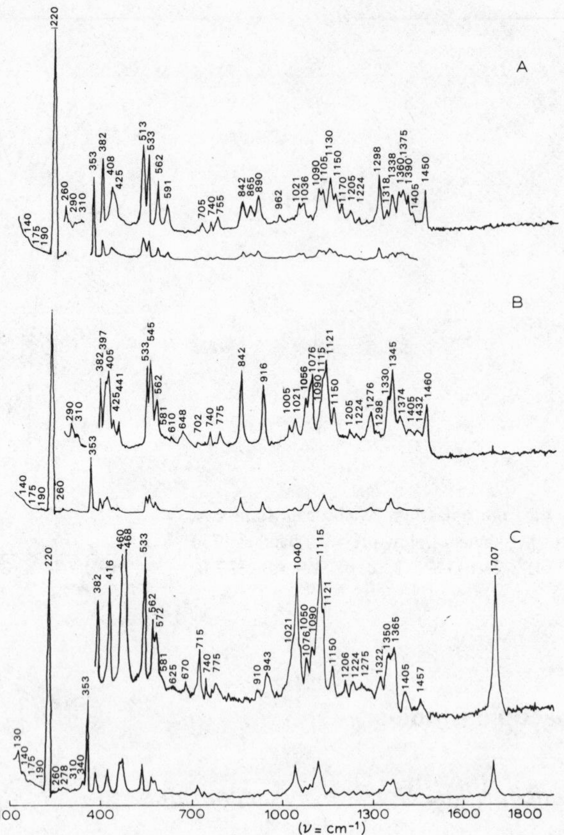

Fig. 8.8. Laser Raman spectra of monosaccharides below 1800 cm^{-1}. (A), deuterated D-glucose; (B), D-glucose; (C), D-glucuronic acid. (She *et al.*, 1974.)

anomeric hydrogen is equatorial in the α-D anomer, and axial in the β-D anomer, in the $^4C_1$(D) conformation.]

Table 8.7 (Tu *et al.*, 1978) presents a comparison between the IR and Raman bands of types 1, 2, or 3. Compounds listed in the table are (1) methyl 1-thio-β-D-galactopyranoside, (2) carboxymethyl 1-thio-β-D-galactopyranoside, (3) carboxymethyl-1-thio-β-D-glucopyranoside, (4) cyanomethyl 2-acetamido-2-deoxy-1-thio-β-D-glucopyranoside, and (5) 2-(acrylamidomethylaminocarbonyl)ethyl 1-thio-β-D-glucopyranoside. The table shows that the equatorial C—H bending for the β-D compounds has a band between 905 and 887 cm^{-1} (Tu *et al.*, 1978). The C—H equatorial bending vibration for α-D compounds in the $^4C_1$(D) conformation lies between 865 and 837 cm^{-1}.

These authors were interested in learning the spectral effect(s) of substitution of a sulfur atom for an oxygen atom in glycosidic linkages, and in complementing what is

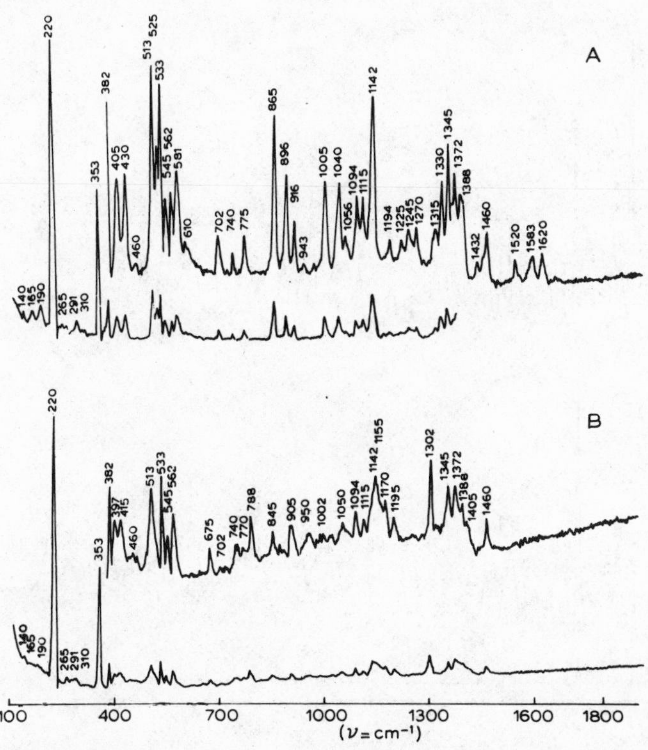

Fig. 8.9. Laser Raman spectra of monosaccharides below 1800 cm^{-1}. A, D-glucosamine · HCl; B, deuterated D-glucosamine · HCl. (She *et al.*, 1974.)

already known from IR spectroscopy about characterizing carbohydrates with such linkages.

Figure 8.11 shows Raman spectra of carboxymethyl 1-thio-β-D-glucopyranoside in the solid state and in aqueous solution. This compound shows a type 1 band at 921 cm^{-1} in the solid state, which is not observed in H_2O; a type 2 band at 895 cm^{-1}, which shifts to 898 cm^{-1} in H_2O; and a type 3 band at 795 cm^{-1}, which is not observed in H_2O.

Methyl 1-thio-β-D-galactopyranoside has a type 1 band at 939 cm^{-1} for the solid and a shoulder at 920 cm^{-1} in H_2O; a type 2 band at 884 cm^{-1} for the solid and a shoulder at 883 cm^{-1} in H_2O; and a weak type 3 band at 775 cm^{-1} for the solid that is not observed in H_2O.

Cyanomethyl 2-acetamido-2-deoxy-1-thio-β-D-glucopyranoside presents no type 1 band for the solid or aqueous state; it has a type 2 band at 884 cm^{-1} for the solid, which becomes a broad band at 887 cm^{-1} in H_2O; and it shows no type 3 bands for the solid or aqueous state.

For all the compounds studied by Tu *et al.* (1978) an anomeric C—H bending vibration (type 2b) was observed at ca. 891 ± 7 cm^{-1}, which permits differentiation of

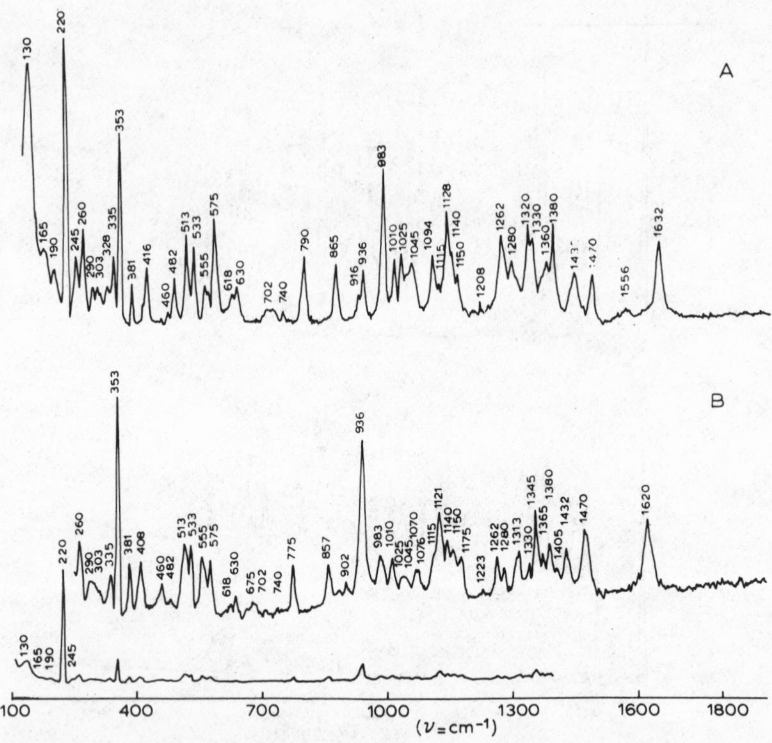

Fig. 8.10. Laser Raman spectra of monosaccharides below 1800 cm^{-1}. (A), *N*-acetyl-D-glucosamine; (B), deuterated *N*-acetyl-D-glucosamine. (She *et al.*, 1974.)

such β-D compounds by Raman spectroscopy. Substitution of S for O in the glycosidic linkage appears to have no effect.

The solid-state Raman spectra (not shown) of the last compound named above presents a band characteristic of the *N*-acetyl group (2-acetamido) at 1659 cm^{-1}, which is obscured in aqueous solution by the water background (broad band around 1648 cm^{-1}) found in all the carbohydrates they examined. She *et al.* (1974) found an *N*-acetyl band at 1632 cm^{-1} in the Raman spectrum of 2-acetamido-2-deoxy-α-D-glucose.

Raman spectra have been recorded of aqueous solutions of D-glucose, cellobiose, maltose, and dextran [(1→6)-α-D-glucan] (Vasko *et al.*, 1971). The use of both water and deuterium oxide showed that the Raman bands at about 1349, 1071, 1021, and 913 cm^{-1} for all these substances were assignable to modes related to the C—O—H group. The Raman spectra of some C-deuterated D-glucoses, when compared to the IR spectrum of α-D-glucose, allowed the identification of some C—H modes. For α-D-glucose, bands assigned to CH_2-related modes were found at 1457, 1337, 1219, and 1011 cm^{-1}; bands at 1404, 1360, 1250, 1076, 1047, 911, and 836 cm^{-1} were assigned to C—H related modes. In the range 700–1500 cm^{-1} the spectra of the four materials listed above are extremely similar, but below 700 cm^{-1} each carbohydrate examined has distinct char-

Table 8.6. Assignments of Vibrations of Monosaccharides Below 1800 cm^{-1}[a]

Glucuronic acid	D-Glucose	Deuterated D-glucose	Glucosamine· HCl	Deuterated glucosamine· HCl	*N*-Acetyl glucosamine	Deuterated *N*-acetyl glucosamine	Assignments
1707	—	—	—	—	—	—	Carbonyl group of C(6)OOH
—	—	—	—	—	1632	1620	Amide I band
—	—	—	1620	—	—	—	NH_3^+
—	—	—	1583	—	—	—	NH_3^+
—	—	—	1520	—	—	—	NH_3^+
—	—	—	—	—	1556	—	Amide II band
1457	1460	1450	1460	1460	1470	1470	CH_3, C(6)-H_2, COH
—	1432	—	1432	—	1432	1432	
1405	1405	1405	—	1405	—	1405	
—	—	1390	1388	1388	—	—	
—	1374	1375	1372	1372	1380	1380	CH_3
1365	—	1360	—	—	1360	1365	
1350	1345	—	1345	1345	—	1345	
—	—	1338	—	—	—	—	
—	1330	—	1330	—	1330	1330	C(6)-H_2, CH_3, COH
1322	—	1318	1315	—	1320	1313	
—	1298	1298	—	1302	—	—	
1275	1276	—	1270	—	1280	1280	COH
—	—	—	—	—	1262	1262	
—	—	—	1245	—	—	—	
1224	1224	1224	1225	—	—	1223	
1206	1205	1205	1194	1105	1208	—	
—	—	1170	—	1170	—	1175	COD
—	—	—	—	—	—	—	ND_3^+
—	—	—	—	1155	—	—	ND_3^+
1150	1150	1150	—	—	1150	1150	
—	—	—	1142	1142	1140	1140	
1121	1121	1130	—	—	1128	1121	
1115	1115	1105	1115	1115	1115	1115	

Continued

Table 8.6. *(Continued)*

Glucuronic acid	D-Glucose	Deuterated D-glucose	Glucosamine·HCl	Deuterated glucosamine·HCl	*N*-Acetyl glucosamine	Deuterated *N*-acetyl glucosamine	Assignments
1090	1090	1090	1094	1094	1094	—	
1076	1076	—	—	—	—	—	C(2)-OH
—	—	—	—	—	—	1076	
—	—	—	—	—	—	1070	
1050	1056	—	1056	1050	1045	1045	
1040	—	—	1040	—	—	—	
—	—	1036	—	—	—	—	
1021	1021	1021	—	—	1025	1025	
—	1005	—	1005	1002	1010	1010	C(6)-H_2,CH_3
—	—	962	—	—	983	983	
943	—	—	943	950	936	936	
910	916	—	916	—	916	—	C(6)-OH
—	—	890	896	905	—	902	
—	—	865	—	—	—	—	C(2)-OD
—	—	—	865	—	865	857	
—	842	842	—	845	—	—	C(1)-H for α-configuration
—	—	—	—	788	790	—	
775	775	—	775	770	—	775	
—	—	755	—	—	—	—	
740	740	740	740	740	740	740	
715	—	—	—	—	—	—	
—	702	705	702	702	702	702	
670	—	—	—	675	—	675	
—	648	—	—	—	—	—	
625	—	—	—	—	630	630	
—	610	—	610	—	618	618	
—	—	591	—	—	—	—	
581	581	—	581	—	—	—	

572	—	—	—	—	575	575	
562	562	562	562	562	555	555	
—	545	—	545	545	—	—	
533	533	533	533	533	533	533	
—	—	—	525	—	—	—	Probably due
—	—	513	513	513	513	513	to skeletal
468	—	—	—	—	482	482	modes
460	—	—	460	460	460	460	
—	441	—	—	—	—	—	
—	425	425	430	—	—	—	
416	—	—	—	415	416	—	
—	405	408	405	397	—	408	
—	397	—	—	—	—	—	
382	382	382	382	382	381	381	
353	353	353	353	353	353	353	Probably due
340	—	—	—	—	335	335	to skeletal
—	—	—	—	—	328	—	modes
310	310	310	310	310	303	303	
278	290	290	291	291	290	290	
260	260	260	265	265	260	260	
—	—	—	—	—	245	245	
220	220	220	220	220	220	220	
190	190	190	190	190	190	190	Probably due
175	175	175	165	165	165	165	to torsional
140	140	140	140	140	—	—	modes
130	—	—	—	—	130	130	

[a] She *et al.* (1974).

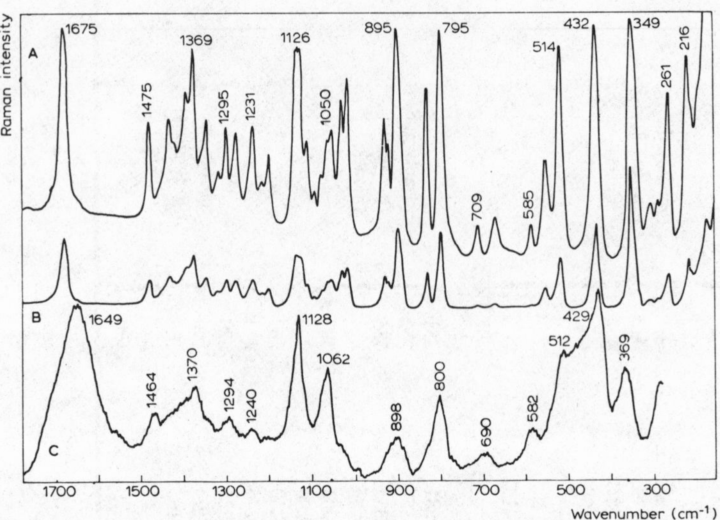

Fig. 8.11. Laser Raman spectra of carboxymethyl 1-thio-β-D-glucopyranoside in the range of 1800 to 200 cm^{-1}. (A) solid phase, radiant power 300 mW, resolution ± 5 cm^{-1}; (B) solid phase, radiant power 250 mW, resolution ± 5 cm^{-1}; (C) in aqueous solution, radiant power 500 mW, resolution ± 6 cm^{-1}; all at integration time 0.2 sec, scanning speed 0.2 cm^{-1} sec^{-1}. (Tu *et al.*, 1978.)

acteristic patterns of bands and their intensities, making this region quite useful for identification of these molecules in solution.

The same investigators (1972) have made a theoretical study of the vibrational spectra of α-D-glucose by normal coordinate analysis, using a program written by Boerio and Koenig (1971). They compared their predicted vibrational frequencies for α-D-glucose with those they had found in the IR and Raman spectra, both in the crystalline state and in aqueous solution. The computed potential-energy distribution for this molecule showed that most of the modes are highly coupled vibrations. Vasko *et al.* (1972) considered that the agreement between the calculated and observed data was reasonably satisfactory for a molecule as large as this one. (See Table III in Vasko *et al.*, 1972, for a detailed comparison of the calculated and observed IR and Raman frequencies, and the approximate percentages of potential-energy distributions.)

Koenig (1972) has recorded Raman spectra of D-glucose, maltose, cellobiose, and dextran in H_2O and D_2O. A band that is present at 847 cm^{-1} for the H_2O solution of D-glucose appears at 846 (strong), 841 (weak), and 847 cm^{-1} (strong) in the spectra of maltose, cellobiose, and dextran, respectively. The weakness of the band for cellobiose is indicative of the β-glycosidic linkage at C1. Koenig assigned the band at ca. 847 cm^{-1} to a vibration at C1 with an α-configuration. Also, he suggested that a strong band in the vicinity of 890 cm^{-1} (890, cellobiose; 898, glucose; 896, maltose) is not unique to the β-configuration at the anomeric C1 position.

The region below 700 cm^{-1} showed distinctive Raman spectra for each of the compounds above, making it a potentially useful region for identifying these carbohydrates in solution.

Table 8.7. Laser Raman Frequencies Corresponding to Infrared-Spectral Characteristics of Anomers[a]

Compounds	Type 1	Type 2 (a or b)	Type 3
Infrared			
α anomers	917 ± 13	844 ± 8	766 ± 10
β anomers	920 ± 5	891 ± 7	744 ± 9
Raman			
3 (β)			
solid	921	895	795
H_2O solution	n.o.[b]	898	n.o.
1 (β)			
solid	939	884	775 (weak)
H_2O solution	920 (sh)[c]	883 (sh)	n.o.
2 (β)			
solid	916	895	795 (weak)
H_2O solution	n.o.	902	n.o.
5 (β)			
solid	n.o.	890	n.o.
4 (β)			
solid	n.o.	884	n.o.
H_2O solution	n.o.	887 (broad)	n.o.

[a] Tu *et al.* (1978).
[b] n.o. = band not observed.
[c] sh = shoulder.

Polysaccharides

Many studies have been done to correlate the conformation of polysaccharides with IR and Raman data. Some of the substances studied have been amylose, amylopectin, dextran, cellulose, β-1,3-xylan, and β-1,4-xylan (Walton and Blackwell, 1973). Cael *et al.* (1973) have shown that Raman lines at 1263 and 946 cm^{-1} shift to 1254 and 936 cm^{-1} when V-amylose is converted to B-amylose. They explained these changes as an extension of the helix and changes in the intramolecular hydrogen bonding when conversion takes place. Walton and Blackwell (1973) reported that the different forms of cellulose could be identified by means of the O—H and C—H stretching modes in infrared spectra.

Koenig (1972) has recorded the Raman spectra of amylose, amylopectin, glycogen, dextran, cellulose, and cellobiose in the solid state. Table 8.8 gives band frequencies and assignments for most of these compounds. He assigned the Raman band at ca. 847 cm^{-1} for D-maltose (848 cm^{-1}), amylose (840 cm^{-1}), amylopectin (850 cm^{-1}), glycogen

Table 8.8. Raman Frequencies and Band Assignments for Amylose, Amylopectin, Glycogen, Dextran, and Cellulose[a,b]

Amylose	Amylopectin	Glycogen	Dextran	Cellulose	Assignment
				1479m	COH bending
1458m	1458s	1456s	1462s	1454m	CH_2 bending
				1432w	
1394sh	1393w	1397w	1408w	1407m	CH bending
1377m	1379s	1382s		1377vs	
			1365s(sh)		
1349sh	1349sh	1350sh	1342s	1359sh	COH bending
1335s	1338s	1338s(sh)		1337s	CH bending
		1326s		1319sh	
1294w	1294w	1293w		1293m	CH bending
			1272m	1277m	CH_2OH group
1267m	1264m	1261m(br)			CH_2OH group
1252m				1249vw	
1224vw				1234vw	
1206w	1198w	1200w	1213w	1204w	CH bending
1152sh	1153sh	1154sh	1150sh	1152s	C—O, C—C, CH bending
1127s	1119s	1122s	1130s	1122s	C—C
1112s(sh)				1109s(sh)	
	1097s	1097s(sh)		1096vs	
1080s	1082s	1084s	1081s	1071s(sh)	COH bending CH bending
1050m(sh)	1050m	1047m	1057w	1057s	C—O, C—C
1034w				1035m	COH bending
	998w	998w	996w	997m	CH_2 rock
				971m	
			953w		
944m	940s	940s			
925m(sh)	921m(sh)		921s		COH bending CH bending
901m	905sh	907sh	896sh	910m	
855m	863m	848m	863sh		
840m(sh)	850m(sh)	840m(sh)	844s		C—O, C—C CH bending α-C_1 confign.
				826w	
		801w			
783vw					
	762m	768w			
755vw			753m		
	715m	714w		726w	
700w			699m		
	672w	678w		674w	
660vw			656vw		
648vw				639m	
			624vw		
	605w	605w		611m	

Continued

Table 8.8. ***(Continued)***

Amylose	Amylopectin	Glycogen	Dextran	Cellulose	Assignment
591w(sh)					
568m	573m	570m	577w	568m	
			538vs		
524w	517w	514w		524s	
			506w	508sh	
481vs	477vs	476s			
				462s	
436w	441m	440m	445w	438vs	
402w	405w	402w	406s		
				394s	
				378s	C—O—C with
			367vw		β confign.
	352m	354m			
				346m	C—O—C with
		329w		330m	β confign.
	307m		299	303	
				277	
	246	242	252		
			211		

[a] Koenig (1972).
[b] s, strong; m, medium; w, weak, v, very, sh, shoulder.

(840 cm^{-1}), and dextran (844 cm^{-1}), to the α-configuration at the anomeric C1 position. Cellulose has no such band and cellobiose shows a weak band at 840 cm^{-1}. Both these compounds have the β-configuration at the anomeric C1 position. Another band occurs in the range 890–910 cm^{-1} for D-cellobiose (891 cm^{-1}), cellulose (910 cm^{-1}), β-D-maltose (902 cm^{-1}), amylose (901 cm^{-1}), amylopectin (905 cm^{-1}), glycogen (907 cm^{-1}), and dextran (896 cm^{-1}). However, spectra–structure correlations showed that this band is not unique to the β-anomeric configuration.

The Raman spectral pattern of each compound below 700 cm^{-1} is distinctive, as is the case for aqueous solution, and can be used for identification purposes.

Atalla and Dimick (1975) have recorded Raman spectra of highly crystalline samples of celluloses I and II. The main feature in the comparison of the two spectra in Fig. 8.12 is the occurrence of significant differences in the low-frequency region. The low-frequency region of the spectra, where most of the changes occur, is associated with skeletal-angle bending vibrations. The explanation offered by these authors for the spectral differences below 500 cm^{-1} is a change in the orientation of the glycosidic linkage between the D-glucose residues. The suggestion of different conformations for celluloses I and II does not preclude the differences in hydrogen-bonding patterns inferred from infrared spectra, but the differences in hydrogen-bonding patterns cannot explain the differences seen in Fig. 8.12 in the low-frequency regions.

Cyclohexaamylose (Fig. 8.13) is a cyclic polysaccharide containing six α-D-glu-

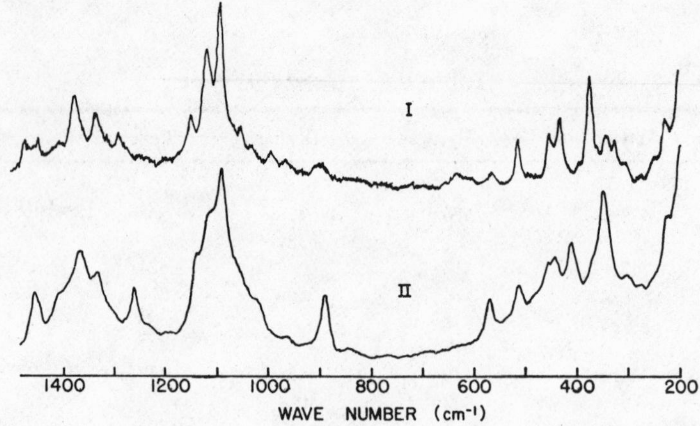

Fig. 8.12. Raman spectra of highly crystalline samples of celluloses I and II. (Attala and Dimick, 1975.)

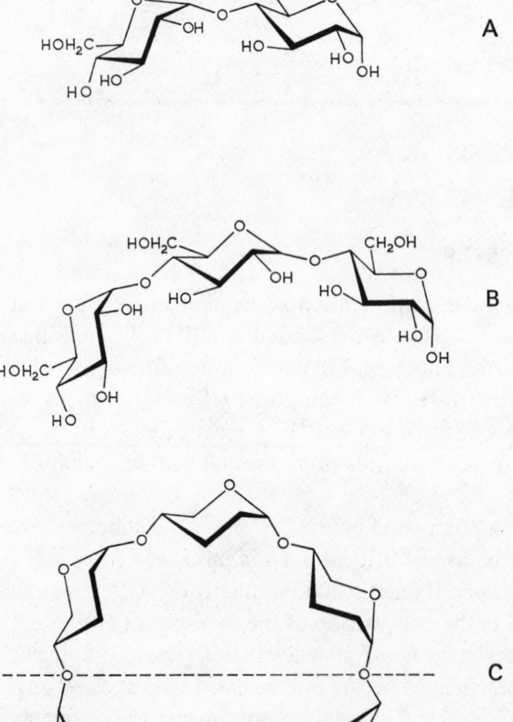

Fig. 8.13. Structure of (A) maltose, (B) maltotriose, and (C) cyclohexaamylose. (Note that a maltotriose molecule is exactly half the size of that of cyclohexaamylose, and it very likely has a conformation similar to that of half of cyclohexaamylose.) (Tu *et al.*, 1979.)

copyranosyl residues and no free hemiacetal hydroxyl group. Tu *et al.* (1979) have compared its Raman spectrum with those of maltose and maltotriose (Fig. 8.14), and have discussed structural implications of these spectra. Spectrum A (maltose) shows an α-anomeric band at 850 cm^{-1}. The band at 905 cm^{-1} indicates β-anomeric maltose. Spectrum B (maltotriose) shows a band at 860–851 cm^{-1} and the absence of one between 890 and 900 cm^{-1}; both of which indicate an α anomer. Spectrum C (cyclohexaamylose) displays a prominent band at 865 cm^{-1}, indicating the α-D-linkage.

Below 1500 cm^{-1}, C—H bending vibrations produce distinct bands at 1457, 1462, and 1454 cm^{-1} for maltose, maltotriose, and cyclohexaamylose, respectively. Mixtures of CH_2 deformation, C—O—H bending, and C—C stretching vibrations occur from 1440 to 1320 cm^{-1}. The maltotriose bands resemble more those of cyclohexaamylose than

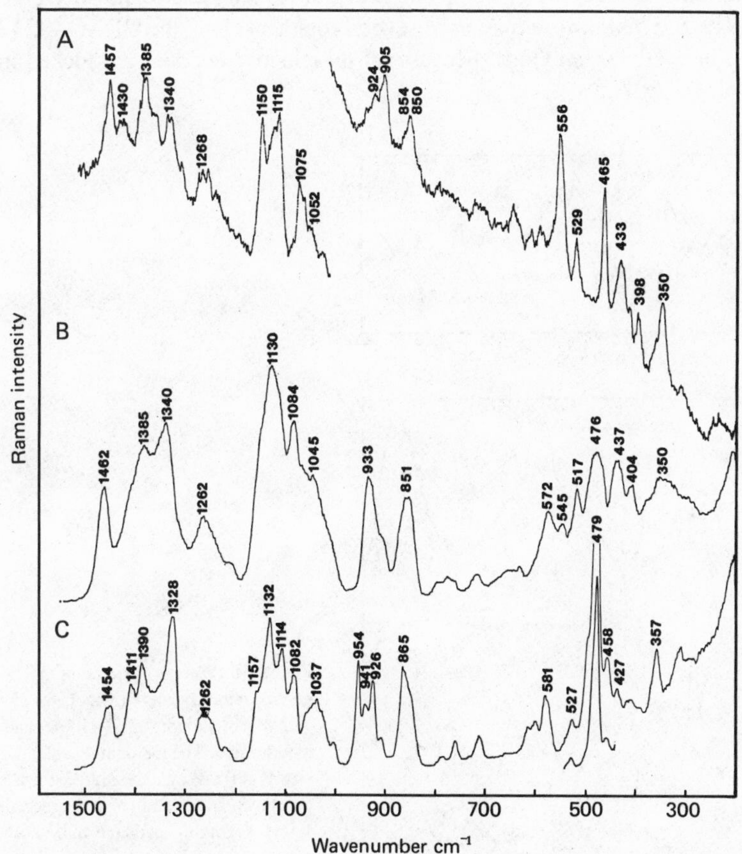

Fig. 8.14. Laser Raman spectra of (A), maltose; (B), maltotriose; (C), cyclohexaamylose in the range of 2000 to 1700 cm^{-1} (radiant power, (A) 220, (B) 100, and (C) 100 mW; slit width, (A) 500, (B) 300, and (C) 500 μm; integration time, 0.2 sec; scanning speed 0.2 cm^{-1}/sec). (Tu *et al.*, 1979.)

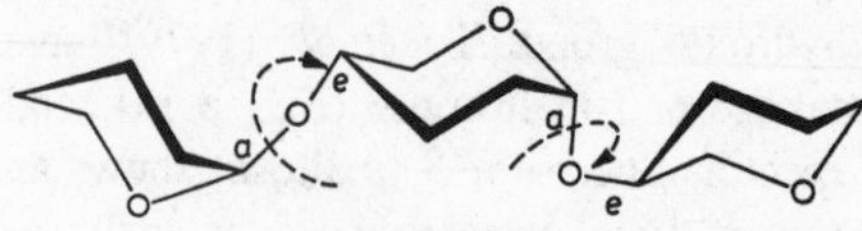

Fig. 8.15. Possible rotational isomers of maltotriose along the C1—O—C4 bonds [α-D-(1→4)-glucosidic linkage]. The symbol *a* denotes an axial bond, and *e* an equatorial bond. (Tu *et al.*, 1979.)

those of maltose. From 1200 to 1000 cm^{-1} (C—O—H stretching and OH deformation) maltotriose and cyclohexaamylose have quite similar spectra, different from that of maltose. The region from 960 to 920 cm^{-1} displays the C—O—C vibration of α-D-(1→4)-bonding. Here, cyclohexaamylose shows bands at 954, 941, and 926 cm^{-1}, but maltose and maltotriose have only one band.

Tu's laboratory has demonstrated (She *et al.*, 1974; Tu *et al.*, 1977; Tu *et al.*, 1978) that monosaccharides display many Raman bands below 600 cm^{-1}. Spectra A, B, and C (Tu *et al.*, 1979) also show complex Raman scattering at the lower frequencies, produced by ring vibrations, made up of complex modes of skeletal and torsional vibrations.

Maltose and maltotriose may have several rotamers along the C1—O—C4 axis (Fig. 8.15), but in cyclohexaamylose, because all the glucose residues are locked into a ring

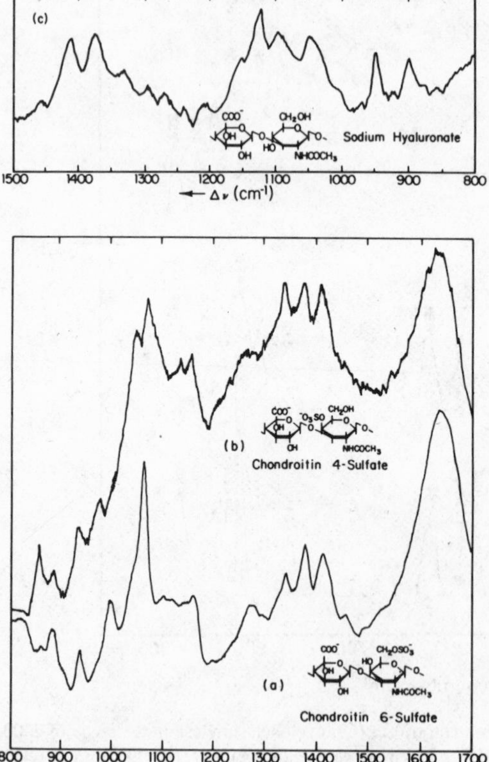

Fig. 8.16. Raman spectra of 5% solutions of glycosaminoglycans in water: (a) chondroitin 6-sulfate, (b) chondroitin 4-sulfate, and (c) sodium hyaluronate. The incident wavelength is 5145 Å, power 100 mW, and resolution 5 cm^{-1}. The spectra shown in (a) and (b) were recorded with the signal averaging package and show 100 scans averaged and Fourier transformed to reduce noise. The spectrum shown in (c) was recorded with the single scan mode of the Spex compudrive at a scan rate of 0.1 cm^{-1}/sec. (Bansil *et al.*, 1978.)

structure, such possibilities are strictly limited. It is interesting to note, therefore, that the spectrum of maltotriose has about the same complexity as that of cyclohexaamylose, and the overall spectrum of maltotriose more resembles that of cyclohexaamylose than that of maltose, indicating that the triose probably has a conformation very like that of the hexose (Tu *et al.*, 1979).

Glycosaminoglycans

Raman spectra of chondroitin 4-sulfate, chondroitin 6-sulfate, and sodium hyaluronate in the solid state and in aqueous solution have been recorded by Bansil *et al.* (1978). IR spectra of these substances were measured in the solid state. Figure 8.16 shows Raman spectra of these compounds in 5% aqueous solutions. Figure 8.17 shows solid state spectra for the chondroitin sulfates. These authors gave a detailed table of Raman frequencies of the above substances and D-glucuronic acid, *N*-acetylgalactosamine, and *N*-acetyl-glucosamine. For example, chondroitin 6-sulfate and chondroitin 4-sulfate show Raman bands at 820 and 853 cm^{-1}, respectively, for the asymmetric vibration of the C—O—S linkages, which have been interpreted to indicate an equatorial OSO_3^- group in the former compound and an axial OSO_3^- arrangement in the latter (Lloyd *et al.*, 1961).

Barrett and Peticolas (1979) have obtained the solvent-subtracted (difference) Raman spectra of hyaluronic acid of >95% purity (Fig. 8.18) at pH values 6.0 and 8.0 and ionic strength 0.1. The differences in the height of the two spectra are due only to differences in percentages of phosphate present. This study reports much better resolution and greater

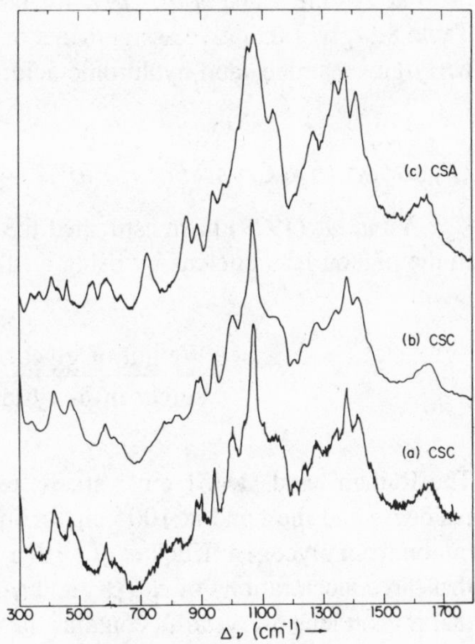

Fig. 8.17. Raman spectra of powders of chondroitin 6-sulfate (a, b) and chondroitin 4-sulfate (c). Incident laser wavelength = 5145 Å, power 20 mW, spectral resolutions, 5 cm^{-1}. The spectrum shown in (a) is the raw data signal averaged for 30 scans, whereas that shown in (b) is Fourier transformed to remove the high-frequency noise. A comparison of the spectra in (a) and (b) shows that data manipulation does not introduce any artifacts into the spectrum: it only reduces the noise, and this makes the weaker peaks show up more clearly. The spectrum shown in (c) is the average of 100 scans and has been Fourier transformed. CSA, chondroitin 4-sulfate, CSC, chondroitin 6-sulfate. (Bansil *et al.*, 1978.)

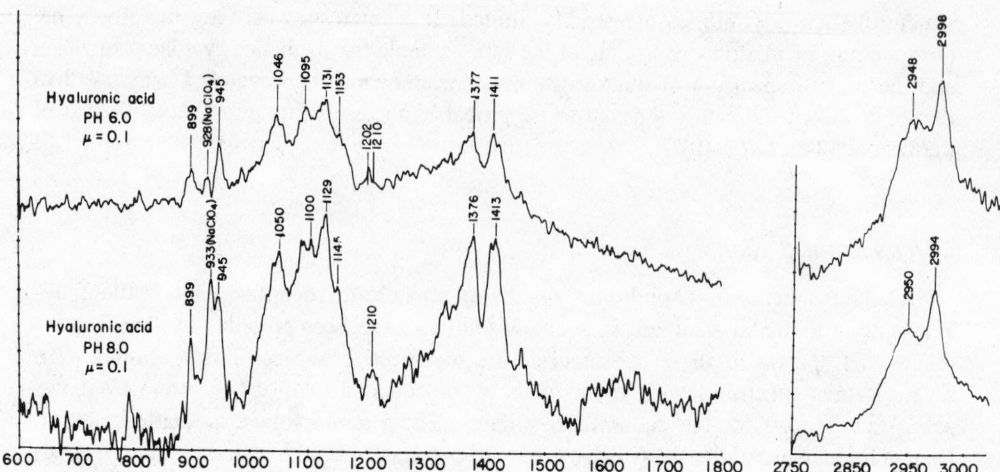

Fig. 8.18. Solvent subtracted Raman spectra of hyaluronic acid at pH 6.0 and 8.0 and constant ionic strength 0.1. Vibrations due to traces of the sodium perchlorate "marker" at ~928 cm^{-1} remain in these spectra. (Barrett and Peticolas, 1979.)

signal-to-noise ratio than a previous study (Tu *et al.*, 1977) on a <90% pure sample for the region 600 to 1800 cm^{-1} and 2750 to 3100 cm^{-1}. Barrett and Peticolas resolved the broad bands of the latter authors at ~1100 and ~1365 cm^{-1} into several peaks at 1050, 1100, 1129, and 1145, and at 1376 and 1413 cm^{-1}, respectively. They also recorded spectra of D-glucuronic acid and *N*-acetyl-D-glucosamine in aqueous 10% solutions. A band at 745 cm^{-1} (Tu *et al.*, 1977) is not observed in the spectra of the present authors. Table 8.9 gives tentative assignments of the Raman spectral lines of glucuronic acid, *N*-acetylglucosamine, and hyaluronic acid.

Glycogen to δ-Crystallin Ratio

Yu *et al.* (1977) have estimated the ratio of glycogen content to δ-crystallin content in the pigeon lens nucleus by using a calibration curve based on the following equation:

$$\frac{\text{Weight of glycogen}}{\text{Weight of }\delta\text{-crystallin}} = \frac{I_{481\ \mathrm{cm}^{-1}}/I_{1004\ \mathrm{cm}^{-1}}}{2.3}$$

The Raman band at 481 cm^{-1} arises from glycogen with no contribution from protein residues, and the band at 1004 cm^{-1} comes from phenylalanine residues with no contribution from glycogen (Kuck *et al.*, 1976). In order to use this equation, Yu *et al.* assumed that the concentrations of α-, β-, and γ-crystallins in the pigeon lens are negligible and that pigeon lens δ-crystallin contains the same mole percentage of phenylalanine as chick lens δ-crystallin, which was the type of crystallin used for the calibration.

Table 8.9. Tentative Assignment of Raman Spectral Lines (cm^{-1})[a]

Glucuronic acid	N-Acetyl-glucosamine	Hyaluronic acid
620		
770		
		899 β-linkages
917		945 C—C stretch
		C—O—C stretch
	953	
1030		
		~1050 C—O, C—C
	1055	
1057		
	1085	
		~1100 C—O—H bend, acetyl group
1117		
	1124	
		~1130 C—C
1154		~1150 C—O, C—C, oxygen bridge
		1210 CH_2 twist
	1325	
		1330 CH bend, amide III
1363		
		~1380 CH_3 ionized carboxyl
	1382 CH_3	
		~1410 CH bend, ionized carboxyl, amide II
	1485 CH_3, C(6)—H_2, COH	
		1460 CH_2 bend
1722 Carboxyl group		
2903		
	2905	
	2942	
		2950 CH_3 stretch
		~3000 NH stretch, CH_2 stretch

[a] Barrett and Peticolas (1979).

References

Atalla, R. H., and Dimick, B. E., *Carbohyd. Res.* **39,** C1-C3 (1975).

Avram, M., and Mateescu, Gh., *Infrared Spectroscopy, Applications in Organic Chemistry,* Wiley-Interscience, New York, 1972.

Bansil, R., Yannas, I. V., and Stanley, H. E., *Biochim. Biophys. Acta* **541,** 535 (1978).

Barker, S. A., and Stephens, R., *J. Chem. Soc.* **1954,** 4550.

Barker, S. A., Bourne, E. J., Stacey, M., and Whiffen, D. H., *J. Chem. Soc.* **1954a,** 171.

Barker, S. A., Bourne, E. J., Stephens, R., and Whiffen, D. H., *J. Chem. Soc.* **1954b,** 3468.

Barker, S. A., Bourne, E. J., Stephens, R., and Whiffen, D. H., *J. Chem. Soc.* **1954c,** 4211.
Barker, S. A., Bourne, E. J., and Whiffen, D. H., *Methods Biochem. Anal.* **3,** 213 (1956).
Barrett, T. W., and Peticolas, W. L., *J. Raman Spectrosc.* **8,** 35 (1979).
Bhavanandan, V. P., and Davidson, E. A., *Carbohyd. Res.* **57,** 173 (1977).
Blackwell, J., Vasko, P. D., and Koenig, J. L., *J. Appl. Phys.* **41,** 4365 (1970).
Boerio, F. J., and Koenig, J. L., *J. Polymer Sci.* **A-2,**9, 1517 (1971).
Brown, C. J., *Acta Cryst.* **13,** 1049 (1960).
Cabassi, F., Casu, B., and Perlin, A. S., *Carbohyd. Res.* **63,** 1 (1978).
Cael, J. J., Koenig, J. L., and Blackwell, J., *Carbohyd. Res.* **29,** 123 (1973).
Cael, J. J., Isaac, D. H., Blackwell, J., Koenig, J. L., Atkins, E. D. T., and Sheehan, J. K., *Carbohyd. Res.* **50,** 169 (1976).
Casu, B., Scovenna, G., Cifonelli, A. J., and Perlin, A. S., *Carbohyd. Res.* **63,** 13 (1978).
Danishefsky, I., and Bella, A., Jr., *J. Biol. Chem.* **241,** 143 (1966).
Eliel, E. L., Allinger, N. L., Angyal, S. J., and Morrison, G. A., *Conformational Analysis,* p. 398, Interscience, New York, 1965.
El Khadem, H. S., and Parker, F. S., in *The Carbohydrates,* 2nd Ed., Vol. IB, pp. 1376–1393 (W. Pigman and D. Horton, eds.) Academic Press, New York, 1980.
Harris, M. J., and Turvey, J. R., *Carbohyd. Res.* **15,** 51 (1970).
Hineno, M., *Carbohyd. Res.* **56,** 219 (1977).
Hoffman, P., Linker, A., and Meyer, K., *Biochim. Biophys. Acta* **30,** 184 (1958).
Hunt, S., and Jevons, F. R., *Biochem. J.* **98,** 522 (1966).
Isbell, H. S., Smith, F. A., Creitz, E. C., Frush, H. L., Moyer, J. D., and Stewart, J. E., *J. Res. Natl. Bur. Std.* **59,** 41 (1957).
Koenig, J. L., *J. Polymer Sci., Part D* **6,** 59 (1972).
Kuck, J. F. R., Jr., East, E. J., and Yu, N.-T., *Exp. Eye Res.* **23,** 9 (1976).
Lee, Y. C., Stowell, C. P., and Krantz, M. J., *Biochemistry* **15,** 3956 (1976).
Liang, C. Y., and Marchessault, R. H., *J. Polymer Sci.* **39,** 269 (1959).
Lloyd, A. G., and Dodgson, K. S., *Biochim. Biophys. Acta* **46,** 116 (1961).
Lloyd, A. G., Dodgson, K. S., Price, R. G., and Rose, F. A., *Biochim. Biophys. Acta* **46,** 108 (1961).
Marbet, R., and Winterstein, A., *Helv. Chim. Acta* **34,** 2311 (1951).
Mathews, M. B., *Nature* **181,** 421 (1958).
Mathews, M. B., and Dorfman, A., *Arch. Biochem. Biophys.* **42,** 41 (1953).
Meyer, K., Davidson, E., Linker, A., and Hoffman, P., *Biochim. Biophys. Acta* **21,** 506 (1956).
Murata, K., *Nature* **193,** 578 (1962).
Nakanishi, K., and Solomon, P. H., *Infrared Absorption Spectroscopy,* 2nd Ed., Holden-Day, San Francisco, 1977.
Nakanishi, K., Takahashi, N., and Egami, F., *Bull. Chem. Soc. Jpn* **29,** 434 (1956).
Onodera, K., Hirano, S., and Kashimura, N., *Carbohyd. Res.* **1,** 208 (1965).
Orr, S. F. D., *Biochim. Biophys. Acta* **14,** 173 (1954).
Parker, F. S., *Applications of Infrared Spectroscopy in Biochemistry, Biology, and Medicine,* Chap. 6, pp. 100–141, Plenum Press, New York, 1971.
Peat, S., Bowker, D. M., and Turvey, J. R., *Carbohyd. Res.* **7,** 225 (1968).
Rees, D. A., *J. Chem. Soc.* **1963,** 1821.
Scheinthal, B. M., and Schubert, M., *J. Biol. Chem.* **238,** 1935 (1963).
She, C. Y., Dinh, N. D., and Tu, A. T., *Biochim. Biophys. Acta* **372,** 345 (1974).
Sheehan, J. K., Gardner, K. H., and Atkins, E. D. T., *J. Mol. Biol.* **117,** 113 (1977).
Suzuki, S., and Strominger, J. L., *J. Biol. Chem.* **235,** 2768 (1960).
Terho, T., Hartiala, K., and Häkkinen, I., *Nature* **211,** 198 (1966).
Tipson, R. S., and Isbell, H. S., *J. Res. Natl. Bur. Std.* **64A,** 405 (1960a).
Tipson, R. S., and Isbell, H. S., *J. Res. Natl. Bur. Std.* **64A,** 239 (1960b).
Tipson, R. S., and Isbell, H. S., *J. Res. Natl. Bur. Std.* **65A,** 31 (1961a).
Tipson, R. S., and Isbell, H. S., *J. Res. Natl. Bur. Std.* **65A,** 249 (1961b).
Tipson, R. S., and Isbell, H. S., *J. Res. Natl. Bur. Std.* **66A,** 31 (1962).

Tipson, R. S., and Parker, F. S., in *The Carbohydrates,* 2nd Ed., Vol. IB (W. Pigman and D. Horton, eds.), pp. 1394–1436, Academic Press, New York, 1980.

Tipson, R. S., Isbell, H. S., and Stewart, J. E., *J. Res. Natl. Bur. Std.* **62,** 257 (1959).

Tipson, R. S., Cerezo, A. S., Deulofeu, V., and Cohen, A., *J. Res. Natl. Bur. Std.* **71A,** 53 (1967).

Tu, A. T., Lee, J., and Lee, Y. C., *Carbohyd. Res.* **67,** 295 (1978).

Tu, A. T., Lee, J., and Milanovich, F. P., *Carbohyd. Res.* **76,** 239 (1979).

Tu, A. T., Dinh, N. D., She, C. Y., and Maxwell, J., *Studia Biophys. (Berlin)* **63,** 115 (1977).

Turvey, J. R., Bowker, D. M., and Harris, M. J., *Chem. Ind. (London)* **1967,** 2081.

Vasko, P. D., Blackwell, J., and Koenig, J. L., *Carbohyd. Res.* **19,** 297 (1971).

Vasko, P. D., Blackwell, J., and Koenig, J. L., *Carbohyd. Res.* **23,** 407 (1972).

Verstraeten, L. M. J., *Anal. Chem.* **36,** 1040 (1964).

Verstraeten, L. M. J., *Carbohyd. Res.* **1,** 481 (1966).

Walton, A. G., and Blackwell, J., *Biopolymers,* Academic Press, New York, 1973.

Yu, N.-T., East, E. J., Chang, R. C. C., and Kuck, J. F. R., Jr., *Exp. Eye Res.* **24,** 321 (1977).

Chapter 9

NUCLEIC ACIDS AND RELATED COMPOUNDS

Introduction

Nucleic acid chemistry, biochemistry, and molecular biology are subjects that continue to hold the attention of scientists everywhere.

The original Watson–Crick model (1953) has since been revised by Wilkins and his co-workers (see Figs. 9.1 and 9.2). The latter figure shows (a) the space-filling model (with van der Waals radii indicated) of the revised structure of DNA in the B-configuration; and (b) a projection of the model. The B-configuration of DNA, which is the one stable at greater than 66% relative humidity, is made up of two right-handed helical polynucleotide chains in which the internucleotide linkage in one strand is $3' \rightarrow 5'$ and in the other is $5' \rightarrow 3'$, that is, the strands are of opposite polarity. The strands are wound around the axis of the molecule so that the double helix they form cannot be separated into two strands unless unwinding occurs. Complementarity of the organic bases is still maintained, that is, the allowed base pairs are adenine and thymine, and guanine and cytosine. Figure 9.3 shows these base pairs hydrogen-bonded. Unusual bases, such as substituted cytosines (for example, 5-methylcytosine) can replace cytosines wherever the former occur. The planes of the aromatic rings are at right angles to the helix axis and the hydrogen-bonded base pairs are stacked on top of one another 3.4 Å apart. The phosphodiester groups are on the outside of the cylinderlike molecule. Since each turn of the double helix contains ten bases and each turn has a pitch of 34 Å, the helix has an exact tenfold screw axis in a right-handed screw.

Wang *et al.* (1979) have described a left-handed double helical form of DNA in a single crystal of a DNA fragment that contains six base pairs with the sequence d(CpGpCpGpCpG), where d is deoxy; C, cytosine; p, phosphate; and G, guanine. The sugar-phosphate backbone of the left-handed double helix has a zigzag path, hence the name Z-DNA has been given to this form. Z-DNA forms a helix which has 12 base pairs per turn of 44.6 Å (Wang *et al.*, 1981). These authors list seven major differences in structure between the double helical forms of Z-DNA and right-handed B-DNA. Also, the GC base pairs of the left-handed double helix are located at one side of the axis near

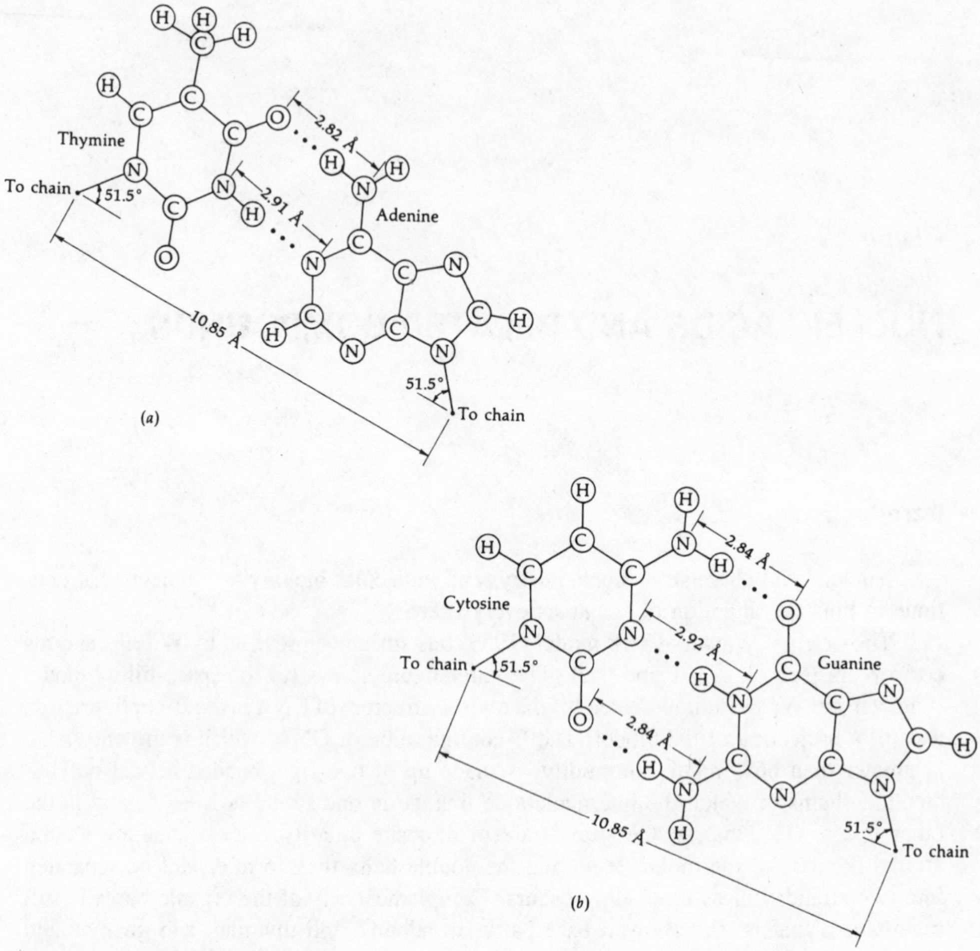

Fig. 9.1. Dimensions and hydrogen bonding of (*a*) thymine to adenine and (*b*) cytosine to guanine in double helix of DNA. (Arnott *et al.*, 1965.)

the periphery of the molecule instead of the center, as in right-handed B-DNA. In addition, right-handed B-DNA presents a major and a minor groove, whereas Z-DNA has only a single groove with phosphate groups located along the opening, the single groove being analogous to the minor groove of B-DNA. The outer convex surface of Z-DNA is analogous to the concave major groove of B-DNA (Wang *et al.*, 1981). These authors have delineated two forms of Z-DNA, Z_I and Z_{II}, the former being *gauche*(−)-*trans* for the phosphodiester conformation, and the latter *gauche*(+)-*trans* (Sundaralingam, 1969). Figure 9.4 shows the space-filling diagrams of the two Z conformations.

Dickerson *et al.* (1982) have compared in detail the structures of A-, B-, and Z-DNA. They presented skeletal and space-filling drawings of each type of DNA as determined from single-crystal X-ray structure analyses of certain base-pair oligomers. Each

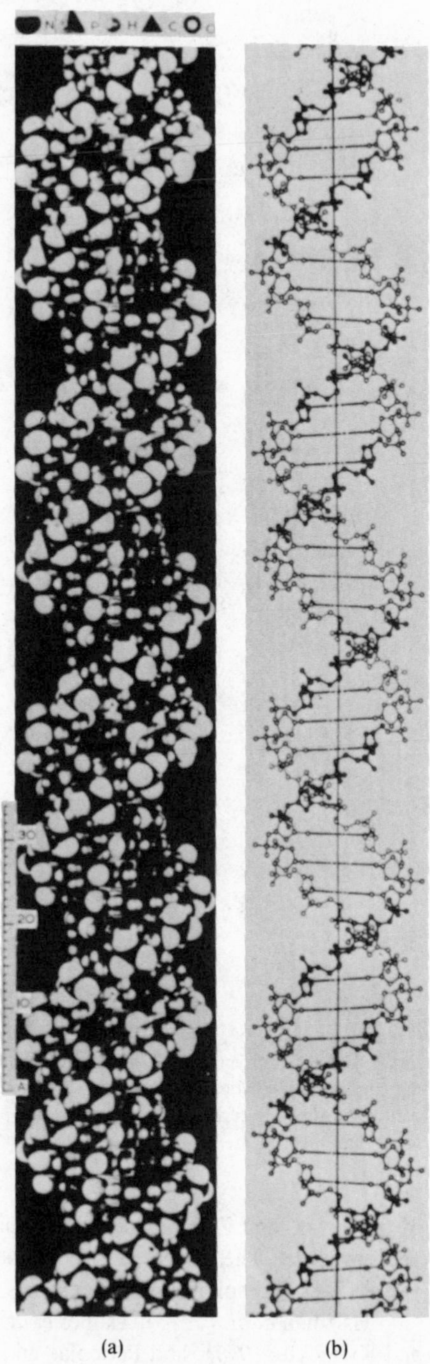

Fig. 9.2. (a) Model of B-configuration DNA (van der Waals radii indicated). (b) A projection of the model (M. H. F. Wilkins). (From Mahler and Cordes, 1971.)

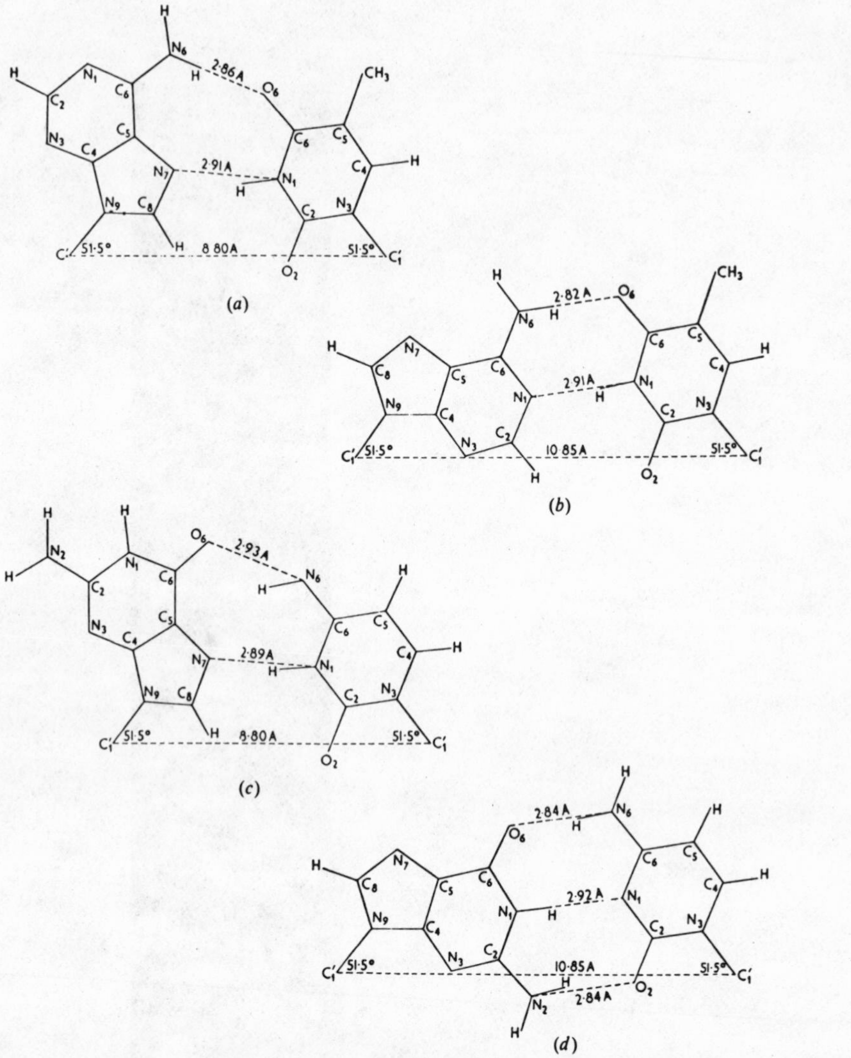

Fig. 9.3. Adenine-thymine hydrogen-bonded pairs: (a) found by Hoogsteen in a 1:1 complex of 3-*N*-methylthymine and 9-*N*-methyladenine; (b) postulated by Watson and Crick (current version refined by Arnott) for part of the structure of DNA. Guanine-cytosine pairs: (c) Hoogsteen type; (d) Watson–Crick type. (Arnott *et al.*, 1965.)

of the A, B, and Z families of DNA has its own intrinsic restrictions on chain folding and structure. The observed solvent positions in these crystal structures corroborated earlier fiber and solution measurements.

Hartman *et al*. (1973), Tsuboi *et al*. (1973), Thomas (1974), Tsuboi (1974), Thomas and Kyogoku (1977), and Peticolas and Tsuboi (1979) have reviewed the application of Raman and resonance Raman spectroscopy to the study of nucleic acids. The reviews by

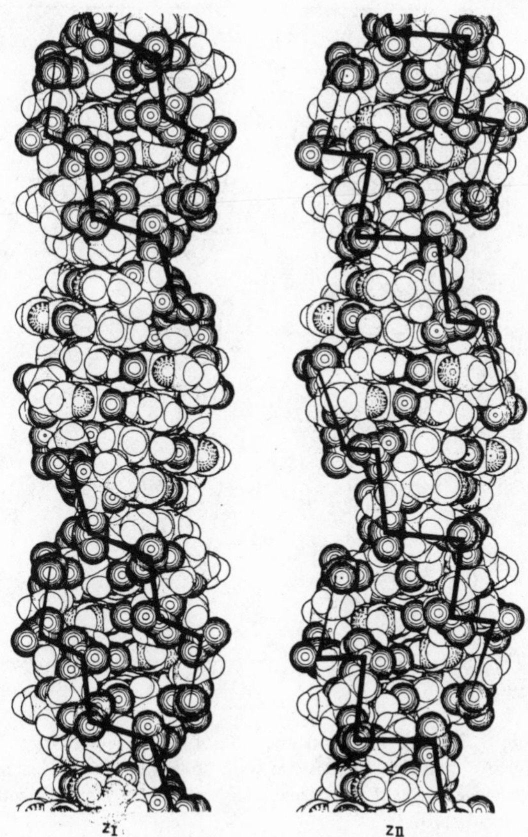

Fig. 9.4. Van der Waals diagrams of the Z_I and Z_{II} conformations. A solid heavy line is drawn from phosphate to phosphate which shows the zigzag course of the sugar phosphate chain. Because the GpC phosphate group is rotated away from the groove, the zigzag is accentuated in Z_{II} relative to Z_I. (Wang *et al.*, 1981.)

Hartman *et al.* (1973), Tsuboi (1974), and Thomas and Kyogoku (1977) also treat the application of IR spectroscopy.

IR Absorption Band Assignments

Figure 9.5 presents IR spectra of films of calf thymus deoxyribonucleic acid (DNA) and yeast transfer ribonucleic acid (tRNA) in the undeuterated, partly deuterated, and deuterated forms. Figure 9.6 shows IR spectra (polarized radiation) of oriented films of the sodium salt of rice dwarf virus RNA in the undeuterated and deuterated forms at 75% relative humidity (r.h.). Figure 9.6 presents many distinct, sharp bands, most of which show strong dichroism. The PO_2^- symmetric and antisymmetric stretching bands display rather different dichroic properties from their counterparts in DNA: in RNA the 1083-cm^{-1} symmetric stretching band shows parallel dichroism and the 1225-cm^{-1} antisymmetric stretching band shows perpendicular dichroism; whereas, in DNA the former displays perpendicular dichroism and the latter is not dichroic. Tsuboi (1969) concluded from dichroic measurements that the PO_2^- group of double-helical RNA of the rice dwarf

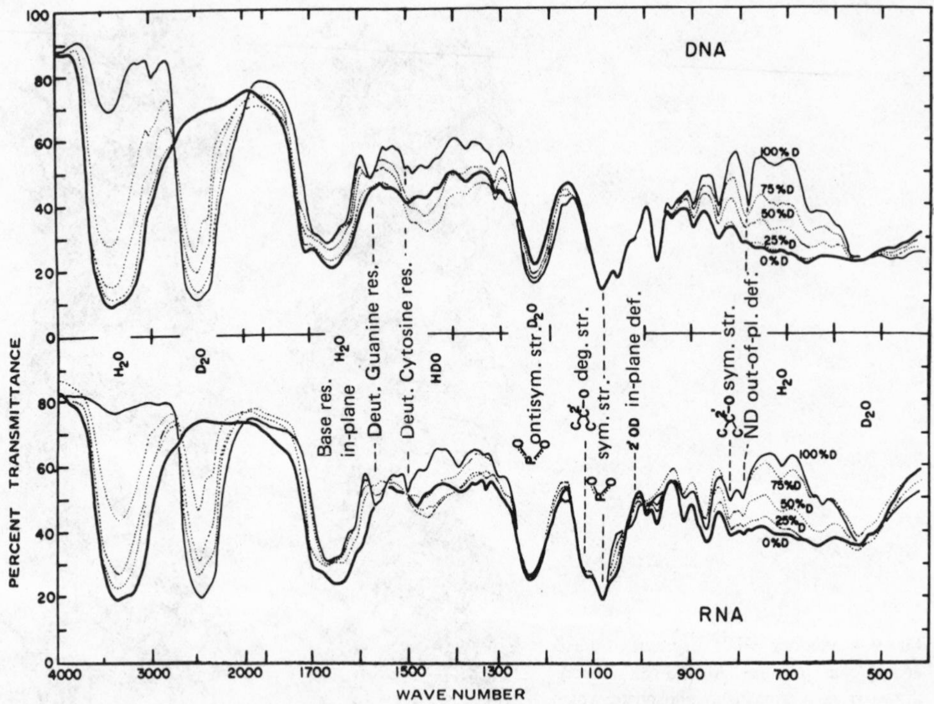

Fig. 9.5. Infrared absorption spectra of calt thymus DNA and yeast tRNA (bulk) films placed in air of 92% relative humidity. The spectra were observed in the undeuterated, partly deuterated, and nearly completely deuterated states. Partial deuteration was achieved by placing the films under partially deuterated water vapor. (Tsuboi, 1969.)

virus is oriented so that its O · · · O line forms an angle of ca. 70° with the polymer axis and the bisector of angle OPO forms an angle of ca. 40° with the axis. (See Beetz *et al.*, 1979, for a discussion of linear IR dichroism and its application to dioxy vibrations of the phosphate group of nucleic acids; see also p. 374 in this book.)

Both DNA and RNA have a broad intense band at 3400 cm^{-1} (or at 2500 cm^{-1} when deuterated), which is assigned to the OH (or OD) stretching vibration of adsorbed water molecules. In the range 1800–1500 cm^{-1} the group of strong bands are due to the C═O stretching, skeletal stretching, and NH bending vibrations of the organic base residues (Sutherland and Tsuboi, 1957). The PO_2^- antisymmetric stretching vibration is found at 1220 cm^{-1} (Sutherland and Tsuboi, 1957; Tsuboi, 1957). Several intense bands in the 1100–1000-cm^{-1} region are assigned to the PO_2^- symmetric stretching vibration of the phosphate group and the C—O stretching vibrations of the ribose or deoxyribose moiety (Sutherland and Tsuboi, 1957; Tsuboi, 1957; Shimanouchi *et al.*, 1964). The 1000–700-cm^{-1} region contains several medium or weak bands, which come from the P—O stretching, C—O stretching, and NH out-of-plane bending vibrations, and also from the librations

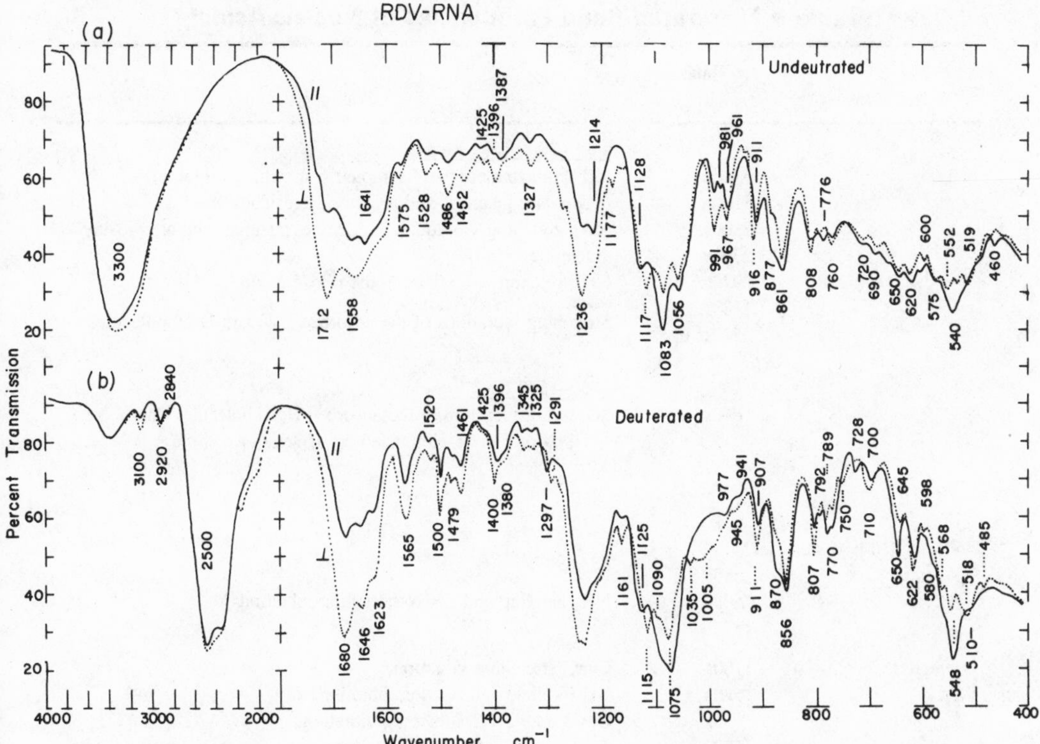

Fig. 9.6. Infrared absorption spectra of oriented films of sodium salt of rice dwarf virus RNA in the region 4000–400 cm^{-1}. (a) Undeuterated, at 75% r.h.; (b) deuterated, at 75% r.h. Full line, electric vector of the incident radiation is parallel to the fiber axis. Dotted line, electric vector of the incident radiation is perpendicular to the fiber axis. (Tsuboi, 1969.)

of adsorbed water molecules. The 600–400-cm^{-1} region should contain the librations of D_2O molecules and the ND out-of-plane vibrations. Table 9.1 (Tsuboi, 1969) presents the frequency, relative intensity, and probable vibrational mode for the IR bands seen in nucleic acids.

The region from 1500 to 1800 cm^{-1} has been used extensively for the IR examination of polynucleotides in ^2H_2O (D_2O) solution (see Miles and Frazier, 1978 and Miles, 1971 for listings of references and discussion of the qualitative identification of polynucleotide-complex forms and the quantitation of their amounts). However, strong absorbance of water at ca. 1645 cm^{-1} has impeded the accurate recording of IR spectra of H_2O solutions in the C═O region. In order to avoid this difficulty when they wanted to measure NH_2 deformations and the effect of coupling of NH and NH_2 bending motions on C═O vibrations, Miles and Frazier (1978) have used 5-μm-thick CaF_2 cells and computer subtraction of the water spectrum to produce accurate spectra that showed resolved, assignable absorption frequencies of nucleotides and polynucleotides from 1500 to 1800 cm^{-1}. Also, they avoided difficulties inherent in the use of such thin cells by employing

Table 9.1. Infrared Band Frequencies of Nucleic Acids[a,b]

Residue	Band (cm^{-1})	Comments
		4000–2000-cm^{-1} Region
	3300m 3100m	NH_2 antisymmetric and symmetric stretching vibrations of adenine, guanine, and/or cytosine residues.[c,d]
	2500w 2360w	ND_2 stretching vibrations of adenine, guanine, and/or cytosine residues.[c,d]
	2920w	CH stretching vibrations in the ribose group.
	3380s, br 2500s, br	Stretching vibrations of the adsorbed H_2O and D_2O molecules.
		1750–1550-cm^{-1} Region
	1640s, br	Scissoring vibration of the adsorbed H_2O molecule. This vibration can be eliminated by deuteration.
Undeuterated base residues		
Adenine	1665s 1605s 1573w	NH_2 bending and C=N stretching vibrations[e]
Uracil	1700s	C=O stretching vibration.
	1680s	NH in-plane deformation vibration.[e]
	1650s	C=O and C=C stretching vibration.
	1620w	C=O and C=C stretching vibration.
Thymine	1720s	C2=O stretching vibration.[d]
	1660s	C4=O stretching vibration.[d]
	1575w	Ring stretching vibration.[d]
Guanine	1690s 1630w	C=O stretching and NH_2 scissoring vibration.[f,g]
Hypoxanthine	1680s,br	In-plane ring vibration.
Cytosine	1700w	
	1670sh	NH_2 scissoring vibration.[e]
	1647s	In-plane ring vibrations.
	1603s	In-plane ring vibrations.
	1585w	In-plane ring vibrations.
Deuterated base residues		
Adenine	1627s 1579w	C=N and C=C stretching vibrations.[e]
Uracil	1692m	C2=O stretching vibration.[e,h]
	1657s	C4=O and C=C stretching vibrations.[e,h]
	1618w	C4=O and C=C stretching vibrations.[e,h]
Pseudouridine[g]	1690m	C2=O stretching vibration.
	1653s	C4=O and C=C stretching vibration.

Continued

Table 9.1. *(Continued)*

Residue	Band (cm^{-1})	Comments
Guanine	1665s	C═O stretching vibration.[i]
	1581s	C═N stretching vibrations.[j]
	1568m	C═N stretching vibrations.[j]
Hypoxanthine[k]	1672s	C═O stretching vibration.[i]
Cytosine	1652s, 1616m, 1585w	In-plane ring vibrations.
		1550–1300-cm^{-1} Region
Deuterated adenine residue	1483w	There are several weak bands in this region. They are assigned to in-plane vibrations of the base residues involving the N—H and C—H in-plane deformation modes. Some examples are given for both deuterated and undeuterated residues.
Undeuterated cytosine residue	1530w	
	1500m	For RNA, a band should appear near 1400 cm^{-1}, which is due to a vibration involving the in-plane C—O—H deformation mode at the 2′ position of the ribose ring.[l]
Deuterated cytosine residue	1525m	
	1503s	
		1300–1000-cm^{-1} Region
Undeuterated cytosine residue[j,l,m]	1296w	
Undeuterated thymine residue[j,m]	1292w	
	1276w	
	1225s	Antisymmetric stretching vibration of the [P(O)(O)] group.[c,n] An absorption of the adsorbed D_2O molecules is superimposed here.
	1135s	Nearly degenerate stretching vibrations of the C(C)–O–C
	1116s	structure in RNA at the 2′ position of the ribose residue.[o]
	1084s	Symmetric stretching of the [P(O)(O)] group.[c,n,p]
	1050s	Some C—O stretching vibration of the ribose or the deoxyribose residue.
	1020m,br	C—O—D in-plane vibration in deuterated RNA at the 2′ position of the ribose residue.[l]

Continued

Table 9.1. *(Continued)*

Residue	Band (cm^{-1})	Comments
		<u>1000–700-cm^{-1} Region</u>
	800–760w	A few weak bands are present. These are due to out-of-plane vibrations of base residues of undeuterated DNA's and RNA's.[d,j,o]
	800–760	Several bands which are stronger than the above are present. These are due to out-of-plane vibrations of base residues of deuterated DNA's and RNA's[d,j,o]
	967s 890m 830w	Every DNA, both deuterated and not. The 967 cm^{-1} band is similar in position and intensity to the 950 cm^{-1} band of diethylphosphate anion, which is assigned to the C—C stretching vibration.[p]
	995m 970m 915s 880m 870s 815s	Many of these appear in all the synthetic polyribonucleotides and in all natural RNA. The 995 and 970 cm^{-1} bands disappear on deuteration; the others remain almost unchanged. All are due to ribose–phosphate chain-backbone vibrations. The 815 cm^{-1} band may be assigned to a nearly symmetrical stretching of the (C)(C)C–O structure at the 2′ position of the ribose residue.[o]
		<u>700–400-cm^{-1} Region</u>
	700–600vbr	Due to adsorbed H_2O molecules.
	550–450br	Due to D_2O molecules.
	650m 645m 622m 598w 580w 568w 548s 518m 510w 485w	These bands are superimposed upon the above broad bands, and are found in deuterated RNA of rice dwarf virus. The 650, 580, and 548 cm^{-1} bands are strongly polarized along the fiber axis. Those at 645, 598, 518, and 485 cm^{-1} are polarized perpendicular to the fiber axis.

[a] Tsuboi (1969).
[b] Key: br, broad; m, medium; s, strong; sh, shoulder; v, very; w, weak.
[c] Sutherland and Tsuboi (1957).
[d] Kyogoku *et al.* (1967b).
[e] Tsuboi *et al.* (1962).
[f] Angell (1955; 1961).
[g] Tsuboi and Kyogoku (1969).
[h] Miles (1964).
[i] Howard and Miles (1965).
[j] Tsuboi *et al.* (1968).
[k] Miles (1961).
[l] Tsuboi (1964).
[m] Tsuboi and Shuto (1966).
[n] Tsuboi (1957).
[o] Tsuboi *et al.* (1963).
[p] Shimanouchi *et al.* (1964).
[q] Sato *et al.* (1966).

mixed H_2O-organic-solvent systems (H_2O–methanol; H_2O–ethylene glycol), which lowered the concentration of water and the absorbance of solvent, thereby allowing thicker cells and standard double-beam compensation.

Poly(adenylic acid) in water solution displays a strong NH_2 deformation mode at 1659 cm^{-1} coupled to an adenine ring vibration at 1605 cm^{-1}. Poly(uridylic acid) shows a wide band at ca. 1695 cm^{-1}, mainly C2=O stretching, an unresolved C4=O band (probably near 1660 cm^{-1}) and a ring vibration at ca. 1636 cm^{-1} (Miles and Frazier, 1978). These workers discussed in great detail and compared the spectra of these polymers and their helical complexes as observed in H_2O and in ^2H_2O.

Tautomeric Forms

The fact that all derivatives of adenine, uracil, cytosine, thymine, and inosine (i.e., the bases and their nucleotides and nucleosides) are found in the keto-amino forms rather than the enol-imino forms in neutral aqueous solutions, has been corroborated by Raman spectroscopy (Lord and Thomas, 1967; Medeiros and Thomas, 1971). This finding is in agreement with earlier studies, e.g., those by Miles and his colleagues with IR studies (Miles, 1956; 1958a,b; 1959; 1961; Miles *et al.*, 1963). Note that guanine is not mentioned in the list above. The support for the keto-amino structure of guanine is on weaker grounds. It is possible that other ring structures may be present in aqueous solutions of guanine in equilibrium with the keto-amino form (Tomasz *et al.*, 1972).

Adenosine Tri-, Di-, and Monophosphates

Rimai *et al.* (1969) have obtained Raman spectra for adenosine tri-, di-, and monophosphate in the pH range 0.5 to 13.5 and between 550 and 1700 cm^{-1}. The spectra were pH dependent, and ATP was easily distinguishable from ADP. Bands at about 960 and 1100 cm^{-1} were good indications of the degree of ionization of the terminal phosphate group.

Heyde and Rimai (1971) have measured the frequency (in the vicinity of 1125 cm^{-1}) of the symmetric P=O stretching band of the phosphate group of ATP as a function of pH in aqueous solutions of ATP alone, and of ATP complexes with Na^+, Mg^{2+}, and Ca^{2+}. From these data they obtained the pK_a of $HATP^{3-}$ and the stability constants of ATP–metal complexes. The logarithms of the stability constants of ATP^{4-} with Ca^{2+} and Mg^{2+} are 3.9 ± 0.4 and 4.5 ± 0.3, respectively. The pK_a' of the secondary proton dissociation of ATP (in concentrations ranging from 0.02 *M* to 0.3 *M*) is 6.50, in good agreement with the value found by other methods.

The logarithms of the stability constants of ATP^{4-} with Mn^{2+} and Zn^{2+} are 3.81 ± 0.09 and 3.40 ± 0.05, respectively (Rimai and Heyde, 1970).

Lewis *et al.* (1975) have used Raman spectroscopy to study a phosphate transfer reaction from adenosine triphosphate to inorganic phosphate or arsenate in dimethyl sulfoxide. This reaction resembles in some respects energy transfer reactions catalyzed by coupling factor 1 from chloroplasts [Nelson *et al.*, (1972)].

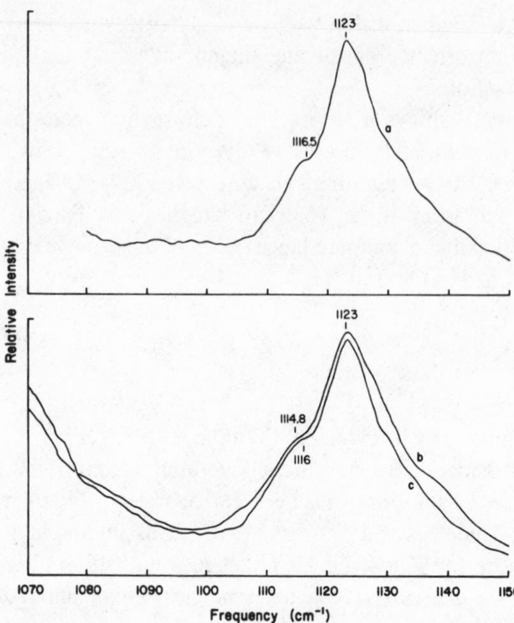

Fig. 9.7. (a) The Raman spectrum of the vibration associated with the triphosphate moiety in a solution containing 20 m*M* ATP, 5 m*M* arsenate, 20 m*M* $MgCl_2$, and 1 ml of dimethyl sulfoxide. (b) Same as (a) except 40 m*M* Tricine-maleate (pH 8) has been added and arsenate has been deleted. (c) Same as (a) except 40 m*M* Tricine-chloride (adjusted to pH 8) has been added. (Lewis *et al.*, 1975.) Reprinted with the permission of the American Chemical Society. Copyright 1975.

In Fig. 9.7 it can be seen that the symmetric stretch of the phosphate group that occurs at 1117 cm^{-1} in an ATP solution free of Mg^{2+} ions (see Fig. 9.8) has shifted and split into two vibrations. The principal band now occurs at 1123 cm^{-1} while a definite shoulder appears at about 1116 cm^{-1} in all the spectra of Fig. 9.7. This change suggested that the Mg^{2+} ions interact with at least some of the phosphate groups. The fact that a shoulder occurs at 1116 cm^{-1} of about half the intensity of the 1123-cm^{-1} band allows the assumption that two of the three phosphate groups are complexed to the Mg^{2+} ions while one remains free. The frequencies of the main band and the shoulder support this explanation. The vibration at 1123 cm^{-1} is higher in frequency than the band at 1117 cm^{-1} obtained for ATP solutions with no Mg^{2+} ions. Rimai *et al.* (1970) had observed earlier such a frequency shift for ATP solutions with Mg^{2+} and with Ca^{2+}. The shoulder at 1116 cm^{-1}, in contrast, is evidence of a free phosphate, so the stretching frequency is rather similar to the 1117-cm^{-1} band for ATP with no Mg^{2+} ions. The spectra in this paper support a mechanism in which Mg^{2+} binds to the α and β phosphates of ATP, leaving the third phosphate free for the transfer reaction.

Nicotinamide Adenine Dinucleotide

Patrick *et al.* (1974) have examined the carbonyl frequencies of the Raman spectra of nicotinamide, 3-acetylpyridine, and nicotinaldehyde as a function of solvent. These compounds displayed similar carbonyl frequencies (~1690 to 1700 cm^{-1}) in solvents that do not hydrogen-bond. In H_2O, D_2O, and CD_3OD, which do hydrogen-bond, the carbonyl frequency of nicotinamide was considerably lower (~30 to 40 cm^{-1}); but that of the other

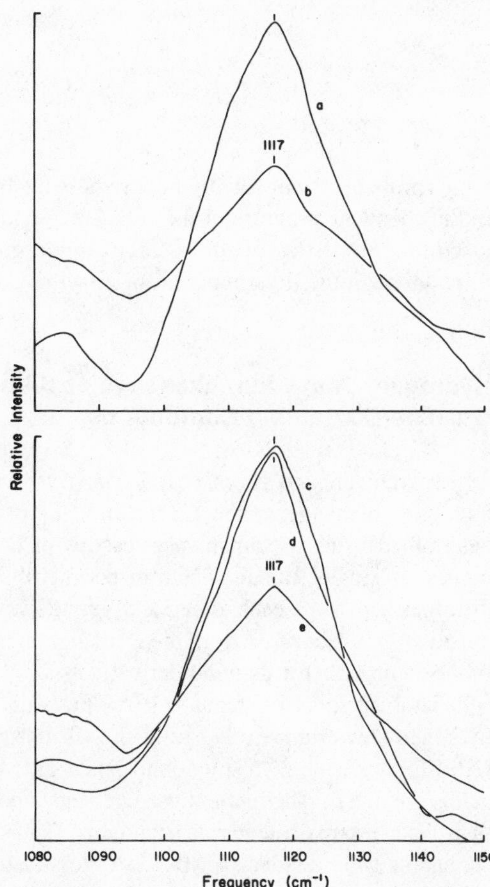

Fig. 9.8. (a) The Raman spectrum of the vibration associated with the triphosphate moiety in a solution containing 20 m*M* ATP, 40 m*M* Tricine-maleate (pH 8), and 5 m*M* arsenate. (b) Same as (a) except 20 m*M* NaCl and 1 ml of dimethyl sulfoxide were added. (c) Same as (a) except Tricine-maleate and arsenate were deleted. (d) Same as (a) except Tricine-maleate was deleted. (e) Same as (a) except 1 ml of dimethyl sulfoxide was added. (Lewis *et al.*, 1975.) Reprinted with the permission of the American Chemical Society. Copyright 1975.

substances changed very little. The authors interpreted their data in terms of preferential stabilization of a resonance form of the carboxamide group, —$(O^-)C{=}NH_2^+$, by hydrogen-bonding solvents producing increased single-bond character (lower vibrational frequency) for the carbonyl of the carboxamide moiety. The *N*-methyl derivatives of these compounds gave similar results, allowing Patrick *et al.* to suggest that the proposed resonance effect was independent of the nature of the *N* substituent and it was therefore functioning in NAD^+. Thus, the carboxamide group of NAD^+ in aqueous solutions would be more polar than it would be, if represented as usual, by the group —$C({=}O)NH_2$.

NADH and its analogs, on the other hand, displayed no solvent effect on the carbonyl frequency in the Raman spectra. Thus, the polar group proposed for the 3-carboxamide group of the oxidized compounds may be largely diminished when the ring is reduced.

The distribution of electrons in the amide group of NAD^+ and NADH could be of major importance in determining interaction for the coenzymes at the binding sites of various pyridine nucleotide dehydrogenases (Patrick *et al.*). This investigation supports the view that, in aqueous solutions, the —$C({=}O)NH_2$ group in the three position of NAD^+ may be better expressed by the structure

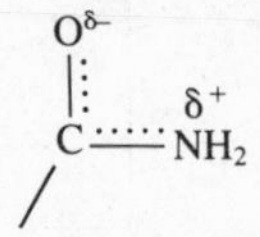

$$O^{\delta-}$$
$$\overset{\delta+}{C \cdots NH_2}$$

which probably is not equally important in the NADH electronic structure. The differences in the chemical structure of the ring (i.e., oxidation vs. reduction state), also change the electronic properties of the 3-carboxamide group—changes which will probably be of importance in the determination of enzyme–coenzyme interactions.

Hydrogen–Deuterium Exchange at C8 of Adenine- and Guanine-Containing Compounds

Livramento and Thomas (1974) have shown that the exchange kinetics for the isotopic hydrogen exchange at the C8 position of purines can be more accurately and easily determined by laser-Raman spectroscopy of D_2O solutions than by tritium labeling techniques. A single Raman spectrum permits the C8-H and C8-D forms of adenine to be distinguished from each other and simultaneously their relative concentrations can be quantitatively measured from the Raman intensities. Another single spectrum can be used in the same way for guanine derivatives.

Thomas and Livramento (1975) have done a comparative study of the kinetics of hydrogen–deuterium exchange at the C8 positions of adenosine 5′-monophosphate (5′-rAMP), adenosine 3′,5′-monophosphate or cyclic AMP (cAMP), and poly(riboadenylic acid) [poly(rA)]. The method used depends upon the use of Raman lines of the nucleotide phosphate groups as internal standards. Thus, reliable assignments and intensity measurements were needed, so these workers also studied a series of Raman spectra of 5′-rAMP and cAMP to evaluate how phosphate-group frequencies and intensities were affected by pH and pH changes in the absence of 8-CH exchange. (The pH and pD values used ranged from five to seven.)

They determined pseudo-first-order rate constants for the deuterium exchange of 8-CH groups in 5′-rAMP, cAMP, and poly(rA) as a function of temperature in the range 20–90°C. In the case of 5′-rAMP, the logarithm of the rate constant displays strictly linear dependence on $1/T$, i.e., $k_\psi = Ae^{-E_a/RT}$, where A is 2.3×10^{14} per h and E_a is 24.2 ± 0.6 kcal/mol. In the case of cAMP, above 50°C, k_ψ is almost identical in magnitude and temperature dependence to that of 5′-rAMP; but below 50°C, deuterium exchange in cAMP is much faster than in 5′-rAMP, accompanied by an activation energy of 17.7 kcal/mol and a frequency factor of 9.6×10^9 per h. The exchange rate for poly(rA) is much lower than that for 5′-rAMP at all temperatures. However, as in the case of cAMP, the plot of ln k vs. $1/T$ is divisible into a high- and a low-temperature section, each of which has different Arrhenius characteristics. Above 60°C, the E_a of poly(rA) is 22.0 kcal/mol and A is 3.2×10^{12} per h. Below 60°C, E_a is 27.7 kcal/mol and A is 1.8×10^{16} per h. Thus, an increase in temperature above 60°C does not decrease the retardation of exchange in poly(rA) compared to 5′-rAMP. The conclusion drawn from these results (Thomas and Livramento, 1975) is that the distribution of electrons in the adenine ring of cAMP is changed when the temperature goes below 50°C, although such a change does not occur in the case of 5′-rAMP. These authors concluded that retardation of exchange in poly(rA) is most probably caused by base stacking at lower temperatures and by steric hindrance from the polyribonucleotide backbone at higher temperatures.

Lane and Thomas (1979) have used Raman spectroscopy to determine pseudo-first-order rate constants as a function of temperature for the deuterium-exchange reaction of C8-H groups in guanosine 5′-monophosphate (5′-rGMP) and guanosine 3′,5′-monophosphate (cGMP). The logarithm of the rate constant (k_ψ) for each compound showed a linear dependence on $1/T$, i.e., $k_\psi = Ae^{-Ea/RT}$, with $A = 8.84 \times 10^{14}$ per h and $E_a = 24.6$ kcal/mol for 5′-rGMP; and $A = 3.33 \times 10^{13}$ per h and $E_a = 22.2$ kcal/mol for cGMP. (A is the frequency factor and E_a is the Arrhenius activation energy.) These authors found that exchange of the C8-H groups in guanine nucleotides is about 2 to 3 times faster than in adenine nucleotides (Thomas and Livramento, 1975).

Lane and Thomas pointed out certain advantages of Raman spectroscopy over the measurement of the rate of tritiation, which is a well-established method for monitoring isotopic hydrogen exchange at the C8 position of purines: (1) the Raman spectrum simultaneously measures *both* the structural integrity of the molecule *and* the concentration of isotopically exchanged molecules; (2) the Raman spectroscopic method can measure the exchange of guanine and adenine moieties separately in a polynucleotide or oligonucleotide; and (3) the Raman spectroscopic method is much simpler compared to labeling and assaying procedures used in the tritium-exchange method. However, the Raman technique as used by these workers needs much larger samples of material than the tritiation method.

Effect of Methylation on Some Nucleosides

Mansy *et al.* (1979) have recorded Raman spectra of methyl derivatives of 5′-AMP, tubercidin (7-deazaadenosine), inosine, uridine, and cytidine; and have discussed the effect of methylation at different sites on the acidity and the spectra of nucleosides.

Titration of a solution of N^6-methyltubercidin in D_2O with acid produces no change in its spectrum until ~pH 7. In the region of pD <7 a Raman band at 1320 cm^{-1} due to I disappears (Mansy *et al.*, 1979), and a band at 1350 cm^{-1} appears characteristic of II caused by protonation (Fig. 9.9 shows the spectra of N^6-methyltubercidin at extremes of pD). The respective decrease and increase in the intensities at 1320 and 1350 cm^{-1} were normalized by dividing the original recorded values by the intensity of ν_1 of an internal ClO_4^- ion reference. These authors used plots of these intensity ratios (Fig. 9.10) to determine a pK for the dissociation of XII of 5.1 ± 0.2.

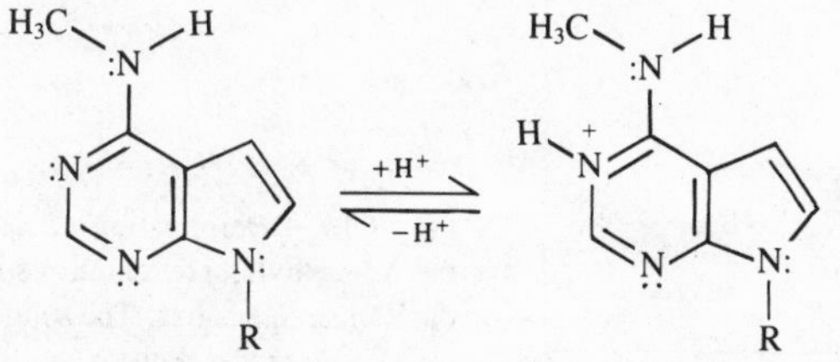

$$\text{I} \underset{-H^+}{\overset{+H^+}{\rightleftharpoons}} \text{II}$$

I

II

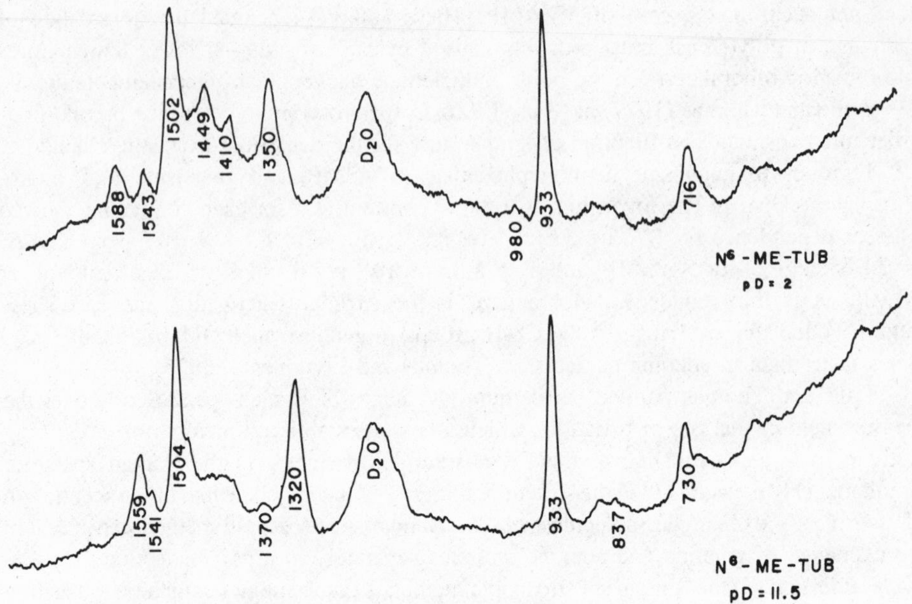

Fig. 9.9. Raman spectra of N^6-methyltubercidin in D_2O at pD 2 and 11.5. (Mansy *et al.*, 1979.)

Metal Binding

Investigators have been studying the binding of metals by nucleosides and nucleotides for many years. Since it was observed several years ago that CH_3Hg^+ is mutagenic (Löfroth, 1972; Mulvihill, 1972) and that specific platinum (II) compounds inhibit mitosis by selective inhibition of DNA synthesis (Harder and Rosenberg, 1970; Gale *et al.*,

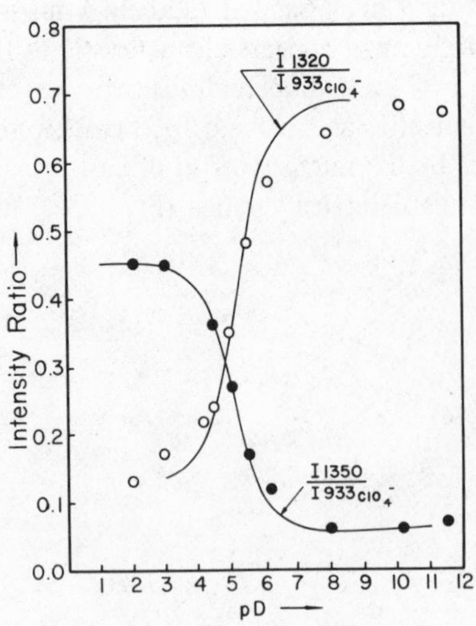

Fig. 9.10. Determination of the dissociation constant for the N^6-methyltubercidinium cation in D_2O at 25° C from the Raman intensities. The solid lines are theoretical curves for a single proton transfer equilibrium with pK 5.2. (Mansy *et al.*, 1979.)

1971), a spurt has occurred in studies of the binding of heavy metals to polynucleotides. Mansy *et al.* (1974) have examined the interaction between CH_3Hg^{II} and pyrimidine nucleotides and nucleosides. They used Raman difference spectroscopy to look for evidence of binding of CH_3Hg^+ to uridine [abbreviated Urd, according to IUPAC-IUB abbreviations for nucleosides, etc.; see *Biochemistry* **9,** 4022 (1970)].

Figure 9.11 shows the difference spectrum for 50 m*M* Urd plus 50 m*M* CH_3Hg^+ versus 50 m*M* Urd, both at pH 7. The vibrations due to group vibrations of CH_3Hg^+ should be displayed as positive peaks in the difference spectrum. The ν(Hg—C) and $\delta_s(CH_3)$ modes are quite apparent at 560 and 1207 cm^{-1}, respectively. Bands that occur due to a metal–nucleoside complex and which are not characteristic of CH_3Hg^{II} or nucleoside are also displayed as positive peaks, examples of which are those at 1291, 1044, and 605 cm^{-1}. A derivative curve, which is centered at 786 cm^{-1}, is caused by a small shift in a Urd band. Such derivative curves allow the detection of very small frequency shifts or intensity changes, or both, in broad bands. When coordination of CH_3Hg^+ takes place, a shift in a Urd band from 780 to 791 cm^{-1} occurs, giving rise to the feature at 786 cm^{-1}.

These authors outlined a procedure for the determination of heavy-metal binding sites on polynucleotides with two or more base moieties in aqueous solution. The methylmercury (II) ion binds to uridine with displacement of a proton and coordination to N3. The ion also binds to cytidine at N3; however, at pH 7, coordination to uridine is the preferred reaction.

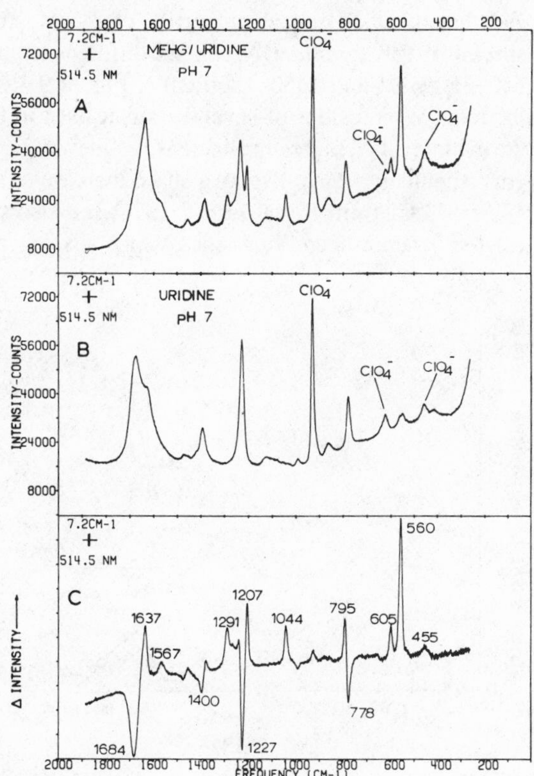

Fig. 9.11. Raman difference spectrum: A, 50 m*M* Urd + 50 m*M* CH_3HgClO_4; B, 50 m*M* Urd (both solutions 0.1 *M* in $NaClO_4$); C, (A − B) difference spectrum. Solution pH 7. Scan conditions: 0.25 Å intervals, 10 sec counting time. (Mansy *et al.*, 1974.) Reprinted with the permission of the American Chemical Society. Copyright 1974.

Mansy and Tobias (1975) have shown by means of Raman difference spectrophotometry that CH_3Hg^{II} binds quantitatively to N1 of inosine at pH 8. As shown by NMR measurements, binding occurs also at N7 when N1 is saturated.

Theophanides (1979a) has reviewed the interaction of metal ions with nucleosides, nucleotides, and nucleic acids; he has presented characteristic properties of the various metal ions, the formation of complexes, and some of the factors that stabilize or disrupt complex formation. He has also given a detailed discussion (1979b) of the IR and Raman spectroscopy of metal nucleic acid systems, e.g., the effect on the $—NH_2$ frequency position when a metal (M) is attached to the $—NH_2$ group of an adenine moiety, thus producing an $—MNH_2^+$ group; or the effects of metal binding by, e.g., GMP and ATP, on spectral bands and their intensities.

Spectra of DNA and Their Temperature Dependence

Table 9.2 shows Raman spectral data for calf thymus DNA obtained by Small and Peticolas (1971b). These, of course, show major differences from data on RNA because of replacements of ribose by deoxyribose and of uracil by thymine.

An early comprehensive review on Raman spectra of nucleic acids has been presented by Koenig (1972). Rimai *et al.* (1974) have investigated the temperature dependence of certain Raman line intensities of double-stranded calf thymus DNA in order to determine correlations that may occur between the temperature dependence of the Raman spectra and that of other optical properties of DNA, particularly the UV absorbance. Raman spectra of calf thymus DNA at six different temperatures were observed for the ranges 650–850 cm^{-1} and 1450–1850 cm^{-1}. Figure 9.12 shows the temperature dependence of the relative intensities of several Raman lines in the DNA. Four classes of Raman lines are evident: (1) temperature-increasing lines, e.g., the 730-cm^{-1} adenine line and the 798-cm^{-1} shoulder, which display a sharp increase of intensity in the region of DNA melting ($T_m \approx 73°C$, from Rimai *et al.*, 1974; see also Gruenwedel and Chi-Hsia Hsu, 1969); (2) temperature-decreasing lines which display a decreasing temperature dependence,

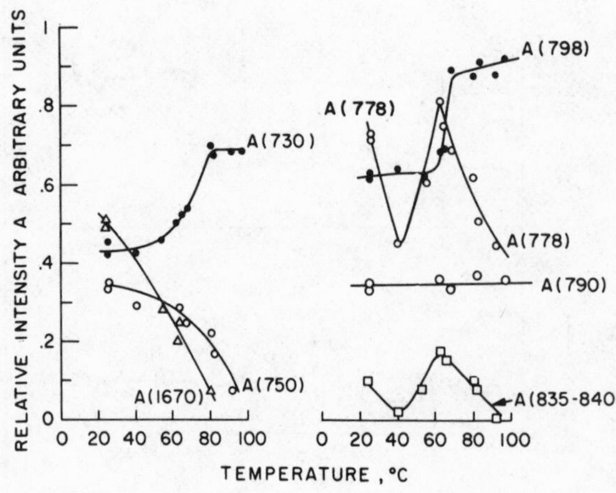

Fig. 9.12. Temperature dependence of the relative intensities of several Raman lines in calf thymus DNA dissolved in 0.015 *M* NaCl, 0.0015 *M* sodium citrate. The Raman intensities were measured relative to the intensity of the 1637-cm^{-1} water bending line. (Rimai *et al.*, 1974.)

Table 9.2. Raman Frequencies for Aqueous Calf Thymus DNA[a]

Frequencies (cm^{-1})		
H_2O solution	D_2O solution	Assignments[b]
	500	Deoxyribose-phosphate
	567	Deoxyribose
672	662	T
683	685	G
730	725	A
752	743	T
	774	C, T
787		O—P—O diester symmetric stretch overlapping C, T
	792	O—P—O diester symmetric stretch
~835	838	Deoxyribose-phosphate
	871	Deoxyribose-phosphate
895	897	Deoxyribose-phosphate
917	921	Deoxyribose
975	977	Deoxyribose
1017	1015	C—O stretch
1058	1053	C—O stretch
1094	1095	$O\cdots P\cdots O^-$ symmetric stretch
1144		Deoxyribose-phosphate
1180		Base external C—N stretch
1214		T
1226		A
1242		T
1259		C, A
1304	1307	A
1320		G
1340	1351	A
1378	1382	T, A, G
1423	1424	A, G
1448	1449	Deoxyribose
1462	1465	Deoxyribose
1491	1486	G, A
	1504	A
1514	1524	A
1534		G, C
1580	1580	G, A
	~1621	
	1672	C=O stretch

[a] Small and Petiolas (1971b).

[b] T, C, A and G indicate vibrations characteristic of the thymine, cytosine, adenine, and guanine bases, respectively, listed in order of their relative contributions with the largest contribution first. Deoxyribose-phosphate indicates probable origin is in the deoxyribose-phosphate chain but cannot be readily assigned specifically to deoxyribose or phosphate.

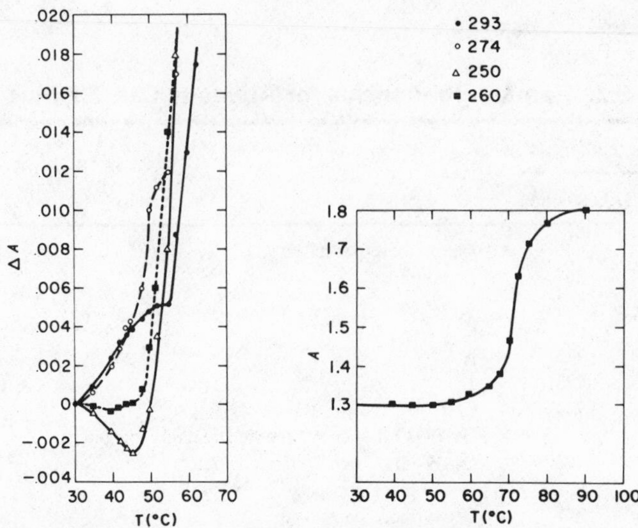

Fig. 9.13. Temperature dependence of the absorbance difference at four different wavelengths in the near-UV taken from difference spectra of a calf thymus DNA solution approximately 65 μg/ml in 0.015 *M* NaCl, 0.0015 *M* sodium citrate. Left: plot of absorbance difference referred to a sample at the reference temperature, 30°C. Right: the complete melting curve at 260 nm. (Rimai *et al.*, 1974.)

spread over a wide range, and have intensities which reach zero in the region of high temperature, e.g., the thymine line at 750 cm^{-1} and the 1674-cm^{-1} carbonyl stretch; (3) lines with an up-and-down temperature dependence, but which display a sharp increase somewhere in the melting region, e.g., 778 and 835 cm^{-1}; and (4) temperature-independent lines such as the 790-cm^{-1} band. (There are other temperature-independent lines in the DNA spectra; see Rimai *et al.*, 1974).

Rimai *et al.* studied the temperature dependence of the UV temperature difference spectrum of calf thymus DNA in the premelting region at several temperatures between 40° and 55°C, with 30°C as the reference temperature. Figure 9.13 shows the temperature dependence of the absorbance difference at 293, 274, 260, and 250 nm taken from the spectra (not shown here). The whole 260-nm melting curve is also given. The data presented by these authors and by Sarocchi and Guschlbauer (1973) show that the 260-nm band is insensitive to temperature changes in the pre-melting region. Figure 9.13 also shows that the bands at 293, 274, and 250 nm are sensitive to temperature changes in the pre-melting region. Using a simple model, Rimai *et al.* were able to account for the variety of Raman temperature dependence and qualitatively correlate these with the temperature dependences seen in the UV spectra.

Hydration of A-, B-, and C-Forms of DNA, RNA, and Other Polynucleotides

The data from IR studies (Falk *et al.*, 1963a) suggest the following sequence of events during the hydration of DNA: the first water molecules adsorb at the PO_2^- groups (site 1), then at the C—O—P (site 2) and C—O—C (site 3) groups of the sugar-phosphate linkages. By the time the relative humidity (r.h.) of 65% is reached, six water molecules are adsorbed per nucleotide, and the sites 1, 2, and 3 are water-saturated. Above 65% relative humidity, water molecules start to bind to adenine, guanine, thymine, and cytosine (these are sites 4), and the DNA structure changes as the bases become ordered.

The major conclusion reached by Falk *et al.* (1963a and 1963b) is that the B (and/or A) helical configuration of sodium DNA changes to a disordered form when the water of hydration is removed from the grooves (sites 4) of the helix. It is interesting to note that the helical structure of DNA appears to collapse *in vivo* during dehydration. When bacteria are dehydrated under controlled humidity conditions (50%–70% r.h.), they are more susceptible to natural death and to damage from X-ray or UV irradiation (Webb, 1967). This is the range of relative humidity in which the DNA structure breaks down.

Hartman *et al.* (1973) have reviewed the subject of the hydration and structure of nucleic acids. Table 9.3 shows the major IR bands observed for films of DNA.

X-ray diffraction is a reliable method for the determination of the structure of nucleic acids and polynucleotide helical chains (Arnott, 1970; Davies, 1967); but the method is not generally useful for work with dilute nucleic acid solutions. Erfurth *et al.* (1972) have tackled the problem of obtaining structural information, both on fibers and on dilute solutions, by Raman spectroscopy. They observed several Raman bands that are produced by the vibration of the sugar-phosphate backbone, in both RNAs and DNAs whose frequencies and intensities are directly related to what form the nucleic acids are in—A, B, or C—according to X-ray crystallographic categories. These Raman bands are inde-

Table 9.3. Major Bands in the Spectrum of Hydrated DNA[a]

Frequency (cm^{-1})	Group and mode of vibration	Information obtained
3400 cm^{-1}	OH stretch of H_2O or of HDO in D_2O	Extent of hydration. Structure of water in hydration layer
1640 cm^{-1}	OH bend of H_2O	Obscured by DNA bands
710 cm^{-1}	H_2O libration	
2520 cm^{-1}	OD stretch of D_2O or HDO in H_2O	Extent of hydration. Structure of water in the hydration layer
1215	OD bending of D_2O	Obscured by DNA bands
3400–3200	NH stretch of amino and imide groups in nucleic acids	Obscured by water but can show rate of exchange with D_2O
1750–1600	In-plane stretching modes of the C=C, C=N and C=O groups of the bases	Observe in films hydrated with D_2O. Shows orientation of the base planes; extent of hydration; extent of base pairing
1240–1220	Antisymmetric stretching motion of the PO_2^- groups	Observe in films hydrated with H_2O. Shows extent of hydration and orientation of PO_2^-
1084	Symmetric stretching mode of the PO_2^- group	Orientation of the PO_2^- group
1054	CO stretch	Hydration
1015	CO stretch	Hydration
960	PO stretch	Hydration

[a] Hartman *et al.* (1973).

pendent of other chemical characteristics, such as base composition, whether or not a 2′-hydroxyl is present, etc.

Several investigators have shown that ribonucleic acid structures always display a Raman band at about 810 to 814 cm^{-1}, when these structures have an ordered or partly ordered form (Small and Peticolas, 1971a,b; Thomas *et al.*, 1971; Aylward and Koenig, 1970). Ribosomal RNA shows a band in this region (Thomas *et al.*, 1971). The spectrum of yeast transfer RNA (Fig. 9.14) shows such a band quite plainly. When a rise in temperature causes the secondary structure to disappear, this band at 814 cm^{-1} characteristically disappears (Small and Peticolas, 1971a,b; Thomas *et al.*, 1971). This band, which is present in all ordered ribo-structures, was assigned to the sugar-phosphate diester stretch, more probably the symmetric stretch than the antisymmetric one. DNA and deoxyhomopolymers do not show a band at 814 cm^{-1}. Calf thymus DNA shows a broader peak at about 835 cm^{-1}.

Erfurth *et al.* hypothesized that the configuration of the phosphate backbone is responsible for the differences in RNA and DNA noted above, i.e., that the conformation of the phosphate group in the A form, characteristic of RNA, causes the strong, highly polarized vibration at 814-cm^{-1}, and the phosphate configuration in the B form, characteristic of DNA in solution, does not. They tested the hypothesis by making DNA fibers in the A form, using low NaCl content and 75% humidity. For this preparation, a strong band much the same as the 814-cm^{-1} band in RNA solutions showed up for the DNA fiber in the A form at 807 cm^{-1}. When they changed the humidity around the fiber to 98%, the DNA went over to the B form when equilibrium was achieved, and the Raman spectrum then had the same identical characteristics of DNA in solution. The disappearance of the 807-cm^{-1} band is completely reversible and humidity dependent.

When the nucleic acid backbone has the A form, the symmetric dioxy-stretch of the phosphate group is found at 1101 cm^{-1}. When it has the B form, it is found at 1094 cm^{-1}, but this small change may not be easily seen.

Erfurth *et al.* (1972) also observed the spectrum of the C form of DNA (low humidity, 32%). For this form, the strong band at 785 cm^{-1} stands alone (i.e., neither the sharp band at 807–814 cm^{-1} nor the broad band at 835 cm^{-1} is present). At high humidity, the C-form spectrum converts back to a spectrum like that of the B form.

The Raman method used here was also applied to short oligoribonucleotides and dimers of ribonucleosides. At low temperature (6°C) and pH 7, the phosphate group takes up the A conformation when the stacking forces between the bases are strong enough.

Erfurth *et al.* (1975) have obtained both Raman spectra and X-ray diffraction patterns from oriented fibers of sodium deoxyribonucleic acid as a function of salt content and relative humidity. For oriented fibers the A-form is always found between 75% and 92% relative humidity. The conformation changes to the B-form at 92% relative humidity only if an excess (3%–5%) of added salt is present. Oriented fibers having low amounts of added salt stay in the A-form at 92% relative humidity and higher.

Correlating X-ray diffraction patterns and Raman spectra, these authors found that a band at 807 cm^{-1} was always present when a fiber displayed the A-type diffraction pattern; this band shifts to 790 cm^{-1} when the nucleic acid is in the B form. Some DNA samples did not lend themselves to X-ray diffraction; these were examined by Raman spectroscopy to study the A $\rightleftharpoons$ B transformation in unoriented fibrous masses of DNA

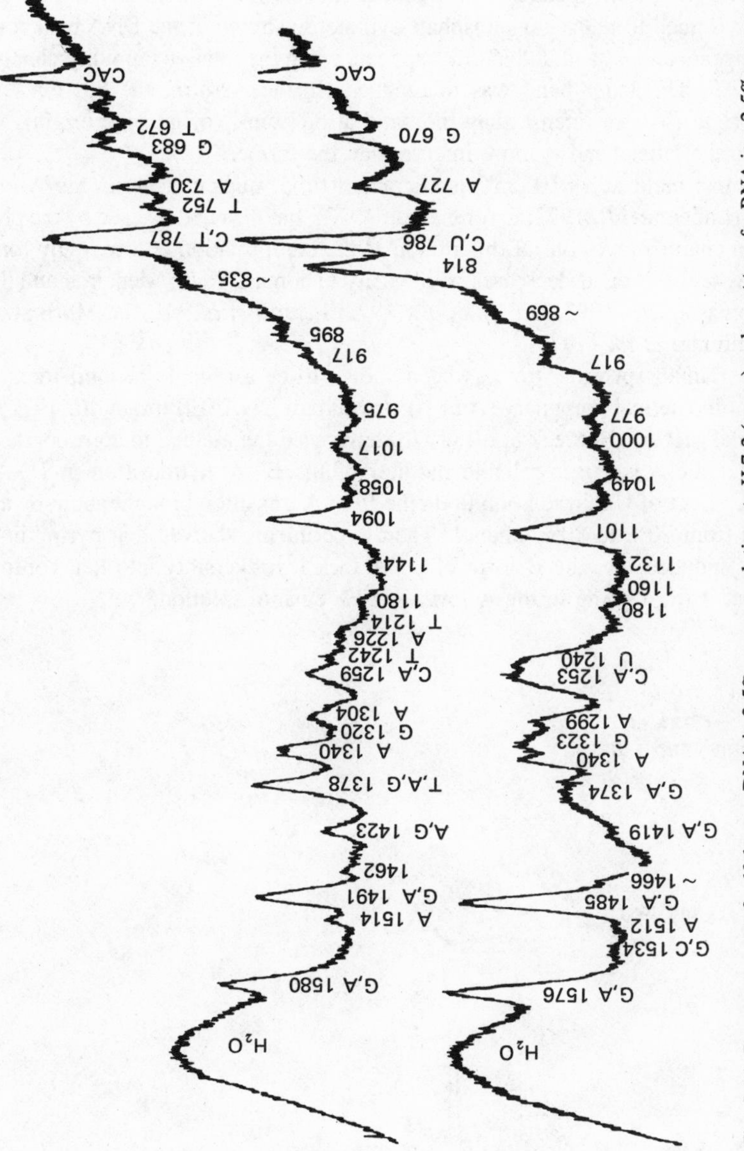

Fig. 9.14. Raman spectra of calf thymus DNA in 2.5% aqueous solution at pH 7.2 (upper) and yeast transfer RNA in a 2.5% aqueous solution at pH 7 (lower) with 0.01 *M* cacodylate buffer. (Erfurth *et al.*, 1972.)

and in concentrated, oriented gels. In the unoriented fibrous masses, the transformation always occurs at 92% relative humidity even at very low salt concentrations; but in oriented DNA gels at low salt concentrations, the A form can persist as a metastable state when the concentration is as low as 20% DNA.

Figure 9.15 shows spectra of an oriented calf thymus sodium DNA. The 807-cm^{-1} band was assigned to the sugar-phosphate symmetric stretch of the DNA backbone. This band disappears and a band at 835 cm^{-1} appears when the relative humidity changes from 75% to 92%. The latter band was unassigned. In the A form, the 665-cm^{-1} band of thymine origin is more intense than the band at 682 cm^{-1} (of guanine origin), while in the B form the latter band is more intense than the former.

A Raman band at ~810 cm^{-1} is characteristic of nucleic acids in the A-type conformation (Lafleur *et al.*, 1972; Erfurth *et al.*, 1972), but the band, caused by the phosphate sugar main chain, moves out of this region as the conformation becomes any form other than the A, and the band decreases in intensity (Thomas, 1970; Medeiros and Thomas, 1971; Thomas *et al.*, 1972; Brown *et al.*, 1972; Erfurth *et al.*, 1975a; Morikawa *et al.*, 1973; Nishimura *et al.*, 1974).

Since Raman spectroscopy has been shown to be a reliable method for the determination of the helical secondary (A or B) structure of DNA (Erfurth *et al.*, 1972; Erfurth and Peticolas, 1975), Herbeck *et al.* (1976) have used the method to corroborate the fact that circular dichroism is a reliable monitor of the B $\rightleftharpoons$ A transition in DNA. They studied the effect of UV irradiation upon the B $\rightleftharpoons$ A transition brought about by a change of solvent from 70% to 80% ethanol. Their experiments showed that pyrimidine-dimer cross-links induced into the B form of DNA lock it irreversibly into that conformation and prevent it from going to the A form in 80% ethanol solution.

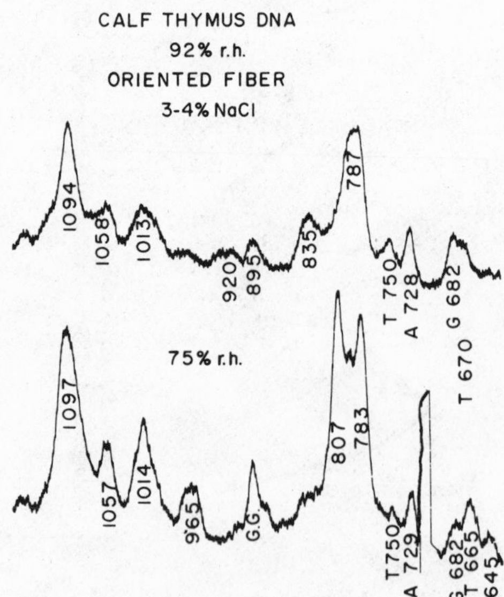

Fig. 9.15. Raman spectra of an oriented fiber of calf thymus Na-DNA with 3%–4% NaCl at 75% and 92% relative humidity. (Erfurth *et al.*, 1975.)

IR Dichroism

Fritzsche *et al.* (1976) have used IR linear dichroism to study oriented films of calf thymus DNA before and after their irradiation with 254-nm UV light. The irradiated DNA films were changed in their conformation. These changes were indicated by a replacement of the B → A transition by a B → C transition when the humidity was lowered, from 90% to 75% relative humidity. C-like structures of DNA in films can be sensitively detected by IR linear dichroism. Pilet and Brahms (1972 and 1973), and Brahms *et al.* (1973) have shown that measurements of the dichroism produce quantitative information about the geometrical arrangement of the phosphate (PO_2^-) group in the backbone of the DNA double helix [however, the reader is advised to consult Beetz *et al.* (1979), who have stated that structural information cannot be obtained from the use of linear IR dichroism of the dioxy vibrations of the phosphate group of nucleic acids. Also, see later, p. 374 in this chapter.]. The antisymmetric PO_2^- stretching vibration at 1230 cm^{-1} has a transition moment in the direction of the (O—O) line of the phosphate group, the symmetric PO_2^- stretching vibration at 1090 cm^{-1} in the direction of the bisector of the OPO angle. The linear dichroism of these two vibrations permits calculation of the two angles $\Theta_{O—O}$ and Θ_{OPO} between the related transition moments and the helix axis (Fraser, 1953 and Brahms *et al.*). Data obtained by Fritzsche *et al.* (1976) for these angles ($\Theta_{O—O}$ at 1230 cm^{-1} and Θ_{OPO} at 1090 cm^{-1}) supported the presence of the B→C transition when the relative humidity was lowered from 90% to 75% (see Table 9.4).

Pilet *et al.* (1975) have used IR linear dichroism measurements (Pilet and Brahms, 1972, 1973) to investigate conformations and structural transitions in several polydeoxy-

Table 9.4. Angles Θ of the Transition Moments Relative to the Helical Axis of DNA Calculated from the Infrared Dichroic Ratio $R = A_\perp/A_\parallel$[a,b]

		$\Theta_{O—O}$ (1230 cm^{-1})			Θ_{OPO} (1090 cm^{-1})		
	Humidity:	high	low		high	low	
Sample	Form:	B	A	C	B	A	C
Calf thymus DNA, nonirradiated		55 ± 2°	65 ± 2°	—	70 ± 3°	45 ± 2°	—
DNA, irradiated at 254 nm, 5.0 × 10^4 J m^{-2} (97% relative humidity)		56 ± 1°	—	52 ± 1°	70 ± 2°	—	69 ± 2°
3 DNA's of different GC content, mean value		56 ± 5°	64 ± 5°	—	70 ± 5°	45 ± 5°	—
C-like forms of DNA, mean value		—	—	50 ± 3°	—	—	66 ± 5°

[a] Fritzsche *et al.* (1976).

[b] The data assigned to the B conformation were measured at relative humidities > 88%; the data assigned both to the A- and the C-like conformations at intermediate relative humidities around 75%.

nucleotides. [See below for definition of dichroic ratio, $R(\perp/\|)$]. Polydeoxynucleotides of different base sequence—the alternating poly[d(A-T)] · poly[d(A-T)], crab *(Cancer pagurus)* satellite DNA, on the one hand, and double-stranded homopolymer complexes poly[d(A)] · poly[d(T)], poly[d(I)] · poly[d(C)], on the other, exhibit significant differences in their conformation and conformational transitions. Poly[d(A-T)] · poly[d(A-T)] takes up at lower humidity an A* form of low stability, i.e., restricted to limited relative humidity. In the A* form the orientation of the bisector of the phosphate OPO group is at 34° ± 7° with respect to the helical axis, which is somewhat lower than that of the classical A form of DNA from *Micrococcus lysodeikticus* (45° ± 3°). In the cases of the homopolynucleotide double-stranded complex poly(dA) · poly(dT) and poly(dI) · poly(dC), the conformational change from B to A is not found. Rather, poly(dA) · poly(dT) at lower humidity is in a stable altered B form. The data of Pilet *et al.* (1975) showed that homo(dA) · homo(dT) double-stranded sequences prevent the change from the B to the A form. All AT-containing polydeoxynucleotides and crab satellite DNA take up at high humidity an altered B form in which the orientation of the OPO bisector at 64° relative to the helical axis is significantly lower than the 68° to 74° seen in DNAs. The data also indicate that the base-pairing geometry in poly(dA) · poly(dT), poly[d(A-T)] · poly[d(A-T)], and in poly (dI) · poly(dC) appears to be of the Watson and Crick type.

The angles referred to above (formed between the transition moment and the polynucleotide axis) are calculated by use of the dichroic ratio $R(\perp/\|)$:

$$R(\perp/\|) = \frac{\sin^2\theta + g}{2\cos^2\theta + g}$$

where g is the parameter characterizing the semicrystalline state of the oriented sample and which can be related to the fraction f of perfectly oriented chains by $f = 1/(1 + {}^3/_2\, g)$.

The observed differences in conformation are not due to different base-pairing schemes. Pilet *et al.* suggested that in DNAs of high AT content the presence of homo(dT) · homo(dA) sequences and the relatively low stability of the A form of d(A-T) alternating sequences may inhibit the change to the A form.

Beetz *et al.* (1979) have stated that structural information cannot be obtained from the use of linear IR dichroism of the dioxy vibrations of the phosphate group of nucleic acids. They have collected data for several types of DNA showing that there are discrepancies between the values calculated for certain angles in these molecules by X-ray diffraction and IR dichroism. For example, in A-type DNA, the angle formed by the bisector of the PO_2^- group (O—P—O) angle and the helix axis is 45° by the IR method and 15° by the X-ray method; the angle formed by the O2–O3 line (Fig. 9.16) and the helix axis is 65° by IR and 79° by X-ray. Such discrepancies appear to arise from misinterpretation of the IR dichroism data. The misinterpretation, these authors say, comes from assigning C_{2v} symmetry to the PO_2^- group in nucleic acids (based on the $H_2PO_2^-$ ion) or C_2 [based on the $(CH_3O)_2PO_2^-$ ion]. The symmetry of the PO_2^- group in the backbone chain is actually C_s because the C3 and C5 carbons linked to the O1 and O4 oxygen atoms are not equivalent (Beetz *et al.*, 1979). They gave an explanation based

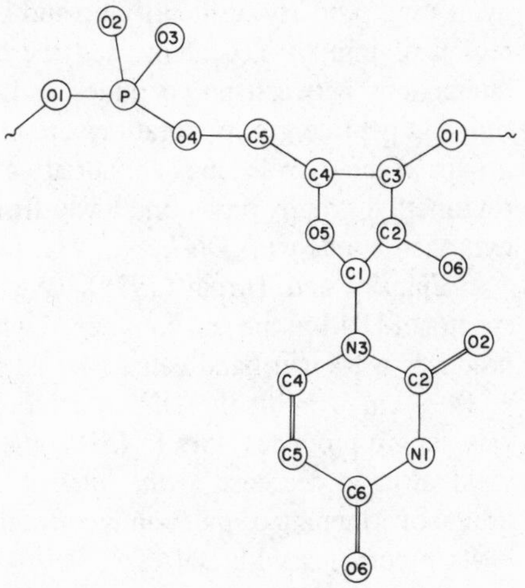

Fig. 9.16. Conformation angles. (Beetz *et al.*, 1979.)

on the $H_2PO_2^-$ vibrational bands and the calculation of normal modes of $HDPO_2^-$, which has C_s symmetry. The usefulness of the linear dichroism data lies in the calculation of transition dipole moments associated with both the vibrations of all the atoms of the nucleotide and their charges.

DNA–Antibiotic Interactions

Raman spectroscopy has been used to study the conformational features of netropsin-DNA and distamycin-DNA complexes (Martin *et al.*, 1978). Netropsin and distamycin (Fig. 9.17) are related oligopeptide antibiotics that bind strongly to adenine-thymine-rich regions of duplex DNA. Netropsin elongates and stiffens DNA, while distamycin may bend the DNA helix (Zimmer *et al.*, 1971; Reinert, 1972). These antibiotics have the

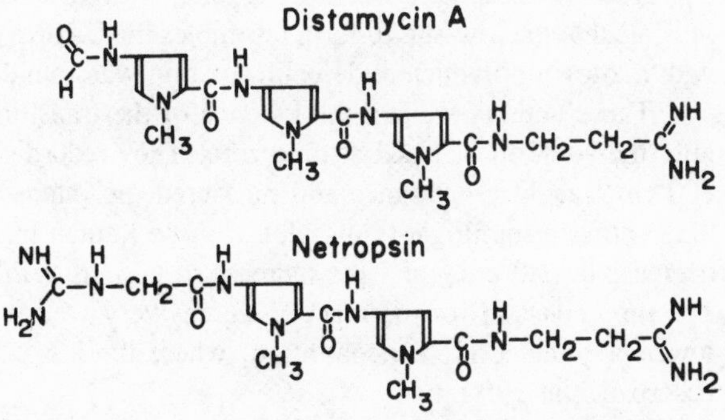

Fig. 9.17. Molecular structures of distamycin A and netropsin. (Zimmer *et al.*, 1971.)

interesting property of identifying and binding to a specific feature of DNA, namely, the region of high A-T content, and thereby may be used as simple models for specific interactions between polypeptides and DNA. Upon drug binding to DNA, the pyrrole ring and peptide group vibrations are changed, although the environments of the methyl groups of the pyrrole rings are not affected; in fact, the binding model shows the pyrrole-ring methyl groups projecting away from the DNA, whereas peptide N—H groups form hydrogen bonds with DNA.

Chinsky and Turpin (1978) have used UV excitation to obtain resonance Raman spectra of DNA alone and complexed with Actinomycin D. With DNA, 300-nm excitation produces a Raman band caused by adenine at 1582 cm^{-1} and one by guanine (mainly) at 1492 cm^{-1}. With the DNA-Actinomycin D complex, the same selective excitation wavelength produces only DNA bands, none from Actinomycin D. Complex formation yields a large decrease in the intensity of the guanine band, which is the result of the orbital overlapping of the 2-amino group of guanine with the ring nitrogen of Actinomycin in its π complex with DNA.

Some of the other assignments made to Raman bands for DNA were as follows: 1665 cm^{-1}, thymine; 1380 and 1256 cm^{-1}, predominantly thymine and to a small extent cytosine; 1309 cm^{-1}, probably cytosine only.

Order–Order Transition in a Synthetic DNA

Pohl and Jovin (1972) have found that synthetic DNA with an alternating base sequence, poly(dG-dC) · poly(dC-dG), shows at neutral pH a salt-dependent, reversible and cooperative transition between two double-helical forms. They called these forms "R" and "L," based on circular dichroism spectra. The circular dichroism spectrum at low salt concentration (0.2 *M* NaCl) is similar to that for DNA of high G-plus-C content and was called the R-form; the CD spectrum at high salt concentration (20% w/w NaCl) was nearly an inversion and was called the L-form. The transition between the R- and L-forms is an example of an "order–order" conformational change of a double-stranded polynucleotide in solution.

Pohl *et al.* (1973) have used Raman spectroscopy to follow the salt-induced "order–order" transition between the R- and L-forms in aqueous solution. The spectral data indicate that the low-salt R-form resembles the B-form of natural DNA, but no correlation with a known polynucleotide conformation was found for the high-salt L-form.

These authors studied the kinetics of the transition from the R-form to the L-form after the addition of NaCl at time zero. They recorded Raman spectra between 600 and 700 cm^{-1} at different times and measured the intensities as a function of time. Figure 9.18 shows a semilogarithmic plot of these Raman intensities as a function of time. The decrease at 680 cm^{-1} and the increase at 625 cm^{-1} follow the same first-order kinetics. The time constant for the reaction agrees very well with one determined at a 1000-fold lower polynucleotide concentration, where the kinetics was followed by the change in absorbance at 295 nm.

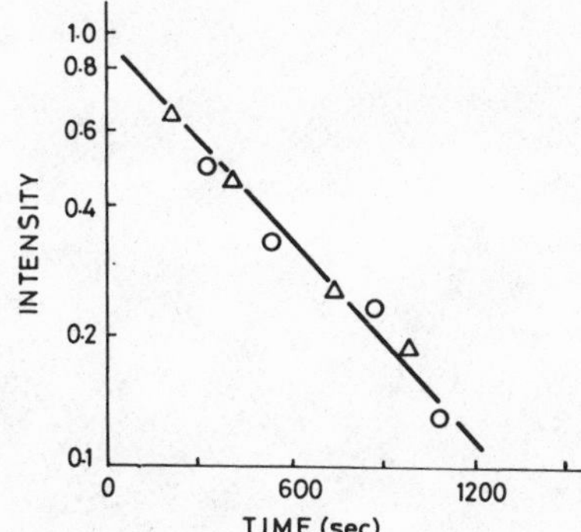

Fig. 9.18. Semilogarithmic plot of Raman intensities during the $R \rightarrow L$ transition as a function of time after changing the salt concentration. The decrease at 680 cm^{-1} (○ – ○) and the increase at 625 cm^{-1} (△ – △) follow the same first-order kinetics. (Pohl *et al.*, 1973.)

Alkylation of DNAs

Mansy and Peticolas (1976) have used Raman spectroscopy to study the alkylation of calf thymus DNA, poly(dG) · poly(dC), and poly(dA) · (dT) by two alkylating agents [*N*,*N*-dimethyl-2-chloroethylamine hydrochloride (HN 1) and *N*-methylbis(2-chloroethyl)amine hydrochloride (HN 2)]. The compound HN 2 is an example of a nitrogen mustard compound which is known to be a potent tumor-inhibiting agent. HN 2 forms interstrand cross-links. An excess of the alkylating agent caused the Raman frequencies due to the guanine ring modes in DNA and poly(dG) · poly(dC) to change essentially quantitatively to those of 7-methylguanosine. This change is evidence that practically all the guanine bases were alkylated in the N7 position. Raman spectra showed no change in bands for any of the other bases in alkylated DNA. Figure 9.19 shows the spectra of double helical poly(dG) · poly(dC) and that of the alkylated compound, both at pH 7.0. The former substance shows strong bands at 686 and 1490 cm^{-1} which have been assigned to the guanine ring modes (Lord and Thomas, 1967; Brown *et al.*, 1972; Lafleur *et al.*, 1972). In alkylated poly(dG) · poly(dC), the 686-cm^{-1} band is absent and the 1488-cm^{-1} band displays a large decrease in intensity. The 1488-cm^{-1} band was assigned to a guanine ring mode which strongly involves the C8—H bond stretch and the N7═C8 double bond stretch.

Raman Normal Modes and Frequencies

The vibrational modes of the purine and pyrimidine bases of nucleic acids have been discussed by Tsuboi *et al.* (1973). These authors have given detailed descriptions of the atomic displacements associated with the vibrations, and their work should be consulted. Table 9.5 presents data for some strong Raman bands of special interest (Peticolas and Tsuboi, 1979).

Uracil displays a strong Raman band near 1690 cm^{-1} in aqueous solution (H_2O) owing to the carbonyl stretching vibration, and a weaker band near 1631 cm^{-1}. These

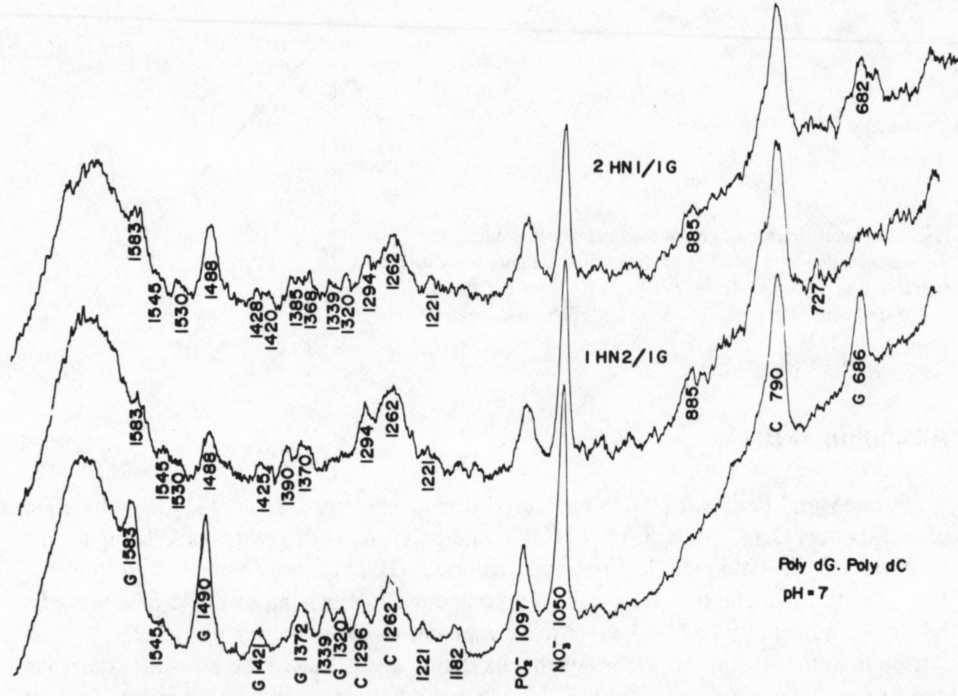

Fig. 9.19. Laser Raman spectrum of (bottom) poly(dG) · poly(dC); (middle) poly(dG) · poly(dC) + one equiv of HN2 per guanine base; (top) poly(dG) · poly(dC) + 2 equiv of HN1 per guanine base. In each case, the concentration of the helical complex is about 20 mg/ml and the pH 7.0. (Mansy and Peticolas, 1976.) Reprinted with the permission of the American Chemical Society. Copyright 1976.

bands show a splitting and frequency shift in D_2O solution. Uracil in a double helix shows a single band at 1680 cm^{-1}. When the double helix is melted, two bands form, apparently the result of a splitting, which was formerly thought to be the result of the rupture of uracil-adenine hydrogen bonding (Small and Peticolas, 1971a,b). It has since been shown, however, that one cannot completely distinguish between the effects on the band split caused by hydrogen bonding or by stacking of bases (Morikawa *et al.*, 1973).

Resonance Raman Effect from Adenine

Blazej and Peticolas (1977) have presented the first resonance Raman excitation profile making use of UV as well as visible radiation. They measured the intensity of the Raman spectrum of AMP as a function of the frequency of the incident laser light in the range from 514.5 to 295 nm. The Raman bands at 1484 and 1583 cm^{-1} (Fig. 9.20) appeared to obtain practically all their intensity from a weak electronic transition at 276

Table 9.5. Some Prominent Raman Lines (cm^{-1}) Due to Base Vibrations in Double Helical Ribonucleic Acid Homopolymers[a]

	Hypochromism			Presonance[b]
Uracil residue	in Poly(AU)	in Poly A · poly U		
1680(s) (C=O str.)	+ +			
1634(m) (C=O str.)	no			
1400(m)	+ +	+ +		265 nm
1235(s)	+ +	+ +		265 nm
785(s) (ring breath.)	+ +	+		210 nm
Cytosine residue	in Poly C	In GpC		
1657(m)				
1607(m)				
1528(m)	+	+		268 nm
1292(s)	+			268 nm
1240(s)				
782(s) (ring breath.)				230 nm
Adenine residue	in Poly A	in Poly AU	in Poly A · poly U	
1580(s)		no		276 nm
1510(m)	+ +	no	+	
1484(m)	no	no	−	276 nm
1379(m)	+	+ +		
1340(s)		no		
1310(s)	+ +	+	+ +	
1255(w)	+	no		
729(s) (ring breath.)	+	+ +	+ +	210 nm
Guanine residue	in 5′ GMP	in GpC		
1582(s)	+			276 nm
1487(s)	+ +	−		276 nm
1375(m)	+			
1328(m)	+ +	−		276 nm
670(s)	− −	−		Far

[a] Peticolas and Tsuboi (1979).
[b] UV absorption bands from which the base vibration appears to obtain its intensity by the preresonance phenomenon.

nm. The group of Raman bands between 1300 and 1400 cm^{-1} appeared to derive at least part of their intensity from an electronic band whose 0—0 transition lies in the 269–259-nm region. These authors presented a schematic diagram of their Raman spectrometer (not given here). Tsuboi *et al.* (1974) have also shown that the Raman vibrations of adenine at ~ 1580 cm^{-1} and 1484 cm^{-1} probably obtain their intensity from the 276-nm absorption band.

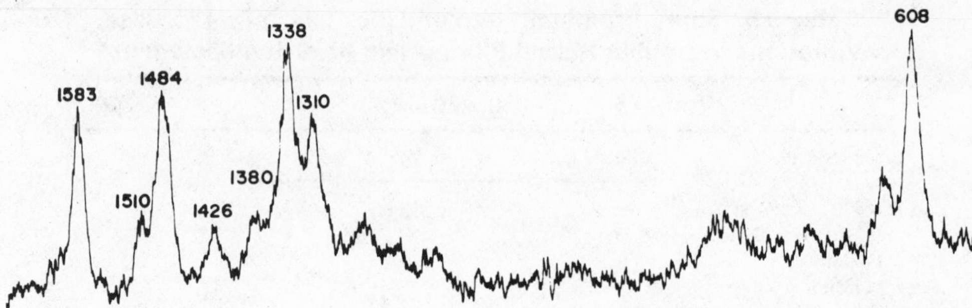

Fig. 9.20. Raman spectrum of 0.01 *M* AMP/0.25 *M* sodium cacodylate, pH 7.1. Excitation wavelength was 295 nm. (Blazej and Peticolas, 1977.)

Resonance Raman Effect from Nucleic Acids

Peticolas and Tsuboi (1979) have discussed theoretical and experimental aspects of the preresonance Raman effect from nucleic acids. According to Tsuboi and his co-workers, if the excited electronic state geometry is similar to that of the normal mode of the ground state, the Raman-active band tends to obtain its intensity from that low-lying electronic state by means of a preresonance effect. Base stacking or base pairing can lead to the following result: electronic transition moments of the absorption bands (of the bases) may interact to change the intensity of the electronic absorption bands (UV hypochromism) and thus the intensities of the Raman bands (Raman hypochromism). The latter hypochromic effect (see Table 9.5) is detectable for each of the types of bases individually, whereas the electronic absorption bands show broad overlap. Thus, the Raman effect can be used more sensitively to measure interactions between specific bases than the intensity of the electronic absorption bands.

Raman Frequencies of the Phosphomonoester and Phosphodiester Groups

At a pH value above seven, all mononucleotides display a prominent Raman band at 980 cm^{-1}, which is strongly polarized. This band is caused by the symmetric PO_3^{2-} stretching vibration (Hartman *et al.*, 1973), and bands near 1170 cm^{-1} arise from the C—N stretching vibration of external amino groups. The 1170-cm^{-1} band of adenylic acid is assigned to the external C6—N stretching vibration of the adenine moiety. Weak Raman bands assigned to the 5′—C—O stretching vibration of the phosphomonoester dianion lie near 1055 cm^{-1}, e.g., for 5′-IMP, 5′-AMP, 5′-CMP, and 5′-GMP.

The phosphomonoester ion $CO(HO)PO_2^-$ displays Raman bands near 1085 and 817 cm^{-1}, at a pH value between the first and second pK_a's, i.e., between 1 and 6.5. The 1085-cm^{-1} band is due to symmetric PO_2^- stretching and the 817 cm^{-1} band to symmetric —O—P—O— stretching in the C—O—P—O—H system (Rimai *et al.*, 1969).

The Raman spectrum of 5′-IMP shows a strong band at 1265 cm^{-1} at a pH value below 1; this band is due to P═O stretching of the $CO(HO)_2PO$ group.

Table 9.6. Approximate Raman Frequencies of the Phosphomonoester and Phosphodiester Groups[a]

Group	Conditions	Frequency (cm^{-1})	Relative intensity[b] (polarization)	Assignment
$COPO_3^{2-}$	pH (pD) > 7	980	m (P)	PO_3^{-2} symmetric stretching
		1055	w (?)	CO stretching
$CO(HO)PO_2^-$	1 < pH (pD) < 6	817	m (P)	OPO symmetric stretching
		1050	w (?)	CO stretching
		1085	m (P)	PO_2^- symmetric stretching
$CO(HO)_2PO$	pH (pD) < 1	1035	w (?)	CO stretching
		1265	s (P)	P=O stretching
$(CO)_2PO_2^-$	pH (pD) > 6.5	795–815[c]	w (P)–s (P)	OPO symmetric stretching
		1050	w (?)	CO stretching
		1100	m (P)	PO_2^- symmetric stretching

[a] Hartman *et al.* (1973).
[b] Abbreviations: w = weak; m = medium; s = strong; P = polarized.
[c] Conformationally sensitive in frequency and intensity: ca. 815 cm^{-1} (s) in ordered polyribonucleotides and RNA; ca. 795 cm^{-1} (w) in disordered polyribonucleotides and RNA; ca. 787 cm^{-1} (s) in double-helical DNA.

A summary of Raman frequencies of the phosphomonoester group appears in Table 9.6. Notable similarities occur in the frequencies of the $CO(HO)PO_2^-$ and $(CO)_2PO_2^-$ groups, showing that the bonding and conformation of phosphate groups are very similar in ordered polyribonucleotides and in mononucleotide monoanions.

Hypochromism and Hyperchromism. Base Stacking and Base Pairing

As with intensity changes in the 260-nm region, the considerable stacking of adenine residues in single-stranded poly A at neutral pH causes notable decreases in the Raman intensities of vibrations of the adenine ring (hypochromism); and the ring vibrations of poly U, poly I, and poly C are also affected in this way (Small and Peticolas, 1971a,b).

When DNA and RNA have a double-helix structure, the planes of the various bases are approximately parallel to one another and nearly perpendicular to the helix axis. Hypochromism (see Mahler and Cordes, 1971) results from such stacking of the bases. The stacking is accompanied by decreases in the intensities of certain Raman bands of the organic bases (Lafleur *et al.*, 1972; Morikawa *et al.*, 1973; Prescott *et al.*, 1974; Nishimura, 1976; Small and Peticolas, 1971a,b). Percentages of hypochromism are given in Table 9.7 in parentheses for some of these bands exhibited by polynucleotides studied by these workers.

Chou *et al.* (1977) have used Raman spectra to show that polyinosinic acid poly(rI) forms an ordered complex in aqueous solutions of high ionic strength. Its T_m is 45°C. The geometry of the structure is A-helical and contains stacked bases which appear to be stabilized by specific hydrogen bonding associated with the C6=O groups of hypoxanthine. Low ionic strength produces poly(rI) of disordered structure lacking base stack-

Table 9.7. Hypochromic Raman Bands[a]

Poly rA · Poly rU	
Uracil	784 cm^{-1} (30), 1231 cm^{-1} (52), 1395 cm^{-1} (36).
Adenine	728 cm^{-1} (41), 1309 cm^{-1} (40), 1339 cm^{-1} (10), 1380 cm^{-1} (30), 1485 cm^{-1} (−25), 1512 cm^{-1} (15), 1583 cm^{-1} (−18).
Poly (rA-rU) · Poly (rA-rU)	
Uracil	783 cm^{-1} (35), 1237 cm^{-1} (49), 1397 cm^{-1} (35).
Adenine	726 cm^{-1} (46), 1302 cm^{-1} (28), 1378 cm^{-1} (35), 1576 cm^{-1} (−14).
Poly rA	725 cm^{-1} (35), 1303 cm^{-1} (39), 1508 cm^{-1} (67).
Poly rC	1296 cm^{-1} (20), 1533 cm^{-1} (10).

[a] Lafleur *et al.*, 1972; Morikawa *et al.*, 1973; Prescott *et al.*, 1974; Nishimura, 1976; Small and Peticolas, 1971a, b.

ing and base pairing interactions. At high salt concentrations aqueous poly(rI) displays a backbone frequency at 815 cm^{-1}, a C═O frequency at 1710 cm^{-1}, and temperature-dependent intensities (hypochromism), particularly at 1495, 1326, and 718 cm^{-1}; these observations show that in solutions of high ionic strength poly(rI) forms at low temperature an ordered A-helical structure, specific hydrogen bonding of C6═O groups, and base stacking interactions. This structure changes, when heated to 70°C, to a disordered form of poly(rI), almost identical to that of the disordered form at low ionic strength.

Using the intensities of bands at 815 cm^{-1} (backbone frequency) and 1100 cm^{-1} in spectra of poly(rI) and triple-stranded poly(rU) · poly(rA) · poly(rU), these workers showed that the correlation found earlier for single- and double-stranded ribopolymer complexes holds true for these multistranded structures also, that is, ordered structures have A-helix geometry and the Raman intensity ratio I_{815}/I_{1100} is directly proportional to the number of ordered nucleotide subgroups (Thomas and Hartman, 1973).

Gramlich *et al.* (1975) have studied the Raman spectra of acidic solutions of polyriboadenylic acid [poly(rA)], in order to estimate secondary structural properties. In the region 600 to 1600 cm^{-1}, the spectra display some well-resolved lines. The lines at 725, 1303, 1336, and 1508 cm^{-1} were used to demonstrate the Raman hypochromic effect at pH values of 5.73 and 5.35, as a function of temperature. The peak heights of these lines were normalized by comparing them to their height at 15°C. The normalized heights were then plotted against temperature. Reference lines were acetate at 925 cm^{-1} and the poly(rA) band at 1100 cm^{-1}.

Figure 9.21 shows spectra of poly(rA) at pH 5.73, observed at 26° and 37.5°C. Figure 9.22 shows a plot of the normalized lines at 1508, 1303, 1336, and 725 cm^{-1} versus temperature. These increases in intensities in the range 25°–40°C are caused by the double helix-coil transition of the polynucleotide, and those above 40°C are caused by the noncooperative unstacking of the bases in the partly ordered single strands. Increasing temperature causes the amount of stacking to decrease.

These authors found that the intensity of the 1336-cm^{-1} line (mainly a C5—N7 stretching mode) is nearly temperature independent at pH 7.0, while at pH values of 5.73 and 5.35 this line is strongly temperature dependent. They, therefore, concluded that the intensity of the 1336-cm^{-1} line has a greater dependence on the double helix-coil transition

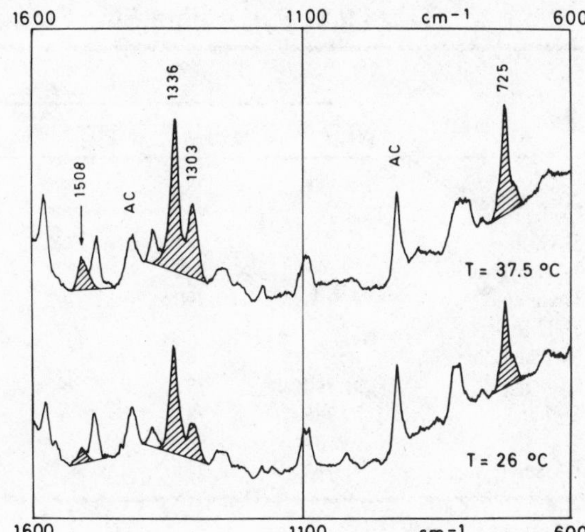

Fig. 9.21. Raman spectra of 2% poly (rA) (w/w) at 26°C and 37.5°C, pH = 5.73. (Gramlich *et al.*, 1975.)

than on the degree of ordering in the single strand. This finding suggested that the C5—N7 bond in adenine has different relative conformations in double helix and single strand.

In Raman spectra of RNA there is an intense line near 780 cm^{-1} which permits the identification of uridine and cytidine residues in the polymer (Thomas, 1970). Table 9.8 (Hartman *et al.*, 1973) presents pyrimidine frequencies that occur at about 780 cm^{-1} under a variety of conditions. The data in the table represent conditions in H_2O, where pH < pK_{ring}, pH 7, and pH > pK_{ring}; and in D_2O, where pD < pK_{ring}, pD7, and

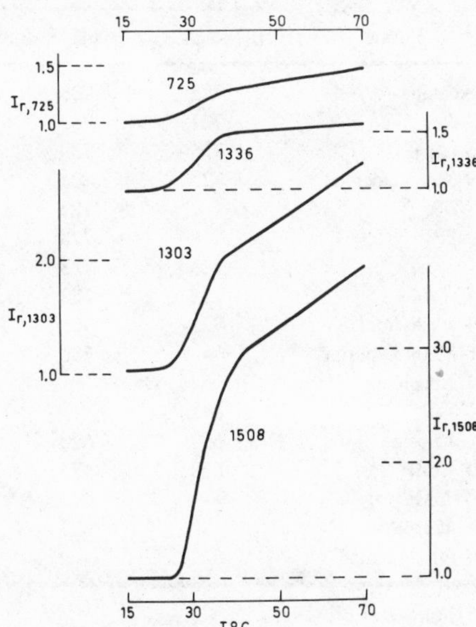

Fig. 9.22. Plot of the intensity of the bands at 725, 1303, 1336, and 1508 cm^{-1} in poly (rA) at pH 5.73 vs. temperature (ratio to intensity at 15°C). (Gramlich *et al.*, 1975.)

Table 9.8. Pyrimidine Ring Frequencies ca. 780 cm^{-1}[a]

	H_2O solutions			D_2O solutions		
	pH < pK_{ring}	pH 7	pH > pK_{ring}	pD < pK_{ring}	pD 7	pD > pK_{ring}
Uracil	788	788	798	783	783	798
Uridine		782	790		782	788
5′-UMP		784	790		783	790
1-Methyluracil		770	778		755	780
Thymidine		790	792		788	790
5′-TMP		785			785	
Pseudouridine					785	
Cytosine	790	787	797	777	777	787
Cytidine	786	785		775	772	
5′-CMP	787	783		776	774	
1-Methylcytosine	795	781		780	775	

[a] Hartman *et al.* (1973).

pD > pK_{ring}. pK_{ring} is ~ 9.5 (see Lord and Thomas, 1967). Table 9.9 shows characteristic purine frequencies that occur in the vicinity of 725 cm^{-1} under the same pH (pD) conditions listed above. The Raman line near this frequency for RNA samples permits the identification of adenine residues in the polymer (Thomas, 1970).

Derivatives of guanine do not show an intense line near 725 cm^{-1}, as do those of adenine. Most guanine derivatives display a line in the range of 625 to 690 cm^{-1} (mainly

Table 9.9. Purine Ring Frequencies ca. 725 cm^{-1}[a]

	H_2O solutions			D_2O solutions		
Purine	pH < pK_{ring}	pH 7	pH > pK_{ring}	pD < pK_{ring}	pD 7	pD > pK_{ring}
Adenine	722	725	725	703	722	722
Adenosine	727	733		720	725	
5′-AMP	725	730		718	721	
9-Methyladenine	724	725		703	715	
ADP		727				
ATP		728				
Inosine	723	723	738	717	721	738
5′-IMP	721	723	738	718	721	738
1-Methylinosine	721	721		715	720	
6-Methoxypurine-riboside		736			736	
Guanine	642		650	625		645
Guanosine	670	670	675	668	660	660
5′-GMP	670	675	680	665	675	680
3′-GMP		665				
5′-dGMP		683				
9-Ethylguanine	629			615		

[a] Hartman *et al.* (1973).

near 670 cm^{-1}), and the details concerning the Raman frequencies of such compounds—under the same conditions as those mentioned above for adenine derivatives—are also given in Table 9.9.

It is of interest to point out that the 670-cm^{-1} line exhibits hyperchromism with respect to base-stacking in 5′-GMP, or transfer RNA, or ribosomal RNA, as opposed to the *hypo*chromism that other pyrimidine and purine ring frequencies display in their Raman spectra.

Ishikawa *et al.* (1973) have prepared poly(2-dimethylaminoadenylic acid), which forms an acid helix of great stability, with a T_m about 50°C higher than that of poly(A) under comparable conditions. They found that the dependencies of different optical properties (IR,UV,CD) upon temperature for this polymer are quite different from one another. Although the changes of the optical properties of neutral poly(2-dimethylaminoadenylic acid) with temperature must be related to structural changes of the polymer, the detailed nature of the relationship could not be defined. The usual assumption that the fraction of bases stacked is directly proportional to change with temperature of an optical property was not valid in the case of this polymer. The temperature profiles varied considerably with the optical property and with the wavelength used for the measurement. The authors stated that, since independent and definitive structural evidence was lacking, the choice of a particular optical property as a linear measurement of stacking is essentially arbitrary. They offered two possible structural interpretations of their optical data.

Chinsky *et al.* (1977) have used UV laser excitation (300 nm) to produce resonance Raman spectra of a DNA containing bromodeoxyuridine (BrdUrd), the poly d(BrU-A). The DNA absorbs at the same wavelength as the pulsed laser excitation, thus allowing only one component to be observed, namely the bromodeoxyuridine. Strong hyperchromism of Raman bands at 1627, 1352, and 1230 cm^{-1} showed that a conformational change takes place when the ordered, base-paired form of poly d(BrU-A) at 25°C is converted to the unordered melted form at 63°C. The intensity of the band at 1627 cm^{-1} was particularly enhanced upon melting of the DNA; this band represents the vibrations of the C4 carbonyl which is hydrogen bonded to the NH_2 group of adenine.

Howard *et al.* (1977) have investigated A-I and A-G polynucleotide pairing by IR spectroscopy. The spectrum of a 1:1 mixture of poly(2-aminoadenylic acid) [poly NH_2A] and polyinosinic acid in the region of double-bond vibrations displays large changes from the spectra of the components (Fig. 9.23), characteristic of base-pairing interactions (see Miles and Frazier, 1964; Howard *et al.*, 1969; Miles, 1971). The inosine carbonyl vibration (see Howard and Miles, 1965) has shifted to 1689 cm^{-1} in the helix ($\Delta\nu =$ 15.5 cm^{-1}) and the half-bandwidth has decreased from 35 to 20 cm^{-1}. The A ring vibration at 1612.5 cm^{-1} (at 50°C) moves to 1622 cm^{-1} in the helix, and the shoulder at ca. 1600 cm^{-1} has lower intensity and occurs as a neatly resolved band at 1595.5 cm^{-1} in the helix.

Melting curves for bands representing only A (1612.5, 1622 cm^{-1}) and I (1673.5, 1689 cm^{-1}) vibrations, respectively, display parallel temperature dependence (Fig. 9.24), indicating that the spectral changes result from specific base-pairing associations rather than from self-structures of either component (for detailed discussions, see Miles and Frazier, 1964, Howard *et al.*, 1969, and Miles, 1971).

In a reinvestigation of poly (A), poly (I) interaction, Howard and Miles (1977) have concluded from UV data that a 1:2 complex is the only interaction product formed by poly (A) and poly (I).

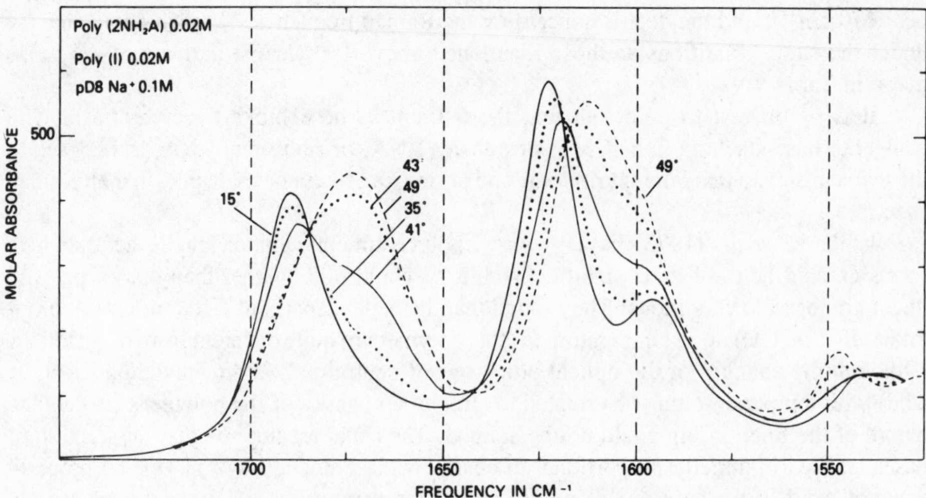

Fig. 9.23. Infrared spectra of poly($2NH_2A$) · poly(I) (15°C) and of mixtures formed by thermal dissociation of the double helix. The large change in I carbonyl-stretching frequency on going to the ordered form (1673 → 1689 cm^{-1}) is similar to that observed in many helices. The isosbestic point in the carbonyl region occurs at 1684 cm^{-1}. The $2NH_2A$ ring vibration (1613 cm^{-1}) shows the same frequency increase observed in the 1:1 complex with poly(U) (Howard *et al.*, 1976) but, in marked contrast to the large loss of intensity in the latter helix, exhibits little change in ϵ_{max}. (Howard *et al.*, 1977.) Reprinted with the permission of the American Chemical Society. Copyright 1977.

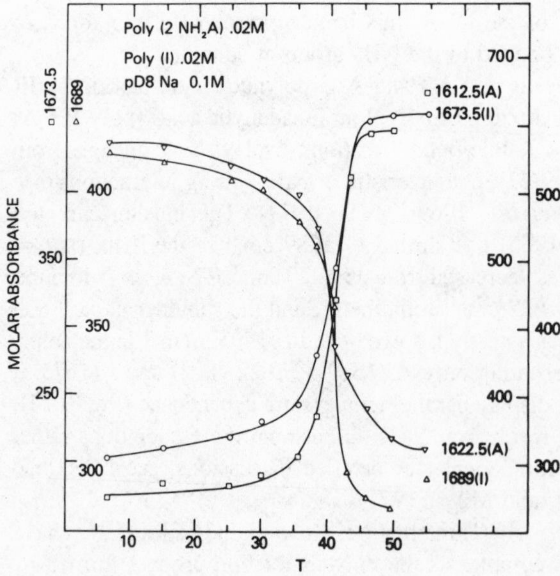

Fig. 9.24. Infrared melting curves of poly($2NH_2A$)·poly(I). Parallel temperature dependence of A and I bands shows that spectroscopic changes are due to a specific A·I pairing interaction. (Howard *et al.*, 1977.) Reprinted with the permission of the American Chemical Society. Copyright 1977.

Transfer and Ribosomal RNAs. Conformations

Tsuboi *et al*. (1971) have recorded the Raman spectra of poly (G), poly (C), poly (A), and poly (U); and that of purified formylmethionine transfer RNA (tRNA) from *E. coli,* in aqueous solutions at pH 7.5. Figure 9.25 shows these spectra. In order to interpret the spectrum of the tRNA, these authors constructed a synthetic spectrum for the tRNA based on normalization of the spectra of the four homopolynucleotides (normalization

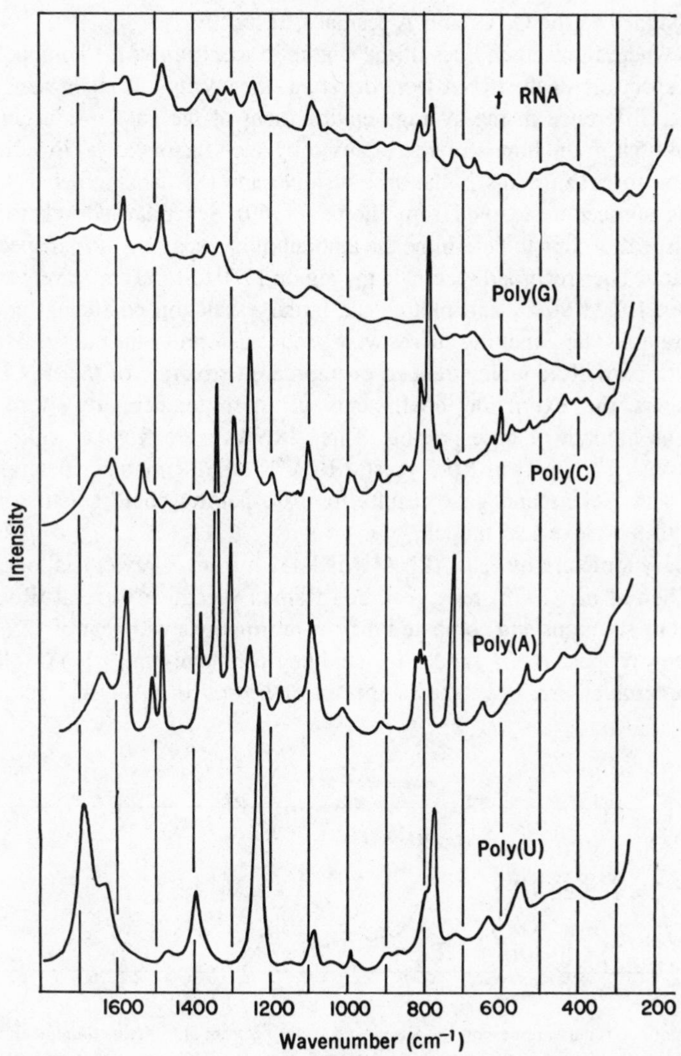

Fig. 9.25. Raman spectra of formylmethionine tRNA of *Escherichia coli*, poly(G), poly(C), poly(A), and poly(U) in aqueous solutions at pH 7.5 at room temperature. (Tsuboi *et al*., 1971.)

with reference to a common band at 1100 cm^{-1}) and on modification of the intensities of the various Raman lines by factors corresponding to the tRNA composition.

The agreement of the synthetic spectrum with the observed tRNA spectrum (Fig. 9.26) allowed assignment of many of the Raman lines of tRNA to be made by comparison of the spectra. The lines at 1572 and 1481 cm^{-1} were assigned to the guanine (G) residue; those at 1530, 1000, 968, 911, 807, and 782 cm^{-1} to the cytosine (C) residue; that at 721 cm^{-1} to the adenine (A) residue; and the broad peak at about 1690 cm^{-1} and a shoulder at about 1230 cm^{-1} to the uracil (U) residue. The peak at 1370 cm^{-1} consists of three unresolved lines due to the U, A, and G residues. The peaks at 1334, 1316, and 1293 cm^{-1} were assigned to two A lines, one G, and one C. The peaks at 667, 594, and 707 cm^{-1} were assigned to the G, C, and A residues, respectively.

In cases where the Raman lines of the synthetic spectrum of the homopolymers and the observed spectrum of the tRNA were different, the spectral changes were interpreted as reflecting a difference in the average environment of the base residue in the tRNA molecule from that in the homopolymer, caused by such factors as (a) lowering of some bond-stretching force constants in the base residue, and (b) stacking between bases.

Applying a procedure adapted from Thomas (1969), Schernau and Ackermann (1977) have used IR spectroscopy to determine the amount of base pairing of four specific tRNAs in ^2H_2O solution. They recorded spectra in the region 1750 to 1550 cm^{-1} over a temperature range of about 15° to 90°C, and plotted the molar extinction coefficient at 1657 cm^{-1} versus temperature. These melting curves were used to determine the ranges of temperature associated with partially double-stranded or unordered structure of the tRNA. For a set of wavenumbers the extinction coefficients at these temperatures were employed to calculate the amount of base pairing. Three tRNAs were from *E. coli:* tRNA$^{f\ Met}$, tRNA$^{Val\ I}$, tRNA$^{Glu\ II}$. One was from yeast: tRNA^{Phe}. The IR-mathematical method used by Schernau and Ackermann gave results for base-pairing content that were in good agreement with the cloverleaf model.

The tertiary structure of yeast tRNA^{Phe} is known from X-ray crystallographic measurements. Chen *et al.* (1978) have recorded Raman spectra of several tRNAs and 5S RNA in aqueous solutions and compared their conformations with that of yeast tRNA^{Phe}, which has been proposed as a model for studying the three-dimensional folding of all tRNAs. For example, Fig. 9.27 shows spectra of *E. coli* tRNA$_1^{Val}$ and yeast tRNA$_2^{Val}$.

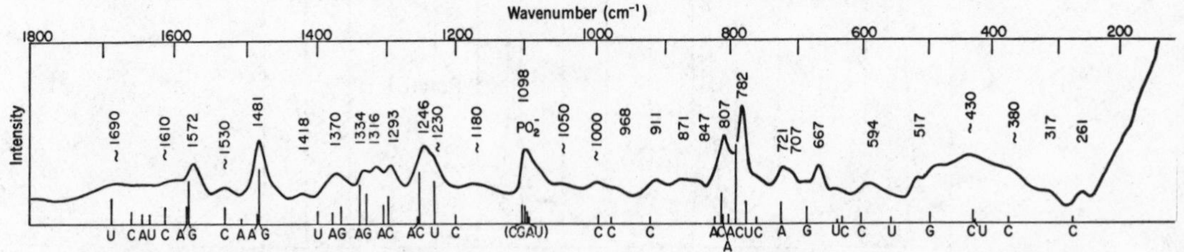

Fig. 9.26. The continuous curve represents the observed Raman spectrum of formylmethionine tRNA (spectrum at top of Fig. 9.25, but with background emission subtracted). The vertical lines are the positions and intensities of the Raman lines in a synthetic spectrum, which is a linear combination of the "normalized" Raman spectra of the four homopolymers. (Tsuboi *et al.*, 1971.)

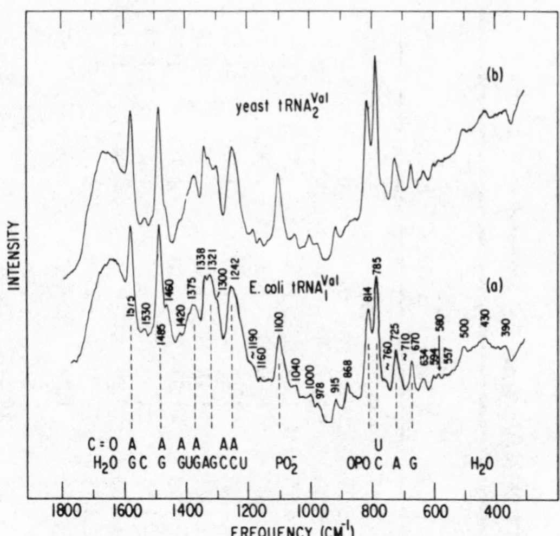

Fig. 9.27. Raman spectra of 3% aqueous tRNAs containing 2 m*M* $MgCl_2$: (a) *E. coli* tRNA$^{Val}_1$; (b) yeast tRNA$^{Val}_2$. The origins of the major Raman lines are indicated by: A, adenine; U, uracil; G, guanine; C, cytosine; OPO, phosphodiester chain mode; PO_2^-, ionized phosphate. When two bases contribute to the intensity of a given line, the symbol of the major contributor is placed above the other. (Chen *et al.*, 1978.) Reprinted with the permission of the American Chemical Society. Copyright 1978.

Table 9.10 shows a summary of spectral data gathered from eleven tRNA's and *E. coli* 5S RNA. The line at 814 cm^{-1} originates from —OPO—; the one at 1100 cm^{-1} from PO_2^-. These authors found that the spectra of the eleven tRNAs displayed a high, rather constant content of order in the ribophosphate backbone, as shown by the ratio I_{814}/I_{1100}, which was 1.73 $\pm$ 0.05 for all eleven.

Chen *et al.* compared their calculated and observed line intensities for adenine (725 cm^{-1}) and guanine (670 cm^{-1}), and found agreement of the calculated and observed values to be within experimental error for the adenine line and possibly for the guanine line. They used such comparisons to determine the degree of stacking of the guanine and adenine moieties. From previous work (Chen *et al.*, 1975) they knew, for example, that in the spectrum of yeast tRNA^{Phe} the 670-cm^{-1} line *decreases* by a factor of 3 when the guanine moieties change from the native (all-stacked) form (Kim *et al.*, 1974; Ladner *et al.*, 1975) to the unstacked configuration, whereas the 725-cm^{-1} line *increases* by 12% when the partially stacked adenine groups in the native configuration become unstacked. Thus, the agreement between expected and observed ratios of the 670- and 725-cm^{-1} lines in, e.g., native tRNA^{Trp} and tRNA^{Phe} shows that the fraction of stacked adenine and guanine in the first is approximately the same as in the second.

From the same type of data for all the tRNAs studied they determined the effectiveness of stacking of guanine and adenine bases in all the samples compared to that in tRNA^{Phe}. There were variations in the amounts of stacking of these bases, but most of the tRNAs studied had amounts of stacking similar to that found for yeast tRNA^{Phe}. The 5S RNA showed a lower stacking efficiency of the adenine bases and a much higher one for the guanine bases than in yeast tRNA^{Phe}.

Many *E. coli* tRNAs contain 4-thiouracil (λ_{max}, 330 nm) at position 8. Adenine, guanine, cytosine, and uracil, the more common bases, absorb near 260 nm (for details, see Beaven *et al.*, 1955). By using either a 351.1- or 363.8-nm Ar^+ laser beam and a rotating sample cell, Tsuboi (1977) has observed the resonance Raman spectrum of the intact 4-thiouracil residue of tRNAs from *E. coli* and *Thermus thermophilus* HB8. Thus,

Table 9.10. Relative Intensities of Principal Raman Lines in 11 tRNA's and 5S RNA[a,b]

Freq (cm^{-1})	Origin	Y Phe[c]	Y Ala	Y Asp	Y fMet	Y Gly	Y Leu	Y Trp	Y Tyr	Y Val	E Val	E Glu	5S RNA
670	G	0.44	0.57	0.57	0.43	0.45	0.38	0.39	0.47	0.37	0.52	0.56	0.69
725	A	0.69	0.62	0.62	0.60	0.54	0.69	0.68	0.77	0.56	0.62	0.55	0.66
785	C,U	2.16	2.21	2.30	2.39	2.08	2.07	2.33	2.40	2.28	2.39	2.44	2.44
814	—OPO—	1.74	1.68	1.79	1.75	1.75	1.78	1.69	1.75	1.69	1.74	1.77	1.54
1100	PO_2^-	1.00	1.00	1.00	1.00	1.00	1.00	1.00	1.00	1.00	1.00	1.00	1.00
1242	U,C,A[d]	1.37	1.36	1.39	1.54	1.38	1.37	1.38	1.43	1.42	1.63	1.63	1.57
1300	A,C	1.12	1.05	1.05	1.22	1.10	1.26	1.00	1.21	1.18	1.19	1.30	1.19
1321	G	1.34	1.34	1.42	1.48	1.34	1.24	1.14	1.48	1.28	1.70	1.35	1.57
1338	A	1.54	1.03	1.13	1.33	1.25	1.56	1.36	1.55	1.43	1.62	1.20	1.59
1375	A,G	1.07	0.96	0.96	1.05	0.98	1.01	0.94	1.01	0.99	0.94	0.94	1.11
1485	A,G	1.90	1.94	2.19	2.08	1.89	1.84	1.75	2.13	1.70	2.26	2.06	2.67
1575	A,G	1.40	1.22	1.36	1.45	1.27	1.44	1.25	1.56	1.28	1.62	1.46	1.67
	composit[b]												
	A	18	8	13	16	13	21	17	17	14	15	14	23
	U	14	13	16	8	15	18	16	10	14	11	11	20
	G	22	27	24	27	23	23	20	23	21	24	22	41
	C	18	22	19	22	20	20	18	21	21	23	27	36
	other	4	6	3	2	3	3	3	7	6	3	2	0
	total	76	76	75	75	74	85	74	78	76	76	76	120

[a] Chen *et al.* (1978).
[b] All solutions were 3%–4% by weight. Intensities are peak intensities near the indicated frequency relative to the PO_2^- line at 1100 cm^{-1}. For shoulders, the relative ordinate value at the indicated frequency is given. All values are averages of at least four spectra. Y = yeast tRNA, E = *E. coli* tRNA. For composition, minor bases are grouped with parent major base. Pseudo-U is included in U but dihydro-U with "other" because its Raman frequencies do not coincide with those of U (Thomas *et al.*, 1973).
[c] Intensities for yeast $tRNA^{Phe}$ were taken from Chen *et al.* (1975).
[d] The peak near 1242 cm^{-1} is a superposition of lines near 1234 due to U and near 1251 due to A and C.

by choosing a beam of proper wavelength, his group singled out the resonance Raman spectrum of the minor base over the more abundant common ones. The Raman lines characteristic of 4-thiouridine are 1450, 1317, 1256, and 1210 cm^{-1}. The importance of the rotating cell is to be noted, since, as soon as the rotation is stopped, a different resonance Raman spectrum is seen, a reflection of photochemical intramolecular cross linking between cytosine at position 13 and 4-thiouracil at position 8. Tsuboi mentions another interesting fact: $tRNA^{Tyr}$ lacks cytosine at position 13, which has guanine, thus precluding cross linking; but the $tRNA^{Tyr}$ displays 4-thiouracil resonance Raman bands even when a stationary cell is used (Nishimura *et al.*, 1976).

Hartman *et al.* (1973) have presented Raman data on unfractionated ribosomal RNA from *E. coli*. These data are shown in Table 9.11, and are based on work by Thomas (1970). Thomas *et al.* (1971) have presented data on fractionated 16S and 23S rRNA, i.e., Raman spectra showing temperature dependence of certain bands: (a) An increase in temperature from 30° to 85°C caused an increase in intensity at 1660 cm^{-1} and a decrease at 1688 cm^{-1}, indicating a conversion of bases from paired moieties to unpaired ones. (b) The 815-cm^{-1} band became much weaker and shifted to lower frequency, thus displaying a net conversion of nucleotides from ordered to disordered states. (c) A band denoting ring frequencies of U and C (780 cm^{-1}) increased in intensity. This also happened with poly U and poly C. These spectral changes were attributed to decreases in stacking with the rise in temperature (Small and Peticolas, 1971a,b). (d) The ring frequencies of G (bands at 1480, 1370, and 670 cm^{-1}) showed unusual changes by decreasing in intensity as tbe temperature rose to 85° C. The explanation for this action (Thomas *et al.*, 1971) is that G moieties are more stacked at 85° C than at 30° C. Hartman *et al.* (1973) have also shown Raman spectra of several conformationally similar transfer RNAs—namely, $tRNA^{fMet}$, $tRNA^{Val}$, and $tRNA^{Phe\,2}$ from *E. coli*. The thermal denaturation of the tRNAs produced dissociation of AU and GC base pairs and the unstacking of pyrimidines. The findings of Thomas *et al.* (1972) strongly supported a clover-leaf type of conformation for each of the tRNAs at low temperature.

A weak shoulder near 710 cm^{-1} in the tRNA Raman spectra was attributed to 4-thiouridine or dihydrouridine, or both (Hartman *et al.*, 1973).

Hartman *et al.* (1973) have discussed determinations of secondary structure in RNAs and have given a summary of quantitative estimates of percentages of RNA bases in paired and unpaired states for a variety of ribosomal RNAs, transfer RNAs, and a bacteriophage RNA (μ2). These percentages were based on IR measurements by various investigators. To obtain such percentages, the IR spectrum of RNA is fitted to a constructed spectrum obtained by summing the proper quantitative reference spectra corresponding to paired and unpaired bases (Thomas, 1969).

Many structures have been proposed for 5S RNA, which, along with 5.8S RNA, is a structural component of the large ribosomal subunit of eucaryotic cells (Monier, 1974; Erdmann, 1976). However, none of those structures satisfies all the known data on physical properties and certain functions of these molecules (Erdmann, 1976; Vigne and Jordan, 1977). Measurements of Raman peak intensities give data especially useful for the examination of RNA secondary structure (Luoma and Marshall, 1978a), by showing how "ordered" the RNA backbone is, what the percentage is of uracil moieties in base-paired regions, and how much stacking of various bases is present (Peticolas, 1972; Thomas

Table 9.11. Raman Frequencies of Aqueous rRNA[a,b]

H_2O solution pH 7	D_2O solution pH 7	Subgroup[c]	Assignments
435 (0)		U, C	?
500 (1)	498 (1)	G	Possibly ring def. and C—O def.
580 (1B)	560 (1B)	A, U, G, C	
635 (0)	625 (0)	A, U, G, C	
670 (2)	668 (2)	G	Ring str.
710 (0S)	705 (0S)	C	Ring str.
725 (3)	718 (3)	A	Ring str.
	755 (0)		
786 (6)	780 (6)	U, C	Ring str.
814 (5)	814 (6)	Phosphate	OPO sym. str.
867 (2)	860 (2B)	Ribose	
918 (1)	915 (1)	Ribose	CO str.; CC str.
975 (0)	990 (1)	Ribose	
1003 (1)		A, U, C	
1049 (2)	1045 (1)	Ribose-phosphate	CO str.
	1090 (S)		
1100 (5)	1100 (4)	Phosphate	PO_2- sym. str.
	1140 (0)	A	
1185 (2)	1185 (1)	A, G, C (U)	External C—N str.
	1235 (S)	A, C	Ring str.
1243 (6)		U, C	Ring str.
1255 (5)	1257 (5)	A, C	Ring str.
1300 (S)		C, U	Ring str.
	1310 (7)	C	Ring str.
1320 (7)	1318 (6)	A, (G)	Ring str.
1340 (7)	1345 (7)	A	Ring str.
	1370 (3B)	G	Ring str.
1380 (5B)		A, G, U, (C)	Ring str.
	1390 (2)	A, (C), (U)	Ring str.
1422 (2)		G, A	Ring str.
1460 (S)	1460 (S)	Ribose	CH def.
1484 (10)	1480 (8)	G, A	
1510 (S)	1503 (S)	C	
1527 (2)	1526 (3)	C, G, (A)	Coupled ring and double-bond str.
	1560 (S)	?	
1575 (8)	1578 (10)	G, A	
1620 (B)	1622 (3)	U, A, C, G	
1650 (B)			
	1658 (4B)	U, G, C	Mainly C═O str.
1692 (4B)	1688 (4B)		

[a] Hartman *et al,* (1973).

[b] Frequencies are accurate to ± 2 cm^{-1} for intense bands and ± 4 cm^{-1} for weak bands. Figures in parentheses refer to relative intensity on 0 to 10 scale. Abbreviations: B = broad; S = shoulder; A = adenine; U = uracil; G = guanine; C = cytosine; sym. str. = symmetric stretching; def. = deformation.

[c] When more than one subgroup contributes appreciably to a Raman band, the larger contributor is listed first. Minor contributors are listed in parentheses.

and Hartman, 1973; Chen and Thomas, 1974; Chen *et al.*, 1975; Thomas, 1975). Luoma and Marshall (1978b) have studied the Raman spectra of 5S RNA from *Saccharomyces cerevisiae* (yeast) in water and ^2H_2O solutions. Their data did not satisfy all the secondary structures proposed earlier; rather, they allowed a highly stable cloverleaf structure which is adaptable to other eukaryotic 5S RNAs. Also, the proposed structure (Fig. 9.28) for the yeast 5S RNA correlated well with a similar secondary structure for yeast 5.8S RNA (Luoma and Marshall, 1978a).

Some of the Raman lines (cm^{-1}) seen in Fig. 9.29 and their origins in yeast 5S RNA and mixed tRNA, are as follows: 670, G; 725, A; 785, C,U; 814, —OPO—; 1100, PO_2^-; 1234, U; 1251, C,A; 1300, C, A; 1321, G; 1338, A; 1375, G,A; and 1485, G, A. The amount of base-stacking can be estimated in RNAs from certain of these lines because their intensities depend on the types and degrees of base-stacking interactions, for example, the lines at 670, 725, and 785 cm^{-1}, and the in-plane vibration lines above 1200 cm^{-1}, particularly the one at 1234 cm^{-1} (uracil). The intensities of these lines all

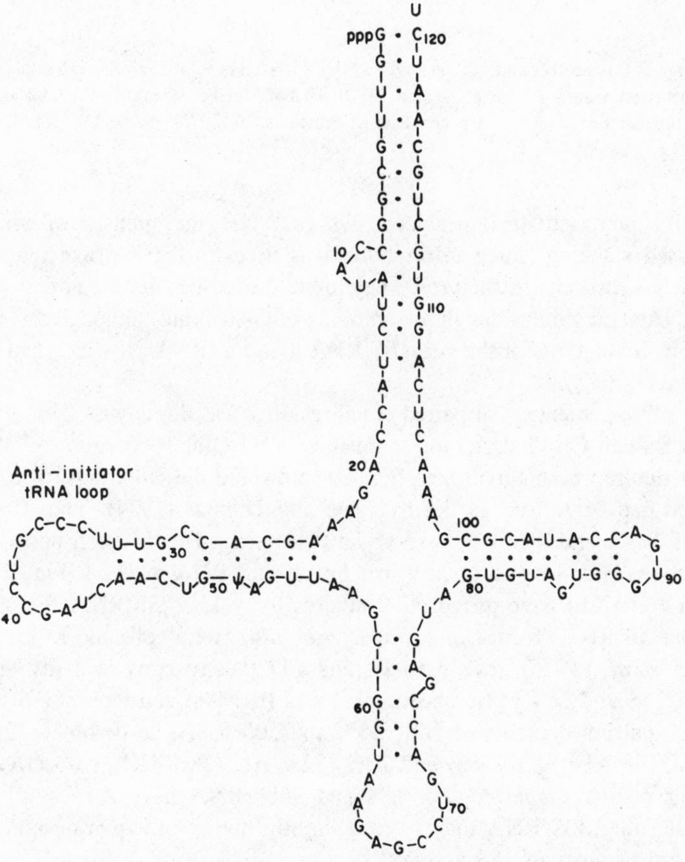

Fig. 9.28. Proposed cloverleaf secondary structure for *S. cerevisiae* 5S RNA. (Luoma and Marshall, 1978b.)

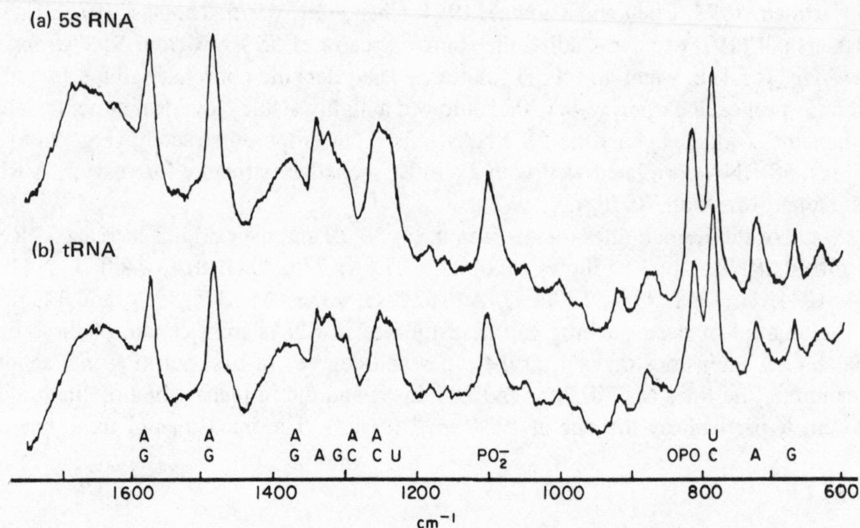

Fig. 9.29. Original Raman spectra of *S. cerevisiae*. (a) 5 S RNA and (b) mixed tRNA. Samples were dissolved (4% w/v) in water containing 10 m*M* phosphate (pH 7), 10 m*M* $MgCl_2$, 100 m*M* NaCl. A background spectrum of the buffer produced only the water peak. (Luoma and Marshall, 1978b.)

decrease with increased stacking except 670 cm^{-1} (G), the intensity of which increases with increased stacking. Such information was used by these workers to compare 5S RNA and tRNA secondary structures. They found similarities in the amounts of A-stacking for these RNAs and similarities in the amounts of G-stacking, although their calculations showed more G-stacking for the yeast 5S RNA than for tRNA^{Phe}, which had been studied by Chen *et al.* (1975).

The relative percentage of paired uracil residues was directly measured by observing the ratio of Raman C═O stretching intensities in ^2H_2O at 1660 cm^{-1} and at 1688 cm^{-1}. The former line represents hydrogen-bonded forms and the latter non-hydrogen-bonded. This method had been used earlier by Chen and Thomas (1974). Thus, the data show that both 5S RNA and tRNA^{Phe} have about two-thirds of their uracil bases paired.

Besides the proposed secondary structure for 5S RNA of yeast (Fig. 9.28), Luoma and Marshall (1978b) gave proposed structures for yeast 5.8S RNA, *E. coli* 5S RNA, and for other 5S RNAs from *Chlorella, Drosophila,* HeLa cell, and *X. laevis*.

Yanagi *et al.* (1975) have used IR and CD spectroscopy to study the secondary structure of 18S and 26S yeast ribosomal RNAs (rRNAs) and their complexes. IR data showed base-pairing contents of 18S, 26S, and 30S RNAs to be 66% (30% AU, 36% GC), 66% (30% AU, 36% GC), and 70% (32% AU, 38% GC), respectively (see Fig. 9.30). Their results suggested that 18S and 26S rRNA have very similar secondary structures and that 30S RNA may have a slightly higher base-pairing content than the estimated sum of those of 18S and 26S rRNAs.

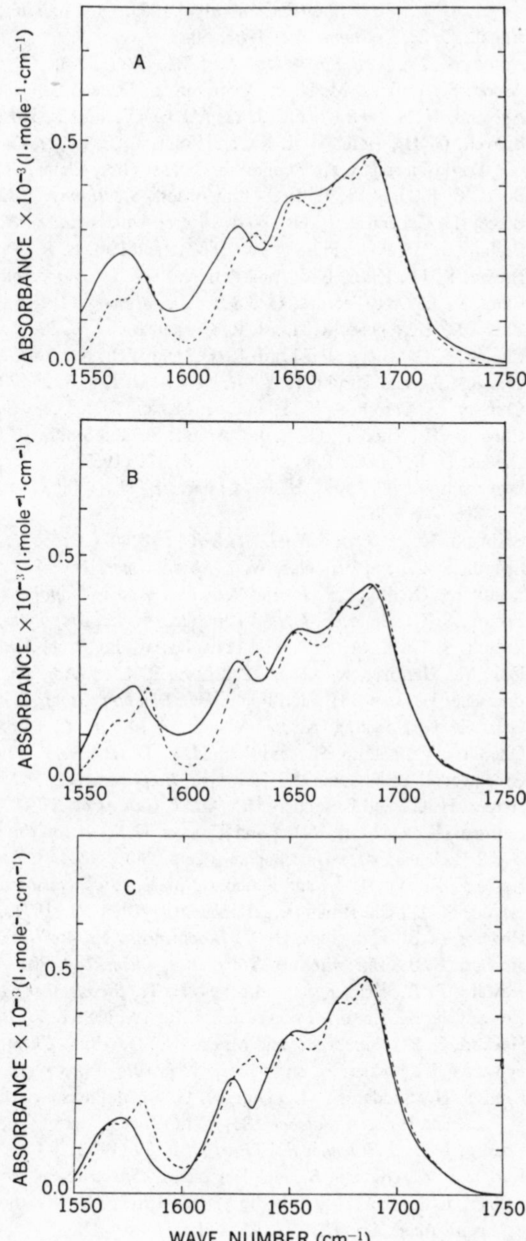

Fig. 9.30. Infrared spectra of yeast rRNA components in 0.1 *M* NaCl, 10 m*M* Tris-DCl (pD = 7.6) at 15°. (A) (——) observed spectrum of 18S rRNA; (- - -) synthetic spectrum corresponding to 0.30 A-U + 0.36 G-C + 0.12 A + 0.14 U + 0.08 G + 0.01 C. (B) (——) observed spectrum of 26S rRNA; (- -) synthetic spectrum corresponding to 0.24 A-U + 0.36 G-C + 0.17 A + 0.04 U + 0.08 G + 0.01 C. (C) (——) observed spectrum of 30S RNA; (- - -) synthetic spectrum corresponding to 0.32 A-U + 0.38 G-C + 0.12 A + 0.11 U + 0.07 G + 0.00 C. (Yanagi *et al.*, 1975.)

References

Angell, C. L., Ph.D. thesis, Cambridge, 1955, cited in Tsuboi (1969).

Angell, C. L., *J. Chem. Soc.* **1961,** 504.

Arnott, S., *Progr. Biophys. Mol. Biol.* **21,** 265 (1970).

Arnott, S., Wilkins, M. H. F., Hamilton, L. D., and Langridge, R., *J. Mol. Biol.* **11,** 391 (1965).

Aylward, N. N., and Koenig, J. L., *Macromolecules* **3,** 590 (1970).

Beaven, G. H., Holiday, E. R., and Johnson, E. A., in *The Nucleic Acids,* Vol. I (E. Chargaff and J. N. Davidson, eds.), Academic Press, New York, Chapter 14, pp. 493–553, 1955.

Beetz, C. P., Jr., Ascarelli, G., and Arnott, S., *Biophys. J.* **28,** 15 (1979).

Blazej, D. C., and Peticolas, W. L., *Proc. Natl. Acad. Sci. USA* **74,** 2639 (1977).

Brahms, J., Pilet, J., Phuong Lan, T.-T., and Hill, L. R., *Proc. Natl. Acad. Sci. USA* **70,** 3352 (1973).

Brown, K. G., Kiser, E. J., and Peticolas, W. L., *Biopolymers* **11,** 1855 (1972).

Chen, M. C., and Thomas, G. J., Jr., *Biopolymers* **13,** 615 (1974).

Chem, M. C., Giegé, R., Lord, R. C., and Rich, A., *Biochemistry* **14,** 4385 (1975).

Chem, M. C., Giegé, R., Lord, R. C., and Rich, A., *Biochemistry* **17,** 3134 (1978).

Chinsky, L., and Turpin, P. Y., *Nucleic Acids Res.* **5,** 2969 (1978).

Chinsky, L., Turpin, P. Y., Duquesne, M., and Brahms, J., *Biochem. Biophys. Res. Commun.* **75,** 766 (1977).

Chou, C. H., Thomas, G. J., Jr., Arnott, S., and Smith, P. J. C., *Nucleic Acids Res.* **4,** 2407 (1977).

Davies, D. R., *Annu. Rev. Biochem.* **36,** 321 (1967).

Dickerson, R. E., Drew, H. R., Conner, B. N., Wing, R. M., Fratini, A. V., and Kopka, M. L., *Science* **216,** 475 (1982).

Erdmann, V. A., *Progr. Nucl. Acids Res.* **18,** 45 (1976).

Erfurth, S. C., and Peticolas, W. L., *Biopolymers* **14,** 247 (1975).

Erfurth, S. C., Bond, P. J., and Peticolas, W. L., *Biopolymers* **14,** 1245 (1975).

Erfurth, S. C., Bond, P. J., and Peticolas, W. L., *Biopolymers* **14,** 247, 1259 (1975a).

Erfurth, S. C., Kiser, E. J., and Peticolas, W. L., *Proc. Natl. Acad. Sci. USA* **69,** 938 (1972).

Falk, M., Hartman, K. A., Jr., and Lord, R. C., *J. Am. Chem. Soc.* **85,** 387 (1963a); **85,** 391 (1963b).

Fritzsche, H., Lang, H., and Pohle, W., *Biochim. Biophys. Acta* **432,** 409 (1976).

Gale, G. R., Howle, J. A., and Walker, E. M., Jr., *Cancer Res.* **31,** 950 (1971).

Gramlich, V., Klump, H., and Schmid, E. D., *Biochem. Biophys. Res. Commun.* **63,** 906 (1975).

Gruenwedel, D. W., and Chi-Hsia Hsu, *Biopolymers* **7,** 557 (1969).

Harder, H. C., and Rosenberg, B., *Int. J. Cancer* **6,** 207 (1970).

Hartman, K. A., Lord, R. C., and Thomas, G. J., Jr., in *Physico-Chemical Properties of Nucleic Acids,* Vol. 2 (J. Duchesne, ed.), Chapter 10, pp. 2–89, 92–143, Academic Press, New York, 1973.

Herbeck, R., Yu, T.-J., and Peticolas, W. L., *Biochemistry* **15,** 2656 (1976).

Heyde, M. E., and Rimai, L., *Biochemistry* **10,** 1121 (1971).

Howard, F. B., and Miles, H. T., *Biochemistry* **16,** 4647 (1977).

Howard, F. B., and Miles, H. T., *J. Biol. Chem.* **240,** 801 (1965).

Howard, F. B., Frazier, J., and Miles, H. T., *Biochemistry* **15,** 3783 (1976).

Howard, F. B., Hattori, M., Frazier, J., and Miles, H. T., *Biochemistry* **16,** 4637 (1977).

Howard, F. B., Frazier, J., and Miles, H. T., *J. Biol. Chem.* **244,** 1291 (1969).

Ishikawa, F., Frazier, J., and Miles, H. T., *Biochemistry* **12,** 4790 (1973).

Kim, S. H., Suddath, F. L., Quigley, G. J., McPherson, A., Sussman, J. L., Wang, A. H. J., Seeman, N. C., and Rich, A. *Science* **185,** 435 (1974).

Koenig, J. L., *J. Polymer Sci., Part D,* **6,** 59 (1972).

Kyogoku, Y., Higuchi, S., and Tsuboi, M., *Spectrochim. Acta* **23A,** 969 (1967b).

Ladner, J. E., Jack, A., Robertus, J. D., Brown, R. S., Rhodes, D., Clark, B. F. C., and Klug, A., *Proc. Natl. Acad. Sci. USA* **72,** 4414 (1975).

Lafleur, L., Rice, J., and Thomas, G. J., Jr., *Biopolymers* **11,** 2423 (1972).

Lane, M. J., and Thomas, G. J., Jr., *Biochemistry* **18,** 3839 (1979).

Lewis, A., Nelson, N., and Racker, E., *Biochemistry* **14,** 1532 (1975).

Livramento, J., and Thomas, G. J., Jr., *J. Am. Chem. Soc.* **96,** 6529 (1974).

Löfroth, G., "Methylmercury," Bulletin No. 4, Swedish National Science Research Council, Stockholm, 1969.
Lord, R. C., and Thomas, G. J., Jr., *Spectrochim. Acta* **23A,** 2551 (1967).
Luoma, G. A., and Marshall, A. G., *J. Mol. Biol.* **125,** 95 (1978b).
Luoma, G A., and Marshall, A. G., *Proc. Natl. Acad. Sci. USA* **75,** 4901 (1978a).
Mahler, H. R., and Cordes, E. H., *Biological Chemistry,* 2nd Ed., Harper and Row, New York, 1971.
Mansy, S., and Peticolas, W. L., *Biochemistry* **15,** 2650 (1976).
Mansy, S., and Tobias, R. S., *Biochemistry* **14,** 2952 (1975).
Mansy, S., Wood, T. E., Sprowles, J. C., and Tobias, R. S., *J. Am. Chem. Soc.* **96,** 1762 (1974).
Mansy, S., Peticolas, W. L., and Tobias, R. S., *Spectrochim. Acta* **35A,** 315 (1979).
Martin, J. C., Wartell, R. M., and O'Shea, D. C., *Proc. Natl. Acad. Sci. USA* **75,** 5483 (1978).
Medeiros, G. C., and Thomas, G. J., Jr., *Biochim. Biophys. Acta* **247,** 449 (1971).
Miles, H. T., *Biochim. Biophys. Acta* **22,** 247 (1956).
Miles, H. T., *Biochim. Biophys. Acta* **27,** 46 (1958a).
Miles, H. T., *Biochim. Biophys. Acta* **30,** 324 (1958b).
Miles, H. T., *Biochim. Biophys. Acta* **35,** 274 (1959).
Miles, H. T., *Proc. Natl. Acad. Sci. USA* **47,** 791 (1961).
Miles, H. T., *Proc. Natl. Acad. Sci. USA* **51,** 1104 (1964).
Miles, H. T., *Proc. Nucleic Acid Res.* **2,** 205 (1971).
Miles, H. T., and Frazier, J., *Biochemistry* **17,** 2920 (1978).
Miles, H. T., and Frazier, J., *Biochem. Biophys. Res. Commun.* **14,** 21 (1964).
Miles, H. T., Howard, F. B., and Frazier, J., *Science* **142,** 1458 (1963).
Monier, R. in *Ribosomes* (M. Nomura, A. Tissières, and P. Lengyel, eds.), p. 141, Cold Spring Harbor Laboratory, New York, 1974.
Morikawa, K., Tsuboi, M., Takahashi, S., Kyogoku, Y., Mitsui, Y., Iitaka, Y., and Thomas, G. J., Jr., *Biopolymers* **12,** 790 (1973).
Mulvihill, J. J., *Science* **176,** 132 (1972).
Nelson, N., Nelson, H., and Racker, E., *J. Biol. Chem.* **247,** 6506 (1972).
Nishimura, Y., Doctoral thesis, University of Tokyo, 1976, quoted in Tsuboi, M., in *Vibrational Spectroscopy—Modern Trends* (A. J. Barnes and W. J. Orville-Thomas, eds.), p. 405, Elsevier, New York, 1977.
Nishimura, Y., Morikawa, K., and Tsuboi, M., *Bull. Chem. Soc. Jpn* **47,** 1043 (1974).
Nishimura, Y., Hirakawa, A. Y., Tsuboi, M., and Nishimura, S., *Nature* **260,** 173 (1976).
Patrick, D. M., II, Wilson, J. E., and Leroi, G. E., *Biochemistry* **13,** 2813 (1974).
Peticolas, W. L. in *Advances in Raman Spectroscopy* (J. P. Mathieu, ed.), p. 285, Heyden and Son, Ltd., New York, 1972.
Peticolas, W. L., and Tsuboi, M., in *Infrared and Raman Spectroscopy of Biological Molecules* (T. M. Theophanides, ed.) NATO Advanced Study Institutes Series, Series C: Mathematical and Physical Sciences, D. Reidel Publishing Company, Dordrecht, Holland, pp. 153–173 (1979).
Pilet, J., and Brahms, J., *Biopolymers* **12,** 387 (1973).
Pilet, J., and Brahms, J., *Nature, New Biol.* **236,** 99 (1972).
Pilet, J., Blicharski, J., and Brahms, J., *Biochemistry* **14,** 1869 (1975).
Pohl, F. M., and Jovin, T. M., *J. Mol. Biol.* **67,** 375 (1972).
Pohl, F. M., Ranade, A., and Stockburger, M., *Biochim. Biophys. Acta* **335,** 85 (1973).
Prescott, B., Gamache, R., Livramento, J., and Thomas, G. J., Jr., *Biopolymers* **13,** 1821 (1974).
Reinert, K.-E., *J. Mol. Biol.* **72,** 593 (1972).
Rimai, L., and Heyde, M. E., *Biochem. Biophys. Res. Commun.* **41,** 313 (1970).
Rimai, L., Heyde, M. E., and Carew, E. B., *Biochem. Biophys. Res. Commun.* **38,** 231 (1970).
Rimai, L., Maher, V. M., Gill, D., Salmeen, I., and McCormick, J. J., *Biochim. Biophys. Acta* **361,** 155 (1974).
Rimai, L., Cole, T., Parsons, J. L., Hickmott, J. T., Jr., and Carew, E. B., *Biophys. J.* **9,** 320 (1969).
Sarocchi, M.-T., and Guschlbauer, W., *Eur. J. Biochem.* **34,** 232 (1973).
Sato, T., Kyogoku, Y., Higuchi, S., Mitsui, Y., Iitaka, Y., Tsuboi, M., and Miura, K., *J. Mol. Biol.* **16,** 180 (1966).
Schernau, U., and Ackermann, Th., *Biopolymers* **16,** 1735 (1977).

Shimanouchi, T., Tsuboi, M., and Kyogoku, Y., in *Advances in Chemical Physics, Vol. VII, The Structure and Properties of Biomolecules and Biological Systems* (J. Duchesne, ed.) pp. 435, 437, Interscience, New York, 1964.

Small, E. W., and Peticolas, W. L., *Biopolymers* **10,** 69 (1971a).

Small, E. W., and Peticolas, W. L., *Biopolymers* **10,** 1377 (1971b).

Sundaralingam, M., *Biopolymers* **7,** 821 (1969).

Sutherland, G. B. B. M., and Tsuboi, M., *Proc. Ry. Soc. (London)* **A239,** 446 (1957).

Theophanides, T. M., in *Infrared and Raman Spectroscopy of Biological Molecules* (T. M. Theophanides, ed.), D. Reidel, Dordrecht, Holland, pp. 187–204, 1979a; pp. 205–223, 1979b.

Thomas, G. J., Jr., *Biochim. Biophys. Acta* **213,** 417 (1970).

Thomas, G. J., Jr., *Biopolymers* **7,** 325 (1969).

Thomas, G. J., Jr., in *Vibrational Spectra and Structure* (J. R. Durig, ed.), Vol. 4 p. 239, Elsevier, New York, 1975.

Thomas, G. J., Jr., in *Vibrational Spectra and Structure* (J. Durig, ed.) Vol. 3, Marcel Dekker, New York, 1974.

Thomas, G. J., Jr., and Hartman, K. A., *Biochim. Biophys. Acta* **312,** 311 (1973).

Thomas, G. J., Jr., and Kyogoku, Y., in *Infrared and Raman Spectroscopy,* Part C (E. G. Brame, Jr. and J. G. Grasselli, eds.), Chap. 11, pp 717–872, Marcel Dekker, New York, 1977.

Thomas, G. J., Jr., and Livramento, J., *Biochemistry* **14,** 5210 (1975).

Thomas, G. J., Jr., Chen, M. C., Lord, R. C., Kotsiopoulos, P. S., Tritton, T. R., and Mohr, S. C., *Biochem. Biophys. Res. Commun.* **54,** 570 (1973b).

Thomas, G. J., Jr., Medeiros, G. C., and Hartman, K. A., *Biochem. Biophys. Res. Commun.* **44,** 587 (1971).

Thomas, G. J., Jr., Medeiros, G. C., and Hartman, K. A., *Biochim. Biophys. Acta* **277,** 71 (1972).

Thomas, G. J., Jr., Chen, M. C., and Hartman, K. A., *Biochim. Biophys. Acta* **324,** 37 (1973a).

Tomasz, M., Olson, J., and Mercado, C. M., *Biochemistry* **11,** 1235 (1972).

Tsuboi, M., *Appl. Spectrosc. Rev.* **3,** 45 (1969).

Tsuboi, M., in *Basic Principles in Nucleic Acid Chemistry,* Vol. 1 (P.O.P. Ts'o, ed.), Chap. 5, pp. 399–452, Academic Press, New York, 1974.

Tsuboi, M., *J. Am. Chem. Soc.* **79,** 1351 (1957).

Tsuboi, M., *J. Polymer Sci. Pt. C* No. 7, 125 (1964).

Tsuboi, M., in *Vibrational Spectroscopy—Modern Trends* (A. J. Barnes and W. J. Orville-Thomas, eds.), Chap. 25, p. 405, Elsevier, New York, 1977.

Tsuboi, M., and Kyogoku, Y., in *Synthetic Procedures in Nucleic Acid Chemistry,* Vol. II (W. W. Zorbach and R. S. Tipson, eds.) Interscience, New York, 1969.

Tsuboi, M., and Shuto, K., *Chem. Pharm. Bull. (Tokyo)* **14,** 784 (1966).

Tsuboi, M., Kyogoku, Y., and Shimanouchi, T., *Biochim. Biophys. Acta* **55,** 1 (1962).

Tsuboi, M., Shuto, K., and Higuchi, S. *Bull. Chem. Soc. Jpn* **41,** 1821 (1968).

Tsuboi, M., Hirakawa, A. Y., Nishimura, Y., and Harada, I., *J. Raman Spectrosc.* **2,** 609 (1974).

Tsuboi, M., Takahashi, S., and Harada, I., in *Physico-Chemical Properties of Nucleic Acids* (J. Duchesne, ed.), Vol. 2, Chap. 11, pp. 91–145, Academic Press, London, 1973.

Tsuboi, M., Takahashi, S., Muraishi, S., Kajiura, T., and Nishimura, S., *Science* **174,** 1142 (1971).

Tsuboi, M., Matsuo, K., Shimanouchi, T., and Kyogoku , Y., *Spectrochim. Acta* **19,** 1617 (1963).

Vigne, R., and Jordan, B. R., *J. Mol. Evol.* **10,** 77 (1977).

Wang, A. H.-J., Quigley, G. J., Kolpak, F. J., Crawford, J. L., van Boom, J. H., van der Marel, G., and Rich, A., *Nature (London)* **282,** 680 (1979).

Wang, A. H.-J., Quigley, G. J., Kolpak, F. J., van der Marel, G., van Boom, J. H., and Rich, A., *Science* **211,** 171 (1981).

Watson, J. D., and Crick, F. H. C., *Nature* **171,** 737 (1953).

Webb, S. J., *Can. J. Microbiol.* **13,** 57 (1967).

Yanagi, K., Katsura, T., and Iso, K., *J. Biochem. (Tokyo)* **78,** 599 (1975).

Zimmer, Ch., Reinert, K. E., Luck, G., Wähnert, U., Löber, G., and Thrum, H., *J. Mol. Biol.* **58,** 329 (1971).

Chapter 10

VIRUSES AND NUCLEOPROTEINS

In Chapter 9 we examined IR and Raman spectroscopic studies of nucleic acids and related substances. In this chapter we discuss spectroscopic studies of nucleic acid-containing substances, viruses and nucleoproteins. Among the viruses and bacteriophages are R17, MS2, Pf1, fd, turnip yellow mosaic, and tobacco mosaic viruses. Examples of the topics that follow these are various chromatins, nucleosomes, and DNA complexes of various types.

R17 Virus. RNA and Protein

Hartman *et al.* (1973) have recorded the Raman spectrum of the RNA phage R17 to obtain information about the structures of RNA and protein components of a virus. Figure 10.1 shows spectra of R17 virus and R17 RNA in aqueous NaCl solutions. Table 10.1 gives frequencies, relative intensities, and assignments for these spectra. The lines of greatest intensity are found between 2800 and 3000 cm^{-1}, caused mostly by aliphatic C—H stretching vibrations of amino acid side chains. Lines caused by proteins, which make up 68.3% by weight of the R17-virion (Kaper, 1968), are prominent between 1600 and 1700 cm^{-1}, due to the fact that amide I vibrations of the peptide groups display greater intensity than the C═O group vibrations of cytosine, uracil, and guanine nucleotides. Between 250 and 1600 cm^{-1} RNA and protein residues display Raman lines of comparable intensities.

These authors compared Raman lines from specific nucleotide vibrations in the phage with their counterparts in the spectrum of protein-free RNA. These comparisons suggested many similarities of RNA structure in the intact phage and in the protein-free condition. A line at 1480 cm^{-1} is much weaker in phage RNA than in the protein-free form, suggesting that a change in RNA secondary structure is imposed by the viral proteins.

Raman frequencies at 500 and 636 cm^{-1} in Fig. 10.1a are due mainly to S—S and C—S stretching vibrations of C—S—S—C linkages between the chains of phage protein

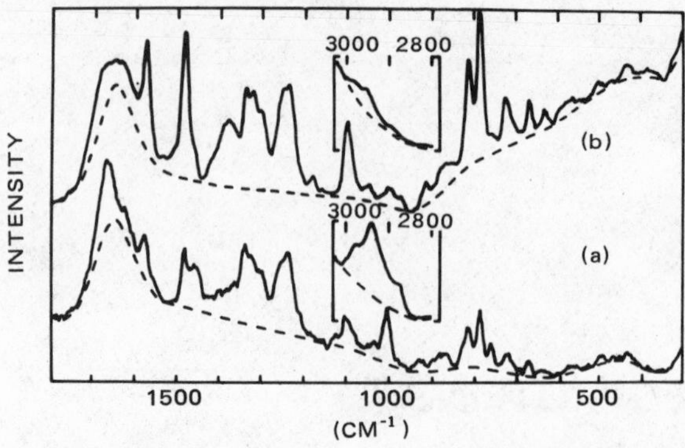

Fig. 10.1. (a) Raman spectrum of R17 virus in aqueous 0.2 *N* NaCl. Conditions: concentration C = 56.3 mg phage/ml (= 17.8 mg RNA/ml), pH = 7, t = 32°C, spectral slit width $\Delta\sigma$ = 10 cm^{-1}, period τ = 20 sec, rate of scan r = 25 cm^{-1}/min, amplification a = 1× (inset a = 1/3×). (b) Raman spectrum of R17 RNA in aqueous 0.2 *N* NaCl. Conditions: C = 37 mg RNA/ml, pH = 7, t = 32PC, $\Delta\sigma$ = 10 cm^{-1}, τ = 20 sec, r = 25 cm^{-1}/min, a = 1× (inset a = 1/3×). The solid curves shown are the spectra as recorded and the dashed curves indicate the background of scattering by liquid H_2O. The volume concentration of RNA in (b) is 2.2 times greater than that in (a). (Hartman *et al.*, 1973.)

Table 10.1. Raman Frequencies, Relative Intensities, and Assignments[a,b]

R17-RNA		R17-Virus		
Frequency	Assignment	Frequency	Assignments	
			RNA	Protein
		~350(0)	r, C	
		385(0)	C	
435(1)	r	435(1)	r	?
502 (0)	G, C	500(1)	G, C	Disulfide
580(1B)	C, G	575(1)	C, G	Trp
635(1)	r	637(1)	r	Disulfide
670(2)	G	670(2)	G	
710(S)	A	710(S)	A	
725(3)	A	722(3)	A	
755(0)	C	760(3)		Trp
787(10)	C, U	787(8)	C, U	
815(7)	P	815(6)	P	
		830(S)		Tyr
860(0)	r	865(3)	r	Trp
880(0)	r	885(S)	r	Trp
915(1)	r			
975(0)	r	930(1)		Trp
1001(1)	A, U, C	1005(7)		Phe
		1015(S)		Trp
1045(1)	r	1035(1)	r	Phe
		1085(S)		Phe
1100(5)	P	1100(5)	P	
		1125(1)		?
1160(0)	r	1162(0)	r	

Continued

Table 10.1. *(Continued)*

R17-RNA		R17-Virus		
Frequency	Assignment	Frequency	Assignments	
1180(1)	C, G, A	1180(0)	C, G, A	
		1210(S)		Tyr, Phe
1238(6)	U, C	1238(10)	U, C	Am III
1250(6)	C, U	1250(8)	C, U	Am III
		1268(S)		Am III
1303(4)	A, C	1300(S)	A, C	
1320(6)	G	1320(S)	G	
1342(6)	A	1343(6)	A	Trp
		1365(2B)	G	Trp
1380(4B)	G, U, A	1400(1)	U, A	
		1450(S) }	C-H def	
1462(0)	C-H def	1462(4) }		
1482(10)	G, A	1480(5)	G, A	
1535(0)	G	1550(S)		Trp
1575(7)	G, A	1575(5)	G, A	Trp
		1600(1)		?
1620(B)	U	1620(2)	U	Trp, Tyr, Phe
~1650(B) }	U, G, C	~1665(B)	U, G, C	Am I
~1685(B) }				
2890(1) }	Aliphatic C-H str	2880(10B) }	Aliphatic C-H str	
2955(6) }		2942(30B) }		
2980(3) }		2975(15B) }		

[a] Hartman *et al.* (1973).
[b] Frequencies in cm^{-1} are accurate to ± 3 cm^{-1} for sharp lines and ± 5 cm^{-1} for weak or broad lines. Relative intensities, in parentheses, are based upon an arbitrary intensity of 10 given to the strongest line in each spectrum below 1600 cm^{-1}. Abbreviations: A, adenine; U, uracil; G, guanine; C, cytosine; r, ribose; P, phosphate; Trp, tryptophan; Tyr, tyrosine; Phe, phenylalanine; Am, amide; str, stretching; def, deformation; S, shoulder, and B, broad.

(Lord and Yu, 1970). The intensities of the lines at 815 and 1100 cm^{-1} (phosphate symmetric PO_2^- stretching) are sensitive to changes in the conformation of the nucleic acid backbone. The intensities of the lines at 722, 787, and 1480 cm^{-1} are proportional to the extent of participation of adenine, uracil plus cytosine, and guanine residues, respectively, in the secondary structure of RNA (Thomas *et al.*, 1971; Thomas *et al.*, 1972; Thomas, 1969; Lafleur *et al.*, 1972; Thomas and Hartman, 1973; Small and Peticolas, 1971). Hartman *et al.* (1973) found a gross change in the molecular configuration of guanine residues between the phage and protein-free states, a finding that showed that guanine is important in the complexing between RNA and protein in the virus.

Bacteriophage MS2. RNA and Capsid Protein

Figure 10.2 shows Raman spectra of the protein-free RNA extracted from the bacteriophage MS2, recorded at limits of high and low temperature (Thomas *et al.*, 1976). From spectra such as these, one can construct RNA melting curves by plotting the intensities of conformationally sensitive Raman lines as a function of the temperature. Figure 10.3 shows some plots of the peak intensities of selected Raman lines from the spectra of MS2 RNA versus temperature. In these plots we find two types of temperature dependence.

Raman lines arising from ring vibrations of the bases (Fig. 10.3, left) exhibit little intensity change with temperatures up to 50°C, then show large changes as the temperature increases beyond 50°C. Previous investigations with model compounds (Thomas and

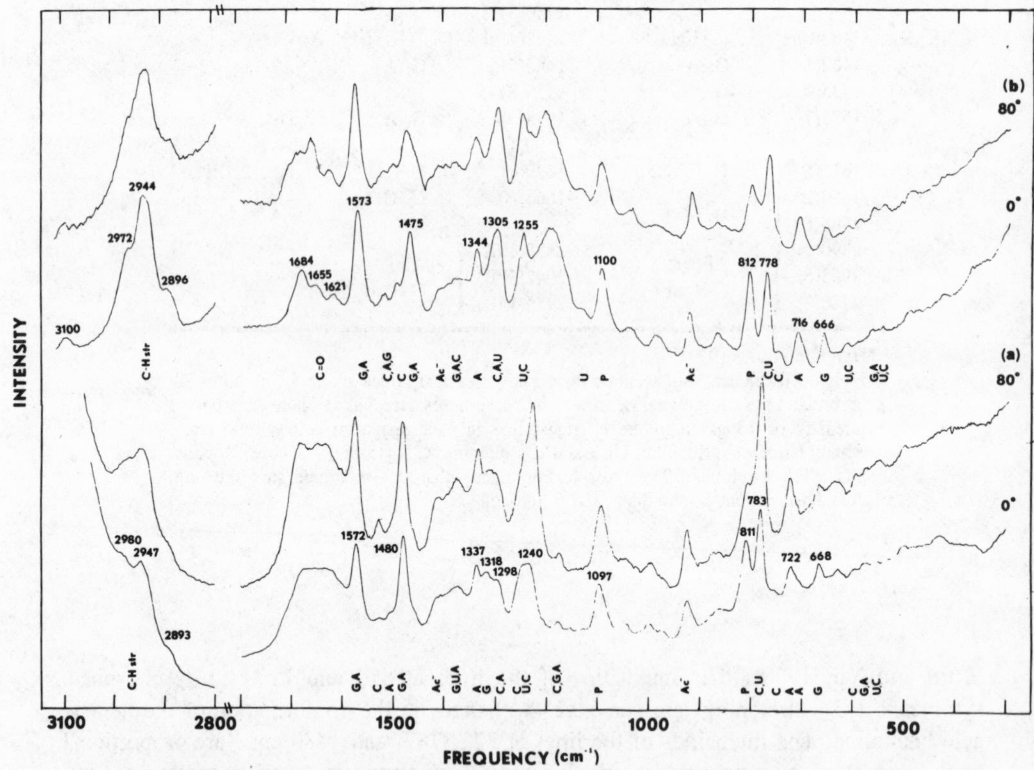

Fig. 10.2. Raman spectra in the regions 300–1800 cm^{-1} (amplification $A = 1$) and 2800–3100 cm^{-1} ($A = 1/3$) of RNA extracted from MS2 virions. (a) H_2O solutions at 0 and 80°C. (b) D_2O solutions at 0 and 80°C. The RNA concentration in each case is 40 μg/μl. Raman frequencies of the prominent lines are listed in cm^{-1}, and assignments are given along the abscissa to RNA base (A, U, G, C) sugar (R), or phosphate (P) groups. The line at 925 cm^{-1}, labeled Ac^-, is due to acetate buffer. Comparison of (a) and (b) shows the advantage in the double-bond region (1600–1700 cm^{-1} interval) of D_2O over H_2O as a solvent for Raman spectroscopy. On the other hand, the conformationally sensitive frequency of the RNA backbone (near 812 cm^{-1}) is unaffected by the change of solvent but greatly altered by the change of temperature. (Thomas *et al.*, 1976.)

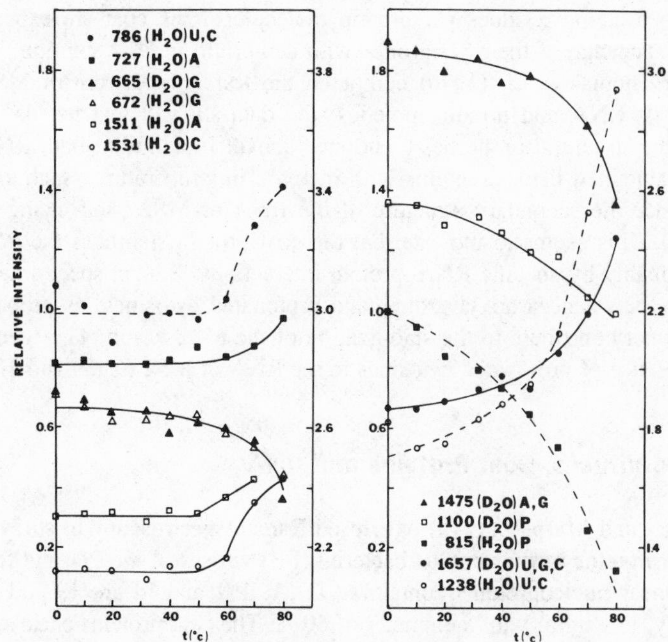

Fig. 10.3. Plots of the peak intensities of selected Raman lines of MS2 RNA vs. temperature. The ordinates (left axis for solid curves and right axis for broken curves) show intensities in arbitrary units after normalization to an internal standard. The group of lines in the left figure display cooperative melting behavior, while those on the right exhibit noncooperative melting. The keys give the Raman frequencies in cm^{-1}, the solvent employed, and the assignments. (Thomas *et al.*, 1976.)

Kyogoku, 1977; Hartman *et al.*, 1973) give evidence that these spectral changes arise from the unstacking of the RNA bases, which is a cooperative event occurring above 50°C. The unstacking of pyrimidines takes place within a smaller temperature range than that of purines.

Raman lines arising from carbonyl and phosphate group vibrations (Fig. 10.3, right) change gradually in intensity over the whole range of temperature, showing that the breakup of hydrogen bonding between paired bases in helical sections of MS2 RNA is a noncooperative structural transition, as is the loss of order in the RNA backbone. In going from 30° to 80°C, the polynucleotide chain of MS2 RNA changes from 85% A-helical structure to less than 30% ordered structure (Thomas *et al.*, 1976).

The MS2 capsids (shells composed of coat protein molecules), freed of RNA, were also studied by Thomas *et al.* (1976). The coat proteins were found to contain 0%, 60%, and 40%, respectively, α, β, and random conformations. The secondary structure of the coat protein molecules withstands temperatures up to 55°C, above which the capsids cave in. In each coat protein molecule four tyrosine side chains act as H-bond donors to negative acceptors (very likely —COO^- groups of glutamic and aspartic acids) of the protein. The capsid protein molecules have no disulfide bridges either within or between them. Two cysteine residues per protein molecule of the coat are exposed enough to allow fast exchange of their SH groups with deuterium in D_2O solution.

When Thomas *et al.* (1976) compared the Raman spectrum of MS2 phage with spectra of its RNA and protein moieties, the data showed that the native virion also preserves the structural properties mentioned above. In addition, both RNA and protein of MS2 maintained their structures with increase in temperature. Such a result is noteworthy, since the secondary structure of the RNA of MS2 itself changes below 50°C (Fig. 10.3). This seems to indicate that the coat protein stabilizes the RNA within the capsid, probably by specific RNA–protein interactions. Raman spectra have shown that interactions between carboxyl groups and protonated cytosines, as proposed by Kaper (1972), do not contribute to the stabilization of the MS2 virion. The spectra have ruled out the presence of protonated cytosines in the RNA of MS2 (Chou and Thomas, 1977).

Pf1 and fd Viruses. Coat Proteins and DNA

Thomas and Murphy (1975) have used Raman spectroscopy to study the structures of the coat proteins of filamentous bacterial (FB) viruses of the Pf1 and fd strains. Both are long linear nucleoproteins containing DNA. Pf1 and fd are 19,500 and 9,000 Å, respectively, in length, with a diameter of 60 Å. The coat proteins encapsulate the DNA molecule, which is believed to be a circular single strand. The amino acid composition differs in these two strains: glutamine and arginine are present in Pf1 but not in fd; tryptophan, phenylalanine, glutamic acid, and proline are present in fd but not in Pf1.

Raman data from several laboratories give the following information concerning proteins and polypeptides in water (Chen and Lord, 1974; Yu *et al.*, 1973; Frushour and Koenig, 1974). For helical structures a strong sharp amide I line appears at 1650 ± 5 cm^{-1} and there is relatively weak Raman scattering in the amide III region, from 1265 to 1300 cm^{-1}. For β-structures a strong sharp amide I line at 1665 ± 5 cm^{-1} and a strong amide III line appear at 1235 ± 10 cm^{-1}. Random, or disordered, structures display a strong broad amide I line near 1665-cm^{-1} and a medium amide III line at 1248 ± 5 cm^{-1}. The amide frequencies and intensities in Fig. 10.4 thus confirm that the conformations of the coat proteins of Pf1 and fd are both α-helix. Also, the lack of additional Raman scattering in the amide I area, which could be attributed to β or random structures, shows that the coat proteins are uniformly α-helix (Thomas and Murphy, 1975). Figure 10.4 also shows that the DNA bases produce weak scattering near 1685 cm^{-1} (Thomas, 1974). This arises from the carbonyl stretching vibrations of the bases in the DNA (Thomas, 1974; Hartman *et al.*, 1973; Thomas and Hartman, 1973; Erfurth *et al.*, 1972).

Further evidence for α-helix structure in Pf1 and fd coat proteins also occurs in the low-frequency area. Examples of this are Raman lines near 377 and 530 cm^{-1} which arise from alanine residues in the right-handed α-helix (Itoh *et al.*, 1974).

The α-helical structures of the coat proteins of both Pf1 and fd respond reversibly to temperature changes. At 32°C, the amide I line of Pf1 is displayed at 1652 cm^{-1} (Fig. 10.4); at 50°C, it moves to 1655 cm^{-1}; and at 75°C, it appears at 1657 cm^{-1} with a constant half-width (line width at half-maximum intensity). When the aqueous Pf1 is cooled to 32°C, its amide I is again found at 1652 cm^{-1}. Coat protein of the fd shows

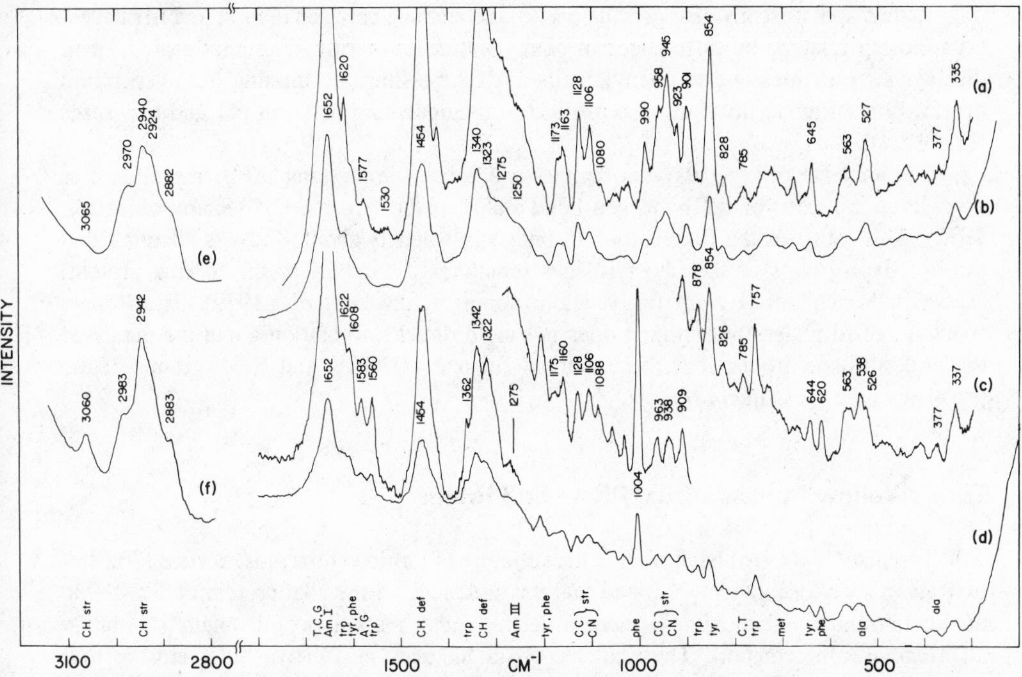

Fig. 10.4. Raman spectra of FB viruses in 0.05 *M* NaCl at 32°C and pH 9. Curve *a*, Pf1: concentration C = 108 μg/μl; excitation wavelength (λ) = 488.0 nm; radiant power (P) = 100 mW; slit width (Δσ) = 5 cm^{-1}; scan rate (r) = 50 cm^{-1}/min; rise time (t) = 1 sec; amplification (A) = 3×. Curve *b*, Pf1: Δσ = 10 cm^{-1}; t = 3 sec; A = 1×; other conditions as in *a*. Curve *c*, fd: C = 147 μg/μl; λ = 514.5 nm; P = 100 mW; Δσ = 5 cm^{-1}; r = 50 cm^{-1}/min; t = 1 sec; A = 3×. Curve *d*, fd: Δσ = 9 cm^{-1}; t = 3 sec; A = 1×; other conditions as in *c*. Curve *e*, Pf1: A = 1/3×; other conditions as in *b*. Curve *f*, fd: A = 1/3×; other conditions as in *d*. Frequencies of prominent lines are given in cm^{-1} units and assignments to molecular subgroups are denoted by standard abbreviations. Abbreviations: str, stretching; def, deformation; CH, carbon–hydrogen bond; CC, carbon–carbon bond; CN, carbon–nitrogen bond; A, T, C, and G, adenine, thymine, cytosine, and guanine; ala, alanine; met, methionine; phe, phenylalanine; trp, tryptophan; tyr, tyrosine; Am, amide. (Thomas and Murphy, 1975.)

the same reversible frequency shift. These frequency reversibilities reflect reversible alterations in the α-helical structures in the temperature range cited.

Figure 10.4 also contains useful information about the DNA structure. DNA with an A-helix displays a Raman line at 810 cm^{-1} of intensity equal to that of the pyrimidine line at 785 cm^{-1} (Erfurth *et al.*, 1972). No Raman line appears between 790 and 825 cm^{-1} in the spectrum of Pf1, although there is a well-resolved pyrimidine line at 785 cm^{-1}. The weak line at 801 cm^{-1} in the fd spectrum is caused by amino acid residues, and is too weak and of too low frequency to result from the A-helix DNA vibrations. Thus, the data of Fig. 10.4 are not consistent with an A-form DNA for the fd and Pf1 virions (Day, 1969). The data indicate a DNA backbone of either the B- or C-form, because neither B-DNA nor C-DNA shows a distinctive Raman line between 790 and 825 cm^{-1} (Erfurth *et al.*, 1972).

Raman spectroscopy also permits one to differentiate between the Pf1 and fd phages, both closely related, by differences in phenylalanine scattering. The coat protein of fd displays several lines due to phenylalanine. All these lines are missing in the spectrum of Pf1. Other Raman lines are also useful for distinguishing between Pf1 and fd viruses (Fig. 10.4).

The intensity ratio I_{850}/I_{830} for the tyrosine doublet in proteins has been proposed as a sensitive indicator of the hydrogen bond status of the tyrosine OH (Siamwiza *et al.*, 1975). This ratio for the filamentous fd phage in water is about 10:2.7, indicating that the hydroxyl groups of the two tyrosines (positions 21 and 24 of the fd coat protein) accept hydrogen bonds from fairly acidic donors (Dunker *et al.*, 1979). The Raman spectrum of fd phage (not shown) does not yield direct information about the nature of the proton donor groups, but these authors suggest —COOH and NH_3^+ groups of the proteins as the most likely donors.

Turnip Yellow Mosaic Virus. RNA and Protein

Turano *et al.* (1976) have studied the structure of turnip yellow mosaic virus (TYMV) by Raman spectroscopy. They found that the molecules of the coat protein of TYMV do not contain much α-helical or β-sheet structure, rather the polypeptide chain has mainly an irregular conformation. There are no S—S linkages in TYMV, but there are four cysteine residues, all exposed to the aqueous solvent. Tryptophan residues are accessible to solvent molecules, but tyrosine moieties apparently do not have access to the solvent. The tyrosine residues form strong hydrogen bonds between the phenolic —OH donor and negative acceptor groups within the virion. Above 54°C, the virus structure collapses, but between 0° and 54°C, the structural features mentioned here remain intact.

The RNA molecule which lies encapsulated in TYMV has only about 60% ordered secondary structure, much lower than the values near 85% found for molecules of transfer RNA, 16S ribosomal RNA, R17 RNA, and MS2 RNA (Hartman *et al.*, 1974; Hartman *et al.*, 1973; Thomas *et al.*, 1976). The encapsulated RNA secondary structure remains essentially intact up to 54°C.

The status of cytosine residues in the RNA of TYMV was also evaluated. Nonprotonated cytosine residues have a very different spectrum in the interval from 1250 to 1600 cm^{-1} than protonated ones, so Turano *et al.* used this information to establish that the cytosine residues of the RNA of TYMV are not protonated at pH 7, either for protein-free RNA or for the encapsulated RNA. They deem it unlikely that specific hydrogen-bonding interaction at pH 7 between cytosine residues of RNA and carboxyl groups of the coat proteins is one of the primary stabilizing factors of the native TYMV virion.

These authors feel that the high resolution and lack of interference from the solvent during Raman detection of cysteine S—H groups or cystine S—S groups provide advantages over such methods as Fourier transform IR spectroscopy (Alben *et al.*, 1974).

Hartman *et al.* (1978) have studied the structure of turnip yellow mosaic virus (TYMV) and its capsids by Raman spectroscopy. Figure 10.5 shows representative spectra of TYMV at several pH values. Previous assignments of the prominent Raman bands to specific groups in RNA and capsid were made by Turano *et al.* (1976) and are shown

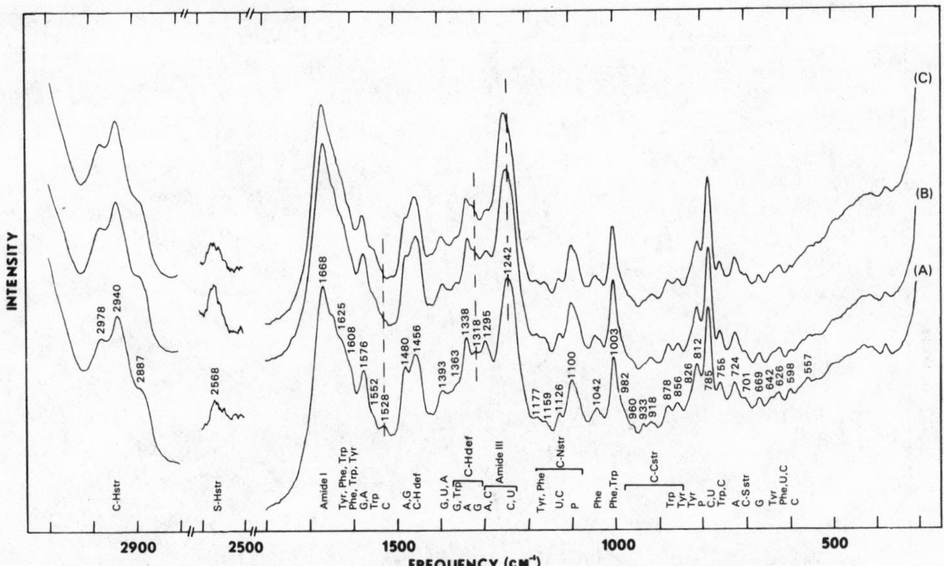

Fig. 10.5. Raman spectra at 32°C of TYMV in 0.3 *M* KCl solutions at (A) pH 7.5, 62 μg/μL; (B) pH 4.8, 75 μg/μL; and (C) pH 4.1, 71 μg/μL. Conditions: excitation wavelength, 488.0 nm; slit width, 10 cm^{-1}; scan speed, 25 cm^{-1}/min; rise time, 10 sec; amplification, $A = 1$ (300–1800 cm^{-1}), $A = 3$ (2500–2600 cm^{-1}), and $A = 1/3$ (2800–3100 cm^{-1}). Broken vertical lines show frequency shifts. Abbreviations: Standard one-letter symbols are used for the RNA bases and three-letter symbols for the amino acids. Also, P = phosphate, str = stretching, def = deformation, and C—S, C—C, C—N, C—H, and S—H denote their respective functional groups. (Hartman *et al.*, 1978.) Reprinted with the permission of the American Chemical Society. Copyright 1978.

in this figure. Figure 10.6 shows spectra of the capsids under various conditions. The coat-protein molecules in TYMV have a secondary structure made up of 48 ± 6% irregular conformation, 43 ± 6% β-sheet, and 9 ± 5% α-helix. When the RNA is removed from the capsid, the secondary structure of the coat-protein molecules is not changed. Also, when as many as 200 breakages per RNA molecule are introduced into the RNA chain, the RNA overall secondary structure in the encapsulated state is not affected. It remains mainly in the A-helix form (77 ± 5%).

Tobacco Mosaic Virus. A-Protein and RNA

Fox *et al.* (1979) have used Raman spectroscopy to study materials isolated from tobacco mosaic virus (TMV)—namely, the virus A-protein [isolated by the method of Fraenkel-Conrat (1957) as described by Knight (1975)], and an infectious RNA [extracted by the method of Gierer and Schramm (1956) as described by Knight (1975)]. The frequency positions of the amide I band (1650- and 1657-cm^{-1} doublet, aqueous spectrum; and 1649, 1659 cm^{-1}, solid spectrum) and amide III (1273 cm^{-1}, aqueous; 1268 cm^{-1}, solid) of the A-protein showed that its peptide backbone contains α-helical structure. (See

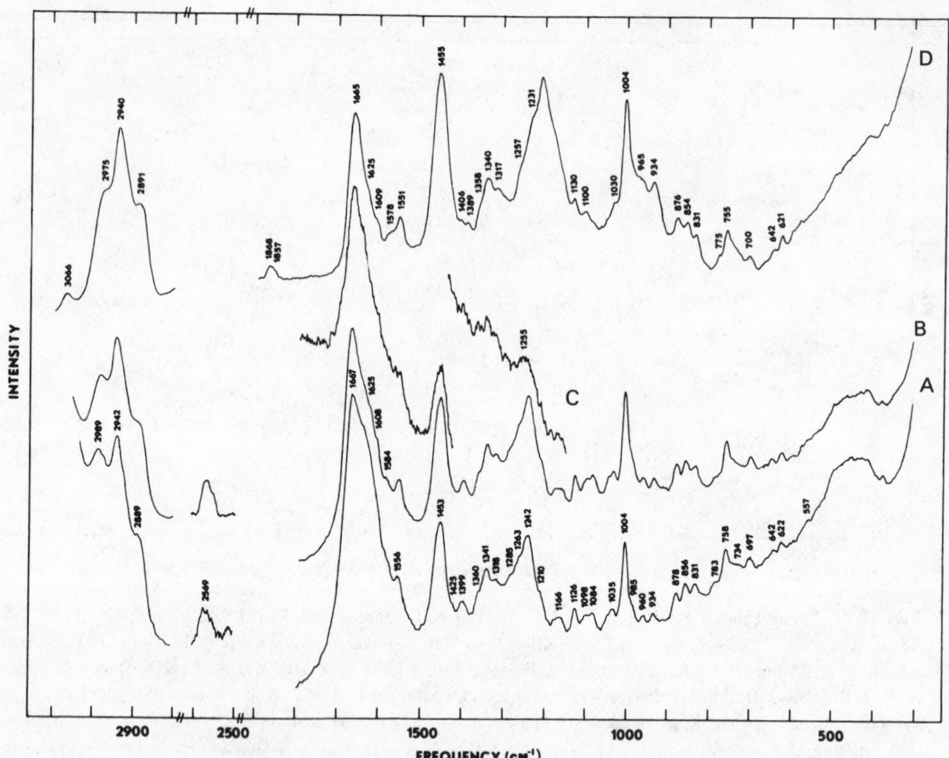

Fig. 10.6. Raman spectra at 32°C of TYMV capsids in H_2O solutions (0.3 *M* KCl) at (A) pH 7.5, 35 μg/μL; (B) pH 4.8, 76 μg/μL; and (C) pH 7.5 (the 1150–1800-cm^{-1} region only), precipitated from solution by denaturation at 70°C; and in D_2O solutions (0.3 *M* KCl) at (D) pD 7.8, 68 μg/μL. Other conditions are as in Fig. 10.5. (Hartman *et al.*, 1978.) Reprinted with the permission of the American Chemical Society. Copyright 1978.

Fig.10.7 for the spectrum of the aqueous solution.) A free —SH group of TMV is detectable as an —SH stretching vibration at 2567 cm^{-1} (not shown here), although many chemical reagents do not react with this sulfhydryl group (Fox *et al.*, 1979). These authors predicted the A-protein structure from its amino acid sequence by the method of Chou and Fasman (1974): α-helix, 32%; β-sheet, 31%; β-turn, 22%; and random coil, 12%. Overall, the structure of the A-protein from the Raman data and the prediction was in agreement with that from circular dichroism (Taniguchi and Taniguchi, 1975; Simmons and Blout, 1960; Schubert and Krafezyk, 1969) and X-ray crystallographic (Stubbs *et al.*, 1977) studies done earlier.

The secondary structure of the viral RNA was estimated from the ratio I_{815}/I_{1080} (Peticolas, 1975; Whitmann-Liebold and Whitmann, 1967; Nozu *et al.*, 1970; Thomas and Hartman, 1973) to have about 50% ordered structure. The 1080-cm^{-1} band, from the symmetrical stretching of the ionic phosphate, is invariant with conformation; while

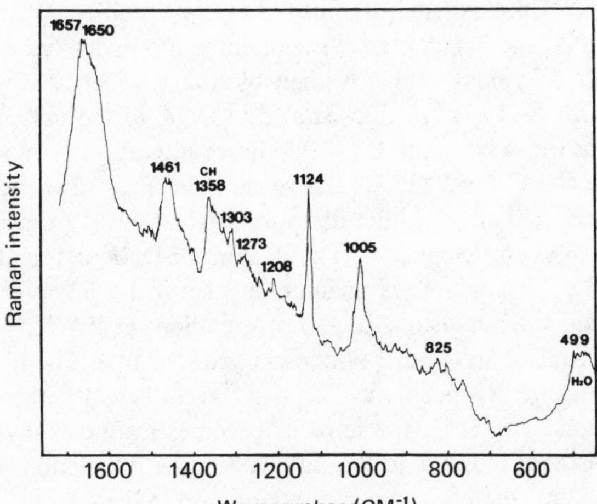

Fig. 10.7. Raman spectrum of isolated protein component (A-protein) from tobacco mosaic virus in aqueous solution. Power: 400 mW, slit width: 400 μm, full scale: 1000, scanning speed: 0.2 cm^{-1}/sec. (Fox *et al.*, 1979.)

the 815 cm^{-1} band is caused by the phosphodiester vibration for Watson–Crick-type base pairing.

Chromatins and Nucleosomes

Thomas *et al.* (1977) have used Raman and circular dichroism spectra to characterize chicken-erythrocyte nu bodies (nucleosomes) and chromatin with respect to the conformational structures of their constituent DNA and histone molecules. The nu bodies are nucleohistone components of eukaryotic chromatin. There are about 140 nucleotide base pairs of DNA in each nu body. These surround a protein core containing two molecules each of the histones H2A, H2B, H3, and H4 (the "inner histones"). A section of DNA connects adjacent nu bodies, and histones H1 or H5 are bound to this DNA, which consists of 30 to 70 base pairs. Raman spectroscopy indicated that the inner histones which are bound to DNA in chromatin or in isolated nu bodies have conformations similar to those of the inner histones separated from DNA in 2 *M* NaCl solutions. The inner histone tetramer contains 50% α-helix and ~50% random structure. Thomas *et al.* proposed that the protein core of the nu body contains about 80% α-helix and they suggested that the basic segments of the inner histones contain little α-helix or β-sheet structure. The Raman spectrum of chicken DNA (not shown) is similar to that of calf thymus DNA (Prescott *et al.*, 1976). Raman bands at 1094 cm^{-1} (PO_2^- dioxy symmetric stretching), 832 cm^{-1} (mainly diester —O—P—O— stretching), and the relative intensities of bands at 671 and 681 cm^{-1}, (thymine and guanine ring vibrations, respectively), indicate that the DNA is in the B-form.

Thomas *et al.* (1977) found that each inner histone tetramer has one cysteine group per 491 amino acid groups (band at 2559 cm^{-1}). Also, each of 15 tyrosine residues is hydrogen-bonded, most probably to $—NH_3^+$ terminal groups of lysine moieties.

Goodwin and Brahms (1978) have studied native chromatins (calf thymus, hepatoma cells, and hepatoma cells containing bromodeoxyuridine-substituted DNA), nucleosomes (calf thymus) and DNA fibers by Raman spectroscopy; and directly correlated their results with X-ray diffraction data. The DNA of the various chromatins and nucleosomes was shown to be in the B (98% relative humidity) or modified-B form. The Raman spectrum of the C form (47% relative humidity), as characterized from fibers, shows a band at 865–870 cm^{-1}, which disappears when the C form is converted to the B form. (A band at 835 cm^{-1} remains.) The A form of DNA is found in fibers at 75% relative humidity. The N7 position of guanine appears to be involved in the interaction between nucleic acid and nonhistone proteins. Vibrations at N7 (N7—C8 stretching) were correlated with a band at 1490 cm^{-1} (Goodwin and Brahms). The histone core does not interact with the guanine N7 position. The band at 1490 cm^{-1} has the same intensity in DNA and in mono-, di-, tri-, and tetranucleosomes from calf thymus. The authors state that the core histones thus cannot be involved in the interaction with the N7 position of guanine and do not, therefore, lie in the large groove.

Goodwin *et al.* (1979) have compared the Raman spectra of nucleosomes from different sources, which contain different amounts of nonhistone proteins (NHP), with the spectrum of free DNA. The nucleosomes were from chicken erythrocytes, calf thymus, rat liver, and Zajdela hepatoma cells. Nucleosomes display Raman spectra in which the intensity of the band at ca. 1490 cm^{-1} is significantly decreased with respect to that seen for free DNA. These workers correlated the size of the reduction with the amount of NHP present in a given nucleosome preparation.

They found that nucleosomes also show a diminished intensity of a band at ca. 1580 cm^{-1} compared to that in free DNA. These two bands, in the case of free DNA, are caused mainly by guanine and adenine vibrations. When DNA is methylated at the guanine N7 and adenine N3 positions, a decrease in the intensities of the 1490- and 1580-cm^{-1} bands occurs. The Raman data support a model for nucleosomes in which the nonhistone protein molecules can become attached in the large DNA groove, whereas the core histones interact with bases in the small groove.

DNA Complexes and Chromatin

Mansy *et al.* (1976) have used Raman spectroscopy to identify an interaction site on DNA for arginine-containing histones in chromatin. They made comparisons of the spectra of DNA, chromatin, and complexes of DNA with poly-L-arginine and *N*-α-acetylarginine. The Raman spectrum of the chromatin from mouse myeloma cells is shown in Fig. 10.8. The amide I frequency at 1655 cm^{-1} shows that the histone proteins are predominantly α-helical (Yu *et al.*, 1973; Frushour and Koenig, 1974; Yu and Liu, 1972). The bands at 835 and 879 cm^{-1} indicate that the DNA is of the B-type conformation (Erfurth *et al.*, 1972; Erfurth and Peticolas, 1975).

One can compare intensities of the Raman bands in chromatin with those in ordinary DNA by using the 1097-cm^{-1} band of the PO_2^- symmetric dioxy stretch as an internal standard for intensity changes. The chromatin spectrum is markedly different from that of ordinary DNA in that the former shows a decrease in the intensity of the 1490 ± 2-

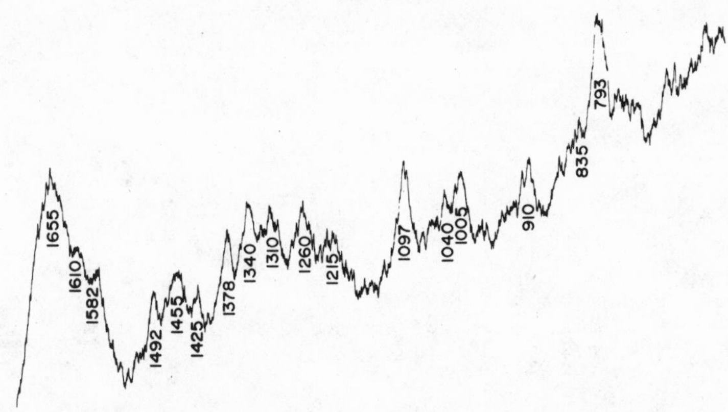

Fig. 10.8. Raman spectrum of mouse myeloma chromatin, cm^{-1}. (Mansy *et al.*, 1976.)

cm^{-1} band, due mainly to guanine and slightly to adenine (Erfurth and Peticolas, 1975). A decrease in the intensity of the 1490 cm^{-1} band was ascribed by Mansy *et al.*, to a hydrogen bond at the N7 position of guanine, and they concluded that a probable interaction of the N-terminal cationic residues of the histone proteins with DNA involves a hydrogen bond with the N7 position of guanine which lies in the major groove of B-form DNA.

In Fig. 10.9, we see the Raman spectrum of calf thymus DNA in (a) 0.1 *M* *N*-α-acetylarginine, (b) mixed with 20% by weight poly-L-arginine, and (c) in water solution. The intensity of the 1490 ± 3-cm^{-1} band is clearly decreased as a result of the interaction of DNA with arginine peptides. The similar decrease in the 1490 ± 3-cm^{-1} band of guanine in chromatin and in arginine–peptide mixtures with DNA supports the view that the N7 position of guanine is a site of attachment for arginine. In a space-filling model of the B-form of DNA these authors observed that the close proximity of the N7 position of guanine (lying in the major groove) to the —PO_2^- group of the backbone allows the terminal ═N—H of the arginine to be bound to the N7 position, while the terminal NH_3^+ group of arginine can form an ionic bond with the PO_2^- group. Three factors tend to stabilize the attachment of arginine side-chains in the major groove: a hydrogen bond at N7 of guanine; a salt bridge between the PO_2^- group of DNA and the NH_3^+ group of arginine; and hydrophobic interaction between the methylene groups and the interior of the major groove.

Studies of DNA-histone complexes provide information that is useful for attempting to understand the structure of chromatin (Van Holde and Isenberg, 1975). Liquier *et al.* (1977) have used IR linear dichroism to study complexes between DNA and histones H2B and H3. They used a broad range of histone: DNA ratios at different relative humidities. Figure 10.10 shows an example of the type of system studied. The dichroic ratios, $R(\perp/\|) = A_{\perp}/A_{\|}$, were calculated from recordings of the spectra of oriented films with the electric field of the incident light parallel and perpendicular to the orientation axis. Specimens were aligned at 45° with respect to the slit, thereby eliminating the polarization due to the gratings of the spectrophotometer. Measurements of the water

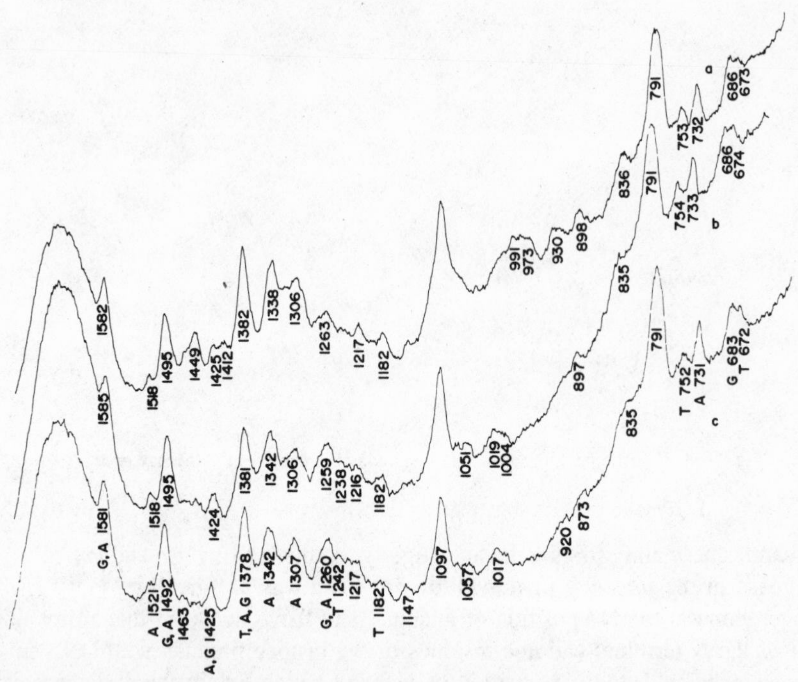

Fig. 10.9. Raman spectra of 2% aqueous solution of calf thymus DNA self-buffered to pH 7.0 (a) in 0.1 *N,N*-α-acetylarginine, (b) with 25% by weight poly-L-arginine and (c) doubly distilled H_2O. (Mansy *et al.*, 1976.)

absorption band at 3400 cm^{-1} allowed control of the relative humidity (r.h.). The top half of Fig. 10.10 shows the spectrum of the DNA:H2B complex (1:0.1 weight ratio) at 100% r.h. Here we can see properties of a B-form type of DNA: (a) a strongly perpendicular absorption at 1710 cm^{-1} (C═O and C═N double-bond stretching), which is characteristic of base pairing (Tsuboi, 1970); (b) a nondichroic band at 1230 cm^{-1} (antisymmetric stretching, ν_{as}, of the OPO group); (c) a strongly perpendicular band at 1090 cm^{-1} [symmetric stretching, ν_s, of the OPO group (Shimanouchi *et al.*, 1964)]. The bottom half of Fig. 10.10 represents the same complex, but at 66% r.h. The 1710-cm^{-1} band is still strongly perpendicular, but the 1230-cm^{-1} nondichroic band changes to perpendicular, and the 1090-cm^{-1} band changes from perpendicular to parallel. This spectrum resembles those recorded from films at low r.h. and is characteristic of DNA in the A form.

The progressive addition of histone H3 or H2B to DNA inhibits the change from B- to A-form and DNA stays in the B-type form at low r.h. The B → A transition is completely inhibited when only one molecule of H2B or H3 histone is bound per about three or four DNA helix turns, respectively. These workers proposed that about four to three turns of DNA helix characterize the "critical length of DNA" for the B → A transition.

Liquier *et al.* (1979) have used IR spectroscopy to study DNA secondary structures in various nucleohistone-DNA complexes. Under conditions of low relative humidity,

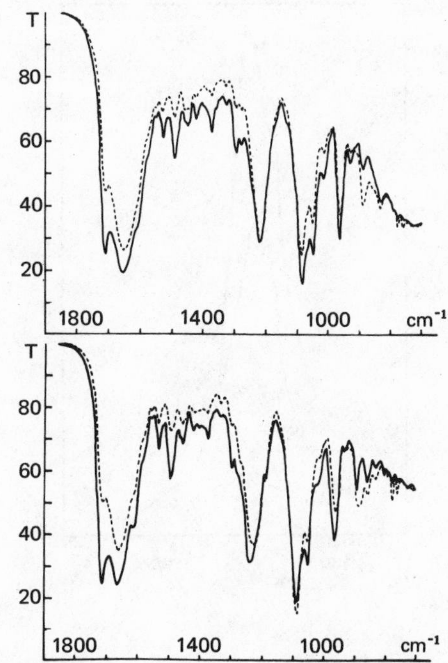

Fig. 10.10. (Top) Infrared spectrum of H2B-DNA 0.1:1 complex, high humidity: (——) electric vector of polarized light perpendicular to the orientation axis; (···) electric vector of polarized light parallel to the orientation axis. (Bottom) Same sample, low relative humidity. (Liquier *et al.*, 1977.) Reprinted with the permission of the American Chemical Society. Copyright 1977.

DNA remains in the B-type conformation in the following DNA-containing systems: chromatin, chromatin extracted by 0.6 *M* NaCl (removal of protein), nucleosomes, and histone-DNA reconstituted complexes. In chromatin extracted by tRNA and in nonhistone protein-DNA reconstituted complexes the DNA can have an A-type conformation. The spectral interval used to study the above systems for different conformations was the 800–900 cm^{-1} region, which shows characteristic bands for the phosphodiester chain.

Cotter and Lilley (1977) have used ^{31}P NMR to compare the environment of DNA phosphate groups in the 140 base-pair chromatin core particle with that of the phosphates in the extracted DNA (i.e., minus histones); and they used IR spectroscopy (Figs. 10.11, 10.12) to compare the secondary structure of core protein *in situ* with that of the complex isolated in 2 *M* NaCl. They concluded that the strain at the phosphate groups of DNA in the core particle is unchanged from that in B-form DNA (in which kinks may occur at intervals of ten base pairs), and that the α-helical and random coil contributions to the isolated core-protein secondary structure are maintained within the chromatin subunit.

DNA-Polylysine and Polyriboadenylic Acid-Polylysine Complexes

Prescott *et al.* (1976) have examined the Raman spectra of DNA–polylysine and polyriboadenylic acid–polylysine complexes. The purpose of this work was to determine whether structural changes take place during complex formation and which molecular subgroups may be involved. Figure 10.13 compares spectra of DNA, polylysine, and

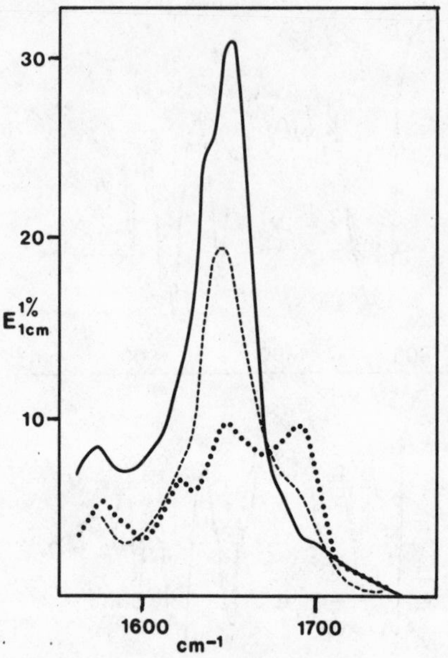

Fig. 10.11. Infrared spectra from core particles and their constituents: (· · ·) DNA in 10 m*M* cacodylate, 0.7 m*M* EDTA, p^2H appt 7.1. (——) Core protein in 2 *M* NaCl, 10 m*M* CHES, p^2H appt 8.6. (- - -) Core particles in 10 m*M* cacodylate, 0.7 m*M* EDTA, p^2H appt 7.1. (Cotter and Lilley, 1977.)

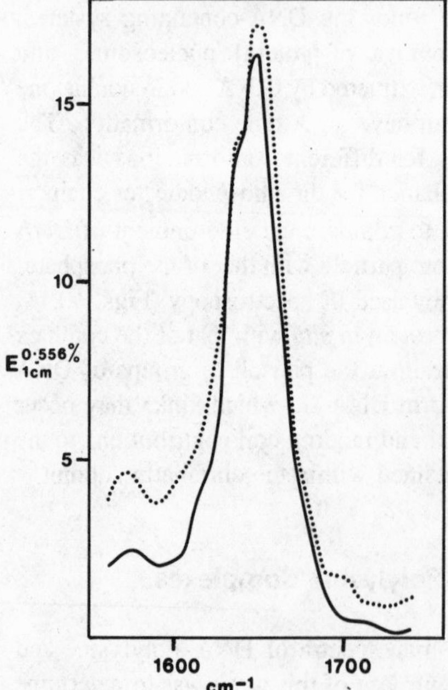

Fig. 10.12. Infrared spectrum of core protein: (· · ·) Experimental spectrum. (——) Spectrum calculated by subtraction of DNA spectrum from core particle spectrum. (Cotter and Lilley, 1977.)

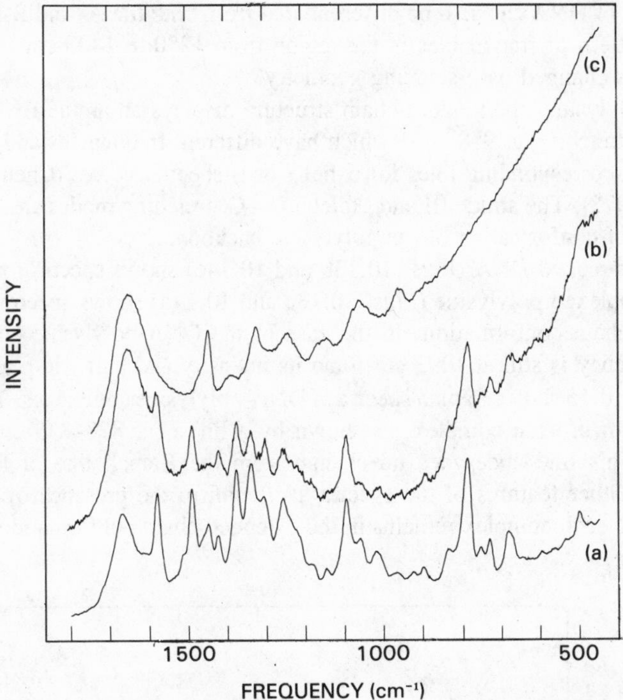

Fig. 10.13. Raman spectra at 32°C of the following in 1.0 *M* NaCl solution (pH 7): (a) DNA-polylysine complex, directly mixed; (b) DNA; (c) polylysine. Excitation wavelength 514.5 nm; radiant power ~300 mW; spectral slit width 10 cm^{-1}; scan speed 25 cm^{-1}/min; rise time 10 sec. (Prescott *et al.*, 1976.) Reprinted with the permission of the American Chemical Society. Copyright 1976.

DNA-polylysine complex in 1.0 *M* NaCl. One can differentiate the A, B, and C forms of DNA by the frequency and intensity of the Raman line near 800 cm^{-1}, a considerable portion of which stems from the O—P—O symmetric stretching vibration of the phosphodiester backbone (Thomas and Hartman, 1973; Erfurth *et al.*, 1972). In the A structure a strong sharp line is present at 807 cm^{-1}; in the B structure a weak broad line occurs at 830 ± 5 cm^{-1} (Erfurth *et al.*, 1972 and 1975; Rimai *et al.*, 1974); in the C structure the corresponding frequency seems to occur outside the span of 790 to 850 cm^{-1}, where it is covered over by frequencies of the DNA bases (Erfurth *et al.*, 1972). The A, B, and C structures of DNA are also distinguishable by other features, particularly in the two lines at 670 and 680 cm^{-1}. For A-structure, the 670-cm^{-1} line is more intense than the 680-cm^{-1} one; for B-structure, the latter exceeds the former; for C-structure, the two lines have equal intensities (Erfurth *et al.*, and Rimai *et al.*, cited above). The lines at 670 and 680 cm^{-1} arise from in-plane ring vibrations of thymine and guanine, respectively (Lord and Thomas, 1967a, 1967b, and 1968), and the reversal of the intensities along with the A → B transition is probably an indication of the different base-stacking geometries in the DNA helices. Also, the B-form of DNA has a strong thymine line at 750 cm^{-1}, the intensity of which is much less in the A-form or C-form (Erfurth *et al.*, 1972).

The C-form of DNA can also be differentiated from both the A- and B-forms by the very different pattern of frequencies in the region from 1250 to 1400 cm^{-1}, a property again reflecting its changed base-stacking geometry.

For polylysine, the random-chain structure displays an amide III line at 1245 cm^{-1} and C—C stretching at 958 cm^{-1} which have different frequencies and intensities, compared to the corresponding lines for α-helix or β-sheet structure (Chen and Lord, 1974; Yu *et al.*, 1973). The amide III and skeletal C—C stretching modes are the most sensitive to changes of conformation of the polylysine backbone.

Noncomplexed DNA (Figs. 10.13b and 10.14c) shows spectra for the B-structure, and noncomplexed polylysine (Figs. 10.13c and 10.14d) shows spectra for the extended or random-chain conformation. In the case of the DNA-polylysine complex, the PO_2^- group frequency is still at 1093 cm^{-1} and its intensity and half-width are unchanged.

Figure 10.15 shows Raman spectra of DNA-polylysine complexes. The DNA remains in the B-form in each complex, as shown by a line near 828–835 cm^{-1} in each DNA spectrum. This line undergoes no change from the Raman line of free DNA at this frequency. Other features of the spectra also confirm the presence of the B-form. The polylysine in each complex remains in the extended chain conformation, which is bound

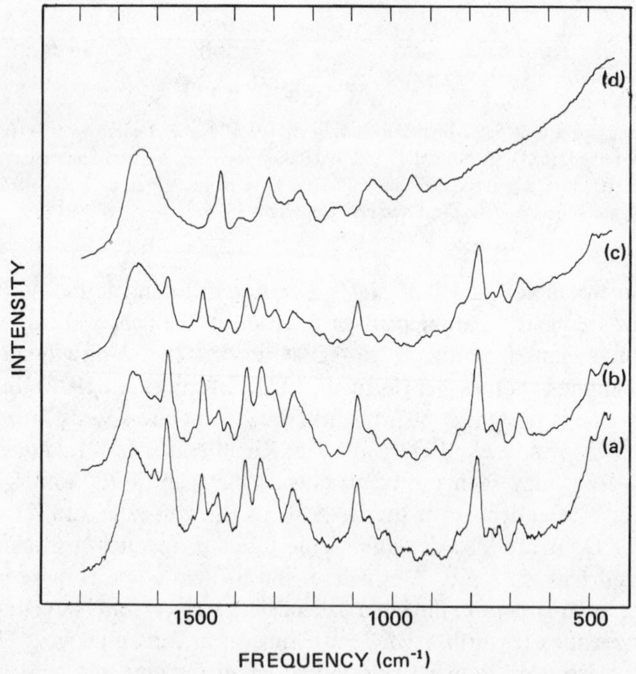

Fig. 10.14. Raman spectra at 32°C of the following in 0.025 *M* NaCl solution (pH 7): (a) DNA-polylysine complex, directly mixed; (b) DNA-polylysine complex, reconstituted; (c) DNA; (d) polylysine. Conditions as in Fig. 10.13. (Prescott *et al.*, 1976.) Reprinted with the permission of the American Chemical Society. Copyright 1976.

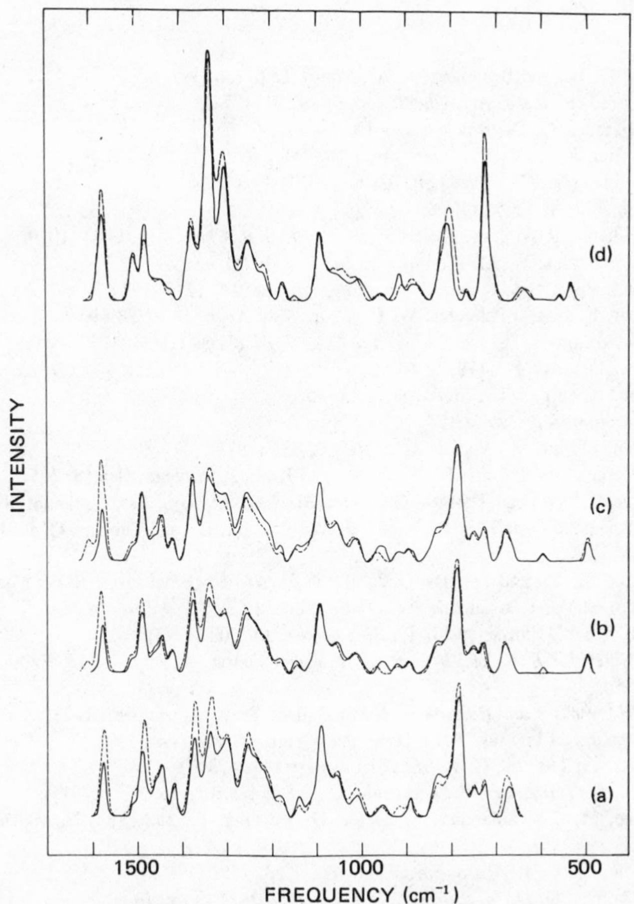

Fig. 10.15. Comparison of the observed Raman spectra of a complex (- - -) with the sum of the spectra of the constituent polymers (——), for each of the following: (a) DNA-polylysine, directly mixed in 1.0 *M* NaCl (pH 7); (b) DNA-polylysine, reconstituted in 0.025 *M* NaCl (pH 7); (c) DNA-polylysine, directly mixed in 0.025 *M* NaCl (pH 7); (d) poly(rA)-polylysine, directly mixed in 0.01 *M* sodium phosphate (pH 7.5). (Prescott *et al.*, 1976.) Reprinted with the permission of the American Chemical Society. Copyright 1976.

by electrostatic interactions between positively charged lysyl side chains and negatively charged phosphate groups of the DNA.

Spectra of the complex between polyriboadenylic acid and polylysine show that both the geometry of the backbone and the mode of interaction between stacked bases of polyriboadenylic acid are altered by complex formation. The extended polylysine chain may also be bound to polyriboadenylic acid, poly r(A), by electrostatic interaction between lysyl and phosphate groups. The *gauche*$^+$–*gauche*$^+$ configuration usually present in the phosphodiester linkages of poly(rA) in aqueous solution is distorted when binding to polylysine occurs.

References

Alben, J. O., Bare, G. H., and Bromberg, P. A., *Nature* **252,** 736 (1974).

Chen, M. C., and Lord, R. C., *J. Am. Chem. Soc.* **96,** 4750 (1974).

Chou, P. Y., and Fasman, G. D., *Biochemistry* **13,** 222 (1974).

Chou, C. H., and Thomas, G. J., Jr., *Biopolymers* **16,** 765 (1977).

Cotter, R. I., and Lilley, D. M., *FEBS Lett.* **82,** 63 (1977).

Day, L. A., *J. Mol. Biol.* **39,** 265 (1969).

Dunker, A. K., Williams, R. W., and Peticolas, W. L., *J. Biol. Chem.* **254,** 6444 (1979).

Erfurth, S. C., and Peticolas, W. L., *Biopolymers* **14,** 247 (1975).

Erfurth, S. C., Bond, P. J., and Peticolas, W. L., *Biopolymers* **14,** 1245 (1975).

Erfurth, S. C., Kiser, E. J., and Peticolas, W. L., *Proc. Natl. Acad. Sci. USA* **69,** 938 (1972).

Fox, J. W., Lee, J., Amorese, D., and Tu, A. T., *J. Appl. Biochem.* **1,** 336 (1979).

Fraenkel-Conrat, H., *Virology* **4,** 1 (1957).

Frushour, B. G., and Koenig, J. L., *Biopolymers* **13,** 455 (1974).

Gierer, A., and Schramm, G., *Nature* **177,** 702 (1956).

Goodwin, D. C., and Brahms, J., *Nucleic Acids Res.* **5,** 835 (1978).

Goodwin, D. C., Vergne, J., Brahms, J., Defer, N., and Kruh, J., *Biochemistry* **18,** 2057 (1979).

Hartman, K. A., Clayton, N., and Thomas, G. J., Jr., *Biochem. Biophys. Res. Commun.* **50,** 942 (1973).

Hartman, K. A., McDonald-Ordzie, P. E., Kaper, J. M., Prescott, B., and Thomas, G. J., Jr., *Biochemistry* **17,** 2118 (1978).

Hartman, K. A., Lord, R. C., and Thomas, G. J., Jr., in *Physico-Chemical Properties of Nucleic Acids,* Vol. 2, (J. Duchesne, ed.) p. 1, Academic Press, New York, 1973.

Itoh, K., Hinomoto, T., and Shimanouchi, T., *Biopolymers* **13,** 307 (1974).

Kaper, J. M., in *Molecular Basis of Virology* (H. Fraenkel-Conrat, ed.) pp. 1–133, Reinhold, New York, 1968.

Kaper, J. M., in *RNA Viruses and Ribosomes,* North Holland Press, Amsterdam, 1972.

Knight, C. A., *Chemistry of Viruses,* 2nd Ed., Springer-Verlag, New York, 1975.

Lafleur, L., Rice, J., and Thomas, G. J., Jr., *Biopolymers* **11,** 2423 (1972).

Liquier, J., Taboury, J., Taillandier, E., and Brahms, J., *Biochemistry* **16,** 3262 (1977).

Liquier, J., Gadenne, M. C., Taillandier, E., Defer, N., Favatier, F., and Kruh, J., *Nucleic Acids Res.* **6,** 1479 (1979).

Lord, R. C., and Thomas, G. J., Jr., *Spectrochim. Acta A* **23,** 2551 (1967a).

Lord, R. C., and Thomas, G. J., Jr., *Biochim. Biophys. Acta* **142,** 1 (1967b).

Lord, R. C., and Thomas, G. J., Jr., *Dev. Appl. Spectrosc* **6,** 179 (1968).

Lord, R. C., and Yu, N.-T., *J. Mol. Biol.* **50,** 509 (1970).

Mansy, S., Engstrom, S. K., and Peticolas, W. L., *Biochem. Biophys. Res. Commun.* **68,** 1242 (1976).

Marvin, D. A., and Hohn, B., *Bacteriol. Rev.* **33,** 172 (1969).

Marvin, D. A., and Wachtel, E. J., *Nature* **253,** 19 (1975).

Nakashima, Y., Wiseman, R. L., Konigsberg, W., and Marvin, D. A., *Nature* **253,** 68 (1975).

Nozu, Y., Ohno, T., and Okada, Y., *J. Biochem. (Jpn)* **68,** 39 (1970).

Peticolas, W. L., *Biochimie* **57,** 417 (1975).

Prescott, B., Chou, C. H., and Thomas, G. J., Jr., *J. Phys. Chem.* **80,** 1164 (1976).

Rimai, L., Maher, V. M., Gill, D., Salmeen, I., and McCormick, J. J., *Biochim. Biophys. Acta* **361,** 155 (1974).

Shimanouchi, T., Tsuboi, M., and Kyogoku, Y., *Adv. Chem. Phys.* **7,** 435 (1964).

Siamwiza, M. N., Lord, R. C., Chen, M. C., Takamatsu, T., Harada, I., Matsuura, H., and Shimanouchi, T., *Biochemistry* **14,** 4870 (1975).

Small, E. W., and Peticolas, W. L., *Biopolymers* **10,** 1377 (1971).

Thomas, G. J., Jr., *Biochim. Biophys. Acta* **213,** 417 (1970).

Thomas, G. J., Jr. in *Vibrational Spectra and Structure,* (J. R. Durig, ed.), Vol. 3, p. 239, Dekker, New York, 1974.

Thomas, G. J., Jr., and Hartman, K. A., *Biochim. Biophys. Acta* **312,** 311 (1973).
Thomas, G. J., Jr., and Kyogoku, Y., in *Infrared and Raman Spectroscopy,* Part C, (E. G. Brame, Jr., and J. G. Grasselli, eds.) pp. 712–872, Marcel Dekker, Inc., New York, 1977.
Thomas, G. J., Jr., and Murphy, P., *Science* **188,** 1205 (1975).
Thomas, G. J., Jr., Medeiros, G. C., and Hartman, K. A., *Biochem. Biophys. Res. Commun.* **44,** 587 (1971).
Thomas, G. J., Jr., Medeiros, G. C., and Hartman, K. A., *Biochim. Biophys. Acta* **277,** 71 (1972).
Thomas, G. J., Jr., Prescott, B., McDonald-Ordzie, P. E., and Hartman, K. A., *J. Mol. Biol.* **102,** 103 (1976).
Thomas, G. J., Jr., Prescott, B., and Olins, D. E., *Science* **197,** 385 (1977).
Tsuboi, M., *Appl. Spectrosc. Rev.* **3,** 45 (1970).
Turano, T. A., Hartman, K. A., and Thomas, G. J., Jr., *J. Phys. Chem.* **80,** 1157 (1976).
Van Holde, K. E., and Isenberg, I., *Accounts Chem. Res.* **8,** 327 (1975).
Whitmann-Liebold, B., and Whitmann, H. G., *Mol. Gen. Gene.* **100,** 358 (1967).
Yu, N.-T. and Liu, C. S., *J. Am. Chem. Soc.* **94,** 5127 (1972).
Yu, T. J., Lippert, J. L., and Peticolas, W. L., *Biopolymers* **12,** 2161 (1973).

Chapter 11

MODEL MEMBRANES, BIOMEMBRANES, AND LIPID-CONTAINING SYSTEMS

Introduction

The structure and properties of membranes and model membranes is a very complex topic being studied in many laboratories, by many different approaches. Raman and infrared spectroscopy have been used to good advantage. This chapter gives some examples of the ingenuity of various investigators in studies of bilayer and multilayer phospholipid systems, deuterated phospholipids, cholesterol-containing systems, protein–lipid and polypeptide–lipid mixtures, and membranes isolated from natural sources.

Andersen (1978) has reviewed various methods for probing biomembrane structure. These include X-ray, neutron, and Raman scattering, fluorescence, NMR, electron spin resonance of spin labels, calorimetry, and freeze-fracture electron microscopy. He has cautioned that arguments employed to deduce structural information from some of these physical methods frequently have hidden structural assumptions that may be incorrect. Many erroneous and/or contradictory statements have been made about membrane structure, and one must, therefore, understand the types of measurements and be careful about accepting the thinking behind some conclusions. Concerning the use of Raman spectroscopy for studying model membrane systems, he says, ". . . it is useful to regard the conclusions drawn from Raman spectroscopy as plausible explanations of the data rather than as necessary and unavoidable consequences of the data."

Conformational Studies with Phospholipid Bi- and Multilayer Systems

Dipalmitoyl Phosphatidylcholine and Others

Raman spectroscopy is quite useful for studying the structural characteristics of phospholipid molecules in model and biological membrane systems (Lippert and Peticolas, 1971, 1972; Larsson and Rand, 1973; Larsson, 1973; Mendelsohn, 1972; Bulkin, 1972;

Bulkin and Krishnamachari, 1972; Brown *et al.*, 1973; Chapman, 1973). One can examine the fluidity of membrane assemblies by paying attention to the sensitivity to conformational changes of spectral characteristics accompanying the vibrations displayed by the fatty acid hydrocarbon chains. If one could study a group of lecithin compounds and make definitive assignments of the transitions involving the choline and phosphate vibrations, one could then possibly look spectroscopically into the polar head groups of the membrane phospholipids and find additional conformational changes originating there.

Inasmuch as the dipalmitoyl phosphatidylcholine–water bilayer system is used in laboratories as a simple model for the structurally more complex cellular membranes, Spiker and Levin (1975) studied the Raman spectra of dipalmitoyl phosphatidylcholine (DPPC) and other structurally related molecules in order to make definitive assignments of the vibrational transitions of conformationally dependent motions.

The 1100-cm^{-1} region shows mainly the characteristics of the hydrocarbon C—C stretching modes and the phosphate O—P—O$^-$ symmetric stretching vibration. Figure 11.1 shows room temperature spectra for this region for polycrystalline samples of some lipid related systems. The calculations of Snyder (1967) and Schachtschneider and Snyder (1963) indicate that the transitions appearing at about 1065, 1100, and 1130 cm^{-1} in systems containing *n*-paraffin chains originate from modes essentially related to C—C stretching motions. Several laboratories have used the changes in relative intensity with temperature of these modes to follow specific conformational changes in the hydrocarbon chains when lipid–water systems are transformed from gel to liquid crystal phase (Lippert and Peticolas, 1971, 1972; Mendelsohn, 1972; Bulkin and Krishnamachari, 1972). Lippert and Peticolas (1972) had assigned the 1100-cm^{-1} transition in the phospholipids to skeletal modes coming from both the hydrocarbon rotamers and the phosphate group. Brown *et al.* (1973) gave a different assignment to it, namely, the PO_2^- symmetric stretching mode.

The data in Fig. 11.1 are useful in making assignments for this region. Palmitic acid and dipalmitoyl glycerine both display transitions characteristic of the hydrocarbon chain at 1100 cm^{-1}, but these molecules do not contain a phosphate group. When a phosphate group is present (lysolecithin), a distinct band appears at 1081 cm^{-1}, which is assigned to the PO_2^- symmetric stretching mode (Spiker and Levin, 1975). The phosphate band near 1081 cm^{-1} is not visually apparent in the spectra (Fig. 11.1) for DPPC and for the cholesterol–DPPC mixture. When one compares the dotted base lines for the two bottom spectra with those of the two top spectra, one sees a large increase in band areas near 1100 cm^{-1}. Spiker and Levin, by measuring band areas and making comparisons of intensity ratios, come to the conclusion that the phosphate mode lies buried within the 1099-cm^{-1} band of DPPC. They emphasize that the 1100-cm^{-1} region of the phospholipid is a superposition of the C—C stretching modes for the all-*trans* hydrocarbon conformations (if the system is in its gel, or crystalline state), the C—C modes of the hydrocarbon sections having *gauche* bonds, and the symmetric PO_2^- stretching mode.

Figure 11.1 also shows an increase in the relative area of the 1100-cm^{-1} feature of the system cholesterol–DPPC compared to that for DPPC. This increase is caused by a further disruption of the crystalline chain lattice by cholesterol, shown by an increase of *gauche* conformations in the hydrocarbon chains (Lippert and Peticolas, 1971). The 1050-cm^{-1} band in the spectrum of lysolecithin is assigned to a C—O stretching vibration.

Spiker and Levin (1975) assigned Raman lines at 1002 cm^{-1} for solid phosphoryl-

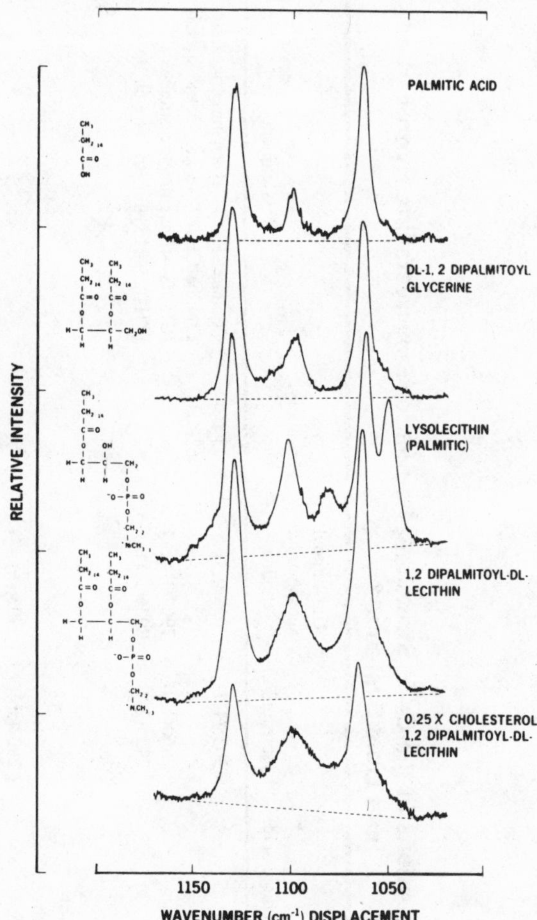

Fig. 11.1. Raman spectra of polycrystalline model lipid compounds in the 1020–1170-cm^{-1} region, with 514.5 nm excitation. χ means mole fraction. (Spiker and Levin, 1975.)

choline (PC) and 987 cm^{-1} for liquid PC to the symmetric PO_3^{2-} stretching mode. The features found at 1094 cm^{-1} for solid PC and 1088 cm^{-1} for liquid PC were assigned to the PO_3^{2-} antisymmetric stretching modes.

The antisymmetric PO_2^- phosphate stretching vibration is displayed as weak lines at 1218 cm^{-1} in liquid glycerophosphorylcholine, 1248 cm^{-1} in solid lysolecithin and 1245 cm^{-1} in solid DPPC.

Table 11.1 summarizes the frequencies and assignments of Raman transitions in the C—H stretching region for polycrystalline hydrocarbon and lipid systems.

Table 11.2 summarizes the frequency assignments for both the C—N and PO_2 diester vibrational modes. The band near 720 cm^{-1}, which is found in the spectra of polycrystalline phosphorylcholine, lysolecithin, and DPPC, had originally been assigned to the PO_2 symmetric diester stretch (Lippert and Peticolas, 1972; Bulkin and Krishnamachari, 1972), but it was later reassigned to the symmetric C—N stretch of the $RN^+(CH_3)_3$ group (Brown *et al.*, 1973).

Table 11.1. Frequencies of Raman Transitions in the C—H Stretching Region for Polycrystalline Hydrocarbon and Lipid Systems[a,b]

Polyethylene[c]	Octacosane	Palmitic acid	Decanedioic acid	Lysolecithin	Dipalmitoyl phosphatidyl-choline	Phosphoryl choline[d]	Assignment
				3035(1.1)	3035(0.5)	3023(10)	Choline(CH_3)C—H asymmetric stretch
						2962(6.1)	Choline(CH_3)C—H symmetric stretch
	2966(0.6)	2966(1.1)		2960(1.9)	2962(1.4)		(CH_3)C—H asymmetric stretch
	2935(0.9)	2938		2935(2.9)	2936(3.1)		(CH_3)C—H symmetric stretch
		(2.7)					
		2926					
≈2932(sh)[e]			2955(sh)				
			2933(sh)				
	2902(sh)	2903(sh)	2911(8.4)	2903(sh)	2903(sh)		
2883(10)	2882(10)	2883(10)	2888(10)	2880(10)	2883(10)		(CH_2)C—H asymmetric stretch
	2862(sh)	2860(sh)	2873(sh)	≈2860(sh)	2861(sh)		
2848(7)	2847(4.3)	2847(5.4)	2848(5.4)	2846(6.1)	2847(7.6)		(CH_2)C—H symmetric stretch

[a] Spiker and Levin (1975).
[b] Relative intensities are shown in parentheses.
[c] Tasumi *et al.* (1962).
[d] Bands in the 2800–3100-cm^{-1} region for phosphorylcholine occurred at 2815(0.4, br), 2861(0.4), 2875(1.0), 2908(1.4), 2931(sh), 2949(sh), 2962(6.1), 2970(sh), 2993(sh), 3023(10), 3040(2.1), and 3065(1.9) cm^{-1}. Only the two bands listed above could be assigned with confidence. (br is broad.)
[e] (sh) refers to shoulder.

Table 11.2. Selected Raman Frequencies of Various Model Phospholipid Compounds[a]

Phosphoryl-choline		L-α-Glycero-phosphoryl choline		Lyso-lecithin (palmitic) (Solid)	Dipal-mitoyl lecithin (Solid)	Palmitic acid	1,2-Dipal-mitoyl glycerine	1,3 Dipal-mitoyl glycerine	Octacosane	Assignments
Solid	Liquid	Solid	Liquid							
722	718 (P)[b]	719	717	718[e]	720					C—N symmetric stretch (choline)
776	766 (P)[c]	779	774	767[e]	760					O—P—O diester symmetric stretch
			820	835	833					O—P—O diester antisymmetric stretch
1002	987 (P)									PO_3^{2-} symmetric stretch
1068, 1077	≈1065[d]		1063	1050	≈1052[f]					C—O stretch
		≈1095	1086	1081	≈1080[g]					PO_2^- symmetric stretch
1094	1088[d]									PO_3^{2-} degenerate stretch
				1062	1064	1064	1064	1063	1064	C—C stretch
				1101	1099	1100	1099	1103	1110	C—C stretch
				1131	1130	1130	1131	1131	1134	C—C stretch
			1218[d]	1227, 1248	1245					PO_2^- antisymmetric stretch

[a] Spiker and Levin (1975).
[b] (P) represents polarized.
[c] P—O stretch for phosphorocholine.
[d] Weak.
[e] Corresponding liquid bands are polarized.
[f] Shoulder of 1064 cm^{-1} band.
[g] Partially obscured by 1098 cm^{-1} band envelope.

The temperature dependence of the Raman spectra of both egg lecithin (phosphatidylcholine) and dipalmitoyl lecithin shows that the region from 1050 to 1150 cm^{-1} is sensitive to *trans-gauche* isomerization of the hydrocarbon chains in phospholipids (Mendelsohn *et al.*, 1976a). For egg lecithin, the intensity ratio of the bands near 1132 and 1104 cm^{-1} can be used to gauge the amount of *gauche* isomer formation.

The all-*trans* chain conformation which is predominant in egg lecithin at −66°C and in dipalmitoyl lecithin at 23°C displays an intensity ratio, I_{1132}/I_{1104}, of about 1.5. Under conditions where *gauche* isomer has formed, the ratio ranges from 0.2 to 0.25. Sonication, which causes vesicles to form, did not cause a change in the relative populations of *trans* and *gauche* isomers in either of these lecithin–vesicle systems.

The spectral region from 2800 to 3100 cm^{-1} is sensitive to both intra- and interchain effects in phospholipid systems. Spectral changes that occur in this region are not due to variation in the relative number of *trans* and *gauche* isomers, but may be related to different interchain interactions of the hydrogen atoms upon sonication, resulting from the change in packing of the chains towards the center of the bilayers of the vesicles.

Spiker *et al.* (1976), searching for an alternative way to determine the IR and Raman spectra of membrane related systems, developed a method to incorporate phospholipid bilayer assemblies in a clay matrix to form ultrathin, self-supporting films. The films contain stabilized bilayers arranged between the silicate layers of the clay, hectorite, and have the shape of disks measuring about 25 μm thick and about 2 cm in diameter. The disks require approximately 2 mg of phospholipid for preparation.

These authors dispersed dipalmitoyl phosphatidylcholine liposomes and erythrocyte ghosts in hectorite, and examined them with both IR and Raman spectroscopic techniques. Several spectral regions below 1100 cm^{-1} are obliterated by the clay, but both head group and acyl chain vibrations are displayed and can be monitored to study conformational changes in phospholipids. Table 11.3 shows a comparison between data collected for 1,2-dipalmitoyl phosphatidylcholine embedded in clay and deposited as a film on a KBr plate.

Gaber *et al.* (1978a) have applied Raman difference spectroscopy to aqueous dispersions of dipalmitoyl phosphatidylcholine. The method required computer subtraction of absolute Raman spectra recorded in each of three different temperature ranges: below the endothermic pretransition at 34 ± 2°C; between 34°C and the melting transition at 42°C; and above the melting temperature. Quantitative and qualitative differences were displayed by the resultant difference spectra, which indicated that a distinct phospholipid conformation occurs in each of the three temperature ranges. The difference spectra displayed greater detail than did the conventional Raman spectra, and the three conformations were described in detail by these authors: (a) the hydrocarbon chains are almost completely in the all-*trans* conformation and are well packed into a crystalline lattice; (b) between the pretransition and the melting temperature the number of *gauche* bonds increases slightly to about one or two *gauche* rotations per chain. *Gauche* bonds are highly restricted in this phase and are present only at the ends of long all-*trans* segments; (c) above the melting transition all the remaining crystalline chain–chain interaction disappears and the number of *gauche* bonds increases dramatically. The restriction on the placement of the *gauche* bonds disappears in this temperature range; they can thus migrate along the chain to locations on adjacent C—C bonds.

Table 11.3. Comparison of Infrared Frequencies and Assignments for 1,2-Dipalmitoyl Phosphatidylcholine Embedded in Hectorite Clay and Deposited as a Film on a Potassium Bromide Plate[a,b]

Dipalmitoyl phosphatidylcholine frequencies (cm^{-1})		
Clay	Film	Assignments
≈2956(sh)	2947(sh)	CH_3 Asymmetric stretch
2926(s)	2919(s)	CH_2 Asymmetric stretch
2853(s)	2851(s)	CH_2 Symmetric stretch
1728(s)	1740(s)	C═O Stretch
≈1488(sh)	≈1490(sh)	CH_2 Deformation
1469(m)	1470(m)	
≈1420(w)	1419(w)	CH_2 Deformation adjacent to C═O[c]
1374(w)	1379(w)	CH_3 Symmetric deformation[c]
1246(m)	1254(m)	PO_2^- Asymmetric stretch
≈1168(m)	≈1168(m)	CH_2 Wag, C—O—C asymmetric stretch
720(w)	721(w)	CH_2 Rock

[a] Spiker *et al.* (1976).
[b] Frequencies calibrated from polystyrene. Abbreviations: sh, shoulder; s, strong; m, medium; w, weak.
[c] Akutsu and Kyogoku (1975).

Fourier transform IR spectroscopy has been used by Cameron and Mantsch (1978) in a study of the gel–liquid-crystal phase transition of 1,2-dipalmitoyl-*sn*-glycero-3-phosphocholine (DPPC) multibilayers. Similar chemical systems have been used by other investigators of model biomembranes (see p. 428). Figure 11.2 shows spectra of DPPC multibilayers at 39°C and at 42°C, and an FTIR difference spectrum of these. The main phase transition of this glycerophospholipid occurs at 41.5°C, so that difference spectra reflect changes in the vibrational pattern of DPPC multibilayers between the two phases. For example, Fig. 11.2 shows a characteristic pattern of intensity changes with minima spaced about 22 cm^{-1} apart at 1285, 1266, 1222, 1199, and 1178 cm^{-1}. These are associated with some weak shoulders in the spectrum of the gel phase. The liquid-crystal phase of DPPC multibilayers does not display such a group of weak bands. Cameron and Mantsch assign it to a $(CH_2)_n$ wagging progression due to the palmitoyl chains, a progression seen for highly ordered aliphatic hydrocarbon chains (Chapman *et al.*, 1967). lecithin–31% water multilayers as a function of the addition of chloroform and of carbon tetrachloride. The data show that a phase transition, which is analogous to a thermally induced one, occurs at constant temperature below T_c (temperature of thermally induced phase transition for a gel–liquid-crystal system) when solute $CHCl_3$ or CCl_4 is added. These authors found a substantial two-phase region, but its molecular nature is unclear. They used the Raman bands of the solute as an internal standard for intensity measurements of various lipid bands occurring in the regions 1025–1350 cm^{-1} and 2800–3000 cm^{-1},

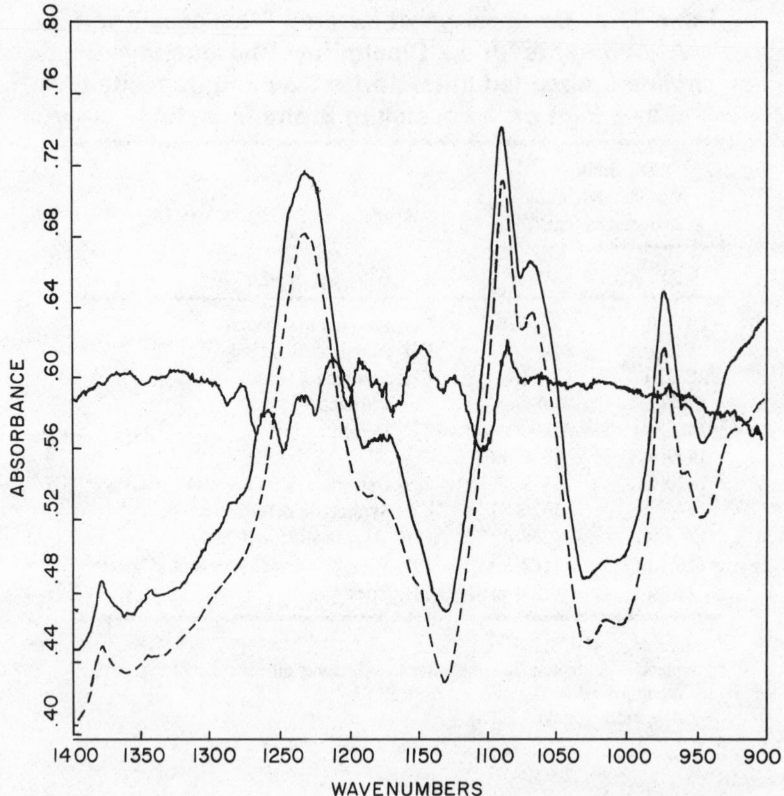

Fig. 11.2. FTIR spectra of DPPC multibilayers in the fingerprint region at 39°C (solid line), at 42°C (broken line) and superimposed on both, the FTIR difference spectrum. (Cameron and Mantsch, 1978.)

e.g., C—H and C—C stretching and CH_2 twisting regions. This referral to an internal standard (e.g., ν_1 of CCl_4 at 459 cm^{-1} and ν_2 of $CHCl_3$ at 669 cm^{-1}) puts the measurement of the intensities of the lipid bands on a somewhat more absolute basis than those previously measured only as relative changes.

These authors found that at a 2.5 mole ratio of CCl_4 to dipalmitoyl lecithin (DPL) or $CHCl_3$ to DPL, 55% of the rotatable carbons are in the *gauche* conformations. They believe that Raman spectra are only useful for obtaining an upper limit on the fraction of *gauche* conformers.

Dimyristoyl Phosphatidylcholine and Others

When a phospholipid bilayer assembly goes through a phase transition from the gel (or crystalline) state to the liquid crystalline form, the hydrocarbon chains undergo both interchain and intrachain disorder, which yields a model in which the covalent bonds nearest the terminal CH_3 groups in the interior of the bilayer exhibit relatively unrestricted conformational mobility (for example, Lee, 1975; Jacobs *et al.*, 1975; Jackson, 1976). Various investigators consider that the hydrocarbon chains in the crystalline form (low-temperature gel) have either all-*trans* configurations or are packed in such a way that the acyl chains are in near-*trans* conformations (Lippert and Peticolas, 1971, 1972; Levine, 1972; Traüble, 1972; Rothman, 1973).

Yellin and Levin (1977), in order to investigate further the intramolecular disorder of the hydrocarbon chains at temperatures below the transition temperature, have studied the activities of some phospholipid–water bilayer systems in the C—C stretching region of their Raman spectra. They followed the temperature dependence of bilayer systems of dimyristoyl phosphatidylcholine (DMPC) and of the dipalmitoyl and distearoyl derivatives (DPPC and DSPC) from slightly above approximately $-180°C$ to a point slightly below the pretransition, or lower transition, temperature. From the Raman data they estimated the enthalpy difference between rotational isomers for each system and interpreted the difference in terms of the number of *gauche* rotamers that formed.

Certain vibrational transitions appear in the 1090–1085-cm^{-1} region as the temperature is increased. These are evidence of *gauche* conformers within the acyl chains. Besides these vibrational transitions, a new feature appears in the three bilayer systems at about 1122 cm^{-1}, and is a C—C stretching mode characteristic of a *gauche* conformation (Fig. 11.3). It is associated with the formation of a *gauche* bond rotation of the terminal methyl group oriented toward the center of the bilayer.

Using the peak height intensities of vibrational features associated with the appropriate rotational isomers, these workers estimated the ΔH values between hydrocarbon chains in all-*trans* conformation and chain configurations containing *gauche* forms. The systems DMPC–H_2O, DPPC–H_2O, and DSPC–H_2O produced Raman data that yielded enthalpy differences of 2.9 ± 0.6, 3.4 ± 0.5, and 9.9 ± 1.2 kcal/mol, respectively. These differences showed that there are approximately two *gauche* bonds per lipid molecule for the DMPC–H_2O and DPPC–H_2O gels and six *gauche* bonds per molecule for the DSPC–H_2O gels.

Lippert and Peticolas (1972) have given Raman spectral assignments of long-chain fatty acids that can be used to determine all-*trans* chain length, as well as the position and configuration of double bonds in homogeneous fatty acid samples.

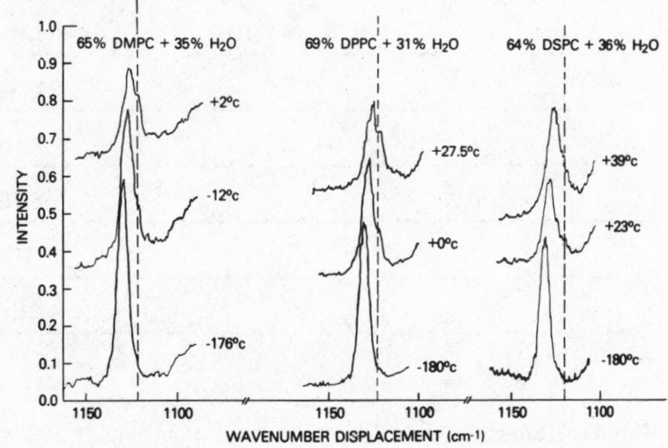

Fig. 11.3. Raman spectra of DMPC–, DPPC–, and DSPC–water gels showing (a) the frequency and intensity changes of the 1130-cm^{-1} C—C all-*trans* stretching mode as a function of temperature and (b) the appearance of the 1122-cm^{-1} C—C *gauche* feature as the temperature increases from $\sim -180°C$. (Yellin and Levin, 1977.) Reprinted with the permission of the American Chemical Society. Copyright 1977.

Figure 11.4 shows eight Raman spectra of the even-number, straight-chain saturated fatty acids between C_8 and C_{22}. One can see regular changes in vibrational frequency with chain length, particularly below 500 cm^{-1} and in the 1030–1130 cm^{-1} range.

Figure 11.5 shows ten spectra of unsaturated fatty acids. The principal lines due to polymethylene chain vibrations are the strongest in the spectrum. Especially the region 1000–1130 cm^{-1} shows relations between vibrational frequency and the length of the separate bound and free hydrocarbon chains (acid terminal and methyl terminal, respectively), e.g., Raman lines appear at 1044 and 1099 cm^{-1} if a C_9 bound chain is present, and at 1112 and 1005 cm^{-1}, if a C_6 free chain is present.

Lippert and Peticolas studied the side-chain melting transition of an L-α-dioleoyl lecithin suspension in water and observed that although the interior of the multilayer is more mobile than the region closer to the polar surface, both regions are in a random configuration at room temperature.

Spiker and Levin (1976) have studied the effect of bilayer curvature on the vibrational Raman spectroscopic behavior of phospholipid–water assemblies. They recorded Raman spectra for dimyristoyl and dipalmitoyl phosphatidylcholine in the gel state for both multilayer and single-wall vesicle assemblies. The vesicle dispersions were prepared by a sonication process. A comparison of intensities between the two classes of bilayer systems shows that a decrease occurs in peak height intensity for the observed hydrocarbon chain transitions in the single-wall vesicle form. No intensity change occurs between bilayer forms for the two observed head group modes.

When these authors examined the peak height intensity ratios for the 1100-cm^{-1} C—C stretching vibrations, they found that trends in the ratios indicated an increase in hydrocarbon chain *trans–gauche* isomerization for the vesicle in comparison to the multilayer arrangements. When they examined various C—C stretching modes, the intensity ratios of these modes for cholesterol-doped dipalmitoyl phosphatidylcholine bilayers indicated that 25 mol % of cholesterol increased the *trans–gauche* acyl chain isomerization in multilayers. The vesicle forms did not display this effect. On the other hand, the ratio of the intensities of methylene twisting and methylene deformation bands, respectively,

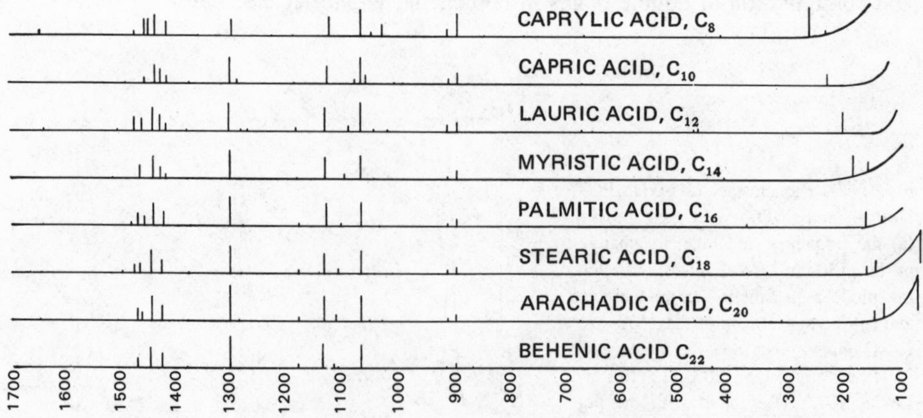

Fig. 11.4. Summary of the Raman spectra of even number, saturated fatty acids. (Lippert and Peticolas, 1972.)

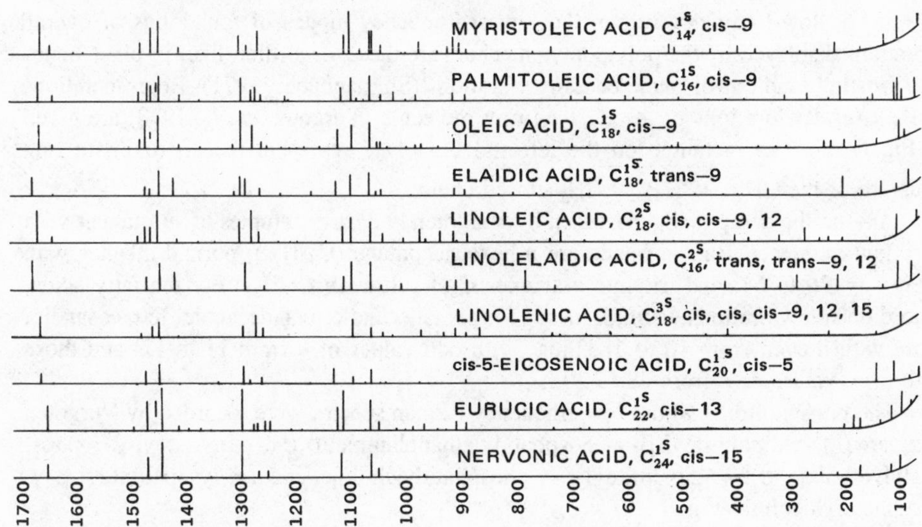

Fig. 11.5 Summary of the Raman spectra of various unsaturated fatty acids, C_na =, i – b: where n is the number of carbons, a is the number of double bonds, i is *cis* and *trans* isomer, and b is the position of the double bond. (Lippert and Peticolas, 1972.)

(I_{1297}/I_{1440}), for the cholesterol-containing systems, suggested that some further type of interchain perturbation occurs in the vesicle aggregations.

Yellin and Levin (1977a) have studied Raman spectra of multilayers of 1,2-dimyristoyl, 1,2-dipalmitoyl, and 1,2-distearoyl phosphatidylcholine as a function of temperature for the gel phase from −180°C to temperatures just below the pretransition value. Temperature profiles for band intensity ratios associated with the acyl chain C—C stretching modes characterize a temperature T_g which defines the onset of *trans–gauche* isomerization for the hydrocarbon chains in the gel. Dimyristoyl (C_{14}), dipalmitoyl (C_{16}), and distearoyl (C_{18}) chain lengths have T_g values of ca. −40°, −40°, and 5°C, respectively. The higher T_g value (5°C) is involved with increasing interactions between the terminal chain regions of the individual monolayers associated in the bilayer unit. The C—H stretching region at 3000–2800 cm^{-1} shows temperature-dependent band intensities which are convenient tools for quantitative assessment of the increase in lateral chain packing disorder during the warming of the gel systems. The temperature profiles for several methylene mode intensity ratios show that both the lateral packing and *trans–gauche* disordering processes are related and are begun at about the same temperatures for each lipid system. Enthalpy increases (due to disordering of the lipid gel matrix by a lattice expansion) for the C_{14}-, C_{16}-, and C_{18}-phospholipids were estimated to be ca. 1.5, 1.2, and 3.2 kcal mol^{-1}, respectively.

The subject of low-frequency vibrations of biological molecules has been treated by Vergoten *et al*. (1978). Among the topics discussed were longitudinal acoustic vibrations (e.g., in polymethylene and polyethylene chains, *n*-alkanes); and lipids (fatty acids, soaps, phospholipids, steroids, cholesterol, and its derivatives).

In the low-frequency region the lowest-frequency modes of four kinds of overall vibrations displayed by the polymethylene chain are these: accordion-like, in-plane bending, twisting, and out-of-plane bending vibrations (Shimanouchi, 1971). Representations of the overall vibrations of the $(CH_2)_{34}$ ring molecule (Vergoten *et al.*, 1978) are given in Fig. 11.6. The rectangle on the left and Fig. 11.7 represent the approximate ring structure of the CH_2—$(CH_2)_{32}$—CH_2 closed chain.

Among the compounds discussed by Vergoten *et al.* as examples of substances with very-low-frequency Raman bands are *n*-hexatriacontane ($C_{36}H_{74}$), normal alkanes with even n = 26 to 64, and triclinic *n*-alkanes C_8H_{18} through $C_{18}H_{38}$. For the fatty acids, accordion-like vibrations were discussed for heptadecanoic, octadecanoic, tetracosanoic, those with n even from 10 to 18, those with odd values of n from 11 to 17, and those with even values of n from 22 to 30.

The phospholipids whose low-frequency Raman spectra were recorded by Vergoten *et al.* are D,L-α-dipalmitoyl-β-γ-phosphatidylethanolamine, D,L-α-dimyristoyl-β-γ-phosphatidylcholine, D,L-α-dipalmitoyl-β-γ-phosphatidylcholine, and L-α-lysopalmitoyl-β-γ-phosphatidylcholine (Fig. 11.8).

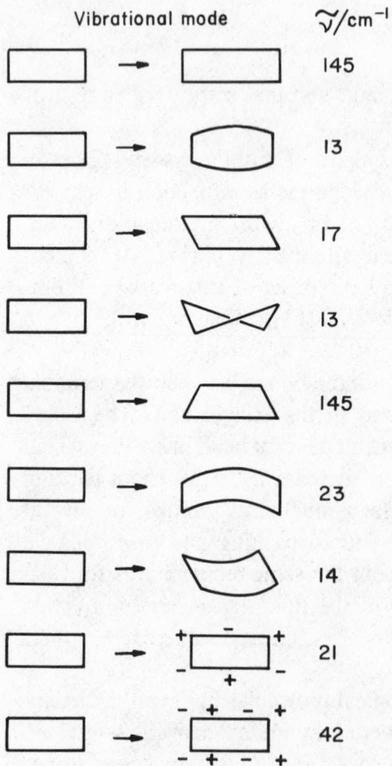

Fig. 11.6. Overall vibrations of the $(CH_2)_{34}$ molecule (Shimanouchi, 1971, quoted in Vergoten *et al.*, 1978).

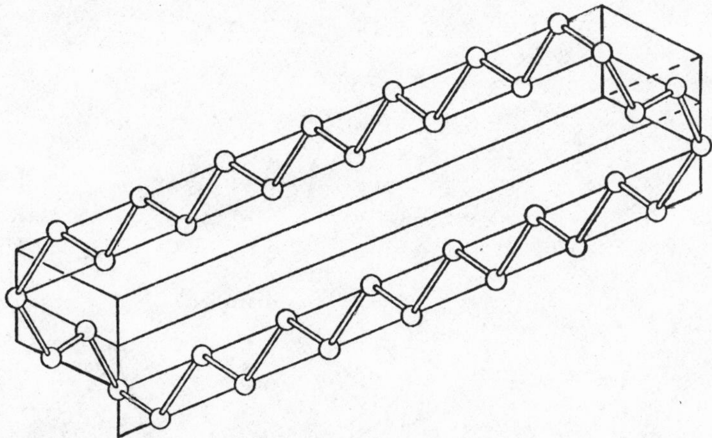

Fig. 11.7. Approximate structure of the $(CH_2)_{34}$ ring molecule in the crystalline state (Shimanouchi, cited in Vergoten, 1978).

Quantitation of the Order of Hydrocarbon Chains

Karvaly and Loshchilova (1977) have discussed the possibility of quantitating information received from Raman spectroscopic examination of aqueous lipid dispersions. An all-*trans* hydrocarbon chain order characteristic, S_T, had been introduced by Gaber and Peticolas (1977) to define the bilayer structure:

$$S_T = \frac{(I_T/I_{\text{ref}})_{\text{sample}}}{(I_T/I_{\text{ref}})_{\text{crystalline lipid}}}$$

The terms were defined thusly: I_T is the Raman line intensity of the A_g vibrational mode of the all-*trans* chain conformation, positioned at ca. 1130 cm^{-1}. I_{ref}, for phospholipids, is the reference-line intensity, I_{722} or I_{1096}. I_{722} represents the symmetric C—N stretch of the choline group; the 1096-cm^{-1} features belongs to the skeletal optical mode (C—C stretch), including PO_2^- symmetric stretching modes and *gauche* rotation. S_T has a value of 1 for crystalline dipalmitoyl phosphatidylcholine, and zero for no order.

Karvaly and Loshchilova showed that the application of S_T for the characterization of biomembrane structure is restricted in its usefulness: S_T may supply enough information if polar head groups are not affected at all by the interaction resulting in *trans-gauche* isomerization.

The intensity of the 1130-cm^{-1} band mentioned above is commonly the reference point in the measurement of conformational disorder in lipid bilayer systems. However, estimates of the concentration of *gauche* bonds may differ by a factor of 6 depending on the model (Gaber and Peticolas, 1977; Pink *et al.*, 1980) used to relate concentration and intensity. Snyder *et al.* (1982), attempting to narrow the degree of variation in these

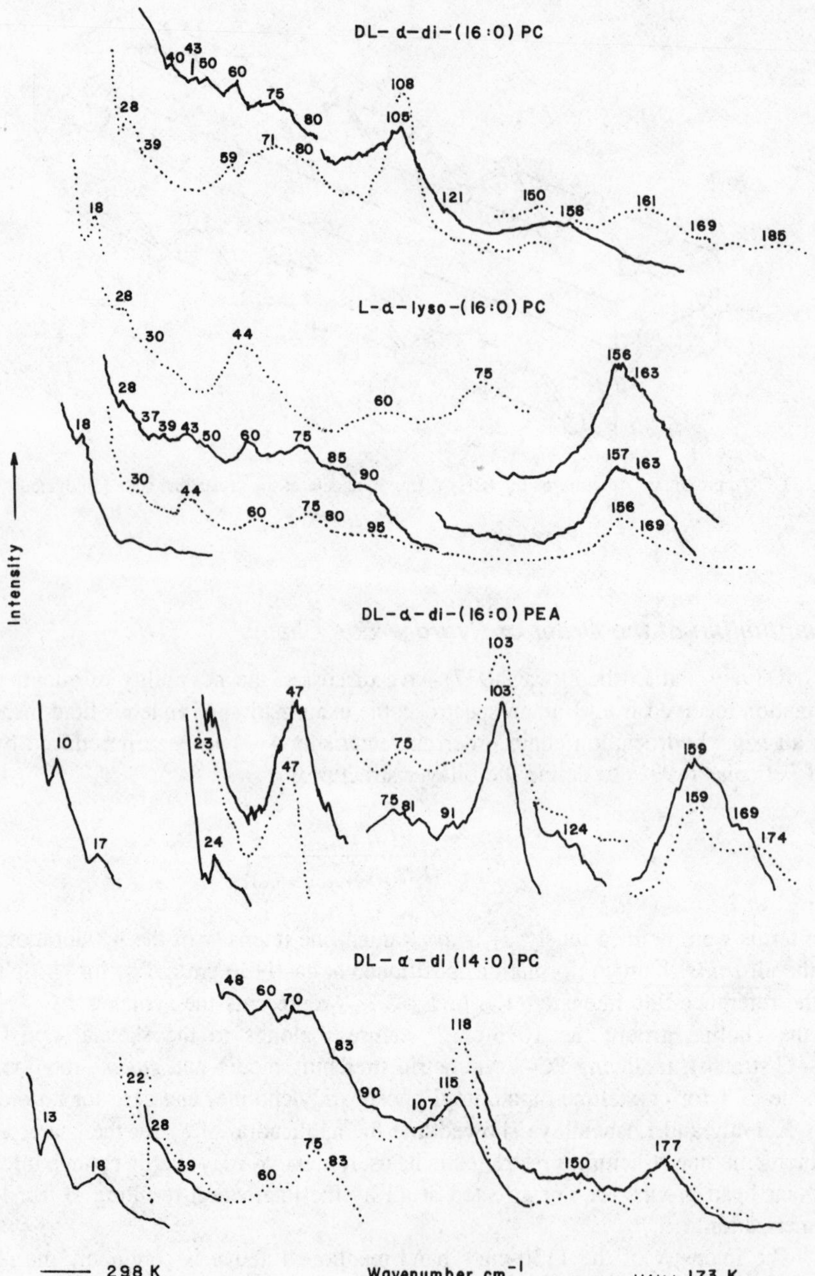

Fig. 11.8. Low-frequency Raman spectra of: D,L-α-dipalmitoyl-β,γ-phosphatidylethanolamine [D,L-α-di(16:0)PEA]; D,L-α-dimyristoyl-β,γ-phosphatidylcholine [D,L-α-di-(14:0)PC]; D,L-α-dipalmitoyl-β,γ-phosphatidylcholine [D,L-α-di-(16:0)PC]; L-α-lysopalmitoyl-β,γ-phosphatidylcholine [L-α-lyso-(16:0)PC]; at room and low temperatures. (Vergoten *et al.*, 1978.)

estimates, have measured the intensity of the 1130-cm^{-1} band of crystalline n-$C_{21}H_{44}$ in its orthorhombic and hexagonal phases. Transition to the hexagonal phase produces a much reduced intensity of the 1130-cm^{-1} band. These workers assumed that the observed intensity reduction is caused by the introduction of *gauche* bonds whose numbers can be independently estimated from other features in the Raman and IR spectra. They concluded from these measurements that the integrated intensity of the 1130-cm^{-1} band is not linearly related to the concentration of *gauche* bonds and that a disproportionately large decrease in the 1130-cm^{-1} band intensity results from the introduction of a low concentration of *gauche* bonds into the otherwise all-*trans* chains during the transition to the hexagonal phase. They further concluded that many earlier estimates of *gauche* bond concentrations based on the assumption of a linear relationship have tended greatly to overestimate the *gauche* bond concentration. Their experimental results agree with those of Pink *et al.* (1980) derived from theory: about one *gauche* bond is present per chain; and not at least six *gauche* bonds, according to the model of Gaber and Peticolas (1977).

The Cooperative Unit Size

Yellin and Levin (1977b) have used Raman spectroscopy to estimate the cooperative unit size ($\approx$11 molecules for the lower limit) in the gel–liquid-crystalline phase transition of lipid multilayers composed of dipalmitoyl phosphatidylcholine (DPPC) and water. They determined the temperature dependence of the Raman spectral transitions assigned to the acyl chain C—C stretching modes of DPPC for the gel, phase transition and liquid crystalline states of the lipid multilayers. They obtained the van't Hoff enthalpy differences ΔH_{vH} between *trans* and *gauche* rotational isomers from the Raman spectral data for the temperature region characteristic of each bilayer, and estimated an average size for the cooperative unit involved in the chain melting process during the phase transition.

An important point noted by these workers is that for the molecular state defined either by the melting of the hydrocarbon chains during the phase transition, or by the liquid crystalline form of the multilayer, the decrease in intensity of the band at ca. 1124 cm^{-1} represents a decrease in chain order, while the increase in intensity of the region from 1090 to 1085 cm^{-1} indicates an increase in chain disorder.

A Dietherlecithin

Sunder *et al.* (1978a) have examined Raman spectra of multilayer dispersions of a dietherlecithin, rather than a diester compound. They used the phospholipid *rac*-1,2-dioctadec-9′-*cis*-enyl-glycero-3-phosphorylcholine in excess water, and examination of the changes in the CH stretching region with change in temperature showed that the multilayer dispersions change from the liquid crystal state to a gel state at -21 ± 4°C. This phase transition temperature is very near that predicted (Barton and Gunstone, 1975) for the corresponding unsaturated lecithins (C═C at position 9) having two ester groups (about -22°C), whereas the lecithin used by Sunder *et al.* had two ether groups. Therefore, changing the ester linkages to ether linkages does not appear to have varied the phase transition temperature much.

Films of Phosphatidylcholine and Phosphatidylethanolamine. Head and Tail Group Interactions

Akutsu and Kyogoku (1975) have recorded the IR and Raman spectra of phosphatidylethanolamine, L-α-glycerophosphorylethanolamine and *O*-phosphorylethanolamine. Table 11.4 presents assignments of the spectra of phosphatidylethanolamine. The spectra of this compound can be interpreted by assuming that the molecule has the dipolar ionic structure both in nonpolar solvents and in the solid form.

The information in reports on the IR spectra of films of phosphatidylcholine (Chapman *et al.*, 1967) and phosphatidylethanolamine (Cherry and Morrison, 1966) has been broadened by Fookson and Wallach (1978). Their work demonstrates the sensitivity of

Table 11.4. Assignments of Vibrational Spectra of Phosphatidylethanolamine[a,b]

Infrared			
Solid (cm^{-1})	Solution (cm^{-1})	Raman solution (cm^{-1})	Assignments
2950(s)	2950(s)		CH_3 Deg. str.
		2940(vs), (p) 2900(s)	CH_3 Sym. str., NH_3^+ sym. deg. str.
2920(vs)	2920(vs)		CH_2 Antisym. str.
2850(s)	2850(s)	2855(vs), (p)	CH_2 Sym. str.
1728(s)	1735(s)	(Solvent)	C=O Str.
1623(w)	1625(w)	(Solvent)	NH_3^+ Deg. def.
1568(w)	1530(w)	(Solvent)	NH_3^+ Sym. def.
1463(m)	1460(m)	1470(sh), (dp) 1440(s), (dp)	CH_2 Sci.
1412(w)	1412(w)		CH_2 Sci. adjacent to C=O
1374(w)	1370(w)		CH_3 Sym. def.
		1310(m), (dp)	
		1263(w), (dp)	
1243(sh) 1215(s)	1228(s)	1230(w), (dp)	PO_2^- Antisym. str.
1172(m) 1140(sh)	1160(m)		CH_2 Wag., C—O—C antisym. str. and others
		1125(w)	C—C Str.
1104(sh) 1090	1090(sh)	1097(m), (p)	PO_2^- Sym. str.
1080(s) 1053(m)	1068(s)	1075(m) 1070(sh)	C—O Str. C—O—C sym. str. and others
1013(s)	1030(m)		C—C—N^+ Antisym. str.
975(w)	980(sh)		
910(m)	890(w)		CH_2 Rocking
		875(w)	C—C—N^+ Sym. str.
805(m)	815(w)		O—P—O Antisym. str.
752(w)	750(w)	755(w)	O—P—O Sym. str.

[a] Akutsu and Kyogoku (1975).
[b] vs: very strong, s: strong, m: medium, w: weak, sh: shoulder, p: polarized, dp: depolarized, deg. str.: degenerated stretching mode, sci.: scissoring mode.

IR spectroscopy to the details of phospholipid head group behavior. They studied multibilayer films of these substances and mixed films (1:1, mol/mol) as a function of fatty acid composition, hydration, and temperature in order to obtain information on the head group interactions among the different phosphatides. Figures 11.9 and 11.10 show differences in the spectra of films of dipalmitoyl phosphatidylcholine (anhydrous and monohydrate) and anhydrous dipalmitoyl phosphatidylethanolamine in the intervals 700–850 cm^{-1} and 1200–3600 cm^{-1}. The band assignments are given in Table 11.5. Fookson and Wallach interpreted the lowered phosphoryl absorption, $\nu(P{=}O)$, in the phosphatidylethanoloamine spectra as caused by N—H→O hydrogen bonding of the ethanolamine head group, which can line up tangentially to the bilayer plane, allowing closer packing of the molecules than can be achieved in phosphatidylcholine. The tighter packing causes strong crystal field effects that result in the complex fine structure of the spectra of phosphatidylethanolamine. These authors also concluded that splitting of the C—H bending fundamental at ca. 1460 cm^{-1} and the terminal CH_2 rocking vibration at ca. 720 cm^{-1} show that the acyl chains of the phosphatidylethanolamine film are orthorhombically packed rather than hexagonally. They attributed the lack of fine structure accompanying the monohydration of phosphatidylcholine to the decreased intermolecular interactions occurring when water is inserted into the lattice of the head group, increasing the intermolecular separations. With complete hydration no further change occurs, suggesting to them that only the first water molecule is bound tightly to the phospholipid.

The work of Fookson and Wallach (1978) was also concerned with phosphate group stretching modes in the interval 800–1200 cm^{-1}. Spectral differences occurred between pure phosphatidylethanolamine (PE) films and films of either pure phosphatidylcholine (PC) or mixed PE–PC systems. They interpreted the two classes of spectra [(a) a doublet

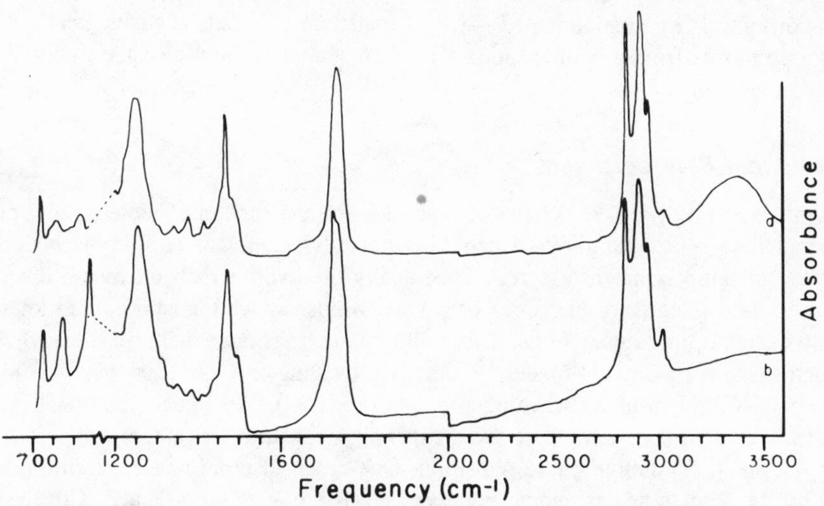

Fig. 11.9. Infrared absorption spectra (at 20°C) from 700–850 and 1200–3600 cm^{-1} of multibilayers of (a) dipalmitoyl phosphatidylcholine monohydrate, and (b) anhydrous dipalmitoyl phosphatidylcholine. (Fookson and Wallach, 1978.)

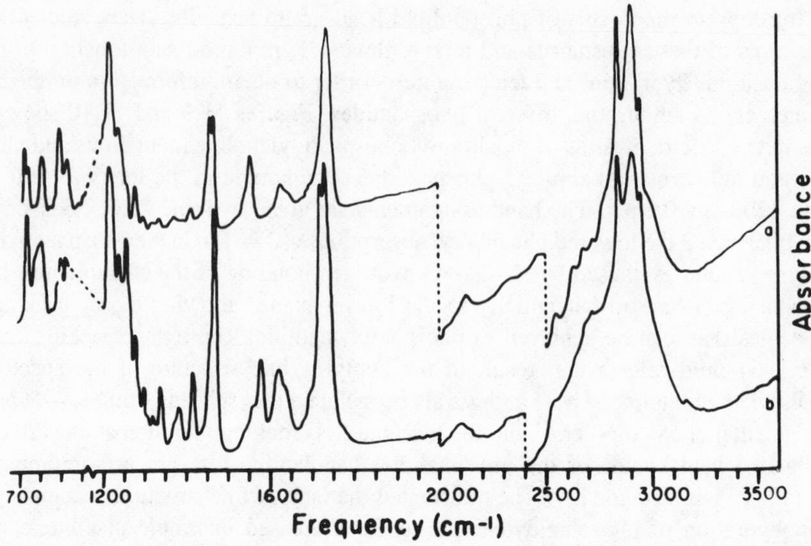

Fig. 11.10. Infrared absorption spectra (at 20°C) from 700 to 850 and 1200 to 3600 cm^{-1} of multibilayers of anhydrous dipalmitoyl phosphatidylethanolamine evaporated from solution in (a) chloroform/methanol (2/1, v/v) and (b) chloroform (Fookson and Wallach, 1978.)

at ca. 1085 and 1055 cm^{-1}, with secondary absorption at ca. 968 cm^{-1} versus (b) individual strong bands at ca. 1080 and 1010 cm^{-1}] as originating from head group orientation tangential versus normal to the bilayer plane. These changes occur because the bands in this interval originate in the P—O—C aliphatic linkage, which always occurs in pairs that share a P atom (yielding coupled modes). They also occur because the relative orientation of the two concomitant P—O—C transition moments changes greatly as the head group moves from a conformation that is tangential to the film plane to one that is perpendicular to it.

Phospholipid Polymorphism

Lapides and Levin (1982) have obtained the Raman spectra of dioleoyl phosphatidylethanolamine and 1-palmitoyl-2-oleoyl phosphatidylethanolamine lipid dispersions as a function of temperature to ascertain specifically the hydrocarbon interchain disorder. They used total integrated intensities of several bands, as well as appropriate intensity ratios to examine lipid interactions. These vibrational transitions included the acyl chain CH_2 deformation modes (1440 cm^{-1}), the CH_2 twisting modes (1296 cm^{-1}), and the double bond C—H in-plane wagging mode (1275 cm^{-1}). All these vibrational probes suggested the occurrence of at least two distinct physical structures at temperatures above the gel ↔ liquid crystalline phase transition. One order–disorder phase transition determined by the Raman spectroscopic parameters was tentatively correlated with the formation of the hexagonal II phase; and additional lipid polymorphic structures could not be ruled out.

Table 11.5. Infrared-Active Bands (in cm^{-1}) of Various Phosphatidylcholines and Phosphatidylethanolamines[a]

Band	DPPC[b] (dry)	DPPC (monohydrate)	PLPC (dry)	Egg PC (dry)	DPPE	DPPE[c]	Egg PE
ν_s CH_2	2848	2849	2848	2853	2848	2848	2848
ν_{as} CH_2	2915	2918	2915	2920	2918	2915	2918
ν_{as} CH_3	2957	2958	2953	2959	2958	2956	2958
δ CH_2	1461	1466	1467	1469	1462	1469	1462
					1471		
ν (C═O)	1724	1733	1734	1740	1739	1720	1740
	1734					1730	
ν (P═O)	1254	1248	1252	1260	1222	1215	1224
					1241s[d]		1240s
Terminal CH_2 rock	722	721	720	720	714	720	720
					723		
$\nu(CH_3)$ of N^+ CH_3	3028	3028	3028	3013			
$\nu(NH_3^+)$[e]					2128	2108	2558
					2328	2328	2638
					2558	2548	2708
					2646		
					2688		
					2728	2728	
δ (NH_3^+)					1550	1580	1559
					1625	1620	1638
ν(OH)		3373					
δ(OH)		1645					

[a] Fookson and Wallach (1978).
[b] All films prepared from solutions in chloroform/methanol (2/1) except DPPE.
[c] Dipalmitoyl phosphatidylethanolamine film prepared from chloroform solution.
[d] s, shoulder.
[e] The ionized, hydrogen-bonded NH_3^+ group displays multiple weak band progressions in the range 2000 to 2700 cm^{-1}.

Modulated Excitation Infrared Spectroscopy (ME-ATR)

Fringeli and Günthard (1976) have used the technique of modulated excitation IR spectroscopy in conjunction with ATR spectroscopy to investigate the hydration sites of egg phosphatidylcholine. The technique is applicable to any system which permits periodic modulation of absorption coefficients by periodic excitation. It allows one to scan selectively those IR absorption bands resulting from molecules (or parts of them) that are involved in the stimulation process. Those absorption bands not affected by excitation are suppressed. (See also Fringeli and Günthard, 1981.)

A germanium ATR plate (50 × 20 × 1 mm) was used and 40–45 internal reflections resulted. The spectrometric apparatus itself is rather complex and descriptions of it and the theory involved are beyond the scope of this book. Newly generated bands (type *a* bands) of significance that resulted from exposure of the evaporated film of phospholipid to nitrogen gas flow (approximately 60% relative humidity), were observed at 3360 and

1650 cm^{-1}. These bands are caused by bound water, O—H stretching and O—H bending vibrations, respectively. Modulated-excitation bands ca. 1730, ca. 1230, and ca. 1090 cm^{-1}, which coincide with the ν(C=O), $\nu_a(PO_2^-)$, and $\nu_s(PO_2^-)$, respectively, give evidence of the main hydration sites of egg phosphatidylcholine, i.e., the phosphate group and the carbonyl group. Another band (type v, due to vanishing molecules or their parts) was observed at ca. 1055 cm^{-1} and was assigned to ν(C—O) from C—O—P stretching in the choline phosphate ester. Thus, the choline group was also found to bind water from the gas phase.

The ME-ATR technique allows the study of model membranes or membranes of viable cells. (See Fringeli and Günthard, 1976; Fringeli and Hofer, 1980; Sherebrin *et al.*, 1972.)

Infrared Dichroic Measurements

Akutsu *et al.* (1975) have built up multilayers of L-α-dipalmitoyl phosphatidylethanolamine on Irtran-2 (ZnS) plates. The type of multilayer obtained was an alternate head-to-head, tail-to-tail structure. These workers used IR dichroic measurements (wire grid polarizer) for the determination of the directions of transition moments in a film sample, in which there is an axis of symmetry perpendicular to the film plane. Such transition moments were determined for the six vibrations assigned to the following: CH_2 antisymmetric stretching (2918 cm^{-1}), CH_2 symmetric stretching (2852 cm^{-1}), CH_2 scissoring (1467 cm^{-1}), carbonyl stretching (1738 cm^{-1}), PO_2^- antisymmetric stretching (1224 cm^{-1}), and C—C—N^+ antisymmetric stretching (1027 cm^{-1}) modes. The data showed that hydrocarbon chains are tilted at ca. 75° to the plane of the film. The transition moments of the C=O stretching, PO_2^- antisymmetric stretching, and C—C—N^+ antisymmetric stretching modes of the polar moiety are oriented nearly parallel to one another and deviate by at most 20° from the plane of the film.

Attenuated total reflection with polarized light has been used to advantage for studying oriented layer assemblies of proteins (Baumeister *et al.*, 1976a; Baumeister *et al.*, 1976b), lipids (Takenaka *et al.*, 1971; Fringeli *et al.*, 1972; Kopp *et al.*, 1975; Baumeister *et al*, 1976c; Fringeli and Günthard, 1976), and lipid–protein systems (Fromherz *et al.*, 1972). Fringeli (1977) studied the structure of dry oriented multilayers of phosphatidylethanolamine, phosphatidylcholine, and other related lipids, e.g., dipalmitoyl phosphatidylethanolamine, by this IR method. Hydrocarbon chains were calculated to have average deviations from the normal to the plane of the bilayer of 20 to 30°; but it could not be decided whether the chains are oriented parallel to each other. Fringeli (1977) stated that the phosphate group of dipalmitoyl phosphatidylethanolamine is probably in the protonated (O=P—OH) form and not in the ionized $>PO_2^-$ form; but he felt that the latter form was to be expected for all the other phospholipids in the series studied.

The Influence of Ions on Lamellar Dispersions of Phosphatidylcholines

Lis *et al.* (1975) have done a Raman study of the influence of ions on lamellar dispersions of phosphatidylcholine by examination of the peak intensity ratio of the all-

trans C—C vibration to the O—P—O stretch plus random *(gauche)* C—C vibration, I_{1064}/I_{1089}. From the ratios of these intensities, it was inferred that dipositive ions decrease the proportion of *gauche* character in the hydrocarbon chains. The relative influences were $Ba^{2+} < Mg^{2+} < Ca^{2+} \approx Cd^{2+}$. Such unipositive ions as Li^+, K^+, and Na^+ produced no changes in the spectrum of the lecithin dispersion. The proportion of *gauche* character of the hydrocarbon chains was nearly independent of the anion for Br^-, Cl^-, acetate, I^-, ClO_4^-, CNS^-, and SO_4^{2-}. Among the substances investigated in this study were the dimyristoyl, dipalmitoyl, and distearoyl derivatives of 1,2 L-α-phosphatidylcholine in powder form and in water dispersions.

Cholesterol–Phospholipid Dispersions

Lippert and Peticolas (1971) have presented evidence indicating that when cholesterol is present with dipalmitoyl lecithin multilayers, its effect on the multilayers is to change the sharp, cooperative gel–liquid-crystal transition to a diffuse, noncooperative event. As a result, the apparent effect of cholesterol on the multilayers is different above and below the transition temperature.

The region 1000–1140 cm^{-1} was of special interest in this study because of peaks assigned to the skeletal optical mode of the hydrocarbon chain with a motion in which alternate carbon atoms move in opposite directions along the chain axis. The Raman spectrum of $CH_3(CH_2)_{n-2}CH_3$ solids with $n > 8$ shows two intense bands at 1064 and 1130 cm^{-1} whose frequencies are independent of n within $\pm$ 2 cm^{-1}, and one very weak band at 1075–1110 cm^{-1} which is highly dependent on n. In hexadecane this band appears at 1081 cm^{-1}, but it is shifted to 1100 cm^{-1} in hexadecanoic acid. These bands had all been assigned by earlier workers to vibrations of the all-*trans* configuration of the chain.

Figure 11.11 shows the 1100 cm^{-1} region of a DL-dipalmitoyl lecithin sonicate (20% by weight in water) at several temperatures. A sharp decrease in the intensity of the bands at 1064–1066 cm^{-1} and 1128–1130 cm^{-1} that are due to a vibration of the extended all-*trans* structure is seen as the temperature goes from 30° to 40°C. Concurrently, a sharp increase in the intensity of the 1089-cm^{-1} band assigned to random liquid-like configurations is seen.

Open triangles in Fig. 11.12 show the change with temperature of the 1066-cm^{-1} band, relative to the height of the 1089-cm^{-1} band. The curve shows a sharp change at 38°–39°C, corresponding to the gel–liquid-crystal phase transition previously observed in the L-isomer at 40°–43°C by other workers using a microscope and differential scanning calorimetry. Lippert and Peticolas concluded that this transition involves a change in the palmitate chains from an all-*trans* to a fluid configuration. The sharply sigmoidal nature of these curves indicates a highly cooperative transition.

The open circles in Fig. 11.12 show the change in the same relative peak heights of dipalmitoyl lecithin in a 1:1 mol/mol cholesterol–lecithin sonicate, 20% by weight in water. An extremely broad transition is seen.

Figure 11.12 also shows a curve (open squares) depicting the temperature dependence of the relative peak heights of 10% (by weight) dipalmitoyl lecithin in chloroform. Here, the paraffin side chains are in a very fluid condition and display no transition characteristics.

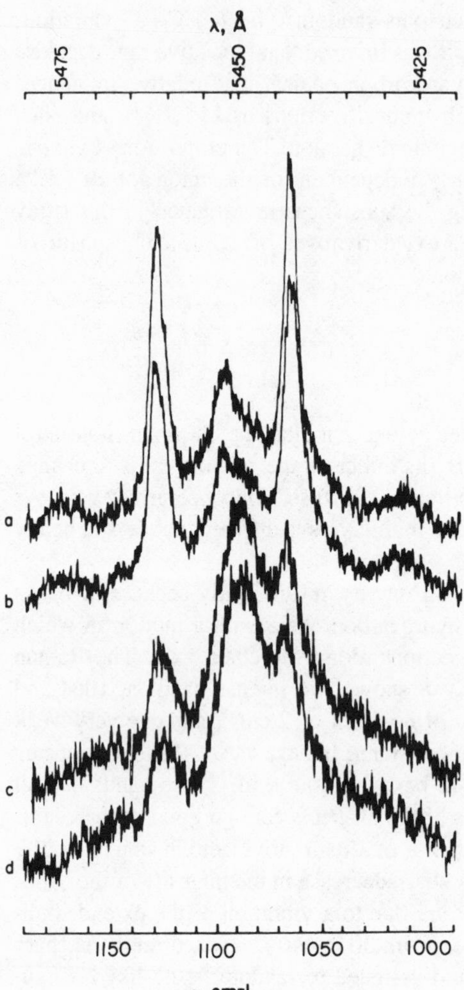

Fig. 11.11. Raman spectra of the 1100-cm^{-1} region of 20% (by weight) D,L-dipalmitoyl lecithin sonicates in water at (a) 20°C, (b) 30°C, (c) 40°C, (d) 50°C. (Lippert and Peticolas, 1971.)

In work with multibilayers, Verma and Wallach (1973) have used polarized IR spectra (wire grid polarizer) to study the effects of cholesterol on the dichroism of phosphatide multibilayers. In the presence of cholesterol the P═O stretching frequency of lecithin (1250 cm^{-1}) shifted to 1230 cm^{-1}, suggesting hydrogen bonding between the phosphatide P═O and the OH of cholesterol. The dichroism study indicated that the addition of cholesterol to lecithin multibilayers induces a configurational rearrangement of the polar head groups. This rearrangement involves changes in the transition vectors for the C═O (ester), P═O, and [C]—O—P (glycerylphosphoryl deformation) bands from 45° to 41°, 49° to 53°, and 60° to 49°, respectively. The electrostatic interaction between the OH of cholesterol and the P═O of phosphatide localizes the steroid ring so

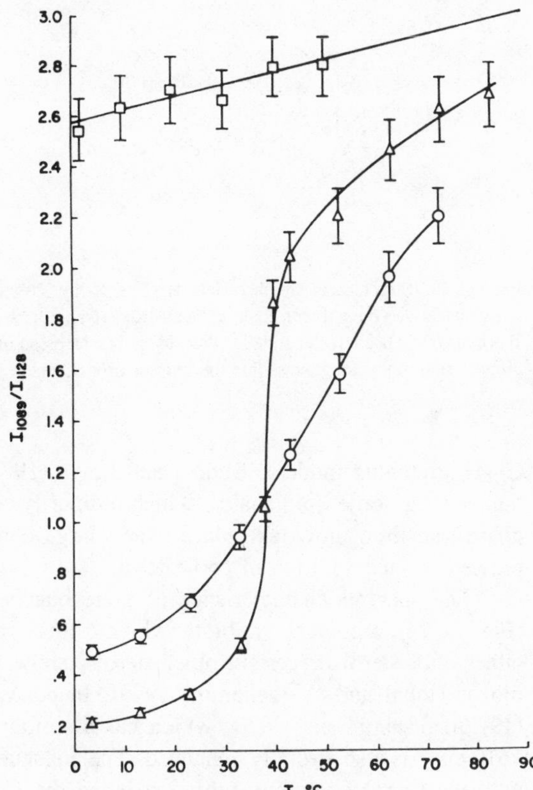

Fig. 11.12. Changes with temperature in the ratio of Raman peak heights of D,L-dipalmitoyl lecithin at 1089 and 1128 cm^{-1} for (a) 20% (by weight) sonicate in water of dipalmitoyl lecithin △, (b) 20% (by weight) sonicate in water of 1:1 cholesterol-dipalmitoyl lecithin ○, and (c) 10% (by weight) solution of dipalmitoyl lecithin in chloroform □. Similar curves are obtained by comparison of the peak heights at 1089 and 1066 cm^{-1}. (Lippert and Peticolas, 1971.)

that it does not extend beyond C6 from the C═O groups of lecithin, thus making rigid the proximal apolar acyl chain segments (Verma and Wallach, 1973).

Bunow and Levin (1977a) have studied the resonance-enhanced and normal vibrational Raman spectra of both single-wall and multilamellar vesicle assemblies of dimyristoyl phosphatidylcholine containing cholesterol and amphotericin B. This antibiotic is believed to form channels through lipid bilayers containing either cholesterol or certain other sterols. Its structure is given in Fig. 11.13.

In the presence of cholesterol, amphotericin B forms hydroxylic channels which penetrate the lipid bilayer. The introduction of the channels into the hydrocarbon region alters the symmetry of the acyl chains, so that the CH_2 asymmetric stretching mode at about 2920 cm^{-1}, which is Raman inactive under local C_{2h} symmetry for the all-*trans* conformation, becomes allowed in the Raman.

The vibrational Raman intensity of the amphotericin B polyene C═C stretching mode is very sensitive to the molecular changes occurring during the phase transition in dimyristoyl lecithin–cholesterol multilayers containing up to 20 mol % of cholesterol. By monitoring the polyene stretching mode (~1559 cm^{-1}) and plotting the ratios of band intensities, $I_{2880/2850}$, for the lipid component (symmetric and asymmetric methylene

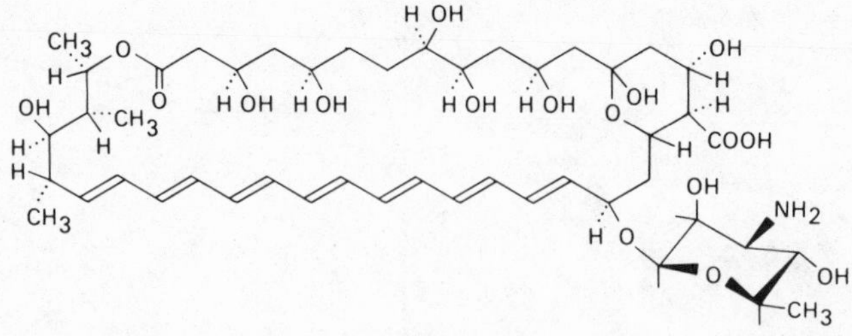

Fig. 11.13. Structure of amphotericin B. The X-ray structure determined for the crystalline *N*-iodoacetyl derivative shows the polyene chain to be all-*trans* (Ganis *et al.*, 1971). One side of the long axis of amphotericin B is hydroxylic in character and the other side polyenic; presumably the hydroxylic sides of an array of molecules, aligned along their long axes, line the interior of a channel in a lipid bilayer. (Bunow and Levin, 1977a.)

C—H stretching modes), Bunow and Levin (1977) confirmed that the large changes in the packing of the lipid chains, which ordinarily occur during the gel-to-liquid crystalline phase transition, now take place over a large temperature range, rather than being suppressed, by the addition of cholesterol.

The spectral characteristics of preresonance-enhanced Raman spectra of nystatin (Fig. 11.14), a polyene antibiotic which forms channels through lipid bilayers containing either cholesterol or certain other sterols, show prominent C═C and C—C stretching modes (Iqbal and Weidekamm, 1979). In contrast to the study of Bunow and Levin (1977a) on amphotericin B, which has a similar structure, a C═O stretching band at 1610 cm^{-1} is also strongly enhanced. The enhancement is due to the mixing of the C═C with the C═O stretching vibration when the C═O group at one end of the nystatin molecule lines up with the polyene section at the other end, and the C═C stretching vibration is modulated by the $\pi \rightarrow \pi^*$ excited state of the polyene structure (Iqbal and Weidekamm, 1979).

The nystatin structure has a largely planar conformation in phospholipid–cholesterol

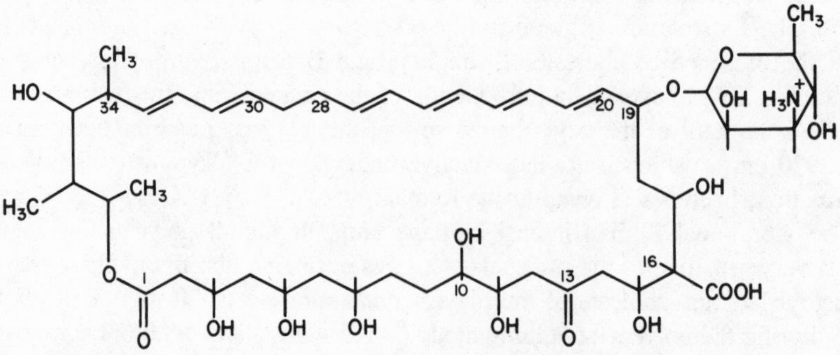

Fig. 11.14. Proposed structure of nystatin (Iqbal and Weidekamm, 1979, from Borowski *et al.*, 1971 and Golding *et al.*, 1966.)

multilayers. Again, in contrast to amphotericin B, the relative spectral intensities of the C═O, C═C, and C—C stretching vibrations are not altered when dimyristoyl and dipalmitoyl phosphatidylcholine–cholesterol multilayers containing nystatin are changed from gel to liquid-crystal phase.

Sphingomyelin Dispersions

The phospholipid sphingomyelin occurs widely in membrane tissue such as brain and nerve cells, in erythrocyte membranes, and in serum lipoproteins. In order to understand more about physicochemical properties of sphingomyelin bilayers, Faiman (1979) has studied the Raman spectroscopy of aqueous dispersions of sphingomyelin during an order-disorder transition over a temperature range of 22.5°–47.0°C. By band profile analysis he identified seven individual bands between 1000 and 1400 cm^{-1}, and followed changes in the intensity of the various components in this region to interpret the structural changes occurring during the order–disorder transition. The relative intensities of various bands were plotted versus temperature. The relative intensities were expressed as the ratio of the total band areas for the bands at 1063, 1087, and 1105 cm^{-1} to the total band areas for the bands at 1128 and 1298 cm^{-1}, to give plots of A_{1063}/A_{1128} and A_{1063}/A_{1298}; A_{1087}/A_{1128} and A_{1087}/A_{1298}; A_{1105}/A_{1128} and A_{1105}/A_{1298} versus temperature. Lippert and Peticolas (1971) have shown that the intensity ratio I_{1087}/I_{1128} plotted against temperature gives information concerning chain conformational changes during a phase transition.

Figure 11.15 (A–C) shows that some form of cooperativity in the melting process can be interpreted from the plots of A_{1063}/A_{1128}, A_{1063}/A_{1298}, A_{1087}/A_{1128}, and A_{1087}/A_{1298} against temperature and to a lesser extent for the plots of A_{1105}/A_{1128} and A_{1105}/A_{1298} against temperature.

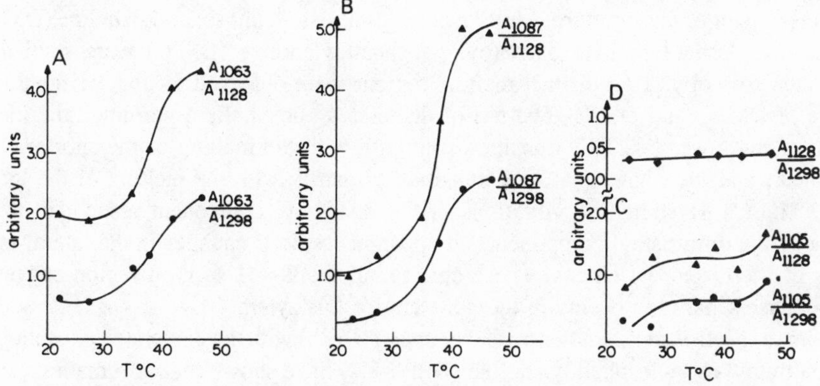

Fig. 11.15. Plots against temperature of (A) A_{1063}/A_{1128} and A_{1063}/A_{1298}; (B) A_{1087}/A_{1128} and A_{1087}/A_{1298}; (C) A_{1105}/A_{1128} and A_{1105}/A_{1298}; and (D) A_{1128}/A_{1298}. All individual points include error bars of ± 0.5°C for the estimated temperature variation at each temperature, and the errors in the relative intensity measurements are obtained from the program. (Faiman, 1979.)

It is of interest that the temperature profile of A_{1063}/A_{1128} in Fig. 11.15A, obtained by the curve-resolving method, is not apparent in similar temperature profiles of I_{1063}/I_{1128} obtained from non-curve-resolving, single-scan techniques.

Deuterated Phospholipid Systems

When Raman spectroscopy is applied as a technique for studying the hydrocarbon chain conformation in biological membranes and related model systems, the two spectral regions most commonly used are the region from 2800 to 3000 cm^{-1} involving the C—H stretching vibrations; and the region from 1000 to 1150 cm^{-1} involving the C—C stretching modes of the hydrocarbon chains and the symmetric P—O stretching vibration of the phosphate group (Mendelsohn *et al.*, 1976b). Of course, biological membranes are complex, and may contain considerable amounts of nonlipid materials that also display Raman lines in these regions. The nonlipid material complicates the matter of determining how much of the spectrum arises from lipids. By introducing a deuterated fatty acid into their model membrane system, Mendelsohn *et al.* examined Raman spectra for the C—D stretching region of stearic acid-d_{35} bound in egg lecithin multilayers. The temperature dependence of the spectra shows that the linewidth of the C—D stretching bands is a sensitive probe of chain conformation of membrane hydrocarbons. The presence of the probe does not perturb the structure of the phospholipid, insofar as the C—H stretching region is a measure of the structure. These workers studied the temperature dependence of the linewidth (full width at half-maximum) of the 2103-cm^{-1} band (symmetric CD_2 stretching) for pure stearic acid–d_{35} and for stearic acid bound in lecithin multilayers. They also studied the intensity ratio of the C—H stretching modes at 2880 and 2850 cm^{-1} for egg lecithin multilayers as a function of temperature. The ratio I_{2880}/I_{2850} was used to measure directly the formation of *gauche* isomers in the hydrocarbon chains of egg lecithin.

Deuterated phospholipids have also been used by Mendelsohn and Maisano (1978) in studies of membrane structure. Multilayers of dimyristoyl phosphatidylcholine (DMPC) (and its -d_{54} derivative) with distearoyl phosphatidylcholine (DSPC) were studied by Raman spectroscopy. Two distinct melting regions were observed for the 1:1 mole ratio mixture of DMPC and DSPC. The use of deuterated phospholipid permitted the identification of the lower (≈22°C) transition primarily with the melting of the shorter chain component, and the higher (≈47°C) transition primarily with the melting of the longer chains. The C—H stretching vibrations of the distearoyl component responded to the melting of the dimyristoyl component, an apparent result of changes in the lateral interactions of the distearoyl chains. These changes in the C—H spectral region suggested that no phase separation occurs in the gel state for this sytem.

Raman spectroscopic studies by Gaber *et al.* (1978b) with the deuterated phospholipid 1,2-dipalmitoyl-d_{62}-phosphatidylcholine (DPPC-d_{62}) have shown that information can be obtained about phospholipid conformation in complex phospholipid and lipid–protein mixtures. Spectra of solid samples of nondeuterated polycrystalline DPPC and polycrystalline DPPC–d_{62} allowed correlations to be made for some of the more important bands, e.g., C—N stretch, O—P—O, CH_2 twist, CH_2 bend, *trans* configuration, and head group

CH_2. These authors made various assignments for groups in DPPC–d_{62} and showed changes of Raman bands when this compound is melted. The set of two compounds together, DPPC and DPPC-d_{62}, provided much detailed information about the total environment of the phospholipid in a bilayer which cannot be obtained from Raman scattering data from either compound alone.

Bunow and Levin (1977b) have recorded the Raman spectra of polycrystalline 1,2-dipalmitoyl phosphatidylcholine–d_9 (fully deuterated choline methyl groups) and 1,2-dipalmitoyl phosphatidylcholine-d_{62} (fully deuterated acyl chains) in the regions 3050–2800, 2250–2050, and 1800–700 cm^{-1} (Fig. 11.16). They examined the dependence of the 2250–2050-cm^{-1} region on the lipid phase for the dipalmitoyl phosphatidylcholine-d_{62} species. The methylene CD_2 deformation and twisting modes at 984 and 919 cm^{-1}, respectively, also display strong isolated vibrational transitions that could be employed to monitor molecular order in mixed deuterated and nondeuterated lipid molecules.

Mendelsohn and Taraschi (1978) have recorded Raman spectra for certain regions that are sensitive to conformational change for dipalmitoyl phosphatidylethanolamine (DPPE) and for chain perdeuterated dipalmitoyl phosphatidylcholine (DPL–d_{62}) alone, as 1:1 binary mixtures, and as a 1:1:1 ternary mixture with cholesterol. Using melting point curves, i.e., plots of relative intensities for certain specific groups versus temperature, they found that DPPE multilayers show noncooperative formation of four to five *gauche* rotamers per chain before the gel–liquid-crystal transition of 66°C. DPL–d_{62}

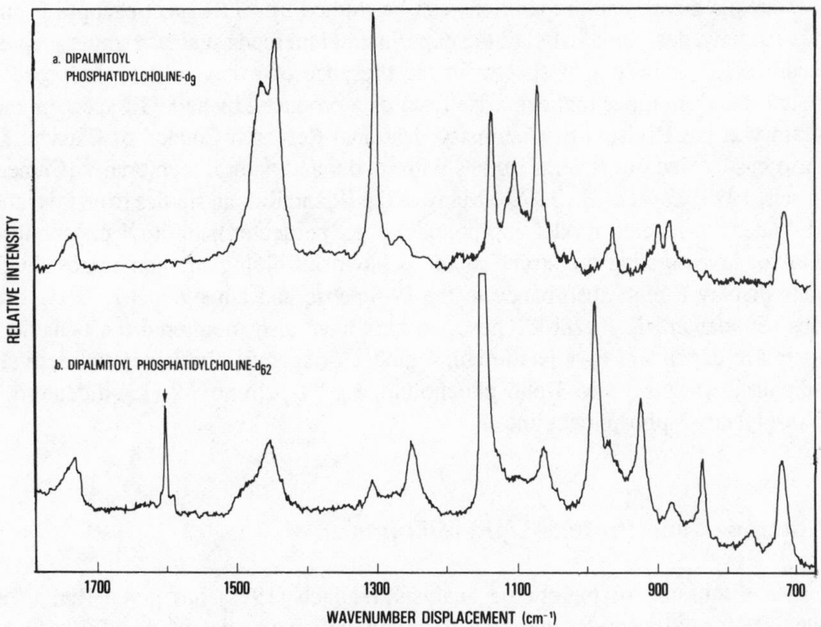

Fig. 11.16. Survey vibrational Raman spectrum of the 1750–700 cm^{-1} region of (a) crystalline dipalmitoyl phosphatidylcholine-d_9 (exposed to room humidity) and (b) crystalline dipalmitoyl phosphatidylcholine-d_{62}. (The peak marked by an arrow represents an atomic argon emission reference line.) (Bunow and Levin, 1977b.)

multilayers had similar properties. In 1:1 DPPE–DPL–d_{62} mixtures the phase separation takes place over a range of 44° to about 63°C, as shown by variation in the band width of the C—D stretching vibrations of the DPL–d_{62}. The stretching vibrations of the DPPE C—H bonds also parallel the phase separation in the 1:1 mixture of DPPE and DPL-d_{62}. The ratio I_{2885}/I_{2850} changes most rapidly near the beginning of phase separation and appears to result from the formation of *gauche* rotamers and from changes in the lateral interactions of the DPPE chains.

When cholesterol is added to form the 1:1:1 ternary system, the beginning and completion temperatures for the phase separation are each reduced by about 15°C. When cholesterol is present, the number of *gauche* rotamers that form during the phase separation is one to two, decreased from about four in the binary mixture DPPE–DPL–d_{62}.

The Raman spectroscopic study of ω-deuterated dipalmitoyl lecithin liposomes allowed Weidekamm *et al.* (1978) to evaluate the assignment of C—H stretching vibrations to methylene and methyl groups. Methyl group vibrations were shifted from the 2800–3000-cm^{-1} region to the 2050–2250 cm^{-1} region by the complete substitution of deuterium in the ω-CH_3 groups; and large intensity changes were produced in the bands at 2883 and 2936 cm^{-1}. These workers concluded that both bands represent vibrations attributable simultaneously to the CH_2 and CH_3 groups of the fatty acid chains. The intensity measurements at 2883 and 2936 cm^{-1} for the CH_2 groups led to the conclusion that the CH_3 groups of the choline moiety do not contribute much to the phase transition curves observed for the liposomes in the presence of valinomycin (a mobile carrier for alkali ions through natural membranes and liposomes).

Aqueous membrane preparations can be studied by FTIR spectroscopy. Cameron *et al.* (1979) have described in detail the experimental methods such as sample preparations, procedures for recording of spectra in the presence of excess water, the processing of data, and the difficulties that can arise from data produced by an FTIR spectrometer. The laboratory at the Division of Chemistry, National Research Council of Canada, Ottawa, Ontario has carried out several studies with model and natural membranes (Cameron and Mantsch, 1978; Casal *et al.*, 1979). Many of the IR and Raman studies from this laboratory used deuterium-labeled model compounds, e.g., perdeuteropalmitoyl phospholipids, as probes for investigating the structure and behavior of biological membranes. Such compounds display a high absorbance in the symmetric and antisymmetric C^2H_2 stretching modes (Sunder *et al.*, 1978b). These workers have also monitored the behavior of the temperature-dependent C—^{2}H stretching modes of specifically deuterated derivatives of 1,2-dipalmitoyl-*sn*-glycero-3-phosphocholine, e.g., 1-palmitoyl-2-(13′-dideutero)-palmitoyl-*sn*-glycero-3-phosphocholine.

Polypeptide– and Protein–Lipid Mixtures

In a discussion of membrane analysis Wallach (1972) has given Raman and IR frequencies for polypeptides and phospholipids, which are presented in Table 11.6. Lord and Mendelsohn (1981) have discussed various aspects of the Raman spectroscopy of membrane constituents and related substances. Their discussion includes amino acids (Lord and Yu, 1970; Koenig, 1972; Simons *et al.*, 1972); and polypeptides and proteins. [See also Chapter 3, this book.]

Table 11.6. Some IR and Raman Frequencies of Polypeptides and Phospholipids[a,b]

	Frequency (cm^{-1})			
	Infrared		Raman	
Assignment[c]	Polypeptide[c]	Phospholipid[c]	Polypeptide[c]	Phospholipid[c]
Skeletal deformation	368s		375s[e]	
Skeletal deformation	528s		527vs	
NH and C═O out-of-plane bending, (amide IV)	625m		662w	
CH_2 rocking		720m		
NH bending + skeletal modes; (amide V)				
C—C stretch and CH rocking		870–1150		
CH_3 stretch + skeletal modes	907m		907vs	
CH═CH bending		970m		
P—O—C stretching		980–1050s		
C═C stretching		1000–1620		
Methyl modes	1049vs		1048vw	
All-*trans* hydrocarbon; extended				1064m
Hydrocarbon chain; random				1089m
Hydrocarbon; all-*trans*				1100m
Methyl-carbon stretching	1105s			
Hydrocarbon chain; all-*trans*				1130s
Methyl rocking	1167s		1168m	
CH_2 twisting		1170–1300		
CH_2 wagging		1180–1380		
P═O stretching[d]		1250–1350		
CN stretching + NH in-plane bending; (amide III)	1300s		1310m	
CH bending	1328m		1331s	
Methyl; symmetric deformation	1376vs		1371m	
Methyl; asymmetric deformation	1453vs		1453vs	
CH_2 stretching and bending		1460		
NH bending in plane + CN stretching; (amide II)	1537–1560s		1549w	
C═O stretching; amide; (amide I)[d]	1635–1655vs		1654s	
C═O stretching; ester		1720–1740s		
CH_2 asymmetrical stretching, (in plane)	2853s	2853s	2874m	
CH_2 asymmetrical stretching, (out of plane)	2926s	2926s	2927vs	
C═C—H stretching		3000		

[a] Wallach (1972).
[b] Selected data from: D. Chapman, *The Structure of Lipids*, Methuen, London, Chap. 4 (1965); J. L. Koenig and P. L. Sutton, *Biopolymers* **8,** 167 (1969). D. F. H. Wallach, J. M. Graham, and B. R. Fernbach, *Arch. Biochem. Biophys.* **131,** 322 (1969); and J. L. Lippert and W. L. Peticolas, *Proc. Natl. Acad. Sci. USA* **68,** 1572 (1971).
[c] Assignments are not all absolute and some frequencies determined for polypeptides also apply to phospholipids.
[d] The P═O stretching and amide I frequencies are very dependent on hydrogen bonding.
[e] vs = very strong; s = strong; m = medium; w = weak; vw = very weak.

Infrared group vibrations of diagnostic importance in the spectroscopic examination of membranes have been summarized by Fringeli and Günthard (1981): phospholipids and amide group vibrations. Parker (1971) has also discussed these topics.

Fringeli and Günthard have included in their discussions several examples of the use of ATR spectroscopy with oriented phospholipids. Examples of the phospholipids studied by them are L-α-DPPE; L-α-DPPC; β,γ-dipalmitoyl-*N*-*N*-dimethyl-L-α-cephalin, and lysolecithin. Fringeli (1981) has recently discovered a new highly ordered phase of L-α-DPPC monohydrate, the existence of which lends support to the postulation of a nonuniform conformation of the polar head of L-α-DPPC in all phases described so far (Powers and Pershan, 1977).

Larsson (1973) has shown that the C—H stretching vibration region in the Raman spectra of lipids can be used to identify different states of order. Larsson and Rand (1973) have also shown that these spectral effects are the result mainly of differences in the environment of the hydrocarbon chains. Lippert and Peticolas (1971,1972) and Mendelsohn (1972) have reported that the region 1000–1200 cm^{-1} contains useful information on the conformation, especially on relations between *trans* and *gauche* conformation of the chains. The 1100- and the 2900-cm^{-1} regions supply complementary information on the physical structure in lipid systems. Larsson and Rand (1973) obtained results that indicated that the 2850- and 2885-cm^{-1} C—H stretching vibrations are more sensitive to chain conformation than are those of the 1100-cm^{-1} C—C stretching region, which was used by Lippert and Peticolas to determine the relative all-*trans* to *gauche* conformation. Thus, changes in the relative intensities of the 2850- and 2885-cm^{-1} lines can be observed in systems where small structural changes do not yield detectable effects in the 1100-cm^{-1} region. The ratio I_{2850}/I_{2885} increases, paralleling the "looseness" of the lateral packing of disordered hydrocarbon chains, in the following order: lamellar liquid crystal < hexagonal or cubic liquid crystal < micellar solution < solution in an organic solvent (Larsson and Rand, 1973).

Larger spectral changes which may be of greater usefulness are found when the CH_2 and CH_3 groups of the lipid chains are surrounded by groups that are not CH_2's or CH_3's. (In the situations mentioned above, the hydrocarbon chains were surrounded by other hydrocarbon chains.) The relative intensity of the 2930-cm^{-1} line relative to the other C—H stretching lines appears to increase when the environment of the hydrocarbon chains becomes more polar. Such changes, examples of which occur in lipid–protein complexes, can be used to indicate the kind of interaction between the components. Table 11.7 gives a summary of the spectral changes related to interaction alternatives. Thus, it is possible to distinguish the case in which hydrocarbon chains of the lipid molecules are associated into separate lipid regions from that in which the chains are located in a protein environment, e.g., in an insulin-long chain phosphate ester complex.

Figure 11.17 shows a Raman spectrum of frog sciatic nerve, a bundle of myelin multilayers. The 2855- and 2885-cm^{-1} lines and the relatively weak shoulder at 2930 cm^{-1} show that most of the lipid hydrocarbon chains are surrounded by other chains, and that the structure determined from the relative heights of the 2855- and 2885-cm^{-1} lines is similar to that of lamellar liquid crystal phases.

FTIR spectroscopy has been used to characterize high-density lipoprotein (HDL) and low-density lipoprotein (LDL) (Gilman *et al.*, 1980) The amide I and II bands are significantly broader and at slightly lower frequencies in LDL compared to HDL.

Table 11.7. Dominating C—H Stretching Vibration Peaks in Lipid–Protein Raman Spectra in Cases of Hydrophobic Interaction and No Interaction[a]

	No hydrophobic interaction		Hydrophobic interaction (lipid hydrocarbon chains in a protein environment)
	Crystalline hydrocarbon chains	Liquid hydrocarbon chains	
Dominating C—H peak due to protein (cm^{-1})	2930	2930	2930
Dominating C—H peak due to lipid (cm^{-1})	2885	2850	2930

[a] Larsson and Rand (1973).

Polypeptide–Lipid Complexes

Glucagon–Dimyristoyllecithin

Taraschi and Mendelsohn (1979) have used Raman spectra to examine the properties of a soluble complex formed between the polypeptide hormone glucagon and dimyristoyllecithin (DMPC) when the latter is in the gel state below 23°C. This complex has characteristics similar to those of serum lipoproteins (Epand *et al.*, 1977; Jones *et al.*, 1978) and is, therefore, useful as a model of lipid–protein interaction. The authors studied DMPC in multibilayer form and in small unilamellar vesicles, as control systems. The complexes were made of either of these two systems plus glucagon (30 to 1 phospholipid-to-protein ratio). Since the ratio of intensities I_{1130}/I_{1100} had been correlated by Gaber and Peticolas (1977) with the number of CH_2 groups in the all-*trans* conformation, Taraschi and Mendelsohn monitored this ratio for the two control systems and the complex during the determination of melting-point curves (from ca. 10° to 40°C). The change in the ratio during melting showed the cooperative loss of four or five *trans* bonds in the DMPC hydrocarbon chains, according to calculations based on Gaber and Peticolas's

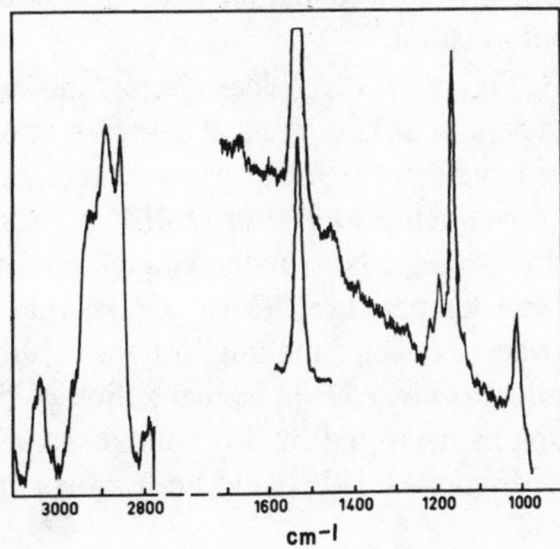

Fig. 11.17. Raman spectrum of a frog sciatic nerve freshly dissected and in Ringer's solution. (Larsson and Rand, 1973.)

method. Also, two or three *gauche* rotamers formed before the melting of the main chain occurred. In the case of the complex of DMPC-glucagon, a 25–30% reduction at 7°C in I_{1130}/I_{1100} (compared to the controls) showed two to three more *gauche* rotations in the chains. Below 20°C there was a small change in the ratio, which suggested that non-cooperative formation of *gauche* rotamers was inhibited.

To investigate the alterations in the lateral packing of the hydrocarbon chains and the formation of *gauche* rotations the authors followed changes in the ratio I_{2885}/I_{2850} (intensities of the antisymmetric to the symmetric C—H stretching vibrations, respectively) over the same temperature range (10°–40°C). In DMPC multilayers, a high degree of order was found in the hydrocarbon chains. At the melting temperature, a loss of *gauche* rotamers occurred. With the formation of single-walled DMPC vesicles lateral packing of the chains was disrupted compared with that in the multilayers at all temperatures. In the case of the complex with glucagon, at temperatures where the complex precipitates, the lateral interactions returned, and the ratio I_{2885}/I_{2850} attained a value almost as large as that of uncomplexed multilayers.

DMPC–Gramicidin A; DMPC–Poly-L-lysine; DMPC–Valinomycin

Susi *et al.* (1979) have done a Raman study of the interactions of several different polypeptides with the bilayers of 1,2-dimyristoyl-L-phosphatidylcholine (DMPC) as a function of temperature. Their interest was to determine what effect the different polypeptides have on liposome structure above and below the main transition temperature, and to examine the cooperativity of the transition.

They studied conformational changes in the phospholipid molecules by observing the intensity of the band at 1062 cm^{-1}, which arises from C—C stretching vibrations of *trans* portions of the hydrocarbon chains (Lippert and Peticolas, 1971; Spiker and Levin, 1975). They made this type of measurement by using as an internal reference the intensity of the C—N stretching vibration at 715 cm^{-1}, since it is constant with temperature changes (Gaber and Peticolas, 1977).

Using the value obtained for the liposomes at liquid-nitrogen temperature as the 100% *trans* standard, they defined a relative intensity characteristic, I_R, as follows: $I_R = (I_{1062}/I_{715})/(I_{1062}/I_{715})_{t(\text{liquid nitrogen})}$. The value of I_R at liquid-nitrogen temperature, where all C—C bonds are *trans,* is unity. If no C—C bonds are *trans* at this temperature, I_R is zero. Thus, these authors used I_R as a measure of the fraction of C—C bonds in a *trans* conformation.

Figure 11.18A (black circles) shows I_R versus temperature for pure DMPC liposomes, and depicts a slow, gradual decrease of orderliness below the main transition temperature. The main transition causes a marked decrease of I_R. Figure 11.18A (open circles) shows I_R versus temperature for DMPC liposomes interacting with gramicidin A, a linear pentadecapeptide. No sudden change appears in the orderliness of the hydrocarbon chains at any temperature. Thus, a relatively small amount of the hydrophobic polypeptide prevents sudden "melting" of the hydrocarbon chains. The Raman data show a more gradual *trans–gauche* isomerization of the hydrocarbon chains than does the phase transition as measured by differential scanning calorimetry.

Figure 11.18B (solid line) shows the change of I_R for DMPC interacting with poly-

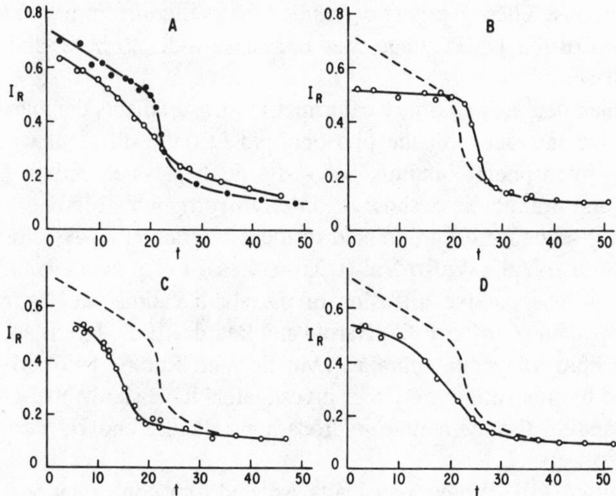

Fig. 11.18. The relative intensity of the 1062-cm^{-1} *trans* band, I_R, as a function of temperature. I_R = 1 at liquid nitrogen temperature. (A) Pure dimyristoylphosphatidylcholine (DMPC) (●——●); DMPC plus 3% w/w gramicidin A (○——○). (B) DMPC plus 25% w/w poly-L-lysine. (C) DMPC plus 25% w/w gramicidin S. (D) DMPC plus 10% w/w valinomycin. All samples were prepared as 1:4 w/w suspensions in water. The dotted lines in B–D represent the I_R values of the pure phospholipid, for comparison. (Susi *et al.*, 1979.) Reprinted with the permission of the American Chemical Society, Copyright 1979.

L-lysine, compared with that for the pure DMPC liposomes (dotted line). These data suggest (Susi *et al.*, 1979) that a certain stabilization of the *trans* structure occurs (Papahadjopoulos *et al.*, 1975), but only in the temperature interval just above the transition temperature of pure DMPC liposomes. Below the transition temperature the hydrocarbon chains seem to conform to a rigid, but disordered structure, showing that poly-L-lysine affects the packing of the hydrocarbon chains in the gel phase.

Figure 11.18C shows a curve for the interaction of DMPC with gramicidin S, whose hydrophobic amino-acid groupings lie outside on the periphery of the molecule. Several features are apparent: a marked lowering of the transition temperature (about 10°C), a large widening of the transition region (lessened cooperativity), and a considerable increase in randomness of structure below the transition range. The relatively small, compact antibiotic molecules appear to be embedded in the hydrophobic regions of the phospholipid in less orderly fashion than are gramicidin A (Fig. 11.18A) and cholesterol. The latter two substances are part of the orderly structure assumed to be present in the hydrocarbon regions of the phospholipid, thus causing a stiffening of the bilayer in the liquid crystalline phase (Chapman *et al.*, 1977; Lippert and Peticolas, 1971; Weidekamm *et al.*, 1977).

Figure 11.18C also shows that the gramicidin S molecules cannot cause order to be established in the liquid crystalline phase.

Figure 11.18D shows I_R versus temperature for the interaction of DMPC with valinomycin; this curve is like the one in Fig. 11.18C, ascribed by Susi *et al.* to hydrophobic binding of gramicidin S to the phospholipid hydrocarbon regions. The curve in Fig. 11.18D differs markedly from that in Fig. 11.18B, which is produced by polar binding to the head groups. These workers concluded that valinomycin protrudes into the hydrocarbon regions of the DMPC, but does not cause order to be established in the liquid crystalline phase.

The Raman data and auxiliary calorimetric measurements demonstrate that the polypeptides studied interact with the phospholipid in quite different ways. Gramicidin A shows orderly hydrophobic binding; poly-L-lysine displays essentially polar binding; and valinomycin and gramicidin S show disorderly hydrophobic binding.

The linear sequence of amino acid residues in gramicidin A is formyl-L-Val-Gly-L-Ala-D-Leu-L-Ala-D-Val-L-Val-D-Val-(L-Trp-D-Leu)$_3$-L-Trp-ethanolamide (Sarges and Witkop, 1965). The passive diffusion of the alkali cations and hydrogen ion through natural (Chappell and Crofts, 1965; Harold and Baarda, 1967; Harris and Pressman, 1967) and artificial lipid bilayer membranes (Mueller and Rudin, 1967; Myers and Haydon, 1972) is aided by this antibiotic. Other investigators have shown that gramicidin A forms a channel spanning the membrane hydrocarbon (Hladky and Haydon, 1972; Krasne *et al.*, 1971).

Veatch *et al.* (1974) have physically isolated four conformational species of gramicidin A and characterized each one by several techniques, including IR spectroscopy. They found two classes of conformations and correlated them with regular dimer models. The first class applied to three of the species, and was helical with predominantly parallel-β hydrogen bonding. The second class applied to the fourth species, and was predominantly antiparallel hydrogen bonding. This fourth species was most likely antiparallel-β double helix. The IR spectra of the first three species had amide I and II bands consistent with parallel-β double helices or π (LD) helical dimers. For the fourth species, the amide I bands at 1633 and 1680 cm^{-1} were within the range of values for antiparallel β-pleated sheet conformation; but the broad amide II band at 1545 cm^{-1} was outside the range (1520–1530 cm^{-1}).

Melittin–Phosphatidylethanolamine

Melittin is a polypeptide from bee venom in which the apolar residues are situated at one end of the polypeptide chain (Dayhoff, 1972). The first twenty residues from the NH_2 terminus are mainly apolar moieties, whereas the COO^- terminus is comprised of the basic sequence -Lys-Arg-Lys-Arg-Glu-Gln. This membrane-active polypeptide might intercalate its apolar residues among the relatively liquid fatty acid chains of a phosphatide bilayer in a random array; or the first 14 residues (up to a proline) might form an apolar α-helical "plug," still permitting the polar section to protrude to the membrane surface (Verma *et al.*, 1974).

Verma *et al.* (1974) have studied the interaction of melittin with phosphatide multibilayers by IR spectroscopy, IR dichroism, and spin label methods. The ratio of the band intensities I_{1460}/I_{1370} increases as the hydrocarbon chain mobility decreases. The ratio increases when melittin is bound, and at the same time the CH_2 rocking band at 720 cm^{-1} sharpens, also showing a decrease in chain mobility.

The IR spectra of phosphatidylethanolamine in the range 1250–800 cm^{-1} depend upon the orientation of the head groups. When melittin interacts, it changes the transition vectors of the following bands (measured by IR dichroism): 1730 cm^{-1} (carbonyl ester stretch), 1240–1220 cm^{-1} (P═O stretch), 1020 cm^{-1} (P—OH) stretch in phosphatidylethanolamine), and 960 cm^{-1} (P—O—C deformation in lecithin). These changes indicate a reorganization of the polar head groups (Verma *et al,* 1974). The amide I band of melittin, when it is incorporated into phosphatide multibilayers, is located at 1650 cm^{-1}. It is not dichroic, thus suggesting that the peptide chain lies in an unordered configuration.

Glycophorin–Dipalmitoyl Phosphatidylcholine

Taraschi and Mendelsohn (1980) have used temperature-dependent Raman spectra to study lipid–protein interaction in the glycophorin–dipalmitoyl phosphatidylcholine (DPPC) model system. Glycophorin is a sialoglycopeptide which spans human red cell membranes. Its amino acid sequence is given by Marchesi *et al.* (1976). It was used for this Raman work because it spans the erythrocyte membrane; it is isolable in rather pure form; it has been efficiently reconstituted into vesicles with DPPC (MacDonald and MacDonald, 1975); and it is not excluded from the solid lipid phase (van Zoelen *et al.*, 1978; Grant and McConnell, 1974), upon which useful studies may be performed during the entire phase transition.

The conformation of the DPPC hydrocarbon chains near the protein is greatly altered from that in pure lipid dispersions. At lipid-to-protein mole ratios of 125 to 1, the melting curve (of chain C—^{2}H stretching vibrations) for complexes of [$^2H_{62}$]-DPPC-glycophorin is broad and noncooperative, and shows a midpoint at 20°C, ~15° below that of the gel–liquid-crystal phase transition for [$^2H_{62}$]-DPPC in multilamellar dispersions. The change in melting behavior induced by the protein is shown in Fig. 11.19. The linewidth function at 2100 cm^{-1} (half-width of the symmetric C—^{2}H stretching vibrations) serves as a probe of phospholipid conformation that responds only to *gauche* rotamer formation. The low-temperature value of 28 cm^{-1} for the linewidth in the 125:1 complex indicates

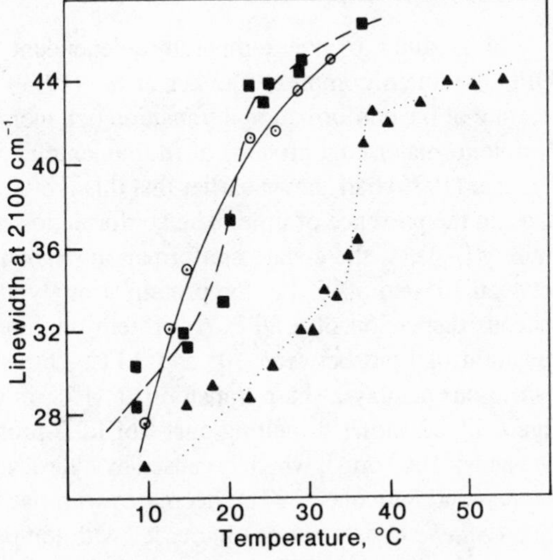

Fig. 11.19. Temperature-induced variation in the linewidth of the C—^{2}H stretching vibrations near 2100 cm^{-1} for [$^2H_{62}$]DPPC in multilayer dispersion (▲ · · · ▲) and complexed 275:1 (■ – – ■) and 125:1 (⊙——⊙) with glycophorin. The gel/liquid crystal phase transition appears as a sharp discontinuity near 35°C in pure [$^2H_{62}$]DPPC. The precision of the linewidth measurement is about ±1–2 cm^{-1}. (Taraschi and Mendelsohn, 1980.)

that no more than one *gauche* rotamer per chain is then present (Taraschi and Mendelsohn, 1980).

Protein–Lipid Complexes

Cytochrome c Oxidase–DMPC and DPPC. Cytochrome c–DMPC and DPPC

Cytochrome *c* oxidase is thought to be a transmembrane intrinsic protein, and cytochrome *c* is thought to act as an extrinsic membrane protein. These and the plasma proteins fibrinogen and albumin interact differently with both dimyristoyl and dipalmitoyl phosphatidylcholine, as shown by differences in the Raman spectra of these phosphatidates in the hydrocarbon C—H stretching region around 2900 cm^{-1} (Lis *et al.*, 1976). An important application of such spectral differences is that they allow one to examine in detail the effects of proteins of unidentified type on the spectra of lipids.

These workers have shown that intrinsic transmembrane protein cytochrome *c* oxidase lowers the intensity ratio, I_{2890}/I_{2850}, of the lipid. Below the lipid liquid crystalline transition this ratio stays constant. Cytochrome *c*, which spreads the lipid layer, lowers the ratio I_{2850}/I_{2930} above and below the liquid crystal transition. Also, when cytochrome *c* was present, the ratio I_{2890}/I_{2850} varied with temperature. When fibrinogen and albumin (extrinsic proteins) were present, decreases occurred at room temperature in the lipid peak ratios I_{2890}/I_{2850} and I_{2850}/I_{2930}. At 5°C both these ratios have the same values as for a standard dimyristoyl phosphatidylcholine–water dispersion. Lis *et al.* interpreted these Raman results to indicate that the dimyristoyl phosphatidylcholine bilayer is hardly affected by fibrinogen and albumin, but cytochrome *c* at low temperatures still has some influence on the environment of the hydrocarbon chains; it does not affect their conformation.

fd Phage B–Protein–DPPC

In a study of the temperature-dependent conformational transitions of a lipid (DPPC)–protein complex, Dunker *et al.* (1979) have observed a lowering of the temperature of the conformational transition of a membrane bilayer by the intrinsic membrane B-protein (major coat protein) of fd filamentous phage. Williams and Dunker (1977) and Wickner (1976) had shown earlier that this protein can attain two different conformational states in the presence of lipid. The conformation of the protein in the membrane is almost entirely β-sheet, the c-state conformation (Raman and CD data), which remains as such between 10° and 50°C; but the protein strongly affects the lipid bilayer organization. An aqueous dispersion of a DPPC/B-protein mixture (1:1) produces a wide conformational transition of lipid between 10° and 30°C. One should note that simple aqueous DPPC dispersions display a sharp transition at 41°C, in marked contrast to the 10°–30°C range. Figure 11.20 shows a melting curve of fd B-protein-DPPC complexes. The intensity of the band at 1062 cm^{-1}, which is caused by hydrocarbon chains in the all-*trans* conformation (Lippert and Peticolas, 1971), decreases with rise in temperature; however, that at ~1100 cm^{-1} (*gauche* conformation) increases with temperature.

Dunker *et al.* (1979) compared their results with those obtained by Boggs *et al.*

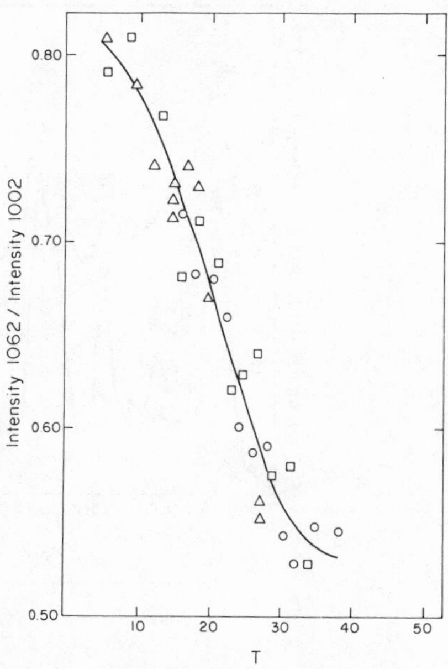

Fig. 11.20. Melting of fd B-protein-DPPC complexes. A 1:1 (mass ratio) lipid/protein complex was heated (○) starting at 15°C, cooled (△), and reheated (□). After 2–3 min equilibration at each temperature, the Raman spectrum over the 950–1200 cm^{-1} region was taken. The phenylalanine peak at 1002 cm^{-1}, which had previously been found to be invariant with temperature, was used as an internal standard. This convenience allows scanning a very narrow frequency range, thus simplifying the experiment. The peak height near 1062 cm^{-1} divided by the phenylalanine peak height provides the given ratios. (Dunker *et al.*, 1979.)

(1976) and Curatolo *et al.* (1977) with the proteolipid isolated from brain white matter. The proteolipid contains much α-helical structure in organic solvents (Sherman and Folch-Pi, 1970) and stabilizes a very large boundary layer of lipid (Curatolo *et al.*, 1977), whereas the β-sheet B-protein appears to destabilize the surrounding lipids. Possibly, hydrophobic β-sheets destabilize boundary lipids, and hydrophobic α-helices stabilize them (Dunker *et al.*, 1979).

Membranes from Natural Sources

Parker (1971) has reviewed the earlier literature on IR examination of membranes. Among the topics included were plasma membranes, endoplasmic reticulum, mitochondrial membranes, nerve and muscle membranes, and erythrocyte and myelin membranes. In this section we discuss more recent Raman experimentation on erythrocyte membranes, lymphoid cell membranes, and the opsin membrane; and Fourier transform IR experimentation on bovine photoreceptor membrane and the purple membrane of *Halobacterium halobium*.

Erythrocyte Membranes

Goheen *et al.* (1977) have compared the Raman spectra of intact erythrocyte ghosts with spectra of membranes from which essentially all peripheral proteins had been stripped (Fig. 11.21). The extraction (stripping) procedure caused much alteration in the environment of peptide bonds and a drastic change in the environment of the tryptophan and

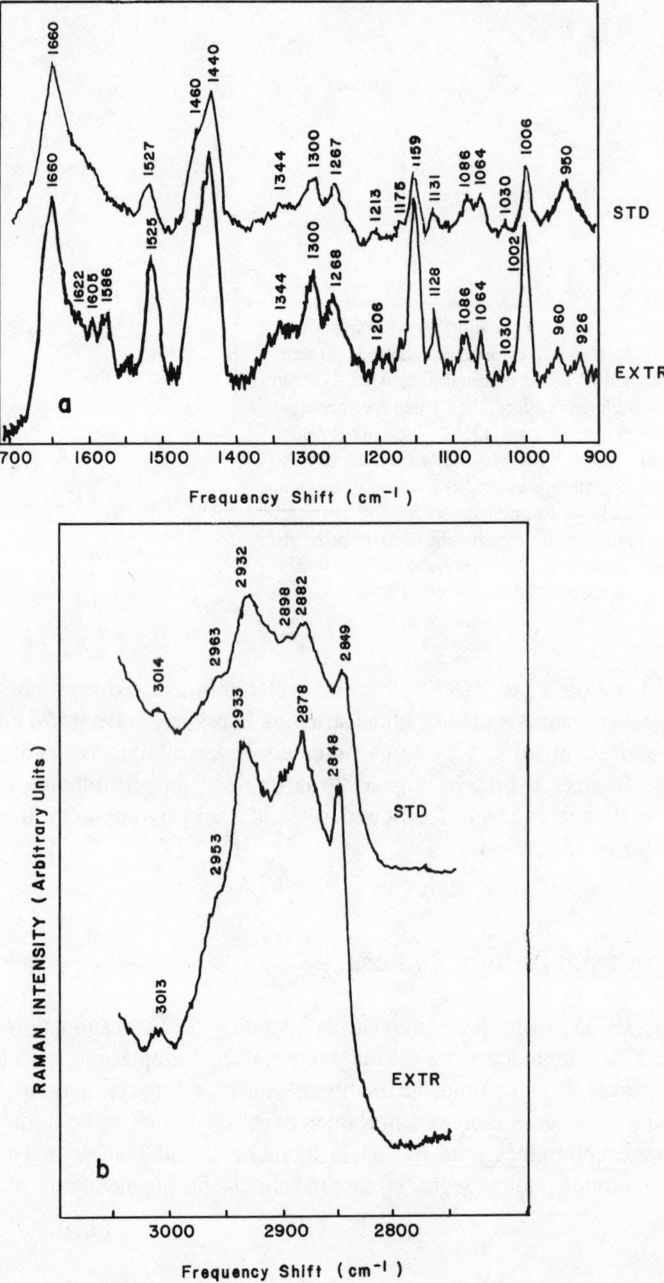

Fig. 11.21. Raman spectra of human erythrocyte vesicles (STD) and vesicles from which peripheral proteins have been removed (EXTR) (a) 900–1700 cm^{-1} (b) 2750–3150 cm^{-1}. Excitation was by argon ion laser with ca. 500 mW power to the sample at 514.5 nm. Resolution is ca. $\pm$ 2 cm^{-1} with an 8.0 cm^{-1} bandpass; time constant, 10 sec; counting rate, 3×10^4 photon counts/sec; scanning speed, 10 cm^{-1}/min. temperature controlled at 5°C. (Goheen *et al.*, 1977.)

phenylalanine rings of the integral membrane proteins. The lipid portion of the bilayer was not affected much by the extraction. These workers noted especially that the total decrease in the heights of the protein-associated peaks of the extracted vesicles was generally less than that expected from the mass of protein removed by extraction.

Verma and Wallach (1976) have examined the effects of pH and temperature variation on the Raman spectra of erythrocyte membrane preparations. In the C—H stretching region there are strong bands at ca. 2930, 2880, and 2850 cm^{-1} at 22°C. When the temperature is raised to above 42°C at pH 7.4, the ≈ 2930-cm^{-1} band comes into prominence, the ≈ 2880-cm^{-1} band is obscured, and the 2850-cm^{-1} band remains unchanged. The ratio I_{2930}/I_{2850} remains stable at pH 7.4 at about 1.2 between 20 and 35°C and then changes rapidly to about 1.4 between 36° and 45°C, a transition which is reversible only up to 39°C. Correction of temperatures due to local heating by the laser beam gives a lower limit of 38°C for the discontinuity and 42°C for the irreversibility. A lowering of the pH to 6.5 brings the transition center to ca. 31°C (corrected) and the lower limit to ca. 22°C (corrected). At pH 6 the range of the transition goes down to 0°–7°C (corrected) and the ratio I_{2930}/I_{2850} ranges from 1.0 to 1.35.

Thus, the change in pH is a significant factor. An increase in hydrogen-ion concentration of about 0.6 μM brings about a 16°C lowering of the transition temperature. Verma and Wallach (1976) had attributed more than 60% of the scattering at 2930 cm^{-1} to protein and had suggested that (i) the apolar residues of penetrating proteins, along with the hydrocarbon chains of phospholipids situated next to these protein residues, constitute a separate phase, and (ii) the high-temperature, pH-sensitive thermotropic step in the ratio I_{2930}/I_{2850} is a cooperative state transition in this phase, associated with a rearrangement of acyl chain configuration and protein residue orientation. They proposed that the pH susceptibility of the heat-sensitive discontinuity may relate to the reversible thermal transitions seen for some globular proteins at pH's below their isoelectric points. The large pH-sensitive change in transition temperature in the range pH 7–6 was thought possibly to result from the low dielectric constant of the membrane core, with small changes in charge having a greater effect on the structure of proteins in membranes than on proteins dissolved in aqueous solvents (Wallach *et al.*, 1979).

Differences have been observed (Verma and Wallach, 1982) between aged normal erythrocytic membranes and Duchenne muscular-dystrophy erythrocytic membranes in certain respects. By studying specific bands in Raman spectra (C—H stretching), these workers observed different thermal transitions of lipid or lipid–protein domains and of proteins in these membranes. The ratios of bands that they studied as a function of temperature were I_{2880}/I_{2850} and I_{2930}/I_{2850}, respectively. Analysis of amide I and amide III bands indicated that the Duchenne muscular dystrophic (DMD) membrane proteins had a greater proportion of β-structure than did the aged normal controls.

Also, in normal membranes the midpoint of protein transition shifted from ~40°C at pH 7.4 to ~30°C at pH 6.5, whereas DMD membranes showed virtually no sensitivity in this range.

The data of Verma and Wallach (1982) suggested a protein anomaly that produces abnormal lipid–protein associations in DMD membranes.

A membrane spectrum with much information has been recorded by Milanovich *et al.* (1976), who used a high signal-to-noise ratio for the study of Dutch Belt rabbit hemoglobin-free erythrocytes. Figures 11.22a and 11.22b present spectra of the erythrocyte ghosts from 300 to 1700 cm^{-1} and 2800 to 3100 cm^{-1}. Table 11.8 gives tentative assignments of the vibrational bands. It should be noted that the bands at 1672 and 1650 cm^{-1} are attributed here predominantly to lipid rather than protein structure as reported in earlier work (Lippert *et al.*, 1975). Also, the ratio of I_{1131} (*trans* conformer C—C stretch) to I_{1087} (*gauche* conformer C—C stretch plus PO_2^- symmetric stretch) gives information on the fluidity of the membrane. Milanovich *et al.* found this ratio to be 0.71 for normal rabbit erythrocytes (close to the value for human erythrocytes), whereas it is ~2.0 in solid dipalmitoyl phosphatidylcholine (Lis *et al.*, 1975), which consists mainly of *trans* conformation. Accordingly, the ghosts contain significant amounts of *gauche* conformation, thus showing much fluidity, which is in accord with current ideas on membrane models (Gulik-Krzywicki, 1975).

The strength of the 549-cm^{-1} band *(trans–gauche–trans)* relative to the 515-cm^{-1} band *(gauche–gauche–gauche)* of the C—C—S—S—C—C structure shows that the —S—S— bonds of the membrane protein have mainly a *trans–gauche–trans* structure, with some *gauche–gauche–gauche* being present.

Mikkelsen *et al.* (1978) have shown that transmembrane ionic gradients across sealed hemoglobin-free, erythrocyte membrane vesicles influence Raman signals originating in the membrane protein. These workers imposed transmembrane cation gradients, and observed that the gradients increased the Raman scattering intensity in the CH_3-stretching region. Imposition of a cation gradient produced changes in the amide I and amide III frequencies characteristic of increased helicity of membrane proteins (Fig. 11.23).

Spectrin-free vesicles also display gradient-sensitive changes of intensity in the CH_3-stretching region, but no gradient-related protein conformational changes were observed in these vesicles.

Figure 11.23A shows three major bands in the CH-stretching region at 2850 cm^{-1} (symmetric CH_2 stretching), ~2890 cm^{-1} (C—CH_3 symmetric stretching), and 2930 cm^{-1} (antisymmetric CH_2 stretching). When a K^+ gradient is imposed, the intensity of the 2930-cm^{-1} band increases relative to that at 2850 cm^{-1} (which is stable, and does not change at temperatures above the acyl chain liquid-crystalline transition). Thus, the intensity ratio I_{2930}/I_{2850} becomes much larger than for vesicles without gradient.

In Fig. 11.23B the amide I region of the native vesicles without a K^+ gradient displays a diffuse band at 1665 cm^{-1} (amide I′, since measurements were made in D_2O solution to lessen the interference at 1640 cm^{-1} by H—O—H bending). When a K^+ gradient is imposed, the diffuse band formerly at 1665 cm^{-1} is resolved into two bands at 1665 and 1640 cm^{-1}. Scattering at 1665 cm^{-1} or 1655 cm^{-1} may originate from polypeptide of unordered or β-structure, but the 1640 cm^{-1} band in D_2O is evidence for α-helix (Lippert *et al.*, 1976; Frushour and Koenig, 1975).

Figure 11.23C shows the amide III region. With no K^+ gradient, the main feature lies at 1267 cm^{-1}, due to α-helix and/or random-coil polypeptide (Lippert *et al.*, 1976; Frushour and Koenig, 1975). When a transmembrane K^+ gradient is imposed, it lessens the intensity of the 1267-cm^{-1} feature and produces a prominent band at 1310 cm^{-1}, which indicates α-helix.

Figure 11.23D shows the amide III′ when no gradient is present (top curve). There is a main band at 960 cm^{-1} (random-coil in D_2O), and a small shoulder at 935-cm^{-1}. In the presence of the gradient, the 935-cm^{-1} band becomes much larger, and a small hint

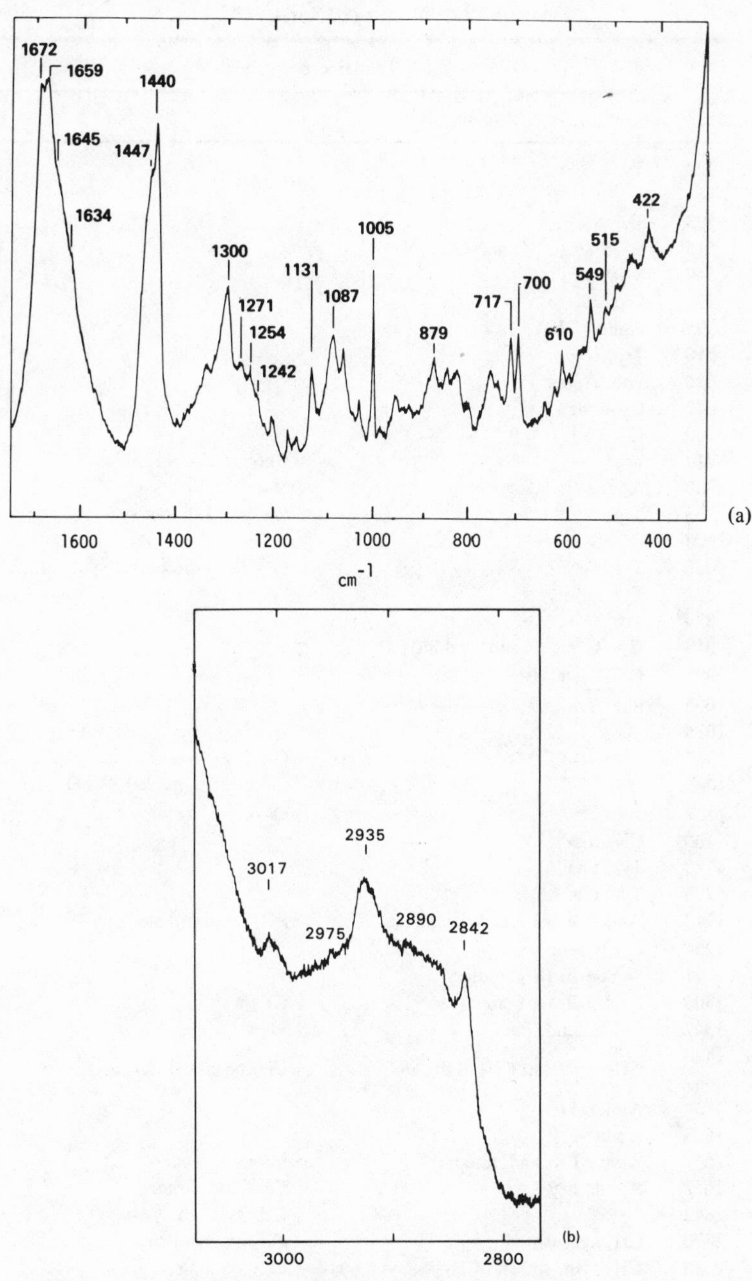

Fig. 11.22 Raman spectrum of Dutch Belt rabbit erythrocyte ghosts in H_2O, (a) 300–1700 cm^{-1}, (b) 2800–3100. Excitation was by argon-ion laser with power at the sample approximately 200 mW at 488.0 nm. The scattered light was viewed transversely with a 75 cm Czerny-Turner double monochromator equipped with holographic gratings. Resolution was 4 cm^{-1}; integration time 10 sec; maximum counting rate, 3000 photons/sec; scanning speed, 4 cm^{-1}/min; temperature 24°C. (Milanovich *et al.* 1976.)

Table 11.8. Frequencies and Assignments of Vibrational Bands in the Raman Spectrum of Dutch Belt Rabbit Erythrocyte Ghosts[a]

	Tentative assignments	
cm^{-1}	Protein	Lipid
422	Tyr	
463		
493	Tyr	
515	S—S str (GGG and GGT)[b]	
549	S—S str (TGT)	
571	Trp	
595	Amide VI	
610	Trp	
626	Phe, Amide IV	
642	C—S str., Tyr	
700	C—S str. (Met)	
717		Choline C—N sym. str.
730	Amide V	
763	Trp	O—P—O diester sym. str.
808		
828	Tyr	O—P—O diester antisym. str.
852	Tyr	
879	Trp	
959	C_α—C str. (α and random)	
991	C_α—C str. (β)	
1005	Phe, Trp	
1035	Phe	
1067		C—C str. *(trans)*
1087		C—C str. *(gauche)* + PO_2^- sym. str.
1131		C—C str. *(trans)*
1160	CH_3 rock	
1180	Tyr, Phe	
1209	Tyr, Phe	
1242	Amide III (β and random)	PO_2^- antisym. str.
1254	C—H bend	
1271	Amide III (α globular)	
1300	Amide III (α fibrous)	CH_2 wag.
1342	C—H bend, Phe	
1440 1447	CH_2 bend and CH_3 asym. def.	CH_2 bend and CH_3 asym. def.
1634	Amide I (α)	
1645	Amide I (β)	
1659	Amide I (α and random)	C=C str. *(cis)*
1672	Amide I (β)	C=C str. *(trans)*
2842		CH_2 sym. str.
2890	CH_2 sym. str.	CH_2 antisym. str.
2935	CH_3 sym. str. and CH_2 antisym. str.	CH_3 sym. str.
2975	CH_3 asym. str.	CH_3 asym. str.
3017		CH olefinic str.

[a] Milanovich *et al.* (1976).
[b] G, *gauche;* T, *trans.*

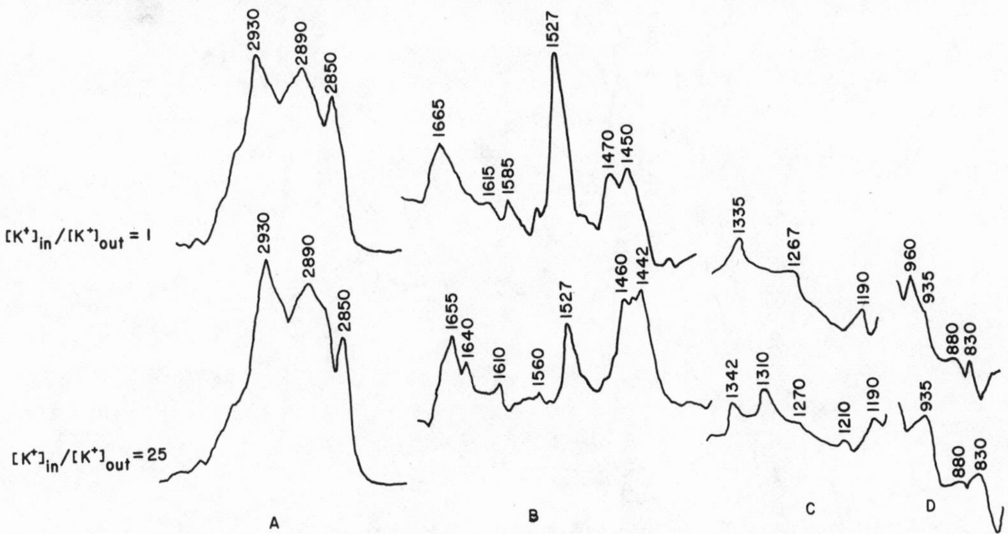

Fig. 11.23. Raman spectra of erythrocyte membrane vesicles without and with transmembrane [K^+] gradients. Intravesicular ionic composition: 10 m*M* Hepes/65 m*M* NaCl/75 m*M* KCl/0.25 m*M* $MgCl_2$. With cation gradients, the extracellular ionic composition is 140 m*M* NaCl/10 m*M* Hepes/0.25 m*M* $MgCl_2$ (pH 7.5). (A) 2800–3100 cm^{-1} in H_2O; (B) 1400–1700 cm^{-1} in ^2H_2O; (C) 1170–1360 cm^{-1} in H_2O (D) 750–1000 cm^{-1} in ^2H_2O. (Mikkelsen *et al.*, 1978.)

of the 960-cm^{-1} band remains. α-Helical polypeptides in D_2O display a band at ca. 930 cm^{-1}, so the changes mentioned here provide evidence for an increase in α-helix content (Frushour and Koenig, 1975).

Imposition of a membrane potential (negative inside) also increased the intensity of Raman bands in the CH_3-stretching region.

Mikkelsen *et al.* concluded that a transmembrane potential (negative inside) and/or a cation gradient can energize membranes by compression of the nonpolar region and transfer of methyl groups of protein into polar regions.

Neoplastically Transformed Lymphoid Cell Membrane

Differences between the structural dynamics of plasma membranes (PM) of normal hamster lymphocytes and lymphoid cells (GD248) that were neoplastically transformed by simian virus 40 (SV40) were revealed by Raman spectroscopy (Verma *et al.*, 1977) in the range of 100–3010 cm^{-1}. Plasma membranes highly purified from the SV40-transformed lymphocytes contain two types of protein that are lacking in the normal membranes, one with an isoelectric point of 4.5, and the other 4.7; and masses of ca. 53,000–58,000 and 90,000–110,000 daltons, respectively.

Verma *et al.* have observed marked differences between the normal and "transformed" membrane types in the thermal response of the C—H stretching (2990–3010 cm^{-1}) (Fig. 11.24) and the acoustical region (400–100 cm^{-1}) (Fig. 11.25). The former

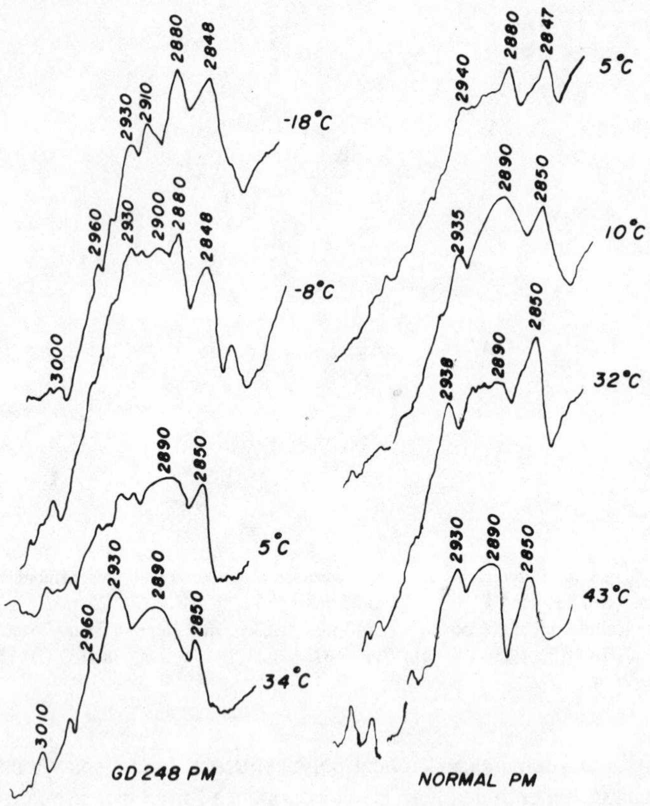

Fig. 11.24. Laser Raman spectra of PM from normal and GD248 cells in the CH-stretching region at various temperatures. (Verma *et al.*, 1977.)

region revealed that the membranes of normal cells display a thermal transition centered at 7°C and ca. 5°C wide. The membranes of GD248 cells, on the other hand, exhibit a lipid transition centered at −5°C and 12°–18°C wide. The acoustical region data produced equivalent results. The proteins of normal membranes display a large thermotropic transition, starting at 39°C (sample temperature). Contrasted with this is the 23°C beginning point for the transition with the GD248 plasma membranes. These workers concluded that the SV 40-specific membrane proteins may have modified the collective thermotropic activity of both normal membrane proteins and membrane lipids. Their laboratory has also used Raman spectroscopy to study highly purified thymocyte plasma membranes (Verma *et al.*, 1975). A plot of the ratio I_{2890}/I_{2850} versus temperature of membrane vesicles displays a discontinuity centered at ca. 21°C. This ratio has a value of 1.3 at 22°C and 2.2 at 10°C, and the discontinuity suggested to Verma *et al.* a lower degree of cooperativity than is observed for phosphatidylcholine liposomes. It may indicate a lipid phase transition or segregation in a region having low content of cholesterol.

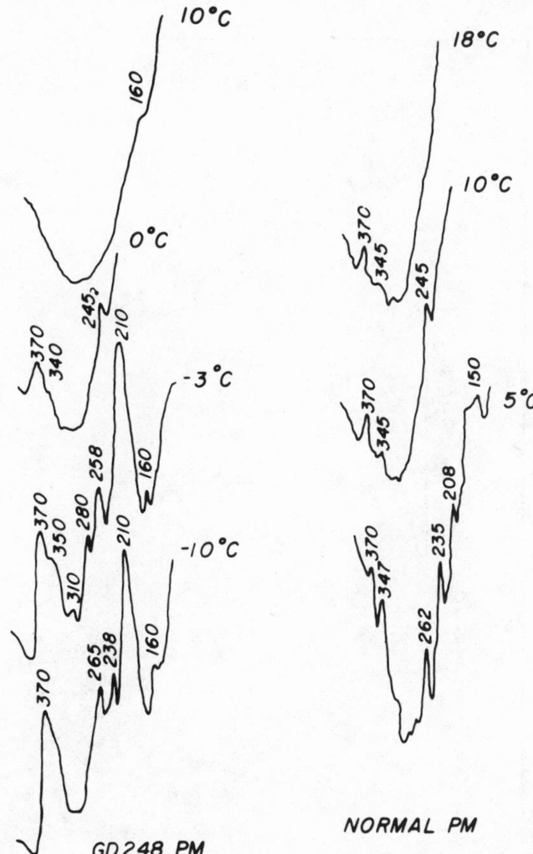

Fig. 11.25. Laser Raman spectra of PM from normal and GD248 PM in the 100 to 400 cm^{-1} region at various set temperatures. (Verma *et al.*, 1977.)

Opsin Membrane

The structure of opsin, the protein of the calf photoreceptor membrane, has been examined by normal (or nonresonance) Raman spectroscopy (Rothschild *et al.*, 1976). It is difficult to obtain normal Raman spectra of photoreceptor membranes because of the highly absorbing retinylidene moiety. Retinal is not easily extractable from the membrane without disruption of the structure, and its optical properties are such that it will completely dominate the Raman spectrum unless one tunes the exciting wavelength far from the chromophore absorption. A way to get around this obstacle is to reduce the chromophore enzymatically to retinol by bleaching with NADPH present, which process activates the retinal oxidoreductase, a protein of the photoreceptor membrane (Futterman, 1963; DePont *et al.*, 1970). The resulting opsin membrane is then isolated from the retinol, and one can record excellent Raman spectra from it.

Most of the bands in Fig. 11.26 arise from lipid and protein group vibrations (Table 11.9). DeGrip (1974) had proposed that opsin contains two S—S bonds, but no band

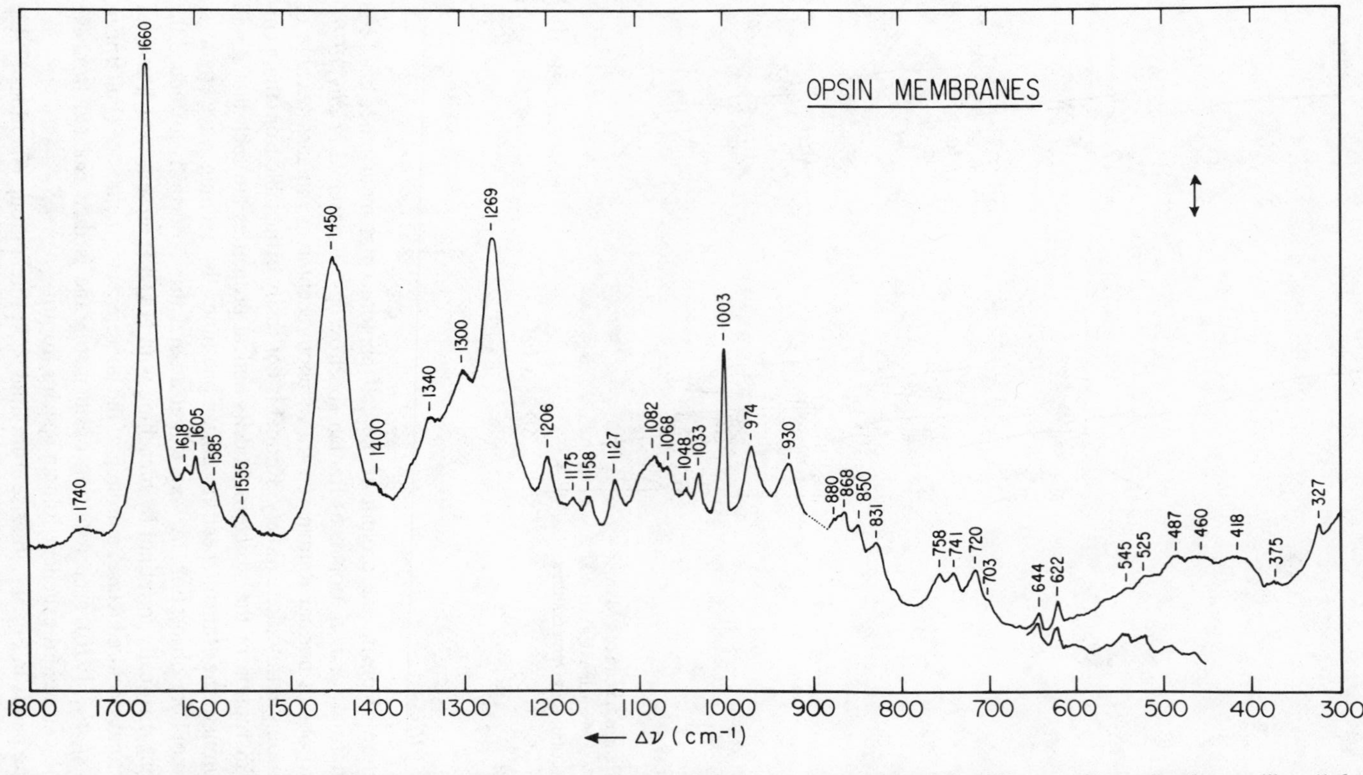

Fig. 11.26. Typical Raman spectrum of calf opsin membranes, recorded at 24°C. The pellet sample is held in a Kimax melting-point glass capillary (1.6 mm diameter) that is in a temperature-controlled sample holder. Spectra were measured with a SPEX *Ramalog* IV system and the 514.5-nm line of an Ar^+ laser (Spectra-Physics 164-03), with incident power 75 mW. Scanning speed was 3 cm^{-1}/min, slit width 200 μm, averaging time 100 sec, and the vertical arrow represents 100 counts/sec. (Rothschild *et al.*, 1976.) Other aspects of the system have been described (Rothschild and Stanley, 1975; Stanley *et al.*, 1975). Spectra were repeated from different membrane preparations, with different laser lines (457.9 and 488.0 nm), and various temperatures. Inset shows 450 to 650 cm^{-1} region recorded with less fluorescent background. A spurious grating ghost was found near 900 cm^{-1} (dotted line). The opsin membranes are estimated to contain 50% lipid and 40% protein by weight. (Daemen, 1973; DeGrip, 1974).

Table 11.9. Principal Peak Frequencies in the Raman Spectrum of Calf Opsin Membranes (Fig. 11.26)[a]

Peak frequency (cm^{-1})	Tentative assignment	Peak frequency (cm^{-1})	Tentative assignment
327		1068	Phospholipid: C—C stretch of *trans* hydrocarbon chain
375	Protein	1082	Phospholipid, O—P—O and C—C stretch of *gauche* hydrocarbon chain
418		1127	Protein, phospholipid: C—C stretch of *trans* hydrocarbon chain
460		1158	Protein, retinal
487		1175	Protein: Tyr, Phe
525	S—S stretch: *trans–gauche* form	1206	Protein: Tyr, Phe
545	S—S stretch: *trans–trans* form, Trp	1269	Phospholipid, protein (amide III)
622	Protein: Phe ring mode	1300	Phospholipid (CH_2), protein (amide III)
644	Protein: Tyr ring mode	1340	Phospholipid, protein: $\gamma(CH_2)$
703		1400	Protein: Asp, Glu $\nu(COO^-)$
720	Phospholipid: choline	1450	Protein, phospholipid: $\delta(CH_2)$
741	Phospholipid	1555	Protein, retinal (—C=C—C=C)
758	Protein	1585	Protein: Phe, Arg
831	Protein: Tyr (hydrogen-bonded OH)	1605	Protein: Tyr, Phe
850	Protein: Tyr (free OH)	1618	Protein: Tyr, Phe
868		1660	Protein (amide I), phospholipid (—C=C—)
880	Protein: Trp	1740	Phospholipid: ester carbonyl
930	Phospholipid	2570	Protein: SH stretch
974	Phospholipid		
1003	Protein: Phe		
1033	Protein: Phe		
1048	Phospholipid		

[a] Rothschild *et al.* (1976).

was observed by Rothschild *et al.* (1976) in the interval 520–500 cm^{-1} where the strong S—S vibration has been seen, e.g., 507 and 516 cm^{-1} in the spectra of lysozyme (Nakanishi *et al.*, 1974) and ribonuclease A (Chen and Lord, 1976). Sugeta *et al.* (1972) stated that the S—S stretching frequency will lie between 500 and 520 cm^{-1} only if the two C—S bonds are in the *gauche–gauche* conformation; will be $\cong$525 cm^{-1} if one C—S is in a *trans* form; and will be $\cong$ 540 cm^{-1} if both are *trans*. Rothschild *et al.* found weak bands at 525 and 540 cm^{-1}, and decided that either the opsin contains no S—S bonds or, if it does, the common *gauche–gauche* conformation is not present.

In opsin membranes $I_{850}/I_{831} < 1$, showing that a sizable fraction of tyrosine side chains is strongly hydrogen bonded (see p. 93). The H-bonded partners with tyrosine may be glutamic and aspartic residues, of which there are 37 and 28 residues, respectively, in opsin; there are 18 tyrosine residues.

The absence of bands below 1240 cm^{-1} shows that a very small amount of β-structure is present, since for antiparallel-pleated β-sheet the amide III band should lie between 1235 and 1229 cm^{-1}; for a random coil, between 1253 and 1243 cm^{-1}; and for an α-helix, between 1300 and 1265 cm^{-1} (see Chen and Lord, 1974). Extensive sections of the native opsin structure are α-helical. Henderson and Unwin's (1975) three-dimensional model for purple membrane shows a protein containing seven closely packed α-helical sections normal to the plane of the membrane.

The opsin membrane phospholipids are in a fluid state at room temperature. This deduction was based on earlier spectroscopic studies of lipids (Lippert *et al.*, 1975; Wallach and Verma, 1975; Verma *et al.*, 1975; Verma and Wallach, 1975) and the fact that the 1082-cm^{-1} band is stronger than the 1068-cm^{-1} band.

Photoreceptor Membrane

Rothschild *et al.* (1980) have used FTIR spectroscopy to study the structure of bovine photoreceptor membrane. They employed FTIR primarily because of its capability to obtain in a few seconds spectra of high signal-to-noise ratio. Figure 11.27A (1800–600 cm^{-1}) and 11.27B (3500–2700 cm^{-1}) present the spectrum of dehydrated membrane. Protein peptide groups appear as the amide II band at 1545 cm^{-1}, the amide I at 1675 cm^{-1}, and the amide A band at 3295 cm^{-1}. The lipid ester C═O stretch vibration lies at 1740 cm^{-1}. (The isopotential spin-dry method mentioned in Fig. 11.27 is discussed in Rothschild *et al.* (1980a,b) and Clark *et al.* (1978). It consists of slowly evaporating solvent from a suspension of membrane fragments while simultaneously ultracentrifuging the fragments onto an isopotential surface.) Table 11.10 lists other assignments.

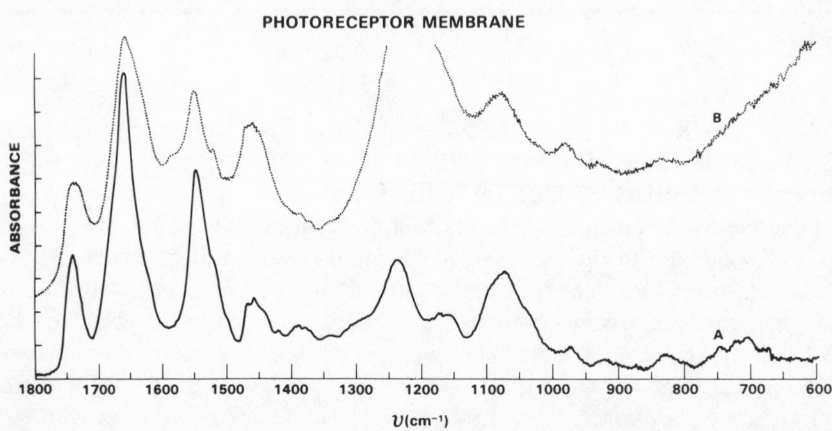

Fig. 11.27. Fourier transform infrared absorption spectrum of cattle photoreceptor membrane from 600 to 1800 cm^{-1} recorded at room temperature with spectral resolution of 4 cm^{-1}. Spectra were obtained with 100 scans of sample and 50 scans of reference beam (~2 min). (A) dehydrated sample prepared on AgCl windows by using the isopotential spin-dry method. Sample had an optical density at 500 nm of 0.5 and at 1657 cm^{-1} of 0.53. (B) sample suspended in ^2H_2O and sealed between 2 AgCl windows. Spectrum was recorded 10 min after mixing with ^2H_2O in a N_2-purged dry-box. The large peak near 1260 cm^{-1} is due to ^2H_2O absorption. (Rothschild *et al.*, 1980.)

Table 11.10. Tentative Assignment of Main Peaks of Dehydrated Photoreceptor Membrane[a]

Wavenumber (cm^{-1})	Assignment
3295	N—H stretch associated (CONH amide A)
3200	N—H stretch associated (CONH amide B)
3015	C—H stretch (in —C=C—)
2960	C—H stretch (CH_3, antisymmetric)
2925	C—H stretch (CH_2, antisymmetric)
2872	C—H stretch (CH_3, symmetric)
2855	C—H stretch (CH_2, symmetric)
1740	C=O stretch (in esters)
	Glu, Asp
1680	Arg guanidinium modes)
1657	C=O stretch (CONH amide D)
	C=C stretch (*cis* only)
1545	N—H bend (CONH amide II)
1520	Tyr (modes of *p*-disubstituted benzene ring)
1500	Lys (NH_3^+ deformation modes)
	Phe (mode of monosubstituted benzene ring)
1467 }	{ C—H scissor (in CH_2)
1457 }	{ C—H bend (CH_3, antisymmetric)
1445	C—H scissor (in CH_2)
	C—H bend (CH_3, antisymmetric)
	Ala
1420	C—H scissor (in CH_2 next to C=O)
1390	CO_2^- stretch (symmetric)
	C—H bend (CH_3 symmetric)
	Asp
1310	C—H wag (in CH_2)
	C—H in-plane deformation (in *trans* —C=C—)
1245	N—H bend (CONH amide III)
1237	P=O stretch (PO_2^-, antisymmetric)
	C—O—C stretch (antisymmetric)
1172	C—O—C stretch (symmetric)
1165	$(CH_3)_2$—C skeletal vibration
	Val, Leu
1157	C—C skeletal stretch
1112	N—H bend (?)
1095	P=O stretch (PO_2^-, symmetric)
1070	P—O—C stretch
1045	C—N stretch
1040	C—O stretch (in hydroxyl groups)
970	C—H bend (in *trans* —C=C—)
	C—C—N$^+$ stretch
955	
920	

Continued

Table 11.10. *(Continued)*

Wavenumber (cm^{-1})	Assignment
875	
825	
775	C—H rock (in CH_2)
742	
720	
700	N—H bend (out of plane; amide V)
	C—H bend (in *cis* —C═C—)

[a] Rothschild *et al.* (1980).

Extensive α-helical structure appears to be contained in rhodopsin and it lies mainly perpendicular to the plane of the membrane. Evidence for this orientation was found in the high ratio of $I_{\text{amide II}}$ to $I_{\text{amide I}}$ (between 0.7 and 1.0) for the highly oriented photoreceptor membrane used to produce spectra in Figs. 11.29, 11.28 and 11.27. Supporting far-UV CD data are given in Rothschild *et al.* (1980b). The content of β-conformation in rhodopsin appears to be quite low, as estimated by Rothschild *et al.* (1980) from the lack of both an amide I band near 1630 cm^{-1} and a weaker one near 1685 cm^{-1}. Deuteration

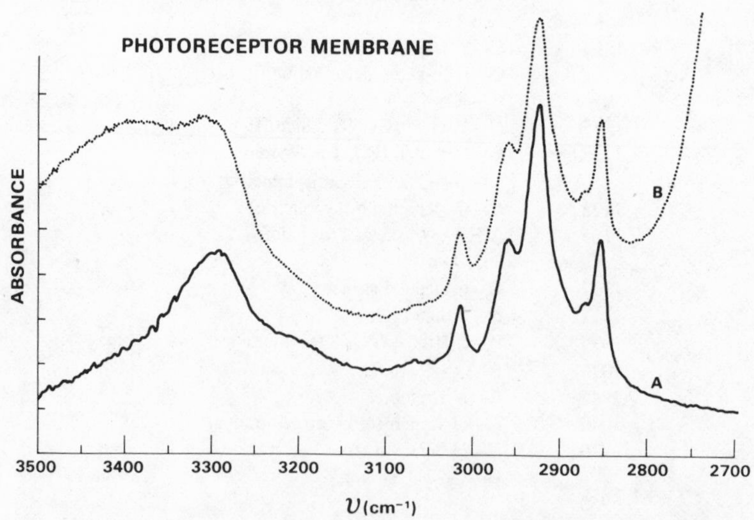

Fig. 11.28. Same as Fig. 11.27 from 2700 to 3500 cm^{-1}. (Rothschild *et al.*, 1980.)

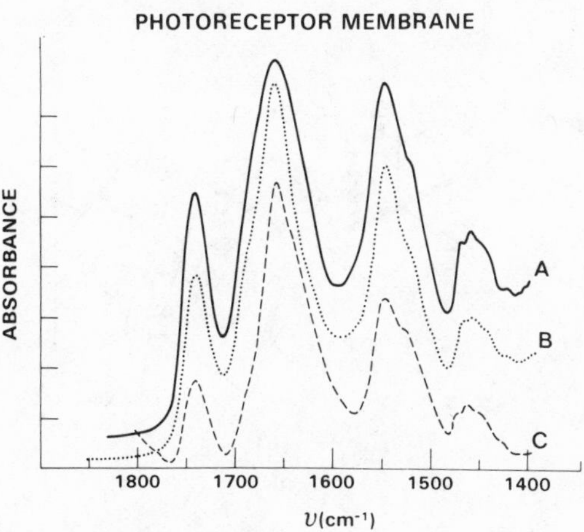

Fig. 11.29. Fourier transform infrared spectra of photoreceptor membrane prepared by using (A and B) the isopotential spin-dry technique and (C) by sedimentation followed by lyophilization. The highest amide II/amide I ratio (~1) is found for sample A. Recording conditions were the same as in Fig. 11.27. The frequency of the amide I and II bands are 1657 and 1545 cm^{-1}, respectively. (Rothschild *et al.*, 1980.)

measurements on bleached photoreceptor membrane exposed to ^2H_2O are consistent with an α-helical conformation for rhodopsin in accord with observations in a Raman scattering paper (Rothschild *et al.*, 1976) and an infrared one (Osborne and Nabedryk-Viala, 1978). Delipidation causes alteration of the rhodopsin structure, which is restored when dioleyl phosphatidylcholine is added to reconstitute delipidated rhodopsin and rhodopsin membranes.

Polarized FTIR and far-UV CD spectroscopy of oriented multilamellar films of photoreceptor membranes also show that the α-helices of rhodopsin are mainly aligned perpendicular to the bilayer plane (Rothschild *et al.*, 1980c). The assumptions made by these authors and the mathematics used for the IR analysis are in this paper.

Michel-Villaz *et al.* (1979) have shown that rhodopsin α-helices in retinal rod (frog) outer segment membranes have an average orientation in the transmembrane direction, and the oriented part of the protein lies in the lipid bilayer.

Purple Membrane (Halobacterium halobium)

Polarized Fourier transform IR spectroscopy has been used by Rothschild and Clark (1979) to study the structure of dried oriented films of the purple membrane (light-adapted form) from *Halobacterium halobium*. The purple membrane contains the protein bacteriorhodopsin, which consists largely of seven rods of high electron density that traverse the bilayer plane of the two-dimensional hexagonal pattern of the crystalline form of the protein (Blaurock and Stoeckenius, 1971; Stoeckenius, 1976; Henderson and Unwin, 1975). The rods are α-helices (Henderson, 1975; Blaurock, 1975).

The specimen was mounted in one beam on a single-axis goniometer, so that it could be tilted to various angles (α_0 = 0, 15, 30, 45, and 60°) around a horizontal axis which was perpendicular to the incident radiation (Fig. 11.30a). A polarizer was inserted in the

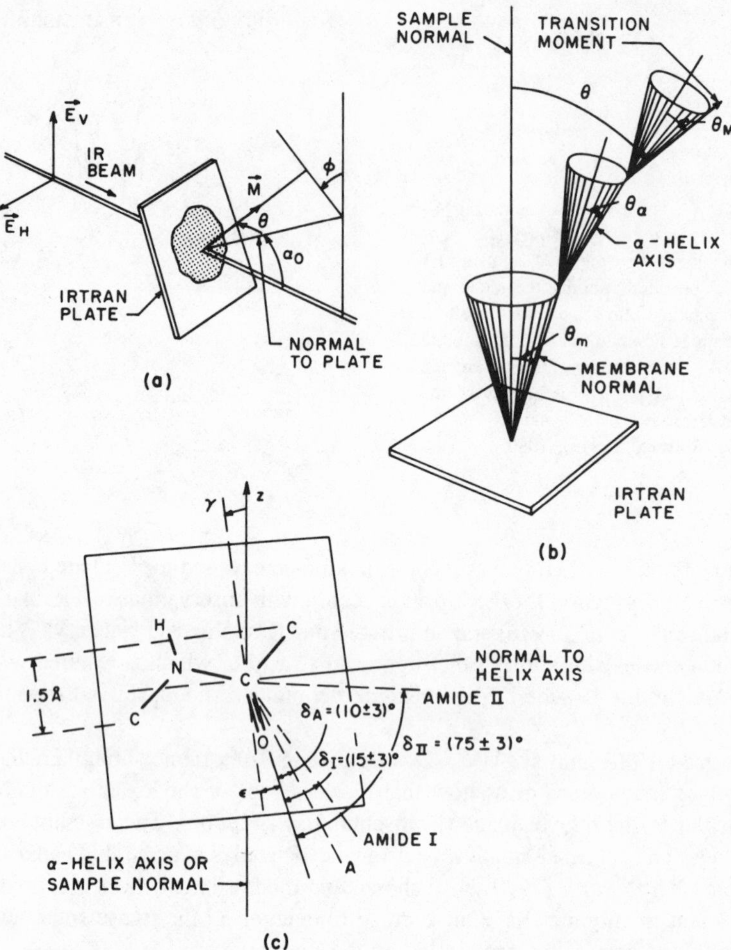

Fig. 11.30. (a) Experimental configuration for polarized Fourier transform IR absorption measurements of purple membrane oriented on Irtran plates. (b) Geometry indicating a set of nested axially symmetric distributions. The moment, M, is distributed about the α-helix axis, which is distributed about the membrane normal that has a mosaic spread distribution about the sample normal. In the case of the purple membrane, the distributions are assumed to be axially symmetric. (c) Peptide plane geometry assumed to be the basic absorbing unit. (Rothschild and Clark, 1979.)

beam before the specimen. Spectra were recorded at the various α_0 values for the horizontal setting of the polarizer, where the incident electric vector was parallel to the plane of the Irtran-4 slide (0.5 mm thickness) for all values of α_0, and in the vertical setting, where the incident electric vector made an angle α_0 with the plane of the slide. Operational procedures were used that eliminated artifacts due to the wavelength-dependent transmission of the polarizer and Irtran in the polarized incident radiation.

The spectra of purple membrane are displayed in Figs. 11.31 and 11.32 for α_0 = 0, 15, 30, 45, and 60° with the polarizer in the vertical and horizontal positions (the latter labeled "H"). The major bands were assigned to protein (bacteriorhodopsin) or membrane lipid vibrations (Table 11.11).

In the case of horizontal polarization, the amide I and II bands display at all goniometer angles a high ratio of the intensities of amide II/amide I of 0.98. Proteins generally give a more typical value of 0.45 (Miyazawa and Blout, 1961). Figures 11.31 and 11.32 (a–e) show that when the polarization is vertical, the amide II/amide I ratio varies from 0.98 when α_0 is 0° to 0.51 when α_0 is 60°. Also, the amide I band moves from 1662 cm^{-1} (α_0 = 0°) to 1667 cm^{-1} as α_0 nears 60°. An amide A band at ca. 3300 cm^{-1} (α_0 = 0°) and 3310 cm^{-1} (α_0 = 60°) increases in absorbance with increase in α_0 (see Fig. 11.32). Rothschild and Clark (1979) attributed these effects to an anisotropic orientational distribution of the transition moments in the purple membrane. In this case the vectors of the amide A, I, and II vibrations make the relative band intensities dependent

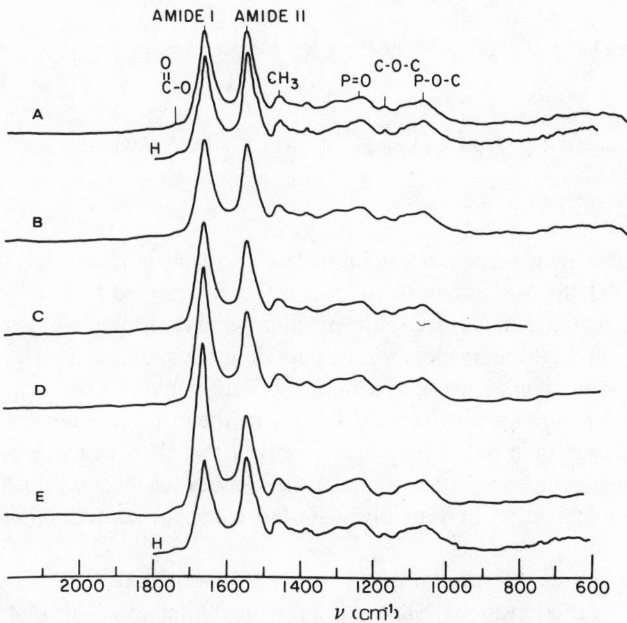

Fig. 11.31. Dichroic IR absorption of dried purple membrane on Irtran-4 slide from 600 to 2200 cm^{-1}. Vertical (absorbance), horizontal axis (wavenumbers, cm^{-1}). Spectra were recorded with 100 scans of sample and 50 scans of reference beam at a resolution of 4 cm^{-1}. Optical densities ranged from 0 to a maximum of 1. A polarizer was placed before the sample in either a vertical or horizontal polarization (denoted *H*) and the sample tilted on a single axis goniometer at an angle α_0, relative to the vertical polarization. Spectra were obtained for α_0 = 0, 15, 30, 45, and 60°, as shown in Fig. 11.31 A–E, respectively. All spectra were made by subtracting the absorption of the polarizer and blank slide. Vertical spectra were smoothed by using a nine-point Lorentzian smoothing factor. (Rothschild and Clark, 1979.)

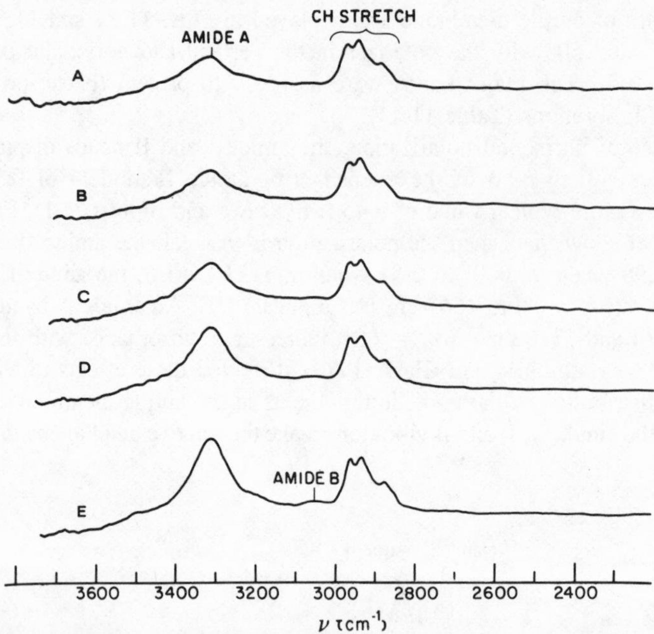

Fig. 11.32. Same as Fig. 11.31. Spectra from 2200 to 3800 cm^{-1}. (Rothschild and Clark, 1979.)

on the angle that the incident electric vector makes with each individual transition moment. (See Tsuboi, 1962, for discussion of such an effect in oriented films of α-helical poly-γ-benzyl-L-glutamate). Having derived a function which relates the observed dichroism to the orientational order parameters for the peptide groups, helical axis distribution, and mosaic spread of the membranes, Rothschild and Clark found that the average orientation of the α-helices is in a range less than 26° away from the membrane normal, in agreement with electron microscopic data. They also observed that the frequency of the amide A and amide I bands is at least 10 cm^{-1} greater than values for most α-helical proteins and polypeptides, an indication of possibly distorted α-helical conformations in bacteriorhodopsin.

Agreement among IR spectral data and several predictions have indicated that the helices in bacteriorhodopsin have the α_{II} structure rather than the more common α_I structure (Krimm and Dwivedi, 1982). The α_{II} structure has the plane of the peptide group tilted with respect to the helix axis with C═O projecting outward, and it may result from additional H-bond stabilization between side chain threonine and serine OH groups and main chain peptide groups. These workers believe that the shorter distance between adjacent amide hydrogens in the α_{II} structure offers the possibility that the helix of NH hydrogens, rather than side chain residues, may be the "proton wire" for translocating protons across the membrane.

Table 11.11. Frequency of Peaks in IR Absorption Spectrum of Purple Membrane from 500 to 3600 cm^{-1}[a,b]

$\nu(cm^{-1})$	Assignment	Dichroism
577	Amide VI (out-of-plane deformation)	
590(s)		∥
605–665(B)		
610		
700	Phenyl B_2	
740	CH (rocking), amide V (?)	
825(B)		∥
937(w)		∥
975(w)	C—O stretching (perpendicular)	⊥
1062	P—O—C stretching	
1100		
1125		
1167	Ester carbonyl stretch, CH_3	
1235	P═O stretch	
1300		
1367(s)	CH_2 wag	
1382		
1415(w)	CO_2 stretch (Glu, Asp)	
1457	CH_3 bending modes	
1515(s)	Tyrosine	
1547	Amide II	⊥
1661	Amide I (α_0 = 0°)	⊥
1667	Amide I (α_0 = 60°)	∥
1725(w)	C(═O)—OH (Glu, Asp)	
1740	Ester carbonyl stretch	
2850(s)	CH stretching modes	
2870	CH stretching modes	
2,929	CH stretching modes	
2958	CH stretching modes	
3060	Amide B	∥
3310	Amide A	∥

[a] Rothschild and Clark (1979).
[b] The type of dichroism found is indicated by ∥ (absorption greatest at α_0 = 90°), ⊥ (absorption greatest at α_0 = 0°). Other symbols referring to the appearance of peaks are s (shoulder), w (weak), and B (broad). Peak frequencies were measured directly from spectra and found to vary no more than ± 1 cm^{-1} for different samples.

References

Akutsu, H., and Kyogoku, Y., *Chem. Phys. Lipids* **14,** 113 (1975).

Akutsu, H., Kyogoku, Y., Nakahara, H., and Fukuda, K., *Chem. Phys. Lipids* **15,** 222 (1975).

Ambrose, E. J., and Elliott, A., *Proc. R. Soc. (London)* **A205,** 47 (1951).

Andersen, H. C., *Annu. Rev. Biochem.* **47,** 359 (1978).

Barton, P. G., and Gunstone, F. G., *J. Biol. Chem.* **250,** 4470 (1975).

Baumeister, W., Hahn, M., and Fringeli, U. P., *Z. Naturforsch.* **31c,** 746 (1976a).

Baumeister, W., Fringeli, U. P., Hahn, M., and Seredynski, J., *Biochim. Biophys. Acta* **453,** 289 (1976b).

Baumeister, W., Fringeli, U. P., Hahn, M., Kopp, F., and Seredynski, J., *Biophys. J.* **16,** 791 (1976c).

Blaurock, A. E., *J. Mol. Biol.* **93,** 139 (1975).

Blaurock, A. E., and Stoeckenius, W., *Nature New Biology* **233,** 152 (1971).

Blout, E. R., de Lozé, C., and Asadourian, A., *J. Am. Chem. Soc.* **83,** 1895 (1961).

Boggs, J. M., Vail, W. J., and Moscarello, M. A. *Biochim. Biophys. Acta* **448,** 517 (1976).

Borowski, E, Zielinski, J., Fakowski, L., Ziminski, T., Golik, J., Kotodziejczyk, E., Jeresczek, E., and Gdulewicz, M., *Tetrahedron Lett.* **1971,** 685.

Brown, K. G., Peticolas, W. L., and Brown, E., *Biochem. Biophys. Res. Commun.* **54,** 358 (1973).

Bulkin, B. J., *Biochim. Biophys. Acta* **274,** 649 (1972).

Bulkin, B. J., and Krishnamachari, N., *J. Am. Chem. Soc.* **94,** 1109 (1972).

Bulkin, B. J., and Yellin, N., *J. Phys. Chem.* **82,** 821 (1978).

Bunow, M. R., and Levin, I. W., *Biochim. Biophys. Acta* **489,** 191 (1977b).

Bunow, M. R., and Levin, I. W., *Biochim. Biophys. Acta* **464,** 202 (1977a).

Cameron, D. G., and Mantsch, H. H., *Biochem. Biophys. Res. Commun.* **83,** 886 (1978).

Cameron, D. G., Casal, H. L., and Mantsch, H. H., *J. Biochem. Biophys. Methods* **1,** 21 (1979).

Casal, H. L., Smith I. C. P., Cameron, D. G., and Mantsch, H. H., *Biochim. Biophys. Acta* **550,** 145 (1979).

Chapman, D., *Biological Membranes* (Chapman, D. and Wallach, D. F. H., eds.), Vol. 2, p. 91, Academic Press, New York, 1973.

Chapman, D., *The Structure of Lipids,* John Wiley and Sons, New York, 1965.

Chapman, D., Williams, R. M., and Ladbrooke, B. D., *Chem. Phys. Lipids* **1,** 445 (1967).

Chapman, D., Cornell, B. A., Eliasz, A. W., and Perry, A., *J. Mol. Biol.* **113,** 517 (1977).

Chapman, D., Kamat, V. B., and Levene, R. J., *Science* **160,** 314 (1968).

Chappell, J. B., and Crofts, A. R., *Biochem. J.* **95,** 393 (1965).

Chen, M. C., and Lord, R. C., *Biochemistry* **15,** 1889 (1976).

Chen, M. C., and Lord, R. C., *J. Am. Chem. Soc.* **96,** 4750 (1974).

Cherry, R. J., and Morrison, A., *Proc. R. Soc. London, Ser. A* **290,** 115 (1966).

Clark, N. A., Luppold, D., Simons L., and Rothschild, K. J., *Bull. Am. Phys. Soc.* **23,** 238 (1978).

Curatolo, W., Sakura, J. D., Small, D. M., and Shipley, G. G., *Biochemistry* **16,** 2313 (1977).

Daemen, F. J. M., *Biochim. Biophys. Acta* **300,** 255 (1973).

Dayhoff, M. O., ed., *Atlas of Protein Sequence and Structure,* Vol. 5, D-223, National Biomedical Research Foundation, Washington, D.C., 1972.

DeGrip, W. J., Thesis, University of Nijmegen, Netherlands, 1974, quoted by Rothschild *et al.* (1976).

DePont, J. J. H. H. M., Daemen, F. J. M., and Bonting, S. L., *Arch. Biochem. Biophys.* **140,** 275 (1970).

Dunker, A. K., Williams, R. W., Gaber, B. P., and Peticolas, W. L., *Biochim. Biophys. Acta* **553,** 351 (1979).

Epand, R. M., Jones, A. J. S., and Sayer, B., *Biochemistry* **16,** 4360 (1977).

Faiman, R., *Chem. Phys. Lipids* **23,** 77 (1979).

Fookson, J. E., and Wallach, D. F. H., *Arch. Biochem. Biophys.* **189,** 195 (1978).

Fraser, R. D. B., and MacRae, T. P., *Conformation in Fibrous Proteins and Related Synthetic Polypeptides,* Academic Press, New York, 1973.

Fraser, R. D. B., and Suzuki, E., in *The Physical Principles and Techniques of Protein Chemistry,* Vol. B (S. J. Leach, ed.) p. 213, Academic Press, New York, 1970.

Fringeli, U. P., *Biophys. J.,* **34,** 173 (1981).

Fringeli, U. P., *Z. Naturforsch.* **32c,** 20 (1977).

Fringeli, U. P., and Günthard, H. H., *Biochim. Biophys. Acta* **450,** 101 (1976).
Fringeli, U. P., and Günthard, Hs. H., "Infrared Membrane Spectroscopy," in *Membrane Spectroscopy* (E. Grell, ed.), pp. 270–332, Springer-Verlag, Berlin, 1981.
Fringeli, U. P., and Hofer, P., *Neurochemistry International* **2,** 185 (1980).
Fringeli, U. P., Müldner, H. G., Günthard, H. H., Gasche, W., and Leuzinger, W., *Z. Naturforsch.* **27b,** 780 (1972).
Fromherz, P., Peters, J., Müldner, H. G., and Otting, W., *Biochim. Biophys. Acta* **274,** 644 (1972).
Frushour, B. G., and Koenig, J. L., in *Advances in Infrared and Raman Spectroscopy,* Vol. 1 (R. J. H. Clark and R. E. Hester, eds.) pp. 35–97, Heyden, London, 1975.
Futterman, S., *J. Biol. Chem.* **238,** 1145 (1963).
Gaber, B. P., and Peticolas, W. L., *Biochim. Biophys. Acta* **465,** 260 (1977).
Gaber, B. P., Yager, P., and Peticolas, W. L., *Biophys. J.* **21,** 161 (1978a).
Gaber, B. P., Yager, P., and Peticolas, W. L., *Biophys. J.* **22,** 191 (1978b).
Ganis, P., Avitabile, G., Mechlinski, W., and Schaffner, C. P., *J. Am. Chem. Soc.* **93,** 4560 (1971).
Gilman, T., Pownall, H., and Kauffman, J. W., *Fed. Proc.* **39,** (6), 1765 (1980).
Goheen, S. C., Gilman, T. H., Kauffman, J. W., and Garvin, J. E., *Biochem. Biophys. Res. Commun.* **79,** 805 (1977).
Golding, B. T., Rickards, R. W., Meyer, W. E., Patrick, J. B., and Barber, M., *Tetrahedron Lett.* **1966,** 3551.
Grant, C. W. M., and McConnell, H. M., *Proc. Natl. Acad. Sci. USA* **71,** 4653 (1974).
Gulik-Krzywicki, T., *Biochim. Biophys. Acta* **415,** 1 (1975).
Harold, F. M., and Baarda, J. R., *J. Bacteriol.* **94,** 53 (1967).
Harris, E. J., and Pressman, B. C., *Nature* **216,** 918 (1967).
Heitz, F., Lotz, B., and Spach, G., *J. Mol. Biol.* **92,** 1 (1975).
Henderson, R., *J. Mol. Biol.* **93,** 123 (1975).
Henderson, R., and Unwin, P. N. T., *Nature* **257,** 28 (1975).
Hladky, S. B., and Haydon, D. A., *Biochim. Biophys. Acta* **274,** 294 (1972).
Iqbal, Z., and Weidekamm, E., *Biochim. Biophys. Acta* **555,** 426 (1979).
Jackson, M. B., *Biochemistry* **15,** 2555 (1976).
Jacobs, R. E., Hudson, B., and Andersen, H. C., *Proc. Natl. Acad. Sci. USA* **72,** 3993 (1975).
Jones, A. J. S., Epand, R. M., Lin, K. F., Walton, D., and Vail, W. J., *Biochemistry* **17,** 2301 (1978).
Karvaly, B., and Loshchilova, E., *Biochim. Biophys. Acta* **470,** 492 (1977).
Koenig, J. L., *J. Polymer Sci. Part D* **6,** 59 (1972).
Kopp, F., Fringeli, U. P., Mühlethaler, K., and Günthard, H. H., *Biophys. Struct. Mechanism* **1,** 75 (1975).
Krasne, S., Eisenman, G., and Szabo, G., *Science* **174,** 412 (1971).
Krimm, S., *J. Mol. Biol.* **4,** 528 (1962).
Krimm, S., and Dwivedi, A. M., Abstracts of the 26th Annual Meeting, *Biophys. J.* **37,** No. 2 Part 2, February (1982).
Lapides, J. R., and Levin, I. W., Abstracts of the 26th Annual Meeting, *Biophys. J.* **37,** No. 2 Part 2, 203a, February (1982).
Larsson, K., *Chem. Phys. Lipids* **10,** 165 (1973).
Larsson, K., and Rand, R. P., *Biochim. Biophys. Acta* **326,** 245 (1973).
Lee, A. G., *Progr. Biophys. Mol. Biol.* **29,** 3 (1975).
Levine, Y. K., *Progr. Biophys. Mol. Biol.* **24,** 1 (1972).
Lippert, J. L., and Peticolas, W. L., *Biochim. Biophys. Acta* **282,** 8 (1972).
Lippert, J. L., and Peticolas, W. L., *Proc. Natl. Acad. Sci. USA* **68,** 1572 (1971).
Lippert, J. L., Gorczyca, L. E., and Meiklejohn, G., *Biochim. Biophys. Acta* **382,** 51 (1975).
Lippert, J. L., Tyminski, D., and Desmeules, P. J., *J. Am. Chem. Soc.* **98,** 7075 (1976).
Lis, L. J., Kauffman, J. W., and Shriver, D. F., *Biochim. Biophys. Acta* **406,** 453 (1975).
Lis, L. J., Goheen, S. C., Kauffman, J. W., and Shriver, D. F., *Biochim. Biophys. Acta* **443,** 331 (1976).
Lord, R. C., and Mendelsohn, R., in *Membrane Spectroscopy,* (E. Grell, ed.), pp. 377–436, Springer-Verlag, Berlin, 1981.
Lord, R. C., and Yu, N.-T., *J. Mol. Biol.* **50,** 509 (1970).
MacDonald, R. I., and MacDonald, R. C., *J. Biol. Chem.* **250,** 9206 (1975).

Marchesi, V. T., Furthmayr, H., and Tomita, M., *Annu. Rev. Biochem.* **45,** 667 (1976).
Mendelsohn, R., *Biochim. Biophys. Acta* **290,** 15 (1972).
Mendelsohn, R., and Maisano, J., *Biochim. Biophys. Acta* **506,** 192 (1978).
Mendelsohn, R., and Taraschi, T., *Biochemistry* **17,** 3944 (1978).
Mendelsohn, R., Sunder, S., and Bernstein, H. J., *Biochim. Biophys. Acta* **419,** 563 (1976a).
Mendelsohn, R., Sunder, S., and Bernstein, H. J., *Biochim. Biophys. Acta* **443,** 613 (1976b).
Michel-Villaz, M., Saibil, H. R., and Chabre, M., *Proc. Natl. Acad. Sci. USA* **76,** 4405 (1979).
Mikkelsen, R. B., Verma, S. P., and Wallach, D. F. H., *Proc. Natl. Acad. Sci. USA* **75,** 5478 (1978).
Milanovich, F. P., Shore, B., Harney, R. C., and Tu, A. T., *Chem. Phys. Lipids* **17,** 79 (1976).
Miyazawa, T., and Blout, E. R., *J. Am. Chem. Soc.* **83,** 712 (1961).
Mueller, P., and Rudin, D. O., *Biochem. Biophys. Res. Commun.* **26,** 398 (1967).
Myers, V. B., and Haydon, D. A., *Biochim. Biophys. Acta* **274,** 313 (1972).
Nakanishi, M., Takesada, H., and Tsuboi, M., *J. Mol. Biol.* **89,** 241 (1974).
Osborne, H. B., and Nabedryk-Viala, E., *Eur. J. Biochem.* **89,** 81 (1978).
Parker, F. S., *Applications of Infrared Spectroscopy in Biochemistry, Biology, and Medicine,* Plenum Press, New York, 1971.
Papahadjopoulos, D., Moscarello, M., Eylar, E. H., and Isac, T., *Biochim. Biophys. Acta* **401,** 317 (1975).
Pink, D. A., Green, T. J., and Chapman, D., *Biochemistry* **19,** 349 (1980).
Powers, L., and Pershan, P. S., *Biophys. J.* **20,** 137 (1977).
Rothman, J. E., *J. Theor. Biol.* **38,** 1 (1973).
Rothschild, K. J., and Clark, N. A., *Biophys. J.* **25,** 473 (1979).
Rothschild, K. J., and Stanley, H. E., *Am. J. Clin. Pathol.* **63,** 695 (1975).
Rothschild, K. J., DeGrip, W. J., and Sanches, R., *Biochim. Biophys. Acta* **596,** 338 (1980).
Rothschild, K. J., Rosen, K. M., and Clark, N. A., *Biophys. J.* **31,** 45 (1980a).
Rothschild, K. J., Clark, N. A., Rosen, K. M., Sanches, R., and Hsiao, T. L., *Biochem. Biophys. Res. Commun.* **92,** 1266 (1980b).
Rothschild, K. J., Sanches, R., Hsiao, T. L., and Clark, N. A., *Biophys. J.* **31,** 53 (1980c).
Rothschild, K. J., Andrew, J. R., DeGrip, W. J., and Stanley, H. E., *Science* **191,** 1176 (1976).
Sarges, R. and Witkop, B., *J. Am. Chem. Soc.* **87,** 2011 (1965).
Schachtschneider, J. H., and Snyder, R. G., *Spectrochim. Acta* **19,** 117 (1963).
Sherebrin, M. H., MacClement, B. A. E., and Franko, A. J., *Biophys. J.* **12,** 977 (1972).
Sherman, G., and Folch-Pi, J., *J. Neurochem.* **17,** 597 (1970).
Shimanouchi, T., *Kagaku no Ryoiki* **25,** 97 (1971).
Siamwiza, M. N., Lord, R. C., Chen, M. C., Takamatsu, T., Harada, I., Matsuura, H. and Shimanouchi, T., *Biochemistry* **14,** 4870 (1975).
Simons, L., Bergström, G., Blomfeldt, G., Forss, S., Stenbäck, H., and Wansen, G., *Comment. Phys.-Math.* **42,** 125 (1972).
Snyder, R. G., *J. Chem. Phys.* **47,** 1316 (1967).
Snyder, R. G., Cameron, D. G., Casal, H. L., Compton, D. A. C., and Mantsch, H. H., *Biochim. Biophys. Acta* **684,** 111 (1982).
Spiker, R. C., Jr., and Levin, I. W., *Biochim. Biophys. Acta* **388,** 361 (1975).
Spiker, R. C., Jr., and Levin, I. W., *Biochim. Biophys. Acta* **455,** 560 (1976).
Spiker, R. C., Jr., Pinnavaia, T. J., and Levin, I. W., *Biochim. Biophys. Acta* **455,** 588 (1976).
Stanley, H. E., Asher, I. M., Rothschild, K. J., Phillies, G. D. J., Carew, E. B., Bansil, R., and Michaels, I. A., in *Peptides: Chemistry, Structure and Biology* (R. Walter and J. Meienhofer, eds.), pp. 227–245, Ann Arbor Science Publications, Ann Arbor, Michigan 1975.
Stoeckenius, W., *Sci. Am.* **234,** (6), 38 (1976).
Sugeta. H., Go, A., and Miyazawa, T., *Chem. Lett.* (1972), 83.
Sunder, S., Cameron, D., Mantsch, H. H., and Bernstein, H. J., *Can. J. Chem.* **56,** 2121 (1978b).
Sunder, S., Bernstein, H. J., and Paltauf, F., *Chem. Phys. Lipids* **22,** 279 (1978a).
Susi, H., in *Structure and Stability of Biological Macromolecules* (S. N. Timasheff and G. D. Fasman, eds.) p. 575, Marcel Dekker, New York, 1969.
Susi, H., Sampugna, J., Hampson, J. W., and Ard, J. S., *Biochemistry* **18,** 297 (1979).
Susi, H., Timasheff, S. N., and Stevens, L., *J. Biol. Chem.* **242,** 5460 (1967).

Takenaka, T., Nogami, K., Gotoh, H., and Gotoh, R., *J. Coll. Interf. Sci.* **35,** 395 (1971).
Taraschi, T., and Mendelsohn, R., *J. Am. Chem. Soc.* **101,** 1050 (1979).
Taraschi, T., and Mendelsohn, R., *Proc. Natl. Acad. Sci. USA* **77,** 2362 (1980).
Tasumi, M., Shimanouchi, T., and Miyazawa, T., *J. Mol. Spectrosc.* **9,** 261 (1962).
Timasheff, S. N., Susi, H., and Stevens, L., *J. Biol. Chem.* **242,** 5467 (1967).
Traüble, H., *Biomembranes* **3,** 197 (1972).
Tsuboi, M., *J. Polymer Sci.,* **59,** 139 (1962).
van Zoelen, E. J. J., van Dijck, P. W. M., de Kruijff, B., Verkleij, A. J., and van Deenen, L. L. M., *Biochim. Biophys. Acta* **514,** 9 (1978).
Veatch, W. R., Fossel, E. T., and Blout, E. R., *Biochemistry* **13,** 5249 (1974).
Vergoten, G., Fleury, G., and Moschetto, Y., in *Advances in Infrared and Raman Spectroscopy,* Vol. 4 (R. J. H. Clark and R. E. Hester, eds.), p. 195, Heyden and Son, Ltd., London, 1978.
Verma, S. P., and Wallach, D. F. H., *Biochim. Biophys. Acta* **330,** 122 (1973).
Verma, S. P., and Wallach, D. F. H., *Biochim. Biophys. Acta* **401,** 168 (1975).
Verma, S. P., and Wallach, D. F. H., *Proc. Natl. Acad. Sci. USA* **73,** 3558 (1976).
Verma, S. P., and Wallach, D. F. H., Abstracts of the 26th Annual Meeting, *Biophys. J.* **37,** No. 2 Part 2, 155a, February (1982).
Verma, S. P., Wallach, D. F. H., and Smith, I. C. P., *Biochim. Biophys. Acta* **345,** 129 (1974).
Verma, S. P., Wallach, D. F. H., Schmidt-Ullrich, R., *Biochim. Biophys. Acta* **394,** 633 (1975).
Verma, S. P., Schmidt-Ullrich, R., Thompson, W. S., and Wallach, D. F. H., *Cancer Res.* **37,** 3490 (1977).
Wallach, D. F. H., *Chem. Phys. Lipids* **8,** 347 (1972).
Wallach, D. F. H., and Verma, S. P., *Biochim. Biophys. Acta* **382,** 542 (1975).
Wallach, D. F. H., and Zahler, P. H., *Biochim. Biophys. Acta* **150,** 186 (1968).
Wallach, D. F. H., Verma, S. P., and Fookson, J., *Biochim. Biophys. Acta* **559,** 153 (1979).
Weidekamm, E., Bamberg, E., Brdiczka, D., Wildermuth, G., Macco, F., Lehmann, W., and Weber, R., *Biochim. Biophys. Acta* **464,** 442 (1977).
Weidekamm, E., Bamberg, E., Janko, K., and Weber, R., *Arch. Biochem. Biophys.* **187,** 339 (1978).
Wickner, W., *Proc. Natl. Acad. Sci. USA* **73,** 1159 (1976).
Williams, R. W., and Dunker, A. K., *J. Biol. Chem.* **252,** 6253 (1977).
Yellin, N., and Levin, I. W., *Biochemistry* **16,** 642 (1977).
Yellin, N., and Levin, I. W., *Biochim. Biophys. Acta* **489,** 177 (1977a).
Yellin, N., and Levin, I. W., *Biochim. Biophys. Acta* **468,** 490 (1977b).

Chapter 12

CAROTENOIDS AND POLYENE AGGREGATES

Introduction

An active area of investigation in photobiology, photosynthesis, and spectroscopy is the transfer of electronic excitation energy from one molecule to another. The exciton model can be used to describe such an energy transfer, e.g., in the case of aggregates of molecules that are connected by weak intermolecular forces (Kasha, 1963; McRae and Kasha, 1964; Hochstrasser and Kasha, 1964; Kasha *et al.*, 1965). Various investigators have used the molecular exciton model to interpret spectra of aggregates of carotenoid aldehydes, a carotenoid complex in *Rhodospirillum rubrum*, dyes, retinal, retinyl polyenes, and lycopene (see Salares *et al.*, 1977). Some of the work discussed in this section shows that the combined use of absorption and resonance Raman spectra permits the separation of electronic ground and excited state effects and furnishes conditions for testing the exciton model.

Gill *et al.* (1970) had hinted at future developments in the use of resonance Raman spectroscopy for biochemical applications when they presented spectra of live carrot and tomato, lycopene, and β-carotene (Fig. 12.1). The reader is also referred to Parker (1975) for a review of some of the earlier work in resonance Raman applications to biochemistry. (See also other sections of this book for resonance Raman studies on rhodopsin, bacteriorhodopsin, heme proteins and porphyrins, chlorophylls, and copper- and other metalloproteins).

Polyene Aggregates of Carotenoids

Salares *et al.* (1977) have studied excited-state effects in model polyene aggregates of the carotenoids astaxanthin, lutein, zeaxanthin, cryptoxanthin, canthaxanthin, and echinenone (Fig. 12.2). These compounds form high-molecular-weight aggregates in aqueous solvent systems. The absorption spectra of the monomer molecules differ from those of the aggregates, the latter showing significant bandwidth narrowing and blue

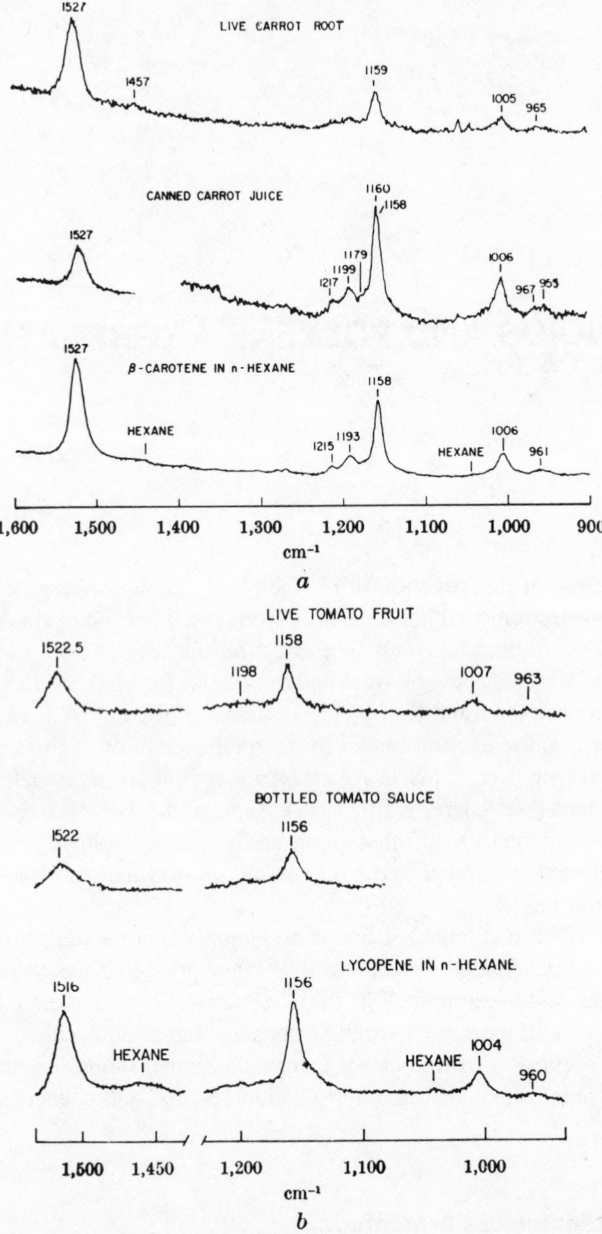

Fig. 12.1. Vibrational fundamentals in the resonance-enhanced Raman spectra of: a, β-carotene in live carrot root, in canned carrot juice, and the *n*-hexane solution of the pure all-*trans* pigment (Sigma) (excited at 488 nm); b, lycopene in live tomato fruit, bottled tomato sauce (the 1527 line of the canned carrot juice spectrum was taken with the gain reduced by a factor of 2), and solutions of lycopene extracted from bottled tomato sauce in an *n*-hexane solution (excited at 514.5 nm). Excitation by argon ion laser, incident power ≈ 100 mW. Grazing angle reflection geometry: the electric vectors of incident and of analyzed scattered light are both perpendicular to the scattering plane. Spectrometer: 75 cm Czerny-Turner double monochromator, digital photon-counting detection, integration time 1 sec, scanning speed 20 cm^{-1}/min. (Gill *et al.*, 1970.)

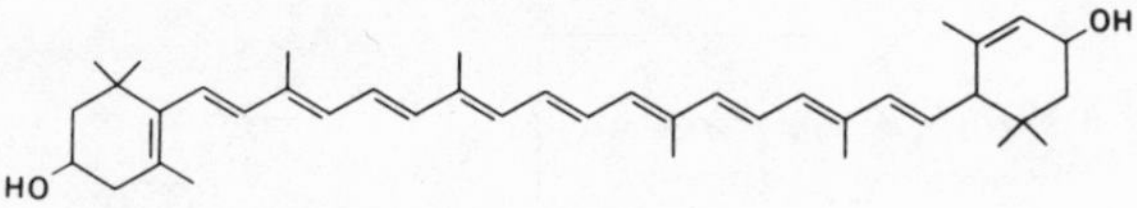

Lutein

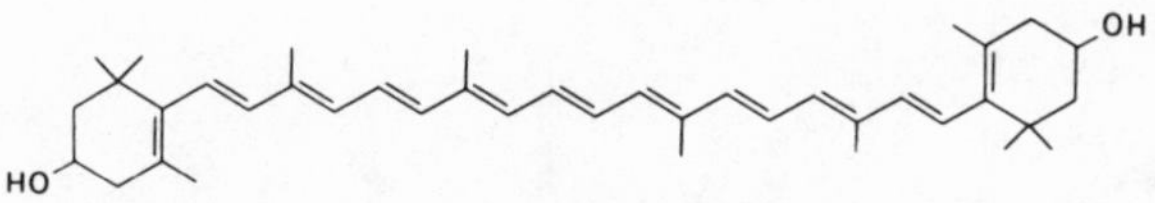

Zeaxanthin

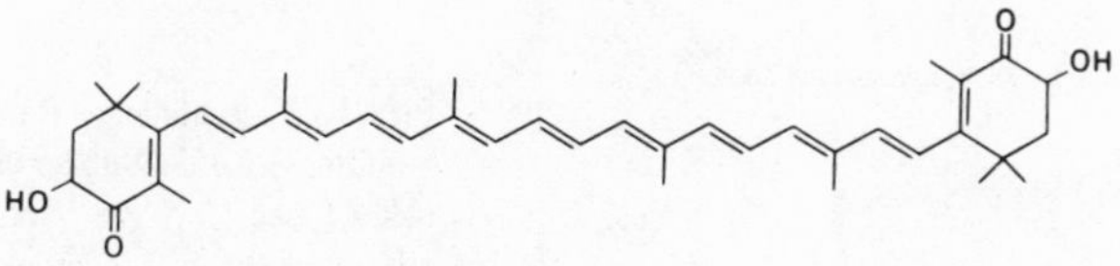

Astaxanthin

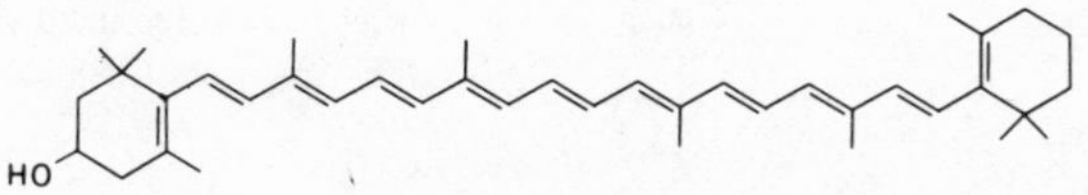

Cryptoxanthin

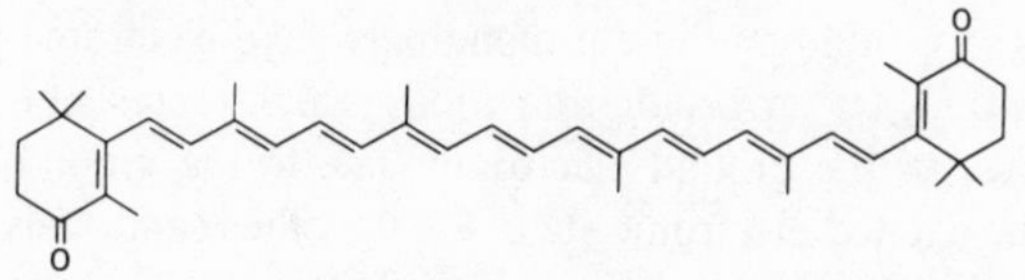

Canthaxanthin

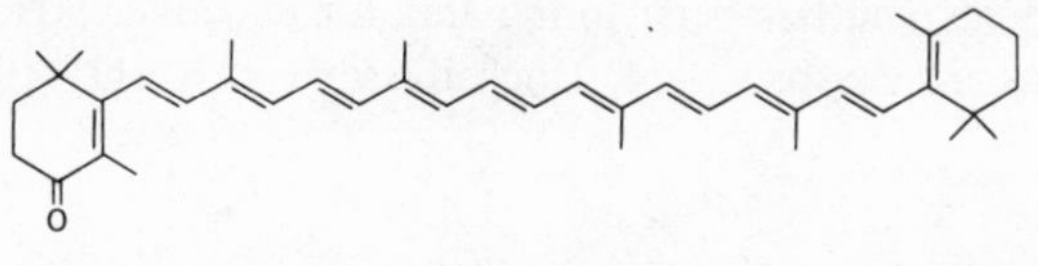

Echinenone

Fig. 12.2. Structural formulas of the carotenoid molecules studied. (Salares *et al.*, 1977.)

shifts of 60–80 nm. The resonance Raman spectra of the aggregated polyenes deviate from the normal (Rimai *et al.*, 1973) in that the position of $\nu_{C=C}$ is near 1518 cm^{-1}, whereas a firmly established correlation of $\nu_{C=C}$ versus $1/\lambda_{max}$ predicts that the C=C stretching band should be near 1560 cm^{-1}. The spectrum of monomeric astaxanthin is presented in Fig. 12.3a. The intense band at 1523 cm^{-1} is essentially a C=C stretching frequency ($\nu_{C=C}$). The spectrum of the aggregate (Fig. 12.3b) shows $\nu_{C=C}$ at 1518 cm^{-1}. Combined data from the absorption and Raman spectra show that strong excited-state (exciton) interactions are occurring. These results are an example of the applicability of the resonance Raman method for identifying excited-state perturbations.

A study of the changes of Raman intensity with wavelength (excitation profile) on aggregated astaxanthin and absorption spectra at -100°C on aggregates of lutein and

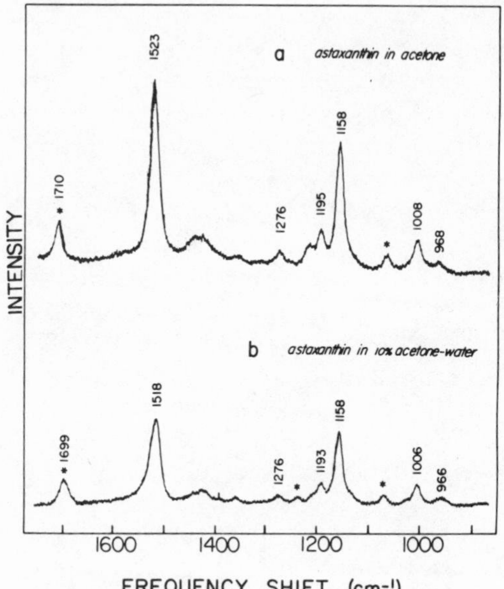

Fig. 12.3. Resonance Raman spectra of astaxanthin with 457.9-nm excitation in (a) acetone (4K, 1 sec, 0.5 cm^{-1}/sec, 9.0-cm^{-1} spectral slits) and (b) 10% acetone–water (4K, 1 sec, 0.5 cm^{-1}/sec, 9.0-cm^{-1} spectral slits). Acetone bands are marked with an asterisk. (Salares *et al.*, 1977a.) Reprinted with the permission of the American Chemical Society. Copyright 1977.

astaxanthin establish that a pronounced change in vibronic structure takes place when aggregation occurs. Astaxanthin and lutein monomers have electronic absorption spectra made up of 0–0, 0–1, 0–2, . . . vibronic transitions which represent transitions from the ground vibrational level of the ground electronic state to the ground, first, and second vibrational levels of the excited electronic state. On the other hand, only the 0–0 transition is seen in the aggregates. These authors explained their results in terms of the molecular exciton model, which is concerned principally with the dipole–dipole coupling of electronic transition moments and has been formulated for noncovalently bound aggregates (Kasha, 1963; McRae and Kasha, 1964; Hochstrasser and Kasha, 1964; Kasha *et al.*, 1965).

Astaxanthin of Carotenoproteins

Salares *et al.* (1977a) have investigated some of the spectral properties of a yellow protein (Buchwald and Jencks, 1968) from the shell of the lobster *Homarus americanus*. This pigment contains ca. 20 astaxanthin molecules per unit of protein. Its absorption and resonance Raman spectral characteristics are identical with those of aggregates of astaxanthin. The maximum in the electronic absorption spectrum of astaxanthin moves from about 480 to 410 nm with aggregation or binding to the yellow protein. Considered together with data on the $\nu_{C=C}$ in resonance Raman spectra (see discussion above), these facts show that in both systems a large perturbation of the electronic excited state occurs, while there is only a minimal perturbation of the ground state. Absorption data recorded at −95°C on the yellow protein and the astaxanthin aggregate display no vibronic structure. Thus, the electronic absorption originates only from the 0–0 or vibrationless transition. The molecular exciton model, mentioned above, adequately explains the spectral observations as the results of interactions of excited-state chromophore molecules. Salares *et al.* (1977a) attribute the shift that occurs in the absorption spectrum when free astaxanthin becomes yellow protein, to interaction of chromophores with chromophores, i.e., exciton coupling, rather than interactions of protein and chromophore. These workers suggested a possible photobiological function for the yellow protein, namely, that it acts as a bulk light gatherer for photons in the deep blue portion of the spectrum.

Salares *et al.* (1976) have attempted to determine if any correlation exists between the resolution obtainable in Raman excitation profiles and in the absorption spectrum of astaxanthin. They recorded excitation profiles for astaxanthin in the interval from 457.9 to 528.7 nm at 23°C and −162°C. At 23°C the visible spectrum was unstructured and at −162°C, structured (i.e., more absorption bands present). A simple mathematical model was used to simulate the experimental data obtained for the profiles, and these data demonstrated that a correlation exists between the development of vibrational structure in the absorption spectrum and in the profile. However, the excitation profiles for astaxanthin at room temperature did not allow resolution of the spacing between the 0–0, 0–1, 0–2, . . . etc. transitions of the broad absorption band, but did demonstrate that the 0–0 transition takes place near 18,850 cm^{-1}.

Continuing their work with astaxanthin, Salares *et al.* (1979) have used resonance Raman spectra to investigate the mechanisms of the absorption spectral shifts that occur for this carotenoid compound when it binds to the lobster carotenoproteins, ovoverdin (from the eggs) and α-, β-, and γ-crustacyanins (from the carapace). These authors recorded resonance Raman spectra of all these proteins. Based on this study and their previous work (Salares *et al.,* 1977a) they identified three classes of astaxanthin binding sites in lobster proteins.

Zeaxanthin–Phospholipid Mixtures

Zeaxanthin (see formula p. 483) has been used by Mendelsohn and Van Holten (1979) as a probe of membrane structure. They noted large changes in the resonance Raman and absorption spectra of this compound when it was mixed with phospholipid dispersions [dipalmitoyl phosphatidylcholine (DPPC) and distearoyl phosphatidylcholine (DSPC), and binary mixtures of these], and the dispersions were heated through their transitions from gel to liquid form.

Having established the validity of the experimental criteria used in their studies of the artificial systems above, these authors studied zeaxanthin aggregation in a phospholipid extract from a *Neurospora crassa* cel$^-$ mutant. They monitored the variation in the zeaxanthin C═C stretching frequency that Salares *et al.* (1977) had found to vary from 1526 cm^{-1} for the monomer to 1520 cm^{-1} for the aggregate. Figure 12.4 shows the temperature-induced changes in $\nu_{C=C}$ for zeaxanthin inserted into multilayers of the lipid extract. The midpoint of the phase separation region, T_m, is at −13°C. The resonance Raman spectral changes are thus seen to reflect conformational changes in the phospholipids of a membrane. How the presence of protein in natural systems will affect the

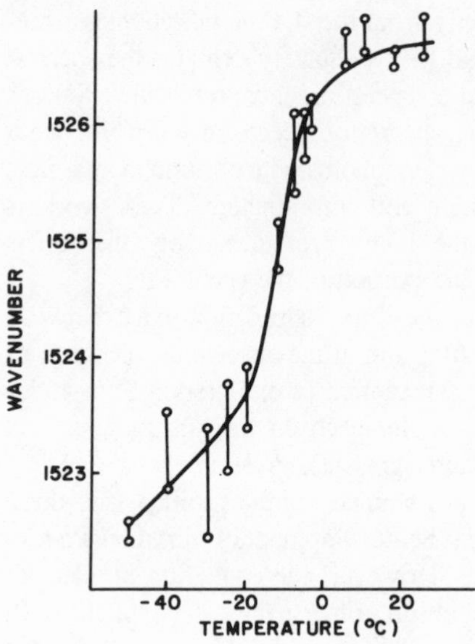

Fig. 12.4. Temperature-induced variation in $\nu_{C=C}$ for zeaxanthin inserted into multilayers of a *Neurospora crassa* (cel⁻ mutant) lipid extract. Phospholipid:carotenoid mole ratio was 55:1. Average precision in frequency data was about 0.3 cm^{-1}. (Mendelsohn and Van Holten, 1979.)

zeaxanthin aggregation mechanism remains to be seen (Mendelsohn and Van Holten, 1979).

Conformation of Bacterial Carotenoids

Lutz *et al.* (1978) have observed that the carotenoids bound to reaction centers of certain photosynthetic bacteria display very similar, but unusual, resonance Raman spectra. By comparing the spectra of these carotenoids with spectra of 15,15′-*cis*-β-carotene, these authors showed that the bacterial carotenoids have *cis* conformations, while the corresponding chromatophores have all-*trans* forms only. Upon extraction from the reaction centers they show their instability by fast isomerization towards the all-*trans* forms. The carotenoids studied were β-carotene; spirilloxanthin, present in reaction centers of *Rhodospirillum rubrum* strain S1; 1,2-dihydrolycopene, in *Rhodopseudomonas viridis;* spheroidene, in *Rps. spheroides* Y; neurosporene, in *Rps. spheroides* GlC; and chloroxanthin, in *Rps. spheroides* Ga.

Identification of Carotene-Containing Organisms

Howard *et al.* (1980) have reported resonance Raman spectra for several types of carotene-containing bacteria and algae in aqueous dispersion. Table 12.1 gives characteristic spectral lines of some of the bacteria produced with incident radiation of 488 nm.

Nelson (1981) has discussed the use of resonance Raman spectroscopy in the study of carotene-containing microorganisms and biological molecules.

Table 12.10. Spectra of Bacteria (488 nm)[a,b]

Flavobacterium aquatile	*Sarcina flava*	*Nocardia corallina*
951(19)	958(23)	961(4)
956(sh)	1006(42)[0.26]	966(4)
1004(57)[0.18]	1157(99)[0.30]	1003(9)
1159(94)[0.33]	1193(37)[0.28]	1154(81)[0.97]
1192(36)[0.29]	1209(25)	1190(sh)
1197(sh)	1530(100)[0.44]	1518(100)[0.92]
1210(22)		
1529(100)[0.42]		
Flavobacterium arborescens	***Micrococcus aurantiaca***	***Rhodopseudomonas palustris***
957(14)	997(13)	1005(35)
992–1009(18) br	1009(21)	1153(81)[0.62]
1023(13)	1028(11)	1157(83)[0.58]
1158(66)[0.87]	1157(59)[1.09]	1176(43)
1193(13)	1192(17)[1.13]	1499(36)
1520(100)[0.94]	1527(100)[1.11]	1514(100)[0.79]
		1519(sh)(89)[0.71]
		1530(61)[small]
(AR—17) *Coccochloris elabens*	**(Pr-6-) *Agmanellum quadruplicatum***	***Flavobacterium suaveolens***
1008(17)	890(w)	964(w)
1060(60)	936(w)	1007(32)
1194(21)	956(w)	1025(w)
1524(100)[0.75]	995(w)	1160(88)
	1023(vw)	1190(32)
	1156(100)[0.29]	1478(w)
	1186(w)	1530(100)
	1524(75)[0.58]br	
Micrococcus luteus	***Micrococcus roseus***	***Rhodospirillum rubrum* (an.)**
960(w)	950(sh)	1000(34)
963(w)	962(8.5)	1146(100)
1006(38)	1004(23)	1184(22)
1017(15)	1155(79)[0.86]	1280(10)
1072(w)	1194(6.5)	1502(94)
1157(100)	1214(6.5)	
1160(sh)	1284(10)	
1190(35)br	1319(4.5)	
1213(25)	1355(5.5)	
1268(14)	1516(100)[0.77]	
1271(25)		
1448(w)		
1529(93)		

[a] Howard *et al*. (1980).

[b] Relative line intensities are in parentheses. Polarization ratios are in brackets.

Carotenoid Spectra of Frog Sciatic Nerves

Szalontai *et al*. (1977) have recorded Raman spectra of frog sciatic nerves in different states of functioning. During excitation they found reversible changes in the C_{40}-carotenoid peaks enhanced by the resonance Raman effect. During the excitatory process the ν(—C═C—) mode (due to β-carotene-like molecules in the membrane) displayed characteristic changes: (a) the ν(—C═C—) band was shifted from 1520 cm^{-1} to 1518 cm^{-1} in high-resolution spectra; (b) the ratio I_{1520}/I_{1157} was decreased and was found to be characteristic of the ν(—C═C—) band alone. These workers explained the change as being due to transient carbon–carbon bond equalization (i.e., rearrangement of π-electrons).

References

Buchwald, M., and Jencks, W. P., *Biochemistry* **7,** 844 (1968).

Gill, D., Kilponen, R. G., and Rimai, L., *Nature* **227,** 743 (1970).

Hochstrasser, R. M., and Kasha, M., *Photochem. Photobiol.* **3,** 317 (1964).

Howard, W. F., Jr., Nelson, W. H., and Sperry, J. F., *Appl. Spectrosc.* **34,** 72 (1980).

Kasha, M., *Radiat. Res.* **20,** 55 (1963).

Kasha, M., Rawls, H. R., and Ashraf El-Bayoumi, M., *Pure Appl. Chem.* **11,** 371 (1965).

Lutz, M., Agalidis, I., Hervo, G., Cogdell, R. J., and Reiss-Husson, F., *Biochim. Biophys. Acta* **503,** 287 (1978).

McRae, E. G., and Kasha, M., in *Physical Processes in Radiation Biology,* (L. Augenstein, ed.) p. 23, Academic Press, New York, 1964.

Mendelsohn, R., and Van Holten, R. W., *Biophys. J.* **27,** 221 (1979).

Nelson, W. H., *Am. Lab.,* p. 94, March 1981.

Parker, F. S., "Biochemical Applications of Infrared and Raman Spectroscopy," in *Appl. Spectrosc.* **29,** 129–148 (1975).

Rimai, L., Heyde, M. E., and Gill, D., *J. Am. Chem. Soc.* **95,** 4493 (1973).

Salares, V. R., Young, N. M., Bernstein, H. J., and Carey, P. R., *Biochemistry* **16,** 4751 (1977a).

Salares, V. R., Young, N. M., Bernstein, H. J., and Carey, P. R., *Biochim. Biophys. Acta* **576,** 176 (1979).

Salares, V. R., Mendelsohn, R., Carey, P. R., and Bernstein, H. J., *J. Phys. Chem.* **80,** 1137 (1976).

Salares, V. R., Young, N. M., Carey, P. R., and Bernstein, H. J., *J. Raman Spectrosc.* **6,** 282 (1977).

Szalontai, B., Bagyinka, Cs., and Horvath, L. I., *Biochem. Biophys. Res. Commun.* **76,** 660 (1977).

Chapter 13

STEROIDS

Introduction

Applications of infrared spectroscopy to studies of steroids and sterols have been previously reviewed by the author (Parker, 1971). Discussions of steroid structures, metabolism, and conformational analysis can be found therein. Also, references to atlases of infrared data and spectra, correlations of spectra with structures, hydrogen bonding of steroids, and the measurement of integrated absorption intensities of steroids, have been discussed therein.

Nes and McKean (1977) have more recently written a comprehensive text giving practical and theoretical treatments of steroid biochemistry. They have discussed in detail steroid structure and nomenclature, the latter having gone through changes over the years. Among the other topics in this excellent book are analytic methods, the biosynthesis of steroids and a chapter concerned with functions of the steroids. This chapter is useful for endocrinologists.

Noone (1973) has described a computerized method for the rapid comparison and retrieval of IR spectral data, particularly for steroid molecules (2000–600 cm^{-1} region). The data base for the time-shared Infrared Information Search program (IRIS) was obtained by extracting and condensing the pertinent information from the data file of the American Society for Testing and Materials (ASTM). It was necessary to condense the data from the comprehensive ASTM coding system (1964) in order to keep computer costs low and data quality optimal.

The Sadtler Research Laboratories (3316 Spring Garden Street, Philadelphia, Pennsylvania 19104) has a collection of 750 infrared steroid spectra.

A very useful reference book concerned with the three-dimensional structures of steroids was published in 1975 (Duax and Norton). It contains a collection of ball-and-stick diagrams that were derived ultimately from X-ray data. Included in this volume are 103 steroids—androstane, pregnane, and estrane derivatives. Top and side views of the steroid molecule are given, as well as atomic coordinates, conformations, bond lengths, angles between bonds, and hydrogen-bonding information.

The book by Görög and Szasz (1978) contains chapters on fundamental steroid chemistry, therapeutic uses of steroid hormones, chromatography of various types (including gas chromatography), and methods of qualitative and quantitative analysis of steroid hormones. These authors have given IR spectra of the following compounds: 5α-androstane; allylestranol; ethynodiol diacetate; quingestanol acetate; ethynylestradiol; mestranol; estrone; estradiol benzoate; estradiol dipropionate; methyl testosterone; methandienone; progesterone; hydrocortisone; prednisolone; depersolone; prednisone; hydrocortisone acetate, hemisuccinate (Na^+), and phosphate (Na^+); triamcinolone acetonide; spironolactone; magestrol acetate; and chlormadinone acetate. This book also contains NMR spectra, including that of cholesterol; and ORD, CD, and mass spectra.

We give here an example of the complementary information available for a steroid from both IR and Raman spectra. Figure 13.1 (Washburn, 1978) shows the contrasting spectra of 17α-ethynyl-5-androstene-3β,17-diol. The Raman spectrum shows a distinct band at 1670 cm^{-1} for the internal C═C stretching vibration, whereas the IR spectrum shows a very weak band. Also, there is in the Raman spectrum a C≡C stretching vibration which is readily observable, but the IR spectrum hardly shows this band at all. On the other hand, the IR spectrum presents strong bands at 3580 and 3420 cm^{-1} due to OH; these are not active in the Raman spectrum. Nor is the 3280 cm^{-1}, due to ≡C—H, active in the Raman. It is, however, strong in the IR spectrum.

Raman and IR Spectra

Jones *et al.* (1956, 1963), using mercury-lamp excitation, have recorded Raman spectra of steroids in solution. Such spectra, along with those from IR spectroscopy, were used for the identification of steroid skeletons. Schrader *et al.* (1959, 1963) have used photographic methods and, then, photoelectric techniques for the recording of Raman spectra of steroid powders. Steigner (1969) has used techniques employing excitation by a low-pressure mercury "point" arc, and has recorded the Raman spectra of many crystalline steroids.

Using laser Raman spectroscopy, Schrader and Steigner (1973) have recorded the Raman spectra of about 80 steroids, and characterized their findings by spectral region. In the region 4000 to 2800 cm^{-1}, the O—H stretching vibrations have low intensity. The N—H vibrations have medium intensity, as do the aliphatic C—H vibrations, comparable to those found in the IR spectra. For a C—H bond on a triple-bonded carbon atom, the vibration around 3300 cm^{-1} is weaker in the Raman spectrum. For a C—H bond on a double-bonded or aromatic carbon atom, the vibration between 3030 and 3070 cm^{-1} is stronger than that in the IR. Steroids having double bonds in the 1-, 1,4-, 4,6-, and 5,16-positions and on aromatic rings in the estrogens, have C—H vibrations stronger than those in which the unsaturation is in the 4- or the 5-position.

In the region 2800 to 1350 cm^{-1} of Raman spectra, the stretching vibrations of central or terminal C≡C bonds (~ 2235 and 2105 cm^{-1}, respectively) and of the isolated C═C bonds (1620–1672 cm^{-1}) are very strong, but in IR spectra, both vibrations display low

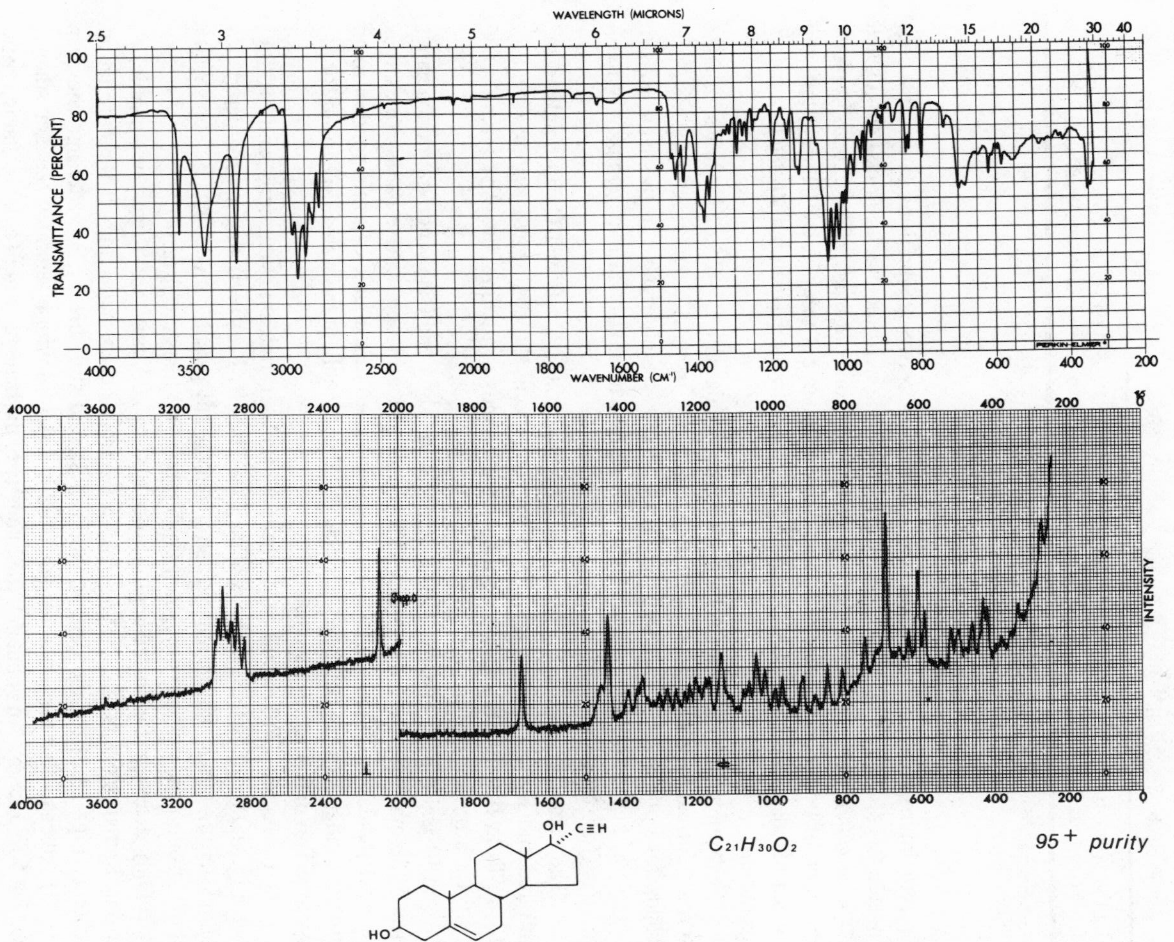

Fig. 13.1. 17α-Ethynyl-5-androstene-3β,17-diol. (Washburn, 1978.)

Table 13.1. Vibrations Caused by Conjugated C—C Double Bonds in Raman and Infrared Spectra[a]

Nature of double bonds	In-phase vibration			Out-of-phase vibration		
	cm^{-1}	Raman	IR	cm^{-1}	Raman	IR
3,5-Diene*(s-trans)*	1670	Very strong	Weak	1640	Weak	Medium
5,7-Diene*(s-cis)*	1596	Strong	Very weak	1650	Weak	Weak

[a] Schrader and Steigner (1973).

or zero intensity. Table 13.1 shows a comparison between vibrations caused by conjugated C—C double bonds in Raman and IR spectra. Homoannular *(s-cis)* conjugated C═C bonds behave differently from heteroannular ones *(s-trans)*. There is an in-phase vibration of both bonds, which is strong in the Raman but weak in the IR spectrum; and an out-of-phase vibration, which is weak or absent in the Raman but stronger in the IR spectrum.

A comparison between isolated carbonyl bonds and C═C bonds shows that the former normally have weaker stretching vibrations in the Raman spectrum. Also, the bands for both isolated carbonyl and C═C bonds are observable in the same Raman spectrum, but in the IR spectrum the very strong carbonyl band may completely mask the C═C absorption.

The characteristic patterns displayed by conjugated C═C and C═O bonds in the infrared and Raman spectra are useful for identifying the conjugated system (see Fig. 13.2). Both IR and Raman spectra display two bands of medium intensity around 1600 cm^{-1} for aromatic rings.

In IR and Raman spectra, CH_3 degenerate bending and CH_2 bending produce bands near 1450 cm^{-1} of medium intensity. For skeletal identification, these bands are used as internal standards. When the CH_3 group is connected to a double or triple bond or to an aromatic ring, a strong band appears near 1380 cm^{-1} only in the Raman spectrum for the symmetrical deformation mode.

In the region 1350 to 800 cm^{-1}, there are skeletal stretching vibrations, and rocking, twisting, and wagging deformations of the C—H bonds. They serve as good characteristics of the steroid, but are not assignable empirically to any possible mode. The Raman spectra show strong bands when conjugated double-bond systems are present.

Only a few bands in the region from 800 to 200 cm^{-1} are directly assignable. A strong Raman band near 725 cm^{-1} may arise from an aromatic ring. If the correlation scheme (Fig. 13.3) for the determination of steroid structure by Schrader and Steigner (1973) indicates that the unknown structure is a steroid with a saturated skeleton, then a band at 500 $\pm$ 5 cm^{-1} of about half the intensity of the band near 1450 cm^{-1} is characteristic for 5β-steroids. 5α-Steroids either have a weak band or it is absent.

It is possible to predict steroid structures by using the spectra in Fig. 13.2 and the correlation scheme given on page 494 (Schrader and Steigner, 1973).

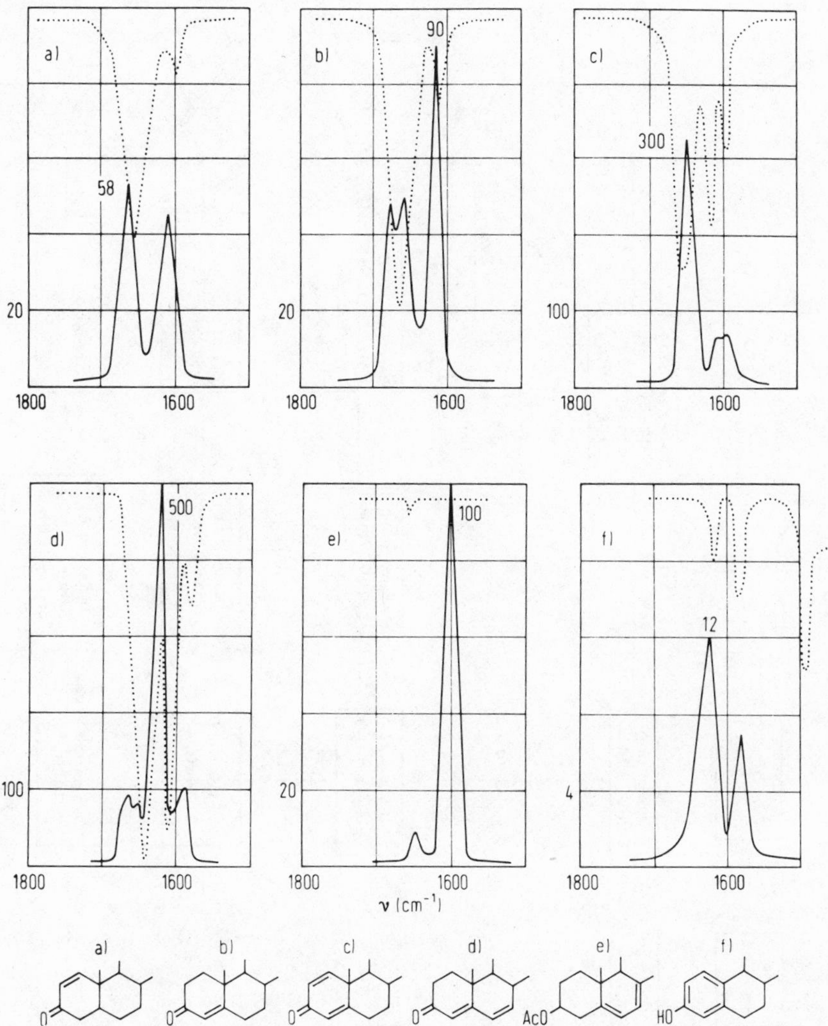

Fig. 13.2. Infrared and Raman spectra in the region 1500–1700 cm^{-1}. (a) 17β-hydroxy-5α-androst-1-en-3-one; (b) 17β-hydroxy-4-androsten-3-one; (c) 17β-hydroxy-1,4-androstadien-3-one; (d) 17β-hydroxy-4,6-androstadien-3-one; (e) 3β-acetoxy-5,7-cholestadiene; (f) 1,3,5(10)-estratriene-3,17β-diol. (· · ·) Infrared spectrum, substances in KBr pellets, linear in absorptivity units (1 unit = ca. 55 liter/ mole cm). (——) Raman spectrum; numbers on ordinate axis or on bands are intensities relative to the intensity at ca. 1450 cm^{-1} ≡ 10 units. (Schrader and Steigner, 1970.)

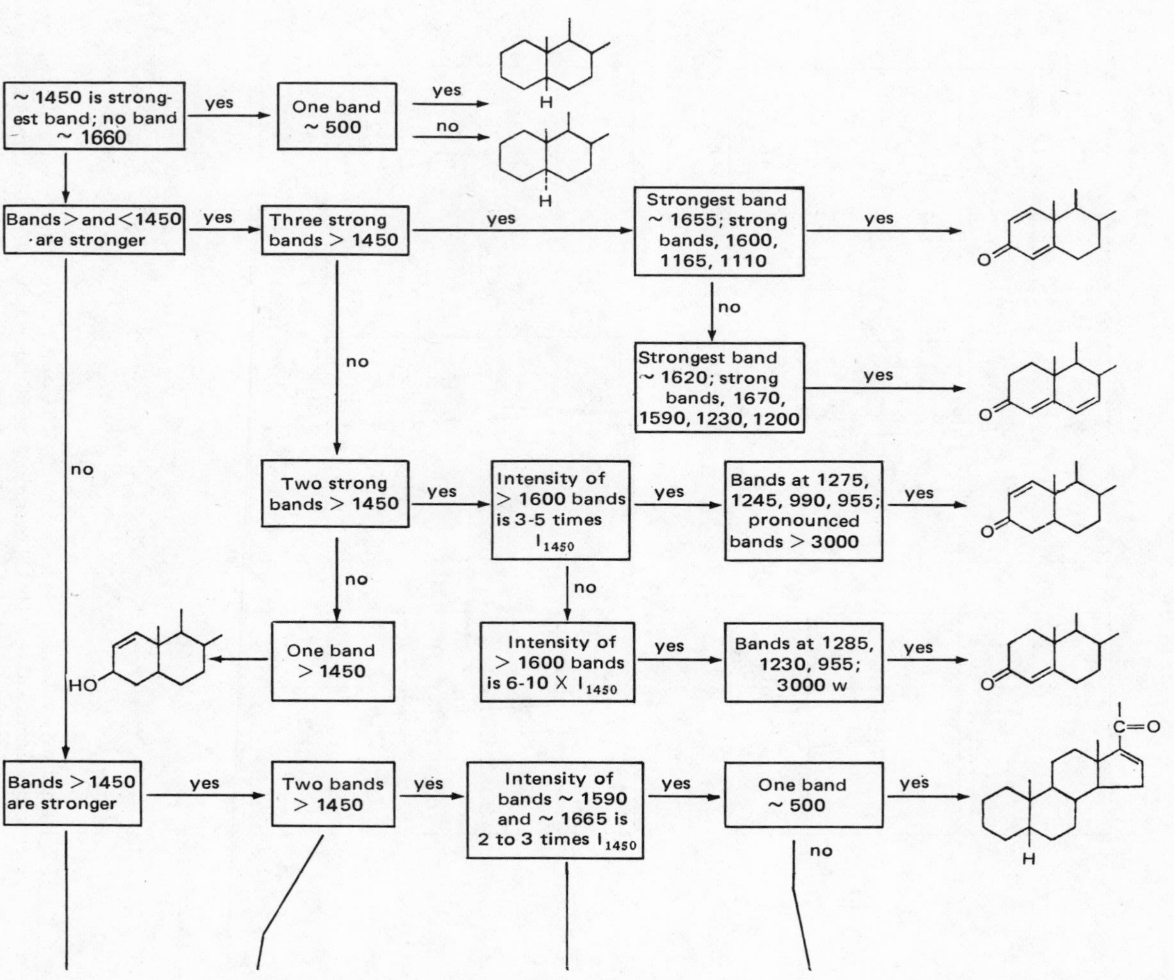

~ 1450 is strongest band; no band ~ 1660
yes
One band ~ 500
yes
no
H
H
Bands > and < 1450 are stronger
yes
Three strong bands > 1450
yes
Strongest band ~ 1655; strong bands, 1600, 1165, 1110
yes
O
no
Strongest band ~ 1620; strong bands, 1670, 1590, 1230, 1200
yes
O
no
no
Two strong bands > 1450
yes
Intensity of > 1600 bands is 3-5 times I1450
yes
Bands at 1275, 1245, 990, 955; pronounced bands > 3000
yes
O
no
no
One band > 1450
HO
Intensity of > 1600 bands is 6-10 X I1450
yes
Bands at 1285, 1230, 955; 3000 w
yes
O
Bands > 1450 are stronger
yes
Two bands > 1450
yes
Intensity of bands ~ 1590 and ~ 1665 is 2 to 3 times I1450
yes
One band ~ 500
yes
C=O
H
no

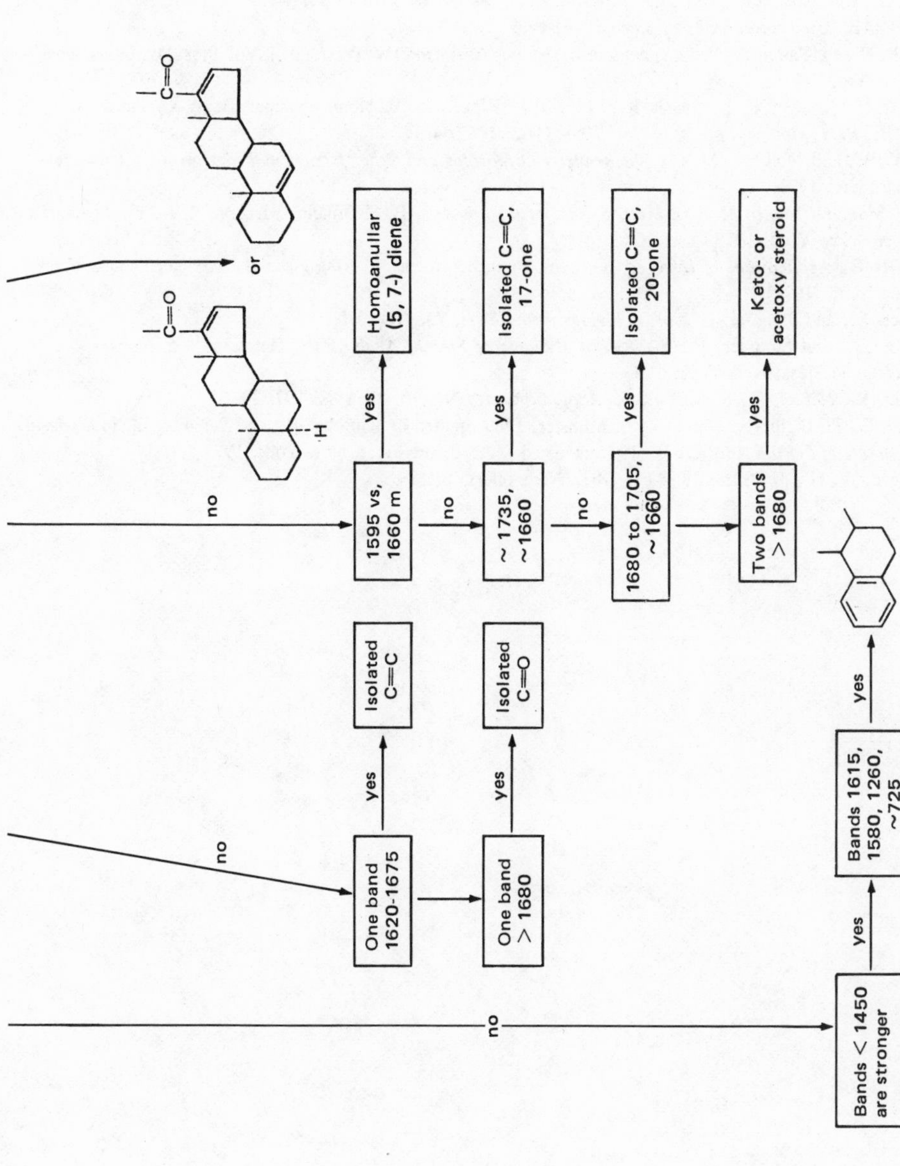

Fig. 13.3. Scheme for the determination of steroid structures by Raman spectroscopy (Schrader and Steigner, 1973). All band numbers are cm^{-1}.

References

ASTM, *Codes and Instructions for Wyandotte—ASTM Punched Cards,* ASTM, Philadelphia, 1964.

Duax, W. L., and Norton, D. A., eds., *Atlas of Steroid Structure*. Vol. I., Plenum Press, New York, 1975.

Görög, S., and Szasz, Gy., *Analysis of Steroid Hormone Drugs,* Elsevier, New York, 1978.

Heftmann, E., *Chromatography of Steroids,* Elsevier, New York, 1976.

Jones, R. N., and Sandorfy, C., in *Technique of Organic Chemistry* (W. West, ed.,), Vol. 9, p. 247, Interscience, New York, 1956.

Jones, R. N., Krueger, P. J., Noack, K., Elliot, J. J., Ripley, R. A., Nonnenmacher, G. A. A., and DiGorgio, J. B., *Proc. Colloq. Spectrosc. Int. 10th, 1962,* **1963,** 461.

Nes, W. R., and McKean, M. L., *Biochemistry of Steroids and Other Isopentenoids,* University Park Press, Baltimore, 1977.

Noone, Margaret M., in *Modern Methods of Steroid Analysis* (E. Heftmann, ed.), pp. 221–230, Academic Press, New York, 1973.

Parker, F. S., *Applications of Infrared Spectroscopy in Biochemistry, Biology, and Medicine,* Plenum Press, New York, 1971.

Schrader, B., and Steigner, E., *Justus Liebigs Ann. Chem.* **735,** 6 (1970).

Schrader, B., and Steigner, E., in *Modern Methods of Steroid Analysis* (E. Heftmann, ed.), pp. 231–243 Academic Press, New York, 1973.

Schrader, B., Nerdel, F., Kresse, G., *Z. Anal. Chem.* **1959,** 170, 43; **1963** 197, 295.

Steigner, E., Ph.D. thesis, University of Münster, 1969, quoted in Schrader, B., and Steigner, E., in *Modern Methods of Steroid Analysis* (E. Heftmann, ed.), Academic Press, New York, 1973.

Washburn, W. H., *Am. Lab.* **10,** 47 (1978). (November)

Chapter 14

USES OF IR SPECTROSCOPY FOR STUDYING GASES

Measurement of Expired $^{13}CO_2$ and $^{12}CO_2$

Hirano *et al.* (1979) have used a simple spectroscopic method to measure continuously the expired $^{13}CO_2$ and $^{12}CO_2$ of a rat that had been injected intraperitoneally with 10 mg of [^{13}C]-sodium bicarbonate containing 49.0 atom % of ^{13}C. Use of their IR analyzer circumvents the employment of radioactive ^{14}C (a potential hazard to be avoided in clinical studies with humans) and a mass spectrometer, which is not convenient for clinical use and maintenance. The IR analyzer measures $^{12}CO_2$ in a short cell at the ~ 2360.2-cm^{-1} band and $^{13}CO_2$ in a cell that is approximately 100 times as long at the ~ 2272.0-cm^{-1} band. The instrument records the volume percentage of $^{12}CO_2$ and the atom percentage excess of $^{13}CO_2$. In the experiment of Hirano *et al.*, which used, e.g., a 210-g normal male rat, a maximum expired $^{13}CO_2$ of 1.7 atom % excess was observed 10 min after the injection of the 10 mg of [^{13}C]-NaHCO$_3$, returning gradually to a value near to the initial natural abundance of $^{13}CO_2$ during a period of about 2 h. The authors reported that since this work was done the analyzer had been improved to the point of being 15 times as sensitive as the one used in this report, and they expected that when the up-dated instrument is marketed, it will have a sensitivity one-fiftieth or less of that described here, thus permitting a much lower dosage of tracer bicarbonate to be administered.

Another development in the use of ^{13}C to replace the use of radioactive ^{14}C in humans comes from the laboratories of Andros, Inc., Berkeley, California, under contract to the National Institute of General Medical Sciences. The nondispersive instrument manufactured by this company uses a "heterodyne principle" for investigating the ratio of $^{13}CO_2$/$^{12}CO_2$ in a sample of expired CO_2 or CO_2 from a combusted organic compound, e.g., a carbon-containing compound from a metabolic pathway. The sensitivity of this instrument for detecting changes in the ratio is reported to be 1 part in 10^4 parts, in a bulletin from the National Institutes of Health (1980). The detector that is used to obtain this sensitivity is InSb, which is cooled by liquid nitrogen.

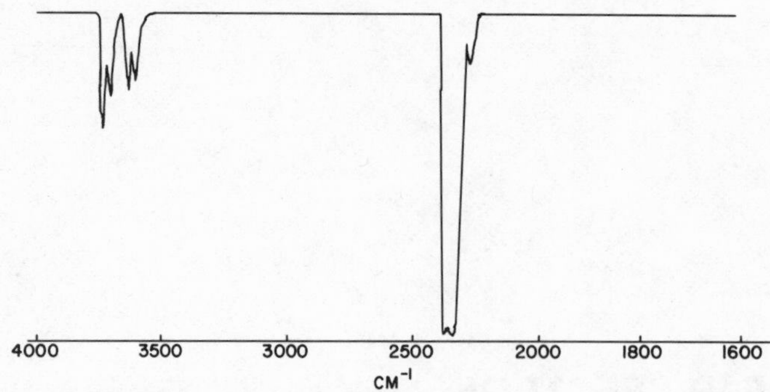

Fig. 14.1. Infrared absorption spectrum of CO_2. (McClatchie; see text.)

The following explanation of the operation of the instrument and the fundamentals involved are taken from McClatchie (see references). Figure 14.1 depicts the IR spectrum of CO_2 from 4000 to 1666 cm^{-1}. Carbon dioxide has three significant bands in this interval, two near about 3704 cm^{-1} and one at 2326 cm^{-1}. The band at 2326 cm^{-1} is the most intense of the three bands. Figure 14.2 curve A displays a more detailed spectrum of this band. The 2326-cm^{-1} band consists of two parts: one caused by $^{12}CO_2$ and a similar absorption due to $^{13}CO_2$, which is shifted to a lower wavenumber because of the larger mass of $^{13}CO_2$. The $^{13}CO_2$ band is weaker because of the low natural abundance of $^{13}CO_2$ compared to $^{12}CO_2$. Figure 14.2 curve B presents the spectrum of CO_2 enriched highly in $^{13}CO_2$, which shows that a large portion of the $^{13}CO_2$ band is completely separated from the $^{12}CO_2$ band. Because of the degree of resolution of the spectrometer used to make Fig. 14.2, the width of the lines is greatly increased and their intensity decreased. The real width of the lines is about 1/10 of the width seen in Fig. 14.2 and the intensity is 10 times greater.

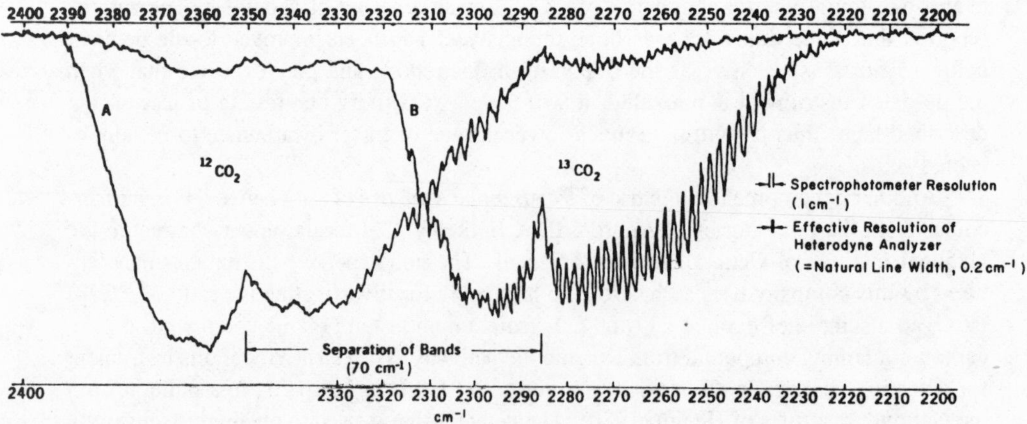

Fig. 14.2. Curve A: Natural CO_2. Curve B: CO_2 enriched in $^{13}CO_2$. (McClatchie; see text.)

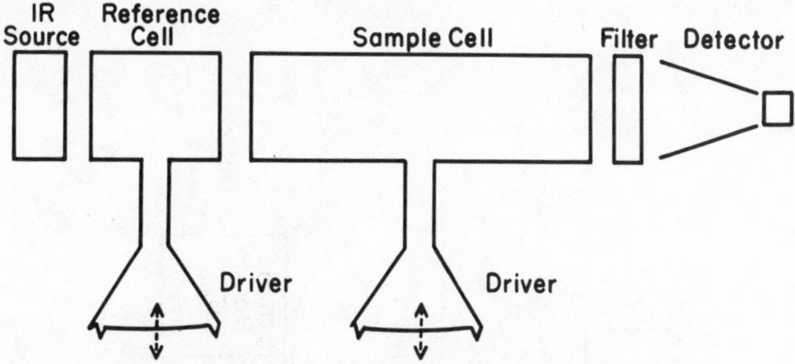

Fig. 14.3. Basic heterodyne analyzer. (McClatchie; see text.)

Figure 14.3 shows the arrangement of the heterodyne gas analyzer. IR radiation from the source passes through the reference cell containing the gas to be measured, e.g., $^{13}CO_2$. The gas will be transparent for wavenumbers corresponding to $^{13}CO_2$. As light passes through the sample cell, additional absorption will occur at the characteristic wavenumbers of the gases in the sample cell.

If the gas in the sample cell is different from the gas in the reference cell, the absorption will be at different wavenumbers, and the total radiation absorbed will be the sum of the radiation absorbed by the references and sample cells. If the gases are the same, the sample cell will absorb radiation in parts of the spectrum that have already been attenuated by the reference cell. In this case the fraction of the radiation transmitted by both cells will be the *product* of the fractions transmitted by both cells. When the gases are different, the absorption is *additive,* whereas for similar gases the transmission (absorption) is *multiplicative*.

McClatchie continues:

> Assume now that the amount of gas in the reference cell is varied sinusoidally at frequency ω (by varying the pressure of the gas, for example) and the sample cell at frequency β. The amount of light falling on the detector will now also vary with time, and a Fourier analysis of the signal on the output of the detector will show large components at frequencies ω and β. In addition, if there is any gas common to both the reference and sample cells, there will be signals at frequencies $\omega \pm \beta$ because of the nonlinear nature of the absorption in this case. The sum and difference frequencies are present *only* if the gases are the same. If the reference cell is filled with $^{13}CO_2$, then a signal at frequencies $\omega \pm \beta$ is present only if there is some $^{13}CO_2$ in the sample cell. A gas with an absorption band overlapping $^{13}CO_2$ will *not* produce a signal unless there is overlap of the individual lines. By changing gases in the reference cell, the system becomes sensitive to different gases. The method is thus capable of producing a highly specific detection of any gas which absorbs infrared radiation.
>
> By adding a $^{12}CO_2$ reference cell to the basic heterodyne instrument, [see Fig. 14.4] we have an instrument capable of measuring both $^{13}CO_2$ and $^{12}CO_2$ simultaneously. By taking the ratio of the $^{13}CO_2$ signal to the $^{12}CO_2$ signal the instrument becomes insensitive to changes in the overall CO_2 level, changes in transmission of the windows or fluctuations in IR source power (all these affect each signal the same way, leaving

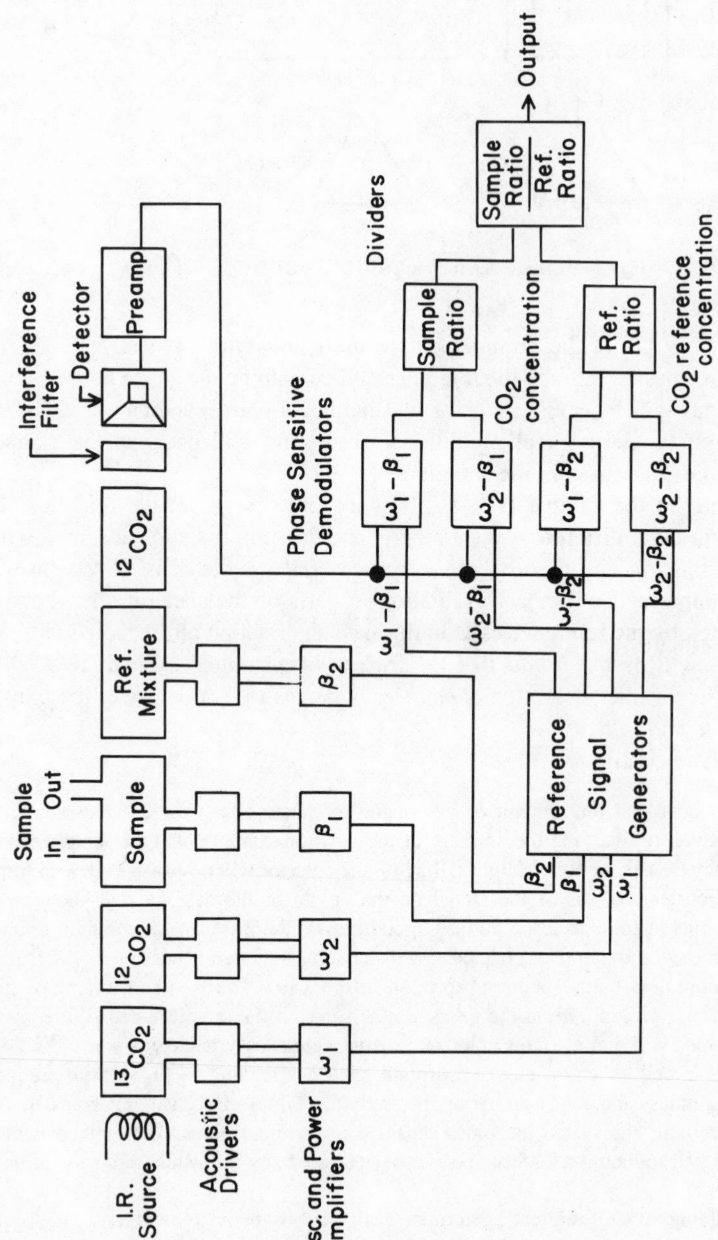

Fig. 14.4. Block diagram of $^{13}CO_2/^{12}CO_2$ ratioing heterodyne analyzer. (McClatchie; see references.)

the ratio unaffected). A further increase in stability is obtained by adding an additional sample cell, filled with CO_2 at approximately the same concentration as the sample cell and having the natural $^{13}C/^{12}C$ ratio [the modified heterodyne configuration is not shown here.] By dividing the $^{13}CO_2/^{12}CO_2$ ratio in the sample cell by the $^{13}CO_2/^{12}CO_2$ ratio in the reference mixture, we obtain an output directly proportional to the enrichment of the $^{13}CO_2$ fraction, starting at 1.0000 and increasing with $^{13}CO_2$ enrichment in the sample cell.

Alveolar Gas Analysis

Continuous IR analysis of end-expired CO_2 concentration yields a good estimate of alveolar or arterial P_{CO_2} in the healthy adult (Cotes, 1975). Evans *et al.* (1977) have examined the performance of an IR analyzer (Beckman LB-2) in sampling alveolar gas at ventilatory frequencies and volumes occurring in the newborn human baby. They found the optimum sampling flow rate to be 300 ml/min; at this flow the 90% response time of the analyzer–catheter system was 140 msec. For measuring CO_2 the IR analyzer is simpler, cheaper, and more convenient than a mass spectrometer.

Blood Gases. Total CO_2 of Plasma

An automated system for IR analysis of total CO_2 of plasma (65 sec per sample) has been described by Rabinow *et al.* (1977). Fifty-microliter samples of plasma are placed in disposable capillaries and are placed into a block, in which the specimens are kept from being exposed to air. A sequencer moves the block forward and "tells" a dispenser to flush the sample into a 10-ml flask with 2.6 ml of 50-mmol/liter H_2SO_4. After incubation for 24 sec, the CO_2 evolved is swept into a 1-meter IR gas cell by nitrogen. The spectrometer (Wilks Miran I analyzer) records the absorbance peak of the gas. It should be noted that the cell of the analyzer is not gas-tight. Normal breathing near the cell raises the CO_2 levels within the cell, causing an appreciable signal. This difficulty is removed by enclosure of the entire IR analyzer in a large plastic bag, secured tightly around the electrical wires and gas lines connected to the instrument. There is excellent correlation between the IR-determined total-CO_2 values and values from the Natelson Microgasometer for plasma samples. The IR method could be adapted to the determination of other blood gases, e.g., CO and NH_3, by the use of specific interference filters in the spectrometer.

CO Binding to Iron (II) Hemeproteins

Maxwell and Caughey (1978) have presented a review on the IR spectroscopy of gases, ligands, and other groups in aqueous solutions and tissue. The metal-bound ligands CO,N_3^-, CN^-, CNO^-, CNS^-, and $CNSe^-$ were discussed as well as the gaseous anesthetic N_2O. (See also pp. 503 and 504 in this chapter.)

Alben *et al.* (1982) have used FTIR spectroscopy to study the binding of carbon monoxide to myoglobin and cytochrome oxidase in isolated rat heart cells. They observed that CO is bound only to myoglobin in the presence of air, thus protecting cytochrome oxidase in heart cells.

Table 14.1 presents IR spectral properties for carbon monoxide bound to iron (II) heme proteins. B (in column 5) = $(\Delta\nu_{1/2}) \times (\epsilon_M) \times 1.24$, and is called the apparent

Table 14.1. Infrared Spectral Properties for Carbon Monoxide Bound to Iron(II) Hemeproteins[a]

	ν_{CO} (cm^{-1})	$\Delta\nu_{1/2}$ (cm^{-1})	ϵ (M^{-1} cm^{-1} $\times 10^3$)	B (M^{-1} cm^{-2} $\times 10^{-4}$)
Hemoglobin A	1951	8	3.7	3.4
Myoglobin (bovine heart)	1944[b]	12	2.0	2.8
Cytochrome *c* oxidase (bovine heart)[c]	1963.5	5.5	4.9	2.8
Cytochrome P 450_{cam} (substrate bound)[d]	1940	12.5	2.3	3.3
Cytochrome P 450_{cam} (substrate free)[d]	1940	~14		3.5
	1962			
Cytochrome P 420_{cam}[d]	1965	~25		
Cytochrome P 450_{LM}(PB)[d]	1948	~22		
Cytochrome P 450_{LM}(3MC)[d]	1954	~26		
Hemoglobin Zurich	1958[e]	8		3.2
	1951[e]	8		
Hemoglobin M_{Emory}[e]	1951	8		
	1970			
Rabbit hemoglobin[f]	1951	8		
	1928	10		
Hemopexin heme carbonyl[g]	1950	25		
Horseradish peroxidase[h]	1933	12		4.4
	1906	14		
Leghemoglobin[i]	1948	7		

[a] Maxwell and Caughey (1978).

[b] S. McCoy and W. S. Caughey. in *Probes of Structure and Function of Macromolecules and Membranes* (B. Chance, T. Yonetani, and A. S. Mildvan, eds.), Vol. 2, p. 289. Academic Press, New York, 1971. A small but significant band at 1933 cm^{-1} is also always present. The origin of this band is still under investigation.

[c] S. Yoshikawa, M. G. Choc, M. C. O'Toole, and W. S. Caughey, *J. Biol. Chem.* **252,** 5498 (1977).

[d] D. H. O'Keeffe, R. E. Ebel, J. A. Peterson, J. C. Maxwell, and W. S. Caughey, unpublished observations. Cytochrome P-450_{cam} was isolated from *Pseudomonas putida* grown with d-camphor. Cytochrome P-450_{LM} was prepared from liver microsomes of rats pretreated with either phenobarbital (PB) or 3-methylcholanthrene (3MC).

[e] W. S. Caughey, J. O. Alben, S. McCoy, S. H. Boyer, S. Charache, and P. Hathaway, *Biochemistry* **8,** 59 (1969).

[f] N. A. Matwiyoff, P. J. Vergamini, T. E. Needham, C. T. Gregg, J. A. Volpe, and W. S. Caughey, *J. Am. Chem. Soc.* **95,** 4429 (1973).

[g] U. Muller-Eberhard, W. T. Morgan, J. C. Maxwell, and W. S. Caughey, unpublished observations.

[h] C. H. Barlow, P.-I. Ohlsson, and K.-G Paul, *Biochemistry* **15,** 2225 (1976).

[i] C. A. Appleby, J. C. Maxwell, and W. S. Caughey, unpublished observations.

integrated intensity. $\Delta\nu_{1/2}$ is the half-bandwidth (width of band in cm^{-1} at one-half peak height in absorbance). ϵ_M is the apparent molar absorptivity. The factor 1.24 accounts for the Lorentzian shape of the band. The broad range of frequencies observed for carbon monoxide in heme proteins reflects the broad differences in structure that are found in the binding sites. (See also, Chapter 6, on porphyrins).

Determination of Carbon Monoxide Hemoglobin in Blood

Maxwell *et al.* (1974) have developed a technically simple IR method for the accurate determination of carbon monoxide hemoglobin in whole blood. The portion of the hemoglobin present as HbCO in whole blood can be directly calculated from the relative band heights for untreated and CO-saturated blood, since the intensity of the CO band for HbCO follows Beer's law. Fig. 14.5A shows IR difference spectra of a nonsmoker's blood and that of a moderate smoker (Fig. 14.5B). The blood of these persons contained 1.5% of hemoglobin and 4.5% of hemoglobin as HbCO, respectively.

These authors have also recorded spectra that permit rough quantitation of the CO-binding heme proteins in heart muscle, e.g., cytochrome oxidase, Hb and myoglobin.

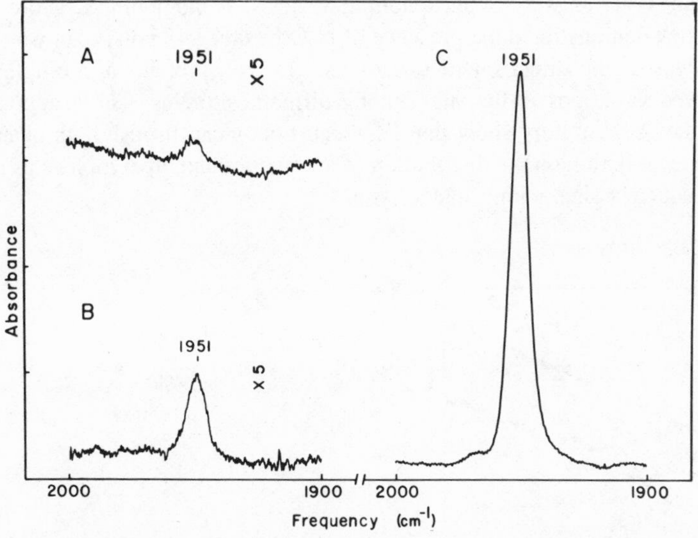

Fig. 14.5. Infrared difference spectra of human whole blood vs. water. (A) CO stretching band of untreated blood from a nonsmoker with 1.5% of hemoglobin as carbonyl hemoglobin. (B) CO stretching band of untreated blood from a moderate cigarette smoker with 4.5% of hemoglobin as carbonyl hemoglobin. (C) Spectrum obtained upon saturation of the blood from (B) with CO. The intensity of the CO band in (B) relative to that in (C) gives directly the fraction of the hemoglobin present as carbonyl hemoglobin in the untreated blood of (B). The ordinates for spectra (A) and (B) are expanded 5 fold compared to spectrum C. (Maxwell *et al.*, 1974.)

These CO-protein complexes show characteristic frequencies at 1963, 1951, and 1944 cm^{-1}, respectively.

IR spectra of CO-saturated rabbit whole blood display two prominent CO stretching bands at 1951 and 1928 cm^{-1}, whereas clinical electrophoresis that can separate hemoglobins A,A_2,F,C, and S could not resolve the second component in rabbit hemoglobin.

Anesthesiology. N_2O Sites in the Brain

Caughey *et al.* (1977) have detected and characterized nitrous oxide sites in the brain of a dog anesthetized by halothane–N_2O. They recorded IR spectra of N_2O in a variety of solvents (CCl_4, $CHCl_3$, CH_2Cl_2, cyclohexane, *n*-hexane, olive oil, H_2O, a concentrated solution of bovine serum albumin in water, glycerol, and methanol) and in the dog brain (Figs. 14.6, 14.7). The appearance or disappearance of nitrous oxide in the brain was easily observed as the N_2O was administered or taken away. These workers found two major types of sites in the brain; one, with $\nu_3 = 2229.8 \pm 0.4$ cm^{-1} and $\Delta\nu_{1/2} = 13.0 \pm 0.6$ cm^{-1}, is rather like the polar site in water; the other, with $\nu_3 = 2216.8 \pm 0.8$ cm^{-1} and $\Delta\nu_{1/2} = 9.6 \pm 1.0$ cm^{-1}, is nonpolar and is probably involved with lipid of the membranes. The methods that were used permit the characterization of *in-tissue* sites occupied by this anesthetic.

Hazzard *et al.* (1982) have extended these studies with N_2O to include red cells and bovine brain tissue as well as phospholipid vesicles, hemoglobin A, and cytochrome *c* oxidase. They demonstrated the presence of N_2O in sites less polar than water and found clear evidence of the direct interaction of the gas with proteins at multiple sites. They also observed variations in the sites among different proteins. The results of this work from Caughey's laboratory show that IR spectroscopy can furnish both quantitative and qualitative information on the distribution of anesthetics and other molecules into protein, lipid, and aqueous sites within intact tissue.

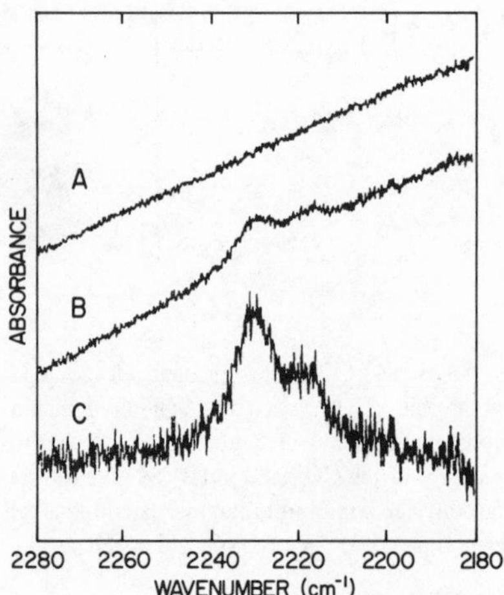

Fig. 14.6. Infrared absorption spectra of dog brain in the region 2280 to 2180 cm^{-1}. Curve *A:* a single scan during administration of halothane–O_2. Curve *B:* a single scan during administration of N_2O–halothane–O_2. Curve *C:* the scan shown in *B* after slope-correction and ordinate scale expansion. (Caughey *et al.*, 1977.)

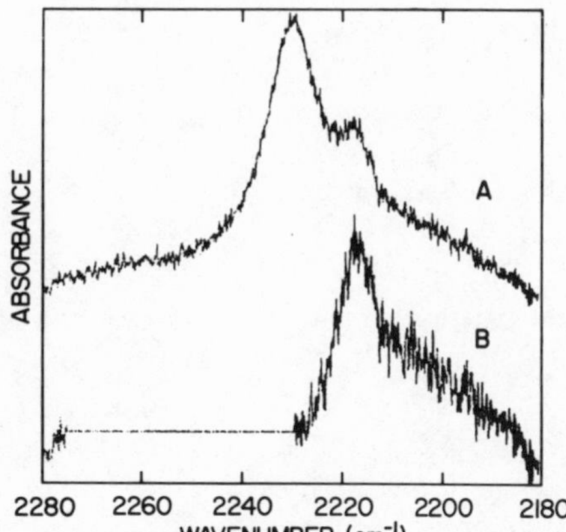

Fig. 14.7. Infrared spectra of dog brain during administration of N_2O–halothane–O_2. Curve *A:* an average of 16 scans, such as Fig. 14.6B, each slope-corrected. Curve *B:* the spectrum of the smaller peak left after subtraction of the larger peak (by taking the mirror image of the left side). (Caughey *et al.*, 1977.)

Metabolism of Dihalomethanes

Dihalomethanes are metabolized to carbon monoxide in the rat. This fact was established by experiments in which ^{13}C-dichloromethane was employed. IR spectra of the blood of rats given the labeled compound displayed absorption bands due to ^{13}CO (Kubic *et al.*, 1974).

References

Alben, J. O., Altschuld, R. A., and Fiamingo, F. G., *Fed. Proc.* **41,** 893 (1982), Abstracts of the 66th Annual Meeting, April, 1982.

Caughey, J. M., Lumb, W. V., and Caughey, W. S., *Biochem. Biophys. Res. Commun.* **78,** 897 (1977).

Cotes, J. E., *Lung Function,* p. 243, Oxford-Blackwell Scientific Publications, England, 1975.

Evans, J. M., Hogg, M. I. J., and Rosen, M., *Brit. J. Anaesth.* **49,** 761 (1977).

Hazzard, J. H., Gorga, J. C., Einarsdottir, O., and Caughey, W. S., Abstracts of the 26th Annual Meeting, *Biophys. J.* **37,** 154a No. 2 Part 2, February (1982).

Hirano, S., Kanamatsu, T., Takagi, Y., and Abei, T., *Anal. Biochem.* **96,** 64 (1979).

Kubic, V. L., Anders, M. W., Engel, R. R., Barlow, C. H., and Caughey, W. S., *Drug Metabolism and Disposition* **2,** 53 (1974).

Maxwell, J. C., and Caughey, W. S., in *Methods in Enzymology* **54,** 302 (1978).

Maxwell, J. C., Barlow, C. H., Spallholz, J. E., and Caughey, W. S., *Biochem. Biophys. Res. Commun.* **61,** 230 (1974).

McClatchie, E. A., "The Development of a $^{13}CO_2/^{12}CO_2$ Ratio Analyzer Using a Nondispersive Infrared Heterodyne Technique," in *Selected Approaches to Gas Chromatography–Mass Spectrometry in Laboratory Medicine* (R. S. Melville and V. F. Dobson, eds.). DHEW Publication No. (NIH) 75-762.

National Institutes of Health Guide for Grants and Contracts, Vol. 9, No. 3, p. 9, February 22, 1980.

Rabinow, B. E., Geisel, A., Webb, L. E., and Natelson, S., *Clin. Chem.* **23,** 180 (1977).

Appendix I

A Table of Reciprocals

	0	1	2	3	4	5	6	7	8	9	Subtract 1	2	3	4	5	6	7	8	9
1.0	1.0000	0.9901	0.9804	0.9709	0.9615	0.9524	0.9434	0.9346	0.9259	0.9174	9	18	27	36	45	55	64	73	82
1.1	0.9091	0.9009	0.8929	0.8950	0.8772	0.8696	0.8621	0.8547	0.8475	0.8403	8	15	23	30	38	45	53	61	68
1.2	0.8333	0.8264	0.8197	0.8130	0.8065	0.8000	0.7937	0.7874	0.7813	0.7752	6	13	19	26	32	38	45	51	58
1.3	0.7692	0.7634	0.7576	0.7519	0.7463	0.7407	0.7353	0.7299	0.7246	0.7194	5	11	16	22	27	33	38	44	49
1.4	0.7143	0.7092	0.7042	0.6993	0.6944	0.6897	0.6849	0.6803	0.6757	0.6711	5	10	14	19	24	29	33	38	43
1.5	0.6667	0.6623	0.6579	0.6536	0.6494	0.6452	0.6410	0.6369	0.6329	0.6289	4	8	13	17	21	25	29	33	38
1.6	0.6250	0.6211	0.6173	0.6135	0.6098	0.6061	0.6024	0.5988	0.5952	0.5917	4	7	11	15	18	22	26	29	33
1.7	0.5882	0.5848	0.5814	0.5780	0.5747	0.5714	0.5682	0.5650	0.5618	0.5587	3	7	10	13	16	20	23	26	30
1.8	0.5556	0.5525	0.5495	0.5464	0.5435	0.5405	0.5376	0.5348	0.5319	0.5291	3	6	9	12	15	18	20	23	26
1.9	0.5263	0.5236	0.5208	0.5181	0.5155	0.5128	0.5102	0.5076	0.5051	0.5025	3	5	8	11	13	16	18	21	24
2.0	0.5000	0.4975	0.4950	0.4926	0.4902	0.4878	0.4854	0.4831	0.4808	0.4785	2	5	7	10	12	14	17	19	21
2.1	0.4762	0.4739	0.4717	0.4695	0.4673	0.4651	0.4630	0.4608	0.4587	0.4566	2	4	7	9	11	13	15	17	20
2.2	0.4545	0.4525	0.4505	0.4484	0.4464	0.4444	0.4425	0.4405	0.4386	0.4367	2	4	6	8	10	12	14	16	18
2.3	0.4348	0.4329	0.4310	0.4292	0.4274	0.4255	0.4237	0.4219	0.4202	0.4184	2	4	5	7	9	11	13	14	16
2.4	0.4167	0.4149	0.4132	0.4115	0.4098	0.4082	0.4065	0.4049	0.4032	0.4016	2	3	5	7	8	10	12	13	15
2.5	0.4000	0.3984	0.3968	0.3953	0.3937	0.3922	0.3906	0.3891	0.3876	0.3861	2	3	5	6	8	9	11	12	14
2.6	0.3846	0.3831	0.3817	0.3802	0.3788	0.3774	0.3759	0.3745	0.3731	0.3717	1	3	4	6	7	8	10	11	13
2.7	0.3704	0.3690	0.3676	0.3663	0.3650	0.3636	0.3623	0.3610	0.3597	0.3584	1	3	4	5	7	8	9	11	12
2.8	0.3571	0.3559	0.3546	0.3534	0.3521	0.3509	0.3497	0.3484	0.3472	0.3460	1	2	4	5	6	7	9	10	11
2.9	0.3448	0.3436	0.3425	0.3413	0.3401	0.3390	0.3378	0.3367	0.3356	0.3344	1	2	3	5	6	7	8	9	10
3.0	0.3333	0.3322	0.3311	0.3300	0.3289	0.3279	0.3268	0.3257	0.3247	0.3236	1	2	3	4	5	6	7	9	10
3.1	0.3226	0.3215	0.3205	0.3195	0.3185	0.3175	0.3165	0.3155	0.3145	0.3135	1	2	3	4	5	6	7	8	9
3.2	0.3125	0.3115	0.3106	0.3096	0.3086	0.3077	0.3067	0.3058	0.3049	0.3040	1	2	3	4	5	6	7	8	9
3.3	0.3030	0.3021	0.3012	0.3003	0.2994	0.2985	0.2976	0.2967	0.2959	0.2950	1	2	3	4	4	5	6	7	8
3.4	0.2941	0.2933	0.2924	0.2915	0.2907	0.2899	0.2890	0.2882	0.2874	0.2865	1	2	3	3	4	5	6	7	8

Continued

Appendix I: A Table of Reciprocals (*Continued*)

											Subtract								
3.5	0.2857	0.2849	0.2841	0.2833	0.2825	0.2817	0.2809	0.2801	0.2793	0.2786	1	2	2	3	4	5	6	6	7
3.6	0.2778	0.2770	0.2762	0.2755	0.2747	0.2740	0.2732	0.2725	0.2717	0.2710	1	2	2	3	4	5	5	6	7
3.7	0.2703	0.2695	0.2688	0.2681	0.2674	0.2667	0.2660	0.2653	0.2646	0.2639	1	1	2	3	4	4	5	6	6
3.8	0.2632	0.2625	0.2618	0.2611	0.2604	0.2597	0.2591	0.2584	0.2577	0.2571	1	1	2	3	3	4	5	5	6
3.9	0.2564	0.2558	0.2551	0.2545	0.2538	0.2532	0.2525	0.2519	0.2513	0.2506	1	1	2	3	3	4	4	5	6
4.0	0.2500	0.2494	0.2488	0.2481	0.2475	0.2469	0.2463	0.2457	0.2451	0.2445	1	1	2	2	3	4	4	5	5
4.1	0.2439	0.2433	0.2427	0.2421	0.2415	0.2410	0.2404	0.2398	0.2392	0.2387	1	1	2	2	3	3	4	5	5
4.2	0.2381	0.2375	0.2370	0.2364	0.2358	0.2353	0.2347	0.2342	0.2336	0.2331	1	1	2	2	3	3	4	4	5
4.3	0.2326	0.2320	0.2315	0.2309	0.2304	0.2299	0.2294	0.2288	0.2283	0.2278	1	1	2	2	3	3	4	4	5
4.4	0.2273	0.2268	0.2262	0.2257	0.2252	0.2247	0.2242	0.2237	0.2232	0.2227	1	1	2	2	3	3	4	4	5
4.5	0.2222	0.2217	0.2212	0.2208	0.2203	0.2198	0.2193	0.2188	0.2183	0.2179	0	1	1	2	2	3	3	4	4
4.6	0.2174	0.2169	0.2165	0.2160	0.2155	0.2151	0.2146	0.2141	0.2137	0.2132	0	1	1	2	2	3	3	4	4
4.7	0.2128	0.2123	0.2119	0.2114	0.2110	0.2105	0.2101	0.2096	0.2092	0.2088	0	1	1	2	2	3	3	4	4
4.8	0.2083	0.2079	0.2075	0.2070	0.2066	0.2062	0.2058	0.2053	0.2049	0.2045	0	1	1	2	2	3	3	3	4
4.9	0.2041	0.2037	0.2033	0.2028	0.2024	0.2020	0.2016	0.2012	0.2008	0.2004	0	1	1	2	2	2	3	3	4
5.0	0.2000	0.1996	0.1992	0.1988	0.1984	0.1980	0.1976	0.1972	0.1969	0.1965	0	1	1	2	2	2	3	3	4
5.1	0.1961	0.1957	0.1953	0.1949	0.1946	0.1942	0.1938	0.1934	0.1931	0.1927	0	1	1	2	2	2	3	3	3
5.2	0.1923	0.1919	0.1916	0.1912	0.1908	0.1905	0.1901	0.1898	0.1894	0.1890	0	1	1	1	2	2	3	3	3
5.3	0.1887	0.1883	0.1880	0.1876	0.1873	0.1869	0.1866	0.1862	0.1859	0.1855	0	1	1	1	2	2	3	3	3
5.4	0.1852	0.1848	0.1845	0.1842	0.1838	0.1835	0.1832	0.1828	0.1825	0.1821	0	1	1	1	2	2	2	3	3
5.5	0.1818	0.1815	0.1812	0.1808	0.1805	0.1802	0.1799	0.1795	0.1792	0.1789	0	1	1	1	2	2	2	3	3
5.6	0.1786	0.1783	0.1779	0.1776	0.1773	0.1770	0.1767	0.1764	0.1761	0.1757	0	1	1	1	2	2	2	3	3
5.7	0.1754	0.1751	0.1748	0.1745	0.1742	0.1739	0.1736	0.1733	0.1730	0.1727	0	1	1	1	2	2	2	2	3
5.8	0.1724	0.1721	0.1718	0.1715	0.1712	0.1709	0.1706	0.1704	0.1701	0.1698	0	1	1	1	1	2	2	2	3
5.9	0.1695	0.1692	0.1689	0.1686	0.1684	0.1681	0.1678	0.1675	0.1672	0.1669	0	1	1	1	1	2	2	2	3

6.0	0.1667	0.1664	0.1661	0.1658	0.1656	0.1653	0.1650	0.1647	0.1645	0.1642	0	1	1	1	1	2	2	2	3
6.1	0.1639	0.1637	0.1634	0.1631	0.1629	0.1626	0.1623	0.1621	0.1618	0.1616	0	1	1	1	1	2	2	2	2
6.2	0.1613	0.1610	0.1608	0.1605	0.1603	0.1600	0.1597	0.1595	0.1592	0.1590	0	1	1	1	1	2	2	2	2
6.3	0.1587	0.1585	0.1582	0.1580	0.1577	0.1575	0.1572	0.1570	0.1567	0.1565	0	0	1	1	1	1	2	2	2
6.4	0.1563	0.1560	0.1558	0.1555	0.1553	0.1550	0.1548	0.1546	0.1543	0.1541	0	0	1	1	1	1	2	2	2
6.5	0.1538	0.1536	0.1534	0.1531	0.1529	0.1527	0.1524	0.1522	0.1520	0.1517	0	0	1	1	1	1	2	2	2
6.6	0.1515	0.1513	0.1511	0.1508	0.1506	0.1504	0.1502	0.1499	0.1497	0.1495	0	0	1	1	1	1	2	2	2
6.7	0.1493	0.1490	0.1488	0.1486	0.1484	0.1481	0.1479	0.1477	0.1475	0.1473	0	0	1	1	1	1	2	2	2
6.8	0.1471	0.1468	0.1466	0.1464	0.1462	0.1460	0.1458	0.1456	0.1453	0.1451	0	0	1	1	1	1	2	2	2
6.9	0.1449	0.1447	0.1445	0.1443	0.1441	0.1439	0.1437	0.1435	0.1433	0.1431	0	0	1	1	1	1	1	2	2
7.0	0.1429	0.1427	0.1425	0.1422	0.1420	0.1418	0.1416	0.1414	0.1412	0.1410	0	0	1	1	1	1	1	2	2
7.1	0.1408	0.1406	0.1404	0.1403	0.1401	0.1399	0.1397	0.1395	0.1393	0.1391	0	0	1	1	1	1	1	2	2
7.2	0.1389	0.1387	0.1385	0.1383	0.1381	0.1379	0.1377	0.1376	0.1374	0.1372	0	0	1	1	1	1	1	2	2
7.3	0.1370	0.1368	0.1366	0.1364	0.1362	0.1361	0.1359	0.1357	0.1355	0.1353	0	0	1	1	1	1	1	2	2
7.4	0.1351	0.1350	0.1348	0.1346	0.1344	0.1342	0.1340	0.1339	0.1337	0.1335	0	0	1	1	1	1	1	1	2
7.5	0.1333	0.1332	0.1330	0.1328	0.1326	0.1325	0.1323	0.1321	0.1319	0.1318	0	0	1	1	1	1	1	1	2
7.6	0.1316	0.1314	0.1312	0.1311	0.1309	0.1307	0.1305	0.1304	0.1302	0.1300	0	0	1	1	1	1	1	1	2
7.7	0.1299	0.1297	0.1295	0.1294	0.1292	0.1290	0.1289	0.1287	0.1285	0.1284	0	0	0	1	1	1	1	1	1
7.8	0.1282	0.1280	0.1279	0.1277	0.1276	0.1274	0.1272	0.1271	0.1269	0.1267	0	0	0	1	1	1	1	1	1
7.9	0.1266	0.1264	0.1263	0.1261	0.1259	0.1258	0.1256	0.1255	0.1253	0.1252	0	0	0	1	1	1	1	1	1
8.0	0.1250	0.1248	0.1247	0.1245	0.1244	0.1242	0.1241	0.1239	0.1238	0.1236	0	0	0	1	1	1	1	1	1
8.1	0.1235	0.1233	0.1232	0.1230	0.1229	0.1227	0.1225	0.1224	0.1222	0.1221	0	0	0	1	1	1	1	1	1
8.2	0.1220	0.1218	0.1217	0.1215	0.1214	0.1212	0.1211	0.1209	0.1208	0.1206	0	0	0	1	1	1	1	1	1
8.3	0.1205	0.1203	0.1202	0.1200	0.1199	0.1198	0.1196	0.1195	0.1193	0.1192	0	0	0	1	1	1	1	1	1
8.4	0.1190	0.1189	0.1188	0.1186	0.1185	0.1183	0.1182	0.1181	0.1179	0.1178	0	0	0	1	1	1	1	1	1
8.5	0.1176	0.1175	0.1174	0.1172	0.1171	0.1170	0.1168	0.1167	0.1166	0.1164	0	0	0	1	1	1	1	1	1
8.6	0.1163	0.1161	0.1160	0.1159	0.1157	0.1156	0.1155	0.1153	0.1152	0.1151	0	0	0	1	1	1	1	1	1
8.7	0.1149	0.1148	0.1147	0.1145	0.1144	0.1143	0.1142	0.1140	0.1139	0.1138	0	0	0	1	1	1	1	1	1
8.8	0.1136	0.1135	0.1134	0.1133	0.1131	0.1130	0.1129	0.1127	0.1126	0.1125	0	0	0	1	1	1	1	1	1
8.9	0.1124	0.1122	0.1121	0.1120	0.1119	0.1117	0.1116	0.1115	0.1114	0.1112	0	0	0	1	1	1	1	1	1

Continued

Appendix I: A Table of Reciprocals (*Continued*)

											Subtract								
9.0	0.1111	0.1110	0.1109	0.1107	0.1106	0.1105	0.1104	0.1103	0.1101	0.1100	0	0	0	1	1	1	1	1	1
9.1	0.1099	0.1098	0.1096	0.1095	0.1094	0.1093	0.1092	0.1091	0.1089	0.1088	0	0	0	0	1	1	1	1	1
9.2	0.1087	0.1086	0.1085	0.1083	0.1082	0.1081	0.1080	0.1079	0.1078	0.1076	0	0	0	0	1	1	1	1	1
9.3	0.1075	0.1074	0.1073	0.1072	0.1071	0.1070	0.1068	0.1067	0.1066	0.1065	0	0	0	0	1	1	1	1	1
9.4	0.1064	0.1063	0.1062	0.1060	0.1059	0.1058	0.1057	0.1056	0.1055	0.1054	0	0	0	0	1	1	1	1	1
9.5	0.1053	0.1052	0.1050	0.1049	0.1048	0.1047	0.1046	0.1045	0.1044	0.1043	0	0	0	0	1	1	1	1	1
9.6	0.1042	0.1041	0.1040	0.1038	0.1037	0.1036	0.1035	0.1034	0.1033	0.1032	0	0	0	0	1	1	1	1	1
9.7	0.1031	0.1030	0.1029	0.1028	0.1027	0.1026	0.1025	0.1024	0.1022	0.1021	0	0	0	0	1	1	1	1	1
9.8	0.1020	0.1019	0.1018	0.1017	0.1016	0.1015	0.1014	0.1013	0.1012	0.1011	0	0	0	0	1	1	1	1	1
9.9	0.1010	0.1009	0.1008	0.1007	0.1006	0.1005	0.1004	0.1003	0.1002	0.1001	0	0	0	0	0	1	1	1	1

Appendix II

SOME USEFUL IR AND RAMAN SOURCES

Infrared Absorption Spectroscopy, 2d ed., by Nakanishi, K., and Solomon, P. H., Holden-Day, San Francisco, California, 1977, 287 pages. Contains a chapter on laser Raman spectroscopy with many examples and interpretations of spectra. Also contains 100 problems in IR spectroscopy with answers worked out.

Advances in Infrared Group Frequencies, by Bellamy, L. J., Chapman and Hall, London, England, 1975, 304 pages.

Atlas of Spectral Data and Physical Constants for Organic Compounds, 2d ed., Grasselli, J. G., and Ritchey, W. M., editors. CRC Press, Cleveland, Ohio, 1975, 4,527 pages.

Ultraviolet and Infrared Spectra of Drugs, No. 1: *Steroids,* Arzamastsev, A. P., and Yashkina, D. S. (Meditsina, Moscow, USSR), 1975, 152 pages. *Chem. Abstr.* **85,** 68365H (1976).

Textbooks for Instrumental Analysis, Vols. 43, 44: *Infrared Spectroscopy. An Introduction,* Guenzler, H., and Boeck, H., Verlag Chemie, Weinheim, Germany, 1975, 363 pages.

Raman/Infrared Atlas of Organic Compounds, Vol. 2 (Schrader, B., and Meier, W., eds.), Verlag Chemie, Weinheim, Germany, 1975, 344 pages.

Practical Spectroscopy Series, Vol. 1, Pt.A. *Infrared and Raman Spectroscopy* (Brame, E. G., Jr., and Grasselli, J. G., eds.), Marcel Dekker, New York, 1977, 345 pages; Part B, 1977, 370 pages; Part C, 1977, 322 pages.

Computers in Chemistry and Instrumentation, Vol. 7, *Infrared, Correlation and Fourier Transform Spectroscopy* (Mattson, J., Mark, H. B., Jr., and MacDonald, H. C., Jr., eds.), Marcel Dekker, New York, 1977.

Vibrational Spectroscopy—Modern Trends (Barnes, A. J., and Orville-Thomas, W. J., eds.) Elsevier Scientific Publishing Co., Amsterdam, The Netherlands, 1977, 442 pages.

Introduction to Infrared and Raman Spectroscopy, 2d ed., Colthup, N. B., Daly, L. H., and Wiberley, S. E., Academic Press, New York, 1975. Gives 624 IR spectra with band assignments. Also gives a set of 24 IR unknowns for interpretation, and 36 selected Raman spectra.

Infrared Spectra of Complex Molecules, 3rd ed., by Bellamy, L. J., Chapman and Hall, London, 1975.

Infrared and Raman Spectra of Inorganic and Coordination Compounds, 3rd edition, by Nakamoto, K., Wiley, New York, 1978.

CHEMICAL COMPOUND INDEX

GENERAL INDEX